# Bioinformatics for Systems Biology

Stephen Krawetz
Editor

# Bioinformatics for Systems Biology

Springer

*Editor*
Stephen Krawetz
Wayne State University
Detroit, MI, USA
steve@compbio.med.wayne.edu



ISBN: 978-1-934115-02-2 e-ISBN: 978-1-59745-440-7
DOI 10.1007/978-1-59745-440-7

Library of Congress Control Number: 2008934788



Printed on acid-free paper

springer.com

# Preface

With the completion of the human genome project, followed by the rise in high-throughput technologies like the various microarray and now high throughput genomic sequencing platforms, we experienced the birth of Systems Biology after its long gestation. This revolution is marked by a change in the research paradigm from the single small-scale experiment, i.e., following the change of a component in a multi component system, to one that attempts to simultaneously monitor the change of tens of thousands of molecules within this body. This clearly necessitates the unparalleled use of project-specific informatic tools, which, to date, requires an unprecedented level of development to collect, manage and mine the data for interesting associations.

To begin to understand this information we now rely on statistical analysis to aid in our selection of the fruit from the tree. However, this often takes us on a journey into a new field for which we are not yet prepared. Samuel Johnson (1709–1784) foreshadowed the dilemma we would face and characterized it as follows: "Knowledge is of two kinds. We know a subject ourselves, or we know where we can find information on it." It is for the latter that we routinely turn to the literature. The rate of growth of the literature parallels that of sequencing data and the array data placing an almost impossible task before each investigator. To partially ease this burden we are again turning towards developing informatic aids that mine the literature and data to develop summaries and associations to directly address the questions posed and the new hypotheses that are to be tested.

Although more clearly articulated, we again face similar challenges as those tackled during the course of the human genome project. It is essential that the training of the biologist and computer-scientist occur in an interdisciplinary environment of cross-fertilization. With this goal in mind the textbook "Bioinformatics for Systems Biology" was undertaken.

We begin this exploration with Part I, to provide the computer scientist with an introduction to the underlying principles of cell biology. This is followed by a brief introduction in Part II as a means for the biologist to become familiar with concepts and the statistical analysis of large datasets. Part III then describes, to date, the best characterized use of the microarray platform that is now moving towards whole genome analysis. With all of this data, how do we begin analysis for common elements guiding the underlying principles? This is discussed in Part IV which leads to Part V and Part VI to test, *in silico,* the relationships on a wide scale in order to assess their applicability. Upon developing the associations, Part VII asks how does this information relate to what was measured? As these basic principles are developed from an "omics" driven biological systems approach, they are applied in Part VIII to translational medicine. An excellent example is the new term "personalized medicine" that is beginning to reverberate in clinical care. It is the culmination of the Systems Biology revolution where technological advances and cross-fertilization have driven the field to mature to the point where it is being incorporated in a true bench-to-bedside manner.

As you read the chapters, you will find that they can stand alone, yet can be combined to emphasize the integral role of informatics in Systems Biology. Most of the figures and tables are in grey scale. I would encourage you to view those that benefit from color on the accompanying CD. The material contained on the CD provides an excellent source of slides for your lectures and presentations.

The chapter-related Glossary and Abbreviations section will assist in familiarizing you with the terms. You will also find the literature and suggested reading sections, including key references, very useful as you delve into the subject matter. Technology, by its very meaning implies refinement and change. The informatics approaches used in systems biology are continually subject to refinement. With this reality, you are encouraged to utilize the web site information provided in various chapters to help access the most current information and resources available. As Systems Biology develops we are able to witness growing pains and milestones. With continued informatic and biological cross-fertilization, advancements in Systems Biology will revolutionize personalized medicine answering questions by integrating information in unexpected ways.

# Contents

# Contributors

**Sachiyo Aburatani** National Institute of Advanced Industrial Science and Technology, Tokyo, Japan

**Dimitra Alexopoulou** Technische Universität Dresden, Dresden, Germany

**Michael R. Alvers** Technische Universität Dresden, Dresden, Germany

**Bill Andreopoulos** Technische Universität Dresden, Dresden, Germany

**Kiyoko F. Aoki-Kinoshita** Soka University, Tokyo, Japan.

**D. Randall Armant** Wayne State University School of Medicine, Detroit, MI

**Richard A. Baldock** MRC Human Genetics Unit, Western General Hospital, Edinburgh, UK

**Jill S. Barnholtz-Sloan** Case Western Reserve University School of Medicine, Cleveland, OH

**Liliana Barrio-Alvers** Technische Universität Dresden, Dresden, Germany

**Thomas L. Beaumont** Wayne State University School of Medicine, Detroit, MI

**Panayiotis V. Benos** University of Pittsburgh, Pittsburgh, PA

**Linda B. Bloom** University of Florida, Gainesville, FL

**David Michael Cherba** Van Andel Research Institute, Grand Rapids, MI

**Jeffrey H. Christiansen** MRC Human Genetics Unit, Western General Hospital, Edinburgh, UK

**Fiona Cunningham** Wellcome Trust Sanger Institute, Hinxton, Cambridge, UK

**Duncan R. Davidson** MRC Human Genetics Unit, Western General Hospital, Edinburgh, UK

**Heiko Dietze** Technische Universität Dresden, Dresden, Germany

**Andreas Doms** Technische Universität Dresden, Dresden, Germany

**K.C. Dukka Bahadur** University of North Carolina at Charlotte, Charlotte, NC

**Anton Epple** Genomatix, Müenchen, Germany

**Myriam Ferro** Université Joseph Fourier, Grenoble, France

**Rivka L. Glaser** Villa Julie College, Stevenson, MD

**Jörg Hakenberg** Technische Universität Dresden, Dresden, Germany

**Robert Hoffmann** Memorial Sloan-Kettering Cancer Center, New York, NY

**Jon Holy** University of Minnesota School of Medicine, Duluth, MN

**Katsuhisa Horimoto** National Institute of Advanced Industrial Science and Technology, Tokyo, Japan

**Jingyi Hui** University of Erlangen, Erlangen, Germany

**Minoru Kanehisa** Kyoto University, Gokasho, Uji, Japan

**Christina Karamboulas** University of Ottawa, Ottawa, ON, Canada

**Amit Khanna** University of Kentucky College of Medicine, Lexington, KY

**Shivendra Kishore** University of Erlangen, Erlangen, Germany

**Stephen A. Krawetz** Wayne State University School of Medicine, Detroit, MI

**Michael L. Kruger** Wayne State University School of Medicine, Detroit, MI

**Gerolamo Lanfranchi** University of Padova, Padova, Italy

**Florian Leitner** Spanish National Cancer Research Centre, Madrid, Spain

**Dennis R. Livesay** University of North Carolina at Charlotte, Charlotte, NC

**Jeffrey A. Loeb** Wayne State University, School of Medicine, Detroit, MI

**Maria Manioudaki** University of Crete, Heraklion, Crete, Greece

**Jan Mönnich** Technische Universität Dresden, Dresden, Germany

**Ian M. Morison** University of Otago, Dunedin, New Zealand

**Anne Parker** Wellcome Trust Sanger Institute, Hinxton, Cambridge, UK

**Ed Perkins** University of Minnesota School of Medicine, Duluth, MN

**Plake Conrad** Technische Universität Dresden, Dresden, Germany

**Adrian E. Platts** Wayne State University School of Medicine, Detroit, MI

**Panayiota Poirazi** Crete, Foundation for Research and Technology-Hellas, Heraklion, Greece

**Daniel A. Rappolee** Wayne State University School of Medicine, Detroit, MI

**Keir T. Reavie** University of California, Davis, Davis, CA

**Martin Reczko** Crete, Foundation for Research and Technology-Hellas, Heraklion, Greece

**Andreas Reischuck** Technische Universität Dresden, Dresden, Germany

**Chiara Romualdi** University of Padova, Padova, Italy

**Sophie Rousseaux** Université Joseph Fourier, Institut Albert Bonniot, Grenoble, France

**Loïc Royer** Technische Universität Dresden, Dresden, Germany

**Shigeru Saito** INFOCOM CORPORATION, Tokyo, Japan

**Michael Schroeder** Technische Universität Dresden, Dresden, Germany

**Christoph W. Sensen** University of Calgary Faculty of Medicine, Calgary, AB, Canada

**Matthias Sherf** Genomatix, Müenchen, Germany

**Gautam B. Singh** Oakland University, Rochester, MI

**Ilona S. Skerjanc** University of Ottawa, Ottawa, ON, Canada

**Stefan Stamm** University of Kentucky College of Medicine, Lexington, KY

**Steven M. Thompson** Florida State University, Tallahassee, FL

**Hemant K. Tiwari** University of Alabama at Birmingham, Birmingham, AL

**Andrei L. Turinsky** Faculty of Medicine, University of Calgary, Calgary, AB, Canada

**Eleftheria Tzamali** University of Crete, Heraklion, Crete, Greece

**Alfonso Valencia** Spanish National Cancer Research Centre, Madrid, Spain

**Thomas Wächter** Technische Universität Dresden, Dresden, Germany

**Craig Paul Webb** Van Andel Research Institute, Grand Rapids, MI

**Thomas Werner** Genomatix, München, Germany

**Nadine Wiper-Bergeron** University of Ottawa, Ottawa, ON, Canada

**David S. Wishart** University of Alberta, Edmonton, AB, Canada

**John J. Wyrick** Washington State University, Pullman, WA

**Matthias Zschunke** Technische Universität Dresden, Dresden, Germany

# Part I
# Life of a Cell and Its Analysis

# Chapter 1
# Structure and Function of the Nucleus and Cell Organelles

**Jon Holy and Ed Perkins**

**Abstract** Living eukaryotic cells must carry out and coordinate an enormous number of biochemical reactions in order to obtain and convert energy to usable forms, break down and interconvert organic molecules to synthesize needed components, sense and respond to environmental and internal stimuli, regulate gene activity, sense and repair damage to structural and genomic elements, and grow and reproduce. This level of complexity requires that biochemical reactions be highly organized and compartmentalized, and this is the major function of cell organelles and the cytoskeleton. Cells have elaborated an elegant cytoplasmic membrane system composed of the nuclear envelope, ER, Golgi apparatus, and associated endocytotic, endosomal, lysososomal, and secretory vesicles and compartments. These membranes serve to both organize and compartmentalize biochemical reactions involved in protein and lipid synthesis, targeting, and secretion. The cytoskeleton not only facilitates cytosolic molecular interactions, but also serves to organize the entire cytoplasmic membrane system. The key to cellular life is organization, and eukaryotic cells display a remarkably rich and elegant architecture to carry out the demands of life.

**Keywords** Cell · Structure · Nucleus · Organelle · Function

## 1.1 Introduction

The myriad of biochemical reactions that comprise life processes are too numerous and complex to be carried out entirely by simple diffusion-mediated interactions between enzymes and substrates. Instead, sequences of biochemical reactions must be efficiently organized and integrated with other sets of reactions by the cell. Two fundamental structural elements are used by eukaryotic cells to organize and integrate these reactions: membranes and a cytoskeletal system. An elaborate system of cellular membranes, in the form of the plasma membrane, membrane-bound organelles, and the nuclear envelope, has evolved to provide reaction surfaces and to organize and compartmentalize molecules involved in specific metabolic pathways. Other cytosolic biochemical reactions, as well as the organization of membranous organelles within the cell, are regulated by interactions with the cytoskeletal system. Consequently, enzymes and proteins involved in biochemical reactions can be located in the cytosol, within membranes, on the surfaces of membranes, within the interior of membrane-bound compartments, or in association with the cytoskeleton. The elaboration of these structural elements has allowed for the sophisticated level of interaction and integration of biochemical reactions that exist in living eukaryotic cells (Fig. 1.1).

Over two hundred different types of cells are found in higher animals, including humans, and the interaction of these diverse cell types is responsible for the formation and functioning of tissues and

J. Holy
Departments of Anatomy, Microbiology & Pathology, University of Minnesota School of Medicine, Duluth
1035 University Avenue, Duluth, MN 55812, USA
e-mail: jholy@d.umn.edu

S. Krawetz (ed.), *Bioinformatics for Systems Biology*, DOI 10.1007/978-1-59745-440-7_1,

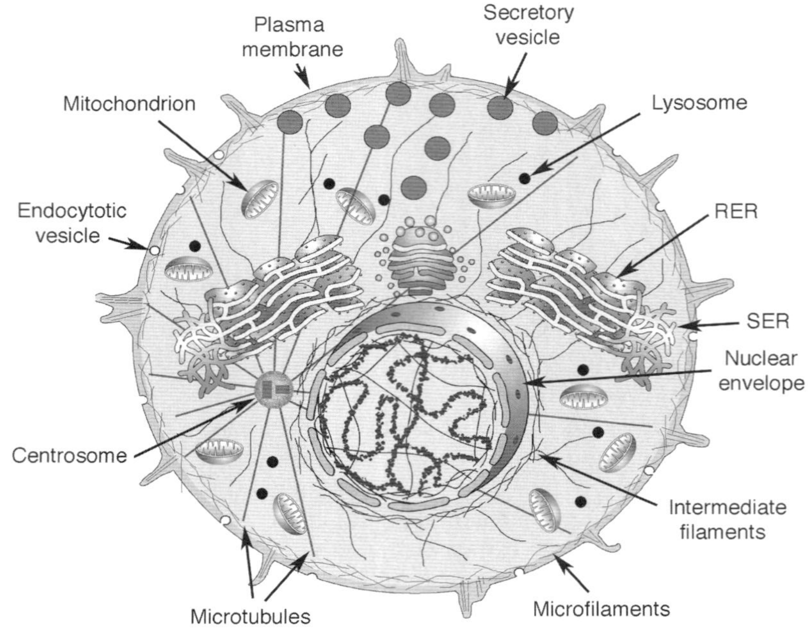


**Fig. 1.1** Diagram of the major cell organelles, including the cytoskeleton and nucleus. This drawing depicts a single idealized cell, and so does not include the cell-cell and cell-ECM interactions. (Copies of figures including color copies, where applicable, are available in the accompanying CD)

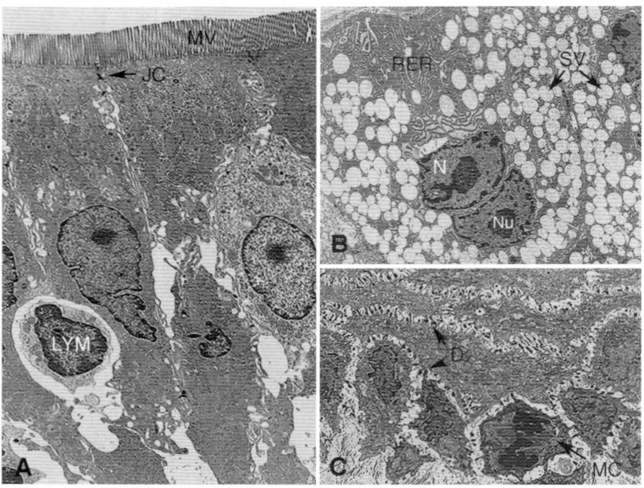


**Fig. 1.2** Examples of low-power electron micrographs of thin sections of rat tissues showing how cell organization reflects cell function. **(A)** Shows intestinal epithelial cells, which are modified to aid in the digestion and absorption of food. The apical membranes of these cells develop highly organized microvilli (MV), which are supported by bundles of microfilaments, to increase the surface area of these cells. The epithelial cells are bound to each other by junctional complexes (JC) consisting of a cluster of tight junctions, zonulae adherens junctions, and desmosomes. The tight junctions prevent material in the intestinal lumen from diffusing between cells into the body cavity, and the adherens and desmosomal junctions firmly anchor cells to each other. A migrating immune cell (LYM, lymphocyte) is also present in this section. **(B)** Shows salivary gland cells, which are specialized to produce and release large amounts of secretory glycoproteins. These cells contain extensive arrays of rough endoplasmic reticulum (RER) filled with secretory vesicles (SV). Two nuclei (N) are present in this section, and both display prominent nucleoli (Nu). **(C)** Shows cells of the esophageal epithelium, which are specialized to accommodate and resist mechanical stresses. These cells are constantly renewed by the mitotic activity of a basal layer of cells (MC, mitotic cell). As cells are produced and differentiate, they move towards the lumen of the esophagus, become flattened, and form extensive desmosomal connections **(D)** with neighboring cells (Copies of figures including color copies, where applicable, are available in the accompanying CD)

organs. Different types of cells carry out specialized functions (Fig. 1.2), but all cells face similar sets of challenges to exist. In general, cells must maintain a barrier against, and sensing mechanisms to interact with their external environment; synthesize and recycle their structural and enzymatic components; repair physical or chemical damage; grow and reproduce; and generate energy for all of these activities. These generalized functions, as well as the more specialized functions of individual cell types, are all performed by cell organelles. Cell organelles perform the basic functions that allow cells to survive and replicate, and are dynamic entities that become modified to help specialized cells carry out specific functions. For example, all cells contain a cytoskeletal filamentous system that functions to maintain cell shape and allows for some degree of movement, but muscle cells contain far greater numbers of these filaments, to carry out the contractile activity of the muscles.

Classically, the phrase "cell organelle" has been used to denote distinct membrane-bound structures that are readily visible by light or electron microscopy and possess characteristic morphological features that make them readily identifiable in essentially all eukaryotic cells. Such structures include the plasma membrane, endoplasmic reticulum (ER), Golgi apparatus, lysosomes, peroxisomes, and mitochondria. The structure and function of these organelles, as well as the cytoskeleton and nucleus, are reviewed in this chapter. Because membrane structure plays fundamental roles in organelle function, the basic features of membrane organization will be considered first.

## 1.2 Membrane Structure

Cell membranes are composed of lipid and protein; the lipid is assembled into two opposing layers, called the lipid bilayer. Although the lipid make-up of membranes is very complex, four major types of phospholipids and cholesterol comprise most of the lipid portion of the bilayer. The phospholipids include the choline-containing lipids phosphatidylcholine and sphingomyelin, and the amine-containing phosphatidylethanolamine and phosphatidylserine. All of these phospholipids possess hydrophilic polar heads, and two hydrophobic fatty acid tails. The membrane bilayer represents an energetically favorable conformation of these lipids in that the tails associate with each other to form a hydrophobic environment in the center of the bilayer, with the polar heads facing outward to interact with the charged aqueous environment of the cytoplasm, organelle lumen, or extracellular space. The hydrophobic region resulting from the association of lipid tails creates a barrier to the passage of charged molecules, and only small uncharged molecules, or lipid-soluble molecules, can freely penetrate the lipid bilayer. Cholesterol, which is shorter and stiffer than phospholipids, comprises approximately 50% of the total membrane lipid. The hydroxyl end of cholesterol interacts with the polar heads of phospholipid molecules, with the rest of the molecule in the same plane as the fatty acid tails of phospholipids. Its presence in membranes is thought to help prevent phase transitions by stiffening membranes at higher temperatures, while also maintaining membrane fluidity at lower temperatures. Although cholesterol is prevalent and equally represented in both bilayers of a membrane, the major phospholipids can be asymmetrically distributed among different membranes, high concentrations of choline-containing phospholipids present in the non-cytoplasmic layer (the layer facing the extracellular matrix), and high concentrations of amine-containing phospholipids in the layer facing the cytoplasm of the plasma membrane, for example.

In addition to lipids, membranes are also composed of proteins. Membrane proteins are either classified as integral membrane proteins, if they penetrate or are anchored in the bilayer, or as peripheral membrane proteins, if they are just associated with the surface of the bilayer. Integral membrane proteins are difficult to remove from membranes, usually requiring disruption of the lipid bilayer (e.g., with detergents), in order to be released. Peripheral proteins are easier to remove from membranes, as they are generally held in place by protein-protein interactions. Integral membrane proteins can penetrate the bilayer completely a single time (single-pass proteins) or multiple times (multi-pass proteins). They can also be anchored in the membrane through covalent attachments to lipid molecules in the bilayer.

A number of the membrane lipids and proteins are glycosylated. Glycosylation of membrane components takes place in the ER and Golgi apparatus. Because glycosylation occurs exclusively within the interior (or lumen) of these organelles, the sugar groups of glycoproteins and glycolipids all face toward the luminal surface of membranes of organelles, and the extracellular matrix (ECM) side of the plasma membrane. Glycosylation of membrane lipids and proteins is thought to help protect membranes, and, in the case of the plasma membrane, to help identify the cell and to assist in the adhesion of cells to the ECM.

Membrane lipids and proteins carry out a number of functions. In addition to serving as the structural framework of the membrane, they mediate the functions of all membranes of the cell. Membrane lipids can form specialized sub domains composed of specific lipid populations (lipid rafts) that are associated with defined functions, and some membrane lipids are intimately involved in signal transduction events. Membrane proteins carry out a wide variety of functions, including serving as membrane channels, carriers, and pumps; transducing cytoplasmic and extracellular signals; targeting membranes to specific locations; and adhering membranes to each other and to the ECM.

## 1.3 The Plasma Membrane

The plasma membrane encloses the cytoplasm of a cell and carries out multiple functions. It forms both a barrier to and an interface with the cellular environment. The plasma membrane is a selectively permeable barrier that, by regulating what enters and exits a cell, is a primary determinant of the composition of the cytoplasm. The plasma membrane is associated with sensing mechanisms that transduce environmental information into a cytoplasmic or nuclear response. The plasma membrane is involved in cell-cell and cell-ECM attachments, and also contains cell-specific molecules that help identify cells, thereby helping to establish the appropriate position and arrangement of each cell in the body.

### *1.3.1 Barrier Functions*

The hydrophobic nature of the central region of the lipid bilayer serves as a barrier to charged and large hydrophilic molecules. Thus, the lipid bilayer is impermeable to small ions (e.g., $Na^+$, $K^+$, $Cl^-$) and proteins. Only small, uncharged molecules (e.g., $CO_2$, $H_2O$), or molecules freely soluble in lipid (e.g., steroid hormones, dioxin) are able to pass directly through the lipid bilayer. In this way the plasma membrane is selectively permeable. However, materials can be transported into and out of the cell by specific transport mechanisms carried out by the plasma membrane (see *Transport Functions*, below). The carbohydrate moieties of glycolipids and glycoproteins contribute to the membrane acting as barriers, impeding the access of molecules to the surface of the plasma membrane, which can also protect the plasma membranes exposed to harsh environments (e.g., the stomach and intestinal lumen).

### *1.3.2 Transport Functions*

Because the lipid bilayer is impermeable to most types of organic molecules, the cell must possess mechanisms to move materials between the cytoplasm and the external environment. Two approaches are used by the cell to move materials into and out of the cytoplasm: (i) transport through the membrane and (ii) transport involving membrane flow.

#### 1.3.2.1 Transport Through the Plasma Membrane

Transport through membranes is mediated by integral membrane proteins, which help conduct materials past the hydrophobic lipid bilayer in a number of ways. Integral membrane proteins can form channels by associating to form pore-like structures in the membrane. Such pores allow for diffusion of molecules small enough to fit through them. This type of transport allows for the flow of molecules down their concentration gradient and an expenditure of energy is not needed if the channel is open. Thus, molecules can move through protein channels by passive diffusion. Examples include ion channels that allow for the passage of ions such as $Na^+$ and $K^+$, and the connexons in gap junctions, which allow for the passage of molecules $<1000$ daltons through the plasma membrane. Whether these channels are open or closed is tightly regulated in order to prevent the constant leakage of small molecules into or out of the cell.

Integral membrane proteins can also act as carriers that bind to specific molecules and help them traverse the lipid bilayer. Binding of the appropriate molecule to carrier proteins results in a conformational change in the structure of the carrier protein, such that the ligand is conveyed across the membrane. Release of the ligand results in reversion of the carrier to its original state, ready to bind another ligand. This type of transport is driven by the concentration gradient of the ligand, and does not require the expenditure of energy by the cell. An example of transport by this method of facilitated diffusion includes glucose transporters in the basolateral membranes of intestinal epithelial cells.

Cells can also transport molecules against their concentration gradients, and this type of transport is carried out by integral membrane proteins that act as pumps and requires expenditure of energy. This is referred to as active transport and it is an essential process in living cells. Examples include a number of different ion pumps, which keep the cytoplasm relatively low in $Na^+$ and high in $K^+$. Ion pumps are vital elements of the plasma membrane, and it has been estimated that as much as one-third or more of the energy consumed by a living cell is used to actively transport $Na^+$ out of the cell. The concentration gradients of certain ions established by these membrane pumps, can themselves serve as motive forces for other transport mechanisms. For example, in addition to moving out of a cell by facilitated diffusion, glucose is actively transported into cells by integral membrane proteins that bind both glucose and $Na^+$. Because these transporters bind both $Na^+$ and glucose, the high concentration of $Na^+$ outside the cell relative to the cytoplasm drives the movement of both $Na^+$ and glucose into the cell, against the concentration gradient of glucose.

Acquisition of glucose from the small intestine represents an example of how active transport can be coupled with facilitated diffusion to move molecules past the epithelia. The apical membrane of intestinal epithelial cells contain $Na^+$ coupled active transporters, which moves glucose against its concentration gradient to accumulate in the cytoplasm. Consequently, the concentration of glucose is higher in the cytoplasm of these cells than in the extracellular spaces underlying them, and carrier proteins in the basolateral membranes of these cells allow for the facilitated diffusion of glucose out of the cell (down its concentration gradient) and into the circulation. It can be seen from this example that directional transport of molecules past epithelial cells require the integral membrane transport proteins to occupy specific locations within the plasma membrane (i.e., either the apical or basolateral membrane). How transporters are organized within the plasma membrane is determined by specific targeting mechanisms acting in conjunction with cell junctions and the cytoskeleton.

#### 1.3.2.2 Transport Involving Membrane Flow

In addition to the movement of materials through membrane channels, carriers, and pumps, the plasma membrane mediates the transport of material into and out of cells by membrane flow. Internalization of extracellular material can occur by entrapment in membrane-bound vesicles that pinch off from the plasma membrane and are transported into the cytoplasm for processing. This process, called endocytosis, can be subdivided into a number of different categories based on the mechanics of how the invagination and formation of vesicles occurs at the plasma membrane, and includes the formation of clathrin-coated vesicles from coated pits, and the formation of nonclathrin-coated vesicles derived from structures called caveoli.

Clathrin-coated vesicles comprise a major pathway in which specific extracellular molecules are recognized and bound to the plasma membrane prior to internalization. This process involves membrane receptors, which are integral membrane proteins of the plasma membrane that recognize specific ligands. One of the best understood examples of this process involves how cholesterol is taken up by cells. In the circulation, cholesterol is packaged in low-density lipoprotein (LDL) particles, which are small particles, composed of protein and cholesterol esters. Specific LDL receptors are present in the plasma membrane that bind and anchor LDL particles to the surface of the cell. LDL bound receptors form clusters in the membrane that recruit adaptor proteins and the cytoplasmic protein clathrin. Clathrin molecules assemble underneath the receptor clusters to form a basketwork that deforms the plasma membrane into an invagination referred to as a coated pit. Continued assembly of the clathrin protein results in the continued invagination, and finally pinching off, and release of a membrane-bound coated vesicle containing LDL receptor and LDL cargo into the cytoplasm. Once the vesicle is formed, the clathrin coating is disassembled and the clathrin recycled to the plasma membrane to assist in the formation of more coated pits. The clathrin-free vesicle then fuses with a membrane-bound compartment called an endosome, which, in addition to receiving vesicles from the plasma membrane, also receives lysosomal vesicles filled with hydrolytic enzymes packaged by the Golgi apparatus. Membrane-bound structures containing a mixture of LDL particles and acid hydrolases then arise from the endosome to form mature lysosomes. During this process, LDL dissociates from the LDL receptor in the acidic endosomal environment, and vesicles enriched with LDL receptors pinch off from the endosome to be recycled back to the plasma membrane. Digestion of LDL particles occurs in the lysosome, followed by the release of cholesterol from the lysosome into the cytoplasm of the cell. This process of receptor-mediated endocytosis is used to concentrate and internalize a number of extracellular molecules. Other common features of receptor-mediated endocytosis include the recycling of both clathrin and receptor; the fusion of internalized vesicles with endosomes; and the formation of lysosomes, which are digestive organelles for materials internalized by this route.

A second endocytotic pathway exists that does not involve clathrin, and may bypass delivery of internalized material to lysosomes. In this pathway, which appears to involve both receptor-mediated endocytosis as well as the non-specific internalization of extracellular fluid, vesicles are created from non-clathrin coated invaginated membrane regions called caveoli. Caveoli may form from specialized membrane domains with distinct phospholipid contents called lipid rafts. Invagination and formation of vesicles in these areas does not require clathrin, and the vesicles formed may be transported directly to the Golgi apparatus or ER instead of the endosomes and lysosomes. Presumably this route is for material that would be damaged or degraded by exposure to lysosomal enzymes. Many cells display a constitutive formation and internalization of these non-coated vesicles in a process sometimes referred to as pinocytosis, or cell drinking (Fig. 1.3J).

### *1.3.3 Signaling Functions*

The plasma membrane serves as the interface with the cell environment and possesses a number of mechanisms to detect and transduce specific extracellular signals. Integral membrane proteins that serve as signal receptors can be categorized into three broad classes: ion channel-linked receptors, G-protein-linked receptors, and enzyme-linked receptors.

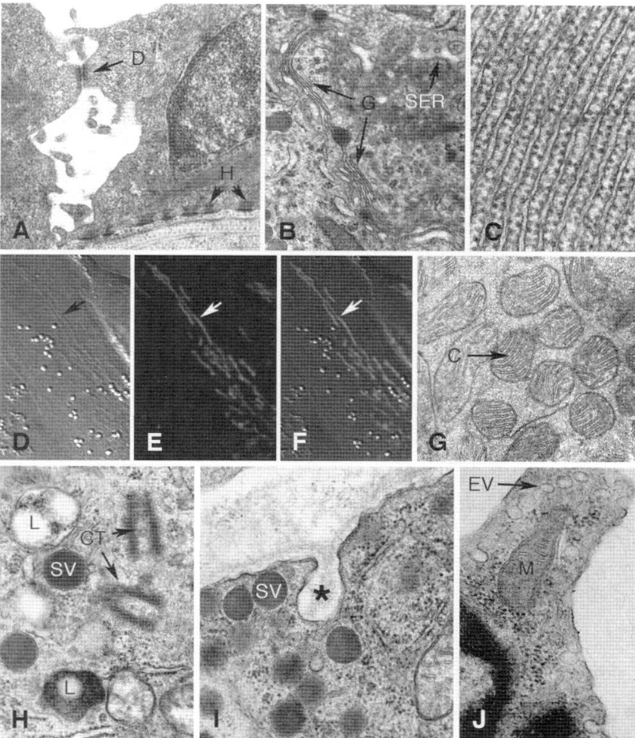


**Fig. 1.3** Micrographs illustrating the structural features of various cell organelles. **(A)** shows epithelial cells from a tadpole (*Rana pipiens*) tail; **(B)**, **(H)** **(I)**, and **(J)** show endocrine cells from a rat pituitary gland; **(C)** and **(G)** show secretory cells from digenetic trematodes (*Halipegus eccentricus* and *Quinqueserialis quinqueserialis*, respectively), and **(D)**, **(E)**, and **(F)** show melanoma cells in culture. **(A)** Low-power electron micrograph of an epidermal cell, showing a number of cell-cell and cell-extracellular matrix junctions. These cells elaborate numerous desmosomes **(D)** and hemidesmosomes **(H)**, and the cytoplasm is filled with prominent bundles of intermediate filaments, which interconnect these junctions. **(B)** Cytoplasm of an endocrine cell, showing smooth endoplasmic reticulum (SER) and flattened cisternae of the Golgi apparatus **(G)**. **(C)** A dense array of RER from a secretory cell; note the high level of organization in the parallel alignment of cisternae. Attached ribosomes appear as small granular bodies. **(D)** Light micrograph of a thin extension of cytoplasm, through which mitochondria (*arrow*) can be seen. Differential interference microscopy. **(E)** Same field of view shown in D, but photographed with fluorescence optics in order to display the elongated mitochondria, which had been labeled with the fluorescent probe tetramethylrhodamine methyl ester. **(F)** Double-exposure (E superimposed on D), showing mitochondrial organization in this region of cytoplasm. The arrows indicate the same mitochondrion in D-F. **(G)** Electron micrograph of a cluster of spherical-to-slightly elongated mitochondria. Their striped appearance is due to the invagination of the inner mitochondrial membrane, which forms the cristae (C). **(H)** Endocrine cell cytoplasm, showing two small lysosomes **(L)**, secretory vesicles (SV), and the centriole pair (CT) of a centrosome. In this section, pericentriolar material and attached microtubules are not clearly displayed. **(i)** Exocytosis in an endocrine cell. This micrograph shows the periphery of the cell, and the deep concavity of the plasma membrane (*) indicates the site where a secretory vesicle has recently undergone exocytosis. **(J)** Pinocytosis in an endocrine cell. The plasma membrane of this cell displays numerous small, smooth invaginations (*arrows*), characteristic of non-clathrin mediated internalization of material (Copies of figures including color copies, where applicable, are available in the accompanying CD)

Ion channel-linked receptors undergo conformational changes upon ligand binding, thereby opening a membrane channel permeable to small ions. Examples of this type of receptors include some types of neurotransmitter receptors. G-protein-linked receptors are integral membrane proteins that, upon ligand binding, activate small GTP-binding proteins (G-proteins), which in turn activate other effector molecules, including ion channel-linked receptors and various enzymes (e.g., adenylate cyclase). Thus, G-protein-linked receptor activity can lead to ion transients across the plasma membrane, or the generation of second messengers such as cAMP. Examples of G-protein-linked receptors include polypeptide-hormone receptors. The third class of membrane-associated signaling molecules is enzyme-linked receptors, which upon ligand binding activate an enzyme activity, which is usually either a protein kinase or guanalyl cyclase. Examples of enzyme-linked receptors are growth factor receptors, whose tyrosine kinase activity is an important regulator of the cell cycle.

In addition to the membrane receptors involved in signal transduction events, cells also possess other types of receptors not associated with the plasma membrane. For example, steroid hormones (e.g., estrogen and testosterone) are lipid soluble and pass directly through the lipid bilayer without the need to bind to proteins exposed on the external face of the plasma membrane. Receptors for these types of signaling molecules are found in the cytoplasm and nucleus.

## *1.3.4 Cell Junctions*

Specializations of the plasma membrane also help cells adhere to each other and to the ECM, form barriers against the diffusion of material between cells of an epithelium, and form channels linking adjacent cells. All of these functions are carried out by cell junctions.

### 1.3.4.1 Cell-Cell Barrier Junctions

It is important for the body to prevent the passage of material between epithelial cells. For example, a major function of intestinal and bladder epithelia is to prevent the direct passage of the contents of the intestine and bladder into the body cavity (Fig. 1.2a). The ability of epithelia to form effective barriers between different compartments is due to the presence of special cell-cell junctions called occluding or tight junctions. Tight junctions are composed of linear arrays of the integral membrane proteins occludin. Occludin binds tightly to itself, and a barrier between cells is formed as the strands of occludin encircling the apical part of adjacent epithelial cells line up and bind to each other. Epithelia that form strong barriers to intercellular leakage contain many tight junction strands, whereas more leaky epithelia generally display fewer strands. Occludin molecules are associated with other cytoplasmic proteins that appear to furnish some linkages with the cytoskeleton, although extensive cytoskeletal interactions, such as those associated with adhesive-type cell junctions (see below), are not readily apparent in tight junctions.

### 1.3.4.2 Cell-Cell Adhesive Junctions

These specializations of the plasma membrane allow cells in an epithelium to bind tightly to each other, and can be subdivided into two categories, adherens junctions and desmosomes based on the type of cytoskeletal system associated with the junctions. Adherens type cell-cell junctions form belt-like arrays encircling the apical part of epithelial cells and are associated with a thick band of microfilament. Zonulae adherens are comprised of integral membrane proteins belonging to the cadherin family of proteins, a number of linking proteins (e.g., vinculin, catenin) that connect cadherins to microfilaments, and associated microfilaments. Cadherins from adjacent cells bind tightly to each other in the presence of $Ca^{++}$, and $Ca^{++}$ chelating agents promote cell dissociation. Interestingly, β-catenin is able to function as an adherens plaque protein, and as a transcription factor in the nucleus. Thus, β-catenin may serve as a sensing device that translates changes in cell-cell adhesion into changes in gene activity.

Desmosomes are punctate cell-cell adhesive junctions that are associated with the intermediate filament (IF) cytoskeleton (Figs. 1.2b and 1.3a). Like adherens junctions, they are also composed of cadherin integral membrane proteins and proteins that link cadherins to IFs. One type of desmosomal cadherin is a protein called desmoglein, and the major linking protein of desmosomes is a member of the plakin family of proteins called desmoplakin. Desmosomal intermediate filaments form dense bundles that interconnect desmosomal plaques, thus strengthening cell-cell attachment and contributing to the mechanical integrity of the epithelium.

#### 1.3.4.3 Cell-Cell Communicating Junctions

Two types of cell-cell junctions, gap junctions and synapses, constitute specializations of the plasma membrane that allow cells to communicate with each other. Gap junctions are punctate structures that electrically couple cells through channels that provide for direct cytoplasmic communication between adjacent cells. Gap junctions are made up of clusters of pore-like structures, called connexons that span the lipid bilayer and allow for the passage of molecules smaller than 1000 daltons. Connexons, made up of hexameric arrays of the integral membrane protein connexin, line up between adjacent cells to form continuous, tightly sealed channels from cell to cell. These connections maintain a barrier against leakage of material to or from the extracellular compartment, but ions and small molecules can diffuse from cell-to-cell to permit electrical coupling. Electrical coupling of gap junctions performs vital functions in propagating the contractions of cardiac muscle. The conformation of connexons is regulated by $Ca^{++}$ such that they remain open in low concentrations of $Ca^{++}$, but close down in the presence of $Ca^{++}$.

In addition to the direct coupling of cells at gap junctions, neurons also communicate with each other at synapses. At synapses, cells release neurotransmitters in quantal fashion through the regulated exocytosis of membrane-bound vesicles. Neurotransmitters rapidly diffuse across a narrow extracellular space to bind to specific receptors on the plasma membrane of the adjacent cell. These receptors are either ion channels or, in some cases, G-protein-linked receptors. When stimulated by a neurotransmitter, the ion channels open, allowing $Na^{+}$ to enter the cell by diffusing down its concentration gradient, thereby depolarizing the plasma membrane. G-protein-linked receptors that bind neurotransmitters release activated G-proteins that may subsequently activate and open other ion channels. The depolarization of the plasma membrane is conducted down the cell body of the stimulated cell, and can trigger the release of neurotransmitters at synaptic junctions between the stimulated cell and other adjacent cells. In this way, signaling activity is propagated between cells connected by synapses.

#### 1.3.4.4 Cell-ECM Adhesive Junctions

Cells elaborate two types of cell-ECM junctions that assist in their adherence to the ECM. Hemidesmosomes anchor epithelial cells to underlying connective tissue and are associated with IF cytoskeletal fibers, whereas focal contacts can be formed by many types of cells and involve microfilament-associated linkages with the ECM. Hemidesmosomes resemble half-desmosomes (Fig. 13a): integrins form the integral membrane component and are linked to IFs by members of the plakin family of proteins. Integrins are also the integral membrane proteins of focal contacts and are connected to bundles of microfilaments by vinculin, talin, and other linking proteins. Focal contacts are associated with protein kinases, called focal adhesion kinases (FAKs). FAKs are thought to help transmit the status of cell-ECM linkage at focal contacts to the cytoplasm and nucleus. Normal (non-cancerous) cells must usually be in contact with a substrate to divide, and FAKs are involved in relaying contact information to the regulatory elements of the cell cycle.

### *1.3.5 Cell Protection and Cell Identity*

Many proteins and lipids of the plasma membrane possess covalently linked sugar groups. These sugar groups are asymmetrically distributed and are present exclusively on the ECM side of the membrane. Plasma membrane glycoproteins and glycolipids serve important roles in both protecting the membrane and in identifying specific cell types. Epithelial cells lining the small intestine elaborate a thick glycocalyx that helps to protect them against the harsh digestive conditions of the intestinal lumen. Protruding sugar groups may help protect the membrane components by limiting access of hydrolytic enzymes to the surface of the membrane. Examples of cell recognition processes involving glycoproteins and glycolipids include the patterns of antigens responsible for blood groupings. Another example involves the initial adhesion of neutrophils to endothelial cells in areas of inflammation. During inflammation, endothelial cells express the integral membrane protein P-selectin, which possesses a lectin domain that recognizes a four sugar group (N-acetylglucosamine, galactose, fucose, and sialic acid) present on glycoproteins and glycolipids of neutrophils. Neutrophils adhere to endothelial cells expressing P-selectin, which facilitates their subsequent migration past the capillary bed to reach the sites of inflammation.

## 1.4 Endoplasmic Reticulum

The endoplasmic reticulum (ER) is a prominent organelle in most cells, and its total membrane area can constitute more than half of all the membrane of a cell. ER membranes delimit enclosed spaces that vary in form from flattened sheets, or cisternae, to branching tubules or distended sacs; the total enclosed luminal space can occupy 10% or more of the cell volume. A number of distinct functions are carried out by the ER. The ER is the primary site of synthesis of membrane lipids and integral membrane proteins for all membranous organelles (ER, Golgi, lysosomes, endosomes, secretory vesicles, and plasma membrane) except mitochondria and peroxisomes. It is also the site of production of secreted proteins and luminal proteins of ER, Golgi, and lysosomes. In addition, the ER functions in lipid synthesis, detoxification and $Ca^{++}$ regulation.

ER can be categorized as either rough (RER) or smooth (SER). RER is so designated due to the presence of numerous ribosomes bound with the cytoplasmic surface of the cisternal membranes (Fig. 1.3c), whereas SER lacks associated ribosomes (Fig. 1.3b). These different forms of ER are specialized for different functions; RER is the site where integral, luminal, and secretory proteins are synthesized, whereas SER is the major site of detoxification and lipid synthesis.

### *1.4.1 Protein Synthesis*

Protein synthesis takes place on ribosomes, which are either present freely in the cytoplasm, or attached to membranes of the ER (forming RER). Cytoplasmic proteins (e.g., cytoskeletal proteins) are synthesized by free ribosomes, whereas proteins associated with membranes (including the plasma membrane) or the luminal compartments of membrane-bound organelles, as well as proteins destined for secretion, are synthesized by RER. The lipid components of membranes are also made by the ER, and both protein and lipid are delivered to the plasma membrane and most organelles by membrane flow involving transport and fusion of membrane-bound vesicles between ER, the Golgi apparatus, and other target organelles. Exceptions to this pattern of membrane biogenesis and renewal include mitochondria and peroxisomes. Interestingly, most proteins of mitochondria, and all peroxisomal proteins, are made by free ribosomes, and subsequently delivered to these organelles via transport mechanisms that move individual proteins into or past their membranes. Membrane lipids are delivered to these organelles by transport proteins that extract lipid from ER membranes and insert them in the membranes of mitochondria and peroxisomes.

Whether ribosomes remain free in the cytoplasm or are bound to ER is determined by the amino acid sequence of the polypeptide chain as it emerges from the ribosome. ER-associated proteins possess a signal sequence that functions in docking the polypeptide to the membranes of the ER.

The signal sequence is recognized and bound by a signal recognition particle, or SRP. The SRP in turn binds to a SRP receptor in the membrane of the ER. A protein translocator apparatus, which forms a pore in the ER membrane through which growing polypeptide chains can pass, also associates with the SRP and SRP receptor, and receives the protein as translation proceeds. Thus, proteins with signal sequences are injected directly into the membrane of the ER as they are synthesized. The signal sequence, which is hydrophobic, remains inserted into the lipid bilayer while the rest of the protein spools past as it elongates. The relative orientation of the signal sequence influences whether the N- or C-terminus of the polypeptide is threaded into the ER lumen. Soluble luminal proteins spool all the way through the bilayer, and a signal peptidase then clips the protein at the signal sequence, liberating the protein into the lumen. Single- and multi-pass membrane proteins are thought to achieve their conformations by internal stop-transfer and start-transfer sequences, which interact with the bilayer to either promote the passage of the growing polypeptide chain through the bilayer (start-transfer sequences), or halt the transmembrane passage (stop-transfer sequences). Multiple start- and stop-transfer sequences therefore can result in a polypeptide chain that doubles back and forth to penetrate the bilayer at multiple points, forming loops in both the cytoplasm and ER lumen.

Because protein function reflects the three-dimensional shape, the polypeptide chain as it folds upon itself adopts a thermodynamically favorable conformation. Protein synthesis in the ER is associated with a robust quality control system that monitors the presence of proteins that fold incorrectly during their production and maturation. Chaperone proteins are proteins whose function is to help newly synthesized proteins fold correctly to adopt their appropriate conformation. Should errors in folding still occur, a number of enzymes function to detect and retain misfolded proteins in the ER, so that subsequent attempts at re-folding can be made. Ultimately, proteins that remain misfolded are exported from the ER and marked for destruction by the enzymatic addition of ubiquitin groups. Proteins with attached ubiquitin molecules are recognized by proteosomes – complexes of proteolytic enzymes that recognize, hydrolyze, and destroy ubiquinated proteins.

A number of post-translational modifications of proteins occur in the ER, as well as in the Golgi apparatus. Disulfide bonds form, and many proteins are glycosylated, or may have glycolipid anchors added. Glycosylation is carried out by the initial assembly of sugars into polymeric structures attached to the membrane lipid dolichol. The assembled carbohyhdrate group is then transferred intact from dolichol to the protein molecule. The glycoprotein may be processed in the ER by trimming of some sugars, and addition of others. Further processing of glycoproteins, and the formation of glycolipids, are the major functions of the Golgi apparatus.

### *1.4.2 Lipid Synthesis*

In addition to the synthesis of proteins for membranes, lysosomes, and secretory vesicles, the ER is also responsible for the synthesis of most membrane lipids for all organelles, including mitochondria and peroxisomes. Enzymes involved in lipid synthesis are embedded in the membrane of the ER, with their active sites facing the cytoplasm. Fatty acids are added to the glycerol phosphate to form phosphatidic acid, which then receives various head groups. Phosphatidylcholine, phosphatidylserine, and phosphatidylethanolamine are formed in this way and initially added to the cytoplasmic leaflet of the lipid bilayer. Phospholipid translocators (flippases and scrambleases) are present in the membrane, and transfer phospholipids between each leaflet of the bilayer. Scrambleases do not exhibit substrate specificity, and help keep the total numbers of phospholipid molecules approximately equal between the two layers. In contrast, some flippases preferentially transfer choline-containing phospholipids from the cytoplasmic half to the luminal half of the lipid bilayer, thereby promoting membrane asymmetry in the distribution of these lipids. Sphingomyelin synthesis is more complex; serine is first attached to fatty acids to form ceramide, which is exported to the Golgi apparatus, where phosphocholine head groups are added. Mitochondria and peroxisomes appear to receive their membrane lipid from the ER through the activity of phospholipid exchange proteins, which transfer phospholipids between membrane systems by extraction and insertion of individual lipid molecules within bilayers. In addition to membrane lipid synthesis, the ER plays important roles in other aspects of lipid metabolism. For example, steroid hormones are synthesized from the cholesterol in the SER.

### *1.4.3 Detoxification*

Harmful substances that are relatively insoluble are difficult to clear from the cell. Such substances can occur either as environmental contaminants or as products of metabolism. SER contains a variety of enzymes that are able to process insoluble toxicants to make them more water soluble and amenable for excretion. The best-studied detoxification enzymes are the members of the cytochrome P450 family of enzymes. Liver hepatocytes are among the most active cells involved in detoxification reactions, and contain large amounts of SER to house the P450 enzymes. The quantity of SER within a cell can fluctuate in response to different levels of exposure to toxic compounds.

### *1.4.4 $Ca^{++}$ Regulation*

The ER membrane contains $Ca^{++}$-ATPases that actively transport cytoplasmic $Ca^{++}$ into the ER lumen. This activity keeps cytoplasmic $Ca^{++}$ levels very low, which is necessary to allow $Ca^{++}$ to effectively function as a signaling molecule. In electrically excitable cells, depolarization of the plasma membrane promotes influx of $Ca^{++}$ from outside the cell; in non-excitable cells, however, most of the $Ca^{++}$ released into the cytoplasm comes from the ER. ER membranes contain $Ca^{++}$ release channels that are activated by inositol triphosphate (IP3), a signaling molecule released by the activation of certain G-protein-linked receptor proteins at the plasma membrane. The contraction of muscle cells is triggered by $Ca^{++}$, and these cells possess an extensive and specialized SER system (the sarcoplasmic reticulum) that contains a second type of $Ca^{++}$-release channel in the SER membrane. After release from the ER, $Ca^{++}$ concentrations are lowered in the cytoplasm by the activity of plasma membrane and ER pumps.

## 1.5 Golgi Apparatus

The Golgi apparatus functions in part as the 'post office' of a cell, packaging and directing different types of cargo from the ER to different organelles and the plasma membrane. In addition to packaging and targeting membrane-associated proteins and lipids, as well as secreted proteins, to their appropriate destinations, the Golgi apparatus modifies certain proteins and lipids received from the ER. In addition, lipid biosynthesis also occurs in the Golgi membranes. Glycolipids are formed in the Golgi by the addition of oligosaccharide chains to ceramide, and processing of glycoproteins continues in the Golgi.

The Golgi apparatus is made up of a set of flattened, membrane-bound cisternae and associated tubulovesicular elements and membrane-bound vesicles in the process of being transported between ER, Golgi, and other locations (Fig. 1.3b). The stacks of Golgi cisternae are biochemically distinct, and the entire stack is polarized, so that a *cis*, or entry face, and a *trans*, or exit face, exist. The *cis* face lies adjacent to ER and is the site of vesicular traffic back and forth between the ER and Golgi. The *trans* face is the site of formation of a number of types of vesicles, which convey material to the plasma membrane, produce secretory vesicles, and form lysosomes. Integral membrane proteins, membrane lipids, and soluble cisternal protein formed by the ER traverse the Golgi and are targeted to their appropriate destinations by this organelle. Three major routes of export from the Golgi occur: (i) constitutive delivery of membrane-bound vesicles to the plasma membrane; (ii) formation of secretory vesicles whose exocytosis is regulated; and (iii) formation of lysosomes. Specific targeting signals are associated with the formation of lysosomal vesicles and secretory vesicles; however, the pathway from the Golgi apparatus to the plasma membrane appears to be largely constitutive and unregulated, forming a default pathway.

Proteins destined to be secreted in a regulated manner (e.g., release of hormones from endocrine cells) are concentrated and packaged in large membrane-bound vesicles in the *trans* Golgi. These secretory vesicles are stored in the cytoplasm until signals to fuse with the plasma membrane are received, thereby resulting in the liberation of their contents outside the cell. Targeting mechanisms exist that direct secretory vesicles to the appropriate cellular location for release. For example, some secretory vesicles are released from the apical plasma membranes of epithelial cells, whereas others fuse with basolateral membranes.

Proteins destined for lysosomes are tagged with mannose-6-phosphate (M6P) groups in the Golgi. M6P receptors are present in Golgi membranes, and vesicles containing lysosomal proteins bound to M6P receptors bud off from the Golgi and fuse with endosomes to form mature lysosomes. During this process, the M6P receptors are recycled for repeated use by vesicular trafficking from endosome to Golgi apparatus.

## 1.6 Lysosomes

Lysosomes (Fig. 1.3h) are the digestive organelles of the cell and are filled with acid hydrolases that are most active at a pH of about 5. Lysosomal vesicles from the Golgi apparatus fuse with endosomes that have received materials from endocytotic vesicles. Endosomes have a moderately acidic pH (about 6) that promote dissociation of ligands from the internalized plasma membrane receptors as well as dissociation of acid hydrolases from M6P receptors. Both types of receptors are recycled by being routed back to the plasma membrane and the Golgi apparatus in membranous vesicles that are pinched off from endosomes. The endosome then matures to form a lysosome by condensing into a spherical or an irregular membrane-bound structure. Proton pumps in the membrane of the maturing lysosome lower the pH inside the organelle to maximally activate the acid proteases to digest the internalized material. Other transporters exist in the lysosomal membrane to allow digested organic molecules to enter the cytoplasm for use by the cell.

In addition to the confluence of lysosomal and endocytotic vesicles at endosomes, materials can be targeted for lysosomal degradation by at least two other mechanisms. Neutrophils and macrophages are cells specialized for the engulfment of bacteria and other large particulate material, which are then internalized by phagocytosis. Lysosomes fuse with these large phagocytotic vesicles to deliver their hydrolases, resulting in the formation of phagosomes. Lysosomes also contribute to the breakdown of cellular material that is not needed or should be turned over. Excess, old, or malfunctioning organelles can be targeted for destruction by being enveloped by cisternae of ER, which then fuse with lysosomal vesicles to form autophagosomes. Recently, evidence has been gathered suggesting that a fourth route to lysosomes may exist that involves the transport of single cytoplasmic molecules through the lysosomal membrane by specific transport proteins.

## 1.7 Membrane Flow

The synthesis, maturation, and ultimate localization of secreted, membrane, and lysosomal proteins, as well as membrane lipid, is reflected in a complex but effective and elegant flow of membranes and organic molecules between the various organelles and cytoplasmic compartments of eukaryotic cells. With the exception of mitochondria and peroxisomes, the membranes of all organelles, vesicles, and the plasma membrane are initially produced by the ER, modified and packaged by the ER and Golgi, then targeted and delivered via the trafficking of membrane-bound vesicles. A common theme in this vesicular trafficking is 'priming' of the donor membrane by the binding of small GTPase proteins, which then recruit proteins that form a polymeric coating on the membrane, which causes the invagination and leads to vesicle formation. The coat proteins usually dissociate from the vesicle shortly after formation, and other membrane-associated proteins mediate the recognition, binding, and fusion of vesicles with the target membranes. The three best understood coat proteins involved in vesicle formation and trafficking include COPI (coat protein I), COPII, and clathrin (described above under *Transport Involving Membrane Flow*). Trafficking between the ER and Golgi is mediated by COPI and COPII, which either direct vesicles from the ER to the Golgi (COPII) or direct vesicles from the Golgi to the ER (COPI). For many ER-resident proteins, there appears to be little restriction on their flow from the ER to the Golgi. However, these ER-resident proteins contain the amino acid sequences KDEL and KKXX (where X is any amino acid) which mark luminal and integral membrane proteins respectively, for return transport from the Golgi to the ER in COPI-coated vesicles. Thus, proteins with these sequences are essentially restricted to the ER by being rapidly returned from the Golgi apparatus. The exact mechanisms of how membrane flows through the Golgi apparatus is still a matter of some debate, but at least part of the flow appears to be carried out by membrane-bound vesicles and COPI. The flow of membrane from Golgi to the plasma membrane includes a constitutive, default pathway that operates in the absence of specific targeting signals. However, delivery of material to secretory vesicles and lysosomes (Figs. 1.3h and I) requires defined targeting information. Interestingly, clathrin is involved in the formation of lysosomal and secretory vesicles from the trans Golgi, but not in the constitutive formation of vesicles destined for the plasma membrane. In addition to the ER and Golgi apparatus, bi-directional membrane flow also occurs between plasma membrane and endosome, and plasma membrane and Golgi. Clathrin-coated endocytotic vesicles from the plasma membrane travel inward to fuse with endosomes, and the receptors subsequently return to the plasma membrane via small vesicles formed from endosomal membranes. Endocytotic vesicles formed from caveolae may bypass lysosomes and fuse directly with Golgi or ER. In general, the outward pathway of membrane flow from ER-to Golgi-to plasma membrane is balanced by the inward flow from plasma membrane-to endosome-to lysosome, and plasma membrane-to Golgi/ER. Both outward and inward trafficking, fusion of vesicles with the appropriate target membrane is carried out through the functioning of SNARE proteins. SNAREs are transmembrane proteins present in both vesicle (v-SNARE) and target (t-SNARE) membranes. The cytoplasmic extensions of v- and t-SNARES recognize and bind each other, and promote the fusion of the vesicle and target membranes.

## 1.8 Mitochondria

The primary function of mitochondria is to convert energy sources into forms that can be used to drive cellular reactions. Not surprisingly, they comprise a significant volume of the cell, normally about 20% of the total cell mass. Mitochondria replicate by a process involving growth and fission of pre-existing mitochondria. Interestingly, mitochondria contain their own DNA in the form of a circular chromosome that resembles a prokaryotic chromosome. Based on this and other lines of evidence, it appears that mitochondria arose by the prokaryotic colonization of eukaryotic cells, early in their evolution. Although mitochondria contain DNA and are able to carry out transcription and translation, they only produce about 5% of their protein, the rest being encoded by nuclear genes, synthesized by cytoplasmic ribosomes, and subsequently transported and inserted into mitochondria. They also appear to obtain most of their membrane lipid from the ER, mediated by phospholipid transfer proteins that shuttle these molecules from the ER to mitochondria.

Cells use ATP as their primary energy source, and the main function of mitochondria is the production of ATP from food sources. Energy is obtained from the oxidation of food material by the sequential transfer of electrons to lower energy states; the released energy from this process is used to drive membrane-bound proton pumps, thus establishing an electrochemical gradient. Protons are then allowed to flow back across the membrane down their concentration gradient, and the released energy is used to drive the synthesis of ATP from ADP and Pi. The electrons are ultimately transferred to $O_2$, and the entire process is therefore referred to as oxidative phosphorylation.

### *1.8.1 Mitochondrial Structure*

Mitochondria are composed of two membranes that enclose distinct compartments. The outer mitochondrial membrane is somewhat permeable to small molecular weight compounds ($< 5000$ daltons); conversely, the inner membrane contains a very high ratio of protein to lipid, and movement of material past this membrane is tightly regulated. The space between the two membranes is called the intermembrane space, and the compartment delimited by the inner membrane is called the mitochondrial matrix. The inner membrane is folded into sheet- or tube-like invaginations into the matrix, thus increasing its surface area (Fig. 1.3G). The increased surface area of this membrane allows mitochondria to house greater numbers of electron transport systems and ATP synthase complexes. The matrix is the site of conversion of pyruvate and fatty acids to acetyl CoA, and the location of the citric acid cycle, where acetyl CoA is oxidized.

### *1.8.2 Chemiosmotic Generation of ATP*

With respect to energy production, the pathways of carbohydrate and lipid metabolism converge in the generation of acetyl CoA in the mitochondrial matrix. Carbohydrate is converted to glucose 6-phosphate, which, as a substrate for glycolysis, gives rise to two pyruvate molecules. Pyruvate is transported to the mitochondrial matrix were it is converted to acetyl CoA by the pyruvate dehydrogenase complex. Fatty acids are oxidized in the mitochondrial matrix, releasing acetyl groups that are then linked to CoA. Acetyl CoA derived from carbohydrate and fatty acid metabolism is fed into the citric acid cycle, resulting in the production of $CO_2$ and NADH. NADH carries electrons to the electron transport chain, which is located in the inner mitochondrial membrane. The electron transport chain is complex, being composed of about 40 different proteins. The actual transfer of electrons is carried out by a number of different heme groups linked to various cytochromes, iron-sulfur centers-containing proteins, ubiquinone, copper atoms, and a flavin. These are organized into three large enzyme complexes, with ubiquinone and cytochrome C serving as carriers of electrons between the complexes. The three complexes are the NADH dehydrogenase complex, the cytochrome b-c1 complex, and the cytochrome oxidase complex. Electrons shuttled across these complexes move to progressively lower energy states, with the released energy used to pump $H^+$ from the matrix to the intermembrane space. Then, $H^+$ is allowed to flow down its concentration gradient (from the intermembrane space to the matrix), through the ATP synthase complex. This is a large transmembrane complex of about 500,000 daltons that contains multiple proteins, and constitutes about 15% of the total inner membrane protein. The energy that is released is used to couple Pi to ADP to make ATP. Finally, the electrons used to drive the $H^+$ pumps are combined with $O_2$ and $H^+$. Thus, the generation of ATP from electron transport in mitochondria by this chemiosmotic mechanism consumes $O_2$ and produces water.

### *1.8.3 Other Mitochondrial Functions*

In addition to converting food energy into ATP, mitochondria are also involved in a number of other functions, including $Ca^{++}$ regulation and apoptotic signaling. Like the ER, mitochondria are able to sequester $Ca^{++}$ to help keep cytoplasmic levels low. In addition to producing ATP, the $H^+$ gradient can be used to import $Ca^{++}$ into the mitochondrial matrix. $Ca^{++}$ is important in regulating the activity of certain mitochondrial enzymes. Deposits of calcium can be formed in mitochondria in response to high cytoplasmic levels of $Ca^{++}$.

In addition to roles in ATP production and $Ca^{++}$ homeostasis, mitochondria are centrally involved in the regulation of programmed cell death, or apoptosis. Programmed cell death is a normal and essential component of embryogenesis, where it helps to sculpt a number of morphogenetic events, such as digit formation and maturation of neuronal circuitry patterns. Furthermore, apoptosis is involved in the maintenance of tissues and organs by the orderly turnover of cells, and in the removal of injured and damaged cells. A central mechanism by which apoptosis is carried out involves the activation of the caspase family of cystine proteases in the cytoplasm of cells. Release of cytochrome C by mitochondria can facilitate caspase activation, and involves the association of this cytochrome with other molecules (e.g., APAF-1) and with pro-caspases to activate the caspase cascade. Release of cytochrome C from mitochondria can occur in response to specific membrane signaling events, as well as from cytoplasmic and nuclear damage. Thus, positive feedback mechanisms exist between the plasma membrane, cytoplasmic, and nuclear signaling, where mitochondria can function in an "amplification loop." Among the triggers that can initiate apoptotic cell death is cellular damage resulting from the generation of reactive oxygen species (ROS). ROS are toxic by-products of the oxidative phosphorylation reactions carried out by mitochondria, and these organelles thus appear to be quite susceptible to ROS-linked damage. It has been proposed that a variety of age-related disorders, including some types of cancer, are fundamentally linked to mitochondrial damage resulting from ROS.

A number of imaging techniques have demonstrated that mitochondria are highly dynamic organelles, continuously undergoing fission and fusion in many types of cells. They can be viewed as forming a network which is in a constant state of flux; what is an individual mitochondrion one moment may fuse with, and become part of, an extended reticular array the next. Similarly, areas of a continuous, branching, mitochondrial network can rapidly fragment to form a number of individual mitochondria of different shapes and sizes (Figs. 1.3D-F). Intriguingly, there is increasing evidence that the ability of mitochondria to undergo fusion and fission is linked to their functional capacity. A number of studies have shown that inhibiting these processes limits the ability of mitochondria to produce ATP and may make the cell more susceptible to the induction of apoptosis.

## 1.9 Peroxisomes

Peroxisomes are membrane-bound vesicular organelles that have been classically understood to be involved in lipid metabolism and a number of oxidative reactions of the cell. This contains high concentrations of enzymes that are able to form $H_2O_2$ by the transfer of hydrogen atoms from substrates to molecular oxygen. In addition, peroxisomes contain catalase, which breaks down $H_2O_2$ to oxidize various substrates, including toxins. Like mitochondria, peroxisomes replicate by fission and growth of pre-existing organelles. All the proteins and lipids of the peroxisome is made in the cytoplasm, and subsequently imported into the peroxisomal membrane and lumen. Interestingly, recent evidence suggests that peroxisomes are also involved in modulating the functioning of signaling lipids that are involved in developmental events, including cell differentiation and morphogenesis.

## 1.10 The Cytoskeleton

The cytoskeleton of eukaryotic cells is composed of a complex system of proteinaceous filaments and filamentous arrays that are present in both cytoplasm and nucleus. Three major cytoskeletal systems are elaborated by cells, and include microfilaments, microtubules, and intermediate filaments. These different cytoskeletal systems are biochemically and functionally distinct.

### *1.10.1 Microfilaments*

Microfilaments are small solid filaments about 6 nm in diameter, composed of the 45 kDa globular protein actin. Actin is one of the most abundant proteins in cells, composing up to 5% or more of the total cell protein. Microfilaments help support and organize the plasma membrane and are involved in cell motility and maintenance of cell shape, serving as the muscle of the cell.

A large number of actin-associated proteins mediate the functions of microfilaments, including regulating actin polymerization (e.g., profilin, WASp, and ARP), cross-linking microfilaments to form organized arrays (e.g., filamin, fimbrin, and villin), interacting with membrane proteins to establish and maintain distinct membrane domains (e.g., vinculin, catenins, and Z0-1 of tight junctions), and functioning as motor proteins (type I and II myosins) to carry out motility events. Much of the actin in cells is present as soluble monomers (g-actin) bound to profilin. This interaction inhibits the polymerization of actin monomers into filaments (f-actin). Signals from the plasma membrane, mediated in large part by small GTP-binding proteins (e.g., rac, rho, and cdc42), activate WASp and ARP proteins, which promote the dissociation of profilin from g-actin and seed the growth of new microfilaments from the sides of pre-existing microfilaments. Microfilament polymerization is controlled by the addition of capping proteins to the end of the growing filaments, and turnover of filaments occurs by the actions of microfilament cutting proteins such as gelsolin, followed by depolymerization of f-actin to g-actin and association of the latter with profilin.

Although some cell movements and membrane activities appear to be driven by the polymerization and depolymerization of actin, many other types of motility events require the interaction of microfilaments with myosin motor molecules. Myosin functions as an actin-activated ATPase, undergoing cycles of microfilament attachment and detachment with associated conformational changes, resulting in power strokes. These events are linked to the binding, hydrolysis, and release of ATP, Pi, and ADP. Myosin activity can slide microfilaments past each other, transport vesicles and other cargo down microfilaments, and deform membranes that are linked to microfilaments.

A number of mutations are known that interfere with microfilament functioning. WASp protein was discovered as the protein mutated in Wiscott-Aldrich syndrome, which results from a deficiency in the ability of actin to polymerize. Dystrophin is a large linking molecule that connects submembrane arrays of microfilaments to integral membrane proteins and ECM proteins of skeletal muscle cells. Mutations in dystrophin that interfere with its ability to link microfilaments to the plasma membrane weaken the plasma membrane, causing the eventual death of the muscle cell and giving rise to some forms of muscular dystrophy.

### *1.10.2 Microtubules*

Microtubules are small proteinaceous tubes about 25 nm in diameter, composed of the protein tubulin. Microtubules function to organize the cytoplasm and mediate intracellular motility events. They are associated with motor proteins that interact with membrane-bound organelles and vesicles to help determine their location and organization within the cytoplasm. They also carry out crucial functions in cell division, forming the spindle apparatus that segregates replicated chromosomes among the daughter cells. Microtubules also support and power cilia and flagella, which are highly motile appendages produced by ciliated epithelial cells and sperm.

Unlike microfilaments and intermediate filaments IFs (see below), microtubules are associated with a distinct organizing center, called the centrosome or MTOC (microtubule-organizing center). The centrosome is composed of a pair of centrioles (Fig. 1.3H) surrounded by an amorphous mass of pericentriolar material. Centrioles themselves are short, barrel-like arrays of microtubules and are associated with the ability of the centrosome to replicate. Pericentriolar material both nucleates microtubule growth and anchors the ends of microtubules. Three types of tubulin genes are expressed in eukaryotic cells, including α-, β-, and γ-tubulin. Microbules are composed of heterodimers of α- and β-tubulin, and γ-tubulin is a component of the pericentriolar material that participates in the nucleation of microtubule growth. Microtubules possess an intrinsic polarity and are all oriented, so one end (the minus end) is anchored in the pericentriolar material, with the free end (the plus end) extending into the cytoplasm. Microtubule polymerization and depolymerization occurs at the plus end, and involves addition or removal of α-β heterodimers in a GTP/GDP-controlled process. Tubulin heterodimers are GTP-binding proteins that also possess GTPase activity. Heterodimers associated with GTP readily polymerize, whereas hydrolysis of tubulin-bound GTP to GDP weakens the heterodimer interconnectivity and favors depolymerization. Prior to assembly, heterodimers of α- and β-tubulin are associated with GTP; at some point after polymerization, the tubulin-bound GTP is hydrolyzed to GDP. As long as the terminal tubulin subunits at the plus end are associated with GTP, the microtubule will grow by the addition of GTP-containing heterodimers. Occasionally, however, the GTPase activity catches up with a growing end of the microtubule, hydrolyzing GTP to GDP in the terminal subunits. At this point, the microtubule rapidly disassembles, shrinking in size back toward the centrosome. Depolymerizing microtubules can be rescued and re-grown if sufficiently high concentrations of GTP-containing tubulin heterodimers are present so that the GTP cap can be re-established. Because of these events, most microtubules in the cell continuously oscillate between slow growth and rapid depolymerization, a feature that has been called dynamic instability.

Like the microfilament cytoskeleton, the organization and functions of microtubules are regulated and carried out by associated proteins. Microtubule-associated proteins are generally categorized as either structural proteins or motor proteins. Structural proteins include higher molecular weight proteins called MAPs (microtubule-associated protein) and lower molecular weight tau proteins. Structural MAPs and tau proteins are thought to help organize microtubule arrays in the cytoplasm. Microtubule-associated motor proteins include dyneins and kinesins, both of which, like myosin, undergo cycles of binding, conformational changes, and dissociation in an ATP-dependent manner to move microtubules past each other, or to move cargo along microtubules. Microtubule-mediated intracellular transport is carried out by multiple members of both the dynein and kinesin family of proteins, but ciliary and flagellar motility selectively utilize dynein. The microtubule bundle, or axoneme, supporting a cilium is anchored in a specialized centrosome called a basal body. Unlike regular centrosomes, axoneme microtubules originate as direct extensions from one of the centrioles in a basal body, and not from associated pericentriolar material. Axoneme microtubules form circular arrays of doublets surrounding a central pair of microtubules. The outer microtubule doublets are associated with dynein, which spans to the adjacent microtubule pairs. Dynein motor activity attempts to slide microtubule pairs past each other, which is converted into a bending of the cilium because the bases of the microtubule pairs are attached to the basal body and not free to slide. In this way, hydrolysis of ATP by dynein powers the rapid, whip-like movements of cilia and flagellae in a microtubule-dependent manner. Other forms of microtuble-mediated motility occur in the cytoplasm, where membrane-bound vesicles, organelles, and other cargo associated with dynein or kinesin move along microtubules. Motor proteins exhibit a directionality, which, combined with the fact that microtubules are polarized, allows for vectorial transport of material within the cell. Dyneins move cargo towards the minus ends of microtubules, and kinesins generally move cargo towards the plus ends of microtubules (although minus end-directed members of the kinesin family are known).

In dividing cells, centrosomes replicate along with DNA in S-phase, and subsequently participate in the formation of the mitotic spindle. Daughter centrosomes move apart and promote the complete reorganization of the microtubular cytoskeleton prior to, and during, nuclear envelope breakdown. The plus ends of microtubules radiating from the centrosomes, now called spindle poles, bind and become stabilized by the kinetochores of chromosomes, forming bundles of kinetochore microtubules. Microtubule-mediated motility events align chromosomes on the metaphase plate and are subsequently responsible for the separation of daughter chromosomes in the anaphase of mitosis. Pinching of the cell into two daughter cells (mediated by bundles of microfilaments in association with the plasma membrane at the cleavage furrow) leads to the inheritance of the correct number of chromosomes and a single spindle pole, which becomes the centrosome in each daughter cell. Interestingly, other organelles such as mitochondria and peroxisomes appear to be randomly apportioned to each daughter cell by virtue of the fact that they are distributed throughout the cytoplasm, whereas the ER and Golgi apparatus vesiculate and disperse throughout the cytoplasm early in mitosis to be inherited in the same way.

A number of drugs interfere with microtubule dynamics; examples include colchicine, which actively promotes tubulin depolymerization, and taxol, which stabilizes microtubules and inhibits depolymerization. Both of these drugs are toxic to cells, indicating that the oscillation between polymerization and depolymerization is crucial to microtubule function and cell health. A number of taxol-based compounds have been developed for use in the chemotherapeutic treatment of cancer, highlighting the importance of microtubule dynamics in cell division.

### *1.10.3 Intermediate Filaments*

Intermediate filaments (IFs) are solid filaments about 10 nm in diameter, made up of one or more of a large family of IF proteins. IFs are found in both the cytoplasm and the nucleus. They function in strengthening the cytoplasm of cells, as well as in mechanically integrating cells of a tissue by interconnecting desmosomes and hemidesmosomes.

The IF family of proteins is the most complex family of cytoskeletal proteins, with over 50 different IF gene products elaborated by cells of higher vertebrates. IF proteins can be divided into five groups: (i) acidic keratins, (ii) neutral/basic keratins, (iii) vimentin-like proteins, (iv) neurofilament proteins, and (v) lamins. IF proteins are expressed in tissue-specific patterns, with epithelial cells containing keratins, cells of mesenchymal origin expressing vimentin-like IF proteins, and neuronal cells expressing neurofilament IF proteins. Lamins are present in essentially all nucleated cells, and form a filamentous network underlying and supporting the inner membrane of the nuclear envelope. There is evidence that lamins help organize chromatin and are involved in some aspects of DNA synthesis. Interestingly, mutations in lamins have been found that give rise to some forms of muscular dystrophy, lipidodystrophies, and progeria (premature aging) diseases.

IF proteins are long, rod-like molecules that contain a central domain, rich in alpha helices. The rod domain of IF proteins coil around each other to form coiled coil dimers, which then associate into higher order structures, much like the weaving together of strands to form a rope. The polymerization state of IF proteins is mediated by phosphorylation, and hyperphosphorylation of IF proteins appears to lead to dissociation of IFs by repulsion of subunits bearing multiple negative phosphate charges. One of the best studied examples is the depolymerization and repolymerization of nuclear lamina filaments during cell division. At the onset of mitosis, lamins are hyperphosphorylated and the nuclear lamina depolymerizes, facilitating nuclear envelope breakdown. At the end of mitosis, lamins are dephosphorylated and a nuclear lamina and nuclear envelope re-forms around each daughter nucleus.

Only a relatively few IF-associated proteins are known, and these appear to help organize intermediate filaments and mediate interactions with other cytoskeletal proteins and organelles.

## 1.11 Nuclear Organization

The largest and most prominent structure in most cells is the nucleus, which serves as a repository for the cell's inheritable genetic material (Figs. 1.4A and E). Within the nucleus several subnuclear bodies have been identified microscopically. The most prominent of these is the nucleolus, a major organizing substructure consisting of a myriad of protein and RNA aggregates responsible for generating and processing ribosomal RNA (rRNA) and the assembly of ribosomes (Figs. 1.4B and C). Portions of a number of chromosomes that contain amplified sequences encoding ribosomal RNA (rRNA) and ribosomal protein mRNA cluster together and associate with a number of other proteinaceous elements in the nucleus to facilitate the formation of the nucleolus. The size of the nucleolus can vary greatly within the nucleus often occupying a significant volume, particularly under conditions when large amounts of protein must be synthesized and there is a high demand for ribosomes by the cell. Within the nucleoli three morphologically distinct regions can be observed (fibrillar center, dense fibrillar component, and granular component) each thought to represent progressive stages of rRNA synthesis, processing and eventual ribosome production (Fig. 1.4D). While much remains to be learned regarding nucleolus organization and function, it is clear that its morphology can alter greatly depending on the particular cell cycle stage. In preparation for mitosis, nucleolar dissociation is observed during cell division with subsequent nucleolar fusion as the resulting daughter cells prepare for a new round of DNA replication.

In addition to the nucleolus, additional subnuclear structures can be observed. Cajal bodies are thought to be sites of ribonucleoprotein (RNP) and small nuclear RNA (snRNA) metabolism, processing, and storage. Often closely associated with Cajal bodies, Gemini of coiled bodies (GEMS) are thought to be functionally and structurally similar to Cajal bodies and together they appear to function as recycling centers for small nuclear ribonucleoproteins (snRNPs) and snRNAs. Another subnuclear domain speckle the interchromatin granule clusters that appear to contain high levels of RNA splicing snRNPs and other mRNA splicing factors. Additional subnuclear domains are being characterized at the genetic and molecular level and their contribution to human disease has only recently begun to be appreciated.

In eukaryotes, the nucleus is the assembly center for packaging and transmitting the genetic material that resides on chromosomes. Mammalian chromosomal DNA has a total linear length of about two meters when fully extended and is packaged into an average nuclear size of about 10 μM (i.e., 1/100 of a millimeter) in diameter. The approximately three billion base pairs that encode the human genome are organized into 24 distinct chromosomes (22 autosomal and one pair of sex chromosomes, X and Y). The chromosomal DNA must be packaged into the nucleus in such a way as to allow access to transcriptional and replication machinery, and the nucleus as a whole must allow for a high level of nuclear-cytoplasmic transport and signaling. For this, DNA fibers complex with a defined set of proteins to form chromatin and facilitate the approximately 200,000-fold compaction necessary to house the entire genome into the nucleus. Packaging of DNA is mediated by histones, which form octameric arrays around which approximately 200 base pairs of DNA are wound. This first order of DNA packing resembles 'beads on a string,' with the beads, or nucleosomes, composed of histone octamers and associated with DNA. Nucleosomes are further coiled to form 30 nm diameter solenoid fibers, which are subsequently more or less condensed to form heterochromatin (tightly compacted and relatively transcriptionally inactive DNA) or euchromatin (loosely compacted and transcriptionally active DNA). Each chromosome usually consists of a mixture of heterochromatin and euchromatin, and occupies more-or-less defined areas within the nucleus. Transitions between heterochromatin to euchromatin and euchromatin to

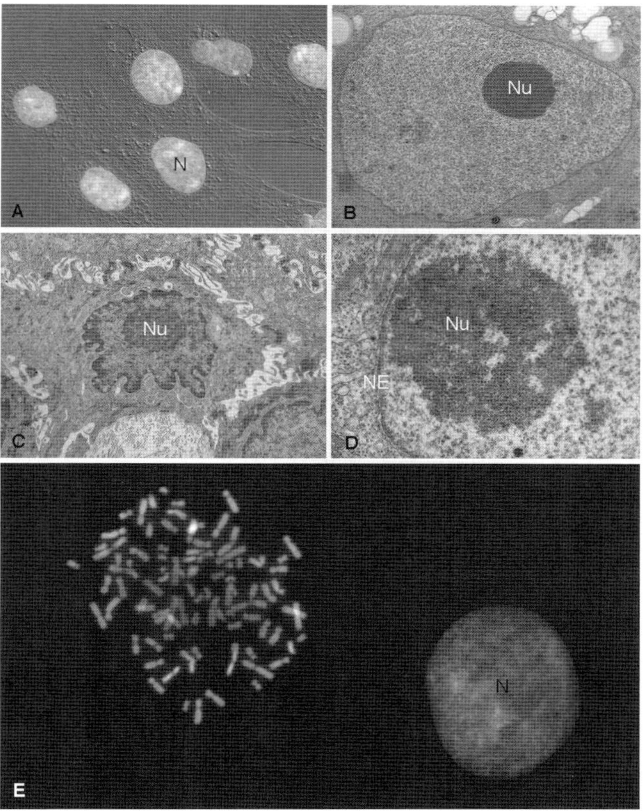


**Fig. 1.4** Light and electron micrographs of nuclear features of cells. **(A)** Light micrograph of cells in culture, stained with a fluorescent dye that binds nuclear DNA. This image is a double exposure showing both differential interference and fluorescence images (similar to Fig. 1.3F). Fluorescent nuclei (N) can be seen, surrounded by the cytoplasm of these cells. **(B)** Electron micrograph of a nucleus of an oocyte from a digenetic trematode. Nuclei from these cells contain mostly euchromatin (loosely compacted, active DNA), and a prominent nucleolus (Nu). **(C)** Electron micrograph of a nucleus of a rat esophageal epithelial cell. These nuclei are smaller than oocytes nuclei, present an irregular border, and contain patches of heterochromatin (more condensed, inactive DNA). A nucleolus (Nu) is also present in this nucleus. **(D)** Higher magnification of a nucleolus (Nu) from a trematode oocyte nucleus. Granular and fibrillar areas can be discerned within the nucleolus, which lies next to the nuclear envelope (NE) in this cell. **(E)** DNA staining and fluorescence microscopy of cell spreads to display chromosome structure. A human fibrosarcoma cell line was treated with colchicine to arrest cells in mitosis, after which the cells were applied to microscope slides. In this image, the nuclear material from two cells are present. To the left are chromosomes from a metaphase-arrested cell; at this stage of the cell cycle, the nuclear envelope has broken down and the replicated chromosomes are highly condensed and visible as distinct structures. To the right is a nucleus (N) from a cell that had not yet reached mitosis; thus, the nucleus is intact and the uncondensed chromosomes fill the entire nuclear volume (Copies of figures including color copies, where applicable, are available in the accompanying CD)

heterochromatin states are greatly influenced by a variety of temporal and physical posttranslational modifications of the histone tails of chromatin. The temporal and spatial pattern of histone modification, referred to as the histone code, together with specific modifications of the DNA sequence itself can greatly influence the whole genome gene expression signatures in specific cell types or tissues as well as in normal and diseased cells. It is becoming increasingly apparent that in addition to local chromatin structure, the physical location of chromosomes within the nucleus can greatly influence gene expression. During interphase, interactions can occur between distal chromatin regions between the same chromosomes or across different chromosomes. Thus, where a chromosome resides within the nucleus, or its chromosome territory, can affect gene expression,

DNA repair in response to damage, and cell development. For example, chromosome regions with few genes (gene poor) or that which are replicated at later stages of DNA replication tend to be located at the nuclear periphery and perinucleolar regions. Also, repositioning of actively transcribed genes with respect to the centromeric heterochromatin can lead to gene silencing.

Mounting evidence suggests that chromosomes may be organized on a protein or protein-RNA based scaffolding. This scaffolding, the nuclear matrix, is biochemically ill-defined, but appears to be composed of filaments that form a three-dimensional meshwork within the nucleus, which is surrounded by the denser filamentous mat of lamin IFs underlying the nuclear envelope. Although the biochemical makeup of the nuclear matrix is not well understood, it is possible that lamin IFs are not restricted to the nuclear periphery, but contribute to at least some of the matrix structure. Certain DNA sequences have been found to bind to the nuclear matrix much more tightly than others, leading to the proposal that distinct matrix attachment regions (MARs) or scaffold attachment regions (SARS), periodically link chromosomes to the matrix, resulting in the formation of large (30–100 kilobase) loops of DNA tethered to the matrix at MAR domains. These DNA sequences do not display a rigid consensus sequence, but have a number of features in common, including being relatively AT rich, histone poor, and possessing multiple topoisomerase II binding sequences. MAR domain DNA may also confer position independent expression of exogenous DNA incorporated into random sites within the genome. Thus, it has been proposed that MAR domains form participate as long-range regulatory elements, helping to control gene expression. Chromosome "insulators" appear to demarcate heterochromatin/euchromatin boundaries and may function by stopping the spreading and expansion of heterochromatin across a chromosomal region.

The nucleus is bounded by the nuclear envelope, which consists of two membranes (inner and outer), the underlying nuclear lamina composed of lamin IFs, and numerous nuclear pores that span the inner and outer membranes. Nuclear pores are multimolecular arrays exhibiting an eight-fold symmetry that are involved in the exchange of material between cytoplasm and nucleus. Material moves through nuclear pores by both passive diffusion and active transport; molecules smaller than 5,000 daltons are freely permeable between nucleus and cytoplasm, but those larger than about 60,000 daltons must be actively transported. Molecules between these sizes can move between nucleus and cytoplasm without active transport, but take longer to equilibrate with increasing size. Proteins actively transported into the nucleus contain nuclear localization sequences that are recognized by the pore complexes. Nuclear localization signals vary in amino acid sequence, but usually contain a number of lysine residues and are positively charged.

Rapidly evolving genomic technologies provide us the unprecedented opportunity to monitor entire transcriptional and proteomic outcomes in response to a variety of cues. Our understanding of global gene regulation will require far more than the simple deciphering and analysis of DNA sequences and must include an appreciation of nuclear architecture and chromosome organization. Towards that end, deduced gene expression signatures must be interpreted with an appreciation of chromosome location within the nucleus, nuclear-cytoplasmic signaling mechanisms, and overall cell structure and function.

## Glossary and Abbreviations

**Actin:** The protein used to form microfilaments. Actin can be either soluble (g-actin) or polymerize to form microfilaments (f-actin).

**Adherens junction:** A type of cell-cell adhesive junction in which bundles of microfilaments are connected to the plasma membrane via linking proteins (e.g., catenin). The linking proteins connect microfilaments to integral membrane proteins called cadherins, which bind to each other in the presence of $Ca^{++}$ to adhere cells together.

**ADP:** Adenosine 5'-diphosphate. A nucleotide associated with cellular energy regulation. The release of one of the three phosphate groups of ATP yields usable energy and ADP (which contains two phosphate groups and is at a lower energy state). ATP can be regenerated from ADP with the input of energy to attach a third phospate group.

**Apoptosis:** A specific way by which cells die, as a result of injury or as a result of programming (e.g., during the maturation of some tissues and organs during embryonic development). Apoptosis includes defined biochemical pathways that, when initiated, activate caspases and result in the destruction of key cytoplasmic and nuclear proteins.

**ATP:** Adenosine 5'-triphosphate. A molecule used as an energy source by the cell, most of which is normally generated by oxidative phosphorylation in mitochondria. ATP is the primary source of cellular energy used to power enzymatic reactions.

**Basal body:** A specialized type of centrosome (or MTOC) that gives rise to a cilium or flagellum.

**Cadherin:** An integral membrane protein found in desmosomes and adherens junctions. In the presence of $Ca^{++}$, cadherens from adjacent cells bind, adhering cells to each other. Cadherins are connected to microfilaments at adherens junctions via linking proteins such as catenin, and to the intermediate filament cytoskeleton at desmosomes via linking proteins such as desmoplakin.

**Cajal bodies:** A nonmembrane-bound sub nuclear domain required for the processing of small nuclear RNAs (snRNA) and small ribonucleoproteins (snRNPs)

**Caspases:** Proteases that are activated during apoptosis. Caspases destroy key cellular components, as well as activate nucleases, thus promoting nuclear disassembly and cell death.

**Catenin:** A type of linking protein found in adherens-type cell-cell junctions. Catenins are particularly interesting in that they can translocate to the nucleus and function as transcription factors, thus transducing events at the cell surface into changes in gene expression.

**Caveolae:** Invaginations of the plasma membrane involved in endocytosis. Caveolae are not associated with clathrin, but possess a distinct lipid makeup, and may constitute a specialized type of lipid raft.

**Cell junction:** Specializations of the plasma membrane that allow for anchorage and communication between cells, and between cells and the extracellular matrix.

**Centriole:** A short, barrel-like structure composed of a cylindrical array of microtubule triplets. Centrioles are associated with centrosomes, basal bodies, and spindle poles, and function in the replication of these microtubule organizing centers.

**Centrosome:** The organizing center for the microtubular cytoskeleton. Centrosomes are composed of centrioles and pericentriolar material. Three forms of centrosomes are found in cells, including the single centrosome of non-dividing cells, the two spindle poles of dividing cells, and the basal bodies of ciliated cells.

**Chromatin:** A complex of DNA, histones and additional proteins that form the chromosomes that reside in the nuclei of cells.

**Chromosome:** The organizational macromolecule of DNA and associated proteins that provides the structural and functional basis for inheritance of the nuclear genetic material.

**Cilia:** Motile, whip-like extensions of the cell, supported by a bundle of microtubules that are connected to basal bodies. Cilia actively beat back and forth as a result of interactions between microtubules and the motor protein dynein, in a ATP-requiring process.

**Clathrin:** A protein involved in receptor-mediated endocytosis. Clathrin molecules assemble onto membranes, forming a coat which leads to an invagination of the plasma membrane. These coated pits subsequently pinch off into the cytoplasm to form coated vesicles.

**Claudin:** An integral membrane protein associated with tight junctions, which are sealing junctions that form transcellular barriers between cells.

**Coated pit:** *see* Clathrin

**Coated vesicle:** *see* Clathrin

**COPs:** Proteins involved in directing membrane-bound vesicles between the ER and the Golgi apparatus.

**Cytochrome P-450:** A group of proteins involved in detoxifications reactions.

**Cytoskeleton:** A system of filaments and tubules in the cytoplasm and nucleus that perform numerous functions, including maintaining cell shape, driving cell motility and cell division, and organizing the cytoplasm. Three major types of fibers comprise the cytoskeleton: microfilaments, microtubules, and intermediate filaments.

**Dalton:** A measure of molecular mass; one Dalton is about the mass of a hydrogen atom. Small proteins are a few thousand Daltons in size, medium and large proteins range from tens of thousands to hundreds of thousands of Daltons.

**Desmosome:** A type of cell-cell adhesive junction in which bundles of intermediate filaments are connected to the plasma membrane via linking proteins of the plakin family (e.g., desmoplakin). The linking proteins connect intermediate filaments to integral membrane proteins of the cadherin family, which bind to other desmosomal cadherins in adjacent cells to adhere cells together. Numerous desmosomes are found in tissues subjected to mechanical stress, such as the epidermis.

**DNA:** Deoxyribonucleic acid.

**ECM:** Extracellular matrix.

**Endocytosis:** A process by which cells internalize material via the formation of plasma membrane invaginations, which pinch off into the cytoplasm to form membrane-bound vesicles containing the engulfed material. Specific extracellular molecules can be concentrated and internalized by this method in a process called receptor-mediated endocytosis.

**Endoplasmic reticulum:** An extensive, membrane-bound cytoplasmic organelle involved in protein and lipid synthesis, as well as in detoxification reactions and $Ca^{++}$ regulation. Membranes of the ER form enclosed compartments that range in morphology from flattened sheets to interconnected tubular network. Protein and lipid products made by the ER can be delivered to other parts of the cell or secreted, via membrane-bound vesicles that traffic between the ER, Golgi apparatus, and plasma membrane. Protein synthesis occurs in ER that possesses attached ribosomes (rough ER, or RER), and lipid synthesis is primarily associated with ER lacking ribosomes (smooth ER, or SER).

**Endosome:** A membrane-bound structure formed by the coalescence of endocytotic vesicles and vesicles containing lysosomal enzymes from the Golgi apparatus. Endosomes can give rise to lysosomes; organelles that efficiently digest internalized material.

**ER:** Endoplasmic reticulum.

**Exocytosis:** A process by which cells secrete material via fusion of membrane-bound secretory vesicles with the plasma membrane.

**Extracellular matrix:** An elaborate system of proteins and polysaccharides that surrounds cells and tissues. Composed of structural elements, as well as soluble factors that influence cell growth, differentiation, and function.

**FAK:** Focal adhesion kinase.

**Flippase:** A membrane-associated enzyme that is able to transfer phospholipids between each layer of the lipid bilayer.

**Focal adhesion kinase:** A protein kinase associated with focal contacts involved in transducing contact information at the cell surface into a cellular response.

**G-protein:** Small proteins involved in signaling functions that are able to bind GTP or GDP. G-proteins cycle between active and inactive states, depending on whether they are associated with GTP or GDP.

**Gap junction:** A type of cell-cell communicating junction that allows for the direct passage of small molecules between cells. Gap junctions are formed by the alignment of membrane-spanning pores, called connexons, between the cells.

**GDP:** Guanosine 5'-diphosphate. A nucleoside formed by the hydrolytic removal of a phosphate group from GTP.

**Hemidesmosome:** A type of adhesive junction that attaches epithelial cells to the extracellular matrix (ECM). Bundles of intermediate filaments (IFs) are connected to integral membrane proteins of the integrin family via linking proteins. Hemidesmosomal integrins bind to proteinaceous elements of the ECM, thereby providing mechanical linkage between IFs and the ECM.

**IF:** Intermediate filament.

**Inositol triphosphate:** A type of membrane-associated lipid molecule involved in cell signaling.

**Integral membrane protein:** A protein that passes through, or is embedded in, at least one layer of the lipid bilayer. Integral membrane proteins are strongly attached to membranes and usually require disruption of the membrane structure to be released.

**Intermediate filament:** One of the three major cytoskeletal groups of proteinaceous fibers found in cells. Associated with desmosomes and hemidesmosomes, they function in strengthening the cytoplasm of cells, as well as mechanically linking cells of an epithelium with each other and with the ECM. Intermediate filaments are also found in the nucleus, where they form a lamina that helps support the nuclear envelope.

**IP3:** Inositol triphosphate.

**LDL:** Low density lipoprotein.

**Lipid bilayer:** The structure phospholipids molecules adopt to form a membrane. Composed of two layers of phospholipids molecules, where the hydrophobic tails face each other and the polar heads face outward, creating a hydrophobic central region sandwiched between charged surfaces.

**Lipid raft:** A region of the plasma membrane exhibiting a specialized phospholipid makeup. It is associated with specific functions, including formation of caveolae and signal transduction.

**Lumen:** An enclosed space, or chamber. Usually refers to the compartment enclosed by a membranous organelle.

**Lysosome:** A digestive organelle formed by the ER and Golgi apparatus. Hydrolytic enzymes synthesized by the ER are concentrated and packaged by the Golgi apparatus into lysosomal vesicles. Lysosomal vesicles fuse with either endosomes or with old cell organelles to digest internalized material, or cellular material to be recycled, respectively.

**M6P:** Mannose-6-phosphate.

**mRNA:** Messenger RNA, the type of RNA that encodes the sequence of amino acids to be assembled into a specific protein in association with a ribosome.

**Mannose-6-phosphate:** A polysaccharide tag attached to hydrolytic enzymes that marks them for packaging into lysosomal vesicles by the Golgi apparatus.

**MAR:** Matrix attachment region.

**Matrix attachment region:** A specialized sequence of DNA that is bound to the nuclear matrix.

**Microfilament:** One of the three major cytoskeletal groups of proteinaceous fibers found in cells. Microfilaments are concentrated underneath the plasma membrane, which they help support and organize. They are also associated with adherens junctions, microvilli, and cleavage furrows. Microfilaments are involved with maintaining cell shape and powering cell motility.

**Microvilli:** Finger-like extensions of the plasma membrane supported by core bundles of microfilaments. Microvilli serve to increase the absorptive area of epithelial cells.

**Microtubule:** One of the three major cytoskeletal groups of proteinaceous fibers found in cells. Microtubules help to organize the cytoplasm, participate in intracellular transport, allow for ciliary and flagellar motility, and organize and segregate chromosomes during mitosis.

**Microtubule organizing center (MTOC):** *see* **Centrosome**.

**Mitochondria:** Double-membraned organelles primarily involved in converting the energy from food molecules into a form usable by the cell. This is largely accomplished by using food energy to create a proton gradient across the inner mitochondrial membrane, which in turn is used to drive the synthesis of ATP. Mitochondria also function in calcium homeostasis and in the regulation of programmed cell death, or apoptosis.

**Mitosis:** The segregation of chromosomes during cell division. Cell division includes mitosis, followed by cytokinesis, or the division of the parental cell into two daughter cells.

**MTOC:** Microtubule organizing center.

**Nuclear envelope:** A double-membraned structure enclosing the nucleus that establishes and maintains a distinct nuclear environment. The nuclear envelope is perforated by nuclear pores, which allow for the regulated transport of material between nucleus and cytoplasm.

**Nuclear matrix:** A protein (and possibly RNA) based scaffolding within the nucleus that is thought to help organize chromatin. The composition of the nuclear matrix is not well understood, but it appears to play important roles in DNA synthesis and regulation of gene activity.

**Nucleolus:** A specialized structure within the nucleus involved in ribosomal RNA synthesis and ribosome assembly.

**Occludin:** An integral membrane protein associated with tight junctions that helps form a transcellular barrier between cells.

**Occluding junction:** Another term for tight junction.

**Organelle:** A readily identifiable cellular inclusion that possesses a characteristic morphology and function. The term is usually used to refer to the major membrane-bound structures within cells, including the nucleus, ER, Golgi apparatus, lysosomes, peroxisomes, and mitochondria.

**Pericentriolar material:** The amorphous material surrounding centrioles in a centrosome. Microtubules associated with centrosomes and spindle poles arise from the pericentriolar material.

**Peripheral membrane protein:** A membrane-associated protein that is not embedded in the lipid bilayer. Peripheral proteins can be associated with the phospholipids heads of the bilayer, or with the portions of integral membrane proteins that extend beyond the bilayer.

**Peroxisome:** An organelle involved in lipid metabolism and oxidation reactions, including the generation and destruction of $H_2O_2$.

**Phagocytosis:** A type of endocytosis where large particulate matter is taken up by a cell (e.g., the engulfment of bacteria by macrophages).

**Pinocytosis:** A type of endocytosis where small vesicles internalize extracellular fluid for uptake by the cell.

**Plasma membrane:** The membrane surrounding a cell. The plasma membrane, sometimes called the plasmalemma, encloses the cytoplasm and protects the cell contents from the environment. It carries out vital functions in protection, transport of material into and out of the cell, sensing and responding to the environment, and in cell identification.

**RER:** Rough endoplasmic reticulum.

**Rough endoplasmic reticulum:** ER that possesses ribosomes (thus presenting a "rough" surface). RER is primarily involved in the synthesis of membrane, lysosomal, and secreted proteins.

**Ribosome:** A multimeric array of protein and ribonucleic acid that is involved in protein synthesis. Ribosomes assemble individual amino acids into a polymer, or polypeptide, in a specific sequence determined by an associated mRNA molecule. Ribosomes can exist either "free" in the cytoplasm or attached to the ER. In the former location, the proteins they express are released into the cytoplasm; in the latter location, the proteins are either inserted into the membrane of the ER or released into the lumen of the ER.

**RNA:** Ribonucleic acid. Includes a number of subtypes, including messenger RNA (mRNA), ribosomal RNA (rRNA), transfer RNA (tRNA), and micro RNA (miRNA).

**SER:** Smooth ER.

**Signal recognition particle:** An assembly of proteins that help dock ribosomes to ER, forming RER.

**Smooth endoplasmic reticulum:** ER that lacks ribosomes (thus presenting a "smooth" surface). SER is involved in lipid synthesis, calcium transport, and detoxification reactions.

**SNAREs:** Proteins that help regulate the trafficking of membrane bound vesicles by mediating their fusion with specific target organelles.

**S-Phase:** The stage of the cell cycle where DNA synthesis occurs.

**Spindle pole:** A microtubule organizing center that assembles the microtubule spindle during cell division. When cells divide, their centrosome duplicates and moves apart to form a bipolar spindle.

**Spliceosome:** An assembly of protein and RNA molecules that processes newly-made mRNA into mature mRNA.

**SRP:** Signal recognition particle.

**Start-transfer sequence:** A specific sequence of amino acids that initiates the penetration of a growing polypeptide chain into the lipid bilayer of the RER.

**Stop-transfer sequence:** A specific sequence of amino acids that stops the insertion of a polypeptide chain into the lipid bilayer of the RER.

**Synapse:** A type of communicating cell-cell junction found between neurons in nervous tissue.

**Tight junction:** A type of cell-cell junction that establishes a transcellular barrier. Also referred to as an occluding junction.

**Trans-Golgi:** The portion of the Golgi apparatus that releases membrane-bound vesicles after their contents have been processed by the Golgi apparatus.

**Tubulin:** The protein used to form microtubules. Most microtubules continually oscillate between growth and disassembly by the addition or removal of tubulin heterodimers at the plus ends of microtubules.

**WASp:** The protein mutated in Wiscott-Aldrich syndrome patients. This protein plays an important role in regulating actin polymerization.

**Zonulae adherens:** Large, belt-like adhesive junctions usually found encircling the apical regions of epithelial cells.

## Suggested Reading

### *Membranes; Plasma membrane; Endoplasmic reticulum; Golgi apparatus; Lysosomes*

Eskelinen, E.L., Tanaka, Y., and Saftig, P. (2003) At the acidic edge: emerging functions for lysosomal membrane proteins, Trends Cell Biol. 13:137–145.

Kroemer, G., and Jaattela, M. (2005) Lysosomes and autophagy in cell death control, Nat. Rev. Cancer 5:886–897.

Kusumi, A., and Suzuki, K. (2005) Toward understanding the dynamics of membrane-raft-based molecular interactions, Biochim. Biophys. Acta 1746:234–251.

Mukherjee, S., and Maxfield, F.R. (2004) Membrane domains, Ann. Rev. Cell Dev. Biol. 20:839–866.

Pjeffer, S. (2007) Unsolved mysteries in membrane traffic, Ann. Rev. Biochem. 76:629–645.

Schroder, M., and Kaufman, R.J. (2005) ER stress and the unfolded protein response, Mutation Res. 569, 29–63.

van Vliet, C., Thomas, E.C., Merino-Trigo, A., Teasdale, R.D., and Gleeson, P.A. (2003) Intracellular sorting and transport of proteins, Prog. Biophys. Mol. Biol. 83:1–45.

Vigh, L., Escriba, P.V., Sonnleitner, A., Sonnleitner, M., Piotto, S., Maresca, B., Horvath, I., and Harwood, J.L. (2005) The significance of lipid composition for membrane activity: new concepts and ways of assessing function. Prog. Lipid Res. 44:303–344.

Watson, P., and Stephens, D.J. (2005) ER-to-Golgi transport: form and formation of vesicular and tubular carriers, Biochim. Biophys. Acta. 1744:304–315.

### *Mitochondria and Peroxisomes*

Chan, D.C. (2006) Mitochondrial fusion and fission in mammals, Ann. Rev. Cell Dev. Biol. 22:79–99.

Fagarasanu, A., Fagarasanu, M., and Rachubinski, R.A. (2007) Maintaining peroxisome populations: a story of division and inheritance, Ann. Rev. Cell Dev. Biol. 23:321–344.

Garrido, C., and Kroemer, G. (2004) Life's smile, death's grin: vital functions of apoptosis-executing proteins, Curr. Op. Cell Biol. 16:639–646.

Kakkar, P., and Singh, B.K. (2007) Mitochondria: a hub of redox activities and cellular distress control, Mol. Cell Biochem. (epub ahead of print)

Ryan, M.T., and Hoogenraad, N.J. (2007) Mitochondrial-nuclear communications, Ann. Rev. Biochem. 76:701.

Schrader, M., and Fahimi, H.D. (2006) Growth and division of peroxisomes, Int. Rev. Cytol. 255:237–290.

Titorenko, V.I., and Rachubinski, R.A. (2004) The peroxisome: orchestrating important developmental decisions from inside the cell, J. Cell Biol. 164:641–645.

Wallace, D.C. (2005) A mitochondrial paradigm of metabolic and degenerative diseases, aging, and cancer: a dawn for evolutionary medicine, Ann. Rev. Genet. 39:359–407.

Schatz, G. (2007) The magic garden, Ann. Rev. Biochem. 76:673–678.

Wanders, R.J.A. and Waterham, H.R. (2006) Biochemistry of mammalian peroxisomes revisited, Ann. Rev. Biochem. 75:295–332.

## *Cytoskeleton*

Ananthakrishnan, R., and Ehrlicher, A. (2007) The forces behind cell movement, Int. J. Biol. Sci. 3:303–317.

Bettencourt-Dias, M., and Glover, D.M. (2007) Centrosome biogenesis and function: centrosomics brings new understanding, Nat. Rev. Mol. Cell Biol. 8:451–463.

Caviston, J.P., and Holzbaur, E.L.F. (2006) Microtubule motors at the intersection of trafficking and transport, Trends Cell Biol. 16:530–537.

Delon, I., and Brown, N.H. (2007) Integrins and the actin cytoskeleton, Curr. Op. Cell Biol. 19:43–50.

Disanza, A., Steffen, A., Hertzog, M., Frittoli, E., Rottner, K., and Scita, G. (2005) Actin polymerization machinery: the finish line of signaling networks, the starting point of cellular movement, Cell. Mol. Life Sci. 62:955–970.

Herrmann, H., Bar, H., Kreplak, L., Strelkov, S.V., and Aebi, U. (2007) Intermediate filaments: from cell architecture to nanomechanics, Nat. Rev. Mol. Cell Biol. 8:562–573.

Honore, S., Pasquier, E., and Braguer, D. (2005) Understanding microtubule dynamics for improved cancer therapy, Cell. Mol. Life Sci. 62:3039–3056.

Houben, F., Ramaekers, F.C., Snoeckx, L.H., and Broers, J.L. (2007) Role of nuclear lamina-cytoskeleton interactions in the maintenance of cellular strength, Biochim Biophys. Acta 1773:675–686.

Kim, S., and Coulombe, P.A. (2007) Intermediate filament scaffolds fulfill mechanical, organizational, and signaling functions in the cytoplasm, Genes Dev. 21:1581–1597.

Magin, T.M., Reichelt, J., and Hatzfeld, M. (2004) Emerging functions: diseases and animal models reshape our view of the cytoskeleton, Exp. Cell Res. 301:91–102.

Mege, R.M., Gavard, J., and Lambert, M. (2006) Regulation of cell-cell junctions by the cytoskeleton, Curr. Opin. Cell Biol. 18:541–548.

Westermann, S., and Weber, K. (2003) Post-translational modifications regulate microtubule function, Nat. Rev. Mol. Cell Biol. 4:938–948.

Yamaguchi, H., and Condeelis, J. (2007) Regulation of the actin cytoskeleton in cancer cell migration and invasion, Biochim. Biophys. Acta 1773:642–652.

## *Nuclear Structure*

Bode J., Goetze S., Heng H., Krawetz S.A, Benham C. (2003) From DNA structure to gene expression: mediators of nuclear compartmentalization and dynamics, Chromosome Res. 11, 435–45.

Branco, M.R. and Pombo, A. (2007) Chromosome organization: new facts, new models.Trends Cell Biol. 17, 127–34.

Boisvert, F.M., van Koningsbruggen, S., Navascues, J., and Lamond, A.I. (2007) The multifunctional nucleolus, Nat Rev Mol Cell Biol. 8, 574–85.

Chakalova, L., Debrand, E., Mitchell, J.A., Osborne, C.S., and Fraser P. (2005)Replication and transcription: shaping the landscape of the genome, Nat Rev Genet. 6, 669–677.

Cioce, M. and Lamond, A.L. (2005) Cajal bodies: a long history of discovery, Annu Rev Cell Dev Biol. 21, 105–131.

Handwerger, K.E. and Gall, J.G. (2006) Subnuclear organelles: new insights into form and function, Trends Cell Biol. 16, 19–26.

Heard, E. and Bickmore, W. (2007) The ins and outs of gene regulation and chromosome territory organization, Curr Opin Cell Biol. 19, 311–316

Pezo, R.C. and Singer, R.H. (2007) Nuclear microenvironment in cancer diagnosis and treatment, J Cell Biochem. 2007 May 3; [Epub]

Zaidi S.K., Young, D.W., Javed, A., Pratap, J., Montecino, M., van Wijnen, A., Lian, J.B., Stein, J.L., and Stein, G.S. (2007) Nuclear microenvironments in biological control and cancer, Nat Rev Cancer. 7, 454–463.

# Chapter 2
# Transcription and the Control of Gene Expression

**Nadine Wiper-Bergeron and Ilona S. Skerjanc**

**Abstract** Transcription, the initial step of gene expression is a tightly regulated process. In addition to variability in the core promoter region of the RNA polymerase II transcribed genes, which can stabilize or destabilize the basal machinery and influence transcription rates, promoters contain enhancer regions which can be far upstream from the gene transcribed. These enhancers and their DNA binding factors are highly variable and can lead to the recruitment of unique co-activator complexes that can influence the initiation and progression of the polymerase through the nucleosomal structure of chromatin. In essence then, every promoter becomes a unique microenvironment, the sum of several enhancer elements, core promoter elements and chromatin structure. The gene's transcription rate is then dependent on the efficiency of these interactions – an average of the effects of each enhancer and co-activator leading to a fine tuning of transcriptional responses according to cellular needs. It is therefore not surprising that transcription is the major checkpoint for gene expression in the cell. The next step for the new mRNA molecule is post-transcriptional modification, which augments the stability of the messenger mRNA export to the cytoplasm (in eukaryotes) and translation into proteins.

**Keywords** Transcription · Initiation factors · Termination · RNA polymerase · Coactivators/Corepressors · Nucleosome

## 2.1 Introduction

Transcription is a process by which genetic information is copied onto a single stranded RNA molecule. This copy, the transcript, can be used to form components of the protein translation machinery that includes the transfer RNA [tRNA], ribosomal RNA [rRNA]) or to direct the production of a specific protein encoded by a messenger RNA (mRNA). Production of these different RNAs is catalyzed by a family of enzymes known as RNA polymerases. RNA polymerases use a single-stranded DNA template and nucleotides to synthesize a complementary RNA strand. Phosphodiester bond formation releases pyrophosphate which is rapidly hydrolyzed to ensure a unidirectional reaction which favors the addition of nucleotides to the nascent chain.

### *2.1.1 Structure of RNA Polymerases*

In prokaryotes, one RNA polymerase exists that catalyzes the production of all cellular RNAs. It is made up of 5 polypeptides, which are assembled into a holoenzyme of approximately 450 kDa with

I.S. Skerjanc
Department of Biochemistry, Microbiology and Immunology, University of Ottawa, 451 Smyth Road, Room 4244, Ottawa, Ontario, KIH 8M5, Canada
e-mail: iskerjan@uottawa.ca

S. Krawetz (ed.), *Bioinformatics for Systems Biology*, DOI 10.1007/978-1-59745-440-7_2,

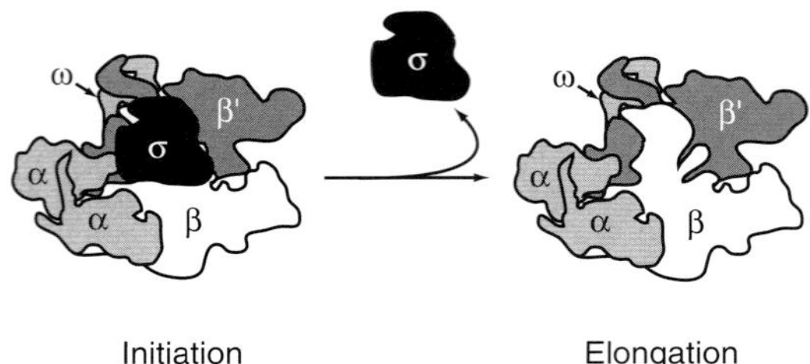


**Fig. 2.1** Schematic representation of the prokaryotic RNA polymerase. Bacterial RNA polymerase is composed of 5 subunits (2 α, β, β', ω and σ). The β and β' subunits form a jaw-like structure into which is fed the DNA template. Upon unwinding of the template, transcription can begin by adding complementary bases forming a DNA:RNA hybrid. Because the σ subunit blocks the exit channel for the nascent RNA, it must dissociate to permit elongation of the chain (Copies of figures including color copies, where applicable, are available in the accompanying CD)

the general structure $\alpha_2\beta\beta'\omega\sigma$ (Fig. 2.1) [1]. The β and β' subunits form a crab-claw structure which positions the DNA template near the $Mg^{2+}$ containing catalytic site found in a tunnel-like structure [2]. The σ subunit, which dissociates from the enzyme once transcription begins, is required to correctly position the polymerase at the initiation site for transcription [3].

In eukaryotes, the process of transcription is more complex and is mediated by three different but related RNA polymerases. RNA polymerase I catalyzes the synthesis of the majority of rRNAs whereas RNA polymerase II synthesizes mRNA. Family member RNA polymerase III synthesizes the 5S rRNA, the tRNAs and other small cellular RNAs.

Eukaryotic polymerases are multi-subunit enzymes that share many structural features with the prokaryotic RNA polymerase and with each other [1]. In particular, RNA polymerase I, II and III share a common catalytic core made up of subunits Rpb5, Rpb6, Rbp8, Rpb10, and Rpb12 (Table 2.1) [4,5]. Outside the core of conserved subunits, there is extensive structural and physical similarities between many of the additional components of the different RNA polymerase holoenzymes [4,5]. For example, subunits Rpa14/43 of RNA polymerase I are thought to be homologous to Rbp4/7 of RNA polymerase II and C17/25 of RNA polymerase III [4].

RNA polymerase II is the enzyme responsible for the catalysis of mRNA formation, and since gene transcription is a major control point for gene expression, the polymerase is tightly regulated.

**Table 2.1** Subunit components of *S. cerevisiae* RNA polymerases. The subunits which are used by all three polymerases are grouped under common subunits. Note that subunits AC40 and AC19 are used by both RNA polymerase I and III

| | RNA pol I | RNA pol II | RNA pol III |
|---|---|---|---|
| Common subunits | Rpb5 | Rpb5 | Rpb5 |
| | Rpb6 | Rpb6 | Rpb6 |
| | Rpb8 | Rpb8 | Rpb8 |
| | Rpb10 | Rpb10 | Rpb10 |
| | Rpb12 | Rpb12 | Rpb12 |
| | AC40 (Rpc5) | | AC40 (Rpc5) |
| | AC19 (Rpc9) | | AC19 (Rpc9) |
| Unique subunits | Rpa1 | Rpb1 | Rpc1 |
| | Rpa2 | Rpb2 | Rpc2 |
| | Rpa3 | Rpb3 | Rpc3 |
| | Rpa4 | Rpb11 | Rpc11 |
| | Rpa5 | Rpb4 | Rpc6 |
| | Rpa8 | Rpb7 | Rpc12 |
| | Rpa9 | Rpb9 | Rpc4 |
| | | | Rpc8 |

(Copies of tables are available in the accompanying CD.)

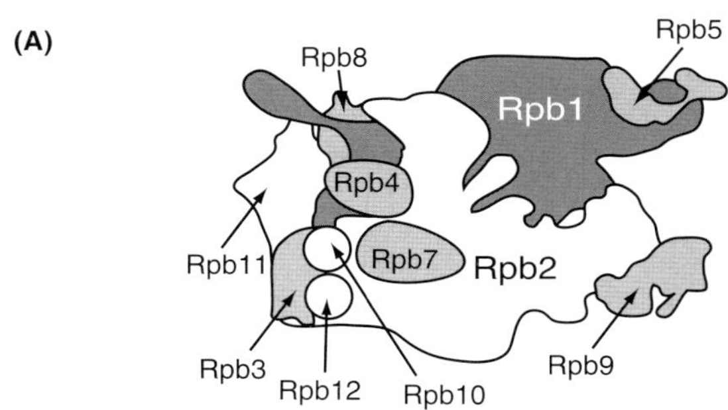


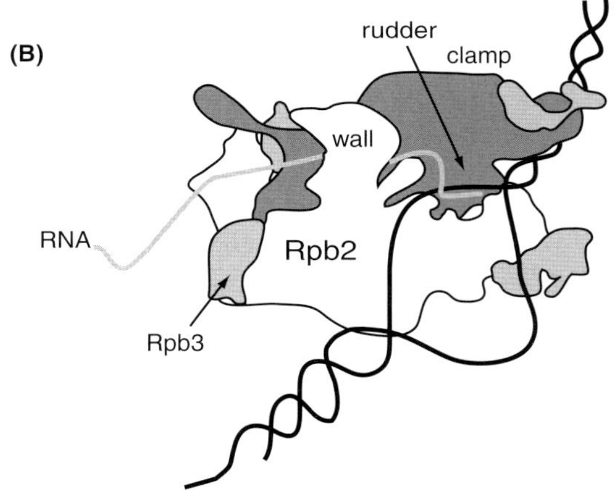


**Fig. 2.2** Schematic representation of eukaryotic RNA polymerase II. **(A)** The structure of RNA polymerase II resembles bacterial RNA polymerase. The jaws are formed by subunits Rpb1 and Rpb2. Note that the C-terminal domain of subunit Rpb1 is unique to RNA polymerase II. **(B)** During transcription, the DNA template is fed into the jaw structure formed by Rpb1 and Rpb2, guided and stabilized by the rudder domain. Unwinding of the template to form the transcription bubble permits the formation of the DNA:RNA hybrid. The nascent transcript exits the polymerase from a site beneath the C-terminal tail or domain (CTD) (Copies of figures including color copies, where applicable, are available in the accompanying CD)

The crystal structure of RNA polymerase II has been solved, both on and off DNA templates, a work that earned Roger Kornberg the Nobel Prize in Chemistry in 2006 [6–8]. These studies have confirmed that the polymerase matures during transcription of a gene through a process involving covalent modifications of the polymerase and conformational changes. RNA polymerase II has many structural features which are linked to both its function and regulation (Fig. 2.2). There exist 2 "jaws", which act to grab and position the DNA template within the catalytic site of the enzyme, which resembles the structure formed by the β and β' subunits of the prokaryotic RNA polymerase (Fig. 2.1). The DNA template is fed into a positively charged cleft within the enzyme, which stabilizes the DNA-polymerase interaction. In addition a "clamp" closes over the DNA template essentially locking it in place (Fig. 2.2B). This region is in an open conformation in the free polymerase molecule, permitting the association of DNA template with the catalytic site. The DNA template is then separated and positioned by 3 loops which originate from the clamp: the rudder, the lid and the zipper (Fig. 2.2, only rudder is visible) [9]. Finally, a bridging helix formed by the Rpb1 subunit of the polymerase makes direct contact with the DNA template and promotes translocation of the template during transcription [3,6,7,10–12].

In addition to these features, RNA polymerase II has a C-terminal tail, or domain (CTD), which acts as a regulatory domain (Fig. 2.2). The CTD is composed of tandem repeats of the amino acid sequence YSPTSPS. The number of repeats varies greatly among species [13]. The CTD is the target of modifying enzymes which control the polymerase's activity (kinases) and acts as a docking site for mRNA processing enzymes [14]. Since the nascent transcript exits the polymerase at a site found beneath the CTD, the mRNA processing enzymes are in close proximity to their substrate, thus effectively coupling transcription to mRNA maturation (discussed chapter 3) [15].

RNA polymerase II functions as part of a multiprotein complex which matures as transcription is initiated and as it progresses along the DNA template [3,10]. RNA polymerase II associates with transcription factors, which are necessary to direct the polymerase to the transcription start site, to stabilize interactions with the DNA, and to unwind the DNA template. These factors, which are required for the transcription of all RNA polymerase II transcribed genes, are called general transcription factors (GTFs) and bear the nomenclature TFIIX, where X is unique [10].

## 2.2 What Drives Transcription – The Promoter Region

To ensure transcription of a region of the DNA, DNA sequences are required upstream (or 5') to signal where transcription should begin and to direct RNA polymerase II positioning. These sequences make up the promoter region of the gene and ensure the induction of transcription from the locus. In prokaryotes, promoter regions are very simple, consisting of a TATAAT element (Pribnow box) located at position –10 relative to the transcription initiation site [16,17]. In addition to the Pribnow box, particularly strong promoters, which drive the transcription of genes that are highly expressed also have an element at position –35 (TTGACA) [18–20]. These elements direct the prokaryotic RNA polymerase to associate with the initiation site, and permit the unwinding of the template DNA, a step necessary for the initiation of transcription.

Similar to prokaryotic promoters, eukaryotic promoter regions for rRNA and tRNA consist of relatively short sequences which direct the correct placement of their respective polymerase [21]. In the case of RNA polymerase I, the promoter region spans approximately 50 base pairs and contains an AT-rich sequence known as the ribosomal initiator element [22]. The RNA polymerase III promoter contains promoter elements found downstream of the transcription initiation site, which direct the initiation of transcription by the polymerase [21]. For genes transcribed by RNA polymerase II, however, the promoter regions are generally more complex and can span to regions distal to the transcription start site. The high variability in RNA polymerase II promoter architecture permits gene-specific transcriptional responses which can be fine-tuned to answer to intra- and extracellular signaling events.

### *2.2.1 The Core Promoter Region*

Although there is high variability, eukaryotic RNA polymerase II transcribed genes have common features in their promoter regions [23,24]. The core promoter region spans from position −35 to +35 and houses the minimal sequence elements required to direct basal transcription of a gene by RNA polymerase II [25–27]. Position +1 is the transcriptional start site.

#### 2.2.1.1 The TATA Box

The TATA box was the first core promoter element identified and consists of a TA rich sequence located at −35 to −25 relative to the transcriptional start site (notable exception is yeast, where the TATA box is located at −120 to −40). The optimal site is 5'–TATATAAG–3', although this is highly variable and it is recognized by the general transcription factor TFIID [25–27]. TFIID is a multi-protein complex which includes the TATA Binding Protein (TBP) which binds to the minor groove of the TATA box in a sequence-specific manner. TBP, with its saddle-like structure, binds DNA in a directional fashion and causes the DNA template to bend almost 90° [28–30]. This interaction is thought to correctly position the RNA polymerase, which is recruited to the DNA after TFIID to the transcription initiation site.

#### 2.2.1.2 The Initiator Element

The Initiator Element is the transcription initiation site (most often adenosine at +1 and cytosine at −1) that is recognized by many general transcription factors including TFIID, TBP Associated Factors (TAFII150, TAFII250; part of TFIID), RNA polymerase II itself, and transcription factor YY-1 [31,32]. The initiator element and the TATA box appear to correctly position the polymerase at the transcriptional start site. When spaced approximately 25 base pairs apart, they act synergistically [25–27].

#### 2.2.1.3 The Downstream Promoter Element

Although many RNA polymerase II transcribed genes have TATA boxes which are recognized by TBP, the presence of a TATA box is not necessary for initiation of transcription. The Downstream Promoter Element (DPE), located in the coding region of the gene (approximately +28 to +32) is typically, though not exclusively found in TATA-less promoters [33]. This element may help correctly position the basal machinery at the promoter in the absence of sequence specific TBP binding.

#### 2.2.1.4 The TFIIB Recognition Element

The TFIIB Recognition Element is a short sequence found upstream of the TATA box [34]. This sequence is recognized and bound by TFIIB, a general transcription factor which helps to stabilize the interaction of TFIID with the TATA box, ensuring correct positioning of the polymerase at the initiation site [34].

#### 2.2.1.5 CpG Islands

CpG islands are core promoter regions with high CG content [35]. They are often found in housekeeping gene promoters, which are expressed ubiquitously at steady levels. Interestingly, the promoters which have CpG islands are often the ones which lack a functional TATA box.

#### 2.2.1.6 The Motif Ten Element (MTE)

The MTE is a highly conserved element found at position +18 to +27 relative to the initiator [36]. While the presence of the initiator element is required, the MTE can function to stimulate transcription in the absence of other core promoter elements such as the TATA box or the DPE.

### *2.2.2 Enhancer Elements*

For a cell to respond/adapt to its environment, it must be able to fine tune the expression of its genome. One way in which this control is achieved is through the induction of transcription rates above that which is achieved by the basal machinery (GTFs and RNA polymerase II). To achieve this induction, DNA sequences called enhancers, located upstream of the core promoter, can direct the DNA binding of sequence specific factors (transcription factors) which can increase or decrease transcription rates via the recruitment or stabilization of the basal transcriptional machinery at the initiation site [37]. In fact, in eukaryotes, the binding of transcription factors to the enhancer region is necessary to stimulate the assembly and association of the basal machinery with the core promoter [37–39].

## 2.3 What Stimulates Transcription – Transcription Factors

Transcription factors are protein molecules that act to control the transcription of genes. They function by binding to specific DNA sequences found in the enhancer region of gene promoters via specialized DNA binding domains. Transcription factors also possess one or more activation domains, which are regions of the protein that allow them to stimulate or repress transcription of the target gene in response to intra- and extracellular signaling pathways. Transcription factors can increase the rate of transcription by (i) recruitment and binding of basal machinery to the core promoter, and (ii) remodeling of the chromatin structure which can present a barrier to the active polymerase [39–41].

## 2.4 The Mechanics of Transcription – Making mRNA

### *2.4.1 Prokaryotic Transcription*

To initiate prokaryotic transcription, the RNA polymerase first binds to the Pribnow box element of the promoter forming the closed complex. The DNA template at the initiation site is then unwound to form the template strand (or anti-sense strand) and is made accessible to the enzyme's active site (open complex). The polymerase then begins to synthesize the RNA strand using base complementarity, but because one of the holoenzyme's subunits ($\sigma$) blocks the exit channel for the nascent RNA strand, the first attempts at transcription produce only truncated abortive transcripts (Fig. 2.1) [42]. Dissociation of the $\sigma$ subunit permits the exit of the growing RNA transcript and productive elongation (Fig. 2.1).

### *2.4.2 Eukaryotic Transcription*

#### 2.4.2.1 Initiation

Once sequence specific transcription factors bind to their respective enhancer elements they can, via their activation domains, stimulate the recruitment and assembly of general transcription factors and RNA polymerase II at the initiation site. One way by which this is achieved is by recruiting the multi-subunit complex, Mediator, a step which appears to be generally required for transcriptional activation by RNA polymerase II [43]. Mediator acts as a bridge between the distal enhancers and the basal transcriptional machinery (Fig. 2.3) and may itself receive regulatory inputs to modify the transcriptional response. For example, the recruitment of Mediator by DNA-bound transcription factors can facilitate the recruitment of the polymerase and the GTFs, thereby promoting the assembly of the pre-initiation complex [41]. In addition, Mediator may directly modulate the activity of RNA polymerase II by stimulating phosphorylation of the CTD [41].

Recruitment of the basal machinery occurs as follows (Fig. 2.3C,D) [44,45]. First, TFIID (TATA binding protein (TBP) and TBP Associated Factors (TAFs)) bind at the TATA box of the core promoter. It is important to note that not all promoters contain TATA elements, and in these instances, TFIID can bind to DNA in a sequence independent fashion, directed by other GTFs. After TFIID DNA binding, TFIIA joins the complex binding to both TFIID and DNA at a position upstream of the TATA box. Subsequently, TFIIB binds to TFIID opposite the TFIIA interaction site, and may serve to correctly position the polymerase which arrives next along with TFIIF. TFIIE recruitment to the complex helps to recruit TFIIH, a large protein complex that contains the CDK7/cyclin H kinase complex and a DNA helicase, to the promoter and to regulate RNA polymerase activity through the stimulation of TFIIH kinase activity (Fig. 2.3) [46]. TFIIH helicase activity ensures that the polymerase transcribes the correct DNA strand by unwinding the template [47] while its kinase activity acts to promote transcriptional elongation through

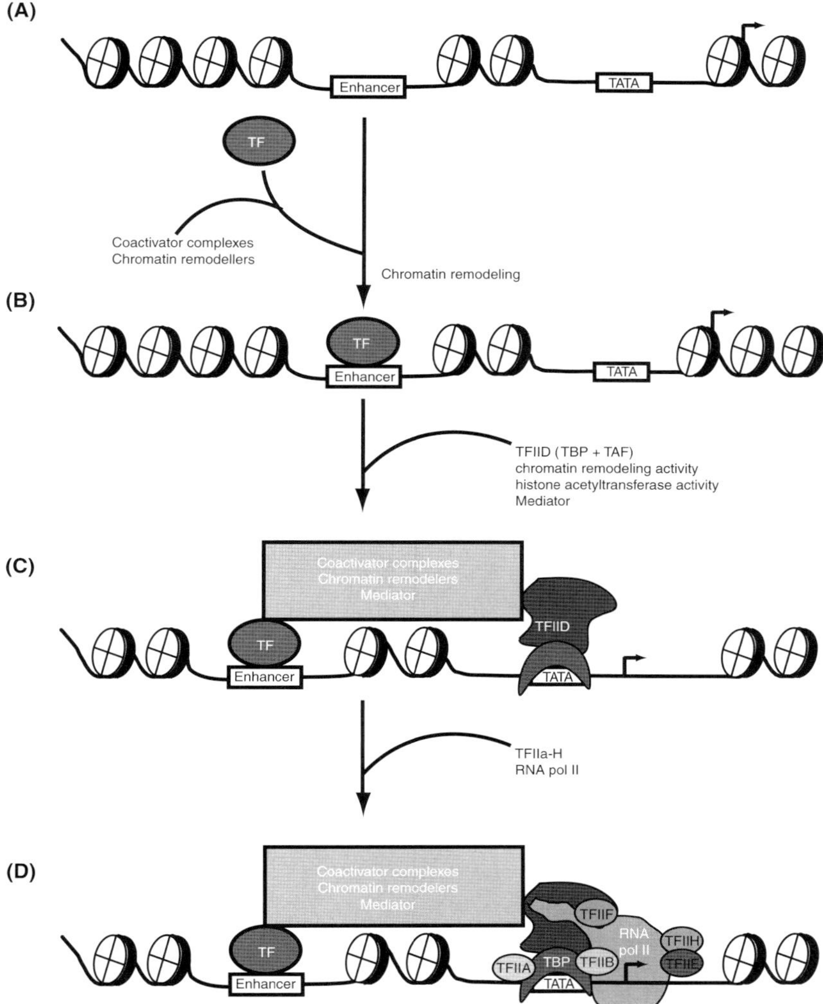


**Fig. 2.3** Assembly of the pre-initiation complex at a eukaryotic RNA polymerase II-transcribed core promoter. **(A)** Structure of a typical eukaryotic RNA polymerase II transcribed promoter consisting of the core promoter (TATA box, initiator element is indicated by the arrow). Note that the initiator element is not accessible due to the nucleosome structure. Upstream is the enhancer element, which can bind to sequence specific transcription factors and due to nucleosome spacing, it is accessible. **(B)** Binding of transcription factors to the accessible enhancer element nucleates the assembly of a co-activator complex which includes Mediator, chromatin remodeling activity and histone acetyltransferase activity. **(C)** Chromatin remodeling activity repositions nucleosomes to make the initiator element accessible to RNA polymerase II. The transcription factor: co-activator complex recruits TFIID (TBP and TAFs) to the TATA box. **(D)** Stable association of TFIID with the TATA box recruits the GTFs and RNA polymerase II to the core promoter (Copies of figures including color copies, where applicable, are available in the accompanying CD)

phosphorylation of the CTD of RNA polymerase II. TFIIH and TFIIE interact strongly, and TFIIE appears to be an important regulator of TFIIH's kinase and ATPase activity [46,48].

Assembly of the GTFs at the core promoter triggers the formation of the transcription bubble, a melted region of DNA which permits the addition of the first nucleotide. Once transcription initiation begins, the pre-initiation complex becomes an initially transcribing complex (ITC). This complex can catalyze the addition of short strands of RNA which are prone to release (abortive transcription) similar to what is observed during prokaryotic transcription.

#### 2.4.2.2 Elongation

The elongation phase of transcription can be divided into 3 steps: promoter clearance, promoter-proximal pausing, and productive elongation (Fig. 2.4B,C) [3,11,49]. Promoter clearance (or escape) is the event in which RNA polymerase II frees itself from the GTFs, thereby liberating it from the core promoter contacts (Fig. 2.4C). This stage is completed once the nascent RNA strand associates stably with the transcription complex, a process stabilized by TFIIB [50,51]. In addition,

**(A) Pre-initiation complex**

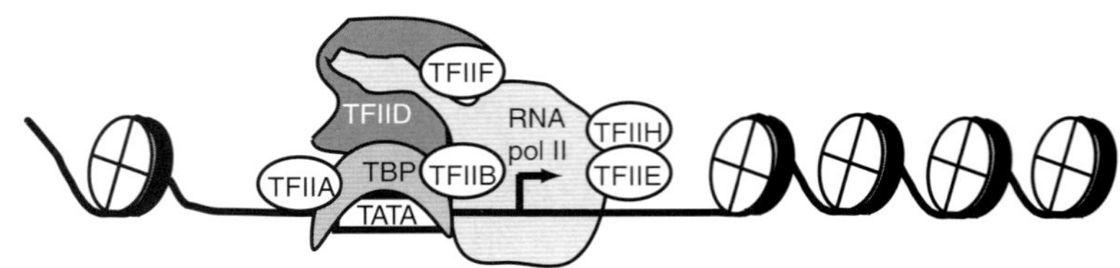


**(B) Initiation/Abortive Transcription**

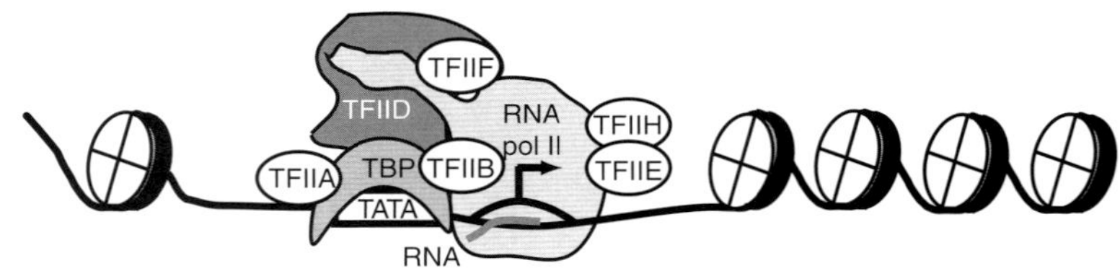


**(C) Productive Elongation**

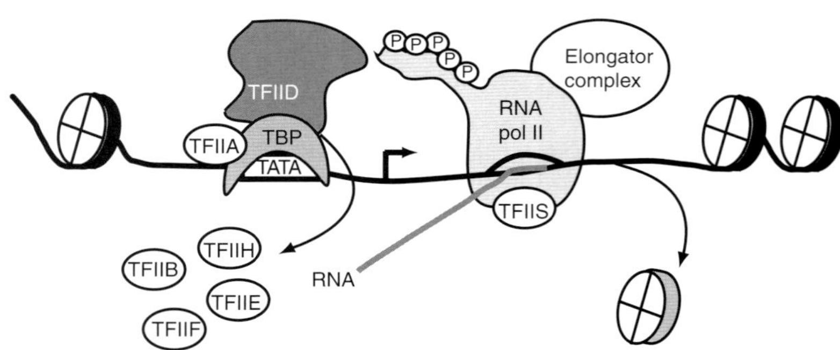


**Fig. 2.4** The conversion of the pre-initiation complex to an elongation complex. **(A)** The pre-initiation complex (PIC) is assembled at the core promoter and the RNA polymerase II is positioned at the initiator element (assembly of the PIC is described in Fig. 2.3). **(B)** Formation of the transcription bubble allows the incorporation of the first few bases of the nascent RNA molecule. At this stage, the polymerase is prone to pausing and reinitiation cycles. **(C)** Phosphorylation of the RNA polymerase C-terminal domain by TFIIH results in the liberation of the polymerase from the GTF complex. Pausing and backtracking are frequent. The paused/arrested polymerase complex can be reinitialized by the elongation factor TFIIS, which cleaves the nascent RNA to reposition it at the catalytic site of RNA polymerase II. The TFIIS complex also possesses chromatin remodeling activity and may promote elongation by removing the nucleosomal barrier to polymerase progression, as indicated by nucleosomal displacement (Copies of figures including color copies, where applicable, are available in the accompanying CD)

TFIIH (which possesses kinase activity) phosphorylates RNA polymerase II on a serine residue within its C-terminal domain (CTD) [48,52,53]. TFIIH phosphorylates the serine residue at position 5 within the repeated sequence, which results in the liberation of the polymerase from the GTF complex, facilitating promoter clearance [52,53].

Promoter clearance converts the ITC into an early elongation complex (EEC) which is prone to frequent pausing and backtracking [49]. This promoter-proximal pausing and 5' movement of the polymerase continues frequently until RNA polymerase II reaches position +30–50 [54]. The backtracked polymerase can spontaneously re-initiate transcription by sliding forward along the template until the active site is in the correct position. However, if the pausing persists, the polymerase will arrest. The arrested EEC can be re-initialized by TFIIS, an elongation factor, which cleaves the nascent transcript to realign the RNA transcript with the active site of the polymerase (Fig. 2.4C) [55]. TFIIS is also part of a larger chromatin remodeling complex (CTEA, chromatin transcription enabling activity), which contains histone acetyltransferase activity and may also help a paused polymerase by remodeling the chromatin structure [56].

Once the polymerase escapes the promoter region, it continues to move in the 3' direction while adding nucleotides to the nascent RNA molecule at a rate of 20–50 nucleotides per second. As the polymerase progresses, it becomes hyperphosphorylated, by accumulating phosphorylated Ser2 in its CTD, a modification catalyzed by the elongation factor p-TEFb, which is composed of Cdk9 and a cyclin (T1, T2a, T2b, or K) [14,57,58]. Modification of the CTD at serine 2 is characteristic of an elongating polymerase [14]. In addition to TFIIS and p-TEFb, both Elongator and Elongin complexes can act to promote RNA polymerase II elongation [59,60].

The choice of which nucleotide to add to the chain depends entirely on base complementarity with the DNA template. Because of its lack of innate error correction, the misincorporation of bases into the nascent mRNA by RNA polymerase II cannot be corrected. Despite the possibility of misincorporation, the fidelity of transcription is quite high. The few errors made by the polymerase are tolerated since RNA molecules are transient copies of the genetic material and do not have an impact on the status of the genome. In addition, since mRNA molecules are produced in large quantities, a few mistakes in the sequence will have little impact on overall functional protein production.

#### 2.4.2.3 Termination

Termination of transcription in prokaryotes and eukaryotes appears to occur via a process of spontaneous dissociation, which is facilitated by a helicase called Rho (Fig. 2.5) [61]. Although there is no "termination" signal in RNA, certain features can lead to the spontaneous dissociation of the nascent RNA from the DNA:RNA hybrid [62]. First, a stretch of 4–10 A/T residues, which base pair less strongly to the DNA template, can trigger the release of the RNA template (Fig. 2.5A). Often, upstream of this A/T rich region is a palindromic sequence that is rich in GC and is able to fold onto itself to form a hairpin. This hairpin, which does not interact with the DNA template, creates a stress that potentiates the detachment of the template at the AT rich region [62].

It has, however, been noted that transcripts produced *in vitro* with recombinant factors were often longer than transcripts produced *in vivo*, leading to the hypothesis that a factor exists *in vivo* that facilitates termination of transcription. Rho, a hexameric protein of identical subunits is a NTP-hydrolyzing helicase which can unwind the RNA molecule from the DNA template, thereby triggering its release from the RNA polymerase [61,63]. Rho appears to recognize a sequence within the RNA molecule which is C-rich and G-poor and contains no secondary structure (Fig. 2.5B). Rho binds the RNA at this site and slides along the nascent RNA in the 3' direction to catch up with the polymerase. When the polymerase pauses at the termination site, Rho catalyzes the unwinding of the RNA:DNA duplex and frees the RNA molecule from the polymerase (Fig. 2.5B) [61].

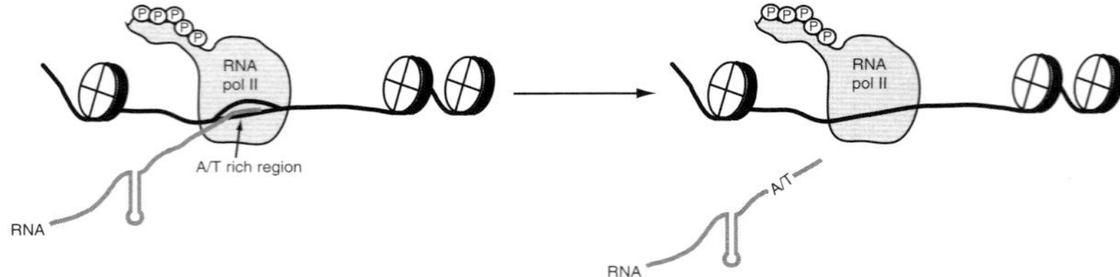


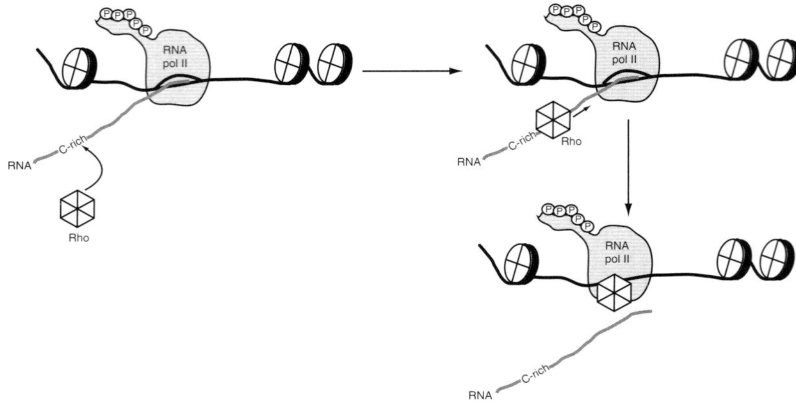


**Fig. 2.5** Termination of transcription can occur by a Rho-dependent or independent mechanism. **(A)** The presence of a palindromic sequence within a nascent transcript can be folded into a hairpin structure in the mRNA molecule. The hairpin, followed by a stretch of 4–10 A/T residues can trigger the spontaneous release of the transcript. The release is believed to be elicited by both the weak base pairing by the A/T rich region and the forces produced by the mRNA secondary structure. **(B)** In the Rho-dependent pathway, the hexameric Rho binds to a C-rich region of the nascent transcript. Once bound to the mRNA, Rho shuttles along the transcript until it reaches the paused polymerase. By unwinding the nascent transcript, Rho catalyzes its release from the DNA template (Copies of figures including color copies, where applicable, are available in the accompanying CD)

## 2.5 DNA Structure and Transcription

*In vivo*, DNA is highly compacted into chromatin. Chromatin is composed of repeating units of protein-DNA complexes called nucleosomes. Nucleosomes are made up of a histone octamer composed of two units each of histone H2A, H2B, H3 and H4 around which a 146 base pair DNA sequence is wrapped [64,65]. This organization of DNA into nucleosomes creates an energetic barrier against transcription, preventing the association of transcription factors and the RNA polymerase with the DNA substrate. The N-terminal tails of the histone molecules appear to be important in the overall compaction of the DNA because they are responsible for the formation of inter-nucleosomal contacts [65]. For efficient transcription to occur on chromatin templates, a transcription factor must be able to direct localized changes in the chromatin structure to allow recruitment of the basal transcriptional machinery [66]. Protein complexes containing enzymatic activities that are correlated with gene activation, such as the ability to acetylate histones (histone acetyltransferase activity), create a permissive state of the DNA that can then be remodeled by ATP-dependent protein complexes [67–69].

## 2.6 Coregulatory Complexes

Co-activator molecules, normally found in large multi-protein complexes, drive transcription by modifying and remodeling the chromatin, and/or by recruiting and stabilizing interactions with the basal transcriptional machinery [70,71]. For a protein to be defined as a co-activator molecule it must be recruited to the DNA via a transcription factor, and it must enhance transcription by that factor but not affect basal transcription rates [72].

### *2.6.1 Histone Acetyltransferases (HAT)*

Acetylation of lysines in the N-termini of histone tails causes the dissociation of higher order nucleosome structures and increases access for the transcription factors and basal transcriptional machinery to DNA. This may be due in part to the charge neutralizing effect of the modification, where ablation of the positive charge of the lysine residues decreases the interaction of the histone tail with the negatively charged DNA.

Histone acetyltransferases have a highly conserved structure, which includes the catalytic domain and a bromodomain [73]. The bromodomain has been implicated in binding acetylated lysine residues and may serve to target the acetylase to the chromatin [73–75]. There are several families of HAT proteins including the p160 family, p300/CBP, and GNAT families.

The p160 co-activator family is a family of evolutionarily related proteins, which were initially characterized by their ability to interact with and potentiate the transcriptional activity of nuclear receptors, a class of transcription factors [76,77]. By virtue of their multiple protein interaction motifs, p160 co-activators can serve as scaffolds onto which larger co-activator complexes can be assembled [78]. Members of the p160 co-activator family are not specific for nuclear receptors, but have been shown to act as co-activators for other transcription factors. For example, the p160 co-activator SRC-1 has been shown to stimulate transcription mediated by AP-1, NF-kB and p53 [79].

p300 was originally purified as the cellular binding protein for the adenoviral protein E1A [80]. The closely related protein CBP was originally purified as a factor associated with the cAMP response element binding protein CREB [81]. p300/CBP share 63% amino acid identity over their length, but have much greater similarity in the domains they share such as the CREB binding site, the E1A binding site and the bromodomain [82]. Although closely related, CBP is not able to rescue the role of p300 in the embryo and the opposite is also true suggesting that the gene dosage of these transcriptional co-activators is essential for development and survival.

Both p300 and CBP possess powerful histone acetyltransferase activities that are able to acetylate histone targets such as histone H3 and H4 as well as many non-histone targets including the basal transcription factor TFIIEβ, p53, GATA-1, NCoA3/p/CIP and HMG-1(Y) [83–89]. The large multi-domain structure of these co-activators also allows them to act as bridges between the DNA-bound transcription factors and the basal transcriptional machinery, as well as providing a link between transcription factors and other co-activator molecules [90].

The GCN5-related N-acetyltransferases (GNAT) family of acetyltransferases includes GCN5, the p300/CBP associated factor PCAF and the more distantly related Hat1, Elp3, and Hpa3 [91]. The family is grouped into sub-families based on sequence similarity over four conserved domains, and all are related to the yeast protein GCN5. Both GCN5 and PCAF are able to interact with p300/CBP and form part of large co-activating complexes associated with nuclear receptors and other transcription factors [92,93]. In addition, PCAF and GCN5 function as part of large distinct multiprotein complexes that share many of the same subunits as the related yeast SAGA complex, including members of the basal transcriptional machinery like the TAFs (part of TFIID) [87,94–96].

**(A)** Histone Displacement

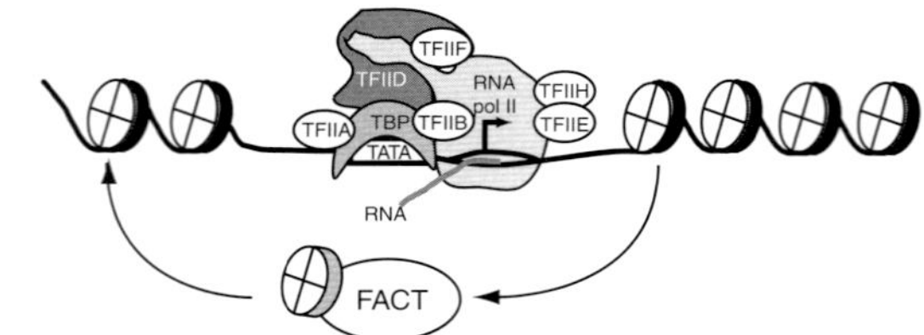


**(B)** Nucleosome Sliding

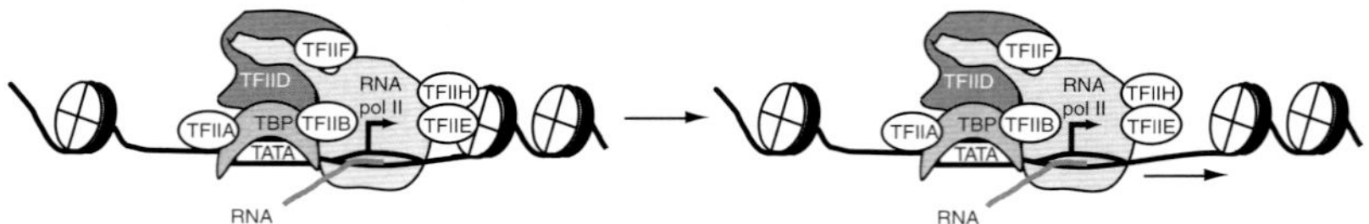


**Fig. 2.6** Nucleosomal reorganization is required for efficient transcriptional elongation. **(A)** Histone chaperone proteins (such as FACT) can act to displace histone proteins from nucleosomes resulting in a reduced number of nucleosomes in the polymerase's path. These nucleosomes can be reassembled behind the transcription bubble. **(B)** Alternatively, the spacing of nucleosomes can be altered by sliding existing nucleosomes along the DNA template, allowing a nucleosome free path ahead of the polymerase (Copies of figures including color copies, where applicable, are available in the accompanying CD)

### 2.6.2 ATP-Dependent Chromatin Remodeling Complexes

To permit the assembly of the pre-initiation complex and the progression of RNA polymerase II along the DNA template during elongation, nucleosomes have to be somehow displaced. This can be accomplished by unwinding the DNA template from around the histone octamer and displacing the histones onto chaperone proteins or by sliding the nucleosome along the DNA template (Fig. 2.6) [97]. A high turnover rate of histone proteins at active genes has been noted, an observation that correlates well with the idea that nucleosomes are removed from the DNA template ahead of the polymerase. Histone chaperone proteins (FACT, Spt6) have also been isolated and may serve to stabilize the displaced histone proteins during elongation [98,99]. To re-position the nucleosomes during transcription, chromatin remodeling factors use ATP to make DNA more accessible to the polymerase and to the transcription factors. There are four families of ATP-dependent chromatin remodelers (ISWI, SNF2, CHD and INO80/SWR) which function as part of large multiprotein complexes [100,101]. It is important to note that not all chromatin remodeling activity is associated with an increase in promoter accessibility and increased transcription rates. Remodeling activity has also been identified in complexes that act to condense the chromatin and shut down active transcription [102].

## 2.7 Covalent Modification of Histones and Transcription Factors

### 2.7.1 Regulation of Transcription Factors by Acetylation and Deacetylation

Acetylation of histone tails increases the accessibility of DNA to transcription factors and to the basal transcriptional machinery. Histone acetylation is associated not with transcription per se, but with a permissive state which would allow transcription to occur. However, modification by acetylation is not reserved to histones alone. Many transcription factors have recently been shown to be targets of acetylation by p300/CBP, PCAF, or GCN5, or combinations thereof, and this covalent modification regulates their transcriptional activation potential. Acetylation can modulate a transcription factor's DNA binding activity (e.g. p53 acetylation by p300/CBP and PCAF), its association with other transcriptional regulators (e.g. acetylation of C/EBPβ by GCN5), or its half-life (e.g. E2F1 acetylation by PCAF) [86,103–106].

### 2.7.2 *Regulation of Transcription by Ubiquitylation*

The ubiquitin-proteasome system is proposed to act to control transcription by recruitment of the 26 S proteasome to the promoters of active genes. Since many acidic transcriptional activation domains have been shown to act as sequences that drive the destruction of the transcription factor (degrons), it has been suggested that the recruitment of members of the ubiquitin pathway to the promoters of active genes would lead to the ubiquitylation of not only the transcription factor, but also the histone targets (such as histone H2B), RNA polymerase II and possibly the recruited co-activator molecules [107–113]. Indeed, the potency of an activation domain has been inversely linked to protein stability, but only when linked to an intact DNA-binding domain [114]. This mechanism would ensure the rapid shutdown of transcription by making the activation of transcription self-limiting.

Both H2A and H2B can be mono-ubiquitylated [115]. These modifications are observed especially at the promoter end of a gene and are correlated with the production of full length transcripts [116]. In addition, ubiquitylation of H2B may direct the post-translational modification of other residues, such as lysine methylation [117,118].

### 2.7.3 *Regulation of Transcription by Other Post-translational Modifications*

In addition to acetylation and ubiquitylation, histones and transcription factors can be methylated (on lysine and arginine residues) and sumoylated (on lysine residues) [67]. Unlike acetylation, which is strongly correlated with a permissive chromatin environment, methylation of histone proteins can be associated with repressed or active gene transcription. For example, the frequency of methylation of lysine 4 on histone H3 decreases from the promoter region and 5' end of the gene to the 3' end, while methylation of lysine 36 increases towards the end of the transcribed region [119,120]. Transcription factors and co-regulatory molecules are also the targets of methylation, especially on arginine residues [121,122].

While mono-ubiquitylation of histones is thought to help the polymerase elongate the RNA strand, sumolation (the addition of SUMO. Small Ubiquitin-like Modifier) to lysine residues prevents acetylation and correlates with a repressed chromatin structure [123]. Correlations between phosphorylation and acetylation have also been noticed, where phosphorylated histone tails act as better templates for acetylases and are less likely to be methylated, producing permissive chromatin [124,125].

## Glossary and Abbreviations

| | |
|---|---|
| CBP | CREB-binding protein |
| CREB | cAMP responsible element binding |
| CTEA | Chromatin transcription enabling activity |
| DPE | Downstream promoter element |

| | |
|---|---|
| EEC | Early elongation complex |
| HAT | Histone acetyltransferase |
| ITC | Initially transcribing complex |
| SUMO | Small ubiquitin-like modifier |
| TAFs | TBP associated factors |
| TBP | TATA binding protein |

## References

### *Structure of RNA Polymerases*

1. Ebright, R.H. (2000) *J Mol Biol* **304**(5), 687–698.
2. Zhang, G., Campbell, E.A., Minakhin, L., Richter, C., Severinov, K., and Darst, S.A. (1999) *Cell* **98**(6), 811–824.
3. Korzheva, N., Mustaev, A., Kozlov, M., Malhotra, A., Nikiforov, V., Goldfarb, A., and Darst, S.A. (2000) *Science (New York, N.Y* **289**(5479), 619–625.
4. Jasiak, A.J., Armache, K.J., Martens, B., Jansen, R.P., and Cramer, P. (2006) *Molecular cell* **23**(1), 71–81.
5. Bischler, N., Brino, L., Carles, C., Riva, M., Tschochner, H., Mallouh, V., and Schultz, P. (2002) *Embo J* **21**(15), 4136–4144.
6. Cramer, P., Bushnell, D.A., and Kornberg, R.D. (2001) *Science (New York, N.Y* **292**(5523), 1863–1876.
7. Gnatt, A.L., Cramer, P., Fu, J., Bushnell, D.A., and Kornberg, R.D. (2001) *Science (New York, N.Y* **292**(5523), 1876–1882.
8. Bushnell, D.A., and Kornberg, R.D. (2003) *Proc Natl Acad Sci U S A* **100**(12), 6969–6973.
9. Kuznedelov, K., Korzheva, N., Mustaev, A., and Severinov, K. (2002) *Embo J* **21**(6), 1369–1378.
10. Asturias, F.J. (2004) *Curr Opin Struct Biol* **14**(2), 121–129.
11. Shilatifard, A., Conaway, R.C., and Conaway, J.W. (2003) *Annu Rev Biochem* **72**, 693–715.
12. Woychik, N.A., and Hampsey, M. (2002) *Cell* **108**(4), 453–463.
13. Phatnani, H.P., and Greenleaf, A.L. (2006) *Genes & development* **20**(21), 2922–2936.
14. Yamada, T., Yamaguchi, Y., Inukai, N., Okamoto, S., Mura, T., and Handa, H. (2006) *Molecular cell* **21**(2), 227–237.
15. de la Mata, M., and Kornblihtt, A.R. (2006) *Nat Struct Mol Biol* **13**(11), 973–980.

### *What Drives Transcription – The Promoter Region*

16. Pribnow, D. (1975) *Proc Natl Acad Sci U S A* **72**(3), 784–788.
17. Schaller, H., Gray, C., and Herrmann, K. (1975) *Proc Natl Acad Sci U S A* **72**(2), 737–741.
18. Siebenlist, U. (1979) *Nature* **279**(5714), 651–652.
19. Siebenlist, U. (1979) *Nucleic Acids Res* **6**(5), 1895–1907.
20. Maniatis, T., Ptashne, M., Backman, K., Kield, D., Flashman, S., Jeffrey, A., and Maurer, R. (1975) *Cell* **5**(2), 109–113.
21. Paule, M.R., and White, R.J. (2000) *Nucleic Acids Res* **28**(6), 1283–1298.
22. Perna, P.J., Harris, G.H., Iida, C.T., Kownin, P., Bugren, S., and Paule, M.R. (1992) *Gene Expr* **2**(1), 71–78.
23. Butler, J.E., and Kadonaga, J.T. (2002) *Genes & development* **16**(20), 2583–2592.
24. Sandelin, A., Carninci, P., Lenhard, B., Ponjavic, J., Hayashizaki, Y., and Hume, D.A. (2007) *Nat Rev Genet* **8**(6), 424–436.
25. Smale, S.T., and Kadonaga, J.T. (2003) *Annu Rev Biochem* **72**, 449–479.
26. Gross, P., and Oelgeschlager, T. (2006) *Biochem Soc Symp* (73), 225–236.
27. Kadonaga, J.T. (2004) *Cell* **116**(2), 247–257.
28. Burley, S.K. (1996) *Curr Opin Struct Biol* **6**(1), 69–75
29. Burley, S.K., and Roeder, R.G. (1996) *Annu Rev Biochem* **65**, 769–799.
30. Andel, F., 3rd, Ladurner, A.G., Inouye, C., Tjian, R., and Nogales, E. (1999) *Science (New York, N.Y* **286**(5447), 2153–2156.
31. Usheva, A., and Shenk, T. (1994) *Cell* **76**(6), 1115–1121.
32. Wieczorek, E., Brand, M., Jacq, X., and Tora, L. (1998) *Nature* **393**(6681), 187–191.

33. Kadonaga, J.T. (2002) *Exp Mol Med* **34**(4), 259–264.
34. Lagrange, T., Kapanidis, A.N., Tang, H., Reinberg, D., and Ebright, R.H. (1998) *Genes & development* **12**(1), 34–44.
35. Saxonov, S., Berg, P., and Brutlag, D.L. (2006) *Proc Natl Acad Sci U S A* **103**(5), 1412–1417.
36. Lim, C.Y., Santoso, B., Boulay, T., Dong, E., Ohler, U., and Kadonaga, J.T. (2004) *Genes & development* **18**(13), 1606–1617.
37. Orphanides, G., and Reinberg, D. (2002) *Cell* **108**(4), 439–451.
38. Roeder, R.G. (1998) *Cold Spring Harb Symp Quant Biol* **63**, 201–218.

## What Stimulates Transcription – Transcription Factors

39. Cosma, M.P. (2002) *Molecular cell* **10**(2), 227–236.
40. Featherstone, M. (2002) *Curr Opin Genet Dev* **12**(2), 149–155.
41. Malik, S., and Roeder, R.G. (2005) *Trends Biochem Sci* **30**(5), 256–263.

## The Mechanics of Transcription – Making mRNA

42. Kuznedelov, K., Minakhin, L., Niedziela-Majka, A., Dove, S.L., Rogulja, D., Nickels, B.E., Hochschild, A., Heyduk, T., and Severinov, K. (2002) *Science (New York, N.Y* **295**(5556), 855–857.
43. Holstege, F.C., Jennings, E.G., Wyrick, J.J., Lee, T.I., Hengartner, C.J., Green, M.R., Golub, T.R., Lander, E.S., and Young, R.A. (1998) *Cell* **95**(5), 717–728.
44. Hampsey, M. (1998) *Microbiol Mol Biol Rev* **62**(2), 465–503.
45. Thomas, M.C., and Chiang, C.M. (2006) *Crit Rev Biochem Mol Biol* **41**(3), 105–178.
46. Sakurai, H., and Fukasawa, T. (1998) *J Biol Chem* **273**(16), 9534–9538.
47. Kim, T.K., Ebright, R.H., and Reinberg, D. (2000) *Science (New York, N.Y* **288**(5470), 1418–1422.
48. Ohkuma, Y., and Roeder, R.G. (1994) *Nature* **368**(6467), 160–163.
49. Saunders, A., Core, L.J., and Lis, J.T. (2006) *Nat Rev Mol Cell Biol* **7**(8), 557–567.
50. Bushnell, D.A., Westover, K.D., Davis, R.E., and Kornberg, R.D. (2004) *Science (New York, N.Y* **303**(5660), 983–988.
51. Chen, B.S., and Hampsey, M. (2004) *Molecular and cellular biology* **24**(9), 3983–3991.
52. Dubois, M.F., Vincent, M., Vigneron, M., Adamczewski, J., Egly, J.M., and Bensaude, O. (1997) *Nucleic Acids Res* **25**(4), 694–700.
53. Trigon, S., Serizawa, H., Conaway, J.W., Conaway, R.C., Jackson, S.P., and Morange, M. (1998) *J Biol Chem* **273**(12), 6769–6775.
54. Ujvari, A., Pal, M., and Luse, D.S. (2002) *J Biol Chem* **277**(36), 32527–32537.
55. Fish, R.N., and Kane, C.M. (2002) *Biochim Biophys Acta* **1577**(2), 287–307.
56. Guermah, M., Palhan, V.B., Tackett, A.J., Chait, B.T., and Roeder, R.G. (2006) *Cell* **125**(2), 275–286.
57. Shim, E.Y., Walker, A.K., Shi, Y., and Blackwell, T.K. (2002) *Genes & development* **16**(16), 2135–2146.
58. Chao, S.H., and Price, D.H. (2001) *J Biol Chem* **276**(34), 31793–31799.
59. Close, P., Hawkes, N., Cornez, I., Creppe, C., Lambert, C.A., Rogister, B., Siebenlist, U., Merville, M.P., Slaugenhaupt, S.A., Bours, V., Svejstrup, J.Q., and Chariot, A. (2006) *Molecular cell* **22**(4), 521–531.
60. Otero, G., Fellows, J., Li, Y., de Bizemont, T., Dirac, A.M., Gustafsson, C.M., Erdjument-Bromage, H., Tempst, P., and Svejstrup, J.Q. (1999) *Molecular cell* **3**(1), 109–118.
61. Richardson, J.P. (2002) *Biochim Biophys Acta* **1577**(2), 251–260.
62. Gusarov, I., and Nudler, E. (1999) *Molecular cell* **3**(4), 495–504.
63. Park, J.S., and Roberts, J.W. (2006) *Proc Natl Acad Sci U S A* **103**(13), 4870–4875.

## DNA Structure

64. Narlikar, G.J., Fan, H.Y., and Kingston, R.E. (2002) *Cell* **108**(4), 475–487.
65. Uberbacher, E.C., and Bunick, G.J. (1989) *J Biomol Struct Dyn* **7**(1), 1–18.
66. Beato, M., and Eisfeld, K. (1997) *Nucleic Acids Res* **25**(18), 3559–3563.
67. Kouzarides, T. (2007) *Cell* **128**(4), 693–705.
68. Marmorstein, R. (2001) *Cell Mol Life Sci* **58**(5–6), 693–703.
69. Mohrmann, L., and Verrijzer, C.P. (2005) *Biochim Biophys Acta* **1681**(2–3), 59–73.

## *Coregulatory Complexes*

70. Xu, J., and Li, Q. (2003) *Mol Endocrinol* **17**(9), 1681–1692.
71. Rachez, C., Lemon, B.D., Suldan, Z., Bromleigh, V., Gamble, M., Naar, A.M., Erdjument-Bromage, H., Tempst, P., and Freedman, L.P. (1999) *Nature* **398**(6730), 824–828.
72. Robyr, D., Wolffe, A.P., and Wahli, W. (2000) *Mol. Endocrinol.* **14**, 329–347.
73. Dhalluin, C., Carlson, J.E., Zeng, L., He, C., Aggarwal, A.K., and Zhou, M.M. (1999) *Nature* **399** (6735), 491–496.
74. Jeanmougin, F., Wurtz, J.M., Le Douarin, B., Chambon, P., and Losson, R. (1997) *Trends Biochem Sci* **22**(5), 151–153.
75. Kouzarides, T. (2000) *Embo J* **19**(6), 1176–1179.
76. Cavailles, V., Dauvois, S., Danielian, P.S., and Parker, M.G. (1994) *Proc Natl Acad Sci U S A* **91**(21), 10009–10013.
77. Halachmi, S., Marden, E., Martin, G., MacKay, H., Abbondanza, C., and Brown, M. (1994) *Science* **264**(5164), 1455–1458.
78. Yao, T.P., Ku, G., Zhou, N., Scully, R., and Livingston, D.M. (1996) *Proc Natl Acad Sci U S A* **93**(20), 10626–10631.
79. Sheppard, K.A., Phelps, K.M., Williams, A.J., Thanos, D., Glass, C.K., Rosenfeld, M.G., Gerritsen, M.E., and Collins, T. (1998) *J Biol Chem* **273**(45), 29291–29294.
80. Eckner, R., Ewen, M.E., Newsome, D., Gerdes, M., DeCaprio, J.A., Lawrence, J.B., and Livingston, D.M. (1994) *Genes Dev* **8**(8), 869–884.
81. Chrivia, J.C., Kwok, R.P., Lamb, N., Hagiwara, M., Montminy, M.R., and Goodman, R.H. (1993) *Nature* **365**(6449), 855–859.
82. Arany, Z., Sellers, W.R., Livingston, D.M., and Eckner, R. (1994) *Cell* **77**(6), 799–800.
83. Bannister, A.J., and Kouzarides, T. (1996) *Nature* **384**(6610), 641–643.
84. Ogryzko, V.V., Schiltz, R.L., Russanova, V., Howard, B.H., and Nakatani, Y. (1996) *Cell* **87**(5), 953–959.
85. Boyes, J., Byfield, P., Nakatani, Y., and Ogryzko, V. (1998) *Nature* **396**(6711), 594–598.
86. Gu, W., and Roeder, R.G. (1997) *Cell* **90**(4), 595–606.
87. Imhof, A., Yang, X.J., Ogryzko, V.V., Nakatani, Y., Wolffe, A.P., and Ge, H. (1997) *Curr Biol* **7**(9), 689–692.
88. Munshi, N., Merika, M., Yie, J., Senger, K., Chen, G., and Thanos, D. (1998) *Mol Cell* **2**(4), 457–467.
89. Chen, H., Lin, R.J., Xie, W., Wilpitz, D., and Evans, R.M. (1999) *Cell* **98**(5), 675–686.
90. Chan, H.M., and La Thangue, N.B. (2001) *J Cell Sci* **114**(Pt 13), 2363–2373.
91. Sterner, D.E., and Berger, S.L. (2000) *Microbiol Mol Biol Rev* **64**(2), 435–459.
92. Yang, X.J., Ogryzko, V.V., Nishikawa, J., Howard, B.H., and Nakatani, Y. (1996) *Nature* **382**(6589), 319–324.
93. Xu, W., Edmondson, D.G., and Roth, S.Y. (1998) *Mol Cell Biol* **18**(10), 5659–5669.
94. Forsberg, E.C., Lam, L.T., Yang, X.J., Nakatani, Y., and Bresnick, E.H. (1997) *Biochemistry* **36**(50), 15918–15924.
95. Bhaumik, S.R., and Green, M.R. (2002) *Mol Cell Biol* **22**(21), 7365–7371.
96. Brand, M., Leurent, C., Mallouh, V., Tora, L., and Schultz, P. (1999) *Science* **286**(5447), 2151–2153.
97. Workman, J.L. (2006) *Genes & development* **20**(15), 2009–2017.
98. Jimeno-Gonzalez, S., Gomez-Herreros, F., Alepuz, P.M., and Chavez, S. (2006) *Molecular and cellular biology* **26**(23), 8710–8721.
99. Adkins, M.W., and Tyler, J.K. (2006) *Molecular cell* **21**(3), 405–416.
100. Corona, D.F., Langst, G., Clapier, C.R., Bonte, E.J., Ferrari, S., Tamkun, J.W., and Becker, P.B. (1999) *Molecular cell* **3**(2), 239–245.
101. Sif, S. (2004) *J Cell Biochem* **91**(6), 1087–1098.
102. Xue, Y., Wong, J., Moreno, G.T., Young, M.K., Cote, J., and Wang, W. (1998) *Molecular cell* **2**(6), 851–861.

## *Covalent Modification of Histones and Transcription Factors*

103. Liu, L., Scolnick, D.M., Trievel, R.C., Zhang, H.B., Marmorstein, R., Halazonetis, T.D., and Berger, S.L. (1999) *Mol Cell Biol* **19**(2), 1202–1209.
104. Wiper-Bergeron, N., Salem, H.A., Tomlinson, J.J., Wu, D., and Hache, R.J. (2007) *Proc Natl Acad Sci U S A* **104**(8), 2703–2708.
105. Narlikar, G.J., Fan, H.Y., and Kingston, R.E. (2002) *Cell* **108**(4), 475–487.
106. Martinez-Balbas, M.A., Bauer, U.M., Nielsen, S.J., Brehm, A., and Kouzarides, T. (2000) *Embo J* **19**(4), 662–671.
107. Kim, T.K., and Maniatis, T. (1996) *Science* **273**(5282), 1717–1719.
108. Lo, R.S., and Massague, J. (1999) *Nat Cell Biol* **1**(8), 472–478.

109. Muratani, M., and Tansey, W.P. (2003) *Nat Rev Mol Cell Biol* **4**(3), 192–201.
110. Brower, C.S., Sato, S., Tomomori-Sato, C., Kamura, T., Pause, A., Stearman, R., Klausner, R.D., Malik, S., Lane, W.S., Sorokina, I., Roeder, R.G., Conaway, J.W., and Conaway, R.C. (2002) *Proc Natl Acad Sci U S A* **99**(16), 10353–10358.
111. Salghetti, S.E., Muratani, M., Wijnen, H., Futcher, B., and Tansey, W.P. (2000) *Proc Natl Acad Sci U S A* **97**(7), 3118–3123.
112. Salghetti, S.E., Caudy, A.A., Chenoweth, J.G., and Tansey, W.P. (2001) *Science* **293**(5535), 1651–1653.
113. O'Connell, B.C., and Harper, J.W. (2007) *Current opinion in cell biology* **19**(2), 206–214.
114. Molinari, E., Gilman, M., and Natesan, S. (1999) *Embo J* **18**(22), 6439–6447.
115. Minsky, N., and Oren, M. (2004) *Molecular cell* **16**(4), 631–639
116. Xiao, T., Kao, C.F., Krogan, N.J., Sun, Z.W., Greenblatt, J.F., Osley, M.A., and Strahl, B.D. (2005) *Molecular and cellular biology* **25**(2), 637–651.
117. Shahbazian, M.D., Zhang, K., and Grunstein, M. (2005) *Molecular cell* **19**(2), 271–277.
118. Ezhkova, E., and Tansey, W.P. (2004) *Molecular cell* **13**(3), 435–442.
119. Rao, B., Shibata, Y., Strahl, B.D., and Lieb, J.D. (2005) *Molecular and cellular biology* **25**(21), 9447–9459.
120. Santos-Rosa, H., Schneider, R., Bannister, A.J., Sherriff, J., Bernstein, B.E., Emre, N.C., Schreiber, S.L., Mellor, J., and Kouzarides, T. (2002) *Nature* **419**(6905), 407–411.
121. Xu, W., Chen, H., Du, K., Asahara, H., Tini, M., Emerson, B.M., Montminy, M., and Evans, R.M. (2001) *Science (New York, N.Y* **294**(5551), 2507–2511.
122. Barrero, M.J., and Malik, S. (2006) *Molecular cell* **24**(2), 233–243.
123. Nathan, D., Ingvarsdottir, K., Sterner, D.E., Bylebyl, G.R., Dokmanovic, M., Dorsey, J.A., Whelan, K.A., Krsmanovic, M., Lane, W.S., Meluh, P.B., Johnson, E.S., and Berger, S.L. (2006) *Genes & development* **20**(8), 966–976.
124. Lo, W.S., Duggan, L., Emre, N.C., Belotserkovskya, R., Lane, W.S., Shiekhattar, R., and Berger, S.L. (2001) *Science (New York, N.Y* **293**(5532), 1142–1146.
125. Lo, W.S., Trievel, R.C., Rojas, J.R., Duggan, L., Hsu, J.Y., Allis, C.D., Marmorstein, R., and Berger, S.L. (2000) *Molecular cell* **5**(6), 917–926.

## Key References

Woychik, N.A., and Hampsey, M. (2002) *Cell* **108**(4), 453–463.

Sandelin, A., Carninci, P., Lenhard, B., Ponjavic, J., Hayashizaki, Y., and Hume, D.A. (2007) *Nat Rev Genet* **8**(6), 424–436.

Malik, S., and Roeder, R.G. (2005) *Trends Biochem Sci* **30**(5), 256–263.

Saunders, A., Core, L.J., and Lis, J.T. (2006) *Nat Rev Mol Cell Biol* **7**(8), 557–567.

Kouzarides, T. (2007) *Cell* **128**(4), 693–705.

Workman, J.L. (2006) *Genes & development* **20**(15), 2009–2017.

# Chapter 3
# RNA Processing and Translation

**Christina Karamboulas, Nadine Wiper-Bergeron, and Ilona S. Skerjanc**

**Abstract** The information that codes for all proteins in a cell is found on specific segments within the DNA. When a cell requires the function of a particular protein, it must initiate the steps involved in the synthesis of this protein. The overall process is termed gene expression. Transcription is the process whereby the cell makes a copy of the genetic information required to build that particular protein. Transcription yields the copy of a particular gene termed the primary transcript, which undergoes several processing events to generate the mature messenger RNA (mRNA). The mRNA molecule is then transported to the cytoplasm where it associates with the ribosome. Here, the information within the transcript is decoded into a polypeptide chain of amino acids to give rise to a particular protein with a specific function. The details of RNA processing and translation are discussed in this chapter.

**Keywords** RNA · Processing · Protein · Translation · Amino acids · Genetic code · Ribosome

## 3.1 Introduction

Once an RNA molecule is transcribed it is referred to as a primary transcript or pre-mRNA because it must be processed before translation can occur. First, the pre-mRNA is modified at its 3' (3' polyadenylation) and 5' (5' capping) ends to increase its stability. Increased stability is important since the mRNA must be shuttled out of the nucleus for translation in eukaryotes.

In addition, the pre-mRNA is cleaved and religated, via a process called RNA splicing, to remove intronic sequences that do not code for any protein (Fig. 3.1A). RNA splicing also permits the production of alternative protein products from a single pre-mRNA species and is therefore a target of regulatory mechanisms.

## 3.2 RNA Processing

### *3.2.1 The 5' Cap*

The 5' cap protects the mRNA from degradation, aids in the transport of the mRNA out of the nucleus and is involved in the initiation of protein translation (reviewed in [1]). The 5' capping of the pre-mRNA occurs during transcriptional elongation once the nascent RNA transcript reaches a length of 25 nucleotides. At this stage, the first 5' phosphate group of the initial nucleotide is removed (Fig. 3.1B). Subsequently, guanosine monophosphate is added in an inverted orientation to yield a unique 5'-5'-triphosphate linkage, a process catalysed by guanylyltransferase, an RNA

I.S. Skerjanc
Department of Biochemistry, Microbiology and Immunology, University of Ottawa, 451 Smyth Road, Ottawa, Ontario, K1H 8M5, Canada
e-mail: iskerjan@uottawa.ca

S. Krawetz (ed.), *Bioinformatics for Systems Biology*, DOI 10.1007/978-1-59745-440-7_3, 

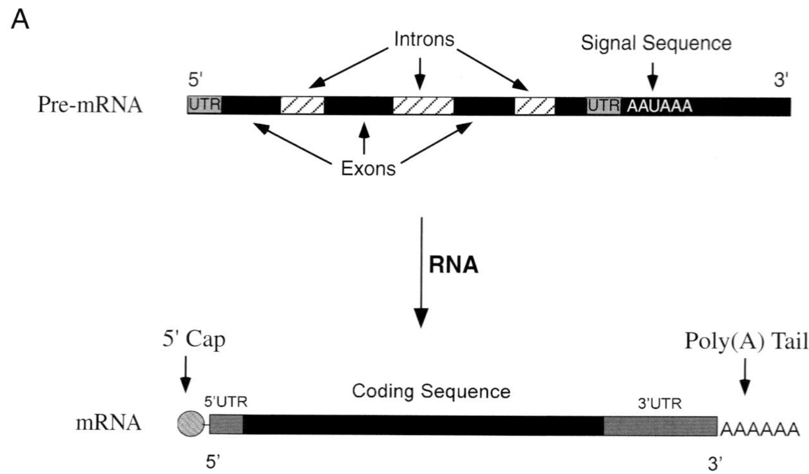


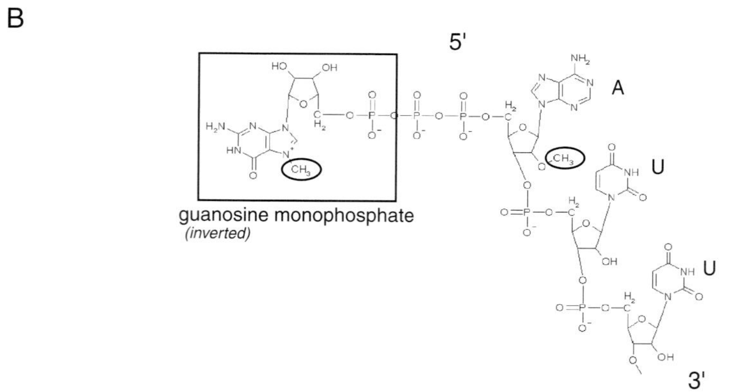


**Fig. 3.1** Schematic of pre-mRNA, mature mRNA and the 5' methylguanosine cap. **(A)** Newly transcribed mRNA or pre-mRNA consists of exons with intervening introns. Pre-mRNA undergoes RNA processing events: 5' capping, splicing (removal of introns) and 3' polyadenylation. The 5' end is modified by capping enzymes with the addition of a methylguanosine cap. The 3' end is modified by initial cleavage of ~20 nucleotides downstream of the polyadenylation signal sequence (AAUAAA). The enzyme poly A polymerase subsequently adds up to 250 adenosines. Mature mRNA also contains 5' and 3' untranslated regions (UTRs). **(B)** Structure of the 5' cap. The 5' nucleotide of an mRNA originally contains three phosphate groups. The first phosphate is removed and a guanosine monophosphate is linked in an inverted orientation, creating a unique 5'-5' triphosphate bridge. Methylation occurs on the newly added guanosine base at the 7 position and on the ribose sugar of the adjacent nucleotide (in this case adenosine) at the 2' position (Copies of figures including color copies, where applicable, are available in the accompanying CD)

polymerase II associated enzyme that catalyzes the transfer of GMP to the nascent mRNA [2]. Next, methylation by methyl transferases takes place on the newly added guanosine base at the 7 position and on the ribose sugar of the initial nucleotide of the mRNA at the 2' position.

### 3.2.2 The 3' Poly(A) Tail

Modification at the 3' end of the RNA transcript includes the cleavage of the transcript downstream of a signal sequence (AAUAAA) and the subsequent addition of a string of adenosine residues (Fig. 3.1A). The mechanism requires the activity of two multi-subunit complexes, that are bound to RNA polymerase II during transcription, termed the Cleavage and Polyadenylation Specificity Factor (CPSF) and the Cleavage Stimulation Factor (CstF) [3,4]. Once the RNA polymerase transcribes the signal sequence, these complexes dissociate from the RNA polymerase and bind to the RNA transcript. CPSF and CstF recruit cleavage factors and stimulate the cleavage of the RNA approximately 20 nucleotides downstream of the signal sequence. Following RNA cleavage, Poly(A) polymerase is recruited to the pre-mRNA, to which it adds ~100 to 250 adenine nucleotides in a template independent fashion. As the poly(A) tail increases in length it recruits Polyadenylate Binding Proteins (PABPs) which act as a molecular ruler, determining the length of the poly(A) tail. Similar to the 5' cap, the poly(A) tail increases the pre-mRNA's stability by protecting it from premature degradation, assisting its transport out of the nucleus, and facilitating the initiation of translation.

### 3.2.3 *RNA Splicing*

The DNA template used to synthesize the pre-mRNA contains sequences, termed introns, which do not code for any protein sequence. Before translation can occur, these non-coding sequences within the RNA must be removed via a process called splicing. In prokaryotes, which have more compact genomes, there are no intronic sequences and so splicing is unnecessary.

Eukaryotic pre-mRNAs can include several intronic sequences that must be cleaved before the initiation of translation and protein production. Intronic sequences are bordered by special sequences, known as splice sites (5' splice site and 3' splice site), which are highly conserved and serve as recognition signals in the splicing process. In addition to the splice sites, other sequences, known as the branch point and a pyrimidine rich region (pyrimidine tract) are required for efficient splicing (Fig. 3.2A). The branch point contains an invariable adenine residue found within a conserved sequence. Sequences located within exons (protein coding regions) called exonic splicing enhancers (ESEs) are also required [5,6]. These sequences ensure accurate recognition and splicing of introns. Precision of splicing is essential as the addition or removal of a single nucleotide can shift the reading frame of the mRNA and can result in the formation of a non-functional protein.

Splicing occurs via two sequential trans-esterification reactions, which are depicted in Fig. 3.2B. At first, the 2'OH group of the ribose sugar of the adenine residue (Fig. 3.2B, see A), also termed the branch point, reacts with the phosphate at the 5' end of the intron. This reaction results in the formation of a 2'-5' phosphodiester bond and a looped structure called a lariat. Secondly, together the free 3'OH group of exon 1 reacts with the 5' end of exon 2 which joins exon 1 and exon 2 and releases the lariat (intron) which is then degraded.

The chemical reactions described above are regulated by a large multi-protein complex called the spliceosome, which assembles in a stepwise manner on the RNA transcript at intron/exon junctions. Five small nuclear ribonucleoproteins (snRNPs) designated U1, U2 and U4-U6 and their associated small nuclear RNAs (snRNAs) are essential for the spliceosome assembly at the splice site. The establishment of an active spliceosome requires the coordinated assembly and disassembly of the 5 snRNPs and several rearrangements of the RNA [7].

The steps in the assembly of the spliceosome are depicted in Fig. 3.3. The U1 snRNP binds to the pre-mRNA at the 5'splice site whereas the U1 snRNA base pairs with the conserved sequence in the 5'splice site. Branch- point binding protein (BBP) and the helper protein U2AF are bound to the branch point A, and recruit U2 snRNP. Base pairing between U2 snRNA and the bases surrounding the A residue displaces BBP and U2AF and causes the A to bulge out of the pre-mRNA. This places the 2'OH group of the branch point adenosine residue in a favorable position to react with the phosphate at the 5' end of the intron. U4,U5, and U6 binding to the 5'splice site results in changes in RNA-RNA interactions within and among the three snRNPs. Consequently, U6 and U2 interact, U1 and U4 are released, and the first trans-esterification reaction occurs. Binding of

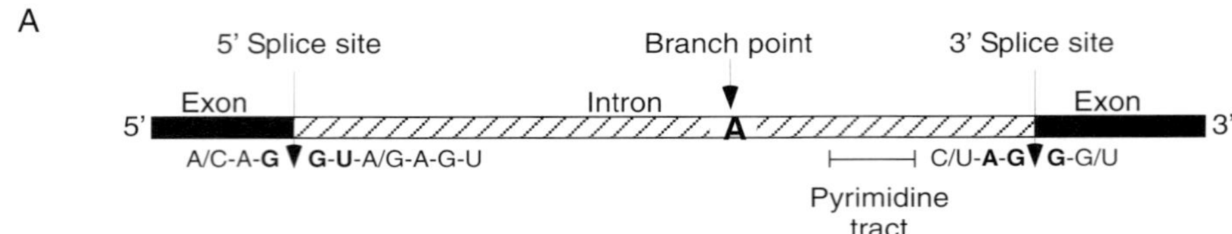


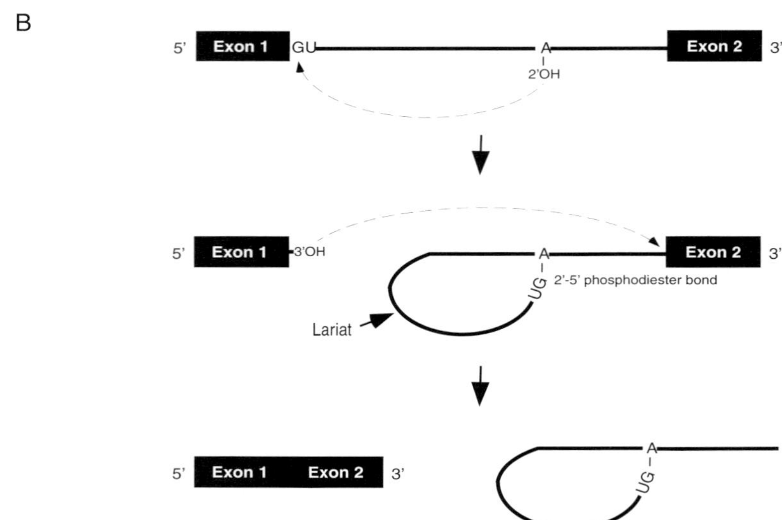


**Fig. 3.2** Splicing of pre-mRNA. **(A)** Eukaryotic pre-mRNAs generally contain conserved sequences at each end of an intron. The 5' (G/GU) splice site and the 3' (AG/G) splice site are located within these conserved sequences. A pyrimidine tract is located near the 3' end of the intron. **(B)** The steps of RNA splicing. The 2' OH group in the ribose sugar of the adenine residue (at the branch point) reacts with the 5' phosphate of the guanosine nucleotide, at the 5' splice site, which results in a 2'-5' phosphodiester bond. The looped structure formed is termed a lariat. The free 3' OH at the end of Exon 1 reacts with the 5' phosphate at the beginning of Exon 2. This results in the joining of Exon 1 with Exon 2 and the release of the lariat, which is degraded (Copies of figures including color copies, where applicable, are available in the accompanying CD)

U5 snRNA with exon 1 stabilizes this site and allows for the interaction with the 3'splice site. Subsequently, U6 catalyzes the second trans-esterification reaction, which involves the reaction of the free 3'OH group of exon 1 with the 5' end of exon 2. Exon 1 and exon 2 are joined together triggering the release of the lariat.

Components that recruit the snRNPs to the splice sites are bound to the CTD of RNA polymerase during transcription, effectively linking the process of transcriptional elongation with splicing. Similar to capping and polyadenylation, RNA polymerase appears to form a platform for the association of the proteins that will be required for the modification of the nascent transcript [8,9].

#### 3.2.3.1 *Alternative Splicing*

It is possible to generate different mature mRNAs from a single transcript by splicing events. For example, consider the splicing of a pre-mRNA containing the exons A, B, C, and D (from 5' to 3').

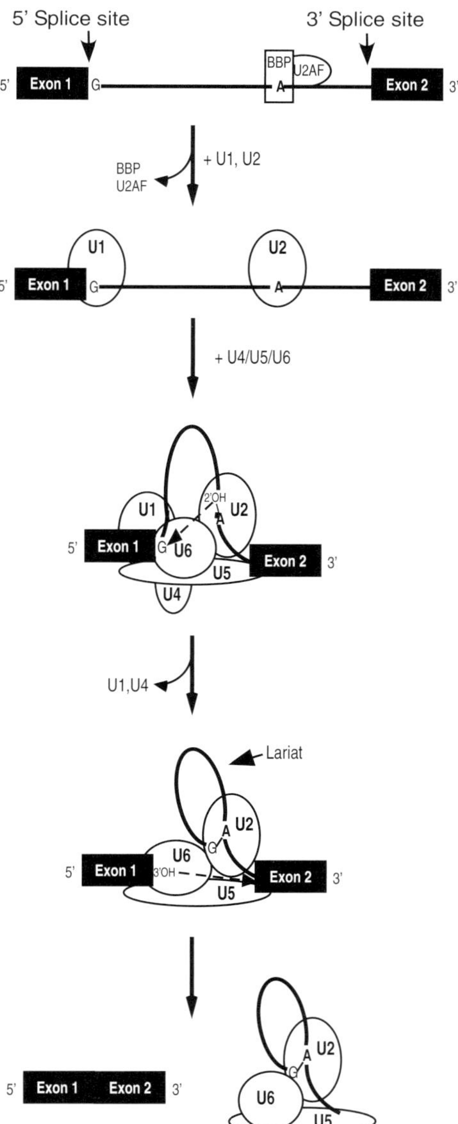


**Fig. 3.3** Assembly of the Splicing Machinery. The removal of introns in pre-mRNA requires the coordinated assembly and disassembly of a multi-protein complex termed the spliceosome. The spliceosome contains five small nuclear ribonucleoproteins (snRNPs) (U1, U2, U4, U5, and U6). See text for details. BBP – Branch-point binding protein (Copies of figures including color copies, where applicable, are available in the accompanying CD)

It is possible that, depending on which splice sites are used, different mRNA species will be produced (for example A-B-C-D, A-C-D, A-B-D among others). In this example, the prediction of some of the mRNA species required the splicing of exons along with introns, producing a unique sequence. The end product includes mature mRNA molecules with different information, which in effect produces different proteins. The significance of alternative splicing lies in the fact that a single gene can give rise to different protein products. Consequently, the alternate proteins produced may differ in binding properties, intracellular localization, enzymatic activity, and protein stability and increase the variability of the cellular response.

Exonic splicing enhancers (ESEs) determine which exons are retained in the mature mRNA and which are spliced out via the recruitment of serine-arginine-rich (SR) proteins (7,10). SR proteins are nuclear phoshoproteins that contain two domains: an N-terminal RNA binding domain and a C-terminal SR-rich domain. Recent studies showed that U1snRNP/SR associates with RNA polymerase II suggesting that the coupling of transcription and splicing results in highly efficient splicing of pre-mRNA transcripts [11,12].

## 3.3 Translation

Translation is the process by which the processed mature mRNA is used as a template for protein synthesis. Translation is a highly regulated process that requires a cell organelle termed ribosome. RNA has proven to play a very important role during translation. Not only is mRNA a copy of the genetic information of DNA, the process of translation requires other RNA molecules such as transfer RNAs and ribosomal RNAs.

### *3.3.1 Messenger RNA*

Messenger RNA (mRNA) serves as the template for protein synthesis. The mRNA molecules are polymers of nucleotides (sugar, phosphate, and base) whose sequence of bases specify the primary structure (amino acid sequence) of a protein. How does a sequence of bases translate into a sequence of amino acids? If each of the four bases coded for 1 amino acid, there would not be sufficient bases to encode all amino acids (20 in total). Using a combination of 2 nucleotides would still not suffice to code for all 20 amino acids ($4^2 = 16$). However, a combination of three nucleotides would consist of more than enough codes to specify all amino acids ($4^3 = 64$). A three-letter code was discovered, termed the genetic code. Each triplet of bases that is translated into an amino acid is called a codon. The majority of codons (61 of the 64), specify individual amino acids, while the remaining three codons (UAA, UAG, and UGA) act as termination signals, which indicate to the translation machinery that the protein synthesis is complete. All the possible codons and the amino acids they code for are shown in Table 3.1. Since the genetic code produces

**Table 3. 1** Genetic Code. All the possible combinations of triplet nucleotides (codons) are shown. Each codon codes for a particular amino acid (except the stop codons UAA, UAG, and UGA)

| 5' | U | | C | | A | | G | | 3' |
|---|---|---|---|---|---|---|---|---|---|
| U | UUU | Phe | UCU | Ser | UAU | Tyr | UGU | Cys | U |
| | UUC | Phe | UCC | | UAC | Tyr | UGC | Cys | |
| | UUA | Leu | UCA | | UAA | Stop | UGA | Stop | |
| | UUG | Leu | UCG | | UAG | Stop | UGG | Trp | |
| C | CUU | Leu | CCU | Pro | CAU | His | CGU | Arg | C |
| | CUC | | CCC | | CAC | His | CGC | | |
| | CUA | | CCA | | CAA | Gln | CGA | | |
| | CUG | | CCG | | CAG | Gln | CGG | | |
| A | AUU | lle | ACU | Thr | AAU | Asn | AGU | Ser | A |
| | AUC | lle | ACC | | AAC | Asn | AGC | Ser | |
| | AUA | lle | ACA | | AAA | Lys | AGA | Arg | |
| | AUG | Met | ACG | | AAG | Lys | AGG | Arg | |
| G | GUU | Val | GCU | Ala | GAU | Asp | GGU | Gly | G |
| | GUC | | GCC | | GAC | Asp | GGC | | |
| | GUA | | GCA | | GAA | Glu | GGA | | |
| | GUG | | GCG | | GAG | Glu | GGG | | |

(Copies of tables are available in the accompanying CD.)

an excess of combinations to code for the amino acids, one amino acid can have more than one codon (e.g. serine, proline, and threonine). Thus, the genetic code is redundant but not ambiguous.

Codons are read by the ribosome in a 5' to 3' direction along the mRNA. It is possible to produce very different proteins depending on where the translation machinery starts the translation of the message. For example, consider the mRNA molecule 5'...UUUCCCAAAGGG...3'. If translation were to start at the first U, the protein sequence would read Phe-Leu-Ile-Gly. However, if the sequence is read from the second U, this sequence of nucleotides would code for the amino acids Phe-Pro-Lys- and if read from the third U, Ser-Gln-Arg. Thus the same mRNA can be read in 3 frames, which produce very different protein products. Indeed, identifying the correct starting point for translation is essential for the production of functional proteins. The first codon to be read by the translation machinery is the 'start' codon. The codon AUG which codes for the amino acid methionine is the first amino acid of almost all proteins. The sequence between the start codon and one of the possible stop codons makes up the open reading frame of the mRNA.

### *3.3.2 Transfer RNA*

During the process of translation, transfer RNA (tRNAs), serve as 'adapter' molecules between the amino acid and the corresponding codon in the template mRNA [13]. tRNAs are small (75–95 nucleotides long) non-coding RNAs that contain complementary regions. These complementary regions base pair and form double helical 'stem' structures, while the non-complementary bases form 'loop' structures. In two-dimensions, tRNAs resemble a cloverleaf structure made up of four base paired stems and three loops: D loop (usually contains dihydrouridine), TΨCG loop (where Ψ is a pseudouridine), and the anticodon loop (Fig. 3.4). The interaction between the D loop and the TΨCG loop causes the tRNA to fold into its classic L-shape structure. The three nucleotides that base pair with the three complementary nucleotides of a codon in the mRNA are called the anticodon, found in the anticodon loop.

The acceptor arm includes a 3' acceptor site that contains the invariable 5'-CCA-3' sequence. The 3'-OH or 2'-OH of this last adenine residue is covalently linked to the carboxyl group of an amino acid by an ester bond. A tRNA molecule linked to an amino acid is referred to as a 'charged' tRNA and is produced by the enzymatic activity of the aminoacyl-tRNA synthetase (aa-tRNA synthetase). There are 20 different aa-tRNA synthetases, one for each amino acid. Aminoacyl synthetases recognize tRNAs by interacting with the anticodon, acceptor arm, and variable pockets and loops in the tRNA molecule [14,15]. Each tRNA is recognized by one aa-tRNA synthetase and is therefore linked to one amino acid, which ensures the specificity of the translation process.

### *3.3.3 Wobble Base Pair*

The genetic code contains 61 codons and 20 amino acids. It was first thought that there would be 61 tRNAs, one corresponding to each codon. However, many cells contain less than 61 tRNAs. Upon closer examination of the genetic code, amino acids with more than one codon, all have the same first two bases, but differ in the last base. For example, CUU, CUC, CUA and CUG all code for Leucine. Furthermore, it was observed that the first base (5') of the tRNA anticodon does not necessarily follow standard Watson-Crick base pair rules with the third (3') base of the mRNA. This is due to the phenomenon known as wobble base pairing [reviewed in 16]. In fact, the first base (5') of the anticodon frequently is ionisine, which can base pair with A, C or U (Fig. 3.4). For example, the anticodon 5' IAG 3', which would carry the leucine amino acid, would recognize all four codons for leucine. Therefore for many amino acids, it is the base pairing between the first two bases of the codon in mRNA, and the last two bases of the anticodon in tRNA that is required for proper recognition. This results in a genetic code that is more tolerant of point mutations.

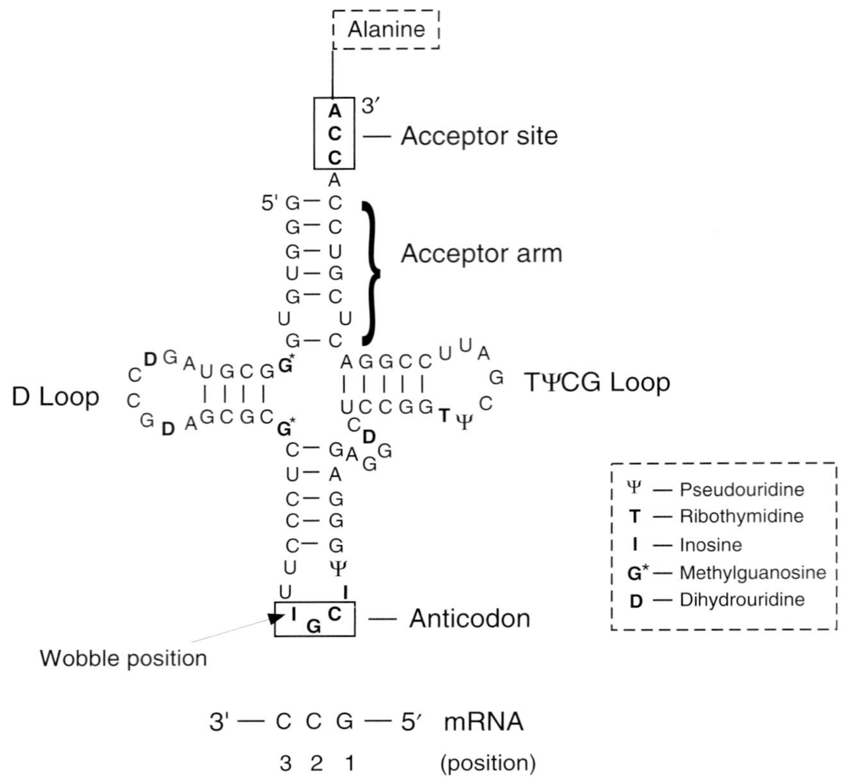


**Fig. 3.4** Schematic Representation of tRNA. Complementary sections along a tRNA molecule will form base pairs together, to produce double helical regions (termed stems). Those portions that do not form base pairs remain single stranded and form loop structures. In the two dimensional state, tRNA molecules resemble a cloverleaf which includes four stems and three loops (D Loop, TΨCG Loop, and the anticodon loop). The acceptor arm contains the acceptor site, the site to which the amino acid is attached. The modification of several bases (bolded and labeled in the dashed box) plays an essential role in the establishment of distinct tertiary structures among the tRNAs (Copies of figures including color copies, where applicable, are available in the accompanying CD)

### 3.3.4 *Ribosomal RNA*

Ribosomal RNA (rRNA) is a non-coding RNA that associates with ribosomes. In eukaryotes, one mRNA transcript codes for three of the four cytoplasmic rRNAs, which is subsequently cleaved into RNA products of varying lengths (18S, 5.8S, and 28S). The fourth rRNA, is transcribed independently (5S rRNA). The rRNA molecules make up ~80% of the total RNA molecules found in a typical cell. Similar to tRNAs, rRNAs form secondary and tertiary structures which are important for their functions. Ribosomal proteins in the cytoplasm are transported into the nucleus where they are assembled with the rRNAs and then transported back into the cytoplasm. As a major constituent of the ribosome, rRNAs play several key roles. rRNAs are important for stabilizing proteins in the ribosome, attracting mRNA molecules, catalyzing peptide-bond formation, recruiting accessory factors, and for proper interaction between the large and small ribosomal subunits during elongation.

### 3.3.5 *Ribosome*

A multi-protein complex associated with a distinct set of rRNAs, make up the small and large subunit of the ribosome (summarized in Table 3.2). In prokaryotes, the small subunit (30S)

**Table 3.2** The constituents of the small and large subunits of the ribosome in prokaryotes and eukaryotes

| | Cell type | | Proteins | | rRNA | |
|---|---|---|---|---|---|---|
| Subunit | Prok. | Euk. | Prok. | Euk. | Prok. | Euk. |
| Small | 30S | 40S | 21 | 33 | 16S | 18S |
| Large | 50S | 60S | 34 | 49 | 5S, 23S | 5S, 28S, 5.8S |

S stands for Svedberg, a measure of the sedimentation rate of suspended particles centrifuged under standard conditions
(Copies of tables are available in the accompanying CD.)

contains 21 proteins (designated S1-S21) and a 16S rRNA. The large subunit (50S) contains 34 proteins (designated L1-L34) and a 5S and 23S rRNA. In eukaryotes, the small subunit (40S) contains 33 proteins and an 18S rRNA and the large subunit (60S) contains 49 proteins and a 5S, 28S and 5.8S rRNA.

The small and large subunits are joined, and they form a space that becomes occupied with the mRNA and tRNAs (17–19). The small subunit is particularly complex in structure, with features called the head, neck, shoulder, body, platform, and spur [20]. The head is predominantly active during translocation. The portion of the small subunit that faces the large subunit forming the space is lined with rRNA. X-ray crystallography of the two ribosomal subunits in prokaryotes, shows the tRNAs bound at three sites termed the A (aminoacyl) site, the P (peptidyl) site, and the E (exit) site (Fig. 3.5) Permission – Cate et al., 1999 [21]. In addition to creating a platform onto which the mRNA meets tRNAs, the ribosome (large subunit) contains what is termed the peptidyl-transfer center (PTC), which catalyzes the formation of peptide-bonds between amino acids. This catalytic site in the large subunit is also lined by rRNA and it is proposed that the rRNA houses the catalytic activity. While the PTC is important for the stabilization of the two aminoacyl-tRNAs, the exact mechanism of catalysis is not clear. The large subunit also contains a tunnel through which the newly synthesized polypeptide chain travels and finally exits at the exit channel.

## 3.4 The Steps of Protein Synthesis

The synthesis of a protein requires the proper assembly of the 'protein factory', which includes ribosomes, tRNAs charged with their corresponding amino acid, mRNA, accessory factors, and GTP. Protein synthesis can be divided into 3 stages: initiation, elongation, and termination.

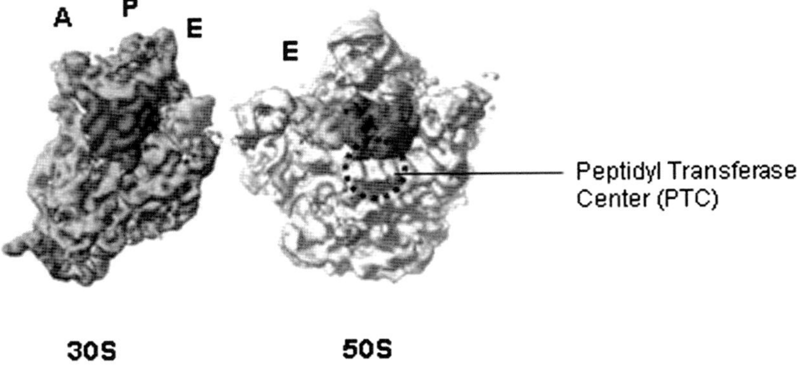


**Fig. 3.5** Model for the bacterial 70S ribosome. Data from X-ray crystal structures of the 70S bacterial ribosome were used to construct the model shown. The small and large subunits are depicted with three tRNAs positioned in the A, P, and E sites. From Cate JH, Yusupov MM, Yusupova GZ., Earnest TN, and Noller HF. (1999) X-ray Crystal Structures of 70S Ribosome Functional Complexes. *Science* 285(5436): 2095–2104. Reprinted with permission from AAAS (Copies of figures including color copies, where applicable, are available in the accompanying CD)

### 3.4.1 Initiation in Prokaryotes

Initiation begins with the binding of the small subunit (30S) to the start site of the mRNA [reviewed in 22,23]. The start site of the messager RNA is specified by the start codon, AUG, which codes for the amino acid methionine. Methionine is generally the amino acid found at the beginning of all proteins, however, methionine is also found within proteins. Therefore, the ribosome must bind at the AUG, which specifies the start site. In prokaryotes, the presence of a sequence (Shine-Delgarno sequence), located upstream of the start codon indicates the correct start codon. Base pairing occurs between the 16S rRNA (in the small subunit) and the Shine-Delgarno sequence (in the mRNA) which allows for enhanced binding of the 30S subunit at the starting AUG codon. The AUG is positioned in the P-site of the ribosome (see below).

The efficient binding of the small subunit (30S) to the mRNA requires accessory factors termed initiation factors (IFs). In prokaryotes, three IFs (IF1, IF2, and IF3) are involved. IF1, assists in the binding of the small subunit to the mRNA and also binds to and blocks the A-site. This directs the initial aminoacyl-tRNA to enter the P-site. IF2 bound to GTP associates with the initial aminoacyl-tRNA and recruits it into the P-site. IF3 is thought to prevent binding of the large subunit (50S) to the small subunit until the small subunit (30S) binds the mRNA and the aminoacyl-tRNA, causing hydrolysis of GTP-IF2 and the release of IF3.

Since methionine residues not only initiate a protein sequence but are also found throughout a protein, there are 2 methionyl-tRNAs. One type is the initiator tRNAs that are only incorporated at the starting AUG site. In prokaryotes, the initiator methionine is $tRNA^{fMet}$, where the methionine is modified to contain a formyl group that is later enzymatically removed. In eukaryotes, the initiator methionine is $tRNAi^{Met}$. For methionines incorporated within the peptide chain, another tRNA is used ($tRNA_m{}^{Met}$ in prokaryotes and $tRNA^{Met}$ in eukaryotes). The same aminoacyl-tRNA synthetase (Met-tRNA synthetase), can add Met to both of these tRNAs, however only the Met-$tRNA_i^{Met}$ can bind to the codon methionine which specifies the start site for translation.

### 3.4.2 Initiation in Eukaryotes

Protein translation in eukaryotes is similar to that of prokaryotes, however, there is a higher degree of complexity [reviewed in 22,23]. Eukaryotic mRNAs contain a 5'methylguanosine cap and a 3' polyadenine tail (poly[A] tail), larger ribosomes and several initiation factors (eIFs, 'e' for eukaryotic). Distinct from prokaryotes, the small subunit (40S) forms a 43S 'pre-initiation complex'. This complex includes the small ribosome (40S) bound with the charged initiator tRNA (Met-$tRNA_i{}^{Met}$)/eIF2-GTP. Thus, the initiator tRNA is bound to the P-site of the ribosome *before* the ribosome interacts with the mRNA (Fig. 3.6). Recall that in prokaryotes, the ribosome binds *directly* to the AUG in the mRNA and thereafter the charged initiator tRNA is recruited.

The initiation factors eIF1, eIF1A, and eIF3 are bound to the small subunit (40S), and they facilitate the binding of the small subunit to the mRNA (Fig. 3.6). Several eIFs are already bound to the mRNA. For instance, eIF4E binds the 5'cap, eIF4F acts as an RNA helicase to remove any double stranded mRNA that would interfere with the movement of the ribosome, and eIF4G links the 5' capped end of the mRNA to the 3' poly (A) end, by interacting with eIF4E and poly(A) binding proteins (PABPs). The eIF3 in the pre-initiation complex interacts with eIF4G, bringing the 40S subunit in association with the mRNA. eIF1A/eIF3 assists in scanning of the pre-initiation complex along the mRNA in search of the AUG codon [24]. The AUG (usually 5'-CCACCAUGC-3') is found by its base pairing with the Met-$tRNA_i{}^{Met}$, after which eIF2-GTP is hydrolyzed and then eIF5 (not shown) aids in the release of eIF2-GDP, eIF3, and eIF4. Finally, the large ribosomal subunit (60S) then binds to complete initiation. Interestingly, rather than using this scanning mechanism for the initiation of translation, in some cases, ribosomes have been shown to

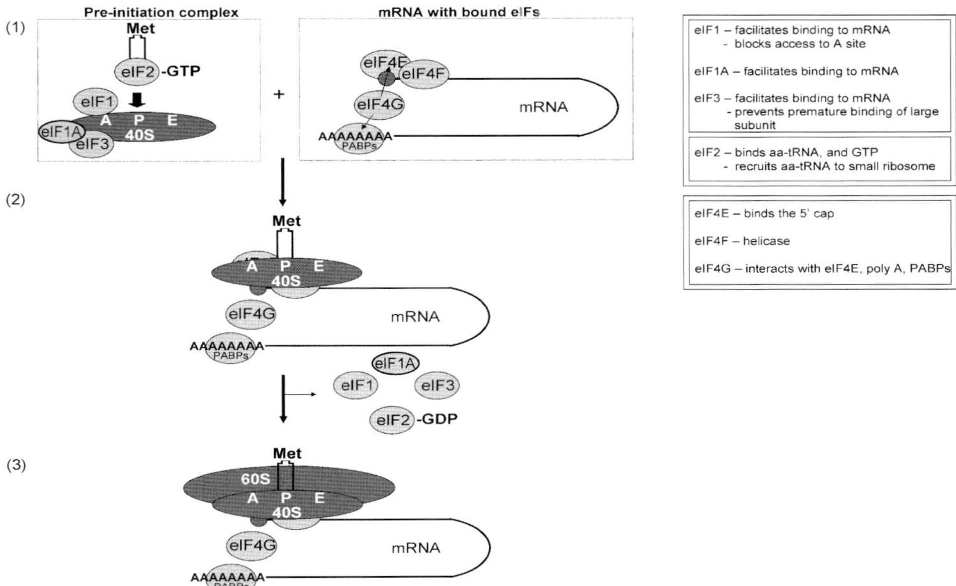


**Fig. 3.6** Initiation of translation in eukaryotes. The initiation of translation requires the binding of the ribosome to the start site of the mRNA. Initiation is facilitated by several initiation factors (eIFs, e = eukaryotic). There are 3 steps: (1) the formation of the pre-initiation complex, (2) the binding of the pre-initiation complex to the 5' end of the mRNA, and (3) the binding of the large (60S) subunit with the release of eIF2-GDP and other eIFs (eIF1, eIF1A, eIF3). The pre-initiation complex contains the 40S subunit bound to eIF1, eIF3, and eIF2, which is itself bound to Met-tRNA$_i^{Met}$. Initiation in prokaryotes is similar, but however, less complex (see text). The eIFs involved are listed to the right, indicating their role and function during initiation (Copies of figures including color copies, where applicable, are available in the accompanying CD)

bind directly at sites within the 5'untranslated region (UTR) called the internal ribosome entry sites (IRES). IRES sequences are many bases in length and form complex secondary structures that can bind the pre-initiation complex (reviewed in [25]).

Initiation of translation requires the proper assembly of all eIFs at the 5' cap along with its interaction with the poly(A) binding proteins. Interference results in the repression of translation. In fact, this is one method the cell utilizes to regulate the amount of protein to be synthesized from a particular mRNA. For instance, in response to various stimuli, active phosphorylated eIF2 is diminished and/or eIF4E is sequestered by other proteins, which inhibits translation. In addition to limiting eIFs, regions in the 5'UTR and 3'UTR of some mRNAs are involved in regulating translation [26]. Secondary stem-loop structures in the 5'UTR block assembly of the pre-initiation complex and/or the scanning of the ribosome [27–29]. In some cases, proteins bind and stabilize these secondary structures. The 3'UTR may also bind proteins that cause the inhibition of translation. Interestingly, small strands of RNA (20–25 nucleotides) termed microRNAs (miRNAs) can also bind to regions of the 3'UTR and prevent translation [30,31].

### *3.4.3 Elongation*

Initiation has now set in place all the required factors to begin decoding the information in the mRNA (Fig. 3.7). The P-site is occupied by aminoacyl-tRNA$_i^{Met}$ and the A-site is empty awaiting the entry of the appropriate aminoacyl-tRNA. The tRNAs travel through the three sites A, P, and E. Similar to initiation, accessory factors are required to enhance the efficiency of the elongation

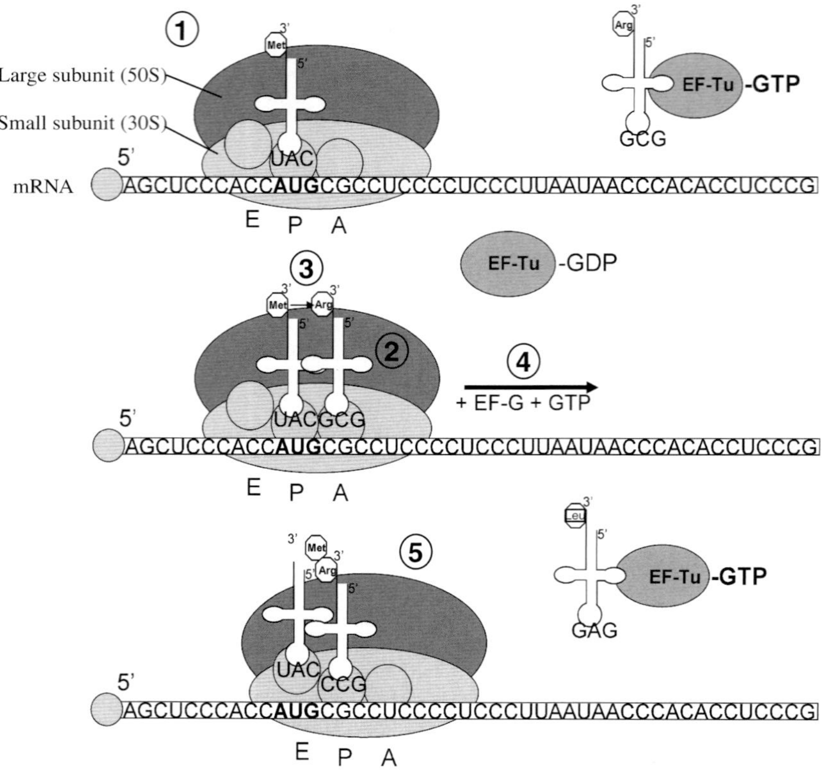


**Fig. 3.7** The stages of elongation. (1) The first step of initiation positions the ribosome at the start codon (AUG) along the mRNA, which is base-paired with the initiator tRNA, which occupies the P-site. The codon exposed to the A site is CGC which codes for arginine. EF-Tu bound to GTP recruits an aminoacyl (arginine) tRNA to the A-site. (2) A correct match between the anticodon of the Arg-tRNA and the codon of the mRNA allows for hydrolysis of GTP and the release of EF-Tu-GDP. (3) A peptide-bond formation between methionine and arginine is catalyzed within the peptidyl transferase center (PTC) of the large subunit. The P-site tRNA is deacylated, and methionine is linked to the arginine on the A-site of the tRNA. (4) The binding of EF-G-GTP and the subsequent hydrolysis of GTP to GDP provides the energy for the shift of the mRNA, with its attached tRNAs through the ribosome. (5) The ribosome moves or translocates forward along the mRNA by three codons in a 5' to 3' direction. Thus, the dipeptidyl-tRNA is shifted from the A-site to the P-site, and the deacylated tRNA is shifted from the P-site to the E-site. The next three codons of the mRNA are now exposed to the A-site (CUC), which codes for leucine. Again, the next tRNA will enter the A-site, the arginine will be linked to the leucine, and the ribosome will shift by three codons. This cycle continues, until the ribosome reaches a stop codon (Copies of figures including color copies, where applicable, are available in the accompanying CD)

process. For example, the elongation factor EF-Tu (eEF1$\alpha$ in eukaryotes) binds to GTP and recruits the aminoacyl-tRNA to the A-site. To ensure that the correct aminoacyl-tRNA interacts stably with the A-site, it must contain the correct anticodon, and interact with the 16S rRNA [19]. If correct matching is in place, the ribosome-tRNA complex is stabilized and translation can occur. EF-Tu hydrolyzes the bound GTP and EF-Tu-GDP is released from the tRNA. In the absence of this proper stabilization of the ribosome, the tRNA will dissociate from the complex.

Recent studies with improved resolution of the 70S ribosome of *Escherichia coli* bound to mRNA and tRNAs, confirm a sharp bend between the codons in the A-site and P-site [19]. This kink is thought to prevent the mRNA from slipping out of the ribosome and becoming out of frame. Furthermore, it was observed that a magnesium ion is responsible for stabilizing this bend

[18]. The tRNA in the P-site is securely bound and surrounded by rRNA and ribosomal proteins from both subunits. This is thought to prevent loss of the peptidyl tRNA and also maintain the reading frame.

The acceptor arms of the tRNAs in the P-site and A-site are located in the PTC of the large subunit. The PTC catalyzes a reaction between the amino acid components of the charged tRNAs. The carbonyl group of the aminoacyl-tRNA in the P-site reacts with the amino group of the aminoacyl-tRNA in the A-site, forming a peptide-bond and releasing water. This reaction occurs spontaneously and the bond produced is extremely stable. The end result is a dipeptidyl-tRNA in the A-site and a deacylated-tRNA in the P-site.

Once the amino acids are linked together, the ribosome shifts in the 5' to 3' direction along the mRNA. Translocation of the ribosome is highly intricate and requires specific interactions between the small and large subunits [18–20]. The small subunit moves against the large subunit in a fashion that 'pulls' the mRNA through the ribosome like a filmstrip. The dipeptide tRNA in the A-site shifts into the P-site and the deacylated tRNA in the P-site shifts into the E-site. The GTP-bound elongation factor EF-G (eEF2 in eukaryotes) binds to the ribosome where hydrolysis of GTP to GDP provides energy for translocation. After translocation, the next codon now occupies the A-site and a new aminoacyl-tRNA is recruited into the A-site. The cycle of aminoacyl-tRNA acceptance, peptide-bond formation and translocation continues until the ribosome reaches a stop codon. The peptide chain elongates one amino acid at a time, translocating through a tunnel within the large subunit and finally leaves the ribosome through an exit channel.

### *3.4.4 Termination*

Termination of translation occurs when one of the three codons (UAA, UAG, or UGA) occupies the A-site. Rather than coding for an amino acid, these codons bind release factors (RFs), a family of factors containing a conserved tripeptide that is proposed to directly interact with the stop codon. There are three RFs in bacteria: RF1 (binds codons UAA and UAG), RF2 (binds UAA and UGA) and the G-protein RF3 which enhances the activity of both RF1 and RF2. In eukaryotes, there are two release factors eRF1 (binds to all stop codons) and the G-protein eRF3 (enhances eRF1 activity).

Once the ribosome reaches a stop codon the release factors enter the A-site. In the P-site, the last amino acid of the peptide is still attached to its tRNA through an ester linkage. The binding of release factors to the A-site causes the transfer of a water molecule to the C-terminal end of the peptide. Hydrolysis allows for the release of the nascent peptide chain and the disassembly of the small and large subunits of the ribosome.

## Glossary and Abbreviations

### *RNA Processing*

**Alternative splicing** refers to the splicing events of pre-mRNA that results in different combinations of exons and gives rise to slightly different forms of proteins.

**Branch point sequence** a conserved sequence located near the 3' end of the intron that contains a conserved adenine required for the first step of splicing.

**BBP** Branch-point binding protein

**C-terminal domain (CTD)** a domain within the large subunit of RNA polymerase II, which is subject to phosphorylation; CTD serves as a major platform for binding of many factors including transcription factors.

**CPSF** Cleavage and polyadenylation specificity factor

**CstF** Cleavage stimulation factor

**Exon** the segments of genes that code for a protein; it is the sequence of bases in the genetic information that will be translated into proteins.

**Exonic splicing enhancers (ESEs)** sequences in exons that act on splice sites to accurately recruit the splicing machinery.

**Intron** the 'non-coding' segments of genes; the sequence of bases in the genetic information that will be removed or spliced from pre-mRNA and not be translated into proteins.

**Messenger RNA (mRNA)** the RNA transcribed from DNA (the message) and refers to the mature RNA transcript that has undergone 5'capping, 3'polyadenylation and splicing.

**PABPs** Polyadenylate binding proteins

**Polyadenylation** the addition of adenine residues (to the 3' end of mRNA).

**Poly(A)** Binding proteins

**Precursor mRNA (pre-mRNA)** describes the mRNA in its nascent form after transcription and before being processed or modified. Also referred to as the primary transcript.

**RFs** Release factors

**Small nuclear ribonucleoproteins (snRNPs)** particles made up of protein and RNA that recognize and bind to splice sites in mRNA and serve as a docking site for the spliceosome.

**Small nuclear RNA (snRNA)** small molecules of RNA that are always associated with proteins forming what is termed small nuclear ribonucleoproteins (snRNPs), and are important in RNA splicing.

**Spliceosome** a complex involved in RNA splicing that contains 5 small nuclear ribonucleoproteins (snRNPs).

**Splice sites** sites within the RNA that contain conserved sequences that are recognized by the splicing machinery, resulting in cleavage at these sites.

**SR** Serine-arginine-rich

**Trans-esterification** the chemical reaction that occurs during splicing where one ester bond is exchanged for another.

**UTRs** Untranslated regions

## *Translation*

**A site or aminoacyl site** site of the ribosome that binds incoming tRNAs charged with an amino acid.

**Anticodon** a region of the tRNA molecule that consists of the three bases that will complementarily base pair with the corresponding codon in mRNA during translation.

**Codon** a triplet sequence of RNA that codes for a specific amino acid.

**E site or exit site** site in the ribosome that occupies the deacylated tRNAs.

**Genetic code** a list of codons that were discovered to code for a particular amino acid, which is universal across all species.

**Internal ribosome entry site (IRES)** are sequence elements located in the 5' untranslated region of eukaryotic mRNAs where ribosomes can bind and initiate translation.

**Peptide-bond** the bond formed after a reaction between an amino group and a carboxyl group resulting in the release of a water molecule:

$$\text{-}\overset{\displaystyle O}{\overset{\|}{C}}\text{-OH} + NH_2 \longrightarrow \text{-}\overset{\displaystyle O}{\overset{\|}{C}}\text{-}\underset{\displaystyle H}{\underset{|}{N}}\text{-} + H_2O$$

**P site or peptidyl site** site in the ribosome that is occupied by the tRNA which is linked to the elongating peptide chain of amino acids.

**Peptidyl transfer center (PTC)** the region in the large subunit of the ribosome that catalyzes the peptide-bond formation between the amino acid linked to the tRNA in the A-site and the amino acid linked to the tRNA in the P-site.

**Ribosome** a complex of proteins made up of two subunits, the small and large subunits, which together function in the binding of mRNA and the synthesis of the peptide chain coded in the particular mRNA.

**Shine-Delgarno sequence** a specific sequence of bases located upstream of the start codon in the mRNA of prokaryotes. This sequence, base pairs with regions of the 16S rRNA found in the small subunit of the ribosome, thereby facilitating the binding of the small ribosome to the mRNA during the initiation of translation.

**Translocation** the movement of the ribosome relative to the mRNA during translation which is catalyzed by the binding of EF-G-GTP and the hydrolysis of the GTP to GDP.

**Wobble base pair** a base pair that does not follow the standard Watson-Crick base pair rules (G-U) or that which is formed between U, A, C, and the modified base I (ionosine). Wobble base pairing occurs frequently in the secondary structures of RNA molecules and plays an important role in the base pairing between the first (5') position in the anticodon of tRNA and the last (3') position in the codon of mRNA.

## References


1. Gu M LC. Processing the message: structural insights into capping and decapping mRNA. Current Opinion in Structural Biology 2005; 15:99–106.
2. Shatkin AJ. Capping of eukaryotic mRNAs. Cell 1976; 9:645–653.
3. Christofori G KW. 3' cleavage and polyadenylation of mRNA precursors in vitro requires a poly(A) polymerase, a cleavage factor, and a snRNP. Cell 1988; 54:875–889.
4. Hirose Y MJ. RNA polymerase II is an essential mRNA polyadenylation factor. Nature 1998; 395:93–96.
5. Lavigueur A LBH, Kornblihtt AR, Chabot B. A splicing enhancer in the human fibronectin alternate ED1 exon interacts with SR proteins and stimulates U2 snRNP binding. Genes and Development 1993; 7:2405–2017.
6. Tsukahara T CC, Helfman DM. Alternative splicing of beta-tropomyosin pre-mRNA: multiple cis-elements can contribute to the use of the 5'- and 3'-splice sites of the nonmuscle/smooth muscle exon 6. Nucleic Acids Research 1994; 22:2318–2325.
7. Jurica MS MM. Pre-mRNA splicing: awash in a sea of proteins. Molecular Cell 2003; 12:5–14.
8. Bentley D. The mRNA assembly line: transcription and processing machines in the same factory. Current Opinion in Cell Biology 2002; 14:336–342.
9. Proudfoot NJ, A Furger, and MJ Dye. Integrating mRNA processing with transcription. Cell 2002; 108:501–512.
10. Graveley BR. Sorting out the complexity of SR protein functions. RNA 2000; 6:1197–1211.
11. Das R YJ, Zhang Z, Gygi MP, Krainer AR, Gygi SP, Reed R. SR Proteins Function in Coupling RNAP II Transcription to Pre-mRNA Splicing. Molecular Cell 2007; 26:867–881.
12. Crispino JD BB, Sharp PA. Complementation by SR proteins of pre-mRNA splicing reactions depleted of U1 snRNP. Science 1994; 265:1866–1869.
13. Raj Bhandary UL KC. Early days of tRNA research: discovery, function, purification and sequence analysis. Journal of Biosciences 2006; 31:439–451.
14. Cavarelli J MD. Recognition of tRNAs by aminoacyl-tRNA synthetases. Federation of American Societies for Experimental Biology 1993; 7:79–86.
15. Vasil'eva IA MN. Interaction of Aminoacyl-tRNA Synthetases with tRNA: General Principles and Distinguishing Characteristics of the High-Molecular-Weight Substrate Recognition. Biochemistry (Mosc) 2007; 72:247–263.
16. Agris PF VF, Graham WD. tRNA's wobble decoding of the genome: 40 years of modification. Journal of Molecular Biology 2007; 366:1–13.
17. Berk V ZW, Pai RD, Cate JH. Structural basis for mRNA and tRNA positioning on the ribosome. Proceedings of the National Academy of Sciences of the United States of America 2006; 103:15830–15834.
18. Selmer M DC, Murphy FV, Weixlbaumer A, Petry S, Kelley AC, Weir JR, Ramakrishnan V. Structure of the 70S ribosome complexed with mRNA and tRNA. Science 2006; 313:1935–1942.
19. Korostelev A TS, Laurberg M, Noller HF. Crystal structure of a 70S ribosome-tRNA complex reveals functional interactions and rearrangements. Cell 2006; 126:1065–1077.

20. Schuwirth BS BM, Hau CW, Zhang W, Vila-Sanjurjo A, Holton JM, Cate JH. Structures of the bacterial ribosome at 3.5 A resolution. Science 2005; 310:827–834.
21. Cate JH YM, Yusupova GZ, Earnest TN, Noller HF. X-ray crystal structures of 70S ribosome functional complexes. Science 1999; 285:2095–2104.
22. Kozak M. Initiation of translation in prokaryotes and eukaryotes. Genes and Development 1999; 234:187–208.
23. Kozak M. Regulation of translation via mRNA structure in prokaryotes and eukaryotes. Genes and Development 2005; 361:13–37.
24. Pestova TV KV. The roles of individual eukaryotic translation initiation factors in ribosomal scanning and initiation codon selection. Genes and Development 2002; 16:2906–2922.
25. Jackson R. Alternative mechanisms of initiating translation of mammalian mRNAs. Biochemical Society transactions 2005; 33:1231–1241.
26. Mignone F GC, Liuni S, Pesole G. Untranslated regions of mRNAs. Genome biology 2002; 3:4.1–4.10.
27. Pelletier J SN. Insertion mutagenesis to increase secondary structure within the 5' noncoding region of a eukaryotic mRNA reduces translational efficiency. Cell 1985; 40:515–526.
28. Kozak M. Circumstances and mechanisms of inhibition of translation by secondary structure in eucaryotic mRNAs. Molecular and cellular biology 1989; 9:5134–5142.
29. van der Velden AW TA. The international journal of biochemistry & cell biology 1999; 31:87–106.
30. Chendrimada TP FK, Ji X, Baillat D, Gregory RI, Liebhaber SA, Pasquinelli AE, Shiekhattar R. MicroRNA silencing through RISC recruitment of eIF6. Nature 2007; 447:823–828.
31. John B SC, Marks DS. Prediction of human microRNA targets. Methods in molecular biology 2006; 342:101–113.

## Suggested Reading

### *RNA Processing*

Gu, M, Lima, CD. (2005) Processing the message: structural insights into capping and decapping mRNA. *Curr. Opin. Struct. Biol.* 15(1): 99–106.

Howe, JK. (2002) RNA polymerase II conducts a symphony of pre-mRNA processing activities. *Biochim Biophys. Acta.* 2(1577): 308–24.

Singh R. (2002) RNA-protein interactions that regulate pre-mRNA splicing. *Gene Expr.* 10(1–2): 79–92.

Soller M. (2006) Pre-messenger RNA processing and its regulation: a genomic perspective. *Cell. Mol.Life Sci.* 63(7–8): 796–781.

### *Translation*

Berk V, Zhang W, Pai RD, and Cate JH. (2006) Structural basis for mRNA and tRNA positioning on the ribosome. *Proc Natl Acad Sci* USA 103(43):5830–5834.

Cate JH, Yusupov MM, Yusupova GZ., Earnest TN, and Noller HF. (1999) X-ray Crystal Structures of 70S Ribosome Functional Complexes. *Science* 285(5436): 2095–2104.

Korostelev A, Trakhanov S, Laurberg M, and Noller HF. (2006) Crystal structure of a 70S ribosome –tRNA complex reveals functional interactions and rearrangements. *Cell* 126(6): 1065–1077.

Liljas A. (2006) Deepening ribosomal insights. *ACS Chem. Biol.* 1(9):567–569.

Schuwirth SB, (2005) Structures of bacterial ribosome at 3.5 Angstrom resolution *Science* 310(827):827–834.

Selmer M, Dunham CM, Murphy FV 4th, Weixlbaumer A, Petry S, Kelley AC, Weir JR, and Ramakrishnan V. (2006) Structure of the 70S ribosome complexed with mRNA and tRNA. *Science.* 313(5795):1935–1942.

# Chapter 4
# DNA Replication, Recombination, and Repair

**Linda B. Bloom**

**Abstract** The instructions needed for producing all the components of a cell and for regulating their functions are encoded in the sequence of DNA. Accurate transmission of this information to the progeny and protection of the genome from chemical degradation are essential to life. Complementary base pairing in duplex DNA provides an elegant means for accurate replication of DNA and repair of DNA damage. Each strand of the duplex provides a template for generating the other strand and in essence acts as a "back-up copy" of the information. The many different proteins and enzymes required to physically manipulate large DNA polymers in replication, recombination, and repair, take advantage of the complementary base pairing between the strands to accomplish their tasks. These enzymes are capable of a sufficient level of accuracy to maintain genetic integrity, yet also allow a low level of mutations to generate genetic diversity ultimately allowing a population to adapt to changing conditions. Understanding how the cellular machinery functions to replicate, recombine, and repair the genome is central to understanding evolution of species and the origin of genetic diseases and cancer.

**Keywords** DNA replication · DNA recombination · DNA repair · Replication fidelity · DNA damage

## 4.1 Introduction

DNA contains an instruction set, which is used by cells, to produce and regulate the components necessary for their survival. This information is encoded in the order, or sequence, of the four possible nucleotide bases that are covalently linked to form a linear DNA polymer. Hydrogen bonding interactions between complementary base pairs bring two linear DNA polymers together to form a double helical structure. James Watson and Francis Crick not only deduced the double helical structure of DNA, but also recognized that hydrogen bonding between complementary DNA bases would provide a mechanism for DNA duplication [1]. Complementary, or Watson-Crick, base pairing of adenine with thymine and guanine with cytosine (Fig. 4.1) allows one strand of DNA to serve as a template for synthesis of the other strand by directing the incorporation of complementary nucleotides into the new polymer. The double-stranded structure of DNA genomes also provides a mechanism for preserving the code. If one strand of DNA is damaged, the other can be used as a template to regenerate the damaged strand and recover information encoded in its sequence.

L.B. Bloom
Department of Biochemistry & Molecular Biology, University of Florida, 1600 SW Archer Rd. JHMHC, room R3-234, Gainesville, FL 32610-0245, USA
e-mail: lbloom@ufl.edu

S. Krawetz (ed.), *Bioinformatics for Systems Biology*, DOI 10.1007/978-1-59745-440-7_4,

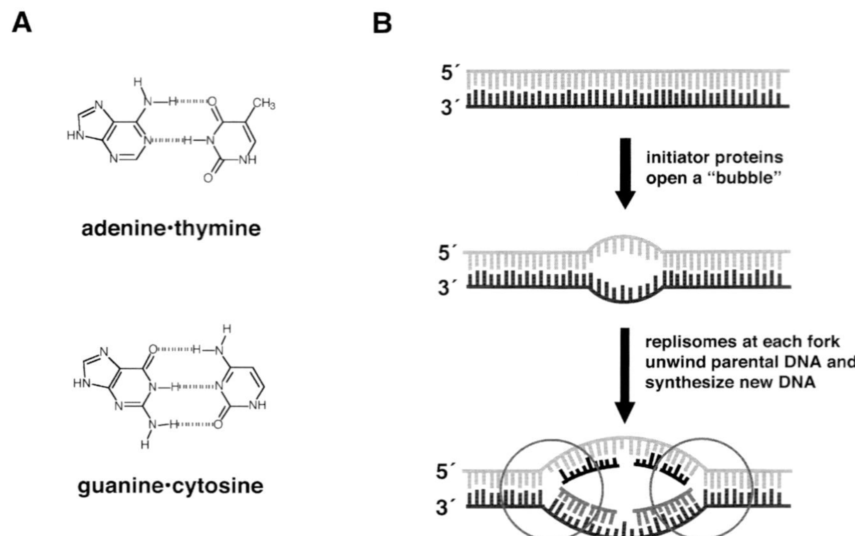


**Fig. 4.1 (A) Chemical structures for Watson-Crick A·T and G·C base pairs.** Adenine and guanine are purines and cytosine and thymine are pyrimidines. For each Watson-Crick base pair, a purine pairs with a pyrimidine to make a base pair that has the same overall size and shape. **(B) Unwinding of the DNA helix for initiation of replication.** Schematic representation of a DNA duplex is shown in the upper panel in which each "rung" of the "ladder" structure represents an individual Watson-Crick base pair as shown in A. A small region of DNA is initially unwound by the activities of the initiator proteins to form a bubble. DNA helicases are loaded at the unwound site that unwinds the DNA duplex further. The rest of the components of the replisome bind and begin DNA synthesis. Ultimately, two replication forks (*circled structures*) form and move away from the origin synthesizing the DNA as they go (Copies of figures including color copies, where applicable, are available in the accompanying CD)

Typically, DNA replication, recombination, and repair are presented as three separate topics, and these processes will be presented one-at-a-time here for clarity. However, at the outset, it should be made clear that these processes are highly interdependent. Like any molecule, DNA can become damaged or it can degrade with time. Damage that is encountered during the process of DNA replication must be repaired either by recombination or other repair mechanisms in order for replication to be completed. DNA recombination and repair that occurs outside the replication phase of the cell cycle requires limited DNA synthesis which is catalyzed by many of the same enzymes that function during replication.

## 4.2 DNA Replication

### *4.2.1 Overview of DNA Replication*

The overall process of DNA replication and many of the key enzymes have been conserved from organisms ranging from bacteria to man. Immediately prior to every cell division, the DNA genome must be duplicated, so that each daughter cell receives an identical copy of the instruction set (Fig. 4.2). Genome replication is not a trivial task. The genome of the bacterium *Escherichia coli* (*E. coli*), a model organism, is comprised of about 5 million base pairs of DNA, organized in a single circular chromosome, or individual DNA molecule. Human cells at mitosis contain about a thousand times more DNA or 6 billion base pairs of DNA which divide into 46 linear chromosomes (see Fig. 4.2 legend). Individual human chromosomes range in size from about 50–250 million base

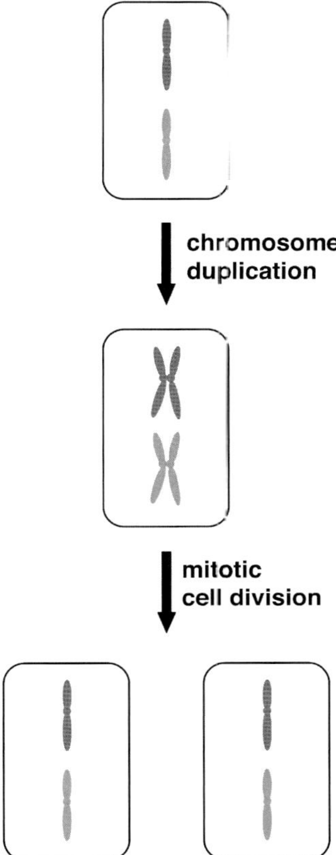


**Fig. 4.2** Mitotic cell division in eukaryotes. Most eukaryotic cells are diploid in that they contain two copies of each chromosome, one derived from each parent. Human cells contain 22 pairs of homologous chromosomes plus two sex chromosomes, two X chromosomes in females or an X and Y chromosome in males, making a total of 46 chromosomes. A single pair of homologous chromosomes is shown in this figure for simplicity. Prior to cell division, chromosomes are duplicated by the process of DNA replication. After replication, the replication products, the sister chromatids, remain tethered together by proteins until the cell divides. This physical linkage is required to help ensure that the chromosomes are properly segregated (divided equally) between the two daughter cells. During cell division, sister chromatids separate and each daughter cell receives one copy of each pair of homologous chromosomes. Genome duplication and cell division is similar in bacteria except that bacterial cells are typically haploid and only contain a single copy of each chromosome (Copies of figures including color copies, where applicable, are available in the accompanying CD)

pairs, and if stretched out, they would range in length from 1.7–8.5 cm. Many enzymes and proteins are required to physically manipulate these large DNA polymers to catalyze the synthesis of new DNA, and to ensure that DNA replication is accurate. In addition, the process of DNA replication must be regulated so that the entire genome is duplicated only once, and each daughter cell receives a complete instruction set.

DNA replication begins at a specific time in the cell cycle and at specific sites, termed as *replication origins*, in the genome. The DNA duplex is unwound at the replication origins to allow the enzymes that synthesize DNA access to the individual DNA strands (Fig. 4.1). Each strand of parental DNA serves as a template for a *DNA polymerase* to synthesize a new strand of DNA. Single nucleotide monomers that form complementary base pairs with the parental template are incorporated into a new DNA polymer by DNA polymerases. As the new DNA grows, the parental duplex is progressively unwound forming *replication forks* that move away from their origin. In *E. coli*, replication forks move at a rate of about 500 nucleotides per second while in eukaryotes they move about 10 times slower. DNA replication is *semi-conservative*, ultimately forming two DNA double helices that contain one strand of parental DNA and one strand of new DNA.

Much of our knowledge of the biochemistry of DNA replication is based on the studies of model systems such as *E. coli*, however, more recent investigations of eukaryotic replication mechanisms are revealing that many of the key features of the bacterial replication machinery are also common to eukaryotes. The process of DNA replication in these model organisms will be compared and contrasted with that of eukaryotes.

### 4.2.2 Initiation of DNA Replication

Initiation of DNA replication requires two elements, a site on the DNA where replication will start or originate and an initiator protein [2, 3]. Origins of DNA replication have been fairly well defined for unicellular organisms such as bacteria and yeast, but not for multicellular eukaryotes. Origins of DNA replication contain two general DNA sequence elements, an element that is recognized by initiator proteins, and an element that is relatively easy to unwind. The *E. coli* genome contains a single 245 base pair (bp) origin of replication, *oriC*, that contains several defined sequence elements. Within the *oriC* are a series of 9-bp sequence elements that are recognized by the *E. coli* initiator protein, DnaA [4], and three 13-bp AT-rich regions of DNA that are relatively easy to unwind. Several DnaA molecules bind to *oriC* through interactions with the 9-bp repeats, and form a large protein-DNA complex that bends the DNA to initiate unwinding of the DNA helix at the 13-bp AT-rich sequence. Many of the steps in DNA replication, recombination, and repair, such as unwinding of DNA at the origin, require energy to manipulate the structures of macromolecules and disrupt noncovalent interactions such as hydrogen bonding between the two complementary DNA strands. Enzymes that catalyze changes in these noncovalent interactions utilize the chemical energy stored in the phosphoryl bonds of adenosine-5'-triphosphate (ATP) to do mechanical work. The DnaA protein utilizes the energy gained from ATP hydrolysis to power the unwinding of DNA at the origin.

Defined origins of replication, or autonomously replicating sequences (ARS), have been identified in the yeast *Saccharomyces cerevisiae (S. cerevisiae)* [5]. Unlike the *E. coli* genome, which contains a single origin, eukaryotic genomes contain multiple origins of replication on a single chromosome. Each origin initiates replication of a segment of the genome. *S. cerevisiae* origins share key features with the *E. coli* origin, a sequence element recognized by the initiator proteins and an A-T rich region that could be easily unwound. Yeast origins are recognized by an origin recognition complex (ORC), which is composed of six different polypeptides. Like the *E. coli* DnaA protein, ORC uses ATP to facilitate the initiation of replication.

The elements that constitute the origin of replication in multicellular eukaryotes remains to be defined [5]. Initiation of replication has been identified at several loci, but conserved DNA sequences such as the 9-bp site to which DnaA binds in *E. coli* have yet to be identified. It is possible that epigenetic elements, or elements that are not defined by the sequence of nucleotides in the DNA, such as the nature of the histone proteins that bind DNA, help to define origins of replication in multicellular eukaryotes. It is estimated that the human genome contains about 30,000 origins of replication, of which each is responsible for replication of about 100,000 base pairs of DNA. Initiator proteins in multicellular eukaryotes are homologous to those in yeast and are comprised of six different ORC proteins [6].

Once the DNA duplex is opened, a *DNA helicase* is loaded to continue the unwinding DNA process. DNA helicases are enzymes that utilize the energy from the hydrolysis of ribonucleoside 5'-triphosphates, most commonly ATP, to break the hydrogen-bonding interactions between complementary DNA strands and unwind the nucleic acid duplexes. In *E. coli*, six molecules of the DnaB protein form a ring-shaped hexamer that encircles single-stranded DNA and functions as a helicase. This hexameric helicase structure is common to other organisms including eukaryotes where the MCM (mini-chromosome maintenance) proteins are believed to perform the function of replicative DNA helicase. In *E. coli*, the DnaB hexamer is assembled around the DNA by the ATP-dependent activity of a helicase loader, DnaC. Likewise in eukaryotic cells, helicase loaders, Cdc6 and Cdt1, are required to assemble the MCM proteins on DNA.

Before DNA synthesis can begin, RNA *primers*, which are short oligonucleotides on the order of 10–15 nucleotides in length must be made. DNA polymerases are unable to synthesize DNA *de novo* and can only extend the RNA (or DNA) primers that are already paired with the template to be copied. *Primases* synthesize these primers using ribonucleoside 5'-triphosphates as building blocks to form a short strand of RNA complementary to the DNA template. The *E. coli* primase interacts with the DnaB helicase and begins the synthesis of RNA primers shortly after DnaB has begun to unwind DNA. In eukaryotes, a hybrid RNA-DNA primer is synthesized by an enzyme complex containing both primase and DNA polymerase $\alpha$. The primase synthesizes a short RNA oligonucleotide which is then extended by DNA polymerase $\alpha$. Replicative DNA polymerases, DNA polymerase III in *E. coli* and DNA polymerases $\delta$ and $\varepsilon$ in eukaryotes, extend these primers to synthesize the bulk of DNA required to duplicate the genome.

Initiation of replication must be regulated, particularly in eukaryotic cells that contain multiple chromosomes and multiple origins of replication, so that the entire genome is replicated only once. The consequences of incomplete or over-replication of parts of the genome can be lethal to eukaryotic cells, or at the very least, result in severe genetic abnormalities and diseases. In *E. coli*, initiation of replication is regulated by controlling the access of DnaA to *OriC*, but it is not as stringent as in eukaryotic cells. When *E. coli* is grown in a rich medium in a laboratory, re-initiation of replication typically occurs prior to cell division. In eukaryotes, initiation of replication is tightly regulated, and occurs in two phases. In the first phase, pre-replication complexes (pre-RCs) composed of the initiator proteins (ORC), helicase loaders (Cdc6 and Cdt1), and helicase (MCM) assemble at the origins. In the second phase, these pre-RCs are activated and the enzymes begin to open the DNA duplex. These two phases are regulated by the activities of cyclin-dependent kinases (CDKs), which are expressed at a specific phase of the cell cycle. Prior to the S phase, a phase of the cell cycle in which DNA is replicated, CDK activity is low and pre-RCs assemble at origins of replication. During the S phase, CDK levels increase and the origins are activated. Importantly, new pre-RCs cannot form during the S-phase when CDK levels are high. This prevents origins that have already been activated and which have initiated replication from being rebound by initiator proteins and being re-replicated.

### *4.2.3 Enzymes at the Replication Fork*

Assembly of the *replisome* is complete when the replicative DNA polymerase and its accessory proteins join the helicase and primase at the replication fork [7] (Fig. 4.3). The replisome synthesizes the new DNA, unwinding the parental duplex as it goes. The chemical synthesis of new DNA is catalyzed by a DNA polymerase contained within the replisome. Cells contain many different DNA polymerases that have different functions in DNA replication and repair. There are 5 known DNA polymerases in *E. coli* and over a dozen in humans. In *E. coli*, DNA polymerase III catalyzes the bulk of DNA synthesis during replication and DNA polymerases $\delta$ and $\varepsilon$ do so in eukaryotes.

All DNA polymerases use 2'-deoxyribonucleoside 5'-triphosphates (dNTP's) as monomeric building blocks for making DNA. DNA polymerases catalyze the attack of the 3' hydroxyl group of the nucleotide at the primer end on the $\alpha$-phosphoryl group of an incoming dNTP displacing pyrophosphate (Fig. 4.4A). Thus, DNA polymerases extend the DNA polymers in the

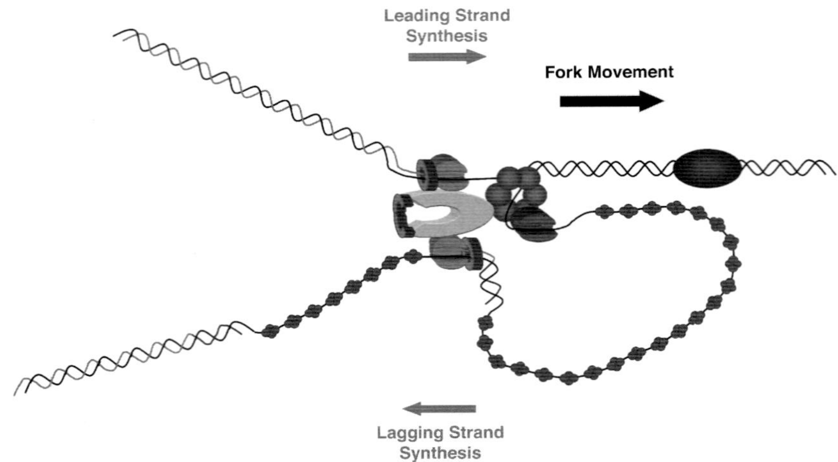


**Fig. 4.3** Proteins at the *E. coli* replication fork. A dimeric polymerase (*golden brown*) complex is capable of coordinated DNA synthesis on the leading and lagging strands. The leading strand polymerase synthesizes new DNA (*gray*) in the direction of the fork movement and the lagging strand polymerase synthesizes DNA (*gray*) in the opposite direction. The hexameric helicase (*six blue spheres*) unwinds the DNA ahead of the polymerase, and primase (*red*) makes RNA primers (*blue lines*) on the lagging strand. Single-stranded DNA that forms as the helix unwinds is coated with single-stranded binding protein (*purple*) to prevent reannealing of strands and to remove secondary structure that may form within a single-strand. Sliding clamps (*green rings*) are assembled on each primer on the lagging strand by the clamp loading complex (*yellow "C"-shaped structure*). Topoisomerase works ahead of the fork to remove superhelical tension that forms in the duplex as a result of unwinding by the polymerases. The same basic proteins are present at the eukaryotic replication fork (Copies of figures including color copies, where applicable, are available in the accompanying CD)

5' to 3' direction by incorporation of 2'-deoxyribonucleoside monophosphates. Watson-Crick base pairing interactions between the incoming dNTP and the next unpaired template base direct the incorporation of correct nucleotides. The frequency of addition of incorrect nucleotides can be less than one in a million, but even with this low error frequency mistakes will be made. To further reduce errors, the DNA polymerases that function in replication contain a 3' to 5' exonuclease activity, that allows them to *proofread* the nucleotides that have been incorporated. This exonuclease activity catalyzes the hydrolysis of the phosphodiester bonds to remove the last nucleotide added to the 3' primer end (Fig. 4.4B). Thus, a nucleotide that has been added incorrectly can be removed.

The overall efficiency of synthesis by DNA polymerases is enhanced by accessory proteins which increase the processivity of the DNA polymerase, or in other words, the number of nucleotides incorporated per DNA binding event. These accessory proteins consist of a ring-shaped sliding clamp and a clamp loader that assembles the clamp on the DNA. Sliding clamps [8, 9], made of identical protein subunits, encircle the DNA and are capable of sliding along the DNA duplex (Fig. 4.5). When a DNA polymerase binds to a sliding clamp on the DNA, it is effectively tethered to the DNA template, so that it is capable of incorporating thousands of nucleotides without dissociating. In the absence of a sliding clamp, DNA synthesis is less efficient because DNA polymerases frequently dissociate from the template and must rebind to continue synthesis. Ring-shaped sliding clamps must be assembled onto the DNA by the activities of clamp loaders. Like DNA helicases, clamp loaders are enzymes that require energy from ATP to catalyze the mechanical reaction of clamp loading. An interface between two clamp monomers must be opened by the clamp loader to allow the DNA to pass through the center of the clamp. After placing the ring on the DNA, this interface must close to prevent the clamp from easily falling off from the DNA. The clamp loader uses energy from ATP to alter noncovalent interactions and open the clamp.

A

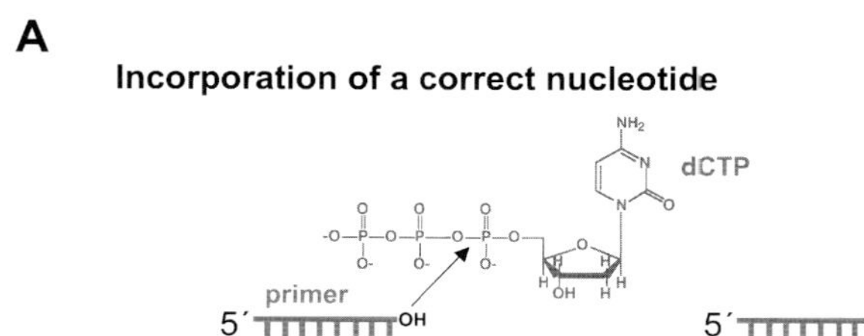


B

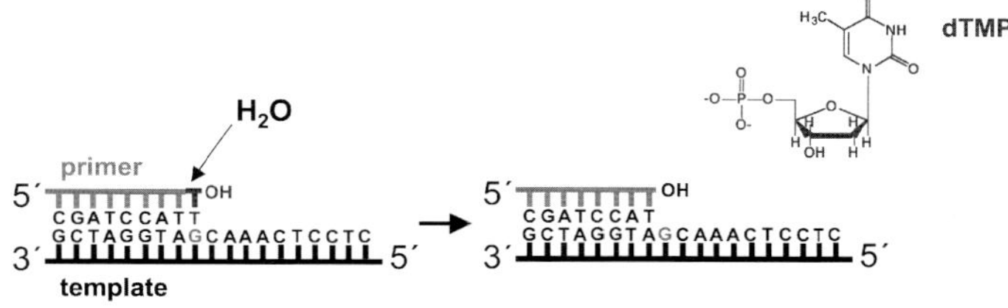


**Fig. 4.4** Reactions catalyzed by DNA polymerases. (**A**) 2'-Deoxyribonucleoside 5'-triphosphates (dCTP in this example) are used as substrates by DNA polymerases to extend the primer in template-directed reactions. The net reaction is incorporation of 2'-deoxyribonucleoside monophosphates onto the 3' hydroxyl of a primer with loss of a pyrophosphate (PP$_i$). (**B**) DNA polymerases can "proofread" newly incorporated nucleotides and excise incorrect nucleotides. This excision reaction removes the last nucleotide that was incorporated by hydrolysis of the phosphodiester bond to release a 2'-deoxyribonucleoside mcnophosphate (dTTP in this example). This is not the reverse of the polymerase reaction because it produces a dNMP rather than a dNTP (Copies of figures including color copies, where applicable, are available in the accompanying CD)

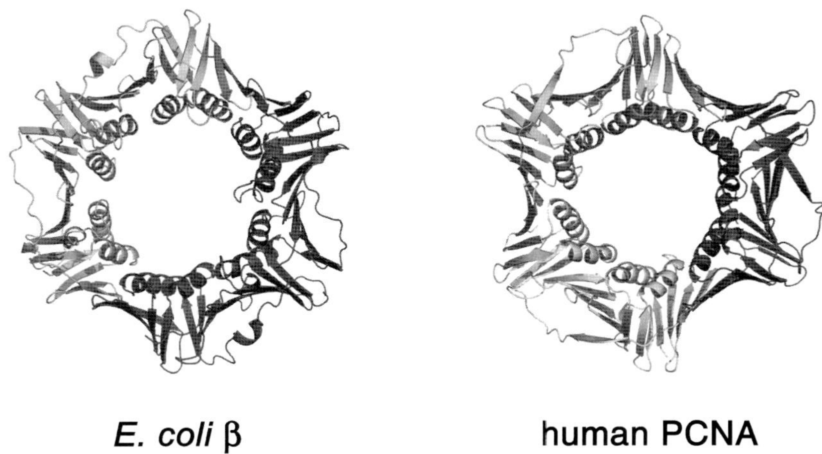


**Fig. 4.5 Structures of sliding clamps from *E. coli* and humans.** (*Left panel*) The *E. coli* β sliding clamp is a head-to-tail dimer of identical monomer subunits. (PDB accession ID: 2pcl [9]) (*Right panel*) The human PCNA sliding clamp is similar in overall structure to the β clamp, but is a trimer of identical monomers. (PDB accession ID: 1axc [8]) Each ring has a central hole that is large enough to encircle the B-DNA (Copies of figures including color copies, where applicable, are available in the accompanying CD)

### 4.2.4 Leading and Lagging Strand Synthesis

In *E. coli*, a complex containing a dimeric DNA polymerase and accessory proteins interacts directly with the helicase and primase to form a replisome. This physical interaction stimulates the activity of the helicase, and increases the rate of fork unwinding. The replisome, which contains two copies of DNA polymerase III, is capable of simultaneously copying both strands of parental DNA at the replication fork. But the two DNA polymerases must work in opposite directions to do this because DNA strands in a duplex are antiparallel and DNA polymerases can only synthesize DNA in the 5' to 3' direction. To accomplish this, one DNA polymerase working on the leading strand, synthesizes the DNA in a single continuous piece moving in the direction of the replication fork. The other DNA polymerase working on the opposite or *lagging strand*, synthesizes the DNA in short fragments called the Okazaki fragments after Reiji Okazaki whose work led to their discovery (Figs. 4.1 and 4.3). As the fork progresses, a loop of single-stranded DNA is created on the lagging strand and a primer is synthesized to begin synthesis of each Okazaki fragment. The clamp loader loads a clamp on each primer and the lagging strand polymerase extends these primers in the direction opposite to fork movement until it encounters a completed Okazaki fragment. Then, the polymerase dissociates and rebinds to a new primer closer to the fork and extends it. Thus, the lagging strand polymerase must repeatedly dissociate from the completed Okazaki fragments and rebind to new primers to continue DNA synthesis in a discontinuous manner. To date, there is no evidence for a direct physical interaction between the leading and lagging strand polymerases in eukaryotes, however, one strand of DNA is synthesized continuously and the other discontinuously in Okazaki fragments. In *E. coli*, Okazaki fragments are 1000–2000 nt in length and in eukaryotes they are 100–200 nt. The overall DNA synthesis is semi-discontinuous because it is synthesized as one continuous strand on the leading strand and as discontinuous fragments on the lagging strand.

The process of unwinding the DNA helix causes the DNA helix ahead of the fork, to become more tightly wound. If this superhelical tension created ahead of the advancing fork were not relieved, it would become increasingly difficult to unwind the DNA at the fork and replication would stop. This problem is overcome by topoisomerases, which are enzymes that function ahead of the fork to remove the tension. Topoisomerases do this by physically breaking one or both strands of DNA, passing DNA through the break, and rejoining the DNA strands.

Each Okazaki fragment starts with an RNA primer. To complete DNA replication, RNA primers must be replaced by DNA and Okazaki fragments must be joined together to form a continuous strand. In *E. coli*, removal of RNA primers and synthesis of DNA can be accomplished by a single enzyme, DNA polymerase I. The 5' to 3' exonuclease activity of DNA polymerase I degrades the RNA primers while the 5' to 3' polymerase activity simultaneously synthesizes DNA to replace the RNA. In eukaryotes, separate enzymes are responsible for degrading the RNA and replacing it with DNA. Finally, the DNA fragments on the lagging strand are joined by a DNA ligase to form one continuous polymer. DNA ligase catalyzes the formation of a phosphodiester bond between the 3' hydroxyl group at the end of one Okazaki fragment and the 5' phosphate at the beginning of the next. Any "nick" in one strand of the DNA duplex that has a 3' hydroxyl on one side and 5' phosphate on the other can be sealed by the activity of a DNA ligase.

### 4.2.5 Fidelity of DNA Replication and Mismatch Repair

DNA replication can be accomplished with less than one mistake, in a billion nucleotides incorporated. This amazing accuracy or fidelity of synthesis is achieved, for the most part, by the DNA polymerase but is further enhanced by a group of mismatch repair enzymes that function to detect and correct replication errors [10, 11]. One main feature of the DNA polymerase that contributes to this fidelity is the geometry of the active site which is optimized for binding Watson-Crick base pairs where the overall shape of both A:T or G:C pairs are the same (Fig. 4.1). Mispairs such as G:T deviate from this ideal geometry so that incorrect nucleotides are incorporated much less efficiently. The frequency of adding an incorrect nucleotide range from 1 in 1000 to 1 in 10,000,000 nucleotides added depending on the nucleotide added and the DNA sequence context. In a rare instance when a mistake is made, DNA polymerases have the ability to remove the incorrect nucleotide using the 3' to 5' exonuclease activity contained in the enzyme. This proofreading capability is further enhanced by a reduced efficiency of adding the next correct nucleotide onto a primer that ends with an incorrect nucleotide. Thus, when a mistake is made, the rate of adding a second nucleotide to the incorrect nucleotide is greatly reduced which gives the exonuclease more time to remove the incorrect nucleotide. Once an incorrect nucleotide is removed, rapid incorporation of correct nucleotides by the DNA polymerase activity resumes. This proofreading activity increases the accuracy of DNA synthesis by a factor of about 10 to 100.

Mispairs that escape proofreading by the DNA polymerase, can be corrected by the post-replicative mismatch repair processes [12, 13] (Fig. 4.6). The net result is the removal of the segment of DNA containing the incorrect nucleotide and resynthesis of DNA to replace the segment that was excised. The key enzymes responsible for mismatch repair are conserved from bacteria to humans. Mispairs are detected in the double-stranded DNA by MutS in *E. coli* or MutS homologs (MSH) in eukaryotes. The MutS proteins recruit other mismatch repair proteins to the site of the mispair to initiate repair. An *endonuclease* cleaves the DNA strand containing the mispair, and an *exonuclease*, degrades the DNA containing the mispair by excising nucleotides starting at the cleavage site created by the endonuclease and working back towards the mispair. Endonucleases are enzymes that hydrolyze a phosphodiester bond at an internal site on one DNA strand to effectively cut the DNA strand into two. Exonucleases are enzymes that progressively degrade the DNA by removing one or a few nucleotides at a time, starting from one end and moving with a

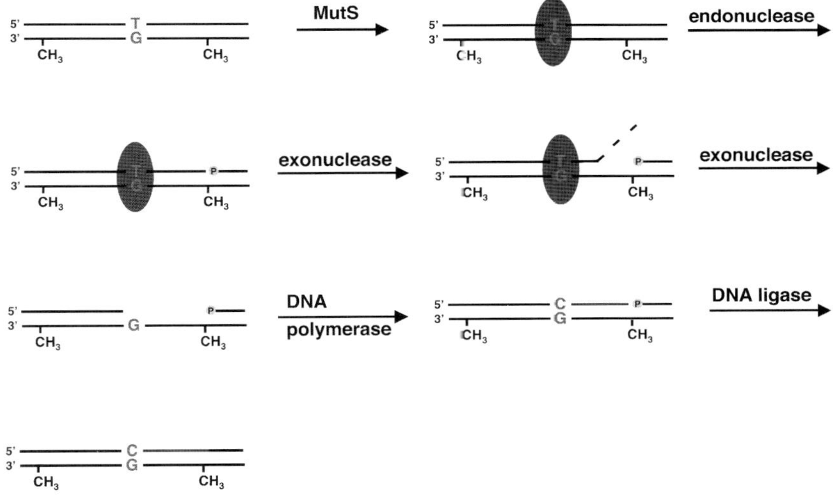


**Fig. 4.6 Methyl-directed mismatch repair in *E. coli*.** The MutS (blue) protein recognizes and binds mismatches such as G·T (red) in the DNA and recruits the rest of the mismatch repair machinery. One of the enzymes recruited to the mispair is an endonuclease which cleaves the newly replicated and unmethylated DNA strand at the GATC sequence, closest to the protein-mismatched DNA complex. The letter P in a gold circle represents a 5' phosphate group. DNA is degraded by an exonuclease until the mismatch is removed The missing segment of DNA is replaced by a DNA polymerase, and the DNA strands are joined together by the activity of a DNA ligase. Mismatch repair in human cells occurs by a similar mechanism except that the mechanism by which the human mismatch repair enzymes recognize which strand of DNA must be degraded is not known. The methylation status of DNA is not used for strand discrimination in human cells (Copies of figures including color copies, where applicable, are available in the accompanying CD)

defined polarity, either in the 5' to 3' or 3' to 5 directions towards the other end. The endonuclease and exonuclease work in concert to remove the segment of DNA containing the incorrect nucleotide by degrading this segment. A DNA polymerase can then resynthesize DNA to replace the segment removed by the mismatch repair enzymes and complete the repair process.

How do the mismatch repair enzymes know which nucleotide of the mispair is incorrect? In *E. coli*, the methylation status of the DNA allows the mismatch repair enzymes to distinguish between the newly synthesized DNA strand and the parental strand. Adenine is methylated in the *E. coli* genome when it appears in the sequence 5'GATC. Methylation of 5'GATC sequences occurs shortly after replication, so for a short period of time, the daughter strand is unmethylated while the parent strand is methylated. These 5'GATC sequences are also recognized by the endonuclease which cuts the unmethylated daughter strand at these sites. These cut sites can be up to 1000–2000 nt away from the mispair so a fairly large segment of the DNA may be removed and replaced. While the overall process of mismatch repair is similar in eukaryotes, it is not yet clear how the eukaryotic enzymes distinguish between the newly synthesized strand and the parental strand. Although eukaryotic genomes can also be methylated, for example on C's at 5'CpG sites in humans, evidence suggests that the methylation status of the DNA strand does not direct repair. The leading theory, at this time, is that the eukaryotic mismatch repair machinery uses some signals from DNA replication such as unligated Okazaki fragments or perhaps the replisome itself to determine which strand is the newly synthesized daughter strand.

## 4.3 DNA Recombination

### *4.3.1 Overview of Recombination*

Through the process of recombination, two DNA duplexes can combine to create hybrid molecules containing sequences from each of the original molecules. Recombination provides mechanisms for generating genetic diversity and repairing DNA strand breaks. Recombination pathways can be grouped into three major classes, homologous, nonhomologous end-joining, and site-specific. Homologous recombination is the most general mechanism for recombination, and it plays a central role in both generation of genetic diversity and DNA repair. Homologous recombination occurs between two regions of DNA with similar or homologous, but not identical, sequences. Crosses between these two regions produce two new molecules that are hybrids of the original molecules. During meiosis (a process of cell division that ultimately produces germ cells containing a single copy of each chromosome), these crosses allow the alleles (alternate forms of the same gene) to be exchanged between homologous chromosomes so that chromosomes passed onto haploid daughter cells are a hybrid of their progenitors (Fig. 4.7A). This process allows a child to inherit traits from all of the grandparents, even though the child only receives a single chromosome from each of his parents. By a similar mechanism, a double-strand break in one chromosome can be repaired without loss of genetic information by recombination with the homologous chromosome.

Double-strand breaks can also be repaired by nonhomologous end-joining to simply reattach the two broken ends. This recombination process "trims" the duplex ends at the break site to remove damaged nucleotides and uncover a small stretch (one to several nucleotides) of homology and then rejoin the two ends. Double-strand break repair by nonhomologous end-joining is not as accurate as homologous recombination because the trimming step can result in loss of genetic information. However, a double-strand break is one of the most lethal forms of DNA damage. A single double-strand break can kill a cell, and nonhomologous end-joining is able to repair these breaks in order to allow the cell to survive at the cost of possible loss of information. In vertebrate cells, nonhomologous end-joining is the predominant pathway for DNA double-strand break repair at times when the cell is not actively replicating DNA.

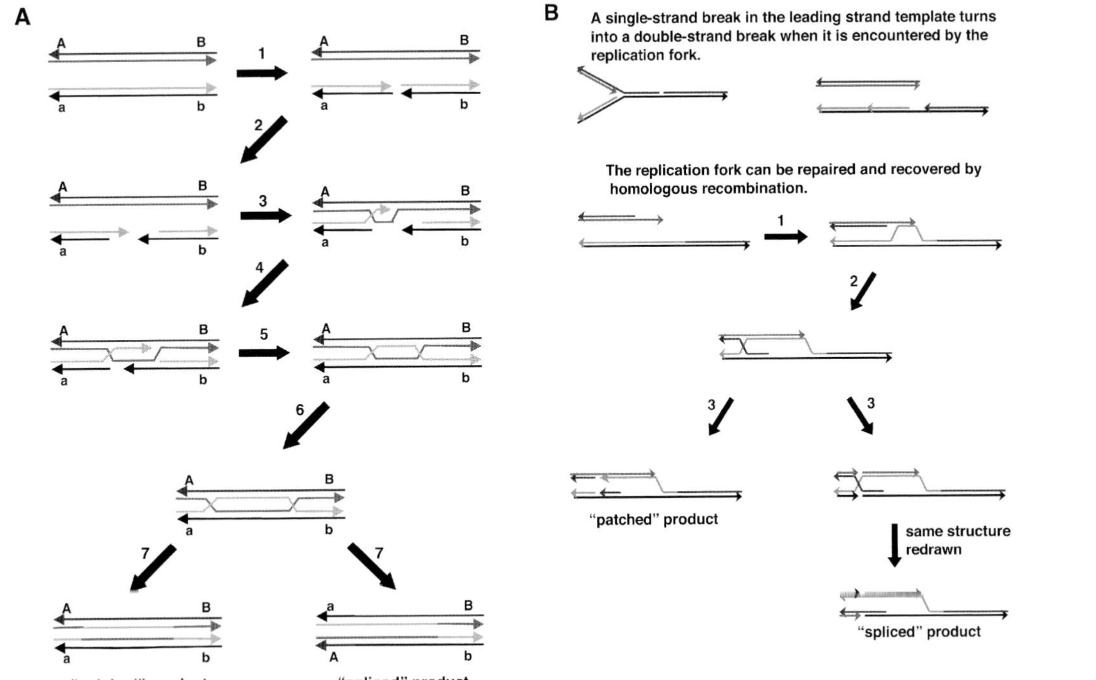


**Fig. 4.7** Homologous recombination. **(A)** Double-strand break model for homologous recombination. DNA duplexes are illustrated by antiparallel arrows. The arrow indicates the polarity of DNA in which the arrowhead is at the 3' end. "A" and "a" represent two alleles of gene A and "B" and "b" represent two alleles of gene B. In this model, recombination is initiated by the formation of a double-stranded break (*step 1*) in one of the homologous duplexes. The broken DNA is then processed by partial degradation, by an exonuclease to generate single-stranded DNA on the 3' ends (*step 2*). One 3' single-stranded end invades the homologous duplex forming a D-loop in the intact duplex (*step 3*). The invading 3' end is extended by a DNA polymerase enlarging the D-loop which can then pair with the remaining 3' single-stranded end (*step 4*). As the D-loop expands, it can displace the 5' end of the broken duplex which is then free to pair with the intact duplex (*step 5*). Branch migration enlarges the regions of heteroduplex DNA by "unzipping" the regions that were originally paired and "zipping" them onto the homologous duplex (*step 6*). Finally, the cross-over points or Holliday junctions are resolved by cleavage of the crossing strands (*step 7*). Two different products, patched and spliced, are formed depending on which of the crossed strands are cleaved. In the patched products the same alleles are maintained on each chromosome, but some intervening sequence comes from the homologous duplex. In the spliced product, the left side of one duplex is spliced onto the right side of the other so that the alleles are swapped. **(B)** Homologous recombination can repair double-strand breaks that form during DNA replication. In step 1, the upper duplex is processed by an exonuclease to generate a 3' end that can strand invade the lower duplex forming a D-loop. Homologous pairing of the light blue and gray strands (*step 2*) displaces the dark blue strand which can then pair with the black strand. Further branch migration can extend the heteroduplex regions. Resolution of the Holliday junction by cutting the "crossed" strands (*dark blue and gray, step 3 left side*) regenerates the replication fork and patched recombination products. Resolution by cutting the uncrossed strands (*light blue and black, step 3 right side*) also regenerates a replication fork and spliced recombination products(Copies of figures including color copies, where applicable, are available in the accompanying CD)

Site-specific recombination does not require homologous sequences between two DNA duplexes. As its name implies, site-specific recombination occurs when one DNA sequence is inserted into a specific site in another DNA duplex. This type of recombination produces a new DNA duplex where information from one is "spliced" into the other duplex. Some viruses use site-specific recombination to integrate their genomes into the genome of a host. Bacteriophage $\lambda$, integrates its viral genome into the genome of its *E. coli* host, and retroviruses such as human immunodeficiency virus (HIV) integrate a double-stranded DNA copy of their viral RNA genome into the host genome. Transposition is an example of site-specific recombination where a genetic element, referred to as the transposon, moves from one location in a genome to another. This repositioning requires a specific nuclease, called the transposase, which is encoded within the transposon.

This section will largely focus on homologous recombination because it provides a mechanism for generating genetic diversity that allows a population to adapt to a changing environment as well as a mechanism for repairing DNA and maintaining the information encoded in the sequence.

## *4.3.2 Homologous Recombination*

### 4.3.2.1 DNA Repair by Homologous Recombination

Extensive homology or significant complementarity in DNA sequence is a prerequisite for homologous recombination between two duplexes. In addition, the physical exchange of information requires breaking and rejoining of the DNA molecules. Modern models for homologous recombination propose that recombination is initiated by formation of a double-stranded break in one of the two duplexes [14–16]. These breaks can result from DNA damage or can be created by specific enzymes as happens during meiosis. The broken duplex is then processed by an exonuclease which partially degrades the ends to generate free single-stranded overhangs, and these processed break sites serve as substrates for initiating recombination. One of the single-stranded ends will invade the intact duplex and pair with the region of homologous sequence to produce a heteroduplex containing strands from two separate DNA duplexes (Fig. 4.7A). The *strand invasion* and homologous pairing reactions are not spontaneous but require the activity of an enzyme, RecA protein in *E. coli* and its homolog, Rad51, in eukaryotes, to catalyze the reactions.

The strand invasion reaction generates a "D-loop" structure which is formed by the complementary strand from the intact duplex that was displaced. The invading strand is extended by a DNA polymerase that increases the size of the single-stranded region in the D-loop. This opens up a region of the DNA homologous to the other single-stranded end of the broken duplex which then pairs with the D-loop. This end can also be extended by a DNA polymerase. Extension of both ends effectively replaces the segments of DNA that were removed by the exonuclease to generate single-stranded ends that initiate recombination. Thus, homologous recombination requires limited DNA synthesis by a DNA polymerase.

Two crossover points between the two homologous duplexes are created following strand invasion by one half of the broken duplex and pairing between the displaced D-loop and the other half of the broken duplex. These crossed strands are named Holliday junctions after Robyn Holliday who first proposed their occurrence in homologous recombination. Holliday junctions can migrate down the DNA duplexes "unzipping" the original duplexes while simultaneously "zipping" the other strands together with new partners thereby extending the heteroduplex regions of the DNA. This process of branch migration can continue as long as homology between the regions of DNA is high. The branch migration reaction is catalyzed by enzymes that are similar to DNA helicases and require ATP.

Following branch migration, the two DNA duplexes are separated or resolved by cleaving the two DNA strands in the Holliday junctions. This cleavage is catalyzed by a nuclease or a resolvase that is specific for the DNA in Holliday structures. Depending on which strands in the junctions are cleaved, two distinct products can be formed that contain different segments of DNA from the original duplexes (Fig. 4.7A). A spliced product can be formed where each end of the hybrid duplex is derived from one of the original duplexes. Alternatively, a patched product can be formed where each new recombinant duplex contains a short patch or segment of DNA from the other original duplex.

All organisms use homologous recombination to repair double-stranded breaks, however, they differ in the extent to which they rely on this mechanism of repair (Fig. 4.7b). In *E. coli*, homologous recombination is the predominate mechanism by which double-stranded breaks are repaired. In eukaryotes, other mechanisms such as nonhomologous end-joining are used in addition to homologous recombination to repair double-stranded breaks. In vertebrate cells, the choice of mechanism depends in part on the phase of the cell-cycle. When cells are actively replicating or have just finished replicating their genomes, sister chromosomes are typically juxtaposed which facilitates strand invasion and the homology search required for homologous recombination [17]. During these phases of the cell cycle, homologous recombination is used extensively to repair double-strand breaks. When cells are not actively replicating, homologous chromosomes may not be located near one another, and regions of the chromosomes required for recombination may be condensed and not accessible to the recombination machinery. During these phases, nonhomologous end-joining is the predominant mechanism for repairing double-strand breaks.

#### 4.3.2.2 Homologous Recombination to Generate Genetic Diversity

Homologous recombination provides a mechanism for generating genetic diversity by recombining genetic material from two different, but homologous, chromosomes [18]. Although bacteria have mechanisms for acquiring genetic material from other bacteria, viruses, and exogenous DNA, the most common mechanism for exchanging genetic information is through sexual reproduction. Individual cells in multi-cellular eukaryotes contain two copies of each chromosome, one homolog from each parent. Normally, these cells divide by a process called mitosis in which the genome is duplicated so that four copies of each chromosome are present, the cell divides into two, and chromosomes are equally divided so that each daughter cell receives two homologous copies of each chromosome (Fig. 4.2). A second mechanism for cell division called meiosis produces germ cells that contain a single copy of each chromosome (Fig. 4.7). During meiosis, DNA is replicated as normally happens prior to cell division, but cells undergo two rounds of division so that each of the four germ cells contains one copy of each chromosome instead of two. Prior to the first cell division, the newly replicated chromosomes, the sister chromatids, remain together and must pair with the homologous chromosomes. This pairing as well as proper segregation of the chromosomes into daughter cells during the first round of cell division requires homologous recombination. The mechanism for homologous recombination during meiosis is the same as that for DNA repair except that an enzyme (Fig. 4.8A), Spo11, initiates the process by making a programmed double-strand break in one of the homologous chromosomes [19]. Following resolution, the two homologous pairs of sister chromatids go to different daughter cells at the first cell division. This differs from the mechanism of chromosome segregation that occurs during mitosis in which each mitotic cell receives a sister chromatid from each of the two homologous chromosomes rather than a pair of chromatids from the same duplicated chromosome. Following meiosis, each of the chromosomes may contain regions derived from the homologous chromosomes.

Homologous recombination requires extensive regions of homology, but not completely identical sequences. As a result, sequence polymorphisms that exist between the two homologous chromosomes result in mispairs in the heteroduplex regions created by strand invasion and branch migration. These mispairs are corrected by the same enzymes required for mismatch repair following DNA replication, however, in this case, there is no "right" or "wrong" nucleotide or strand. Repair can occur on either strand, randomly, and depending on which strand is corrected,

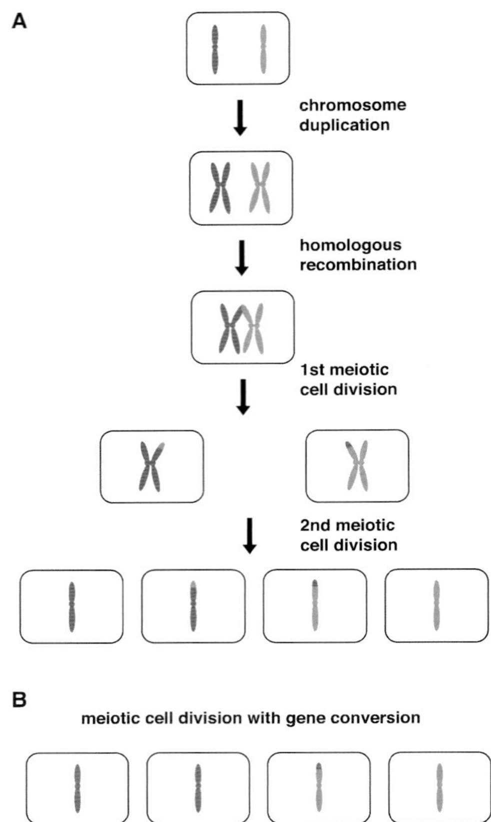


**Fig. 4.8** Meiotic cell division in eukaryotes. **(A)** Meoisis is a specialized pathway for cell division that produces haploid germ cells (e.g., eggs and sperm). Chromosomes are duplicated normally by DNA replication. Prior to the first cell division, homologous chromosomes pair and undergo homologous recombination, which is required for proper chromosome segregation. During the first cell division the sister chromosomes remain paired and homologous chromosomes go to different daughter cells. This differs from mitotic cell division (Fig. 4.2) in which the sisters separate and a homologous pair goes to each daughter cell. In the second cell division, the sister chromatids are separated and each daughter cell receives a single chromosome. **(B)** Meiotic cell division with gene conversion. Repair of the base mispairs resulting from homologous recombination during meiosis can result in a gene conversion event. If homologous recombination without gene conversion occurred as illustrated in panel A, alleles present on the regions of genes that crossed over, the upper tip of the chromosome in this example, would be equally represented among the daughter cells. That is there would be two copies of the paternal (*blue*) and two of the maternal (*pink*) alleles. A gene conversion event would result in an unequal number of alleles as illustrated in the example in panel B in which there are three copies of the paternal alleles (*blue*) in the upper tip of the chromosomes, and single daughter cell containing the maternal allele (Copies of figures including color copies, where applicable, are available in the accompanying CD)

the sequence will either match that of the chromosome that was originally broken, or the chromosome that was invaded. This process can lead to a gene conversion event in which alleles of a given gene are not equally represented among the four daughter cells following meiosis (Fig. 4.8B). For example, gene conversion could result in three daughter cells with a paternal allele and one with a maternal allele rather than the expected pattern of two cells with maternal and two with paternal alleles. Gene conversion events can also result from the DNA synthesis that occurs following the initial strand invasion event. If a gene was located near the site of a double-strand break, DNA synthesis will make a copy of the allele that was present on the intact duplex.

## 4.4 DNA Repair

### *4.4.1 Overview of DNA Repair*

DNA, like any other molecule, can spontaneously decompose with time, react with chemicals including those naturally present in a cell, or be damaged by UV or ionizing radiation [20]. Damage to DNA can have many deleterious effects on the cell. Some types of damage alter the structure of DNA bases so that the sequence is misread by DNA polymerases causing it to incorporate erroneous bases and generate mutations. Others are so severe that DNA synthesis by DNA polymerases is completely blocked at the site of damage. DNA damage occurs with a frequency high enough that it would be lethal to a cell if it were not repaired. Recombination, as discussed above, provides a mechanism for repairing DNA strand breaks. This section describes the mechanisms responsible for repairing damaged bases in DNA so that the code contained in its sequence is preserved.

There are two general strategies that can be taken up by a cell to fix DNA damage, (1) direct reversal of the damage, and (2) replacement of a segment of DNA containing the damage. Direct reversal of DNA damage requires enzymes capable of catalyzing different chemical reactions to undo different types of damage. There are fewer examples of enzymes that repair DNA by direct reversal of damage perhaps because each type of damage would require a different enzyme and chemistry. The majority of DNA damage is repaired using an excision/resynthesis strategy in which a group of enzymes removes a segment of DNA containing damage and replaces it with a new DNA. For this type of process, the same basic set of enzymes can remove many types of DNA damage because they are catalyzing the same basic reaction that involves removing a segment of the damaged DNA.

### *4.4.2 Direct Reversal of DNA Damage*

#### 4.4.2.1 DNA Photolyase Catalyzed Splitting of Thymine Cyclobutane Dimers

When exposed to UV light, two neighboring pyrimidine bases, C or T, in the same DNA strand can react with each other to become covalently linked. One reaction that occurs is the formation of thymine-cyclobutane dimers (Fig. 4.9). The covalent linkage between the two T residues affects the twist of the DNA helix at the site thus distorting the local structure of DNA. Cyclobutane dimers are no longer recognized as two consecutive T's by DNA polymerases, and in general, DNA polymerases can incorporate an A opposite the first T, but are blocked and cannot synthesize past the second T. In bacteria, the enzyme, DNA photolyase, reverses the formation of thymine dimers. Enzyme-bound cofactors absorb light and initiate an electron transfer reaction that catalyzes the splitting of the thymine dimer to regenerate two intact thymine bases. There is no known homolog of photolyase in mammalian cells where cyclobutane dimers are repaired predominantly by nucleotide excision repair.

#### 4.4.2.2 Removal of Methyl Groups by $O^6$-methylguanine Methyltransferase

Chemicals that are naturally present in the cell, as well as chemicals in the environment, can react with DNA to methylate DNA bases. One product of this reaction, $O^6$-methylguanine ($O^6$-MeG), induces mutations because a DNA polymerase has the tendency to incorrectly incorporate a T opposite $O^6$-MeG rather than a C that normally pairs with G (Fig. 4.9). $O^6$-MeG is repaired by $O^6$-methylguanine methyltransferase (MGMT) by directly removing the methyl group to regenerate a normal guanine residue. MGMT is not an enzyme in the true sense of the word because it does not act catalytically;

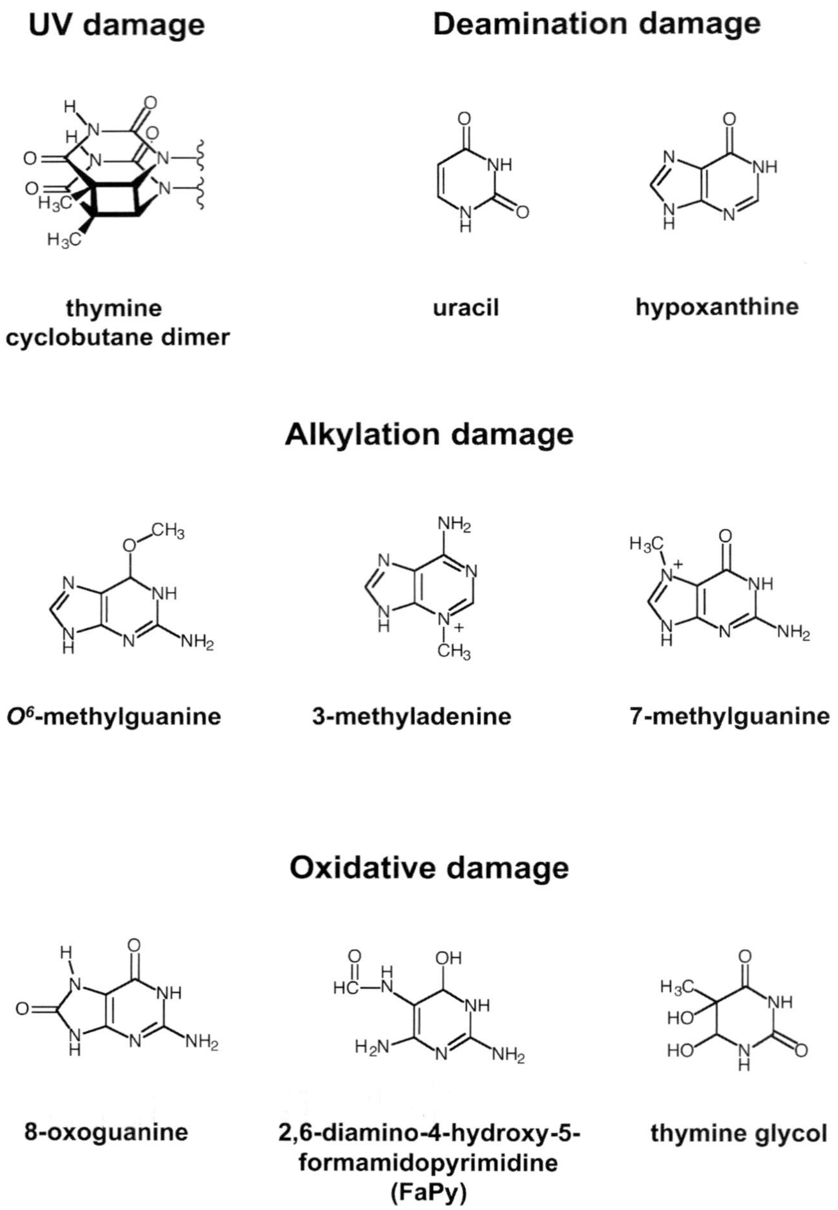

**UV damage**

**Deamination damage**

**thymine cyclobutane dimer**

**uracil**

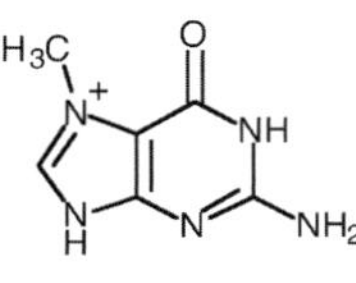


**hypoxanthine**

**Alkylation damage**

**$O^6$-methylguanine**

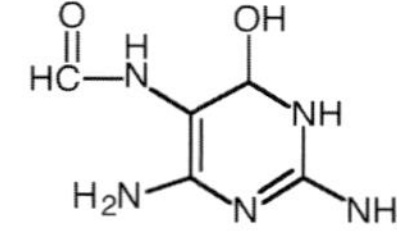


**3-methyladenine**

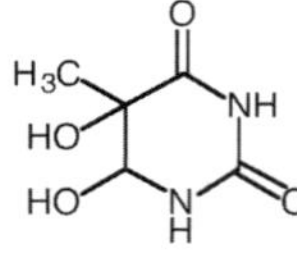


**7-methylguanine**

**Oxidative damage**

**8-oxoguanine**

**2,6-diamino-4-hydroxy-5-formamidopyrimidine (FaPy)**

**thymine glycol**

**Fig. 4.9** Some examples of damaged DNA bases. Thymine cyclobutane dimers are created by exposure to UV light and are repaired in *E. coli* by photolyase, and in mammals by nucleotide excision repair. Deamination is the result of a reaction with water and converts cytosine to uracil and adenine to hypoxanthine. Alkylation damage produces $O^6$-methylguanine which is repaired by methylguanine methyl transferase and also produces 3-methyladenine and 7-methylguanine which are repaired by base excision repair. Reactive oxygen species that generate oxidative damage like, 8-oxoguanine, formamidopyrimidines (FaPy), and thymine glycol are repaired by base excision repair (Copies of figures including color copies, where applicable, are available in the accompanying CD)

instead it is a suicide enzyme. The methyl group is transferred from guanine to a cysteine residue on the protein where a covalent bond is formed. Because MGMT cannot be demethylated, it is incapable of catalyzing the removal of other methyl groups and becomes inactive. MGMT is the only protein known to repair $O^6$-MeG and is found in both bacteria and eukaryotes.

## *4.4.3 Repair by Excision/Resynthesis*

The process of excision of a section of damaged DNA followed by resynthesis allows cells to use the same basic chemical reaction and a common set of enzymes to repair many different types of DNA damage. The two main pathways responsible for excision repair are base excision repair that usually replaces a single damaged base and nucleotide excision repair that replaces a short segment of DNA containing the damaged nucleotide.

### 4.4.3.1 Base Excision Repair

The base excision repair pathway primarily repairs damage to DNA bases much of which occurs spontaneously in the cell without the influence of environmental hazards [21]. Examples include oxidation, deamination, and alkylation of DNA bases. These damaged bases are recognized by damage-specific *DNA glycosylases* that initiate the base excision repair pathway. DNA glycosylases bind to the damaged bases and cleave the C1'-N glycosylic bond between the base and sugar. This leaves a baseless sugar, or AP (for apurinic or apyrimidinic) site, in DNA that is removed by other enzymes in the pathway (Fig. 4.10A). Several different DNA glycosylases are present in cells, and each recognizes a specific damaged base or a class of damaged bases. For example, uracil DNA glycosylase recognizes and excises only uracil which can be formed in DNA by deamination of cytosine. Formamidopyrimidine (FaPy) DNA glycosylase recognizes several different bases damaged by oxidation including 8-oxoguanine and formamidopyrimidines (Fig. 4.9).

Once a DNA glycosylase has removed a damaged base, the AP site that it leaves must be repaired. An AP endonuclease starts this process by cleaving the phosphodiester bond on the 5' side of the AP site to generate a 3'hydroxyl and a 5'deoxyribose phosphate. A deoxyribophosphate lyase removes the 5'deoxyribose phosphate and leaves a one nucleotide gap in the DNA. A DNA polymerase extends the 3' end to replace the missing nucleotide and the resulting nick is sealed by the activity of a DNA ligase. In *E. coli*, separate enzymes catalyze the removal of the baseless sugar and incorporation of the missing nucleotide. In humans, DNA polymerase β contains two enzymatic activities and is capable of both incorporating the missing nucleotide and removing the baseless sugar residue.

### 4.4.3.2 Nucleotide Excision Repair

The nucleotide excision repair pathway recognizes and repairs damage that generates larger more bulky lesions and local distortions in the DNA structure, such as a thymine- cyclobutane dimer. In the nucleotide excision repair pathway, damage is recognized by a protein complex capable of identifying many different types of damage [22, 23]. It is believed that this complex recognizes distortions in the overall DNA structure at sites of damage, or an increase in the ease of unwinding the duplex in the region of the damage. Regardless of the mechanism, this complex identifies sites of damage and recruits the rest of the repair machinery to these sites (Fig. 4.10B). The DNA duplex is then locally unwound at the site of damage, and endonuclease activities within the complex cleave the DNA backbone both at 5' and 3' to the site of damage. These endonucleases are structure-specific and cleave single-stranded DNA at the single-strand/double-strand junction that has been

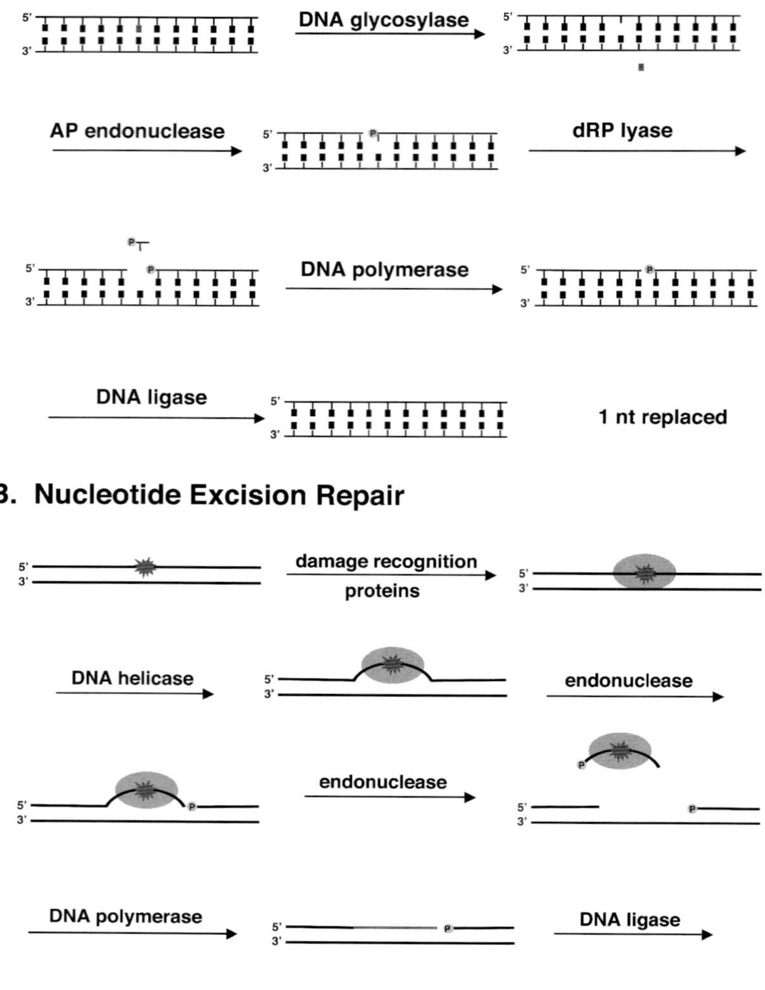


**Fig. 4.10** Repair of DNA by excision of the damage and resynthesis of DNA. (**A**) The base excision repair pathway begins with the removal of a damaged base by a DNA glycosylase. In this scheme undamaged DNA bases are indicated by black squares and the damaged base is indicated by a red square. The C1'-N glycosylic bond between the base and the sugar is cleaved leaving a baseless sugar residue (AP site) in DNA. The DNA strand is cut 5' to the AP site creating a 3' hydroxyl on one side of the cut and a 5'phosphate ("P" in gold circle) on the other. Deoxyribose phosphate lyase (dRP lyase) activity is required to excise the sugar-phosphate residue to create a one nucleotide gap that can be filled in by a DNA polymerase. Repair is complete when the strands are ligated by a DNA ligase. (**B**) The nucleotide excision repair pathway removes a short segment of DNA containing a helix-distorting DNA damage (*red starburst*). The damage is recognized and bound by a protein complex. This protein complex serves to direct the other proteins to the site of damage so that it can be repaired. A DNA helicase separates the DNA strands on either side of the damaged nucleotide. Specific endonucleases recognize the forked single-stranded/double-stranded DNA junctions at these sites, and cleave the DNA at the junctions. The DNA strand is cleaved 3' to the damaged nucleotide, followed by cleavage on the 5' side. The gap created by excision of the damaged DNA segment is filled in by a DNA polymerase and the two strands are joined by a DNA ligase (Copies of figures including color copies, where applicable, are available in the accompanying CD)

created by the DNA unwinding activity of the repair complex. DNA cleavage creates a short oligonucleotide segment, 12–13 nt long in *E. coli* and 24–32 nt long in eukaryotes, that is displaced by a DNA helicase. The resulting gap in DNA can be filled in by a DNA polymerase to leave a nick that is sealed by DNA ligase.

## Glossary and Abbreviations

**Allele** alternate forms of the same gene that differ in primary sequence at some sites.

**AP endonuclease** an enzyme that cleaves the DNA backbone on the 5' side of an AP site to create a 3' hydroxyl and 5' phosphate.

**AP site** a baseless sugar in DNA. AP is an abbreviation for apurinic or apyrimidinic.

**Base excision repair** a pathway that repairs damage to DNA bases by excising the damaged base and replacing it with an undamaged nucleotide.

**Branch migration** movement of a Holliday junction that unwinds DNA strands ahead of the junction and pairs them with homologous strands behind the junction.

**CDKs** Cyclin-dependent kinases

**DNA glycosylase** an enzyme that catalyzes the excision of damaged bases by cleaving the C1'-N glycosylic bond between the damaged base and the sugar.

**DNA helicase** an enzyme that catalyzes the unwinding of the duplex DNA.

**DNA ligase** an enzyme that joins two strands of the DNA at a nick by ligating a 3' hydroxyl end to a 5' phosphate end.

**DNA polymerase** an enzyme that catalyzes the extension of a DNA polymer by incorporating 2'-deoxynucleoside monophosphates in a template directed reaction.

**Endonuclease** an enzyme which cleaves DNA at an internal site by hydrolysis of the phosphodiester bond between the two nucleotides.

**Epigenetic** changes in DNA function that are not due to changes in the primary sequence of DNA.

**Exonuclease** an enzyme that catalyzes the excision or removal of nucleotides from a DNA strand by progressively removing one or a few nucleotides from one end of the DNA strand. Exonucleases have defined polarity and degrade in either a 5' to 3' or 3' to 5' direction.

**Fidelity** refers to the accuracy of synthesis by DNA polymerases.

**Holliday junction** a four-way DNA junction formed when strands from homologous duplexes crossover during recombination.

**Initiator protein** a protein that recognizes origins of replication, and helps in the initial unwinding of DNA and recruiting of replication proteins. The term comes from the original replicon model proposed by François Jacob, Sydney Brenner, and Jacques Cuzin to explain the initiation of replication.

**Lagging strand** the strand synthesized as discontinuous segments called Okazaki fragments.

**Leading strand** the strand of DNA synthesized as one continuous piece in the direction of replication fork progression.

**MCM proteins** Mini-chromosome maintenance proteins form the eukaryotic replicative DNA helicase.

**Meiosis** process of cell division that generates eukaryotic germ cells.

**Mitosis** process of cell division in eukaryotic somatic cells.

**MSH** MutS homologs

**Nick** a single-stranded break in a DNA duplex.

**Nucleotide excision repair** a pathway that repairs DNA damage by removing a short segment of DNA containing the damage and resynthesizing that segment.

**Okazaki fragments** discontinuous segments of DNA that are synthesized on the lagging strand.

**Origin of replication** a site in the genome where DNA synthesis is initiated during replication.

**ORC** Origin recognition complex

**Primase** an enzyme that synthesizes short segments of RNA to prime synthesis by DNA polymerases.

**Primer** a short oligonucleotide complementary to and paired with a larger DNA template.

**Processivity** refers to the number of nucleotides which a DNA polymerase can incorporate in a single DNA binding event.

**Proofreading** the process of removing incorrectly paired nucleotides that is catalyzed by the 3' to 5' exonuclease activity of DNA polymerases.

**Replication fork** the branched DNA structure formed, when a DNA helicase separates the two complementary strands of DNA during replication.

**Replisome** the ensemble of enzymes that function at the replication fork to duplicate the DNA.

**S-phase** the phase of the eukaryotic cell cycle in which DNA is replicated.

**Semi-conservative** refers to the replication of DNA which produces duplexes composed of one newly synthesized and one original strand.

**Semi-discontinuous** refers to DNA synthesis that produces one continuous segment of DNA on the leading strand and Okazaki fragments or discontinuous segments on the lagging strand.

**Strand invasion** pairing of a single-stranded DNA with a homologous region of duplex DNA to form a D-loop.

**Topoisomerase** enzymes that relax the superhelical tension which builds up ahead of the replication fork as the result of unwinding the DNA strands during replication.

## Suggested Reading

### *DNA Replication*

1. Watson, J. D., and Crick, F. H. C. (1953) Genetical implications of the structure of deoxyribonucleic acid. Nature 171, 964–967.
2. Machida, Y. J., Hamlin, J. L., and Dutta, A. (2005) Right place, right time, and only once: replication initiation in metazoans. Cell 123(1), 13–24.
3. Mott, M. L., and Berger, J. M. (2007) DNA replication initiation: mechanisms and regulation in bacteria. Nature Reviews 5(5), 343–354.
4. Kaguni, J. M. (2006) DnaA: controlling the initiation of bacterial DNA replication and more. Annual Review of Microbiology 60, 351–375.
5. Costa, S., and Blow, J. J. (2007) The elusive determinants of replication origins. EMBO Reports 8(4), 332–334.
6. Kelly, T. J., and Brown, G. W. (2000) Regulation of chromosome replication. Annu. Rev. Biochem. 69, 829–880.
7. Johnson, A., and O'Donnell, M. (2005) Cellular DNA replicases: components and dynamics at the replication fork. Annu Rev Biochem 74, 283–315.
8. Gulbis, J. M., Kelman, Z., Hurwitz, J., O'Donnell, M., and Kuriyan, J. (1996) Structure of the C-terminal region of $p21^{WAF1/CIP1}$ complexed with human PCNA. Cell 87, 297–306.
9. Kong, X.-P., Onrust, R., O'Donnell, M., and Kuriyan, J. (1992) Three-dimensional structure of the b subunit of *E. coli* DNA polymerase III holoenzyme: A sliding DNA clamp. Cell 69, 425–437.
10. Goodman, M. F. (1988) DNA replication fidelity: kinetics and thermodynamics. Mutat. Res. 200, 11–20.
11. Kunkel, T. A., and Bebenek, K. (2000) DNA replication fidelity. Annu. Rev. Biochem. 69, 497–529.
12. Iyer, R. R., Pluciennik, A., Burdett, V., and Modrich, P. L. (2006) DNA mismatch repair: functions and mechanisms. Chem. Rev. 106(2), 302–323.
13. Yang, W. (2007) Human MutLalpha: the jack of all trades in MMR is also an endonuclease. DNA Repair (Amst) 6(1), 135–139.

### *DNA Recombination*

14. Haber, J. E. (2000) Partners and pathways repairing a double-strand break. Trends Genet. 6, 259–264.
15. Kowalczykowski, S. C. (2000) Initiation of genetic recombination and recombination-dependent replication. Trends Biochem. Sci. 25, 156–165.

16. West, S. C. (2003) Molecular views of recombination proteins and their control. Nat. Rev. Mol. Cell. Biol. 4(6), 435–445.
17. Heller, R. C., and Marians, K. J. (2006) Replisome assembly and the direct restart of stalled replication forks. Nat. Rev. Mol. Cell. Biol. 7(12), 932–943.
18. Whitby, M. C. (2005) Making crossovers during meiosis. Biochemical Society transactions 33(Pt 6), 1451–1455.
19. Keeney, S., and Neale, M. J. (2006) Initiation of meiotic recombination by formation of DNA double-strand breaks: mechanism and regulation. Biochemical Society Transactions 34(Pt 4), 523–525.

## *DNA Repair*

20. Lindahl, T. (1993) Instability and decay of the primary structure of DNA. Nature 362, 709–715.
21. Lindahl, T. (2000) Suppression of spontaneous mutagenesis in human cells by DNA base excision-repair. Mutat. Res. 462, 129–135.
22. Batty, D. P., and Wood, R. D. (2000) Damage recognition in nucleotide excision repair of DNA. Gene 241, 193–204.
23. Reardon, J. T., and Sancar, A. (2005) Nucleotide excision repair. Prog Nucleic Acid Res Mol Biol 79, 183–235.

# Chapter 5
# Cell Signaling

**Daniel A. Rappolee and D. Randall Armant**

**Abstract** Signal transduction is the molecular process whereby a cell receives, transmits, amplifies, and integrates extracellular stimuli. Key parameters of the signaling pathways are duration and magnitude of the input and subsequent biological responses. Key features of signaling processes include the time- and dose-dependent responses mediated by downstream molecular effectors. The precision of signaling is accomplished through negative feedback and branching between the pathways that mediate cross-talk. Since rate-limiting enzymes in different pathways have different inputs and outputs, integration of signal transduction pathways leads to the optimal biological responses that have been selected during evolution.

**Keywords** Signal transduction · Pathway · Communication cascade

## 5.1 Introduction

The intent of this chapter is to present a broad overview of the functions of signal transduction mechanisms, the basic pathways mediating these functions, and factors that affect the initiation, completion and control of signal transduction pathways. We define signal transduction as the molecular mechanism used by a cell to sense and transduce stimuli from one place to another over a period of time. In this chapter, we focus on signal transduction from the cell surface to the cytoplasm and nucleus. In a broader sense, signal transduction can also occur between cells and from one point to another within a cell, but this is not the focus here. To facilitate the assimilation of a large body of data, there is a section with acronyms and abbreviations at the end of the chapter. To should facilitate future information assimilation. A table is provided that includes electronic resources that aid in attending up to date information on fast-moving research in the field of signal transduction.

## 5.2 Basic Concepts

### *5.2.1 Types of Intercellular Signaling Molecules: The Ligand*

Intercellular communication is essential for development and homeostatic function in multicellular organisms. The language of intercellular communication is via protein growth factors (e.g., local paracrine FGF or blood-borne endocrine growth hormone), hydrophobic steroids (e.g., estrogen) and lipid mediators (prostaglandins, leukotrienes), modified amino acids (e.g., neurotransmitters

D.A. Rappolee
Wayne State University Medical School, Departments of Ob/Gyn and Anatomy, 275 East Hancock St, Detroit, MI 48201, USA
e-mail: drappole@med.wayne.edu

S. Krawetz (ed.), *Bioinformatics for Systems Biology*, DOI 10.1007/978-1-59745-440-7_5,

such as adrenaline), other metabolites and precursors to second messengers (see below) not present in the previous groups (e.g., NO or **n**itric **o**xide, $Ca^{++}$, cyclic AMP). Some of these ligands are present at low concentrations in the nanomolar range (FGF and insulin), whereas others can be as high as the millimolar range (amino acids that can be modified to make neurotransmitters or $Ca^{++}$). This chapter will focus largely on small protein growth factors that modulate growth, differentiation, apoptosis, and steady state function during development and in the adult.

### *5.2.2 General Function and Speed of Intercellular Communication*

An essential role of signal transduction is to coordinate functions of identical or diverse cell types within an organ to synchronize the cellular activity in multicellular organisms. The speed of intercellular communication is dependent upon distance and the mode of delivery of the intercellular signal. Local or paracrine intercellular communication acts within milliseconds over distances less than 10–20 cell diameters (about 200 microns), but endocrine or blood-borne signaling acts over distances up to meters, and requires seconds or minutes [1, 2]. Specialized short distance signaling, like that mediated by gap junctions, and allows linked cells to share small intracellular signal transduction intermediates (second messenger molecules) downstream of cell surface receptors or metabolic intermediates. Specialized long distance signaling mechanisms can span meters within milliseconds due to the increased conductance speed accomplished through mechanisms such as saltatory movement of signals in neurons coupled with the fast action of neurotransmitters at post-synaptic membranes [3].

### *5.2.3 Speed and Magnitude of Intracellular Communication Dictates the Nature of the Signal*

Intracellular signal transduction is initiated once a ligand has bound to its receptor. In assessing the mechanisms that govern intracellular signaling, it is useful to consider the speed and magnitude of the signaling. It is important to keep in mind the vast concentration range of extracellular signals with respect to intracellular signaling. Whereas most growth factors are in the nanomolar range, some neurotransmitter precursors (amino acids) are near the millimolar level, and $Ca^{++}$ is at about 1 millimolar concentration outside the cell. There is an intrinsic requirement to amplify the growth factor signal inside the cell, that requires energy at every amplification step. The signaling intermediates in the growth factor pathways are inherently at comparatively low concentrations and this limits the number of downstream effector molecules that can be regulated. Therefore, these pathways serve as servomechanisms. However, signaling by some second messengers, exemplified by $Ca^{++}$, is intrinsically different. The high potential energy of 1 millimolar extracellular $Ca^{++}$ enables the cell to use this signaling molecule for regulating a very high number of effector outputs. Therefore, growth factors initiate pathways that control transcription through a hierarchy of downstream networks. Downstream networks with large number of effector proteins require stoichiometric control. These networks, exemplified by the contractile sarcomere of heart muscle or cortical granules in the oocyte are best controlled by the highly abundant second messengers (see below), such as $Ca^{++}$.

Another intrinsic property of signaling mechanisms is speed. After the receptor is activated by phosphorylation or some conformational change, receptor binding by non-enzymatic docking molecules peaks within approximately one minute, as in the case of insulin receptor substrate (IRS)-1 [4]. After transiting the cytoplasm, these signal transduction pathways climax in the nucleus where, for example, phosphorylation of AP-1 transcription factors results in peak transcription of c-fos within 15 minutes and initial transcription within one minute [5]. Other hormones, such as estrogen, activate transcription over slower time periods, approaching one hour. In some signal transduction processes, such as in vision chemistry, the complete process of photons activating the rhodopsin is finished in one second [3]. Kinase cascades such as MAPK, and estrogen signaling, operate over minutes to hours, whereas second messengers tend to mediate rapid signaling lasting seconds to minutes. The speed of signal transduction mediates very rapid organismal survival responses, rapid cellular metabolic and homeostatic responses, and slower, permanent changes exemplified by those occurring during embryonic development.

### *5.2.4 Second Messengers*

Second messengers are small molecules that direct and amplify signals within the cell. They are smaller than proteins and can range in size from ions like $Ca^{++}$ to lipids like diacylglycerol (DAG). Other important second messengers include cyclic AMP, cyclic GMP, inositol 1,4,5-trisphosphate, ($IP_3$), and NO. Induced changes in the cellular concentration of second messengers amplify the signal and send it to specific cellular compartments that are accessible to these small molecules [3]. In most cases, second messengers are generated directly or indirectly by enzymes that are activated upon activation of receptor molecules by primary messengers, such as growth factors. For example, phospholipase C becomes enzymatically activated either by binding to phosphorylated tyrosines on receptor tyrosine kinases (phospholipase C$\gamma$ isoforms) or by G proteins coupled to seven pass transmembrane receptors (phospholipase C$\beta$ isoforms) [6]. Both $IP_3$ and DAG are produced as a result of hydrolysis by phospholipase C of phosphatidylinositol 4,5-bisphosphate, a phospholipid residing in the plasma membrane. $IP_3$ is hydrophilic and can enter the cytoplasm where it activates the $IP_3$ receptor, an ion channel located in the membranes of several cellular organelles. The opened channel releases $Ca^{++}$ into the cytoplasm where many $Ca^{++}$-dependent regulatory proteins reside. DAG is hydrophobic and partitions within the lipid bilayer of the plasma membrane where it activates protein kinase C, which phosphorylates numerous protein substrates that influence cell adhesion and other plasma membrane functions. These three second messengers significantly alter the cellular state through their ability to impact a large number of biochemical reactions within specific cellular compartments.

Second messengers are produced by a burst of synthetic or regulatory activity, which subsides and becomes overtaken by constitutive degradative activity, generating a biphasic concentration profile. Phospholipase C generates $IP_3$ within seconds of its activation by a growth factor, but the cytoplasmic $IP_3$ concentration returns to basal levels within minutes [6]. The opening of the $IP_3$ receptor is inhibited by cytoplasmic $Ca^{++}$, providing a negative feedback that limits the maximal concentration of cytoplasmic $Ca^{++}$ and allowing $Ca^{++}$ pumps to begin moving $Ca^{++}$ back into storage organelles. Because second messenger signaling is transient, it is possible for different receptors to use a common second messenger as a switch to turn on downstream signaling cascades. The pathways downstream of a second messenger can change when cells alter their production of target proteins. As a result, the same second messenger could have widely differing effects as a cell differentiates or matures. Indeed, signaling downstream of a second messenger could alter expression of its target proteins so that a repeated burst of signaling produces different outcomes. It is, therefore, possible for a single second messenger to transit a cell over time through sequential steps of differentiation by temporally activating different downstream events. For example, $Ca^{++}$ transients occurring after fertilization are required to advance the early embryo through a series of developmental processes, from completion of meiosis to exocytosis of cortical granules, to mitosis, to embryonic genome activation, and then to morphogenesis [7]. The biochemical basis of each event differs widely, but all are similarly initiated by a transient elevation of cytoplasmic $Ca^{++}$.

In summary, second messengers are generated transiently downstream of receptor-induced enzymes and target large numbers of signaling proteins within specific cellular compartments. They operate as generic switches for the upstream signaling pathways to recruit downstream targets that can shift with the physiologic state of the cell. Second messengers have unique capabilities and functions in contrast to protein and kinase cascades such as the MAPK pathway. Although second messengers can influence long-term decision-making such as cell cycle commitment, they specialize in rapid regulation of housekeeping functions. Since second messengers are small and have special properties, they have specialized capabilities in transducing large signals with relatively low energy requirements through gap junctions, cell membranes, and nuclear pores.

### *5.2.5 The Nature of the Receptor*

If the ligands are the 'verbs' of intercellular language, the receptors are the 'ears'. Trans-membrane plasmalemma receptors are of several types: (1) receptor tyrosine kinases such as the FGF receptor, (2) serine-threonine kinase receptors such as the TGF-beta receptor, (3) non-enzymatic transmembrane receptors (e.g., integrins) that are linked to intracellular tyrosine kinases (e.g., FAK –focal adhesion kinase- and PYK2-proline rich tyrosine kinases), (4) transmembrane-seven pass G-proteincoupled receptors (e.g., the Wnt receptor; frizzled), (5) non-enzymatic receptors that are linked to signaling pathways which are derepressed by allosteric-conformational changes (e.g., smoothened receptor for hedgehog ligand), and (6) receptors that are proteolytically converted to ligand (e.g., the Notch intracellular C terminal domain) [1].

Another class of signaling receptors is located in the cytoplasm. They bind hydrophobic hormones such as steroids (e.g., estrogen), thyroid hormones and retinoic acid ligands, and then translocate to the nucleus where they act as transcription factors [3].

Whether positively activated or if its function is regulated by derepressing signaling enzymes, receptors initiate and amplify signals from outside the cell. Receptors can typically be activated for minutes and tend to undergo negative feedback regulation to limit signaling, even if the ligand continues to be present (see below).

### *5.2.6 Nature of the Signal Cascade Within the Cell: Aspects of Pathway Function – Amplification, Timing, Branching, and Negative Feedback*

Signaling cascades begin as allosteric changes in the receptors or non-enzymatic docking proteins (IRS-1 family and FRS2). These events convey the signal through conformational changes, and by becoming phosphorylation targets of receptor kinases. In either case, the function of the signal transduction pathway is to quickly amplify and directionally conduct information that must reach the cell by transmitting it through a series of tyrosine and serine-threonine kinases. The directionality, speed, and magnitude of the signaling are based on the functions of the cell type and history of signaling by the cell.

The majority of the transmembrane receptors are in the 'off' state until induced by an extracellular ligand. Receptors act like allosterically regulated enzymes with the active site in the cytoplasmic domain. Many of the tyrosine kinases require oligomerization, as there is no sufficient flexibility in the transmembrane alpha-helix to mechanically transduce the ligand–induced conformational change [8]. Oligomerization brings together cytoplasmic kinase domains that cross-activate and then signal to downstream docking and enzymatic signal transduction proteins.

The receptor is activated for a period of time before it is internalized and degraded, desensitized through phosphorylation by a receptor induced kinase such as β-adrenergic receptor kinase/BARK; [9]), or dephosphorylated [3].

During the activation period, a single activation event can become highly amplified. For example, a single quantum of light activating the photoreceptor rhodopsin, leads to the hydrolysis of one hundred thousand cGMP signaling molecules for a duration of one second [3].

The activated receptor has multiple phosphorylation sites available in its cytoplasmic domain. These phosphorylated sites are capable of interacting with a large number of signaling

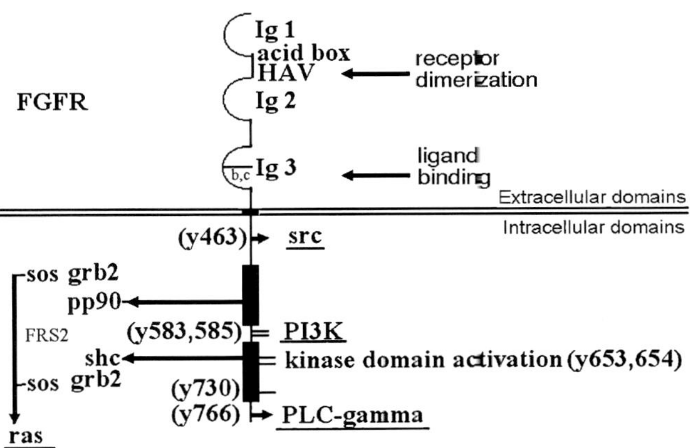


**Fig. 5.1** The FGF receptor is an allosteric enzyme with the allosteric sites in the ectodomain and the enzymatic tyrosine kinase in the cytoplasm. Many mapped potentially functional and functionally active tyrosine activation sites are known (Copies of figures including color copies, where applicable, are available in the accompanying CD)

intermediates that bind to the activated receptor through src homology domains-2 (SH2) which recognize phosphorylated tyrosines (Fig. 5.1). If the activated receptor is the hub of activity, the docking proteins provide the spokes that radiate out to multiple downstream targets. Docking proteins such as IRS-1 and FRS-2 (insulin and FGF receptor substrate) are receptor-binding proteins that can initiate many branches of the signal transduction pathway. This branching of pathway choices immediately downstream of the FGF receptor is considered in the next section of this chapter (Figs. 5.1, and 5.2; Receptor tyrosine kinases, mitogenic signal). Although receptors

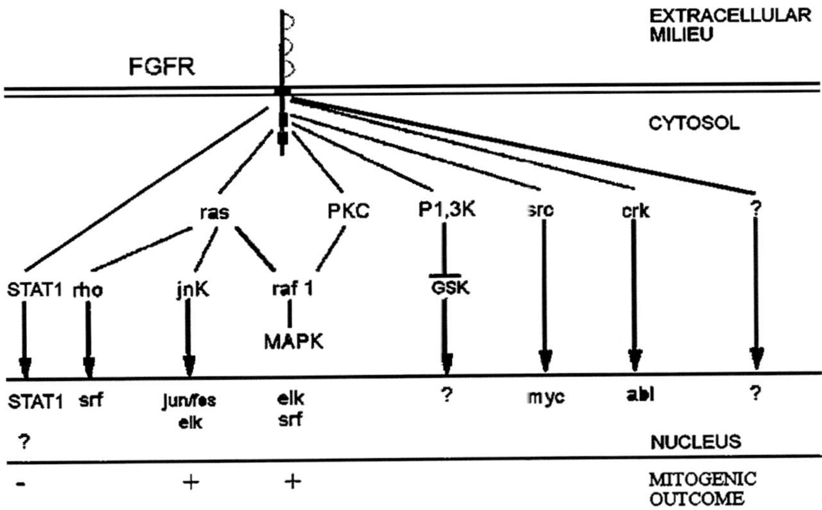


**Fig. 5.2** shows the known mitogenic and transcription activating signal transduction pathways downstream of FGFRs. Note that the preponderance of evidence in cell lines suggests that the *ras*-MAP kinase pathway mediates the mitogenic signal of FGFR. src, crk are only indirectly implicated in FGFR signal transduction due to sequence homology with other receptors and possible binding sites in the cytosolic domain. Jun kinase and P1, 3 Kinase can be mitogenic in certain circumstances, but mutation of the P1, 3 Kinase activating site in an FGFR *in vitro* did not prevent mitogenic response to FGF. STAT1 pathway has recently been shown to be anti-mitogenic in FGFR-3 mediated chondrocyte cell division cessation. The + indicate the most likely pathways for FGFR cell division control (Copies of figures including color copies, where applicable, are available in the accompanying CD)

can share docking proteins such as Grb2 and GAB1, some receptors have more complex sets of docking proteins as is the case for FGF receptor campared with EGF receptor [48]. An array of these docking proteins distinguishes different cell types, which may each express the same receptor to facilitate different signaling pathways. The importance of the additive effect of branched pathways, is indicated by studies analyzing the effects of mutations in the phosphorylation/docking sites of the PDGF receptor on activation of sets of newly transcribed genes (reviewed in [10]. Sets of newly transcribed genes were analyzed for quality and magnitude of induced transcription by cDNA array. The results suggest that more than one phosphorylation/docking site on the PDGF receptor is needed for the full and proper magnitude and breadth of the transcriptional response.

For the three MAPK families (Figs. 5.2, 5.3, Table 5.1), there are three non-enzymatic signal transduction intermediates before the ras GTPase becomes activated. After ras, which is a key component in mitogenic signal transduction [11], there are 4 sequential vertical tiers of serine-threonine kinases (Table 5.1), that each can have different 'horizontal' interacting components. In the three MAPK signal transduction pathways, the first two serine-threonine kinases have only or largely cytoplasmic targets, whereas the last two tiers have both cytoplasmic and nuclear targets. Therefore, the activation of these last two tiers of serine-threonine kinases (MAPKs and MAPKAPS), can lead to nuclear localization, transcription factor phosphorylation (e.g., Elk1 and ATF2 for ERK1, 2), and gene transcription. Other non-tyrosine kinase initiated pathways also have homologous cytoplasmic signal transducers. For example, although it is thought that Raf1

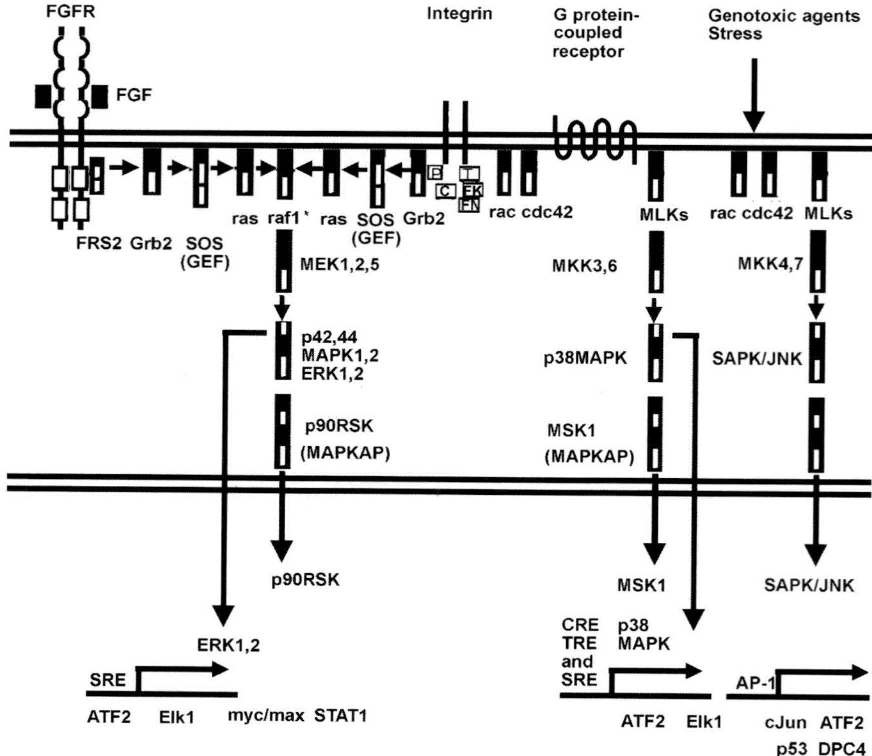


**Fig. 5.3** MAPK Families the three MAPK families are embedded in enzymatic cascades of intracellular serine-threonine kinases regulated by tyrosine kinases and allosteric docking proteins. The pathways initiate at the plasmalemma by receptor tyrosine kinases, ECM-binding integrins, and G protein-coupled receptors. See text for details (Copies of figures including color copies, where applicable, are available in the accompanying CD)

**Table 5.1** Extracellular to intracellular signal transduction for three MAPK families (example FGF ligand receptor) Levels of signal transduction in the cascades of the three MAPK families. FGF signaling is used as an example. For detail, see text

| Ligand ↓ | | FGF1-20 | | | Other ligands | | | | | |
|---|---|---|---|---|---|---|---|---|---|---|
| Receptor ↓ | | FGFR1-4 | | | Other receptors | | | | | |
| Non-enzymatic<br>Docking molecule 1<br>RTK binding ↓ | | FRS2 | | | GAB1,2 | | IRS1-4 | | | |
| Non-enzymatic<br>Docking molecule 2<br>RTK binding<br>Docking molecule<br>1 binding ↓ | | Grb2 | | | shc | | shb | | | |
| Ras superfamily<br>Regulatory molecules ↓ | | SOS (GEF family) | | | GRF family<br>Rate limiting | | GAP family | | vav | |
| Ras superfamilyras<br>G protein family<br>Enzyme ↓ | | | | Rac | Rho | cdc42 | Rap1 | | | |
| Kinase 1 (MEKK)<br>Enzyme ↓ | | Raf-1 | | | RafB | KSR | Tak1 | | | |
| Kinase 2<br>Enzyme ↓ | | MEK1, 2, 5 | | | MEK3,6 | | MEK4,7 | | | |
| Kinase 3 ↓ | | MAPK family<br>One<br>p42, 44 ERK1,2<br>MAPK1,2, MAPK5 | | | MAPK family<br>Two<br>p38MAPK | | MAPK family<br>Three<br>SAPK/JUNK enzyme | | | |
| Kinase 4<br>Enzyme ↓ | | p90RSK | | | MAPKAP | | MNK1,2 | | MSK1 | |
| Resident or Trans-<br>located nuclear<br>Factors | Elk1 | myc/max | pRSK90 | | MAPK family | | ATF-2 | STAT1 | MSK1 | Jun |

(Copies of tables are available in the accompanying CD.)

mediates the effects of the FGF receptor (Fig. 5.2, Table 5.1), Raf1 homologs also have a function in the TGF-beta pathway (TAK1; Fig. 5.3, Table 5.1), and in G protein-coupled receptor pathways (RafB; Fig. 5.3, Table 5.1). Branching of the pathway can occur at any tier of the vertical cascade (Table 5.1). Regulation of branching also distinguishes cell types and prior extrinsic interactions of the cell [12].

### 5.2.7 *The Nature of Signal Transduction, Thinking About What Different Time-and Dose-Dependent Responses Means*

Signal transduction mechanisms can be studied in many ways. In order to understand, and prior to the cause-and-effect testing for enzyme function in transduction pathways, common foundation studies include elucidating the time- and dose-dependent responses of the enzymes to stimuli.

There are key common themes and interpretations for studies of time-dependence. The most obvious one is that the enzyme must be activated before it can be considered as a candidate to mediate downstream molecular or biological effects. Two common mechanisms for positive stimuli are a biphasic response or a plateau response, where enzyme activity in unstimulated cells is arbitrarily set as the baseline.

As an example, a biphasic peak enzyme activity reached at 30 minutes may be 10-fold over the baseline and may return to baseline by 2 hrs. There are two major interpretations of this. The first mechanistic interpretation is that the timing of the peak activity for an enzyme is a function of how many molecular steps are there, between the reception of the stimulus and the enzyme. For example, a docking protein at step two in the pathway has transient peak activity well before the peak activity of an enzyme at step eight. Also, mechanistically the downward slope of the biphasic activity is mediated by negative feedback, but can also be a product of toxicity. Secondly, the functional interpretation is that the enzyme with a biphasic peak activates substrates that ultimately lead to effector pathways that set into motion a biological response. This response outlasts the enzymatic peak, and the enzyme is not directly responsible for this long-term response. An additional functional significance is that, resetting the reception pathway to zero allows the cell to accurately sense fresh stimuli. The significance of this is intuitively understood, as if one has driven for several miles behind a car with an uncancelled turn signal. Once the cell has set a new homeostatic equilibria, or the car has changed lanes, it is important to reset the sensing mechanism so that new inputs can be interpreted and sent to downstream effectors.

The plateau response for time-dependent enzyme activation suggests that the enzyme may participate in the effector mechanism as well as being part of the signal transduction pathway. This is in contrast to the biphasic response. Shared with the biphasic response is an initial lag phase whose length is a function of the pathway length prior to the enzyme.

There are also biphasic and plateau responses observed when measuring dose-dependence. As an example, a biphasic response starting at a baseline might increase to 10-fold over background at 400 mM hyperosmolar sorbitol and return to baseline at 1,000 mM [13]. This suggests that high doses of stimulant activate a negative feedback or that there is toxicity. A plateau response suggests that the enzyme itself has become saturated. Alternately, effectors downstream may be saturated and put up a feedback response on the upstream enzyme to maintain it at constant, plateau levels of activity.

### 5.2.8 *Turning Off the Response*

Signaling requires expeditious and tight control to maintain homeostasis and to ensure proper regulation. Control is exerted at all tiers of the pathway and at various levels of production and activation of the signal transduction proteins. As mentioned above, the receptor is activated for a period of time before it is: (i) destroyed, (ii) desensitized, or (iii) dephosphorylated. Other tiers of the pathway are regulated in similar ways. Ras is inactivated by GTPase activating proteins (GAPs), MEK family members are dephosphorylated and inactivated by protein phosphatase (PP)1 and PP2A. ERK family members are dephosphorylated and inactivated by MAP kinase phosphatases

(MKP)3 and MKP6 [14], and JNK is inactivated by MKP3/6 [15] and references therein). The mRNA transcripts for many of the signal transduction genes have a consensus destruction sequence in the 3' untranslated region that confers a short half-life on them [16]. At the levels of protein and mRNA stability, and protein activation, signal transduction component levels can be regulated rapidly.

## 5.3 Signal Transduction Pathways

This section with will focus on FGF receptors and their activation of the three MAPK families (Figs. 5.2 and 5.3) emphasis on recent progress in illuminating signal transduction pathways. Some models will be touched upon only briefly. Although these are very important pathways and will make up a bulk of future research, the space limitations in this short review prohibit their discussion. They are mentioned here for the sake of completeness and to alert the reader to other sources (Table 5.2). These include activation of PKA and PKC, the mechanism of Calcium-calmodulin signaling, prostaglandins and leukotrienes, and nitric oxide. Other cytostatic pathways mediated by JAK-STAT receptors for the IFNγ receptor, and apoptosis pathways through TNFα are also not covered in depth. Non-canonical pathways mediated by serine-threonine receptor kinases (TGFβ receptor), and novel pathways for signaling by derepressing G protein-coupled receptors (Hedgehog-ligand derepression of patched receptor by smoothened) are covered elsewhere. A novel signaling mechanism important in development, signaling by proteolytically cleaved ligand-activated receptor (Delta ligand activation of Notch protein to cleave and translocate to the nucleus) is also described in signal transduction knowledge environment (STKE). The Wnt-frizzled-GSK3-β-catenin pathways interaction with Ecadherin-βcatenin are not covered here. These pathways are very important in embryonic development and in pathology (especially carcinogenesis), and their elucidation and interaction with other pathways will provide important future areas of research.

**Table 5.2** Signal transduction websites. Electronic resources for signal transduction reagents and information

| | |
|---|---|
| http://stke.sciencemag.org/ | Signal transduction knowledge environment (STKE) Excellent resource for broad and focused signal transduction electronic and archival published literature PDFs and full text articles with JPG figures are available Requires AAAS membership and an STKE users fee |
| http://kinase.oci.utoronto.ca/signallingmap.html | Very good focus with map of the 3 MAPK families and clickable short to long descriptions of molecules on the map |
| http://www-personal.umich.edu/%7Eino/List/ | Good outline of signal transduction pathways with links to PubMed discovery articles |
| http://www.copewithcytokines.de/cope.cgi | Extracellular signaling |
| http://www.grt.kyushu-u.ac.jp/spad/index.html | Good, clickable diagrams, but not recently updated |
| http://vlib.org/Science/Cell_Biology/signal_transduction.shtml | Good cross-referenced site for information about function and sequence references for signal transduction genes |
| http://www.clontech.com/ | Company site. Short description of signaling intermediates in literature on antibodies. Also, check sections under pathway diagrams check other web resources links |
| http://www.clontech.com/ | Company site for signal transduction expression transgenes |
| http://www.scbt.com/ | Company site. Short description of signaling intermediates in literature on antibodies. Also, check sections under pathway diagrams check other web resources links. |
| http://www.ihop-net.org/UniPub/iHOP/ | Information hyperlinked over proteins (IHOP). Good site for all proteins, their structure, function, expression, and interaction. See Chapter 22. |

(Copies of tables are available in the accompanying CD.)

## 5.4 Receptor Tyrosine Kinases (FGF Receptor) and Mitogenesis FGF Receptor as an Example of the Selection of Possible Signaling Pathways that Mediate Distinct Functions

The FGF signaling system includes 22 FGF ligands and 4 receptors [17,18]. Ligand-dependent autophosphorylation and activation of the FGFRs leads to signaling through four proven pathways that lead to new transcription (Figs. 5.1 and 5.2). Two of the downstream targets, p38MAPK and Jun kinase, are generally not mitogenic cell lines ([19] and references therein). The major FGFR mitogenic signaling pathway is the *ras*-Raf-1-MEK-ERK pathway, also known as the 'universal cassette' because of the weight of evidence for its mitogenic role in diverse cell lines. FGF activation of ERK1/ERK2 is necessary and sufficient to activate transcription factors elk and SRF, leading to new transcription and a strong mitogenic responses [8,20].

A second pathway that mediates mitogenesis through the FGFRs leads to binding of phospholipase C (PLC)γ and PI3kinase activation. Phosphoinositol turnover by PLCγ generates inositol trisphosphate, leading to intracellular $Ca^{++}$ mobilization, and diacylglycerol, leading to the activation of PKC [21,19]. There are three groups of PKC noted: conventional (α, β, γ), novel (ξ, η, ν, θ), and atypical (λ). The ξ and atypical families are not mitogenic and are brain-specific [22,23]. Activation of PKCα, β, and γ leads to an increased mitogenic response, mainly through Raf1 and MAPK, although this appears to be less important than the *ras*-dependent MAPK pathway [24,19]. Substitution of tyrosine on FGFR-1 responsible for PLC-γ binding and activation of PKC does not diminish FGF-dependent mitogenesis in two cell lines, suggesting that PKC is not necessary for mitogenesis [25,26]. However, in studies with the related PDGF receptor, mutation of all tyrosine phosphorylation sites was rescued from the mitogenic effects, by inclusion of only the tyrosine that activates the PLCγ-PKC pathway, suggesting that the PKC pathway can be sufficient for mitogenesis. It is important to note that PKC and ras activate Raf1 by separate mechanisms and that Raf1 activates mitogenesis via MAPK activity. Recent analysis of the r*af1* null mutant has suggested that ras is more important, but that the RafB, not Raf1 is necessary for growth factor mediated mitogenesis in embryonic fibroblasts (reviewed in [27]).

Activation of src is a third possible pathway of FGFR activation of cell cycle. The FGFR also activates src (Fig. 5.1, [21,28,19]). This src activation is not mitogenic for FGF, but src itself is mitogenic in other cell lines and mediates functions like cell scattering and activation of nuclear transcription during PC-12 differentiation [29] and references therein). However, these studies were done in cell lines, not in animals.

A fourth pathway has arisen in which FGF activity suppresses cell division through STAT1. In the gain-of-function mutation leading to sporadic non-familial dwarfism, a single change in a transmembrane amino acid in FGFR-3 results in a gain-of-function enzymatic activity that leads to cessation of cell cycle in chondrocytes [30]. Recently, suppression of chondrocyte division was shown to require STAT1 activity, but the mode of activation of STAT1 is not understood [31]. However, STAT1 should be considered when interpreting results after perturbing FGF receptors, but its perturbation will not be considered as a major area of effort until more is understood about its mechanisms.

## 5.5 Signal Transduction Cascades, the 'Universal Cassette', and the Three MAPK Families

In their biochemical pathways of docking proteins and kinases, the three MAPK family pathways (Table 5.1, Fig. 5.3) are very similar. However, functionally, the ERK pathway tends to be more mitogenic and the p38MAPK and SAPK/JNK pathways are more homeostatic and cytostatic. The mechanistic basis for the separation of function is not clear, but is based on the quality and quantity of each type of transcription factor that is activated, more than their differences in enzymatic rates. Each MAPK family activates a largely overlapping group of transcription factors (Fig. 5.3), but overexpression of receptors (increasing pathway strength), can change an outcome from mitogenesis to differentiation within a single cell type (reviewed in [10]). Recent studies using cDNA arrays suggest that different factors induce transcriptional activation of similar sets of genes. The strength of transcription is an important difference. Others have concluded that the choice of activation of individual monomers and ensuing heterodimeric pairs can have different effects. For example, in the AP-1 heterodimeric transcription factor, junB inclusion leads to an anti-growth inflammatory response, whereas junC inclusion is primarily mitogenic [32]. However, early studies looked at functionally similar receptors (mitogenic FGF and PDGF) and did not compare receptors that mediated more diverse biological outcomes. Also, less work has been done on the analysis of the second wave of transcription after the immediate early response. The use of cDNA arrays to analyze intermediate transcriptional outcomes (primary and secondary waves of induced transcription) with respect to upstream receptor signaling capacity and downstream biological outcome, will yield strong insights into the function of signaling pathways.

The MAPK pathway governs cellular functions that lead to outcomes as diverse as S phase and cell cycle commitment [33], and determining cell fate during embryonic development [34]. SAPK and p38MAPK pathways are structurally similar, but tend to mediate homeostatic responses generated during stress in adult somatic cells as exemplified by inflammatory reactions [35]. Emerging evidence suggests that p38MAPK and SAPK pathways also mediate homeostatic responses during embryonic development, but may play a role in normal development, and also intervene in long-term decision-making by "taking over" the developmental decision making during transient stress [36,37].

In summary, MAPK cassette pathways typically amplify nanomolar extracellular signals and transduce these signals over minutes to hours. The biological effects include long-term nuclear decision-making that determine the cellular efforts for periods of time lasting hours or days (cell cycle commitment) or years (in determining cellular fate).

## 5.6 Insights into the Functions Between Proliferation-Promoting and Inhibiting Subfamilies of the MAPK Superfamily and How Mechanisms Underly These Functions

Using the tools and current knowledge discussed in this chapter, biological responses can be understood in terms of enzymatic function. Two contrasting models with different biological functions serve to illustrate key differences in how signal transduction pathways accomplish biological effects. During development, the embryo and fetus in humans must produce several trillion cells in the nine months between the one cell zygote and the newborn baby. In contrast, the 100 trillion cells in adult humans exist for years without net increases in accumulation, although a few of the approximate 200 different cell types maintain stem cells with proliferative capacity [3].

For the MAPK pathway, negative feedback leads to a canonical biphasic response during *in vitro* growth factor stimulation of somatic cells derived from adults. MKP can lead to a shortened MAPK response, that may lead to failure of cells to sustain an anabolic response leading to G1-S phase commitment and cell proliferation [38,33]. In contrast, activated MAPK persists in FGFR response fields in the embryo and fetus for many consecutive days [39,40], enhances cell accumulation and has other consequences. Thus embryos may prolong MAPK signaling to enhance cell accumulation.

Observations also imply that signal transduction pathways in embryos are more sensitive than those in adults. The concentration producing half maximal rates of anabolic and other functions, for insulin [41] and EGF [42] can be 1,000–10,000 fold less in embryos than it is in adult somatic cells. The Kd of the EGF receptor in the embryo is not known, but the Kd for the insulin receptor in embryos is similar to adult somatic cells [43]. Taken together, these data suggest that embryonic cells are more sensitive to growth factors than adult somatic cells, and this enhanced sensitivity is downstream of the receptor. The enhanced sensitivity may be a product of signal transduction.

## 5.7 Cross-Talk Among Receptors

Cross-talk occurs when downstream signaling of a receptor, leads to activation of another receptor-mediated pathway. Cross talk increases the complexity of signaling and permits greater fine-tuning of the regulatory machinery than possible by individual pathways. Of the many signaling pathways available within the cell, multiple pathways can be used in various combinations to transduce signals downstream of a single ligand-receptor interaction. Shared second messengers can lead to the activation of a common downstream pathway by different intercellular signals. Many G protein-coupled receptors, for example, activate phospholipase C downstream of their associated G protein, resulting in elevation of DAG and intracellular $Ca^{++}$ (Fig. 5.4). Therefore, exposure to any of the ligands would result in activation of calmodulin and protein kinase C, as well as other common pathways, and produce similar biological outcomes. During early pregnancy, the uterus secretes calcitonin and lysophosphatidic acid, which both accelerate preimplantation embryonic development to a similar extent through a mechanism dependent on intracellular $Ca^{++}$ [44,45]. Calcitonin and lysophosphatidic acid each bind to a specific receptor, but both receptors are coupled to G proteins that activate phospholipase C. This type of cross-talk provides biological systems with redundancy to ensure the success of critical processes such as blastocyst implantation.

Another form of cross-talk occurs when completing the biological output of a signaling pathway, downstream of one receptor that requires the participation of a signaling pathway downstream of another class of receptor. In the example of signaling in the uterus discussed above, lysophosphatidic acid was not able to accelerate embryonic development if tyrosine kinase activity was inhibited, suggesting the participation of a protein tyrosine kinase. In this case, intracellular $Ca^{++}$ signaling was found to induce the secretion of a member of the epidermal growth factor (EGF) family, heparin-binding EGF-like growth factor (HBEGF). HBEGF then binds to its

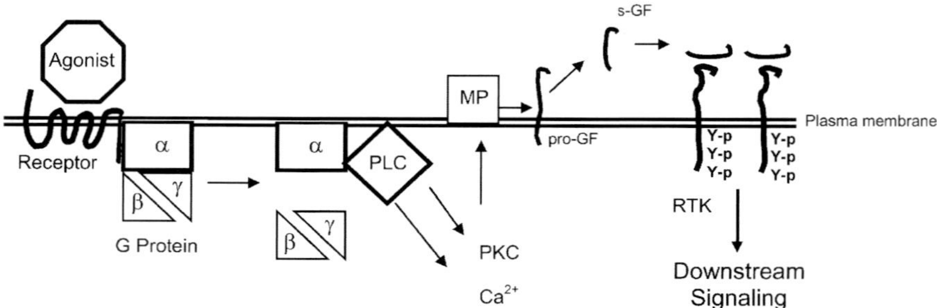


**Fig. 5.4** Receptor cross-talk as a result of EGF receptor transactivation by a G protein-coupled receptor agonist. The trimeric (a, b and g subunits) G protein coupled to a seven pass transmembrane receptor becomes dissociated upon agonist binding, permitting the subunit to activate phospholipase C (PLC). PLC activity leads to the elevation of intracellular $Ca^{++}$ concentration and activation of protein kinase C (PKC), which are responsible for increased metalloproteinase (MP) activity at the cell surface. The transmembrane form of EGF family growth factors (pro-GF) become secreted by the action of MP on their exodomain and are free to bind to their receptor tyrosine kinase (RTK) on the same or neighboring cells. Transphosphorylation of tyrosine (Y) residues in the cytoplasmic domains of dimerized RTKs, create sites for docking proteins with SH-2 domains and downstream signaling (Copies of figures including color copies, where applicable, are available in the accompanig CD)

receptors, EGF receptor and ErbB4. Treatment with HBEGF alone is sufficient to cause a similar stimulation of embryonic development [46]. This type of receptor cross-talk has been described in many cellular systems [47], where the signal passes through the plasma membrane three times; once when the G protein-coupled receptor is ligated, or when the EGF family growth factor is secreted and a third time as the receptor tyrosine kinase is activated. There are numerous examples of EGF receptor transactivation by agonists of G protein-coupled receptors, as well as after exposure to stress or radiation. In each case, it has been shown that the stimulus activates a metalloproteinase by either raising intracellular $Ca^{++}$ or inducing protein kinase C activity [47]. The metalloproteinase activity is required for the secretion of EGF-like growth factors, which are initially produced as transmembrane proteins which are shed upon cleavage of their extracellular domain. The released growth factor is then free to bind to its receptor in an autocrine signaling loop (Fig. 5.4). Signaling downstream of the EGF receptor can then orchestrate additional biochemical changes, such as MAPK activation, as a result of the original stimulus.

## 5.8 Conclusions and Future Directions

Previous research in the field of signal transduction has focused upon identifying novel signaling intermediates and their major roles in the pathways. The three MAPK families illustrate the following common features of related and interacting signal transduction pathways: speed of transduction, amplification via cascading enzymatic activity, branching, interaction with other pathways, and distinct and shared biologic functions mediated through distinct and shared transcription factors.

The next phases of research will focus on the functions of all members within each family of signal transduction genes as the human and mouse genome projects provide complete sets of family members. Large-scale approaches will give a broad picture of the responses of cells to ligands that induce distinct biological outcomes, such as cell death, mitosis, differentiation or motility. Proteomics and cDNA arrays will be used to detect broad changes in signaling quantity and quality of related transcription factors between wild type cells and null mutants or cells with receptor mutants with differential signal transduction capacities [10]. Arrays will also be used to detect differences in transcriptional quantity and quality between functionally different receptors that activate mitogenesis through FGF receptor-ERK signaling, that block mitogenesis through TGFβ receptor–SMAD signaling, or induce apoptosis through TNFα receptor–Fas (FADD) signaling.

Deep and narrow research will still focus on the function of single-signal transduction genes. A central principal in studying the functions of these families will be the use of loss-of-function of single genes (and crosses to knockout out multiple genes in one family) through null mutations *in vivo*. Key focus will be on the unique properties of signal transduction multi-functional enzymes with overlapping ranges of inputs and outputs.

## Glossary and Abbreviations

| | |
|---|---|
| AP1 | Activating protein |
| ATF2 | Activating transcription factor 2 |
| CRE-BP1, CREB2 | Cas Crk–associated substrate, p130CAS (**C in integrin signaling complex); cRaf-Raf proto-oncogene S/T protein kinase |
| DAG | Diacylglycerol |
| DPC-4 | Deleted in pancreatic cancer locus 4, SMAD4 |
| EGF | Epidermal growth factor |
| ELK1 | Ets domain transcription factor |

| | |
|---|---|
| ERK | Extracellular signal-regulated kinase, MAPK |
| FAK | Focal adhesion kinase (**FK in integrin signaling complex) |
| FGF | Fibroblast growth factor |
| FGFR | Fibroblast growth factor receptor |
| FRS2 | FGF receptor stimulated, lipid-anchored Grb2 binding protein |
| Fyn | Src family tyrosine kinase (**Fn in integrin signaling complex) |
| GEF | Guanine nucleotide exchange factor (example is SOS son-of-sevenless) |
| GRB2 | Growth factor receptor-bound protein 2 |
| GSK3 | Glycogen synthase kinase |
| HB | EGF-heparin binding-EGF |
| IFN | Interferon; JNK-Jun N-terminal kinase; Jun-transcription factor |
| MAPK | Mitogen-activated protein kinase |
| MAPKAP | MAP kinase-activated protein kinase 2 |
| MEK | MAPK/ERK kinase, MAPKK; MKK MEK kinase |
| MKP | MAPK phosphatase |
| MLK | Mixed lineage kinase |
| MSK | 1-Mitogen and stress-activated kinase 1 |
| p53 | Tumour suppressor protein that protects from DNA damage; |
| Paxilin | P in integrin signaling complex; |
| PKA and PKC | Protein kinase A and C, respectively |
| PLC | Phospholipase C |
| PP | Protein phosphatase |
| PYK2 | Proline-rich tyrosine kinase-2 (**P in integrin signaling complex) |
| rac | G-protein |
| ras | G-protein |
| RSK | Ribosomal S6 kinase |
| SAPK | Stress-activated protein kinase |
| Sos | Son of sevenless guanine nucleotide exchange factor |
| SRF | Serum response factor |
| STAT | Signal transducer and activator of transcription |
| TAK | TGFβ-activated kinase |
| Talin | (**T in integrin signaling complex) |
| TGF | Tumor growth factor |
| TNF | Tumor necrosis factor |

## References


Rappolee, D. (1998). Growth factors in the mammalian pre- and post-implantation embryo. Growth factors and hormones in mammalian development. In *Hormones and Growth Factors in Development and Neoplasia.*, (ed. D. S. R Dickson), pp. 93–115. NYC: Wiley.

Rappolee, D. and Werb, Z. (1994). The Role of Growth Factors in Early Mammalian development. *Advances in Developmental Biology* 3:41–71.

Alberts, B., Bray, D., Lewis, J., Raff, M., Roberts, K. and Watson, J. D. (1994). *Molecular biology of the cell.* NY: Garland Publishing.

Myers, M. G., Jr., Wang, L. M., Sun, X. J., Zhang, Y., Yenush, L., Schlessinger, J., Pierce, J. H. and White, M. F. (1994). Role of IRS-1-GRB-2 complexes in insulin signaling. *Mol Cell Biol* 14, 3577–3587.

Angel, P. and Karin, M. (1991). The role of Jun, Fos and the AP-1 complex in cell-proliferation and transformation. *Biochim Biophys Acta* 1072, 129–157.

Berridge, M. J., Lipp, P. and Bootman, M. D. (2000). The versatility and universality of calcium signalling. *Nat Rev Mol Cell Biol* 1, 11–21.

Whitake, M. (2006). Calcium at fertilization and in early development. *Physiol Rev* 86, 25–88.
Partanen, J., Vainikka, S., Korhonen, J., Armstrong, E. and Alitalo, K. (1992). Diverse receptors for fibroblast growth factors. *Prog Growth Factor Res* 4, 69–83.
Hausdorff, W. P., Lohse, M. J., Bouvier, M., Liggett, S. B., Caron, M. G. and Lefkowitz, R. J. (1990). Two kinases mediate agonist-dependent phosphorylation and desensitization of the beta 2-adrenergic receptor. *Symp Soc Exp Biol* 44, 225–240.
Hill, C. S. and Treisman, R. (1999). Growth factors and gene expression: fresh insights from arrays. *Sci STKE* 1999, PE1.
Feramisco, J. R., Clark, R., Wong, G., Arnheim, N., Milley, R. and McCormick, F. (1985). Transient reversion of ras oncogene-induced cell transformation by antibodies specific for amino acid 12 of ras protein. *Nature* 314, 639–642.
Jun, T., Gjoerup, O. and Roberts, T. M. (1999). Tangled webs: evidence of cross-talk between c-Raf-1 and Akt. *Sci STKE* 1999, PE1.
Xie, Y., Zhang, W., Wang, Y., Trostinskaia, A., Wang, F., Puscheck, E. E. and Rappolee, D. A. (2007). Using hyperosmolar stress to measure biologic and stress-activated protein kinase responses in preimplantation embryos. *Mol Hum Reprod* 13, 473–481.
Sun, H., Charles, C. H., Lau, L. F. and Tonks, N. K. (1993). MKP-1 (3CH134), an immediate early gene product, is a dual specificity phosphatase that dephosphorylates MAP kinase in vivo. *Cell* 75, 487–493.
Mourey, R. J., Vega, Q. C., Campbell, J. S., Wenderoth, M. P., Hauschka, S. D., Krebs, E. G. and Dixon, J. E. (1996). A novel cytoplasmic dual specificity protein tyrosine phosphatase implicated in muscle and neuronal differentiation. *J Biol Chem* 271, 3795–3802.
Shaw, G. and Kamen, R. (1986). A conserved AU sequence from the 3' untranslated region of GM-CSF mRNA mediates selective mRNA degradation. *Cell* 46, 659–667.
Zhang, X., Ibrahimi, O. A., Olsen, S. K., Umemori, H., Mohammadi, M. and Ornitz, D. M. (2006a). Receptor specificity of the fibroblast growth factor family, part II. *J Biol Chem.*
Zhang, X., Ibrahimi, O. A., Olsen, S. K., Umemori, H., Mohammadi, M. and Ornitz, D. M. (2006b). Receptor specificity of the fibroblast growth factor family. The complete mammalian FGF family. *J Biol Chem* 281, 15694–15700.
Mohammadi, M., Dikic, I., Sorokin, A., Burgess, W. H., Jaye, M. and Schlessinger, J. (1996). Identification of six novel autophosphorylation sites on fibroblast growth factor receptor 1 and elucidation of their importance in receptor activation and signal transduction. *Mol Cell Biol* 16, 977–989.
Tsang, M. and Dawid, I. B. (2004). Promotion and attenuation of FGF signaling through the Ras-MAPK pathway. *Sci STKE* 2004, pe17.
Huang, J., Mohammadi, M., Rodrigues, G. A. and Schlessinger, J. (1995). Reduced activation of RAF-1 and MAP kinase by a fibroblast growth factor receptor mutant deficient in stimulation of phosphatidylinositol hydrolysis. *J Biol Chem* 270, 5065–5072.
Dekker, L. V. and Parker, P. J. (1994). Protein kinase C—a question of specificity. *Trends Biochem Sci* 19, 73–77.
Newton, A. C. (1997). Regulation of protein kinase. C. *Curr Opin Cell Biol* 9, 161–167.
Lavoie, J. N., L'Allemain, G., Brunet, A., Muller, R. and Pouyssegur, J. (1996). Cyclin D1 expression is regulated positively by the p42/p44MAPK and negatively by the p38/HOGMAPK pathway. *J Biol Chem* 271, 20608–20616.
Mohammadi, M., Dionne, C. A., Li, W., Li, N., Spivak, T., Honegger, A. M., Jaye, M. and Schlessinger, J. (1992). Point mutation in FGF receptor eliminates phosphatidylinositol hydrolysis without affecting mitogenesis. *Nature* 358, 681–684.
Peters, K. G., Marie, J., Wilson, E., Ives, H. E., Escobedo, J., Del Rosario, M., Mirda, D. and Williams, L. T. (1992). Point mutation of an FGF receptor abolishes phosphatidylinositol turnover and $Ca^{++}$ flux but not mitogenesis. *Nature* 358, 678–681.
Murakami, M. S. and Morrison, D. K. (2001). Raf-1 without MEK? *Sci STKE* 2001, PE30.
Landgren, E., Blume-Jensen, P., Courtneidge, S. A. and Claesson-Welsh, L. (1995). Fibroblast growth factor receptor-1 regulation of Src family kinases. *Oncogene* 10, 2027–2035.
Spivak-Kroizman, T., Mohammadi, M., Hu, P., Jaye, M., Schlessinger, J. and Lax, I. (1994). Point mutation in the fibroblast growth factor receptor eliminates phosphatidylinositol hydrolysis without affecting neuronal differentiation of PC12 cells. *J Biol Chem* 269, 14419–14423.
Deng, C., Wynshaw-Boris, A., Zhou, F., Kuo, A. and Leder, P. (1996). Fibroblast growth factor receptor 3 is a negative regulator of bone growth. *Cell* 84, 911–921.
Sahni, M., Ambrosetti, D. C., Mansukhani, A., Gertner, R., Levy, D. and Basilico, C. (1999). FGF signaling inhibits chondrocyte proliferation and regulates bone development through the STAT-1 pathway. *Genes Dev* 13, 1361–1366.
Shaulian, E. and Karin, M. (2002). AP-1 as a regulator of cell life and death. *Nat Cell Biol* 4, E131–6.
Roovers, K. and Assoian, R. K. (2000). Integrating the MAP kinase signal into the G1 phase cell cycle machinery. *Bioessays* 22, 818–826.

Chazaud, C., Yamanaka, Y., Pawson, T. and Rossant, J. (2006). Early Lineage Segregation between Epiblast and Primitive Endoderm in Mouse Blastocysts through the Grb2-MAPK Pathway. *Dev Cell* 10, 615–624.

Kyriakis, J. M. and Avruch, J. (1996). Protein kinase cascades activated by stress and inflammatory cytokines. *Bioessays* 18, 567–577.

Rappolee, D. A. (2007). Impact of transient stress and stress enzymes on development. *Dev Biol* 304, 1–8.

Xie, Y., Liu, J., Proteasa, S., Proteasa, G., Zhong, W., Wang, Y., Wang, F., Puscheck, E. and Rappolee, D. (2008). Transient stress and stress enzyme responses have practical impacts on parameters of embryo development, from IVF to directed differentiation of stem cells. *Mol Repro Dev*, 75, 689–697.

Bhalla, U. S., Ram, P. T. and Iyengar, R. (2002). MAP kinase phosphatase as a locus of flexibility in a mitogen-activated protein kinase signaling network. *Science* 297, 1018–1023.

Corson, L. B., Yamanaka, Y., Lai, K. M. and Rossant, J. (2003). Spatial and temporal patterns of ERK signaling during mouse embryogenesis. *Development* 130, 4527–4537.

Wang, Y., Wang, F., Sun, T., Trostinskaia, A., Wygle, D., Puscheck, E. and Rappolee, D. A. (2004). Entire mitogen activated protein kinase (MAPK) pathway is present in preimplantation mouse embryos. *Dev Dyn* 231, 72–87.

Harvey, M. B. and Kaye, P. L. (1991). Mouse blastocysts respond metabolically to short-term stimulation by insulin and IGF-1 through the insulin receptor. *Mol Reprod Dev* 29, 253–258.

Dardik, A. and Schultz, R. M. (1991). Blastocoel expansion in the preimplantation mouse embryo: stimulatory effect of TGF-alpha and EGF. *Development* 113, 919–930.

Mattson, B. A., Rosenblum, I. Y., Smith, R. M. and Heyner, S. (1988). Autoradiographic evidence for insulin and insulin-like growth factor binding to early mouse embryos. *Diabetes* 37, 585–589.

Liu, Z. and Armant, D. R. (2004). Lysophosphatidic acid regulates murine blastocyst development by transactivation of receptors for heparin-binding EGF-like growth factor. *Exp Cell Res* 296, 317–326.

Wang, J., Rout, U. K., Bagchi, I. C. and Armant, D. R. (1998). Expression of calcitonin receptors in mouse preimplantation embryos and their function in the regulation of blastocyst differentiation by calcitonin. *Development* 125, 4293–4302.

Wang, J., Mayernik, L., Schultz, J. F. and Armant, D. R. (2000). Acceleration of trophoblast differentiation by heparin-binding EGF-like growth factor is dependent on the stage-specific activation of calcium influx by ErbB receptors in developing mouse blastocysts. *Development* 127, 33–44.

Prenzel, N., Zwick, E., Daub, H., Leserer, M., Abraham, R., Wallasch, C. and Ullrich, A. (1999). EGF receptor transactivation by G-protein-coupled receptors requires metalloproteinase cleavage of proHB-EGF. *Nature* 402, 884–888.

Schlessinger, J. (2004). Common and distinct elements in cellular signaling via EGF and FGF receptors. *Science* 306, 1506–1507.

# Chapter 6
# Epigenetics of Spermiogenesis

## Combining *In Silico* and Proteomic Approaches in the Mouse Model

**Sophie Rousseaux and Myriam Ferro**

**Abstract** One of the most dramatic chromatin remodeling and genome reorganization ever observed takes place during the post-meiotic maturation of male germ cells. Indeed, after meiosis, early male haploid cells, or spermatids, inherit a somatic-like chromatin-based genome organization. In the following stages of their maturation, histones are removed and replaced by testis-specific basic proteins, while their genome undergoes a dramatic compaction. Despite the fundamental nature of these events, the molecular basis of the driving elements controlling them is not known. Moreover, the resulting sperm specific genome structure could be crucial for the epigenetic information as welll as move conveyed to the embryo, but this structure is completely unknown.

**Keywords** Spermatogenesis · Proteome · Histone · Protamine · *In silico* analysis

## 6.1 Introduction

In order to characterize the molecular events and identify structures and factors involved in the post-meiotic chromatin re-organization of the male genome, we initially characterized post-translational modifications affecting the histones before their removal. These include a global acetylation of the core histones H2A and H4, which dramatically increase in early elongating spermatids. Our hypothesis is that this global histone acetylation could act as a signal for recruiting factors important for the chromatin re-organization events occurring later on.

Facing a complete lack of data concerning the molecular basis of post-meiotic chromatin remodeling we then undertook two complementary approaches for the systematic identification of factors involved in this process.

Our first approach was a comprehensive *in silico* analysis of known and unknown chromatin-binding factors expressed in the testis. A particular focus was made on bromodomain-containing factors, since bromodomains (Brd) are known to specifically bind acetylated histones. A list was established and interesting factors were selected for further studies.

On the parallel front, our second approach was based on a global proteomic strategy to identify additional chromatin-associated factors in the post-meiotic cells. This proteomic approach has led to the identification of approximately 200 factors, including new histone variants, which could specifically be involved in the pericentric heterochromatin reprogramming in late condensing spermatids, as well as chaperones, which might have a role in histone/testis-specific protein exchanges.

The functional data obtained for selected factors provide a basis to build new working hypotheses and shed light on one of the most dramatic and less known process in biology. In addition to genetic information, the spermatozoon conveys an epigenetic message to the future embryo.

---

S. Rousseaux
Inserm U823, 38706 La Tronche cedex, France and Université Joseph Fourier, Institut Albert Bonniot, Grenoble F-38706, France
e-mail: sophie.rousseaux@ujf-grenoble.fr

S. Krawetz (ed.), *Bioinformatics for Systems Biology*, DOI 10.1007/978-1-59745-440-7_6,

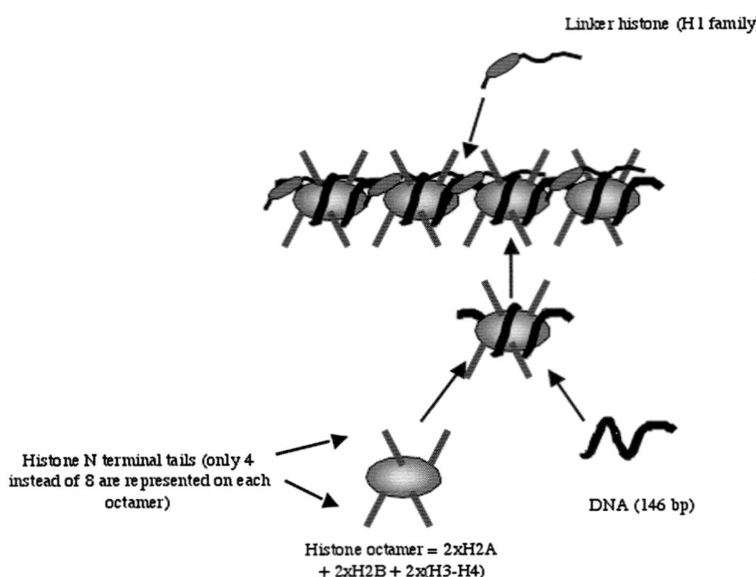


**Fig. 6.1** Chromatin in somatic cells. Schematic view of the nucleosome. The basic unit of chromatin is the nucleosome, which is constituted of an octamere of core histones, around which 146 bp DNA is wrapped. Each core histone is comprised of a well-structured C-terminal part as well as an N-terminal extremity, or "tail", which can be modified on specific amino-acid residues (Copies of figures including color copies, where applicable, are available in the accompanying CD)

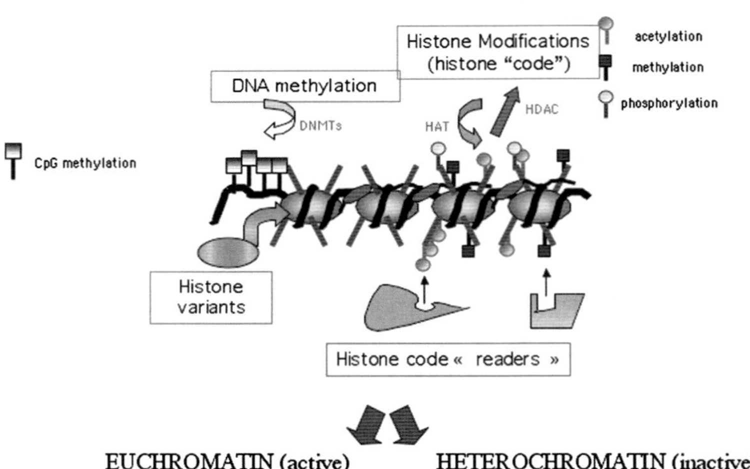


**Fig. 6.2** Epigenomic modifications in somatic cells. Modifications include: (i) methylation of Cytosine residues of the DNA molecule (catalysed by DNMTs – DNA Methyl Transferases) (ii) post-translational modifications of histones (mainly affecting residues of the N-terminal tails): acetylation of lysines, methylation of lysines or arginines, phosphorylation of serines, etc... these modifications are thought to interact and combine with one another in order to define a specific code, known as the "histone code". Enzymatic activities catalyse these modifications or remove them, for example the histone acetyl transferases (HAT) acetylate histones, whereas histone deacetylases (HDACs) remove the acetyl residues. (iii) These histone modifications can be recognized by factors or complexes, which are recruited on chromatin, via the binding of specific "domains" on specific modifications. For example, bromodomain-containing proteins bind acetylated histones, whereas chromodomain-containing proteins are recruited on histone H3 when is tri-methylated Lysine 9. These factors are named as "histone code readers". (iv) The incorporation of histone variants also modify the nucleosome and can play a role in specific epigenomic modifications. (v) Other modifications of the nucleosome structure are likely to be involved, which are not described here. These chromatin modifications help in defining specific chromatin regions, such as euchromatin and heterochromatin (Copies of figures including color copies, where applicable, are available in the accompanying CD)

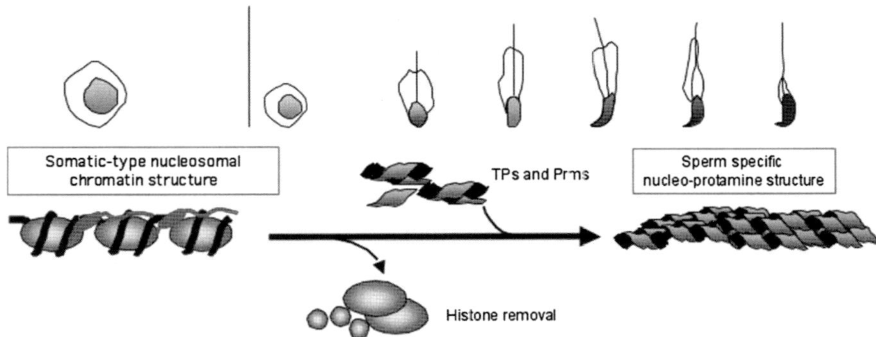


**Fig. 6.3** Global chromatin re-organization during the male post-meiotic cell maturation. In pre-meiotic, meiotic and early post-meiotic cells (round spermatids), the chromatin has a nucleosomal somatic-like structure. During elongation and condensation of the genome this structure is completely re-organized since most histones are removed and replaced by transition proteins (TPs) and protamines (Prms), the latter being responsible for the very compact sperm-specific nucleoprotein structure (DNA associated with protamines), ensuring a protective environment for the male genome (Copies of figures including color copies, where applicable, are available in the accompanying CD)

In the somatic cells, the molecular basis of the epigenetic information is the "epigenome" and involves DNA methylation as well as modifications of the genome packaging structure, or chromatin. The basic unit of chromatin is the nucleosome, which is a 146 bp DNA wrapped around an octamere of core histones, H2A, H2B, H3 and H4 (Fig. 6.1) [1]. Well-studied modifications of this structure include histone covalent modifications, the incorporation of histone variants, the recruitment of chromatin remodeling factors and the expression of non-coding RNAs (Fig. 6.2) (Bernstein et al. 2007).

Spermatogenesis is the differentiation of male germinal cells from spermatogonia to spermatozoa. Beginning at puberty, spermatogenesis is a continuous process, characterized by three major stages: pre-meiotic, meiotic and post-meiotic (or spermiogenesis) (for review see [2]). Spermatogonia divide by mitosis. They then enter meiosis by the formation of primary spermatocytes, which replicate their DNA, undergo meiotic division I, yielding secondary spermatocytes, which then rapidly go through meiotic division II, and generate haploid round spermatids.

During the post-meiotic maturation of the male germinal cells, or spermiogenesis, the spermatid undergoes a global remodeling of its nucleus, which elongates and compacts itself into the very unique nucleus structure of the spermatozoon. During this process, a drastic reorganization of the chromatin structure takes place where most histones are removed and replaced by sperm specific nuclear proteins, transition proteins (TPs) and protamines (Prms), the latter being responsible for the very tight compaction the sperm genome (Fig. 6.3).

Despite the fundamental nature of these events, the molecular basis of the driving elements controlling them is not known. Moreover, the resulting sperm specific genome structure could be crucial for the epigenetic information conveyed to the embryo, but this structure remains completely unknown.

## 6.2 Objective and Strategies

Our objective was to shed light on the events and factors involved in the post-meiotic reorganization of the male genome, one of the last "black boxes" of modern biology, and more specifically to find answers to the following specific questions: what are the molecular mechanisms and factors controlling the general histone removal and the assembly of new chromatin structures? what are the structures replacing nucleosomes? are there genomic regions specifically marked by these structures, and if so which ones? and do sperm specific genomic structures convey male-specific epigenetic information?

Due to a complete lack of data on factors involved in the male post-meiotic genome reprogramming, we have established strategies for the massive identification of factors potentially important in this process (specifically regarding the issues raised above). We then proceeded to their functional studies.

## 6.3 What Was Initially Known About Chromatin Re-Organization During Spermatogenesis?

Prior to histone removal, two specific events affect the nucleosome: the incorporation of histone variants and an increased global acetylation of the core histone tails. It has been postulated that these modifications could result in a looser nucleosomal structure, and facilitate the recruitment of factors and complexes further involved in histone replacement. For review see [3,4].

Our initial working hypothesis was that this global histone acetylation could be a signal for the recruitment of factors and complexes re-organising chromatin. We therefore initiated this work by characterizing the nature and the timing of histone acetylation during spermatogenesis in the mouse as well as in humans. Both published, and yet unpublished, investigations showed the occurrence of a massive increase in global histone acetylation in elongating spermatids in both species, which mostly affected H4 and H2A [5,6].

In order to establish a molecular link between histone acetylation and their replacement, we focused our attention on factors potentially capable of reading and functionally interpreting these specific histone acetylation marks, i.e., factors bearing the acetyl-lysine binding module, the bromodomain (Brd).

Our first approach was a comprehensive *in silico* analysis of known and unknown bromodomain-containing factors expressed in the testis. A list was established and interesting factors were selected for further studies. In parallel a global proteomic approach was undertaken to identify additional factors.

## 6.4 Characterization of Histone Modifications During Spermiogenesis

The analysis of the best-characterized histone modifications rapidly pointed to acetylation of H4 and H2A, as well as to phosphorylation of the H4 serine 1 (S1), as potentially important events in the genome post-meiotic reprogramming. The analysis of the other histone modifications did not allow highlighting any other remarkable event.

Our observations showed that a genome-wide histone H4 hyperacetylation occurs at the beginning of spermatid elongation and is accompanied by a dramatic reprogramming of the pericentric heterochromatin [7].

Indeed, after the completion of meiosis, in the early round spermatids, the pericentric heterochromatin regions become grouped and organized in a unique round chromocenter. Our investigations have shown that, in these cells, the chromocenter bears all the expected characteristics of somatic-cell heterochromatin, i.e., association with HP1 proteins and enrichment in H3 tri-methylated in K9 (H3K9me). Interestingly, in elongating spermatids, we showed that the occurrence of the global histone H4 hyperacetylation corresponds to a dramatic de-compaction of this unique chromocenter, followed by a complete reprogramming of these pericentric regions. Indeed, initially devoid of acetylated histones, these regions gain histone H4 acetylation, and loose the associated HP1 proteins. However, despite the spreading of histone acetylation, H3K9me remains associated with these pericentric regions (Fig. 6.4). A detailed analysis shows that, at this stage, pericentric heterochromatin is composed of a mixture of nucleosomes individually bearing, either the usual H3K9me mark or acetylated H4, but not both. Moreover, as the elongation/condensation of the

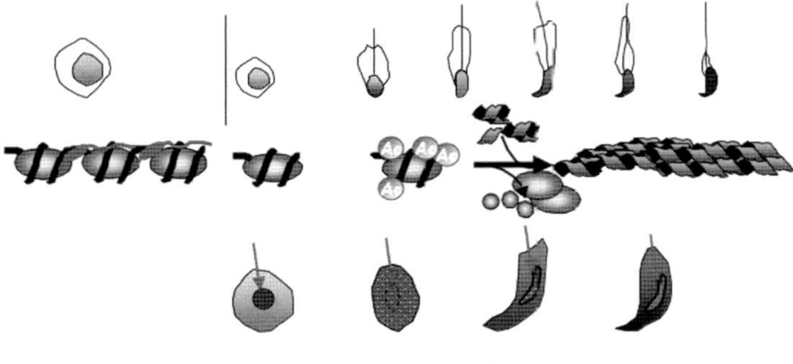


**Fig. 6.4** Dynamics of H4 acetylation during mouse spermatogenesis. Before removal, histones are massively and globally hyperacetylated in the elongating spermatids. By immunofluorescence with an anti-acetylated H4 antibody, we have shown that in early elongating spermatids there is a global increase of acetylated H4, which is present in the whole volume of the nucleus. As the spermatid elongates and starts to condens, the acetylated H4 is removed, but remains present in a restricted area of the genome, mainly corresponding to heterochromatin structures (arrows). It then completely disappears in condensed spermatids (Copies of figures including color copies, where applicable, are available in the accompanying CD)

spermatids proceeds, histone acetylation persists in the pericentric regions, while it disappears from the other regions, reflecting global histone removal in the latter.

Additionally, we observed the occurrence of H4S1 phosphorylation in meiotic cells, that persists during the post-meiotic stages and precedes histone H4 acetylation [8].

This meticulous observation and thorough description of chromatin-related events during the post-meiotic stages of mouse spermatogenesis was used as a guide to select appropriate factors for further functional studies. For instance the detailed knowledge of histone hyperacetylation during spermiogenesis was a determinant in the selection of the appropriate Brd-proteins to be studied: a potentially interesting factor needs to be expressed in cells when hyperacetylation occurs.

## 6.5 *In Silico* Identification of Factors Involved in Post-meiotic Chromatin Re-organization

The data described above suggested that histone acetylation could be associated with either histone replacement by transition proteins (in most of the genome) and/or with constitutive heterochromatin reprogramming.

Logically, factors capable of catalysing these events should bear bromodomains, which would allow for a specific interaction with acetylated histones. Indeed these domains are the best-characterized structural motifes capable of binding acetylated lysines.

### *6.5.1 Principle of In Silico Approaches*

This approach is based on the identification of transcripts from the testis in various databanks, encoding important factors potentially capable of acting on chromatin. For the reasons expressed above we focused on bromodomain-containing factors capable of binding acetylated histones.

### *6.5.2 Search for Bromodomain-Containing Testis-Specific Proteins*

Originally, dbEST and dbEST Cumulative Updates were searched using the bromodomain region of the factor named GCN5 as a query. In order to rapidly identify interestingly expressed sequence tags (ESTs) (those presenting the best similarity the query), we designed a computer program to treat the raw data and to set up a schematic representation of homologous ESTs. The ESTs were sorted according to their origin and only the ESTs present in libraries for testis were selected (Fig 6.5). For details, see [9].

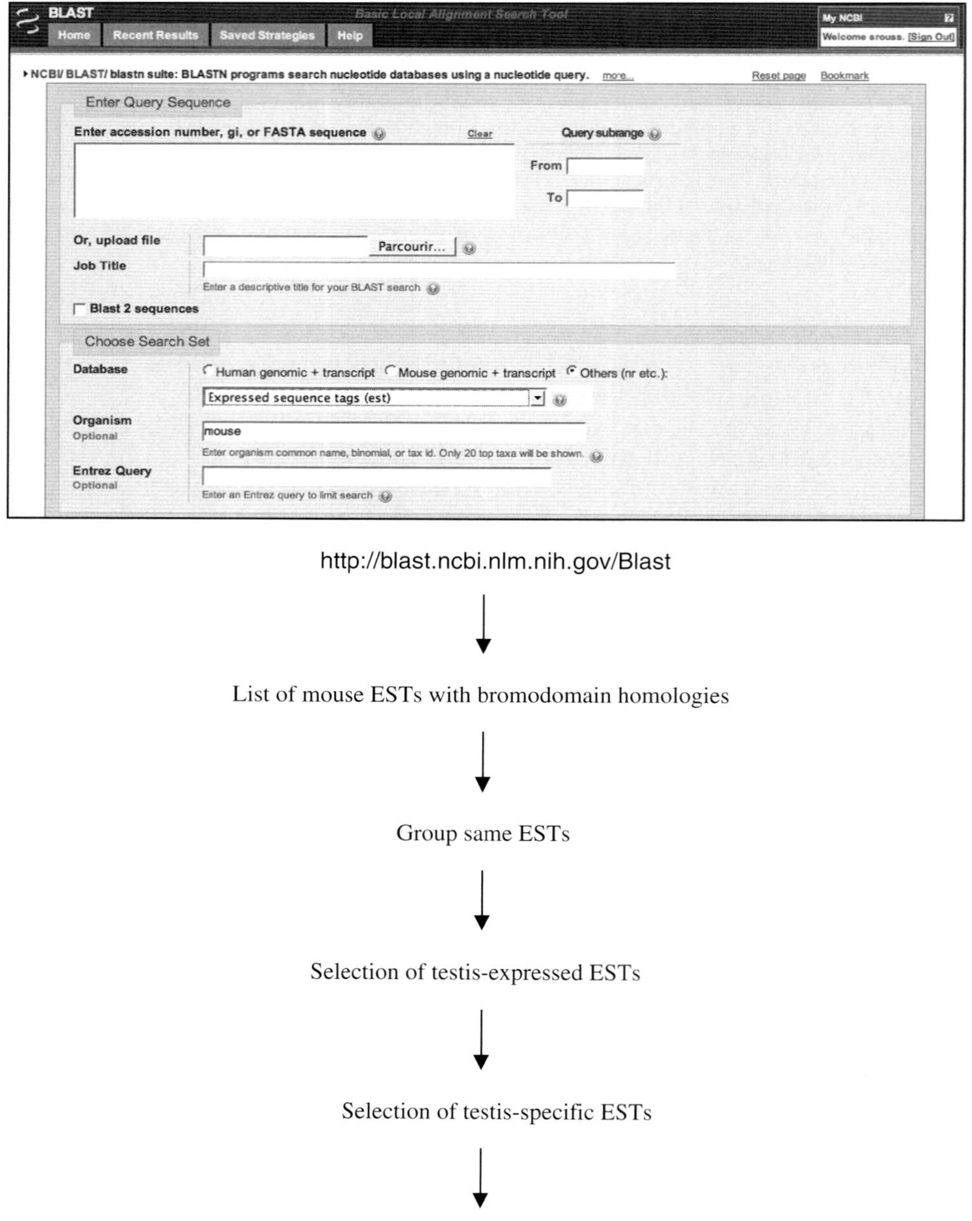


**Fig. 6.5** *In silico* strategy for the identification of testis-specific bromodomain containing factors (Copies of figures including color copies, where applicable, are available in the accompanying CD)

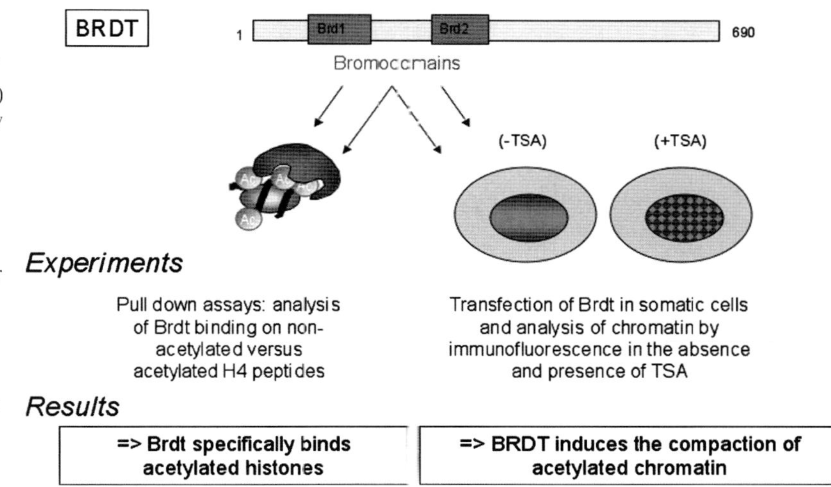


**Fig. 6.6** The functional studies of Brdt show that it binds acetylated histones (pull down assay – left panel) and has the unique property of compacting acetylated chromatin when overexpressed in somatic cells (when transfected cells are treated with TSA (Trichlora statin A) in order to increase histone acetylation, Brdt induces the compaction of chromatin-right panel) (Copies of figures including color copies, where applicable, are available in the accompanying CD)

### *6.5.3 Expression and Functional Studies*

Functional studies were immediately started on a testis-specific double bromodomain-containing protein of completely unknown functions, named Brdt [9].

During this work, we also gathered information on other chromatin-related factors with a link to acetylation and selected Cdyl, a chromodomain containing protein, for functional studies.

#### 6.5.3.1 BRDT

BRDT, a double bromodomain containing protein expressed in the testis and present during the critical post-meiotic stages, appeared as a very interesting candidate. Our early investigations showed that Brdt has the extraordinary capacity for compacting chromatin, *in vivo* and *in vitro* in a strictly acetylation-dependent manner [9] (Fig. 6.6). Later we obtained an antibody, which allowed us to show that Brdt co-localized very precisely with acetylated genomic regions all through spermiogenesis [7]. These results suggested that Brdt could indeed be involved in an early global compaction of acetylated chromatin occurring in condensing spermatids.

#### 6.5.3.2 CDYL

CDYL, another testis-expressed chromatin-related factor, also attracted our attention because of its potential link to acetylation. Indeed, CDYL possesses a N-terminal chromodomain fused to a domain showing significant homology to the Coenzyme A (CoA) binding pocket of enoyl-CoA hydratase. We have shown that CDYL can bind chromatin via its chromodomain, and is likely to be involved in the regulation of histone acetylation during spermiogenesis by recruiting histone deacetylases via its CoA binding pocket [10].

## 6.6 Proteomic Identification of Chromatin-Associated Factors in Male Germinal Cells

### *6.6.1 Principle of Proteomic Approaches*

This approach is based on a global proteomic analysis of extracts obtained from staged post-meiotic cells (i.e. round, elongating and condensing spermatids representing various levels of genome reorganization).

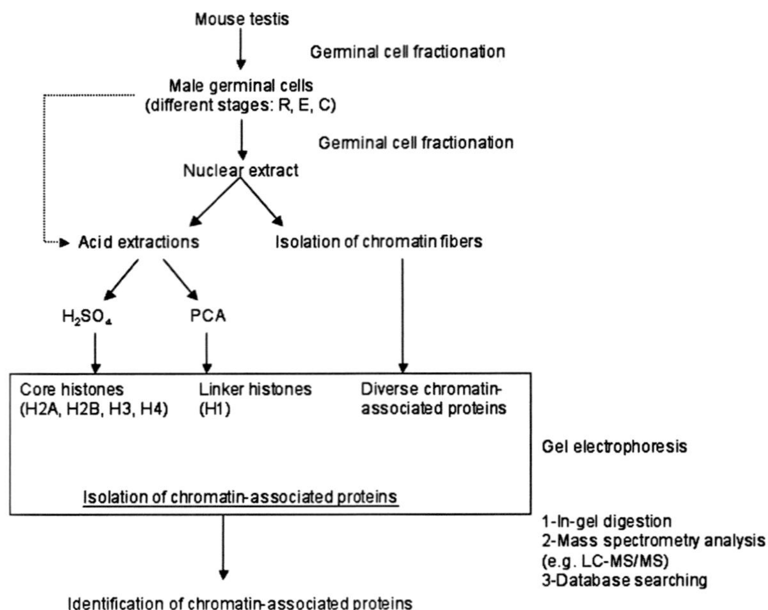


**Fig. 6.7** Strategies for proteomic identification of post-meiotic chromatin re-organising factors. R = round spermatids, E = elongating spermatids, C = condensing spermatids, PCA = perchloric acid, LC = liquid chromatography, MS/MS = tandem mass spectrometry (Copies of figures including color copies, where applicable, are available in the accompanying CD)

Two types of extracts were used for proteomic analyses (Fig. 6.7).

- Proteins associated with chromatin fibres (proteins released by micrococcal nuclease digestion), isolated from staged spermatogenic cells were individually or collectively identified. The selected stages include spermatocytes, as well as round and elongating spermatids.
- Acid-soluble nuclear proteins (mostly DNA-interacting proteins), extracted from condensing spermatids were collectively identified. We focused our attention on these basic proteins, because most of the known DNA organizing proteins, including histones, Transition Proteins (TPs) and Protamines (Prms), are highly basic. The objective was to identify new and yet unknown factors that specifically organize the genome in the spermatids after histone removal. This approach yielded very exciting data. Indeed it led to the discovery of new histone variants [7], and chaperones involved in the assembly of yet unknown spermatid-specific DNA-packaging structures [11].

Proteins from the extracts prepared as above were then analysed by electrophoresis. As some specific chromatin proteins are highly basic (e.g., histones and protamines), one dimension SDS-PAGE electrophoresis was preferred to 2-dimensional electrophoresis (2DE). Indeed standard 2DE allows efficient protein separation for proteins within a 4–7 pI range. Once the proteins were separated by SDS-PAGE, *in*-gel trypsin digestion was carried out directly on the excised gel bands. The so-generated peptides were then analyzed by mass spectrometry either by peptide mass fingerprinting (PMF) or by tandem mass spectrometry (MS/MS) sequencing. The PMF method allows the MW(molecular weight) of the peptides corresponding to a same gel band to be measured. This is generally achieved with a MALDI-TOF instrument. For database searching purposes the measured masses of the peptides, are compared to the theoretical peptide masses retrieved from the *in silico* digestion of proteins present in a protein database (e.g., UniProtKB). The PMF method is well-adapted to 2DE-separated proteins but shows some limitations when more complex samples have to be analyzed. In this case MS/MS analyses is preferred to the peptide sequence information that is used for database searching purposes. In context of MS/MS data, the fragmentation pattern of a given peptide is compared to the theoretical fragmentation patterns of peptides derived from the *in silico* digestion of proteins present in a protein database (for review see). Overall, the proteomic analyses carried out with these two types of extracts have led to the identification of about 200 proteins (unpublished data, [11, 7]). Among these proteins, several may yield new concepts.

### 6.6.2 *Proteomic Identification*

The proteomic strategy described in the previous section was applied to the acid extracts obtained from condensing spermatid (steps 12–16) nuclei in order to gain an insight into the nature of genome-organizing proteins in these cells. About 50 proteins were identified, including, as expected, many DNA-packaging proteins [11]. Among the linker histones, the newly identified H1t2 [13] was present, and within the high mobility group proteins, testis-specific as well as uncharacterized members could be identified. In addition to canonical core histones, many variants were found including new H2A and H2B variants, that were called H2AL1, H2AL2, and H2BL1 and were further characterized [7]. Other proteins not related directly to chromatin remodeling were also found.

Characteristic features, such as the isoelectric point (pI) and the molecular weight (MW), were examined for the 50 proteins soluble in $H_2SO_4$ (sulphuric acid). Fig. 6.7A shows the distribution of those proteins according to their pI, MW and function. Three functional classes were pointed out: (i) chromatin proteins that include histones (core, linker and variant), HMG proteins, transition proteins and protamines, (ii) chaperones, such as nucleoplasmin and HSP70.2, and (iii) a miscellaneous class that gathers elongation factors, proteasome-related proteins and others. As expected, chromatin proteins, most of which are basic, remain clustered in the high pI range and the low MW range on the diagram presented in Fig. 6.8A. More surprisingly acidic proteins, and especially known chaperones with a pI between 4.5 and 6, were identified from acid extracts of condensing spermatids. Based on these data, we hypothesized that a tight association between these acidic proteins and their potential basic partners, such as the chromatin proteins, may have induced their solubility in $H_2SO_4$. Accordingly, these chaperones might have a specific function in the context of chromatin remodeling. Thus, further investigations were undertaken regarding the HSP70.2 protein, which was known to play a critical role in the completion of meiosis during male germ cell differentiation but whose function in post-meiotic cells was unknown. Our recent data strongly suggested that, after meiosis, HSP70.2 acquires new functions, such as being the chaperone of TP1 and TP2, and contributes to the dramatic spermatid-specific genome-wide reorganization [11]. The work described by Govin et al. points out how a global proteomic approach can open up perspectives to investigate yet unknown processes involved in the genome-condensing structure assembly in the course of spermiogenesis.

### 6.6.3 *Expression and Functional Studies of Factors Selected from the Proteomic Approaches*

#### 6.6.3.1 Histone Variants

The persistence of nucleosomes associated with the constitutive heterochromatin of condensing spermatids, despite their replacement in the other regions, attracted our attention to a group of yet unknown histone variants, which were found among the acid-soluble proteins extracted from the condensing spermatids.

We reasoned that these surviving nucleosomes could constitute a privileged site for their assembly.

(A)

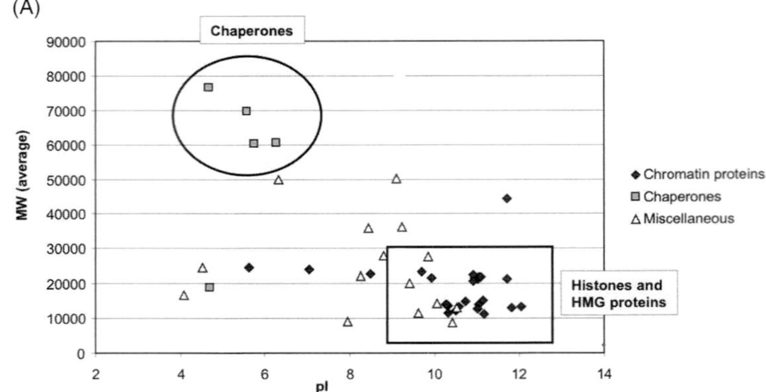


(B)

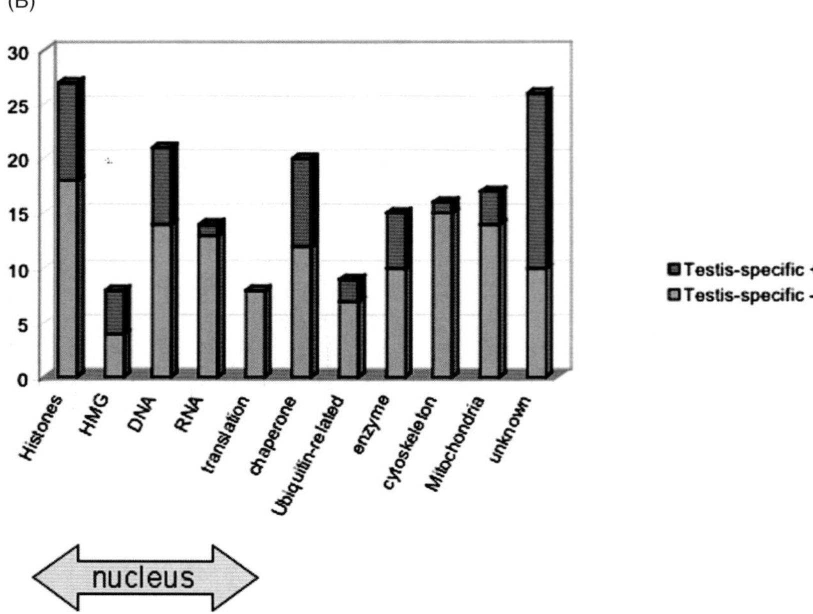


**Fig. 6.8** Factors identified by the proteomic strategy. **(A)** A subset of the protein identified by $H_2SO_4$ acid extraction in late spermatids: distribution according to their isoeclectric point (pI) and molecular weight (MW). **(B)** All proteins (approximately 200) identified from acid extractions and isolated chromatin fibres during spermiogenesis: distribution according to their function and/or their known localization. The proportions of testis-specific proteins versus non-testis specific factors are shown (Copies of figures including color copies, where applicable, are available in the accompanying CD)

Antibodies were generated against these new histone variants in order to study their expression and localization during spermatogenesis. The data obtained showed that they are late expressing histones, which accumulate at the same time or even later than TPs and become specifically associated with the pericentric heterochromatin (Fig. 6.9). These findings are a very important contribution to the understanding of spermiogenesis, since for the first time they evidence elements of a differential genome organization with potentially important implications in the transmission of male-specific epigenetic information.

More detailed investigations have shown that TH2B, an already identified testis-specific H2B variant of unknown function, could provide a platform for the structural transitions accompanying the incorporation of these new histone variants [7].

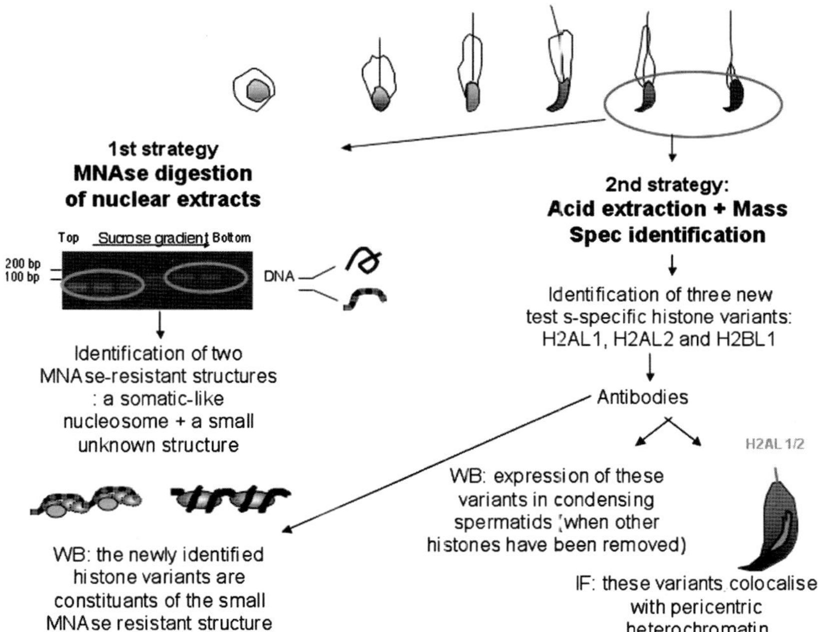


**Fig. 6.9** Late expressing histone variants accumulate in pericentric heterochromatin in the late condensing spermatids, contributing to the existence of new nucleosome-like structures. Right Panel: acid extraction of nuclear proteins from late condensing spermatids and mass spectrometry has allowed for the identification of three new histone variants, H2AL1, H2AL2 and H2BL1. Antibodies raised against these variants show that they are present at the very late stages of spermiogenesis, and specifically localize in pericentric heterochromatin regions (Western Blots and IF). Left Panel: MNAse digestion of late condensing spermatids has led to the identification of a regular nucleosomal structure (left circle) as well as an unknown MNAse resistant smaller structure (right circle). The histone variants H2AL1/2 and H2BL1 are specifically detected in the new MNAse resistant structure (Copies of figures including color copies, where applicable, are available in the accompanying CD)

#### 6.6.3.2 Chaperones with Unexpected Functions

The examination of the list of acid-soluble proteins (in principal basic, positively charged proteins (see Fig. 6.8) from condensing spermatids revealed the unexpected presence of a group of acidic proteins. The analysis of the identity of these proteins showed that they were mostly chaperones. We reasoned that these chaperones could have been solubilized because of their acid-resistant association with the basic (DNA-binding) proteins. We became very interested in these proteins since there is a considerable lack of information on the chaperone systems involved in the assembly of spermatid-specific DNA packaging structures. The analysis of testis-specific HSP70.2, revealed a specific post-meiotic role for the protein in the assembly of TP-containing structures [11].

## 6.7 Conclusion

The combination of *in silico* and proteomic strategies has led to the identification of several hundred of protein potentially involved in post-meiotic genome reorganization in male germ cells. A detailed analysis of chromatin modifications during spermiogenesis was followed by the functional studies of factors carefully selected because of their direct link with these chromatin re-modeling events. The resulting data have enabled us to establish a general scheme of post-meiotic genome reprogramming during mouse spermatogenesis, which is as follows (Fig. 6.10).

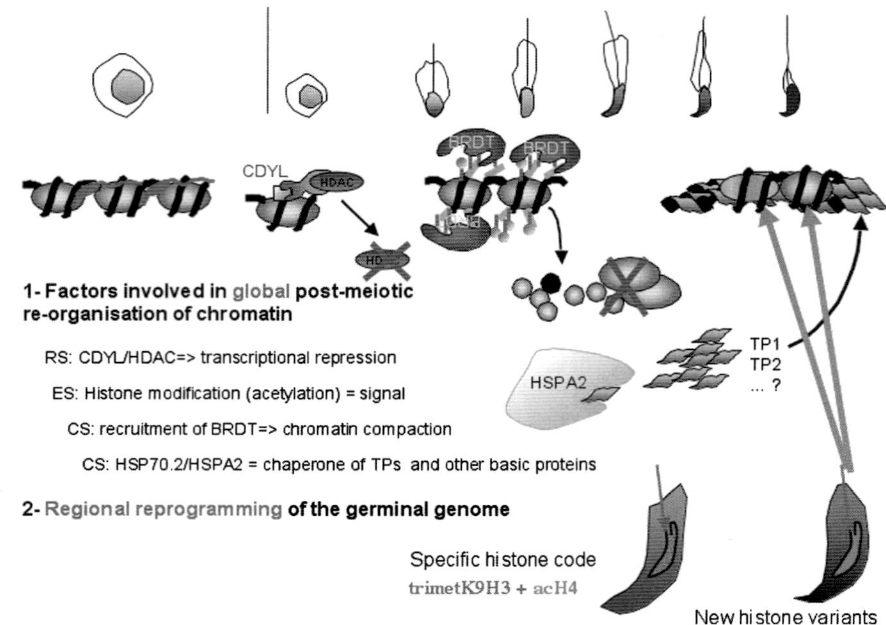


**Fig. 6.10** Our current working model for the molecular basis of post-meiotic reprogramming of the male genome (see text for comment) Rs = round spermatids; ES = elongating spermatids; CS = condensing spermatidss copies, where aplicable, are available in the accompanying CD)

After meiosis, the induction of a massive and genome-wide histone hyperacetylation triggers reprogramming. This is associated with histone replacement in most of the genome and specific reprogramming of pericentric heterochromatin.

Our functional investigations, suggests the following working hypotheses.

- In round spermatids, CDYL associates with HDACs, maintaining a low level of histone acetylation in these cells, which could contribute to the general repression of transcription.
- HDACs are then massively degraded, which induces a global increase in histone acetylation in elongating spermatids.
- Histone acetylation signal their removal and degradation.
- BRDT would link histone acetylation to a first step of genome compaction and massive histone removal in most of the genome, as well as to nucleosome survival and histone exchange, with incorporation of new histone variants, in pericentric regions.
- The assembly of these structures would allow a regional differentiation of the male genome to be established in mature spermatozoa and would mediate the transmission of male-specific epigenetic information to the embryo.

All these events should somehow be linked together. Future efforts will focus on establishing these links and on detailing the molecular basis of major transitions occurring during spermiogenesis.

## Glossary and Abbreviations

**2DE** 2-dimensional electrophoresis
**Brd** Bromodomain
**CoA** Coenzyme A
**ESTs** Expressed sequence tags

**HDACs** Histone deacetylases
**MS/MS** Tandem mass spectrometry
**PMF** Peptide mass fingerprinting

## References


Bernstein BE,Meissner A, Lander ES (2007) The mammalian epigenome. Cell 128: 669–681.

Wolffe A (1995) Chromatin – Structure and function., 2nd edition edn Academic Press, London.

Holstein AF, Schulze W, Davidoff M (2003) Understanding spermatogenesis is a prerequisite for treatment. Reprod Biol Endocrinol 1: 107.

Caron C, Govin J, Rousseaux S, Khochbin S (2005) How to pack the genome for a safe trip. Prog Mol Subcell Biol 38: 65–89.

Govin J, Caron C, Lestrat C, Rousseaux S, Khochbin S (2004) The role of histones in chromatin remodeling during mammalian spermiogenesis. Eur J Biochem 271: 3459–3469.

Faure AK, Pivot-Pajot C, Kerjean A, Hazzouri M, Pelletier R, Peoc'h M, Sele B, Khochbin S, Rousseaux S (2003) Misregulation of histone acetylation in Sertoli cell-only syndrome and testicular cancer. Mol Hum Reprod 9: 757–763.

Hazzouri M, Pivot-Pajot C, Faure AK, Usson Y, Pelletier R, Sele B, Khochbin S, Rousseaux S (2000) Regulated hyperacetylation of core histones during mouse spermatogenesis: involvement of histone deacetylases. [In Process Citation]. Eur J Cell Biol 79: 950–960.

Govin J, Escoffier E, Rousseaux S, Kuhn L, Ferro M, Thevenon J, Catena R, Davidson I, Garin J, Khochbin S, Caron C (2007) Pericentric heterochromatin reprogramming by new histone variants during mouse spermiogenesis. J Cell Biol 176: 283–294.

Krishnamoorthy T, Chen X, Govin J, Cheung WL, Dorsey J, Schindler K, Winter E, Allis CD, Guacci V, Khochbin S, Fuller MT, Berger SL (2006) Phosphorylation of histone H4 Ser1 regulates sporulation in yeast and is conserved in fly and mouse spermatogenesis. Genes Dev 20: 2580–2592.

Pivot-Pajot C, Caron C, Govin J, Vion A, Rousseaux S, Khochbin S (2003) Acetylation-Dependent Chromatin Reorganization by BRDT, a Testis-Specific Bromodomain-Containing Protein. Mol Cell Biol 23: 5354–5365.

Caron C, Pivot-Pajot C, Van Grunsven LA, Col E, Lestrat C, Rousseaux S, Khochbin S (2003) Cdyl: a new transcriptional co-repressor. EMBO Rep 4: 877–882.

Govin J, Lestrat C, Caron C, Pivot-Pajot C, Rousseaux S, Khochbin S (2006) Histone acetylation-mediated chromatin compaction during mouse spermatogenesis. In: Berger SL, Nakanishi O, Haendler B (eds.) The Histone Code and Beyond. Springer-Verlag, Berlin Heidelberg, pp. 155–172.

Aebersold R, Mann M (2003) Mass spectrometry-based proteomics. Nature 422: 198–207.

Govin J, Caron C, Escoffier E, Ferro M, Kuhn L, Rousseaux S, Eddy EM, Garin J, Khochbin S (2006) Post-meiotic shifts in HSPA2/HSP70.2 chaperone activity during mouse spermatogenesis. J Biol Chem 281: 37888–37892.

Martianov I, Brancorsini S, Catena R, Gansmuller A, Kotaja N, Parvinen M, Sassone-Corsi P, Davidson I (2005) Polar nuclear localization of H1T2, a histone H1 variant, required for spermatid elongation and DNA condensation during spermiogenesis. Proc Natl Acad Sci U S A 102: 2808–2813.


## Web Resource

www.expasy.uniprot.org

# Chapter 7
# Genomic Tools for Analyzing Transcriptional Regulatory Networks

**John J. Wyrick**

**Abstract** In this chapter, we described how genomic tools can be used to construct and analyze transcription networks. We focus on the ChIP-chip technique as this genomic tool has proven to be a powerful method for identifying the genomic targets of transcription factors. We discuss bioinformatics methods that can be used to analyze ChIP-chip data and discover the DNA motifs bound by a transcription factor. Finally, we describe bioinformatics tools, such as Athena, that can be used to visualize and analyze transcription factor binding sites. We have shown that integrating the analysis of transcription factor binding data with information about gene function or expression can be a powerful tool for deciphering transcriptional networks.

**Keywords** Chip-chip · Transcription · Network · Athena

## 7.1 Introduction

Dynamic regulation of gene transcription is critical for the development and adaptation of eukaryotic organisms to their environment. The central regulators of gene transcription are the transcription factor proteins. Transcription factors identify their target genes by binding to short, specific DNA sequences, which are typically located in promoter regions upstream of genes. Once bound, transcription factors can either activate or repress the transcription of adjacent genes.

This simple picture of transcription regulation is complicated by a number of biological factors. First, even the simplest eukaryote species contains a large number of transcription factor proteins. For example, more than 200 transcription factors have been identified in the unicellular yeast *Saccharomyces cerevisiae*. The human genome is thought to encode even more—at least 1300 genes encode transcription factor proteins, according to recent tallies. Past studies have focused primarily on individual transcription factors and individual target genes. While they are important, these studies are insufficient to decipher how the global *network* of transcription factors regulates the transcription of the genome.

Secondly, the mechanisms by which transcription factors regulate gene expression are complex. In the simplest case, a transcription factor regulates the transcription of a gene independent of other factors that may be bound to the gene's promoter. In reality, many transcription factors act synergistically or redundantly with one another to specify a gene's transcription frequency. Hence, it can be difficult to untangle the individual contributions of each factor in the transcription network.

Thirdly, transcription factor proteins often bind to short DNA sequences with relatively low sequence specificity. For this reason, it can be difficult to predict which DNA sites in the genome will

J.J. Wyrick
School of Molecular Biosciences, Washington State University, Fulmer Hall 675, Pullman, WA 99164-4660, USA
e-mail: jwyrick@wsu.edu

S. Krawetz (ed.), *Bioinformatics for Systems Biology*, DOI 10.1007/978-1-59745-440-7_7,

be bound by a transcription factor even if its binding specificity has been extensively characterized using biochemical methods. While computational methods are being used to predict transcription factor targets based on their DNA binding specificity, many of the predicted binding sites are inaccurate (i.e., the sites are not bound by the transcription factor protein *in vivo*). Hence, constructing networks that accurately identify a transcription factor's target genes is a nontrivial problem.

Fortunately, many of these limitations can be addressed using a new generation of genomic and computational tools. In this chapter, we will discuss a method known as chromatin immunoprecipitation microarray analysis or *ChIP-chip*. This experimental method can be used to identify the global set of genes bound by a transcription factor. By using ChIP-chip analysis to identify transcription factor targets, a transcriptional network can be rapidly constructed. We will then discuss how DNA binding motifs can be defined using ChIP-chip data, and how networks of these binding motifs can be analyzed using bioinformatics tools.

## 7.2 Genome-Wide Profiling of Transcription Factor Binding Targets

Transcription factors regulate transcription by binding to specific DNA sequences in the genome. Hence, the first step in deciphering a transcription network is to identify the genomic DNA targets that are bound by each transcription factor protein. This is a non-trivial problem. To date, one of the most successful methods of identifying transcription factor targets is by chromatin immunoprecipitation (ChIP). A ChIP experiment results in the purification of the DNA fragments that are bound by a protein of interest, typically a transcription factor. The purified DNA fragments are subsequently identified by PCR amplification, DNA microarray analysis, or high-throughput DNA sequencing.

In this section, we will focus on the ChIP experiments that are coupled with DNA microarray analysis, as this method has been proven to be very successful in mapping the genome-wide binding sites of transcription factors. Our focus will be on the general concepts and practice of the ChIP-chip method. Readers interested in detailed experimental protocols should consult the references listed at the end of chapter.

### *7.2.1 Chromatin Immunoprecipitation*

Chromatin immunoprecipitation (ChIP) consists of four essential steps: (A) In vivo chemical cross-linking of the protein to DNA; (B) Fragmentation of the DNA; (C) Purification of the DNA-protein complexes by immunoprecipitation; and (D) Reversal of protein-DNA cross-links. Fig. 7.1 shows a schematic describing the ChIP protocol. Each of these steps will be described in detail below (Copies of figures including color copies, where applicable, are available in the accompanying CD).

#### 7.2.1.1 Chemical Cross-linking of Protein to DNA

The power of the ChIP technique is that it measures *in vivo* DNA-binding events. This is accomplished through chemical cross-linking of intact cells. Essentially, the addition of a cross-linker "freezes" the transcription factor proteins to their *in vivo* DNA targets. This enables the researcher to capture the DNA binding events that may be transient or dependent upon particular cellular or environmental conditions (e.g., transcription factor binding during heat shock or at a particular phase of the cell cycle). The cross-linking step also lends confidence that the detected DNA binding events were present in the cell, and not caused by experimental artifacts during subsequent steps of the ChIP protocol.

The most commonly used chemical cross-linker in the ChIP experiments is formaldehyde. Formaldehyde reacts with the amino groups of lysine residues in proteins and the amino groups

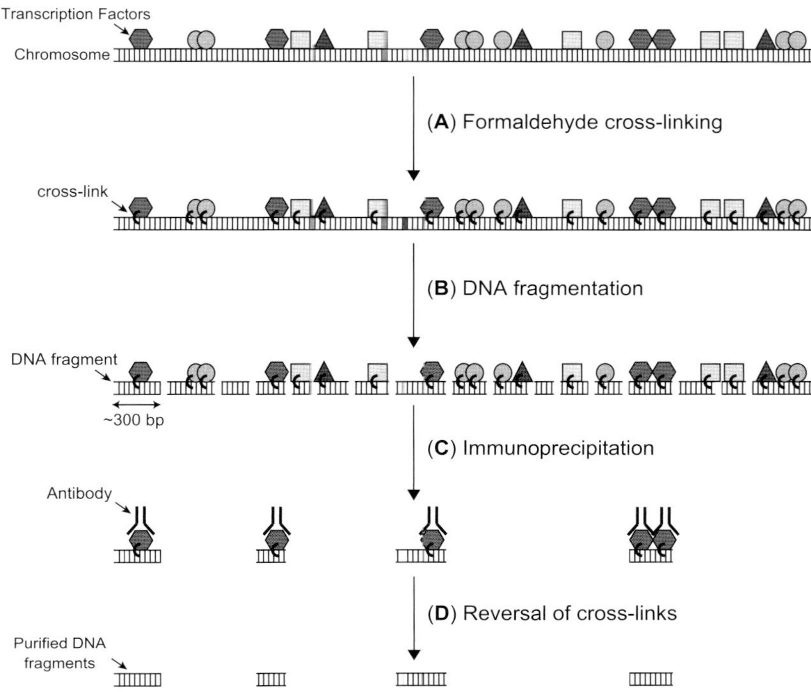


**Fig. 7.1** A schematic diagram illustrating the experimental steps involved in the chromatin immunoprecipitation (ChIP) technique (Copies of figures including color copies, where applicable, are available in the accompanying CD)

present in DNA bases. If a protein is associated with the DNA, a molecule of formaldehyde can react both with the DNA base and the protein to form a stable molecular link between the DNA and its bound protein (see Fig. 7.1A). Formaldehyde is advantageous because it can readily diffuse into the nucleus of a cell to form protein-DNA cross-links, and these cross-links can be easily reversed at subsequent steps by heat-treatment (see below).

#### 7.2.1.2 DNA Fragmentation

Ideally, one would want to cross-link the transcription factor only to the DNA that comprises its actual binding site, and not to the surrounding DNA. In order to accomplish this goal, the cross-linked DNA must be broken into very short fragments, approximating the size of a transcription factor binding site (5–30 bp). In practice, it is not feasible to fragment the DNA to this degree. Instead, one must make do with longer DNA fragments. In the conventional ChIP protocol, the cross-linked DNA is fragmented by exposure to high frequency sound waves emitted by a sonicator. This generates a range of DNA fragments with sizes ranging from 200–1000 bp (Fig. 7.1B).

The DNA fragmentation step is an important determinant of the resolution and sensitivity with which one can measure DNA binding events. For example, if the average size of the DNA fragments is relative large (>1 kb), it will be difficult to narrow down the transcription factor's actual DNA binding site. In addition, a large DNA fragment size can also influence the sensitivity of the ChIP experiment due to the confounding influence of adjacent DNA fragments.

DNA can also be fragmented through digestion by a DNA endonuclease. For example, digestion of DNA with micrococcal nuclease (MNase) has been frequently used in ChIP experiments for histone associated proteins or histone post-translational modifications. Digestion with MNase typically results in a smaller average DNA fragment size (~150 bp) than that achieved by DNA sonication. Unlike DNA sonication, however, the MNase cleavage pattern is not random, but is dependent upon the positioning of nucleosomes in the genomic DNA.

#### 7.2.1.3 Purification of Protein-DNA Complexes by Immunoprecipitation

The crucial step in the ChIP protocol is the purification of transcription factor-bound DNA fragments from the overall pool of genomic DNA. This is typically accomplished by immunoprecipitating the protein-DNA complexes using an antibody that binds with high affinity to the transcription factor of interest (Fig. 7.1C). The affinity and specificity of the antibody for the transcription factor is critical. Antibodies lacking these qualities will not adequately purify the cross-linked DNA, or worse will purify the DNA that is bound by other, nonspecific proteins. Frequently, even antibodies that work well in other experimental applications (e.g., western blot or immunofluorescence) will not work well in ChIP experiments. Hence, it is essential that the antibody used is of ChIP-grade.

A common strategy is to fuse the transcription factor with a short protein sequence that encodes an *epitope tag*. Epitope tags are recognized by high affinity commercially available antibodies. Examples of commonly used epitope tags include the Hemagglutinin (HA) tag, Myc tag, or FLAG tag. The use of an epitope tag obviates the need to raise a ChIP-grade antibody against the transcription factor of interest, as the high affinity anti-tag antibody can be used to immunoprecipitate the transcription factor-DNA complexes. The effectiveness of this strategy was highlighted in a recent study, in which the authors fused a Myc epitope tag to each of 203 yeast transcription factors, in order to systematically perform the ChIP-chip analysis on yeast transcription factors [1].

#### 7.2.1.4 Reversal of Protein-DNA Cross-Links

In order to identify the DNA fragments bound by the transcription factor, the immunoprecipitated DNA must be released from its protein cross-links. Fortunately, formaldehyde cross-links can be easily reversed by heat treatment (Fig. 7.1D). This frees up the immunoprecipitated DNA for subsequent enzymatic steps, which are the topic of the next section.

### 7.2.2 DNA Microarray Analysis of ChIP DNA

Once the DNA fragments bound by the transcription factor have been isolated by chromatin immunoprecipitation, the identity of each of these DNA fragments must be determined. This can be accomplished by analyzing the immunoprecipitated DNA using DNA microarray analysis. This involves two steps: (i) Amplification and labeling of the immunoprecipitated DNA; and (ii) Hybridization of the labeled samples to a DNA microarray.

#### 7.2.2.1 Amplification and Labeling of Immunoprecipitated DNA

ChIP experiments typically yield a rather small amount of DNA. Hence, the DNA must be amplified prior to DNA microarray analysis. A number of protocols have been developed to amplify the DNA from ChIP experiments: ligation mediated PCR (LM-PCR), random primer PCR amplification, or T7 RNA polymerase-based amplification. In the LM-PCR method (see Fig. 7.2), the immunoprecipitated DNA is ligated to short linker oligonucleotides (Fig. 7.2A). The ligated DNA fragments are then PCR amplified using primers that anneal to the linker oligonucleotides. To label the DNA for subsequent DNA microarray analysis, the PCR amplification mix includes modified nucleotides that are conjugated to a fluorescent dye (e.g., Cy5). These modified nucleotides are incorporated during PCR amplification (Fig. 7.2B), and thus the amplified ChIP DNA can be subsequently detected using a microarray scanner.

Typically, a control sample of DNA is also PCR amplified and labeled in parallel with the ChIP DNA sample. The choice of the control sample is important for the success of the ChIP-chip

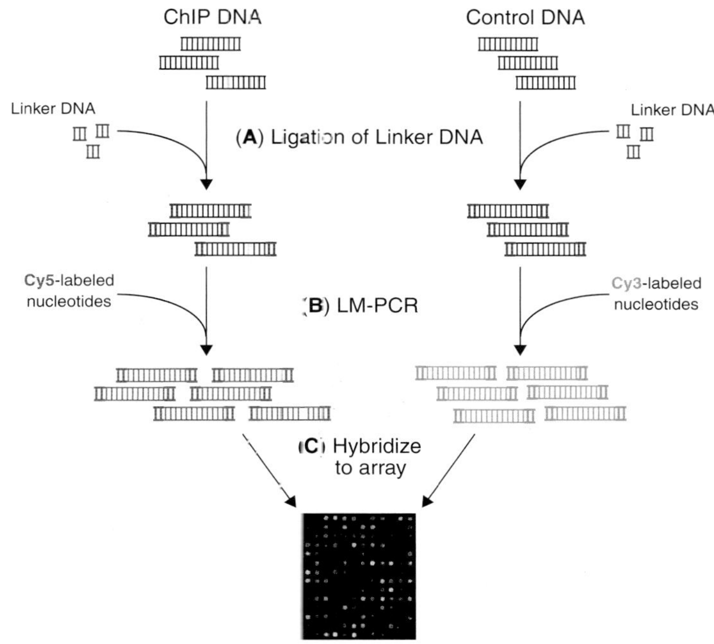


**Fig. 7.2** A schematic diagram illustrating the experimental steps involved in the amplification, labeling, and hybridization of ChIP DNA to DNA microarrays. The experimental protocol for the Ligation Mediated Polymerase Chain Reaction (LM-PCR) method is shown (Copies of figures including color copies, where applicable, are available in the accompanying CD)

experiment. Common controls include fragmented genomic DNA that has not been immunoprecipitated (otherwise known as whole cell extract DNA), or DNA that has been immunoprecipitated using a control antibody. If an epitope tagged transcription factor is used, then DNA immunoprecipitated from an untagged cell line is often the best control. The control DNA sample is commonly labeled with a different fluorescent dye (e.g., Cy3), so that a two-color microarray hybridization can be performed (see below).

#### 7.2.2.2 Hybridization of Labeled Samples to DNA Microarrays

The labeled immunoprecipitated (IP) and control DNA samples are mixed and hybridized to the same DNA microarray (Fig. 7.2C). The DNA microarray is then washed and scanned, and the relative signal for the IP and control DNA is quantified for each probe on the microarray. An example of a scanned ChIP-microarray is shown in Fig. 7.3. The DNA microarray experimental methodology is covered in more detail in the other chapters in this book.

The choice of microarray depends upon the organism being studied, but usually a promoter microarray is used. This is because it is assumed that transcription factors generally bind to promoter sequences. A promoter microarray contains probes for each promoter region in the genome. For the yeast *S. cerevisiae* promoter microarrays contain a single DNA probe for each promoter. This is due to the small size of the yeast promoter regions (~0.5 kb on an average). For the human genome, which contains much longer promoter regions, multiple probes for each promoter sequence are included in a promoter microarray.

Tiling microarrays are a recent innovation that greatly expands the genome coverage and resolution of DNA microarrays. Tiling microarrays contain millions of oligonucleotide probes spaced at short intervals from one another (e.g., every 35 bp). The large number and density of microarray probes allows the user to interrogate the genome at very fine sequence resolution. In particular, the use of tiling microarrays to analyze ChIP DNA provides high resolution detection of transcription factor binding sites.

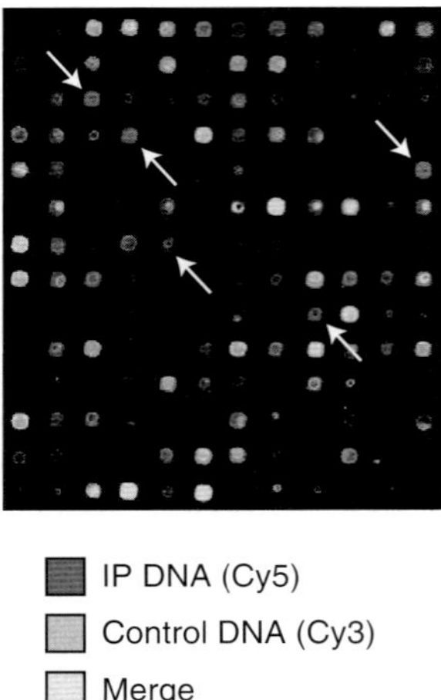


**Fig. 7.3** An example of a scanned promoter microarray following hybridization of amplified and labeled ChIP DNA. Each spot corresponds to a yeast promoter region. The fluorescence signal for the experimental ChIP DNA is shown in red (see color image provided in supplemental CD); the fluorescence signal for the control DNA (whole cell extract DNA in this case) is shown in green. The ChIP DNA was immunoprecipitated using an antibody against the yeast Gcn4 transcription factor. Spots that show enrichment for the IP DNA are indicated by white arrows, and are likely Gcn4 binding sites (Copies of figures including color copies, where applicable, are available in the accompanying CD)

## 7.3 Bioinformatics Analysis of ChIP-chip Microarray Data

The process of chromatin immunoprecipitation selectively purifies DNA fragments that are bound by the transcription factor of interest. It is important to note, however, that the ChIP protocol does not completely remove the unbound genomic DNA fragments. Instead, the ChIP protocol enriches the sample for the bound DNA fragments relative to the unbound DNA fragments. Hence, when analyzing the DNA microarray data, one looks for promoter regions whose spots show significant enrichment in the IP DNA sample relative to the control. Fig. 7.3 illustrates this process. The spots highlighted with arrows indicate promoter regions that show significant enrichment in the IP sample (colored red in the color image provided in the supplementary CD) compared to the control sample (colored green). Promoters that display this level of enrichment in the ChIP sample are likely to be the transcription factor binding sites.

### *7.3.1 Error Model Analysis of ChIP-Chip Data*

A common statistical approach to identify the enriched microarray probes/promoter regions is to use *error model analysis*. The error model calculates the statistical significance of the enrichment for each microarray probe and assigns a confidence measure (*P*-value) for the likelihood of protein binding. The confidence measure for each probe is calculated using the following formula:

$$X = \frac{a_2 - a_1}{\left(\sigma_1^2 + \sigma_2^2 + f^2\left[a_1^2 + a_2^2\right]\right)^{1/2}} \tag{7.1}$$

$$P = 1 - \mathrm{Erf}(|X|) \tag{7.2}$$

where $P$ is the confidence measure, $a_1$ and $a_2$ are the Cy3 (control) and Cy5 (IP) fluorescent intensities, respectively; $a_1$ and $a_2$ are the uncertainties associated with the Cy3 and Cy5 measurements; and $f$ is an empirically derived scaling factor. The scaling factor, $f$, is chosen so that the calculated $X$ values fit a Gaussian distribution, which allows a $P$-value to be estimated using equation 7.2.

Typically, a $P$-value threshold of 0.005 or 0.001 is chosen to identify enriched microarray probes. Promoter regions with a $P$-value less than the threshold are considered to be bound by the transcription factor. An example of error model analysis of ChIP-chip data for the yeast Gcn4 transcription factor is shown in Fig. 7.4. Each point in the graph corresponds to the fluorescent intensities of a single microarray probe representing a yeast promoter region. Points in red (see color image provided in supplemental CD) indicate promoter regions that are significantly bound by Gcn4 ($P < 0.005$) according to the error model.

### *7.3.2 Identifying Transcription Factor Binding Motifs from ChIP-Chip Data*

When one has identified the promoter regions that are bound by a transcription factor, the next step is to identify the short DNA motif present in the promoter sequence that comprises the actual binding site. This search process can be aided if tiling microarrays are used for the ChIP-chip experiment. The high density of the probes in a tiling microarray will allow one to localize the DNA motif bound by the transcription factor with high resolution (Fig. 7.5). When using tiling microarrays, one typically finds that multiple consecutive probes will show enrichment for ChIP DNA. This phenomenon is due to the large size of the DNA fragments relative to the high density of

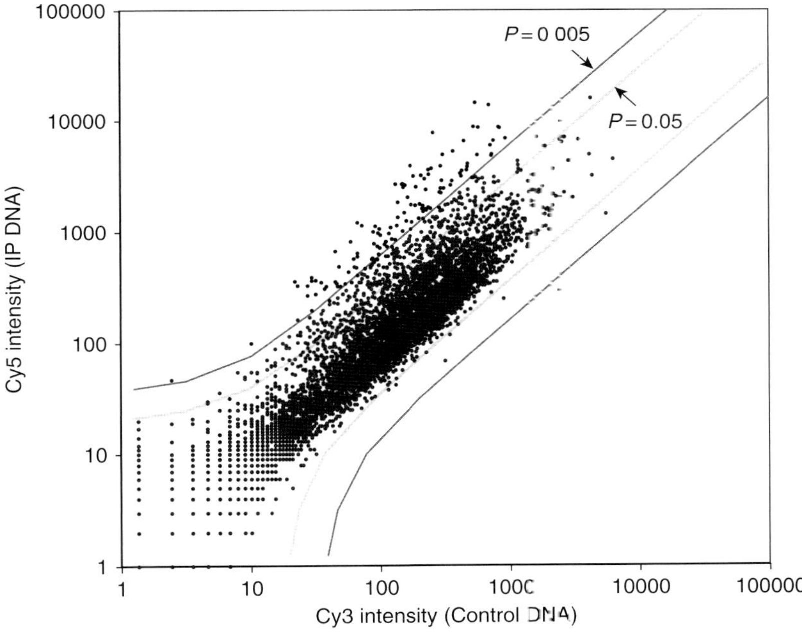


**Fig. 7.4** Error model analysis of Gcn4 ChIP-chip data. Each spot represents the signal from a yeast promoter region on the microarray. The y-axis plots the Cy5 intensity (IP DNA); the x-axis plots the Cy3 intensity (control DNA). The error model cutoffs for $P$-values equal to 0.005 (red line) and 0.05 (yellow line) are displayed. Spots in red (see color image provided in supplemental CD) are significantly bound by Gcn4 according to the error model criteria. Spots in blue are not bound by Gcn4 (Copies of figures including color copies, where applicable, are available in the accompanying CD)

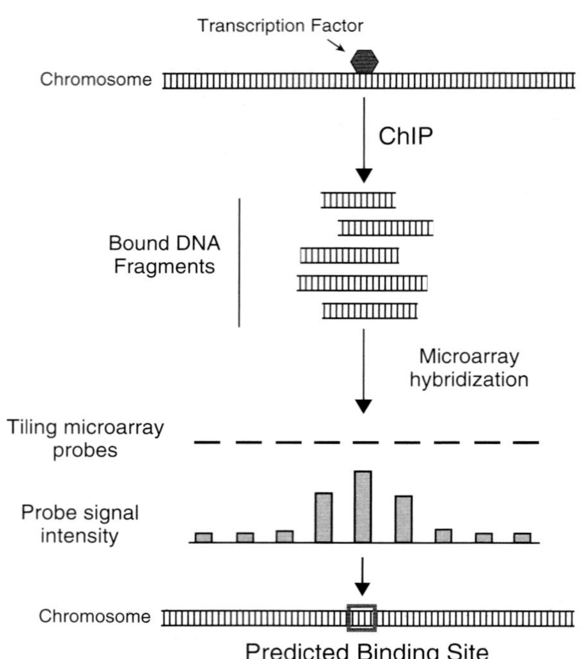


**Fig. 7.5** Schematic diagram showing how the peak signal from high density tiling microarray data can be used to predict a transcription factor binding site with sequence resolution and accuracy (Copies of figures including color copies, where applicable, are available in the accompanying CD)

microarray probes on the tiling microarray. The location of the bound DNA motif can be inferred by finding the probe(s) with the peak signal (see Fig. 7.5), as these probes are likely to correspond to the actual transcription factor binding site.

Most current ChIP-chip experiments are conducted using low-resolution promoter microarrays that contain only a few (or one) microarray probes per promoter region. In such cases, computational methods are necessary to define the DNA motif that constitutes the actual transcription factor binding site. A typical approach is to compile the complete set of promoter regions bound with high affinity by the transcription factor (based on the ChIP-chip data) and search these promoter regions for DNA sequence commonalities that constitute the DNA binding site. A number of software tools have been developed to identify conserved DNA sequences from promoters identified by ChIP-chip analysis, including MDscan, MEME, CONVERGE, and AlignACE.

#### 7.3.2.1 Computational Representations of DNA Motifs

Analysis of ChIP-chip data using the methods described above will typically yield a series of short DNA sequences that comprise the direct binding targets of the transcription factor. These sequences can be used to generate a sequence motif that expresses the commonalities present in this pool of high affinity binding sequences.

One common method is to write the motif as a *pattern*. A pattern represents the consensus sequence of a binding motif. To construct a motif pattern, one typically performs a multiple sequence alignment of the identified DNA sequences bound by the transcription factor. The motif pattern is written using the alignment, and the conserved sequence at each position in the motif is expressed using IUPAC symbols. IUPAC symbols can readily handle ambiguities in the DNA pattern. For example, the IUPAC symbol W indicates that the nucleotide at that position in the pattern can be either A or T. The definition of each DNA IUPAC symbol is given in Table 7.1.

**Table 7.1** IUPAC codes for nucleotide ambiguities

| IUPAC Code | Nucleotides Indicated | IUPAC Code | Nucleotides Indicated |
|---|---|---|---|
| A | A | N | A, C, G, T |
| B | C, G, T | R | A or G |
| C | C | S | C or G |
| D | A, G, T | T | T |
| G | G | U | U |
| H | A, C, T | V | A, C, or G |
| K | G or T | W | A or T |
| M | A or C | Y | C or T |

(Copies of tables are available in the accompanying CD.)

Alternatively, the binding motif can be represented as a *position-specific weight matrix* (PSWM). A PSWM gives the quantitative likelihood of observing each nucleotide at each position in the motif. This is commonly expressed as the log-likelihood ratio. The calculated log-likelihood ratios comprise the elements of the PSWM. It must be noted that because the PSWM provides a quantitative model of the nucleotide propensities, it provides a more accurate representation of the binding motif. More details about PSWM can be found in [2].

#### 7.3.2.2 Example of the Gcn4 DNA Binding Pattern and Profile

As an example, the alignment of high affinity binding sequences for the yeast transcription factor Gcn4 is shown in Fig. 7.6A. These binding sequences were identified by computational analysis of the Gcn4 ChIP-chip data [1]. The motif pattern derived from this alignment is shown in Fig. 7.6B. In column 2 in this example (k = 2), the IUPAC symbol G represents that Guanine nucleotides are predominately found at this position in the motif. In column 4 the IUPAC symbol S is used to indicate that only G or C nucleotides are found at this position.

The PSWM of the Gcn4 binding motif is shown in Fig. 7.6C. In column 4 in this example (k = 4), we observe high log-likelihood scores for the nucleotides G (0.5) and C (1.2) and low likelihood scores for the nucleotides A (–2.6) and T (–2.6). The PSWM also more accurately describes the greater

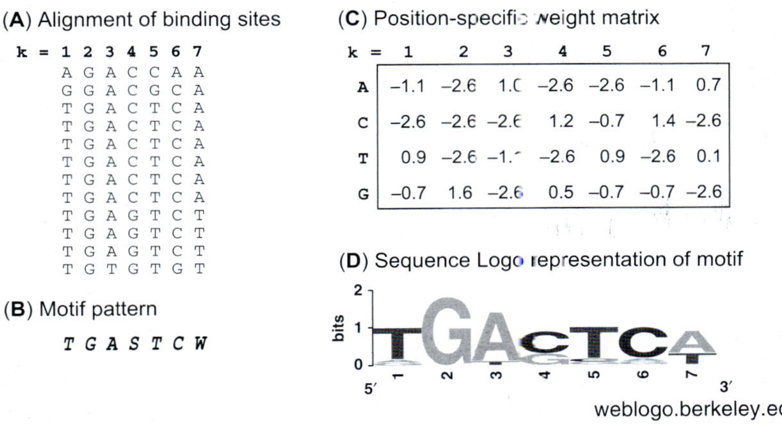


**Fig. 7.6** Computational analysis and representation of the Gcn4 transcription factor binding motif. (**A**) Shows the sequence alignment of DNA sequences bound with high affinity by Gcn4. 'K' indicates the column numbers of the alignment. (**B**) Shows the motif pattern that can be derived from the sequence alignment shown in A. (**C**) Shows the position-specific weight matrix for the Gcn4 binding sites. Each term in the matrix is the log-odds score for that nucleotide at each motif position. (**D**) Sequence logo representation of the Gcn4 binding motif. The sequence logo was generated using the WebLogo software [3] (Copies of figures including color copies, where applicable, are available in the accompanying CD)

propensity of C nucleotide to G nucleotide at position 4 of the motif. This quantitative information is lost in the pattern representation, as the S symbol gives equal weight to the G and C nucleotides.

A common method for visually representing a DNA motif is to use sequence logos. An example of a sequence logo, which in this case represents the Gcn4 DNA binding motif, is shown in Fig. 7.6D. The x-axis indicates each position in the motif. The y-axis measures the information content (in bits of information) for each position. Nucleotide symbols indicate the predominant nucleotide at each position. The smaller symbols indicate nucleotides that are present in a minority of sequences at that position. The sequence logo is a simple but powerful method to graphically convey the sequence similarities of transcription factor binding sites.

## *7.3.3 Assessment and Interpretation of ChIP-Chip Data*

To decipher a transcriptional network one must identify the target genes regulated by each transcription factor. The ChIP-chip approach is a powerful method for determining the genome-wide binding sites of a transcription factor. Unlike *in vitro* or computational methods for predicting binding sites the ChIP-chip approach should, in principle, reveal the DNA sites that are truly bound in the cell. Hence, the ChIP-chip data can be used to construct a network of binding targets for each transcription factor. Indeed, ChIP-chip studies have deciphered the transcriptional networks controlling the cell cycle [4] and stem cell identity [5], for example. An ambitious study used the ChIP-chip method to survey the binding targets of all the 203 yeast transcription factors, which allowed them to construct a comprehensive transcriptional network for a simple eukaryote cell [1].

### 7.3.3.1 Constructing the Gcn4 Transcriptional Network

An example of a transcriptional network that has been constructed from this comprehensive ChIP-chip data set is shown in Fig. 7.7. In this diagram, each of the circles represents a transcription

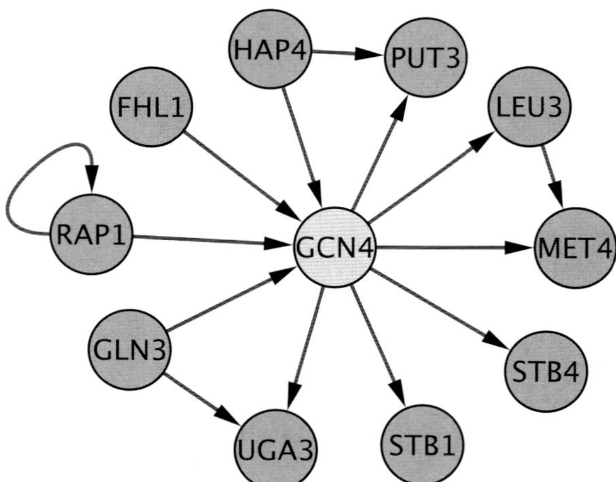


**Fig. 7.7** Diagram showing the transcriptional network associated with the Gcn4 transcription factor. Each circle (node) indicates a transcription factor gene. The arrows connecting the circles denote transcriptional regulation. The direction of the arrow indicates who regulates whom. For example the arrow pointing from *GLN3* to *GCN4* indicates that Gln3 binds to the promoter of the *GCN4* gene. The network was constructed using Cytoscape [6], based on transcription factor binding data from [1] (YPD data only; $P$-value $< 0.005$). Only genes encoding transcription factors are shown in the network (Copies of figures including color copies, where applicable, are available in the accompanying CD)

factor gene that is functionally related to the Gcn4 transcription factor. The arrows indicate transcription regulation as measured by promoter binding. The direction of the arrow indicates who regulates whom, for example the arrow pointing from *GCN4* to *MET4* indicates that the Gcn4 transcription factor binds to the promoter of the *MET4* gene and presumably regulates its transcription. Hence, some of the transcription factors regulate *GCN4* expression (i.e., *FHL1*, *GLN3*, *HAP4*, and *RAP1*), while other factors are themselves regulated by Gcn4 (i.e., *LEU3*, *MET4*, *PUT3*, *STB1*, *STB4*, and *UGA3*). Gcn4 is a yeast transcription factor that regulates genes that function in amino acid biosynthesis. Intriguingly, many of the transcription factors that are connected to Gcn4 in this transcriptional network are also known to regulate genes involved in amino acid biosynthesis (i.e., *LEU3*, *MET4*, *PUT3*) or protein synthesis (i.e., *FHL1*, *RAP1*). Hence, this simple network shows the striking interconnectedness of the protein and amino acid biosynthetic pathways.

#### 7.3.3.2 Potential Pitfalls in Interpreting ChIP-Chip Data

A common assumption when interpreting ChIP-chip data is that gene regulation can be directly inferred from promoter binding. It is important to note, however, that the binding of a transcription factor to a gene's promoter does not necessarily imply that the transcription factor actually regulates that gene. There are numerous examples where a transcription factor binds to a promoter, but does not fully or even partially control the transcription of the adjacent gene. For this reason, one must be cautious in inferring functional regulation from binding data alone.

An alternative approach to identify transcription factor target genes is to use expression microarrays to determine the genes whose mRNA levels are altered when the transcription factor of interest is genetically perturbed (either mutated or over-expressed). A change in mRNA levels caused by a perturbation in a transcription factor provides strong evidence that the gene is functionally regulated by the transcription factor. Again, however, caution must be taken when inferring direct functional relationships from expression data alone. Transcription factor mutants, for example, can affect the expression of large sets of genes indirectly, either by influencing the expression of other transcription factors (which is common), or by altering the expression of genes in important signaling pathways. A recent study of the yeast Leu3 transcription factor found that nearly 300 genes showed altered mRNA levels in *LEU3* mutant cells [7]. However, only 9 of these genes (3%) were directly bound by the Leu3 transcription factor. These results indicate that 97% of the genes identified by expression profiling were influenced indirectly by the *LEU3* mutant.

Because of the caveats discussed above, the most powerful approach is to combine ChIP-chip analysis with expression profiling of transcription factor mutants. The intersection of these data sets typically yields high confidence target genes for each transcription factor. These target genes can then be used to model the transcriptional network.

## 7.4 Bioinformatics Analysis of Transcription Factor Binding Sites

Once the DNA binding sites for each transcription factor have been identified; the next step is to analyze the resulting transcription factor network in biologically meaningful ways. The hope is that through this analysis, one can begin to connect transcription factors to biological pathways. For example, one might test whether a transcription factor regulates genes that function in glycolysis or other metabolic pathways, or if a transcription factor functions to regulate genes expressed periodically during the cell cycle. Identifying such connections enables us to better understand how the transcriptional network functions to control key biological processes such as metabolism and cell division.

For the purposes of this chapter, we will describe analysis methods using the Athena database and software tool [8]. Athena is a promoter database for the model plant *Arabidopsis thaliana*.

Athena contains predicted genomic binding sites for 105 distinct transcription factors. These binding sites have been predicted in the promoter sequences for more than 30,000 *Arabidopsis* genes. While these transcription factor binding sites have not been experimentally determined—by ChIP-chip experiments, for example—similar analysis tools can be used in both cases.

### 7.4.1 Visualization of Transcription Factor Binding Sites

Perhaps the simplest means of studying transcription factor binding sites is through promoter visualization. Promoter visualization allows one to rapidly inspect a set of promoters for shared binding sites. The Athena promoter visualization tool graphically represents the relative positions of transcription factor binding sites. In this example, Athena was used to visualize a 300 bp promoter sequence of the gene *BRCA2A* (systematic *Arabidopsis* gene name: At4g00020), which encodes the *Arabidopsis* homologue of the human breast cancer susceptibility protein 2a. The *BRCA2A* gene is required for meiosis in *Arabidopsis*. A portion of the resulting display is shown in Fig. 7.8A. The display uses color-coded boxes to indicate the locations of the predicted transcription factor binding sites in the promoter sequence (see color image provided in supplemental CD). In this example, the identity of each transcription factor binding site is labeled in Fig. 7.8A. The transcription start site of the *BRCA2A* gene is indicated with a black arrow.

Athena also displays a transcription factor table beneath the promoter image. This table provides a key to the color-coded transcription factor binding sites, lists the number of instances of each binding site in the promoter (#S column in Fig. 7.8B), and calculates the significance of enrichment (*P*-value) for each binding site in the promoter sequence (for more details about enrichment analysis, see section below). An example of a transcription factor table is shown in Fig. 7.8B.

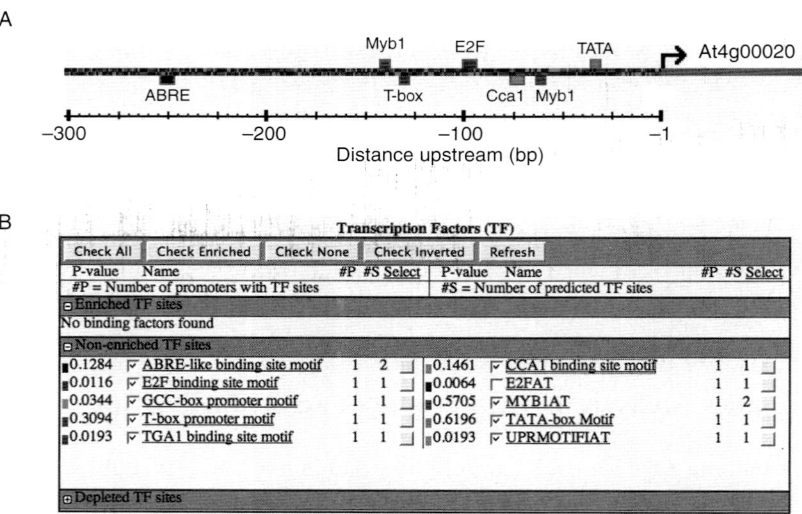


| P-value | Name | #P | #S | Select | P-value | Name | #P | #S | Select |
|---|---|---|---|---|---|---|---|---|---|
| 0.1284 | ABRE-like binding site motif | 1 | 2 | | 0.1461 | CCA1 binding site motif | 1 | 1 | |
| 0.0116 | E2F binding site motif | 1 | 1 | | 0.0064 | E2FAT | 1 | 1 | |
| 0.0344 | GCC-box promoter motif | 1 | 1 | | 0.5705 | MYB1AT | 1 | 2 | |
| 0.3094 | T-box promoter motif | 1 | 1 | | 0.6196 | TATA-box Motif | 1 | 1 | |
| 0.0193 | TGA1 binding site motif | 1 | 1 | | 0.0193 | UPRMOTIFIAT | 1 | 1 | |

**Fig. 7.8** Visualization of the promoter region of the plant *BRCA2A* (At4g00020) gene. (**A**) The promoter sequence of BRCA2A was visualized using the Athena *Arabidopsis* promoter database (www.bioinformatics2.wsu.edu/Athena). The locations of predicted transcription factor binding sites are denoted by color-coded boxes (see color image provided in supplemental CD), which match those in the transcription factor table (B). The arrow indicates the start of transcription. The DNA sequence can be read off of the image (A => red, T => brown, C => blue, G => green). (**B**) The transcription factor binding site table lists the binding sites that are present in the first 500 bp of the At4g00020 promoter (Copies of figures including color copies, where applicable, are available in the accompanying CD)

### 7.4.2 Enrichment Analysis of Transcription Factor Binding Sites

Often one wishes to discover whether transcription factor binding sites are enriched or over-represented in a selected set of genes or promoters. Over-represented binding sites are indicative that such factors may function to regulate the set of selected genes. A common method is to use the hyper-geometric probability model to calculate the statistical significance of enrichment. The following equation can be used to calculate the resulting p-values:

$$P = 1 - \sum_{x=0}^{k-1} \frac{\binom{m}{x}\binom{N-m}{n-x}}{\binom{N}{n}} \tag{7.3}$$

where $N$ is the total number of promoters in the genome, $n$ is the number of promoters in the genome containing the specified transcription factor binding site, $m$ is the size of the selected set of promoters, and $k$ is the number of selected promoters with the specified binding site (i.e., the overlap between $m$ and $n$).

This sort of analysis can be automated to systematically investigate connections between transcription factors and gene function. *Arabidopsis* genes have been classified into functional categories based on their known or predicted function. Such categories are known as gene ontology (GO). An example is the *DNA replication* GO category. Sixty-four *Arabidopsis* genes have been annotated as functioning in this cellular process, including many DNA polymerase subunits and replication origin recognition proteins (e.g., *ORC1A* and *ORC2*).

Figure 7.9 shows the results of systematically mining the set of genes in a variety of GO categories for enriched transcription factor binding sites. In this example, the calculated p-values were used to hierarchically cluster the functional categories and transcription factor binding sites. The clustering grouped together and binding sites that were similarly enriched in GO terms, GO terms that showed similar transcription factor binding site enrichment. Hence, this display shows the relationship between transcription regulatory sites and gene functions. Note, for example, that the three E2F family transcription factor binding sites are enriched in the *regulation of cell cycle*, *DNA repair*, *DNA replication*, *DNA replication initiation*, and *chromosome segregation* GO categories (Fig. 7.9). This connection between E2F family transcription factors and DNA replication has been previously identified in both animal and plant cells.

### 7.4.3 Spatial Analysis of Transcription Factor Binding Sites

Some transcription factor binding sites show a strong position bias in the promoter regions. For example, in higher eukaryotes the *TATA box* site is often located at position –35 bp upstream of the transcription start site. The Athena visualization tools was used to examine whether the E2F transcription factor binding site showed a position bias in the promoters of genes that function in DNA replication. The E2F transcription factor binding site is defined by the following sequence pattern: *TTTCCCGC*. To visualize these binding sites in multiple promoters, Athena's compact promoter visualization tool was used (Fig 7.10). Only the E2F binding site is visualized in this display; all other transcription factor binding sites are hidden. The position of each E2F binding site is indicated by a vertical line. Inspection of Fig 7.10 indicates that the E2F binding site may indeed show a position bias near the transcription start site.

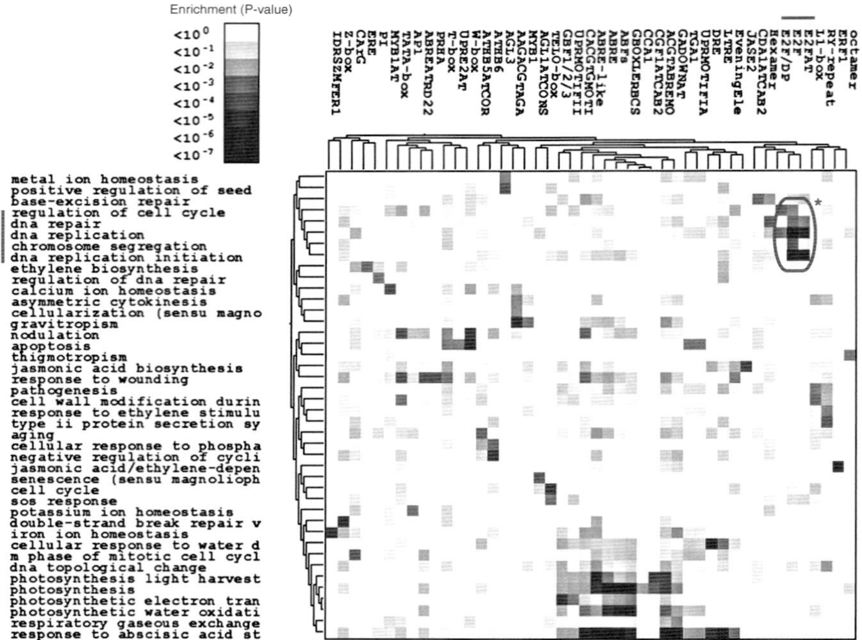


**Fig. 7.9** Transcription factor binding site enrichment analysis for Gene Ontology (GO) terms is present in 'other physiological processes' category. Only those binding sites that showed at least marginal enrichment ($P < 10^{-2}$) in at least one GO category are included in the display. The intensity of the blue spots at the intersection of each GO term and binding site indicates the magnitude of the p-value of the enrichment, with the darker blue being more significant (see color image provided in supplemental CD). The significant enrichment of E2F binding sites in the DNA replication and cell cycle GO categories is indicated in red (Copies of figures including color copies, where applicable, are available in the accompanying CD)

To quantify this bias, Athena's histogram tool was used to plot the spatial distribution of E2F binding sites in the promoters of its functional target genes. The resulting plot is shown in Fig 7.11. A significant peak of binding sites is observed between positions –200 and –1 bp upstream of the transcription start site. This display confirms that the E2F binding sites show a strong position bias near the transcription start site. The location of a binding site in the promoter sequence can be critical to its proper function in transcription regulation, and thus this observed position bias could be important for the proper functioning of the E2F transcription factor.

### *7.4.4 Integrated Analysis of Transcriptional Networks and Gene Expression Patterns*

Genes that contain the same transcription factor binding sites in their promoter sequence frequently have similar patterns of expression. Indeed, often one wishes to test whether gene expression patterns fit the predictions of the transcriptional regulatory network. To test these predictions, one needs tools that integrate gene expression data from expression microarrays with transcription networks.

Athena expression profile tool enables one to conduct such analyses. This tool displays the gene expression patterns for a user-selected set of genes. These expression patterns are derived from a comprehensive microarray data set, which profiled the changes in gene expression in nearly one hundred distinct *Arabidopsis* tissue types and developmental stages.

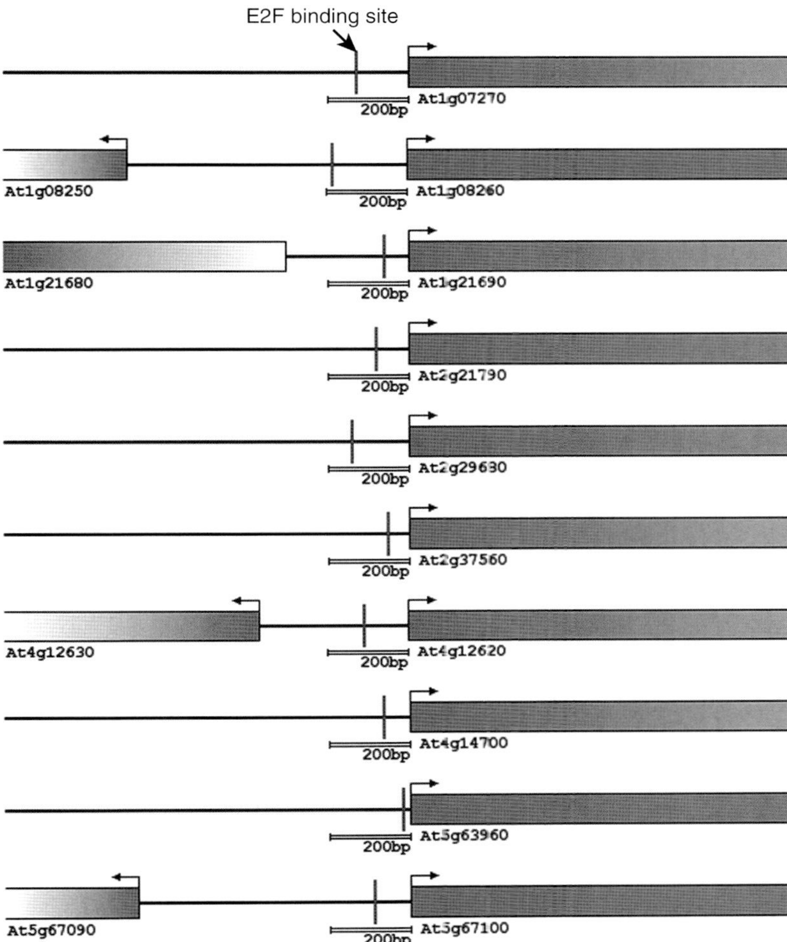


**Fig. 7.10** Compact visualization of E2F binding sites in the promoters of genes implicated in DNA replication. The Athena database was used to visualize the promoters of 10 genes that function in DNA replication. Only the E2F binding sites, which are indicated by a vertical aqua line or hash mark, are displayed in the visualization (Copies of figures including color copies, where applicable, are available in the accompanying CD)

The Athena expression profile tool was used to display the gene expression patterns for DNA replication genes that contained E2F binding sites in their promoter sequence. The resulting display is shown in Fig 7.12. The x-axis plots the different microarray experiments, which are organized according to the tissue type from which the mRNA was extracted. The y-axis plots the change in mRNA levels represented in this case as a modified Z-score. A positive Z-score indicates that a gene's mRNA level (or expression) is increased in that experiment; a negative Z-score indicates that a gene's mRNA level is decreased in that experiment. Each black line corresponds to the expression pattern of a single gene. The red line (see color image provided in supplemental CD) shows the average expression pattern for all the selected genes for each tissue type and developmental stage.

Inspection of the data indicates that the genes regulated by the E2F transcription factor show peak expression in the *Apex* tissue (see Fig 7.12). The shoot apex is the site of rapid cell division and DNA replication in plants. Hence, it is not surprising that E2F regulated genes are highly expressed in apex tissue.

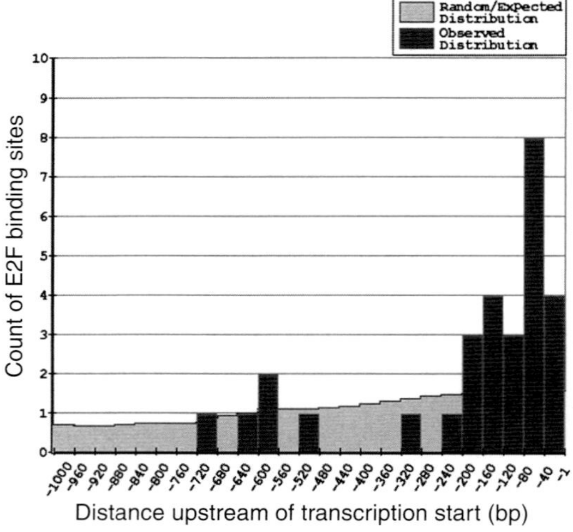


**Fig. 7.11** Histogram plot of the distribution of E2F binding sites in the promoters of genes involved in DNA replication or cell cycle. The bins indicate the number of E2F binding sites present in each 40 bp window of promoter sequences. The gray distribution backdrop is the expected distribution if the E2F binding sites are randomly distributed (Copies of figures including color copies, where applicable, are available in the accompanying CD)

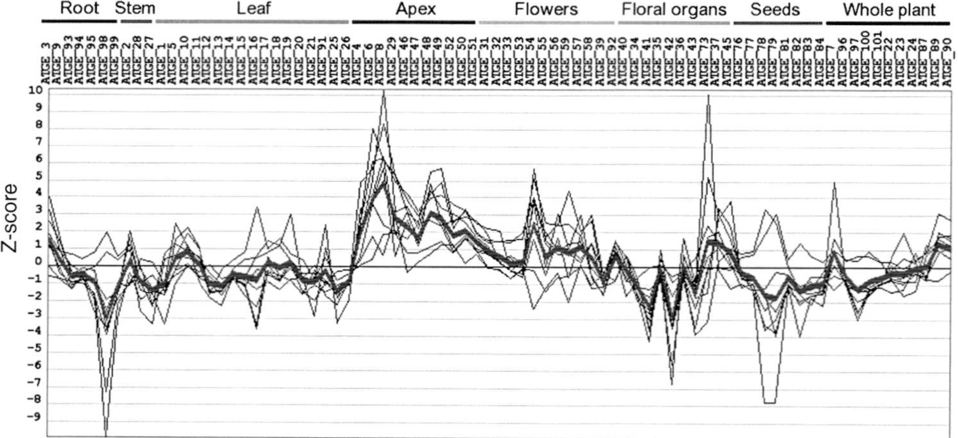


**Fig. 7.12** Genome expression profile of DNA replication genes that contain an E2F binding site in their promoter region (10 genes total). Each black line represents the expression pattern of a single gene; the heavy line indicates the average expression pattern of the selected genes. The x-axis lists the microarray experiments (e.g., ATGE3) organized according to the tissues from which the mRNA samples were collected. The y-axis indicates the change in mRNA levels for that gene in that experiment, in this case represented as a modified Z-score. Expression data from (Schmid, et al., 2005) (Copies of figures including color copies, where applicable, are available in the accompanying CD)

## Glossary and Abbreviations

| | |
|---|---|
| ChIP-chip | Chromatin immunoprecipitation microarray analysis |
| GO | Gene ontology |
| LM-PCR | Ligation mediated polymerase chain reaction |
| PSWM | Position-specific weight matrix |

## References

1. Harbison CT, Gordon DB, Lee TI, Rinaldi NJ, Macisaac KD, Danford TW, Hannett NM, Tagne JB, Reynolds DB, Yoo J, Jennings EG, Zeitlinger J, Pokholok DK, Kellis M, Rolfe PA, Takusagawa KT, Lander ES, Gifford DK, Fraenkel E, Young RA. Transcriptional regulatory code of a eukaryotic genome. Nature 2004; 431:99–104.
2. Hertz GZ, Stormo GD. Identifying DNA and protein patterns with statistically significant alignments of multiple sequences. Bioinformatics 1999; 15:563–577.
3. Crooks GE, Hon G, Chandonia JM, Brenner SE. WebLogo: a sequence logo generator. Genome Res 2004; 14:1188–1190.
4. Simon I, Barnett J, Hannett N, Harbison CT, Rinaldi NJ, Volkert TL, Wyrick JJ, Zeitlinger J, Gifford DK, Jaakkola TS, Young RA. Serial regulation of transcriptional regulators in the yeast cell cycle. Cell 2001; 106:697–708.
5. Boyer LA, Lee TI, Cole MF, Johnstone SE, Levine SS, Zucker JP, Guenther MG, Kumar RM, Murray HL, Jenner RG, Gifford DK, Melton DA, Jaenisch R, Young RA. Core transcriptional regulatory circuitry in human embryonic stem cells. Cell 2005; 122:947–956.
6. Ideker T, Ozier O, Schwikowski B, Siegel AF. Discovering regulatory and signalling circuits in molecular interaction networks. Bioinformatics 2002; 18 Suppl 1:S233–240.
7. Tang L, Liu X, Clarke ND. Inferring direct regulatory targets from expression and genome location analyses: a comparison of transcription factor deletion and overexpression. BMC Genomics 2006; 7:215.
8. O'Connor TR, Dyreson C, Wyrick JJ. Athena: a resource for rapid visualization and systematic analysis of Arabidopsis promoter sequences. Bioinformatics 2005; 21:4411–4413.

## Key References

Lee TI, Johnstone SE, Young RA. Chromatin immunoprecipitation and microarray-based analysis of protein location. Nat Protoc 2006; 1:729–748.

Che D, Jensen S, Cai L, Liu JS. BEST: binding-site estimation suite of tools. Bioinformatics 2005; 21:2909–2911.

O'Connor TR, Dyreson C, Wyrick JJ. Athena: a resource for rapid visualization and systematic analysis of Arabidopsis promoter sequences. Bioinformatics 2005; 21:4411–4413.

## Suggested Reading

### *Genome-Wide Profiling of Transcription Factor Binding Targets*

1. Buck MJ, Lieb JD. ChIP-chip: considerations for the design, analysis, and application of genome-wide chromatin immunoprecipitation experiments. Genomics 2004; 83:349–60.
2. Harbison CT, Gordon DB, Lee TI, Rinaldi NJ, Macisaac KD, Danford TW, Hannett NM, Tagne JB, Reynolds DB, Yoo J, Jennings EG, Zeitlinger J, Pokholok DK, Kellis M, Rolfe PA, Takusagawa KT, Lander ES, Gifford DK, Fraenkel E, Young RA. Transcriptional regulatory code of a eukaryotic genome. Nature 2004; 431:99–104.
3. Iyer VR, Horak CE, Scafe CS, Botstein D, Snyder M, Brown PO. Genomic binding sites of the yeast cell-cycle transcription factors SBF and MBF. Nature 2001; 409:533–5338.
4. Kurdistani SK, Grunstein M. In vivo protein-protein and protein-DNA crosslinking for genomewide binding microarray. Methods 2003; 31:90–95.
5. Lee TI, Rinaldi NJ, Robert F, Odom DT, Bar-Joseph Z, Gerber GK, Hannett NM, Harbison CT, Thompson CM, Simon I, Zeitlinger J, Jennings EG, Murray HL, Gordon DB, Ren B, Wyrick JJ, Tagne JB, Volkert TL, Fraenkel E, Gifford DK, Young RA. Transcriptional regulatory networks in Saccharomyces cerevisiae. Science 2002; 298:799–804.
6. Lee TI, Johnstone SE, Young RA. Chromatin immunoprecipitation and microarray-based analysis of protein location. Nat Protoc 2006; 1:729–748.
7. Liu CL, Schreiber SL, Bernstein BE. Development and validation of a T7 based linear amplification for genomic DNA. BMC Genomics 2003; 4:19.

8. Ren B, Robert F, Wyrick JJ, Aparicio O, Jennings EG, Simon I, Zeitlinger J, Schreiber J, Hannett N, Kanin E, Volkert TL, Wilson CJ, Bell SP, Young RA. Genome-wide location and function of DNA binding proteins. Science 2000; 290:2306–2309.
9. Shivaswamy S, Iyer VR. Genome-wide analysis of chromatin status using tiling microarrays. Methods 2007; 41:304–311.
10. Che D, Jensen S, Cai L, Liu JS. BEST: binding-site estimation suite of tools. Bioinformatics 2005; 21:2909–2911.
11. Liu XS, Brutlag DL, Liu JS. An algorithm for finding protein-DNA binding sites with applications to chromatin-immunoprecipitation microarray experiments. Nat Biotechnol 2002; 20:835–839.
12. Luscombe NM, Babu MM, Yu H, Snyder M, Teichmann SA, Gerstein M. Genomic analysis of regulatory network dynamics reveals large topological changes. Nature 2004; 431:308–312.
13. Macisaac KD, Gordon DB, Nekludova L, Odom DT, Schreiber J, Gifford DK, Young RA, Fraenkel E. A hypothesis-based approach for identifying the binding specificity of regulatory proteins from chromatin immunoprecipitation data. Bioinformatics 2006; 22:423–429.
14. Qi Y, Rolfe A, Macisaac KD, Gerber GK, Pokholok D, Zeitlinger J, Danford T, Dowell RD, Fraenkel E, Jaakkola TS, Young RA, Gifford DK. High-resolution computational models of genome binding events. Nat Biotechnol 2006; 24:963–970.
15. Zhu X, Gerstein M, Snyder M. Getting connected: analysis and principles of biological networks. Genes Dev 2007; 21:1010–1024.
16. Guha, Thakurta D. Computational identification of transcriptional regulatory elements in DNA sequence. Nucleic Acids Res 2006; 34:3585–3598.
17. Li H, Wang W. Dissecting the transcription networks of a cell using computational genomics. Curr Opin Genet Dev 2003; 13:611–616.
18. Riechmann JL, Heard J, Martin G, Reuber L, Jiang C, Keddie J, Adam L, Pineda O, Ratcliffe OJ, Samaha RR, Creelman R, Pilgrim M, Broun P, Zhang JZ, Ghandehari D, Sherman BK, Yu G. Arabidopsis transcription factors: genome-wide comparative analysis among eukaryotes. Science 2000; 290:2105–2110.
19. Xie X, Lu J, Kulbokas EJ, Golub TR, Mootha V, Lindblad-Toh K, Lander ES, Kellis M. Systematic discovery of regulatory motifs in human promoters and 3' UTRs by comparison of several mammals. Nature 2005; 434:338–345.

# Part II
# Statistical Tools and Their Application

# Chapter 8
# Probability and Hypothesis Testing

**Michael L. Kruger**

**Abstract** This chapter provides an elementary introduction to the concept and characteristics of probability and the use of probability theory in hypothesis testing. Examples of the use of the binomial theorem and Fisher's exact in associating gene expression with observed outcomes (i.e., disease) are presented. A brief introduction to the concepts of hypothesis testing, power, ROC (receiver-operating curves) and sensitivity and specificity are also presented.

**Keywords** Probability · Binomial probability distribution · Power · Hypothesis testing · Experimental design

## 8.1 Introduction

In order to determine whether an association exists between genes and disease we need to first identify candidate genes or groups of genes, and then determine if the observed association is significantly greater than what we would expect by chance. This chapter is designed to provide an introduction to probability theory and hypothesis testing. Although the material presented here does not directly refer to the area of bioinformatics, the principles and concepts can be applied to all areas of scientific research.

Probability can be simply defined as the likelihood of the occurrence of an event, where an event is simply defined as the outcome of a trial or an experiment. These events are said to be *independent* if the occurrence or non-occurrence of the event has no effect on the probability of occurrence of other events. A simple example would be a coin toss. Each coin toss outcome is completely independent of the outcome of previous coin tosses. The outcome of heads or tails each has equal probability (0.5 or 50%) of occurrence. This is an example of a *binary* or *dichotomous* outcome. An example of an independent event that has more than two outcomes is the rolling of a six-sided dice. Each outcome, 1 through 6, has equal probability of occurrence (0.166 or 16.6%), and each roll of the dice is completely independent of the outcome of previous rolls of the dice. A common example of an event where the outcomes are not independent is the drawing of a single card from a standard deck of 52 playing cards. Since there are four suits (clubs, diamonds, hearts and spades) and there are thirteen cards in each suit, the probability of drawing a heart is 0.25 (13/52), but on the next drawing of a card the probability of drawing a heart is 0.235 (12/51) since there is one less heart in the deck of cards following the drawing of the initial card. The probability of drawing subsequent cards is therefore *dependent* on the outcome of the prior drawing of cards.

M.L. Kruger
Wayne State University School of Medicine, Department of OB/GYN, C.S. Mott Center, 275 E. Hancock St., Detroit, MI, 48201, USA
e-mail: mkruger@med.wayne.edu

S. Krawetz (ed.), *Bioinformatics for Systems Biology*, DOI 10.1007/978-1-59745-440-7_8,

Events are said to be *mutually exclusive* if they cannot occur simultaneously. An individual cannot simultaneously have both blue and brown eyes, or be less than 21 and greater than 21 years of age, or both a mammal and lay eggs.

Using these two characteristics of probability, we can define probability in the following manner: If an event can occur in $N$ mutually exclusive and independent ways and if $m$ of these possesses a specific characteristic, $E$, the probability of the occurrence is equal to $m/N$.

$$P(E) = \frac{m}{N} \tag{8.1}$$

For example, if we sample 100 people for a particular gene linked to hypertension and 25 of them have the gene, then the probability of having the gene is 0.25 or 25%. Probability is measured on a continuous scale of values between 0 and 1 inclusive. An event that is impossible has a probability of occurrence of 0, and an event that is certain to occur has a probability equal to 1. An event that is equally likely to occur or not occur has a probability of 0.5 (e.g., getting a head or a tail on a coin toss).

Sickle cell anemia was the first genetic disease to be characterized at the molecular level. The mutation responsible is just a single nucleotide of the DNA. It is the result of a point mutation in the β-globin gene. As a result of this mutation, valine is inserted into the β-globin chain instead of glutamic acid. The mutation causes red blood cells (RBCs) to become stiff when they release their load of oxygen and become 'sticky' resulting in their sickle shape. These cells tend to get stuck in narrow blood vessels which may cause severe pain and damage to the heart, lungs or kidney. The recessive gene tends to be maintained in the gene pool because individuals who are heterozygous do not develop sickle cell anemia and are less likely to contract malaria. They are able to survive and reproduce in malaria-infected regions of the world such as Africa.

Below is some hypothetical outcome data of the inheritance patterns of the sickle cell gene for forty (40) offspring of heterozygous mothers and fathers. This data will be used to illustrate some of the basic properties of probability.

**Table 8.1** Human sickle cell anemia: (observed frequencies)

| | Father | | |
|---|---|---|---|
| Mother | B ( + ) | b ( − ) | TOTAL |
| B ( + ) | BB (20) | Bb (6) | 26 |
| b ( − ) | Bb (8) | Bb (6) | 14 |
| TOTAL | 28 | 12 | 40 |

B: dominant normal globin gene (+)
b: recessive sickle cell gene (−)
(Copies of tables are available in the accompanying CD.)

## 8.2 Basic Rules of Probability

1. All events have a probability greater than or equal to zero. The probability of any event is never greater than 1 or less than 0.

$$P(E) \geq 0 \tag{8.2}$$

2. *Addition rule*: If an event is satisfied by any one of a group of mutually exclusive outcomes, the probability of the event is the *sum* of all of the mutually exclusive probabilities in the group.

$$P(E_1 \text{ or } E_2) = P(E_1) + \mathrm{P}(E_2) \tag{8.3}$$

For example, the probability of drawing either an Ace or a Jack is equal to (1/13 + 1/13) = 2/13 (15.4%), or using the data from Table 8.1, of having an offspring with a non-sickle cell phenotype (BB, Bb) is equal to (20/40 + 6/40 + 8/40) = 34/40 (85%).

3. The sum of all mutually exclusive probabilities is equal to one (exhaustive).

$$P(E_1) + P(E_2) + \ldots P(E_n) = 1 \tag{8.4}$$

From Table 8.1, the sum of the probabilities of homozygous normal (BB) and homozygous sickle cell (bb) and heterozygous (Bb) equals 20/40 + 6/40 + 14/40 = 40/40 = 1.0 (100%). Each of the three possible sickle cell outcomes is referred to as a *simple probability*.

A *conditional probability* is the chance of a particular event happening that depends on the outcome of some other event. It is the probability that event *A* occurs given that event *B* has already occurred. The probability is calculated with a subset of the total group as the denominator:

$$P(A/B) = \frac{A \cap B}{B} \tag{8.5}$$

An example would be the probability of drawing a face card (King, Queen or Jack) that is a heart = 3/13 = 0.231 or using the data from Table 8.1, if the father contributes a normal sickle cell gene (*B*) the probability of having a heterozygous child is 8/28 = 0.286.

*A Joint probability* is the probability that a subject picked at random will possess two characteristics simultaneously:

$$P(\mathrm{A} \cap \mathrm{B}) = (P_A \times P_B). \tag{8.6}$$

For example, the probability of a homozygous normal offspring (BB) with both the mother and father contributing a B sickle cell gene if they each had equal probability of donating a normal or recessive sickle cell gene is 0.5 × 0.5 = 0.25.

*Independent* probability is when the probability of event *A* has no effect on the probability of event *B*. If events *A* and *B* are independent then their joint probability is:

$$P(A \cap B) = P(B) \times P(A) \text{ if } \mathrm{P(A)} \neq 0 \text{ and } \mathrm{P(B)} \neq 0 \tag{8.7}$$

For example, the probability of picking an ace of spades is (4/52) × (1/4) = 1/52. The probability of picking an Ace has no effect on the probability of picking a spade. They are statistically *independent* probabilities.

## 8.3 Counting Rules

In order to determine probabilities we need to define the number of all possible events that can occur and their frequency. This requires a discussion of what are generally referred to as *counting rules*. In many cases an event is a sequence of observations, and the outcomes of a series of trials can be thought of as a particular sequence. We can calculate the number of *permutations*, or ordered sequences, by N!. This is the number of different ways that N distinct things may be arranged in order. If we are interested in how the four nucleotide bases, Thymine, Guanine, Cytosine and

Adenosine can be arranged with respect to order; 4! = 24. If we are interested in determining the number of ways to select '$r$' objects from among $N$ distinct objects, the following formula is used:

$$_{\mathrm{n}}P_{\mathrm{r}} = \frac{N!}{(n-r)} \tag{8.8}$$

Using the example of four nucleotide bases arranged into ordered pairs:

$$_{\mathrm{n}}P_{\mathrm{r}} = \frac{4!}{(4-2)} = \quad 12 \quad \text{ordered pairs (TG TC TA GT GC GA CT CG CA AT AG AC)}$$

If the order of the elements is not important, we are then interested in determining the number of possible *combinations,* or arrangements of objects without regard to the order. We use the following formula to calculate the number of possible combinations of $N$ things taken '$r$' at a time:

$$_{\mathrm{n}}C_{\mathrm{r}} = \frac{_{\mathrm{n}}P_{\mathrm{r}}}{r!} = \binom{N}{r} \frac{N!}{\frac{(N-r)!}{\mathrm{r}!}} = \frac{N!}{r!(\mathrm{N}-\mathrm{r})!} \tag{8.9}$$

Using the four nucleotide bases TGAC taken two at a time:

$$\binom{4}{2} = \frac{4!}{2!(4-2)!} = \frac{4 \times 3 \times 2 \times 1}{2 \times 1(2 \times 1)} = \frac{24}{4} = 6 \text{ (TG TC TA GC GA CA)}$$

This formula can be used to determine how many possible unique sequences a group of genes can be arranged into in order to calculate the probability of the occurrence of a particular sequence. For example, if we have 20 genes available to make a pattern of 5 genes and we want to determine the number of unique patterns of the five genes:

$$_{20}C_5 = \frac{20!}{5! \times 15!} = \frac{20 \times 19 \times 18 \times 17 \times 16}{5 \times 4 \times 3 \times 2 \times 1} = \frac{1860480}{120} = 15,504$$

When a process or trial can result in only one of the two mutually exclusive outcomes, such as a coin toss, it is called a *Bernoulli trial.* The characteristics of a Bernoulli trial are:

1. Each trial results in one of two possible mutually exclusive outcomes (success or failure).
2. Probability of success, p, is constant across all trials and the probability of failure $(1-p)$ is $q$.
3. Each trial is independent. The outcome of one trial is not affected by the outcome of any other trial.

When $N$ Bernoulli trials are conducted, the number of successes forms a distribution called the *binomial distribution.* This distribution represents the probability of '$r$' successes in $N$ trials and can be represented by the following formula:

$$_{\mathrm{N}}C_{\mathrm{r}} \times p^{\mathrm{r}} \times q^{(\mathrm{N}-\mathrm{r})} \tag{8.10}$$

For large samples and where order is not important we use the combination rule to calculate the number of combinations of $n$ objects that can be formed by taking $r$ at a time and $p$ as the probability of a 'success' and $q$ as the probability of 'failure'. The binomial formula can be used to compute the probability of each possible outcome of $r$ 'successes' in $N$ trials. This method can be used to determine the expected probability of each number of possible successes and whether an observed outcome is significantly different than that expected by chance. For example, if the

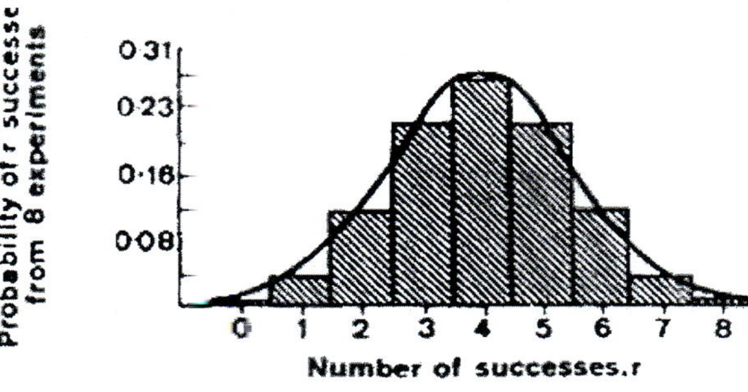


**Fig. 8.1.** Comparison of binomial (*histogram*) and normal (*curved line*) distributions for $n = 8$ (Copies of figures including color copies, where applicable, are available in the accompanying CD)

probability of either a male or female birth is 0.5 ($p = 0.5$ and $q = 0.5$), we can use the binomial formula to obtain the probability of getting exactly 3 male births in 8 pregnancies.

$$_{N}C_{r} = \frac{N!}{r!(N-r)!} \quad \text{The number of combinations of N things taken r at a time.} \qquad (8.11)$$

$$_{5}C_{3} = \frac{8!}{3!(8-3)} = \frac{8 \times 7 \times 6}{3 \times 2 \times 1} = 56 \text{ possible combinations}$$

The binomial probability for 3 male babies in 8 births is:

$$56 \times (0.5)^3 \times (0.5)^5 = 56 \times (0.125)(0.03125) = 0.2188 = 21.9\% \text{ probability of 3 male births.}$$

Therefore we would conclude that this observed probability is not significantly different than chance if we were to use the traditional scientific significance level of 5% (more on this later in the chapter).

Examining Fig. 8.1 we can see that the binomial distribution is an approximation of the normal distribution, and as the number of trials ($N$) increases the binomial distribution becomes a closer approximation of the normal distribution.

## 8.4 Hypothesis Testing using the Binomial Formula

The data in Table 8.2 represents the association of two gene patterns (ABC and abc) that are thought to be associated with male infertility, with ABC representing the pattern associated with the abnormal condition of infertility and 'abc' with normal male fertility in 19 males ($N = 19$). The question to be addressed is whether the observed outcomes are significantly different than what we would expect from chance.

To use the binomial formula to determine the probability of obtaining the observed results we use the column marginal totals to obtain the expected probability of infertility $(9/19) = 0.474$ and

**Table 8.2** Hypothetical outcome for two hypothetical 'gene patterns' associated with male infertility

| | Male Fertility | | |
|---|---|---|---|
| Gene Pattern | Infertile | Fertile | Total |
| ABC | 7 | 3 | 10 |
| abc | 2 | 7 | 9 |
| TOTAL | 9 | 10 | 19 |

(Copies of tables are available in the accompanying CD )

of fertility (10/19) = 0.526 for each of the gene patterns. If there is no difference between the gene patterns then the percentage of fertile and infertile should be the same for each pattern.

We calculate the probability of obtaining the outcome seen in Table 8.2 using the ABC frequencies:

$$ {}_nC_r \times \frac{10!}{7!(3!)} = \frac{10 \times 9 \times 8}{3 \times 2 \times 1} = \frac{720}{6} = 120 $$

$$ 120 \times p^7 q^3 = 120(.474)^7 \times (.526)^3 = 120 \times .0054 \times .1455 = .0939 \text{ (2-tailed) or } .0469 \text{ (1-tailed)} $$

We would use a 1-tailed or directional hypothesis since we are asking whether the probability of the observed association of infertility with the 'ABC' pattern is significantly greater than with the 'abc' pattern. If we did not have a hypothesis about which pattern was associated with infertility, then we would use a two-tailed hypothesis test to establish that there is simply a difference in the proportion of infertile cases between the two gene patterns. Using a significance level of $\alpha = 0.05$, we can conclude that the observed results (Table 8.3) are significantly different than what we could expect from chance since the one-tailed observed probability is (0.0939/2) = 0.047. This use of the binomial formula is a statistical test called the Fisher's Exact test. As the name implies, the Fisher's Exact Test involves the calculation of exact probabilities for the observed frequencies in any 2×2 table. We can calculate the exact probability of any pair of dichotomous outcome variables such as disease state (+ or −) versus treatment (control, treatment) or gene state (present, absent).

**Table 8.3** All possible outcomes for hypothetical male infertility and 'gene pattern' based on the observed marginal totals

| Outcome 1 | | | | Outcome 6 | | | |
|---|---|---|---|---|---|---|---|
| Pattern | Infertile | Fertile | Total | Pattern | Infertile | Fertile | Total |
| ABC | 0 | 10 | 10 | ABC | 5 | 5 | 10 |
| abc | 9 | 0 | 9 | abc | 4 | 5 | 9 |
| Total | 9 | 10 | 19 | Total | 9 | 10 | 19 |
| Outcome 2 | | | | Outcome 7 | | | |
| Pattern | Infertile | Fertile | Total | Pattern | Infertile | Fertile | Total |
| ABC | 1 | 9 | 10 | ABC | 6 | 4 | 10 |
| abc | 8 | 1 | 9 | abc | 3 | 6 | 9 |
| Total | 9 | 10 | 19 | Total | 9 | 10 | 19 |
| Outcome 3 | | | | Outcome 8 | | | |
| Pattern | Infertile | Fertile | Total | Pattern | Infertile | Fertile | Total |
| ABC | 2 | 8 | 10 | ABC | 7 | 3 | 10 |
| abc | 7 | 2 | 9 | abc | 2 | 7 | 9 |
| Total | 9 | 10 | 19 | Total | 9 | 10 | 19 |
| Outcome 4 | | | | Outcome 9 | | | |
| Pattern | Infertile | Fertile | Total | Pattern | Infertile | Fertile | Total |
| ABC | 3 | 7 | 10 | ABC | 8 | 2 | 10 |
| abc | 6 | 3 | 9 | abc | 1 | 8 | 9 |
| Total | 9 | 10 | 19 | Total | 9 | 10 | 19 |
| Outcome 5 | | | | Outcome 10 | | | |
| Pattern | Infertile | Fertile | Total | Pattern | Infertile | Fertile | Total |
| ABC | 4 | 6 | 10 | ABC | 9 | 1 | 10 |
| abc | 5 | 4 | 9 | abc | 0 | 9 | 9 |
| Total | 9 | 10 | 19 | Total | 9 | 10 | 19 |

(Copies of tables are available in the accompanying CD.)

**Table 8.4** The probabilities of observing each of the outcomes shown in Table 8.3 if the null hypothesis of equal proportions is true ($H_o p_1 = p_2$)

| Outcome | Probability | Outcome | Probability |
|---|---|---|---|
| 1 | 0.00001 | 6 | 0.3437 |
| 2 | 0.0009 | 7 | 0.1910 |
| 3 | 0.0175 | 8 | 0.0468 |
| 4 | 0.1091 | 9 | 0.0044 |
| 5 | 0.2864 | 10 | 0.00019 |

(Copies of tables are available in the accompanying CD.)

Table 8.4 provides the probabilities of all possible outcomes based on our row and column marginal values. These are the probabilities of observing each of the tables if the null hypothesis is true. The total of all of these probabilities is 100% since the observed outcomes are exhaustive of all possible outcomes for the observed data. With increasingly large sample sizes ($N$) the probability distribution will approach a normal distribution.

The Fisher's Exact Test is an example of a statistical test. Statistical tests are procedures by which one determines the degree to which collected data are consistent with a specific hypothesis. A statistical hypothesis is a statement about a population, which is based on the observed data that one seeks to either support or refute the observed data. A statistical test is a set of rules by which a decision about the hypothesis is reached. The measure of the statistical test's accuracy is a probability statement about making the correct decision when certain conditions are true for the population being studied.

## 8.5 The Null Hypothesis and the Alternative Hypothesis ($H_0$ and $H_1$)

The *null hypothesis* ($H_0$) is a statement that there is *no* relationship between the variables or factors being examined. The *alternative hypothesis* ($H_1$) states that there is a relationship between the variables or factors being examined. Using our hypothetical data from Table 8.2 we could state the null and alternative hypotheses as:

$H_0$ – there is *no difference* in the proportion of infertile cases between gene patterns 'ABC' and 'abc'or there is NO genetic basis for male infertility.
$H_1$ – there is *a difference* in the proportion of infertile cases between gene patterns 'ABC' and 'abc' or there IS a genetic basis for male infertility.

The statement of the alternative hypothesis is *non-directional* or *two-sided*. It states that there is a difference between the gene patterns without specifying which of the gene patterns are associated with male infertility. In our example we made a *directional* or *one-sided* hypothesis that the 'ABC' pattern is associated with male infertility. The outcome of a research study can be classified in the following decision table (Fig. 8.2). The condition of the null hypothesis may be thought of as the 'true state of affairs' in the population.

| | | Condition of Null Hypothesis | |
|---|---|---|---|
| | | True | False |
| Possible Action | Fail to reject $H_0$ | Correct action | Type II error |
| | Reject $H_0$ | Type I error | Correct action |

**Fig. 8.2** Null hypothesis decision table (Copies of figures including color copies, where applicable, are available in the accompanying CD)

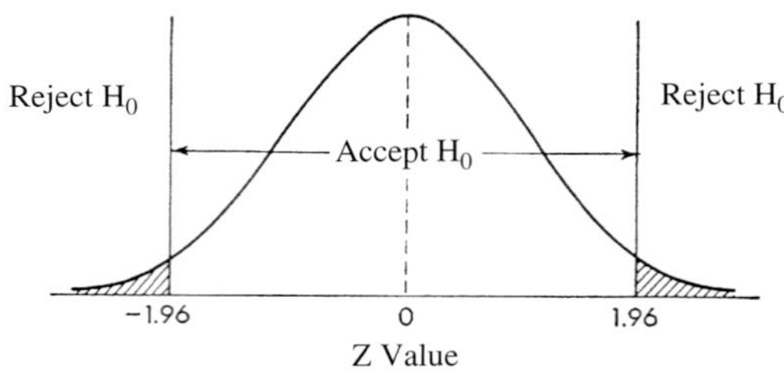


**Fig. 8.3** Region of rejecting $H_o$ when $p < 0.05$ (Copies of figures including color copies, where applicable, are available in the accompanying CD)

With the collection of data we are obtaining a single sample from a population and then making a decision based on this sample data. If we choose to reject the null hypothesis based on our sample data and it is actually true that there is no relationship between the variables being studied, we have made a *Type I* or $\alpha$ error. The probability of making this kind of an error is almost equal to the level of significance of our statistical test (*the p value*). Traditionally, this value has been set by scientific convention to 0.05 or 5%. This means that we are accepting a 5% risk of erroneously rejecting the null hypothesis when it is actually true. If the probability of the observed data is smaller than the significance level ($p < 0.05$) then the data is said to contradict the null hypothesis and the null is rejected. Rejection of the null hypothesis is equivalent to supporting or accepting the alternative hypothesis.

Figure 8.3 represents the two-tailed regions of rejection ($p < 0.05$) for a normally distributed variable. If the point marked by 0 on the X-axis represents the population mean, then any sample mean obtained that falls either 1.96 standard deviations above or below the population mean would lead to a rejection of the null hypothesis. The shaded regions are referred to as the regions of rejection since mean values obtained that fall into these areas lead to the rejection of $H_0$. If we were to use a directional alternative hypothesis our region of rejection would be at one end of the normal distribution, (Fig. 8.4) depending on whether we expected our alternative mean to be greater or smaller than that of the null hypothesis.

If the decision rules fail to reject the null ($H_0$), when in fact it is not true (false), we have made a *Type II* error. The magnitude of the type II error depends upon the level of significance and which of the alternative hypotheses is actually true. In the design of an experiment, the $\alpha$ and $\beta$ error are not independent. Decreasing the probability of a Type I error (alpha) increases the probability of a Type II error (beta).

If instead of simply labeling our male subjects as fertile or infertile, we were to use sperm motility values obtained from semen analysis, we would then state our null and alternative hypotheses with respect to the mean sperm motility values for the two gene patterns.

$$H_0 : \mu_{ABC} = \mu_{abc}$$

$$H_1 : \mu_{ABC} < \mu_{abc}$$

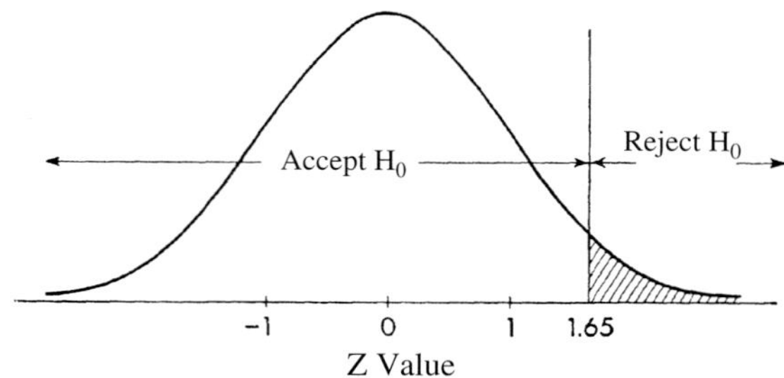


**Fig. 8.4** Region of rejecting $H_o$ when $p < 0.05$ and $H_1$: $p_1 > p_2$ (Copies of figures including color copies, where applicable, are available in the accompanying CD)

The null hypothesis would be that the mean sperm motility is not different for the two gene patterns and the alternative hypothesis would be that the mean sperm motility is lower for the ABC gene pattern.

If this observed difference is not 'real' and is a result of sampling bias or confounding with another variable, then we have committed a Type I error. We rejected the null hypothesis of no difference when it is actually true. If we had used another measure of fertility, such as sperm concentration data, and concluded that there was no significant difference between the two gene patterns, when in reality there is, we would have committed a Type II error. *Type II error*, or β error, is the failure to reject the null hypothesis when it is false. By convention, β error is set to 0.20 or 20%. The complement of Type II error (1 – β) is referred to as the *power* of a statistical test.

The power of a test with respect to a specified alternative hypothesis is equal to 1 minus the probability of Type II error (β). This is the area of the sampling distribution when $H_1$, the alternative hypothesis, is true that falls in the region of rejection of $H_0$. Power is the probability that the decision rule rejects the null when the alternative is true. By convention we use 0.80 as our desired level of power. Each alternative hypothesis has its own power since the region of rejection is different for each alternative hypothesis. The closer the alternative hypothesis is to the null hypothesis the lower the power of the test will be with respect to that specific alternative hypothesis. In Fig. 8.5, we can see that by reducing our level of Type I (α) error from 0.05 to 0.01 results in a large reduction of power since it increases the risk of Type II (β) error with all other factors being equal. A well-designed experiment will have relatively high power. The bigger the difference one expects between groups the more powerful the test. The desired power and significance levels selected should be based upon issues other than statistical principles. The cost of the various types of error and the questions being studied should be part of the decision process. The risks of harmful side effects, the failure to detect an effective treatment, its costs (life or death) as well as practical issues such as the cost of conducting the study, the number of patients and length of time required to complete the study, and all the other factors must be considered in the design of a study. For example, in the treatment of infertility, there may be long-term harmful effects or the cost of the treatment may outweigh the benefits that it provides.

The power of a statistical test is determined by the effect size, the sample size and the alpha (Type I error) level. Effect size is the magnitude of the treatment effect and it is influenced by several factors such as the variability of the sample population, the differences in the treatment variable (i.e., dosage, duration and other protocol differences).

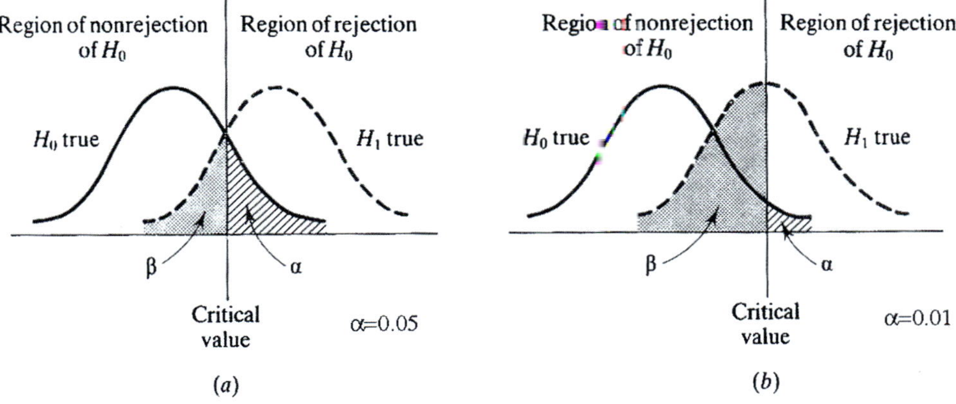


**Fig. 8.5** Regions of rejection and non-rejection of $H_0$ for $\alpha = .05$ and $\alpha = 0.01$ (Copies of figures including color copies, where applicable, are available in the accompanying CD)

Regardless of the setting of the $\alpha$ level, a test may be made more powerful against any given specific alternative hypothesis by increasing the sample size. Large sample sizes aren't always possible and may be costly. So ideally, we determine the minimum sample size required to obtain the desired power for a certain expected effect size. Careful experimental design and control will result in smaller population variability resulting in greater precision and thus more power.

## 8.6 Sensitivity and Specificity

In our example of using the binomial distribution for hypothesis testing we used gene pattern, 'ABC' and 'abc' as a dichotomous or binary variable and associated it with sperm motility. In practice, gene expression is actually a continuous measure of the gene's activity level in the production of specific proteins. With the use of ROC (Receive Operator Curve e.g., Fig. 8.6) we can determine the best cut-point value for categorizing gene expression as 'positive' or 'negative'.

Values of gene expression that are above the cut-point, 'positives', are associated with the presence of disease (e.g., male infertility) and those values below the cut-point, 'negatives', with the absence of the disease. From this we can construct a 2×2 table of the 'Predicted Condition' of positive or negative based on the cut-point determined by the ROC method, and the 'True Condition' of the presence or absence of disease in the observed sample:

*Sensitivity* is the proportion of cases with the disease that are predicted positive on a test (true positives).

$$\text{Sensitivity} = \frac{\text{cases with true positive results}}{\text{total number of cases with disease}}$$

*Specificity* is the proportion of cases without the disease that are predicted negative on a test (true negatives).

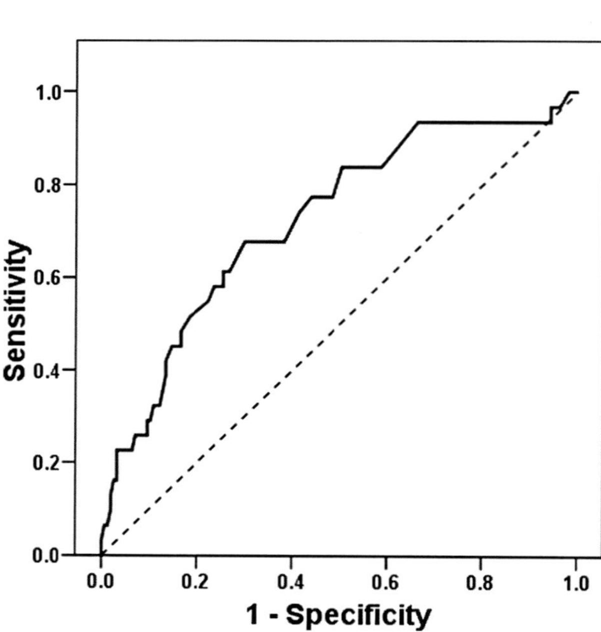


**Fig. 8.6** ROC Curve in which the diagonal dotted line represents chance prediction. Curve represents corresponding sensitivity (*y-axis*) and 1-specificity (*x-axis*) for various cut-points in the observed dependent measure (Copies of figures including color copies, where applicable, are available in the accompanying CD)

**Table 8.5** Sensitivity and specificity table disease present

| Test Result | Yes | No | Total |
|---|---|---|---|
| Positive | True Positive (TP) | False positive (FP) | Subjects positive on the test (TP + FP) |
| Negative | False negative (FN) | True negative (TN) | Subjects negative on the test (FN + TN) |
| Total | Subjects with disease | Subjects without disease | All subjects |

(Copies of tables are available in the accompanying CD.)

$$\text{Specificity} = \frac{\text{cases with true negative results}}{\text{total number of cases without disease}}$$

Using the coordinates obtained from the ROC curve we can obtain measures of accuracy of our classification using the gene expression value. The ROC curve provides measures of sensitivity on the Y-axis and 1-specificity on the X-axis. From these values we can calculate the number of cases in each cell of the sensitivity-specificity table seen in Table 8.5. Other commonly used measures of diagnostic accuracy that can be obtained from this are the *positive predictive and negative predictive values* of the test. The positive predictive value of a test is the probability that the patient has the disease when restricted to those patients who test positive. This term is sometimes abbreviated as PPV. You can compute the positive predictive value as:

PPV = True Positives (TP) / Number of subjects who test positive on the test (TP + FP).

The negative predictive value of a test is the probability that the patient will not have the disease when restricted to all patients who test negative. This term is sometimes abbreviated as NPV. You can compute the negative predictive value as

NPV = True Negatives (TN) / Number of subjects who test negative on the test (TN + FN).

Note that all four of these diagnostic measures sensitivity, specificity, PPV and NPV, are examples of conditional probabilities indicating a patient's status given another condition is met. Sensitivity is the conditional probability of a positive test when the disease or condition is present, $P(T^+ / D^+)$, while the positive predictive value is the conditional probability of having the disease when the test is positive, $P(D^+ / T^+)$. These values can be utilized to determine the usefulness of particular genes or sets of genes in predicting disease. They also provide an objective method of measuring the effectiveness of various treatments as well as the cost-benefit tradeoff of both missed diagnoses (false negatives) and incorrect diagnoses of the disease (false positives).

## 8.7 Concluding Remarks

In this chapter we have tried to present the basic concepts of probability and its' uses in testing the hypotheses about the association of gene pattern and expression with disease. An introduction to hypotheses testing is provided to guide the reader in examining gene association and disease. The concept of statistical power provides a way of thinking about how to best ask questions and design research for revealing these relationships between genetics and human characteristics.

The diagnostic tools of ROC and the indices of sensitivity, specificity, and positive and negative predictive value have been widely used to demonstrate the utility of a variety of screening tools and are likely to be useful in demonstrating the extent of associations between genetic patterns and disease conditions and other human characteristics.

## Suggested Reading

1. Matthews, D.E. and Farewell, V.T. (2007) *Using and Understanding Medical Statistics*, Karger, Basel.
2. Andy Field (2005) *Discovering Statistics Using SPSS* (2nd ed.), Sage, London.

# Chapter 9
# Stochastic Models for Biological Patterns

**Gautam B. Singh**

**Abstract** This chapter reviews the prevalent mathematical formalisms for modeling biological sequences and patterns. Underlying theoretical principles for computationally simple IID models is followed by a discussion on the Markov models for sequences. Similarly, discussions on pattern models focused on the theoretical background of Position Specific Scoring Matrices and Profiles for representation of biological sequence patterns. A detailed discussion and illustration of modeling patterns using Hidden Markov Models is presented.

**Keywords** HMM · Hidden markov model · PSSM · Position specific scoring matrix · Patterns

## 9.1 Introduction

Mathematical formulations for representing bio-sequence patterns are of great significance for detection of patterns and functional motifs in biological sequences. Patterns at various levels of abstractions are the drivers of genomics and proteomics research. Starting at the fine level of granularity, the patterns are comprised of the splice sites, binding sites, and domains. These are subsequently utilized for the definition of patterns at a higher level of abstraction such as introns, exons, repetitive DNA, and locus control regions.

DNA sequences are often modeled as probabilistic phenomena, with the patterns of interest being defined as samples drawn from the underlying random process. For example, the underlying DNA sequence is modeled as a Markov chain of random variables taking on the values $\{A, C, T, G\}$. Given this underlying assumption, one may next model a splice site as a function $P$ that assigns a sample sequence, $S$, of 8–10 nucleotides a value equivalent to their probability of being a splice site.

A pattern detection algorithm would consider each substring from a DNA sequence that could potentially be a splice site, and assign a probability to each candidate. The substrings scoring a high value may be further investigated using the appropriate wet-bench approaches. While this is a simple illustration, it does bring forth an important point. The underlying models that define the DNA sequences and their accuracy are ultimately a determinant of the accuracy that the patterns are subsequently detected. The modeling of the sequences and that of patterns are two complementary objectives. Sequence models provide a basis for establishing the significance of the patterns observed, while the pattern models help us look for specific motifs that are of functional significance. We therefore must consider both of these issues

G.B. Singh
Department of Computer Science and Engineering, Advanced Software and Information Engineering Laboratory, Oakland University, Rochester, MI 48309, USA
e-mail: singh@oakland.edu

S. Krawetz (ed.), *Bioinformatics for Systems Biology*, DOI 10.1007/978-1-59745-440-7_9,

## 9.2 Sequence Models

The two main sequence models are the Independent Identically Distribution and Markov Chain. Sequence models are needed to represent the background stochastic processes in a manner that enables one to analytically justify the significance of the observation. To provide an analogy, the determination of the sequence model is similar to determining the probability of obtaining a *head* (H) while tossing a coin. (For a fair coin, this probability would be 1/2). In general, however, we may estimate this probability by studying the strings of the heads and tail sequences that a given coin has produced in the past. Similarly, given the DNA sequence(s), we may induce the underlying model that represents the maximally likely automaton that produced the sequence.

Let us continue our analogy further. After the coin model has been induced, it would be possible to predict the probability of observing coin tossing pattern such as "three heads in a row", etc. Similarly, after inducing a DNA sequence model, it would be possible to deduce the expected frequency of occurrence of a DNA sequence pattern. This is helpful in classifying the patterns in terms of their relative abundance in a sequence specific manner.

### *9.2.1 Independent Identically Distribution (IID)*

The simplest of all the sequence models is the Independent Identically Distribution or IID model. In this model, each of the four nucleotides is considered to occur independent of each other. Furthermore, the probability of occurrence of a given nucleotide at a given location is identical to the probability of its occurrence at another location. Thus, for example, assume that the sequence is defined using an IID random variable which can take on the possible values defined by the DNA alphabet $\Sigma = \{A, C, T, G\}$. In this case, defining the individual probability values ($p_A$, $p_C$, $p_T$, and $p_G$) specifies the complete model for the sequence. The values may in turn be computed simply by considering the prevalence of each base in the given sequence. In statistical terminology, the maximally likely or ML estimator for probability of occurrence of a given base, $X$, is simply $\frac{n_X}{L}$ where $n_x$ is the frequency of occurrence of the base $X$ in a sequence of length $L$.

In general, the maximal likely estimator for the parameters may be used. Using the ML estimation, the probability of each base $\alpha$ may be estimated as:

$$\hat{P}(\alpha) = \frac{n_\alpha(L)}{|L|} \tag{9.1}$$

This simply counts the relative frequency of the nucleotide $\alpha$ in a sequence of length $L$. This estimator has the advantage of simplicity, and usually works well when $|L|$ is large. It may not work well when $|L|$ is small.

Given the Model $M_{IID}$ has been induced from the sequence data, the probability of the occurrence of a pattern $x$ may be computed using the following:

$$P(x|M_{IID}) = \prod_{i=1,\ldots,n(x)} P(x_i) \tag{9.2}$$

where $P(x_i)$ is the probability of nucleotide $x_i$, at position $i$, along the pattern. The model assumes that the parameters (probability of each of the four nucleotides) are independent of the position along the pattern.

*Example 1:* Consider the following DNA sequence of length = 25:

SEQ = AACGT CTCTA TCATG CCAGG ATCTG

In this case the IID model derived from the sequence given the alphabet $\Sigma = \{A, C, T, G\}$, the sequence model parameters are $\{\frac{6}{25}, \frac{7}{25}, \frac{7}{25}, \frac{5}{25}\}$, corresponding to the maximally likely estimation of the occurrence of each of the four bases. These are thus the IID parameters for the background sequence. The probability of finding the pattern CAAT on this sequence would be equal to $p_C{\cdot}p_A{\cdot}p_A{\cdot}p_T$ or $(\frac{7}{25})\cdot(\frac{6}{25})\cdot(\frac{6}{25})\cdot(\frac{7}{25}) = 0.0045$.

## 9.3 Markov Chain Models

In a Markov chain the value taken by a random variable is dependent upon the value(s) taken by the random variable in a previous state(s). The number of historical states that influence the value of the random variable at a given location along the sequence is also known as the degree of the Markov process. The first-degree Markov chain model has $|\Sigma| + |\Sigma|^2$ parameters, corresponding to the individual nucleotide frequencies as well as dinucleotide frequencies. In this manner, this model permits a position to be dependent on the previous position. However, the frequencies are modeled in a position-invariant manner, and thus may be unsuitable for modeling signals.

This sequence model $M$ is defined on the sample space $\Sigma^*$ and assigns to every sequence $x$ of length $n(x)$ on $\Sigma^*$ a probability:

$$P(x|M) = P_1(x_1) \prod_{i=2,\ldots,n(x)} P_2(x_i|x_{i-1}) \tag{9.3}$$

where $P_1$ is a probability function on $\Sigma$ that models the distribution of $\alpha$'s at the first position in the sequence and $P_2$ is the conditional probability function on $\Sigma \times \Sigma$ that models the distribution of $\beta$'s at position $i > 1$ on the alphabet symbol $\alpha$ at position $i$-$1$.

The parameter estimation using the Maximally Likely estimator proceeds in a manner analogous to the IID model estimation. The transition probabilities are, however, estimated using Bayes theorem. Using Laplace's rule all the frequency counts are incremented by one to eliminate the effect of zeros in the frequency table.

$$P_2(\beta|\alpha) = \frac{P(\alpha\ \beta)}{P(\alpha)} = \frac{freq(\alpha\ \beta) - 1}{\sum_{\beta}(freq(\alpha\ \beta) + 1)} \tag{9.4}$$

In this manner, the conditional transitional probabilities of finding a base $\beta$ at position $(i)$ given that the base $\alpha$ was found at position $(i$–$1)$ is computed by finding the abundance of the dinucleotide $\alpha\beta$ as a fraction of the abundance of the nucleotide $\alpha$.

*Example 2:* Consider once again the same 25-nucleotide sequence as above:

SEQ = AACGT CTCTA TCATG CCAGG ATCTG

While considering the first-degree Markov chain models, the 4-parameters corresponding to individual nucleotide frequencies, and the $4^2$ parameters corresponding to the dinucleotide frequencies need to be computed. The $\Sigma$ parameters are the same as those computed before: $\{p_A, p_C, p_T, p_G\} = \{6/25, 7/25, 7/25, 5/25\}$.

In order to compute $P_2$, the $\Sigma \times \Sigma$ conditional probability values, the dinucleotide frequencies and probabilities are computed from the sequence data. The dinucleotide frequencies and the probabilities are shown below (*with the parenthesized numbers representing the probabilities*):

| | | | |
|---|---|---|---|
| *freq* (AA) = 1 | *freq* (AC) = 1 | *freq* (AT) = 3 | *freq* (AG) = 1 |
| *freq* (CA) = 2 | *freq* (CC) = 1 | *freq* (CT) = 3 | *freq* (CG) = 1 |
| *freq* (TA) = 1 | *freq* (TC) = 4 | *freq* (TT) = 0 | *freq* (TG) = 2 |
| *freq* (GA) = 1 | *freq* (GC) = 1 | *freq* (GT) = 1 | *freq* (GG) = 1 |

**Table 9.1** Conditional nucleotide probabilities for the 25-nt example sequence

| $\downarrow S_{i-1}S_i \rightarrow$ | A | C | T | G |
|---|---|---|---|---|
| A | 2/10 | 2/10 | 4/10 | 2/10 |
| C | 3/11 | 2/11 | 4/11 | 2/11 |
| T | 2/11 | 5/11 | 1/11 | 3/11 |
| G | 2/8 | 2/8 | 2/8 | 2/8 |

(Copies of tables are available in the accompanying CD.)

The condition probabilities are next computed using the Bayes theorem equation 9.4. For example, the probability of finding "C" at position $(i+1)$ given that an "A" has been found at position (i) is $P(C|A) = \frac{p_{AC}}{p_A} = \frac{freq(AC)+1}{\sum\limits_{X\in\{A,C,T,G\}} (freq(AX)+1)} = \frac{2}{10}$. The conditional probabilities for the example sequence are shown in Table 9.1.

Using these model parameters, the probability of finding the pattern CAAT on this sequence using the first order Markov model of the underlying sequence would be equal to $P(C).P(A|C).P(A|A).P(T|A)$ or $(\frac{7}{25}) \cdot (\frac{3}{11}) \cdot (\frac{2}{10}) \cdot (\frac{4}{10}) = 0.0061$. This probability is higher than the ID sequence model probability of the same pattern computer earlier as 0.0045.

### *9.3.1 Higher Order Markov Models*

Higher order Markov chains have been described. For example, the *nth* order Markov process has a memory of $n$, and thus the occurrence of a nucleotide depends on the previous $n$ nucleotides. The probability of observing a sequence $x$ is defined in equation 9.5, in a manner similar to the first order Markov chains.

$$P(x|M) = P_1(x_1) \prod_{i=2,\dots,n(x)} P_2(x_i|x_{i-1}, \dots, x_{i-n}) \tag{9.5}$$

An *nth* order Markov chain over some alphabet $A$ is equivalent to a first order Markov chain over the alphabet $A^n$ of $n$-tuples. This follows from calculating the probability of $A$ and $B$, given $B$ is the probability of $A$ given $B$, i.e., $P(x_k|x_{k-1}, \dots, x_{k-n}) = P(x_k, x_{k-1} \dots x_{k-n+1}|x_{k-1} \dots, x_{k-n})$. In other words, the probability of $x_k$ given in the $n$-tuple ending in $x_{k-1}$ is equal to the probability of the $n$-tuple ending in $x_k$ given the $n$-tuple ending in $x_{k-1}$.

## 9.4 Pattern Models

This section describes the statistical modeling procedures for DNA patterns. Thus, our attention is now focused somewhat more on modeling the motifs that are associated with certain biological functions. In the previous section our goal was to characterize the "sea" of data in which these biological "nuggets" of information are hidden. In contrast, our goal in this section is to model the nuggets themselves.

There are a growing number of well-established patterns that we may wish to model and search for in DNA sequences. Often these patterns, which are of functional significance, are brought forth after an alignment of the sequences belonging to a particular family. Such a multiple sequence alignment is often interspersed with gaps of varying sizes. However, there are sections in the final alignment that are free of gaps in all of the sequences. This fixed size, ungapped aligned regions represent the patterns we would like to model for detection in DNA sequences. The statistical

**Table 9.2** Weight Matrix for TATAA box

| | | | | | | | | |
|---|---|---|---|---|---|---|---|---|
| T | 6 | 49 | 1 | 56 | 6 | 22 | 6 | 20 |
| C | 14 | 6 | 0 | 0 | 3 | 0 | 1 | 2 |
| A | 8 | 4 | 58 | 4 | 51 | 38 | 53 | 30 |
| G | 32 | 1 | 1 | 0 | 0 | 0 | 0 | 8 |

(Copies of tables are available in the accompanying CD.)

techniques that may be employed for developing such a closed form representation of a set of patterns are described below.

### 9.4.1 *Weight Matrices*

A DNA sequence matrix is a set of fixed-length ,DNA sequence segments aligned with respect to an experimentally determined biologically significant site. The columns of a DNA sequence matrix are numbered with respect to the biological site, usually starting with a negative number. A DNA sequence motif can be defined as a matrix of depth 4, utilizing a *cut-off* value. The 4–column/mononucleotide matrix description of a genetic signal is based on the assumptions that the motif is of fixed length, and that each nucleotide is independently recognized by a *trans*-acting mechanism. For example, the frequency matrix in Table 9.2 has been reported for the TATAA box:

If a set of aligned signal sequences, of length "L", corresponding to the functional signal under consideration, then $F = [f_{bj}], (b \in \Sigma), (j = 1..L)$ is the nucleotide frequency matrix, where $f_{bi}$ is the absolute frequency of occurrence of the *b-th* type of the nucleotide out of the set $\Sigma = \{A, C, G, T\}$ at the *j-th* position along the functional site.

The frequency matrix may be utilized for developing an ungapped score model when searching for the sites in a sequence. Typically a log-odds scoring scheme is utilized for this purpose of searching for pattern $x$ of length $L$ as shown in equation 9.6. The quantity $e_i(b)$ that specifies the probability of observing the base $b$ at position $i$, is defined using the frequency matrix such as the one shown above. The quantity $q(b)$ represents the background probability for the base $b$.

$$S = \sum_{i=1}^{L} \log \frac{e_i(x_i)}{q(x_i)} \tag{9.6}$$

The elements of $\log \frac{e_i(x_i)}{q(x_i)}$ behave like a scoring matrix similar to the PAM and BLOSUM matrices. The term Position Specific Scoring Matrix (PSSM) is often used to define the pattern search with the matrix. A PSSM can be used to search for a match in a longer sequence by evaluating a score $S_j$, for each starting point $j$ in the sequence from position 1 to $(N-L+1)$, where $L$ is the length of the PSSM.

A method for converting the frequency matrix into a weight matrix has been proposed by Bucher (1990). The weights at a given position are proportional to the logarithm of the observed base frequencies. These are increased by a small term that prevents the logarithm of zero and minimizes sampling errors. The weight matrix is computed as shown in equation 9.7. The term $e_{bi}$ represents the expected frequency of the base $b$ at position $i$, $c_i$ is a column specific constant, and $s$, a smoothing percentage.

$$W(b,i) = \ln\left(\frac{f_{bi}}{e_{bi}} + \frac{s}{100}\right) + c_i \tag{9.7}$$

These optimized weight matrices can be used to search for functional signals in the nucleotide sequences. Any nucleotide fragment of length $L$ is analyzed and tested for assignment to the proper functional signal. A matching score of $\sum_{i=1}^{L} W(b_i, i)$ is assigned to the nucleotide position being examined along the sequence. In the search formulation, $b_i$ is the base at position $i$ along the oligonuclotide sequence, and $W(b_i, i)$ represents the corresponding weight matrix entry for base $b_i$ occurring along the $i$th position in the motif.

Profiles are similarly defined for modeling functional motifs in amino-acid sequences. A profile is a scoring matrix $M(p, a)$ comprised of 21 columns and $N$ rows, where $N$ is the length of the motif. The first 20 scores represent the weight for each individual amino acid, and the 21st column specifies the cost associated with an insertion or deletion at that position. The value of the profile for amino acid "$a$" defined for position $p$ is $M(a,p) = \sum_{b=1}^{20} W(b,p) \times Y(b,a)$, where $Y(b, a)$ is Dayhoff's matrix and $W(b, p)$ is the weight for the appearance of amino acid $b$ at position $p$. The position specific weight is defined by $\log[f(b,p)/N]$, or as the frequency of occurrence of the amino acid as $b$, as a fraction of the total $N$ sequences utilized for construction of the profile, with a frequency of 1 being used for any amino acid that does not appear at position $p$.

### 9.4.2 Position Dependent Markov Models

The Markov models have been considered as a means to define the background DNA sequence. This model enabled us to define the probability of a nucleotide conditioned upon the nucleotides occurring in the previous position. However, the modeled dependency is position-invariant. A position-dependent Markov model may be utilized for the representation of a sequence signal or motif. This model is defined on the sample space $\Sigma^n$ and assigns to every sequence $x$ on $\Sigma^n$, a probability:

$$P(x|M) = P_1(x_1) \prod_{i=2,\ldots,n} P_{2,i}(x_i|x_{i-1}) \tag{9.8}$$

This model has $|\Sigma| + (n-1) * |\Sigma|^2$ parameters. This model permits position-specific dependencies on the previous position by allowing a unique set of transition probabilities to be associated with each position along the signal. This model assumes that sufficient training data is available to induce position specific Markov probabilities.

### 9.4.3 Hidden Markov Models

There are several extensions to the classical Markov chains, and the hidden Markov models (HMM) are one such extension. The rationale for building a hidden Markov model comes from the observation that as we search a sequence, our observations could arise from a model characterizing a pattern, or from a model that characterizes the background. Hidden Markov DNA sequence models are developed to characterize a model as an *island* within the *sea* of non-island DNA, And the Markov chain characterizing both of these, needs to be present within the same model, with the ability to switch from one chain to the other. In this manner, a HMM utilizes a set of hidden states with an emission of the symbols associated with each state.

From a symbol generation perspective, the state sequence executed by the model is not observed. Thus, the state sequence must be estimated from the observed symbols generated by the model. From a mathematical perspective, the HMM is characterized by the following parameters:

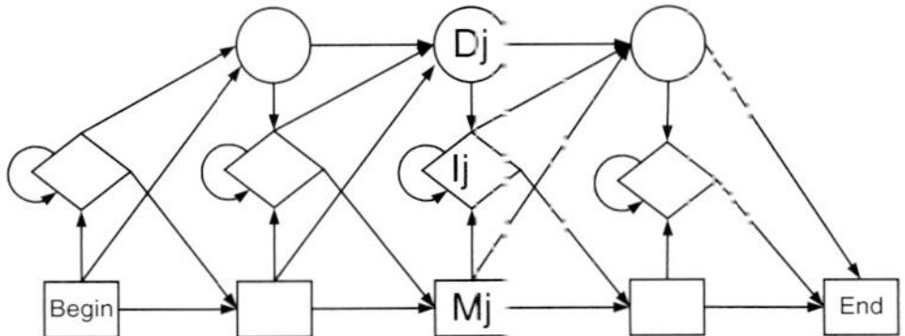


**Fig. 9.1** A Profile HMM utilizes the insert (*diamond*) and delete (*circle*) states. The delete states are *silent* and are not associated with the emissions of any symbols. The parameters of a profile HMM are learned from a multiple sequence alignment (Copies of figures including color copies, where applicable, are available in the accompanying CD)

- $\Sigma$ is an alphabet of symbols.
- $Q$ is a set of states that emit symbols from the alphabet $\Sigma$.
- $A = (a_{kl})$ is $|Q| \times |Q|$ matrix of state transition probabilities.
- $E = (e_k[b])$ is a $|Q| \times |\Sigma|$ matrix of emission probabilities.

Although a general topology of a fully connected HMM allows the state transitions from any state to any other, this structure is almost never used. This is primarily due to the inadequacy of the available data for training a model with the large number of parameters needed for a fully connected HMM developed for any practical problem. Often, the over-generalized model produces sub-optimal results due to the lack of training data. Consequently, more restrictive HMMs that rely on the problem characteristics to suitably reduce the model complexity and the number of model parameters that are needed, are utilized. One such model is defined to be the *profile-HMM*, which is induced from a multiple sequence alignment. The structure of a profile HMM is shown in Fig. 9.1.

The parameters of a profile HMM are estimated using the sample alignments of the sequences used for training. The transitions and the emissions of the symbols in the alignment are used to derive the Maximally Likely (ML) estimator of the HMM parameters. These values are assigned as shown in equation 9.9. The actual transition and emission frequencies $A_{kl}$ and $E_k(a)$ respectively, are used to define the transition and emission probabilities, $a_{kl}$ and $e_k(a)$. Furthermore, pseudo-counts are added to the observed frequencies to avoid zero probabilities. The simplest pseudo-count method is the Laplace's rule, which requires adding one to each frequency.

$$a_{kl} = \frac{A_{kl}}{\sum_X A_{kX}} \quad \text{and} \quad e_k(a) = \frac{E_k(a)}{\sum_Y E_k(Y)} \tag{9.9}$$

*Example 3:* Consider the following multiple sequence alignment defined on the amino acid residues for five globin sequences:

```
HBA_HUMAN       ...VGA--HAGEY ...
HBB_HUMAN       ...V----NVDEV  ...
GLB3_CHITP      ...VKG------D  ...
LGB2_LUPLU      ... FNA--NIPKH ...
GLB1_GLYDI      ... IAGADNGAGV ...
Match States    *** *****
```

A HMM with 8 match states may be constructed based on this alignment. The residues AD in GLB1_GLYD1 are treated as insertions, with respect to the consensus. In match state 1, the emission probabilities are (using Laplace's rule):

$$e_{M1}(V) = \frac{4}{25},\ e_{M1}(F) = \frac{2}{25},\ e_{M1}(I) = \frac{2}{25},\quad \text{and}\ e_{M1}(a) = \frac{1}{25}\quad \textit{for all others}$$

The transition probabilities from match state 1 are as follows:
$a_{M1,M2} = \frac{5}{8}$, $a_{M1,D2} = \frac{2}{8}$, $a_{M1,I1} = \frac{1}{8}$, corresponding to the one deletion in HBB_HUMAN, and no insertions. The emission probabilities for state $I_1$ are all equal to (1/20). The Viterbi algorithm yields the optimal path through the HMM, as well as the log-odds score for observing a given sequence from a HMM. It is commonly used to match a profile HMM to a sequence. The Viterbi algorithm is a recursively defined optimization procedure that is quite similar to the dynamic programming algorithm used for sequence alignment.

The various scores for matching a sequence to the profile HMM are defined in equation 9.10. In this formulation, $V_j^M(i)$ represents the log-odds score of the best path matching subsequence $x_{1\ldots i}$ to the submodel up to state j, ending with $x_i$ being emitted by state $M_j$. Similarly $V_j^I(i)$ is the score of the best path ending in $x_i$ being emitted by $I_j$, and $V_j^D(i)$ for the best path ending in state $D_j$.

$$V_j^M(i) = \log\frac{e_{M_j}(x_i)}{q_{x_i}} + \max\begin{cases} V_{j-1}^M(i-1) + \log a_{M_{j-1}M_j} \\ V_{j-1}^I(i-1) + \log a_{I_{j-1}M_j} \\ V_{j-1}^D(i-1) + \log a_{D_{j-1}M_j} \end{cases}$$

$$V_j^I(i) = \log\frac{e_{I_j}(x_i)}{q_{x_i}} + \max\begin{cases} V_j^M(i-1) + \log a_{M_jI_j} \\ V_j^I(i-1) + \log a_{I_jI_j} \\ V_j^D(i-1) + \log a_{D_jI_j} \end{cases} \tag{9.10}$$

$$V_j^D(i) = \max\begin{cases} V_{j-1}^M(i) + \log a_{M_jD_j} \\ V_{j-1}^I(i) + \log a_{I_{j-1}D_j} \\ V_{j-1}^D(i) + \log a_{D_{j-1}D_j} \end{cases}$$

Generally, there is no emission score $e_{I_j}(x_i)$ in the equation for $V_j^I(i)$, as it is often assumed that the emission distribution from the insert states, $I_j$, is the same as the background distribution. Also, the D→I and I→D transition terms may not be present, and those transition probabilities may be very close to zero.

A detailed example showing the procedure for building a HMM from a multiple sequence alignment and using the HMM to search a sequence is presented next. DNA sequence models are chosen for simplicity.

*Example 4:* Consider the following multiple sequence alignment that needs to be formalized as an HMM.

The alignment of six sequences shows that there are three states where there is an agreement of 3 or more sequences. These are labeled as M1, M2, and M3, and represent the core match states of the model. States M0 and M4 are dummy match states representing the beginning and end of the model Table 9.3.

The multiple sequence alignment corresponds to the HMM topology shown below Fig. 9.2.

| *seq1* | G | A | C | C | A | ϕ | |
|---|---|---|---|---|---|---|---|
| *seq2* | A | G | ϕ | ϕ | ϕ | C | |
| *seq3* | A | G | A | G | ϕ | C | |
| *seq4* | ϕ | ϕ | A | A | A | C | |
| *seq5* | A | G | ϕ | ϕ | ϕ | C | |
| **Consensus** | * | * | | | | * | |
| **M0** | **M1** | **M2** | | | | **M3** | **M4** |
| **Begin** | | | | | | | **End** |

(Copies of tables are available in the accompanying CD.)

**Table 9.3** Simple multiple sequence alignment

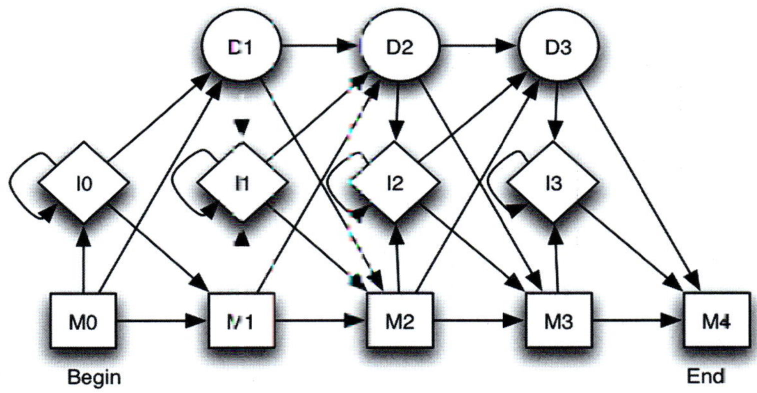


**Fig. 9.2** HMM topology of MSA from Table 9.3 (Copies of figures including color copies, where applicable, are available in the accompanying CD)

**Table 9.4** Frequency of math emissions, insert emisson and frequenceis in MSA example in Table 9.3

**Frequencies**

| Match Emission Frequencies | | M0 | M1 | M2 | M3 | M4 |
|---|---|---|---|---|---|---|
| | A | | 3 | 1 | 0 | |
| | C | | 0 | 0 | 4 | |
| | G | | 1 | 3 | 0 | |
| | T | | 0 | 0 | 0 | |

| Insert Emission Frequencies | | I0 | I1 | I2 | I3 |
|---|---|---|---|---|---|
| | A | 0 | 0 | 5 | 0 |
| | C | 0 | 0 | 2 | 0 |
| | G | 0 | 0 | 1 | 0 |
| | T | 0 | 0 | 0 | 0 |

| Transition Frequencies | | 0→1 | 1→2 | 2→3 | 3→4 |
|---|---|---|---|---|---|
| | MM | 4 | 4 | 2 | 4 |
| | MD | 1 | 0 | 0 | |
| | IM | 0 | 0 | 2 | 0 |
| | ID | 0 | 0 | 1 | |
| | DM | | 0 | 0 | 1 |
| | DD | | 1 | 0 | |

| | 0→0 | 1→1 | 2→2 | 3→3 |
|---|---|---|---|---|
| MI | 0 | 0 | 2 | 0 |
| DI | | 0 | 1 | 0 |
| II | 0 | 0 | 5 | 0 |

(Copies of tables are available in the accompanying CD.)

Using the topology of the HMM, the number of transitions from each state to the next, generated by the training sequences is counted as shown in Table 9.4.

Consider, for example, the alignment for *seq 1*. The given sequence GACCA goes through the states: M0 M1 M2 I2 I2 I2 D3 M4. While in M1, it emits symbol G, in M2 it emits symbol A, then it emits symbols C, another C and A in state I2. From I2, it moves to state D3 and then to M4. Corresponding transition frequencies in the above table are also incremented. Specifically, transition frequency counts corresponding to the following, M0→M1, M1→M2, M2→I2, I2→I2, I2→I2, I2→D3, D3→M4, are all incremented by 1.

These frequency counts are next converted to probabilities using Laplace rule where all the frequency terms are incremented by one to avoid zero values from occurring in the probability table. Further, the probability values are converted into likelihood by assuming that

**Table 9.5** Probability values for frequencies observed in Table 9.4

**PROBABILITIES (using Laplace Rule)**

| Match Emission Probabilies log2 (e/q)* | | M0 | M1 | M2 | M3 | M4 |
|---|---|---|---|---|---|---|
| | A | | 1 | 0 | -1 | |
| | C | | -1 | -1 | 1.3219 | |
| | G | | 0 | 1 | -1 | |
| | T | | -1 | -1 | -1 | |

| Insert Emission Probabilites log2(e/q)* | | I0 | I1 | I2 | I3 |
|---|---|---|---|---|---|
| | A | 0 | 0 | 1 | 0 |
| | C | 0 | 0 | 0 | 0 |
| | G | 0 | 0 | -0.585 | 0 |
| | T | 0 | 0 | -1.585 | 0 |

| Transition Probabilites log2 (Aij) | | 0→1 | 1→2 | 2→3 | 3→4 |
|---|---|---|---|---|---|
| | MM | -0.68 | -0.49 | -1.222 | -0.263 |
| | MD | -2 | -2.81 | -2.807 | |
| | IM | -1.58 | -1.58 | -1.874 | -1 |
| | ID | -1.58 | -1.58 | -2.459 | |
| | DM | | -2 | -2 | -0.585 |
| | DD | | -1 | -2 | |

| | | 0→0 | 1→1 | 2→2 | 3→3 |
|---|---|---|---|---|---|
| | MI | -3 | -2.81 | -1.222 | -2.585 |
| | DI | | -2 | -1 | -1.585 |
| | II | -1.58 | -1.58 | -0.874 | -1 |

(Copies of tables are available in the accompanying CD.)

$q(A) = q(C) = q(T) = q(G) = \frac{1}{4}$. That is, all the bases are equally probable under the background model. The probability values listed below are thus obtained from Table 9.5

Searching for the pattern represented by the HMM in a sequence is illustrated in the table below. In this case, a sequence AGGAC is matched to the HMM. Viterbi equations are utilized to complete the cells. The maximally likely state sequence is M0 M1 M2 I2 I2 M3 M4 (Table 9.6).

**Table 9.6** Patterns presents in MSA example

| | | A | G | G | A | C | |
|---|---|---|---|---|---|---|---|
| | | | | | | | |
| **M0** | **0.00** | -100.00 | -100.00 | -100.00 | -100.00 | -100.00 | |
| **M1** | -100.00 | **0.32** | -4.58 | -6.17 | -6.75 | -10.34 | |
| **M2** | -100.00 | -100.00 | **0.84** | -3.07 | -5.66 | -8.24 | |
| **M3** | -100.00 | -100.00 | -0.68 | -1.39 | -3.85 | **-1.40** | |
| **M4** | -100.00 | -100.00 | -100.00 | -100.00 | -100.00 | -100.00 | **-1.66** |
| | | | | | | | |
| **I0** | -100.00 | -3.00 | -4.58 | -6.17 | -7.75 | -9.34 | |
| **I1** | -100.00 | -4.58 | -2.49 | -4.07 | -5.66 | -7.24 | |
| **I2** | -100.00 | -100.00 | -0.26 | **-0.97** | **-0.85** | -1.72 | |
| **I3** | -100.00 | -100.00 | 0.32 | 0.32 | -3.97 | -4.97 | |
| | | | | | | | |
| **D1** | -100.00 | -4.58 | -6.17 | -7.75 | -9.34 | -10.92 | |
| **D2** | -100.00 | 0.32 | -4.07 | -5.66 | -7.24 | -8.83 | |
| **D3** | -100.00 | 0.32 | -1.97 | -3.43 | -3.30 | -4.18 | |

(Copies of tables are available in the accompanying CD.)

### 9.4.4 Mixture Models

Mixture models are defined in relation to the sample space $\Sigma^n$. The mixture model $M$, is a mixture of component models $M_i, i = 1, ..., k$, and assigns to every sequence $x$ on $\Sigma^n$, a probability defined in equation (9.11).

$$P(x|M) = \prod_{i=1,...,k} P(x|M_i)P(M_i) \tag{9.11}$$

where $P(M_i)$ is the weight of the component model $M_i$ in the mixture. Any probability model may be used as a component model. Furthermore, the mixture model may have component models of a different type.

A mixture model is best suited for modeling data that is comprised of sub-groups. In this manner, an observed data set may be assigned to a class, if a high probability is assigned to it by at least one component model of sufficient weight. This may be considered to be a stochastic analog of a weighted OR function.

Consider for example a set $D$ of short aligned sequences corresponding to some functional site. Further, assume that we have a reason to believe that the observed sequences are characterized to belong to two categories. The goal, which is to establish if an observed sequence $x$ is similar to set $D$, may be achieved by developing a mixture model for the set $D$. The mixture model is comprised of two constituent sub-models that are represented in the set.

### 9.4.5 Goodness of Fit

As described in the previous sections, there are often several methodologies for developing a model for a pattern. Consequently, the natural question to ask is whether there is a systematic methodology that may be utilized to evaluate which of the possible models is best suited for the data at hand. We may utilize the *goodness of fit* measure described below to estimate how well a given model represents the observed dataset. Generally, all data items, $d$, in the training, or another data set, $D$, are considered to be independent.

Under the independence assumption, the likelihood of the dataset $D$ may be estimated as a product of all the probabilities of the individual observations $d$.

$$P(D|M) = \prod_{d \in D} P(d|M) \tag{9.12}$$

Rewriting equation 9.11 in its log-likelihood form, yields equation 9.13.

$$\log P(D|M) = \prod_{d \in D} \log P(d|M) \tag{9.13}$$

This is often desired to prevent numeric underflows that are likely to occur when small numbers are multiplied. In this manner, one may choose a model instance $M^*$ that best fits the dataset $D$ according to the maximum likelihood principle defined below.

$$M^* = \arg\max_M (P(D|M)) = \arg\max_M (\log P(D|M)) \tag{9.14}$$

## Suggested Reading

HMMER: http://www.genetics.wustl.edu/eddy
SAM: http://www.cse.ucsc.edu/research/compbio/sam.html

Besemer, J. & Borodovsky, M. Heuristic approach to deriving models for gene finding, *Nucleic Acids Research*, 27(19), 3911–3920, 1999.

Bucher, P. Weight matrix descriptions of four eukaryotic RNA polymerase II promoter elements derived from 502 unrelated promoter sequences. *Journal of Molecular Biology*. 212:563–578, 1990.

Dayhoff, M.A., Schwartz, R.M. and Orcutt., B.C. A model for evolutionary change in proteins. In *Atlas of Protein Sequence and Structure,* chapter 5, pp. 345–352. 1978.

Feller, W. *An introduction to probability theory and applications*, Volume II. John Wiley and Sons, 1971.

Gribskov, M. and Veretnik, S. Identification of sequence patterns with profile analysis. *Methods of Enzymology,* 266:198–212, 1996.

Gribskov, M., McLachlan, A., and Eisenberg, D. Profile analysis: Detection of distantly related protein, *Proceedings of the National Academy of Science USA,* 87:4355–4358, 1987.

Haussler, D., Krogh, A., Mian, I.S. and Sjolander, K. Protein modeling using hidden Markov models: analysis of globins. *Proceedings of 26th. Annual Hawaii International Conference on System Sciences,* volume 1, 792–802. IEEE Computer Society Press.

Hayes, W.S. Hayes and M. Borodovsky, M. How to interpret an anonymous bacterial genome: machine learning approach to gene identification. *Genome Research*, 8(11), 1154–1171, 1998.

Henikoff, S. and J. G. Henikoff, J.G. Amino acid substitution matrices from protein blocks. *Proceedings of the National Academy of Science USA,* 89:10915–10919, 1992.

Krogh, A. An introduction to hidden Markov models for biological sequences. In Salzberg, S., Searls, D. and Kasif, S., eds. *Computational Biology: Pattern analysis and machine learning methods*. Elsevier Press.

Ohler U., S. Harbeck, S., H. Niemann, H., E. Noth, E. and M. G. Reese, M.G.. Interpolated Markov chains for eukaryotic promoter recognition. *Bioinformatics*, 15(5), 362–369, 1999.

Smith, T. and Waterman, M. Identification of common molecular subsequences. *Journal of Molecular Biology*. 147:195–197, 1981.

Staden, R. Methods for calculating the probabilities of finding patterns in sequences. *Comput. Applic. Biosci.*, 5(2):89–96, 1988.

Staden, R. Methods to define and locate patterns of motifs in sequences. *Comput. Applic. Biosci.*, 4:53–60, 1988.

# Chapter 10
# Population Genetics

**Jill S. Barnholtz-Sloan and Hemant K. Tiwari**

**Abstract** *Population genetics* is the study of evolutionary genetics at the population level focusing on the exchange of alleles and genes within and between populations as well as the forces that cause or maintain these exchanges. This exchange of genes and alleles causes changes in the specific allele and hence genotype frequencies within and between populations. Studying this evolution helps us to better understand how to use human populations as a data set to clarify genetic predisposition to disease.

Even with all these discoveries in the field of population genetics and in the characteristics that cause population-based changes and their consequences (e.g., how genetics can affect human disease susceptibility), until recently, the advances in molecular biology and genetics had only enabled genotypes to be assessed one at a time by a technician in a laboratory. Now with the advent of the gene chips or microarrays, these methods can be automated and carried out at a much larger scale, i.e., 1 million genotypes per person on a single gene chip. The faster techniques will allow all genes to be tested for polymorphism within and between populations for many individuals in a population at a time and many populations at a time. The new technology will allow an even greater insight into the relationship between evolutionary forces and genetic changes in human populations.

**Keywords** Hardy-Weinberg equilibrium · SNPchip genotype · Disease · Populations

## 10.1 Introduction

A species is comprised of many populations of individuals who breed with each other, each with a unique set of genes (or loci) and alleles. Even so, the population as a whole shares a pool of all genes and alleles. *Evolution* is the change of frequencies of alleles in the total gene pool. In order to predict changes in the incidence of disease, some genotypes, such as those associated with rare and sometimes deadly human diseases must be understood from a population genetics perspective. Most individuals do not carry the genotypes that cause an extreme *phenotype* (or trait), but of the ones that do, their phenotype varies greatly from the average person in the population.

J.S. Barnholtz-Sloan
Case Comprehensive Cancer Center, Case Western Reserve University School of Medicine, Cleveland, OH, 44106, USA
e-mail: jsb42@case.edu

S. Krawetz (ed.), *Bioinformatics for Systems Biology*, DOI 10.1007/978-1-59745-440-7_10,

## 10.2 Hardy-Weinberg and Linkage Equilibrium (HWE and LE)

A basic understanding of equilibrium in populations is needed in order to begin to understand these changes in allele frequencies over time and across generations. The two types of equilibrium assumed in populations are *Hardy-Weinberg Equilibrium* (*HWE*) within loci and *Linkage Equilibrium* (*LE*) between loci.

Fundamental to understanding these equilibria is the understanding that genotype frequencies are determined by mating patterns, with the ideal being *random mating*. *Random mating* assumes that mating takes place by chance with respect to the locus under consideration. With random mating, the chance that an individual mates with another having a specific genotype is equal to the population frequency of that genotype. In human populations, mating seems to be random with respect to some traits, such as blood group type, and non-random with respect to certain physical and cultural characteristics such as height, race/ethnicity, and age.

One's *genotype* at any *locus* is made up of two *alleles* (one of two or more forms that can exist at a locus), one allele from the mother, and one allele from the father. A *gene* (*or locus*) is the fundamental physical and functional unit of heredity and will carry information from one generation to the next (NOTE: gene and locus will be used interchangeably). By definition, a gene encodes for some gene product, like an enzyme or a protein. A gene can have exactly two alleles in a *di-allelic locus* or a large number of alleles in a *multi-allelic locus*. A genotype can either be *homozygous* (both alleles received from the mother and father are the same allele), or *heterozygous* (the alleles received from the mother and the father are different).

Independently in 1908, G.H. Hardy [1] and W. Weinberg [2, 3], both published results showing that in a very large population without overlapping generations, the frequency of particular genotypes in a sexually reproducing diploid population reaches equilibrium after one generation of random mating, assuming there is no selection, mutation, or migration. They then showed that the frequencies of the genotypes in the "equilibrium" population are simple products of the allele frequencies. Suppose that a locus, $L$, has alleles $A$ and $a$, with $p$ = frequency $(A) = f(A)$ and $q = f(a) = 1-p$. Then, if members of the population select their mates at random, without regard to their genotype at the $L$ locus, the frequencies of the three genotypes, $AA$, $Aa$, and $aa$, in the population can be expressed in terms of the allele frequencies: $f(AA) = p^2$; $f(Aa) = f(Aa) + f(aA) = 2pq$; $f(aa) = q^2$ (see Table 10.1).

Statistical tests for the presence or absence of HWE for each of the loci of interest are performed either using a *chi-square goodness of fit test* or an *exact test*, to test the null hypothesis that the locus is in HWE versus the alternative hypothesis that the locus is not in HWE. Both tests examine the fit between the observed and expected numbers. If the observed numbers are similar enough to the expected numbers, then we accept the null hypothesis of the locus being in HWE. Exact tests are used when sample sizes are small and/or when they involve multi-allelic loci.

Exact tests for HWE are much more computationally extensive and complicated than the chi-square goodness of fit test. However, an exact test is more powerful for multi-allelic loci than the goodness of fit chi-square test because it does not depend on large sample asymptotic theory. The exact test for HWE is permutation based and is founded on the theory proposed by Zaykin et al.

**Table 10.1** Observed and expected frequencies of the genotypes $AA$, $Aa$ and $aa$ at Hardy-Weinberg equilibrium after one mating of female and male gametes, for a total sample of n individuals

| | | Female gametes | |
|---|---|---|---|
| | | Observed $= p = f(A)$ | Observed $= q = f(a)$ |
| Male Gametes | Observed $= p = f(A)$ | Observed $= p^2 = f(AA)$<br>Expected $= np^2$ | Observed $= pq = f(Aa)$<br>Expected $= npq$ |
| | Observed $= q = f(a)$ | Observed $= pq = f(Aa)$<br>Expected $= npq$ | Observed $= q^2 = f(aa)$<br>Expected $= nq^2$ |

(Copies of tables are available in the accompanying CD.)

[4] and Guo and Thompson [5]. Zaykin et al. [4] proposed an algorithm that performs an exact test for disequilibrium between alleles within or between loci by calculating the probability of the set of multi-locus genotypes in a sample, conditional on the allelic counts, from the multinomial theory under the null hypothesis of equilibrium being present. Alleles are then permuted and the conditional probability is calculated for the permuted genotype array. In order to permute the arrays, they employ a Monte-Carlo method, as suggested by Guo and Thompson [5]. The proportion of permuted arrays that are no more probable than the original sample provides the significance level for the test. The complexity of this testing procedure makes a using of computer program essential.

In short, HWE refers to the concept that there are no changes in genotypic proportions in a population from generation to generation. This equilibrium will remain constant unless the frequencies of alleles in the population are disrupted. Any selection, migration, non-random (assortative) mating and inbreeding, population substructure or subpopulations, mutation or genetic drift (e.g., [6, 7]) can disrupt the equilibrium.

When we look at a single locus, we find two important random-mating principles: (1) genotype frequencies are simple products of the allele frequencies, and (2) HWE is reached after one generation of random mating. However, when we look at two or more loci, the first principle still applies for each locus independently, whereas the second may not because the alleles of one locus may not be in random association with the alleles of the second locus. The *linkage equilibrium* (*LE*) between loci is defined as a random allelic association between alleles at the loci. In other words, considering any two loci, the probability of the combination of alleles, one from each locus, is the same, if the loci are in the same individual or in different individuals. This state of LE will be reached given enough time, but the approach to equilibrium is slow and highly dependent on the *recombination fraction*, $\theta$. Note that HWE is defined within a single locus and LE is defined between two or more loci.

The genetic distance in centimorgans (cM) between two loci can be estimated from the *recombination fraction*, $\theta$, between these loci, where $\theta$ can be calculated as the probability that the gamete transmitted by an individual is recombinant, i.e., an individual whose genotype was produced by recombination after DNA duplication in which adjacent chromosomes can change parts, or "recombine" (see Fig. 10.1). It is this genetic shuffling that enables species to have a rich diversity of phenotypical expressions from generation to generation in a given population. If the loci are extremely close on the same chromosome, then there will hardly ever be a crossover between them and the recombination fraction will approach zero. By contrast, if the loci are far apart or are located on different chromosomes, then recombination will occur by chance in 50% of meioses, and the alleles at one locus will be inherited at random with respect to the alleles at the other locus. Loci with recombination fractions close to zero are tightly linked to each other whereas loci with recombination fractions of 0.5 are completely unlinked to each other.

Thus, allelic *linkage disequilibrium* (*LD*) is measured by a statistic $D$, which is defined as $D = P_{ij} - p_i q_j$, where $P_{ij}$ is the frequency of the gamete carrying the $i^{th}$ allele of one locus and the

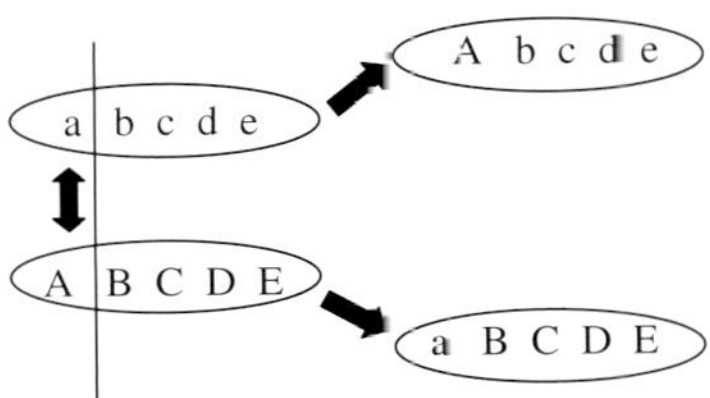


**Fig. 10.1** A pictorial representation of recombination or crossing over between chromosomal segments (or loci) (Copies of figures including color copies, where applicable, are available in the accompanying CD)

$j^{th}$ allele of another and $p_i$ and $q_j$ are the frequencies of the $i^{th}$ and $j^{th}$ alleles of the two loci. $D$ can be a positive or negative number or zero. Another commonly used measure is the standardized LD statistic $D'$, which is $D$ divided by $D_{max}$ therefore making $D'$ bounded by 0 and 1. In random mating populations with no selection, LD is reduced in every generation at a rate of $\theta$ (recombination fraction), $0 \leq \theta \leq 0.5$; leading to $D(t) = (1 - \theta)^t D(0)$, where $D(t)$ is the disequilibrium at generation $t$ and $D(0)$ is the disequilibrium at generation zero.

Tests for significant allelic LD for different combinations of alleles from two and three loci adjacent on a chromosome are performed using a chi-square procedure to test the null hypothesis that $D = 0$, versus the alternative that $D \neq 0$. In order to calculate an allelic disequilibrium measure, allele and genotype/haplotype frequencies must first be calculated which could be done very simply by directly counting the possible genotypes/haplotypes present in the population of interest. This total can then be used to assess allele frequencies. A *haplotype* is the combination of several alleles from multiple loci.

Linkage disequilibrium can result from many different circumstances. LD could have been present in the founding population, and because of a very small $\theta$, this LD would not have had ample time (in generations of random mating) to disappear. Alternatively, the loci could be tightly linked, making recombinants rare and slowing the approach to equilibrium. Interaction between the loci of interest can cause LD when the loci are closely linked. Additionally, the selection of specific heterozygotes can also overcome the natural tendency for $D$ to go to zero. *Population stratification* (*PS*), or the matings of different subpopulations with different allele frequencies, can also cause LD. LD can also be caused purely by chance, in that some loci may present themselves at a higher frequency in a population and stay at that frequency. For further information on estimating and testing of LD refer [8–10].

### *10.2.1 Linkage Disequilibrium Mapping*

The recent completions of the human genome sequence [11] and the HapMap project [12] have renewed interest in the area of LD mapping. Also called allelic association, LD mapping was proposed many years ago [13, 14]. LD mapping is based on the premise that regions adjacent to the mutation in a putative disease gene are transmitted through generations along with that mutation because of the presence of a strong LD. The idea behind LD mapping is to exploit LD between the putative disease locus and single nucleotide polymorphisms (SNPs) which are abundant throughout the genome. This strategy is useful for localizing genes for complex traits which show non-Mendelian patterns of inheritance and are most likely affected by multiple genes acting together and/or by environmental factors. LD mapping uses population-based samples, such as case-control designs, instead of family-based samples and provides a greater statistical power for detection of genes for complex traits. However, there are a few disadvantages to LD mapping; for example (i) when the extent of LD is small in population data, a very dense set of SNPs as well as a large number of cases and controls would be necessary to have a reasonable power to detect the gene of interest (i.e., 10,000 or more SNPs genotyped on at least 1000 cases and 1000 controls); or (ii) when LD is a result of population stratification, additional markers need to be genotyped to adjust for this stratification in order to avoid any false results, false positive or false negative, in the association analysis [15–17].

### *10.2.2 Population Stratification*

Allele frequencies are known to vary widely within and between populations and these differences are widespread throughout the genome [12, 18]. When cases and controls have different allele frequencies attributable to variation in genetic ancestry within or between race/ethnicity groups, population stratification (PS) is said to be present. Essentially ancestry is now a confounding variable. PS is present in recently admixed populations like African- Americans and Latinos [19–21], and also in European-American populations [22–25] and historically isolated populations including Icelanders [26].

A consequence of PS is the potential for bias in the estimate of allelic associations because of deviations from Hardy-Weinberg equilibrium and induction of linkage disequilibrium [27]. In order for the bias due to PS to exist, both of the following must be true: (i) the frequency of the marker variant of interest varies significantly by race/ethnicity, and (ii) the background disease prevalence varies significantly by race/ethnicity [28]. If either of these is not fulfilled, bias due to PS cannot occur. Bias due to PS can induce both false positive and false negative associations. This bias has been shown in some studies to be small in magnitude [28–30], and bounded by the magnitude of the differences in background disease rates across the populations being compared [31]. Simulation studies have also shown that the adverse effects of PS increase with increasing sample size [32, 33].

No true consensus has been reached as to how to test for and/or adjust for population stratification [29, 34], although many methods have been developed [15–17, 35, 36]. Controlling for self reported race has generally been thought to suffice [37], but recent data shows that matching based on ancestry is more robust although in many populations, whether recent admixed or not, individuals are not aware of their precise ancestry [38, 39]. Genomic control [15, 36] and structured association [16, 17, 35] are two techniques commonly used to control and adjust for possible PS in association studies. Genomic control uses a set of non-candidate, unlinked loci to estimate an inflation factor, $\lambda$, which was caused by the PS present and then corrects the standard Chi-square test statistic for this inflation factor. The structured association method utilizes Bayesian techniques to assign individuals to "clusters" or subpopulation classes using information from a set of non-candidate, unlinked loci and then tests for an association within each "cluster" or subpopulation class. A newer, less widely used technique involves the estimation of ancestral proportions through the genotyping of ancestry informative markers (AIMs), which are markers that show large allele frequency differences between ancestral populations and have been found throughout the genome [22, 40–42]. These estimates of either individual or group-specific ancestry can then be used to delineate associations between genetic variants and traits of interest by using genetic ancestry instead of "race/ethnicity" to measure stratification within a study sample [43–47].

## 10.3 Darwin and Natural Selection

The assumptions of equilibrium, both HWE and LE, and hence, allele and genotype frequencies, are all directly affected by the forces of evolution that exist around us, such as natural selection, mutation, genetic drift, inbreeding, and non-random mating. The mechanism of evolution was speculated by a number of people in the early nineteenth century, but it was Charles Darwin [48] who brought this problem to the forefront. He proposed that the cause of evolution was *natural selection* in the presence of variation, which is the process by which the environment limits the population size.

Darwin based this theory on three key observations: (i) When conditions allow individuals in a population to survive and reproduce they will have more offspring than can possibly survive (population size increases exponentially), (ii) lack of resources threatens the individuals' ability to survive and reproduce, and (iii) No two individuals are the same because of variation in inherited characteristics, and, accordingly, all will vary in their ability to reproduce and survive. From these observations he deduced the following: (i) There is competition for survival and successful reproduction, (ii) Heritable differences that are favorable will allow those individuals to survive and reproduce more efficiently than individuals with unfavorable characteristics; i.e., elimination is selective, and (iii) Subsequent generations of a population will have a higher proportion of the favorable alleles than previous generations. With the increase of these favorable alleles in the population, comes an increase of the favorable genotype(s), so that the population gradually changes and becomes better adapted to the environment. This is the definition of *fitness*; genotypes with greater fitness produce more offspring than less fit genotypes. Fitness of a gene or allele is directly related to its ability to be transmitted from one generation to the next.

Because individuals must compete for resources in order to stay alive and reproduce successfully, genotypes determining more favorable characteristics in a population will become more common than those that determine less favorable characteristics. Group interactions, environmental factors, and the relative frequency of particular genotypes can affect a genotype's likelihood of success. Sexual selection is a key component to the likelihood of success can be affected both by direct competition between individuals of the same sex for a mate and by mating success, which is a direct result of the choice of a mate [49].

There are three modes of natural selection (Fig. 10.2): (1) *Stabilizing selection* – this removes individuals who deviate too far from the average and maintains an optimal population, i.e., selection for the average individual, (2) *Directional selection* – this favors individuals who are at one extreme of the population, i.e., selection of either of the extreme individuals, and (3) *Disruptive selection* – this favors individuals at both extremes of the population, which can cause the population to break into two separate populations.

Darwin's theory was transformed into Neo-Darwinism when variation was recognized to be a direct result of spontaneous mutation. The theoretical basis of Neo-Darwinism is based on a mathematical framework (e.g., [*50–52*]) and has become essential to our understanding of molecular evolution. From the 1930s to the 1950s, researchers to tried to better understand the empirical basis of neo-Darwinism, but this was met with great difficulty since a human's lifetime is generally not long enough to be able to observe substantial changes in populations [6].

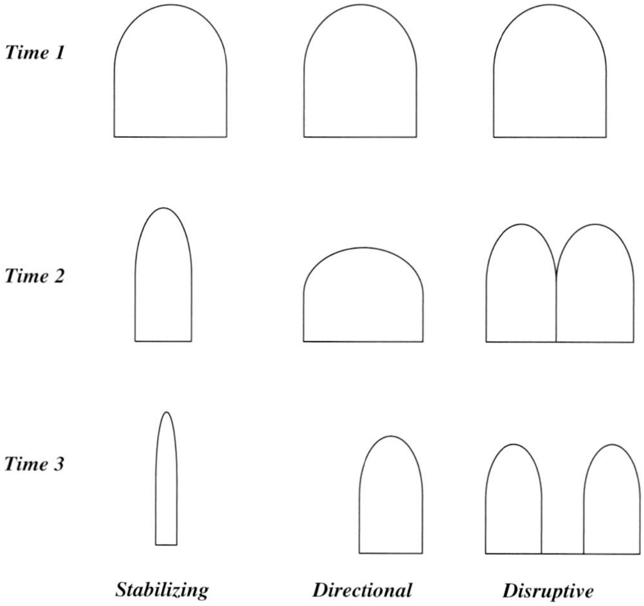

**Fig. 10.2** Three types of natural selection, stabilizing, directional and disruptive, over the course of three different time periods 1, 2 and 3 (three subsequent generations of mating) and their effects on the normally distributed initial population in time period 1 (Copies of figures including color copies, where applicable, are available in the accompanying CD)

**Table 10.2** The effect of the different forces of evolution on variation within and between populations. (Either an increase or decrease of variation within and/or between populations is shown for each force)

| Evolutionary component | Within populations | Between populations |
|---|---|---|
| Inbreeding and non-random mating | Decrease | Increase |
| Genetic drift | Decrease | Increase |
| Mutation | Increase | Decrease |
| Migration | Increase | Decrease |
| Selection: | | |
| - Stabilizing | Increase | Decrease |
| - Directional | Decrease | Increase and Decrease |
| - Disruptive | Decrease | Increase |

(Copies of tables are available in the accompanying CD)

## 10.4 Types of Variation

Darwin's work and the work of the Neo-Darwinists helped us to better understand that the variation within and between populations is caused and maintained by mutation, genetic drift, migration, inbreeding and non-random mating, as well as by the types of natural selection that were discussed in the previous section. Table 10.2 summarizes whether each of the components of evolution increases or decreases variation within and between populations.

In 1953, the Watson-Crick model [53] of DNA (deoxyribonucleic acid) was put forward, opening doors to the application of various molecular techniques in population genetics research. Because DNA is the chemical substance that encodes for all genes, relationships between and within populations could now be characterized through the study of DNA. Now researchers could study the variation within a species instead of having to study the species as a whole. Researchers began by studying amino acid changes, that was accelerated in the mid 1960s, with the advent of electrophoresis (a simpler method of studying protein variation), they switched to studying genetic polymorphisms within populations. Many other technical breakthroughs have emerged such as gene cloning, rapid DNA sequencing and restriction enzyme methods that have uncovered many unexpected properties of the structure and organization of genes.

### *10.4.1 Mutation*

These new methods of molecular study led researchers to discover that all new variations begin with a *mutation* or a change in the sequence of the bases in the DNA. A mutation in the DNA sequence caused by nucleotide substitution (e.g., sickle cell hemoglobin production [54]), by insertions/deletions (e.g., hemophilia A and B, cystic fibrosis [55], *cru-du-chat* syndrome [56]), by triplet expansion (e.g., Huntington's disease [55]), by translocation (e.g., Down syndrome [55]), etc., may be spread through the population by genetic drift and/or natural selection (e.g., [6,57]) and eventually become fixed in a species. If this mutant gene produces a new phenotype, this new characteristic or trait will be inherited by all subsequent generations unless the gene mutates again. Some mutations, called silent mutations, will not affect the protein product and some mutations will occur in non-coding regions, which may or may not have regulatory roles. Clearly, the most interesting variations in terms of effects on the population are those that occur in regions where the sequence is functionally significant.

Spontaneous mutation rates are appreciably small, $10^{-4}$ to $10^{-6}$ mutations per gene per generation, so it is the cumulative effects of mutation over long periods of time that become important (see Fig. 10.3). The simplest kind of mutation is when one nucleotide is replaced by another (a base substitution). These substitutions can be transitions, A to G or C to T, or transversions (all other types of substitutions). If a base substitution results in the replacement of an amino acid in the

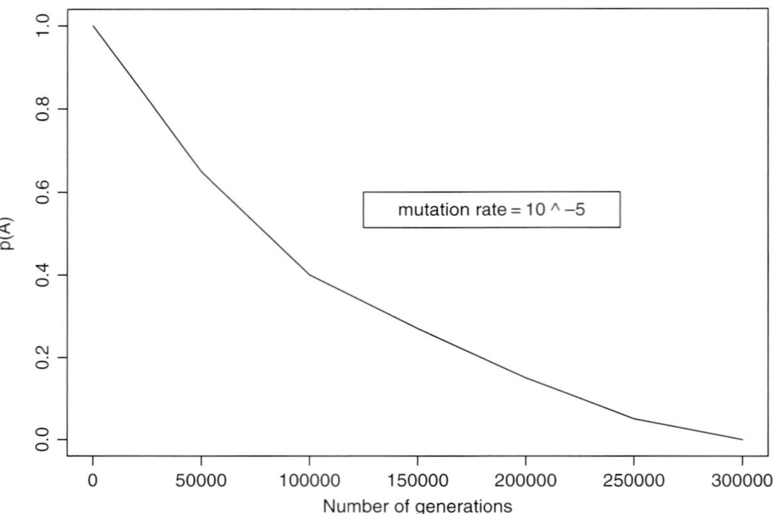


**Fig. 10.3** The cumulative effect of mutation over generations of mating (over time), on the change in frequency of allele *A*, where the mutation rate for *A* to become *a*, is maintained at a constant rate of $10^{-5}$ (Copies of figures including color copies, where applicable, are available in the accompanying CD)

protein product, this is a missense mutation. Mutations can also result in a loss or gain of genetic material, deletion or insertion, which can result in frame-shift mutations. Genetic material can also be rearranged, as with a translocation in which pieces of different chromosomes change places with one another. Mutations due to gene conversion come from the misalignment of DNA, which is associated with the unequal crossing over of parts of adjacent chromosomes.

Fully grasping the concept of mutation requires a basic understanding of polymorphisms, heterozygosity, and gene diversity. The ability to correctly tabulate allele and genotype frequencies for a given population is also imperative.

#### 10.4.1.1 Allele and Genotype Frequencies

The most fundamental quantitative variable in the study of population genetics is allele frequency which is determined as follows. In a population of *N* diploid individuals, we have *2N* alleles present. If the number of alleles, *i*, in the population is $n_i$, then the frequency of that allele in the population is defined as $p_i = n_i/2N$. There is no limitation to the number of alleles that may exist at a single locus but their frequencies must always sum to one. When a locus has only two alleles we denote their frequencies as *p* and $q = 1-p$. A bi-allelic locus, *L*, with alleles *A* and *a*, has three possible genotypes, *AA*, *Aa* and *aa*. However, not all genotypes are necessarily present at all times in a population.

The genotype frequencies at a particular locus are similarly defined; the frequency of a particular genotype is the number of that genotype present in the population divided by the total number of genotypes present. Like the allele frequencies, genotype frequencies must sum to one over all the genotypes present in the study population. However, the number of genotypes is constrained and equals $[m(m+1)]/2$, if there are *m* alleles at the locus with *m* homozygotes and $[m(m-1)]/2$ heterozygotes [54].

As an example of how to count alleles and genotypes in a population, the *MN* blood group will be used. Assume that a population consists of 543 *MM* (*phenotype M*), 419 *MN* (*phenotype MN*), and 457 *NN* (*phenotype NN*) individuals (total = 1419 individuals). We first need to determine the values of $p = f(M)$ and $q = f(N)$ and the genotype frequencies in the population. In this simple example, the values of *p* and *q* and the genotype frequencies can be determined by the straightforward counting of genotypes and alleles. To determine the genotypic frequencies, we simply divide the number of each genotype present in the population by the total number of individuals present in

**Table 10.3** Genotypic frequencies of the three genotypes present in the MN blood group; *MM*, *MN* and *NN*, calculated from a total of 1419 individuals

| Genotype | Number of individuals | Genotypic frequencies |
|---|---|---|
| *MM* | 543 | 543/1419 = 0.38 |
| *MN* | 419 | 419/1419 = 0.30 |
| *NN* | 457 | 457/1419 = 0.32 |
| Total | 1419 | 1.0 |

(Copies of tables are available in the accompanying CD.)

the population (see Table 10.3), where the genotypic frequencies will add up to one. To determine the allelic frequencies we simply count the number of *M* or *N* alleles and divide by the total number of alleles. The 543 *MM* individuals will contribute $543 \times 2 = 1086$ *M* alleles. The 419 *MN* individuals will contribute 419 *M* alleles and 419 *N* alleles, and the 457 *NN* individuals will contribute $457 \times 2 = 914$ *N* alleles. Therefore, there are $1086 + 419 = 1505$ *M* alleles total in the population and $419 + 914 = 1333$ *N* alleles total in the population. So, $p = f(M) = 1505/[2(1419)] = 0.53$ and $q = f(N) = 1333/[2(1419)] = 0.47 = 1-p$.

#### 10.4.1.2 Polymorphism

When a locus has many variants, or alleles, it is referred to as being polymorphic. *Polymorphism* is defined as the existence of two or more alleles with large relative frequencies in a population (occurrence of no less than 1–2%). The limiting frequency of the most common allele, and thus for the polymorphism, is set at 99% [58]. Mutation(s) at a locus generate these multiple alleles, most of which are eliminated from the population by genetic drift or purifying selection. Only a small number of them are incorporated into the population by chance or selection. The first human polymorphism discovered was the ABO blood group identified by Landsteiner [59]. Most polymorphisms are genetically straightforward, with two alleles directly determining two versions of the same protein. Some, however, can be highly complex, with multiple, related loci engaged in a complex system on a chromosome.

There are four primary ways to determine polymorphisms: Restriction fragment length polymorphisms (RFLPs), "Minisatellites" or Variable number of tandem repeats (VNTRs), "Microsatellites" or Short tandem repeats (STRs) or/and Single nucleotide polymorphisms (SNPs). RFLPs are DNA segments of different lengths generated by restriction enzyme cuts, which depend on specific base sequences at a potential cut site. The different sized DNA fragments can be separated using electrophoresis. Since RFLPs are based on single nucleotide changes, they are not very polymorphic in the population and usually have heterozygosities of less than 50%. Minisatellites or VNTRS are repeats of a relatively short oligonucleotide sequence that vary in number from one person to another. They are much more polymorphic than RFLPs. Microsatellites or STRS are multiple (often 100 or more) repeats of very short sequences (2–4 nucleotides), e.g., (CA)n repeats that are amplified by PCR and electrophoresed to score allele sizes. These are highly polymorphic in the population, with most individuals being heterozygous. Thousands of such markers are available, conveniently located throughout the genome. Tri- and tetranucleotide repeats are often easier to score than dinucleotide repeats. Microsatellites are often the markers of choice for genetic linkage studies. SNPs, abundantly available throughout the genome, are a class of recently identified markers characterized by variation at a specific nucleotide. Only 2 alleles exist for a given SNP in the population, so they are less polymorphic than microsatellites.

#### 10.4.1.3 HapMap Project

There are approximately 10–15 million SNPs in the human genome. Alleles of SNPs that are close to each other on the same chromosome tend to segregate together resulting in a non-random association between alleles, or LD. A set of associated SNP alleles in a region of a chromosome is called a *haplotype*. Closely located SNPs inherited together are also referred to as *haplotype blocks*. A haplotype block is a discrete (does not overlap another block) chromosome region of high LD and low haplotype diversity. Haplotype blocks are determined by SNPs in strong LD, thus a few SNPs can provide most of the information on the pattern of genetic variation in the chromosomal region of interest. The SNPs in LD within the block are called haplotype tagging SNPs, or tag SNPs. Due to historical recombination events, LD is strong among SNPs within the haplotype blocks but weak between haplotype blocks. Therefore, chromosome regions corresponding to these haplotype blocks have only few common haplotypes (i.e., frequency of at least 5%), accounting for most of the variation among individuals, since haplotype frequencies vary widely across different world-wide populations. The whole genome can be parsed in haplotype blocks of variable lengths [18, 60, 61].

Hence, gaining a complete understanding of SNP diversity and hence haplotype block diversity in world-wide populations could facilitate gene discovery for complex disease. Begun in October 2002, with its first full release in 2005 [12], the HapMap project has developed a haplotype map of the human genome which describes common patterns of genetic variation among different populations world-wide. The HapMap project, a collaborative project, including scientists from Canada, China, Japan, Nigeria, the United Kingdom, and the United States of America, has become an essential resource assisting scientists world-wide in discovering genes predisposing to diseases and drug response. The complete description of the HapMap project, including the populations from which the samples were selected for generating haplotypes, the Ethical, Legal, and Social Implications (ELSI), and the list of participating scientists and planning groups can be found at http://www.hapmap.org. All data generated by the HapMap project is freely available to the scientific community.

#### 10.4.1.4 Gene Diversity and Heterozygosity

When examining a large number of loci, the amount of variation is usually measured by the proportion of polymorphic loci, which can be reported for a single locus, as an average over several loci, or as the *average heterozygosity* per locus (or *gene diversity*). The *heterozygosity* (the proportion of heterozygotes or polymorphic loci) is defined purely in terms of the genotype frequencies in the population. If $n_{ij}$ is the observed count of heterozygotes $i$, $j$, at locus $L$, where $i$ and $j$ are different alleles in a sample of size $n$, then the sample heterozygosity for that locus $L$ is given by

$$H_L = \sum_i \sum_{i \neq j} \frac{n_{ij}}{n}. \tag{10.1}$$

Heterozygosity is calculated separately for each locus under study and then averaged over all loci under consideration ($m$), to give

$$\bar{H} = \frac{1}{m} \sum_{l=1}^{m} H_L. \tag{10.2}$$

Average heterozygosity or gene diversity is a more useful measure of variation than the proportion of heterozygotes (heterozygosity) because it is not subject to bias caused by sample size, whether it be the size of the study sample or the number of loci being examined. Also, the average heterozygosity is calculated from allele frequencies and not genotype frequencies. Assume that $p_j$ is the frequency of the $j^{th}$ allele at the $l^{th}$ locus, then the gene diversity at this locus, $L$, is

$$D_L = 1 - \sum_j p_j^2, \tag{10.3}$$

and as an average over $m$ loci,

$$\bar{D} = 1 - \frac{1}{m}\sum_l \sum_j p_{jl}^2, \tag{10.4}$$

where $l = 1,\ldots,m$. In a randomly mating population, $\bar{D}$ is equal to the average proportion of heterozygotes per locus. Hence, a very polymorphic locus will have higher gene diversity, because with more alleles present at a locus, more heterozygotes are possible.

However, it is not just mutation that is responsible for sustaining variation. Natural selection, genetic drift, and migration also play key roles in maintaining variation, as do inbreeding and non-random mating and the genetic structure of a population. Selection acts against the dysfunctional alleles that are continuously being created by mutation such that, at equilibrium, the number of new dysfunctional alleles equals the number lost by selection. Selection, in fact, favors heterozygotes, because they maintain two different alleles in their genotypes, and rare alleles are more common in a heterozygous individual. Hence, the heterozygous individual can carry more genetic information than the homozygous individual.

### *10.4.2 Genetic Drift and Migration*

*Genetic drift* is the change in allele frequency that results from the chance difference in the transmission of alleles between generations. The gene pool changes at each generation, so the frequencies of individual alleles will change (drift) through time, and these frequencies can go up or down, accumulating with time. Drift's largest effects are seen on small populations (because larger samples will be closer to the average), and on rare alleles (the more individuals carrying the allele, the higher the transmission frequency). Drift is important since it has a greater effect on transmission of rare alleles than selection, because the latter mechanism helps remove or promote very rare alleles.

In small populations, drift can cause certain allele frequencies to be much larger or smaller than would likely occur in a large population. From this process emerges the *founder effect* which occurs when a small, under-represented group forms a new colony. The Amish in the United States are a good example of this because the roots of this population can be traced to a small number of immigrant families. Occasionally some environmental factor like disease reduces a population to a small number of individuals who subsequently become the parents of a new large population. This process is called *bottle-necking*. Drift can also cause small isolated populations to be very different from the norm, and this can in turn lead to the formation of new species and races.

The basic calibrator of genetic drift is *effective population size*, $N_e$. This is the size of a homogeneous population of breeding individuals, half of which are male and half female, that would generate the same rate of allelic fixation as observed in the real population of total size $N$ [54]. Thus, in a population of size $N$, with random mating, the variance of the random deviation of allele frequencies is $[p(1-p)]/2N$ and the rate of decay is $1/2N$. But a real human population is structured in many different ways: its individuals are of different sexes, ages, and geographical and social groups, for example. Because an actual population does not match the "ideal" population, $N_e$ is estimated indirectly to be $N/2$ to $N/3$ [54]. $N_e$ can be estimated directly if one knows the heterozygosity or the inbreeding coefficient of the population being studied.

*Migration* also causes variation within a population, because of the possibility of mixing many populations together. Geographically defined populations generally show variation between each other, and such variation can have an effect on the fate of that population. Through migration, these populations subdivide and mix with new individuals to form new, sustainable populations. Population subdivision (or population stratification) causes a decrease in homozygous individuals, known as *Wahlund's principle* [62], because it increases variation in the newly formed population. In human populations, the main effect of this fusion of populations is a decrease in the overall frequency of children born with genetic defects resulting from homozygous recessive genes that have high frequency in one of the mixing populations.

#### 10.4.2.1 Wright's Fixation Indices

The genetic structure of a species is characterized by the number of populations within it, the frequencies of the different alleles in each population and the degree of genetic isolation of the populations. The evolutionary forces previously discussed will cause a differentiation within and between subpopulations within a larger species population. Wright [51, 63, 64] showed that any species population has three levels of complexity: $I$, the individual, $S$, the various subpopulations within the total population and $T$, the total population. In order to assess this population substructure and test for allelic correlation within subpopulations, Wright defined three measurements called *fixation indices* that have correlational interpretations for genetic structure and are a function of heterozygosity. $F_{IT}$ is the correlation of alleles within individuals over all subpopulations; $F_{ST}$ is the correlation of alleles of different individuals in the same subpopulation; and $F_{IS}$ is the correlation of alleles within individuals within one subpopulation. Cockerham [65, 66] later showed analogous measures for these three fixation indices, which he called the overall inbreeding coefficient, $F$, and the coancestry, $\theta$ and $f$, respectively.

In order to calculate the three fixation indices, we must first calculate the heterozygosities. $H_I$ is the heterozygosity of an individual in a subpopulation and can be interpreted as the average heterozygosity of all the genes in an individual. $H_S$ is the expected heterozygosity of an individual in another subpopulation and can be interpreted as the amount of heterozygosity in any subpopulation if it were undergoing random mating. And $H_T$ is the expected heterozygosity of an individual in an equivalent random mating total population and can be interpreted as the amount of heterozygosity in a total population where all subpopulations were pooled together and mated randomly.

If $H_i$ is the heterozygosity in subpopulation $i$, and if we have $k$ subpopulations in total, then

$$H_I = \sum_{i=1}^{k} \frac{H_i}{k}. \tag{10.5}$$

If $p_{js}$ is the frequency of the $j^{th}$ allele in the subpopulation $s$, then $H_S$ is the expected heterozygosity in subpopulation $s$, for a total of $h$ alleles at that locus.

$$H_S = 1 - \sum_{i=1}^{h} p_{js}^2, \tag{10.6}$$

$\bar{H}_S$, is the average taken over all subpopulations. Finally if $\bar{p}_j$ is the frequency of the $j^{th}$ allele averaged over all subpopulations,

$$H_T = 1 - \sum_{i=1}^{k} \bar{p}_j^2, \tag{10.7}$$

for all $k$ subpopulations.

Thus, Wright's $F$ statistics are

$$F_{IS} = \frac{\bar{H}_S - H_I}{\bar{H}_S}; \; F_{ST} = \frac{H_T - \bar{H}_S}{H_T}; \; F_{IT} = \frac{H_T - H_I}{H_T} \tag{10.8}$$

and these three equations are related by the following identity, $(1 - F_{IS})(1 - F_{ST}) = (1 - F_{IT})$.

#### 10.4.2.2 Genetic Distance

The degree of genetic isolation of one subpopulation from another can be measured by *genetic distance*, which can be interpreted as the time since the subpopulations that are under comparison diverged from their original ancestral population. Nei proposed the most widely used measure of genetic distance between subpopulations in 1972 [67], even though the concept of genetic distance was first used by Fisher [68] and Mahalanobis [69] and later refined by Sanghvi [70]. Thus, Nei's standard genetic distance is given by $D = -ln(I)$ where $I$ (called the *genetic identity* and corrected for bias) is calculated through the following equation:

$$I = \frac{(2n - 1)\sum_{l}\sum_{j} p_{j1}p_{j2}}{\sqrt{\sum_{l}(2n\sum_{j} p_{j1}^2 - 1)\sum_{l}(2n\sum_{j} p_{j2}^2 - 1)}}, \tag{10.9}$$

where we are examining the $j^{th}$ allele at the $l^{th}$ locus for populations 1 and 2 and $p_{j1}$ is the frequency of the $j^{th}$ allele at the $l^{th}$ locus for population 1, and $p_{j2}$ is the frequency of the $j^{th}$ allele at the $l^{th}$ locus for population 2, from a total sample of $n$ individuals.

### *10.4.3 Inbreeding and Non-random Mating (Assortative Mating)*

*Inbreeding* and other forms of *non-random mating, or assortative mating*, can also have a profound effect on variation within a population. Inbreeding refers to mating between related individuals, or, more precisely, mating in which genetically similar (related) individuals mate more frequently than would be expected in a randomly mating population. Inbreeding mainly causes departures from Hardy Weinberg equilibrium (HWE) and as a consequence of this departure from equilibrium, an increase in homozygotes. This occurs because inbreeding can cause the offspring to have replicates of specific alleles present in the shared ancestor of the parents. Thus, inbred individuals may carry two copies of an allele at a locus that are *identical by descent* (*IBD*) from a common ancestor. The measure of how frequently two offsprings share copies of the same parental (ancestral) allele is referred to as the proportion of IBD. While inbreeding alone cannot change allele frequencies, it can change how the alleles come together to form genotypes. The amount of inbreeding in a population can therefore be measured by comparing the actual proportion of heterozygous genotypes in the population that is inbreeding to the proportion one would expect in a randomly mating population.

In order to illustrate this change in genotype frequencies, we can examine a simple case of inbreeding, where the reference population will be the preceding generation so that the inbreeding coefficient, $F$, measures the increase in IBD from one generation to the next. If allele $A$ has a frequency of $p$ and allele $a$ has a frequency of $q$, $q = 1-p$, then the frequency of $AA$ genotypes in an inbred gamete will be $f(AA) = p^2(1-F) + pF = p^2 + pqF$. An individual of genotype $AA$ can be formed in one of the two ways, either by independent origin, which has a probability of $p(1-F)$, or by identical by descent, which has a probability of $F$. Therefore, if $F > 0$, there will be an excess of $AA$ homozygotes relative to what would be expected by HWE. If $F = 0$ then the frequency of the $AA$

**Table 10.4** Frequencies of genotypes *AA*, *Aa* and *aa* after one generation of inbreeding where the reference population is the preceding generation before inbreeding. The inbreeding coefficient, *F* and the two different types of origin, independent and identical by descent, are incorporated in the calculations

| | Origin | | | |
|---|---|---|---|---|
| Genotype | Independent | Identical by descent | Original frequencies | Frequency change after inbreeding |
| *AA* | $p^2(1\text{-}F)$ | $+ pF$ | $= p^2$ | $+ pqF$ |
| *Aa* | $2pq(1\text{-}F)$ | | $= 2pq$ | $- 2pqF$ |
| *Aa* | $q^2(1\text{-}F)$ | $+ qF$ | $= q^2$ | $+ pqF$ |

(Copies of tables are available in the accompanying CD.)

homozygotes is what would be expected by HWE. Similarly, the frequency of the homozygote *aa* would be.

$$f(aa) = q^2(1 - F) + qF = q^2 + pqF, \tag{10.10}$$

and the same rules would hold for the relationship between the value of *F* and HWE. The probability of the heterozygote *Aa* is more complicated to calculate. But, we know that the frequencies of the genotypes must sum to one so,

$$f(Aa) = 1 - f(AA) - f(aa) = 2pq(1 - F) = 2pq - 2pqF. \tag{10.11}$$

Table 10.4 shows a summary of the changes in genotype frequencies after one generation of inbreeding.

If natural selection and inbreeding act together, they can have a profound effect on the course of evolution because of the increase in the frequency of homozygous genotypes. Inbreeding in human populations can result in a much higher frequency of recessive disease homozygotes, since recessive disease alleles generally have low frequencies in humans. Inbreeding affects all alleles and genes, in inbred individuals, thereby exposing rare recessive disorders that may not have presented themselves if no inbreeding had occurred.

*Non-random mating or assortative mating* occurs when a mate is chosen based on a certain phenotype. In other words, it is a situation when mates are more similar (or dissimilar) to each other than would be expected by chance in randomly mating population. In positive assortative mating, a mate chooses a mate that phenotypically resembles himself or herself. In negative assortative mating, a mate chooses a mate that is phenotypically very different from himself or herself. Assortative mating will only affect the alleles that are responsible for the phenotypes affecting mating frequencies. The genetic variance (or variability) of the trait that is associated with the mating increases with more generations of assortative mating for that trait. In humans, positive assortative mating occurs for traits like intelligence (IQ score), height, or certain socio-economic variables. Negative assortative mating occurs mostly in plants.

## Glossary and Abbreviations

**Allele** one of two or more forms that can exist at a locus (variants of a locus).

**Average Heterozygosity** (*or Gene Diversity*) the average proportion of heterozygotes in the population and the expected proportion of heterozygous loci in a randomly chosen individual.

**Bottle-Necking** when a population is reduced to a small number, possibly because of disease, and later becomes the parents of a new large population.

**Chi-Square Goodness of Fit Test** a statistical test used to test the fit between the observed and expected numbers in a population; the test statistic has a chi-square distribution.

**Degrees of Freedom** the number of possible classes of data minus the numbers of parameters estimated from the data minus 1.

**Directional Selection** favors individuals who are at one extreme of the population, i.e., selection of either of the extreme individuals.

**Disruptive Selection** favors individuals at both extremes of the population, which can cause the population to break into two separate populations.

**Effective Population Size, N$_e$** the size of a homogeneous population of breeding individuals, half of which are male and half are female, that would generate the same rate of allelic fixation as is observed in the real population of total size $N$.

**Evolution** the change of frequencies of alleles in the total gene pool of a given species.

**Exact Test** a statistical test used when the sample sizes are small or the locus under study is multi-allelic because it is more powerful than the chi-square goodness of fit test.

**Fitness** the ability of a gene or locus to be transmitted from one generation to the next; genotypes with greater fitness produce more offspring than the less fit genotypes.

**Founder Effect** when a small, underrepresented group forms a new colony.

**Gene** (*or locus*) the fundamental physical and functional unit of heredity, which will carry information from one generation to the next; generally encodes for some gene product, like an enzyme or protein. It can have anywhere from two alleles, a bi-allelic locus, to a large number of alleles, a multi-allelic locus. (Note: gene and locus are used interchangeably).

**Genetic Distance** the degree of genetic isolation of one subpopulation from another; interpreted as the time since the subpopulations that are under comparison diverged from their original ancestral population.

**Genetic Drift** the change in allele frequency that results from the chance difference in transmission of alleles between generations.

**Genotype** made up of two alleles at a particular locus, one from the mother and one from the father; homogeneous genotype, i.e., both alleles received from the mother and father are the same allele or heterogeneous genotype, i.e., the alleles received from the mother and the father are different.

**Haplotype** the combination of several alleles from multiple loci.

**Hardy-Weinberg Equilibrium (*HWE*)** the frequencies of the genotypes in the "equilibrium" population are just simple products of the allele frequencies.

**Heterozygosity** the proportion of heterozygotes or polymorphic loci in a population.

**Identical by Descent (*IBD*)** how frequently two offspring share copies of the same parental (ancestral) allele.

**Inbreeding** when genetically similar (related) individuals mate more frequently than would be expected in a randomly mating population.

**Linkage Equilibrium (*LE*)** random allelic association between alleles at any loci, i.e., considering any two loci, the probability of the combination of alleles, one from each locus, is the same if the loci are in the same individual or in different individuals

**Migration** the movement of individuals within and between populations.

**Mutation** a change in the sequence of the bases in the DNA.

**Natural Selection** the process by which the environment limits population size.

**Non-random Mating (*or Assortative Mating*)** mating does not take place at random with respect to some traits; for example, mates can choose each other based on such physical and cultural characteristics as height, ethnicity, age and etc.

**Phenotype** trait or characteristic of interest.

**Polymorphism** the existence of two or more alleles with large relative frequencies in a population (occurrence of no less than 1–2%); limiting frequency is 99%; four ways to determine them include: Restriction fragment length polymorphisms (RFLPs), "Minisatellites" or Variable number of tandem repeats (VNTRs), "Microsatellites" or Short tandem repeats (STRs) or and Single nucleotide polymorphisms (SNPs).

**Population Stratification** the matings of different subpopulations with different allele frequencies.

**Population Genetics** the study of evolutionary genetics at the population level.

**Random Mating** mating takes place by chance with respect to the locus under consideration; the chance that an individual mates with another having a specific genotype is equal to the population frequency of that genotype.

**Recombination Fraction** ($\theta$) the probability that the gamete transmitted by an individual is a recombinant, i.e., an individual whose genotype was produced by recombination

**Stabilizing Selection** removes individuals who deviate too far from the average and maintains an optimal population, i.e., selection for the average individual.

**Wahlund's Principle** a decrease in homozygous individuals in a population, caused by population subdivision.

## References

1. Hardy GH. Mendelian proportions in a mixed population. Science 1908;28:449–450.
2. Weinberg W. Uber den Nachweis der Vererbung biem Menschen. Jh. Verein f. vaterl. Naturk. in Wurttemberg 1908;64:368–382.
3. Weinberg W. Uber Verebungsgestze beim Menschen. Ztschr. Abst. U. Vererb. 1909;1:277–330.
4. Zaykin D, Zhivotovsky L, Weir BS. Exact tests for association between alleles at arbitrary numbers of loci. Genetica 1995;96(1–2):169–178.
5. Guo SW, Thompson EA. Performing the exact test of Hardy-Weinberg proportion for multiple alleles. Biometrics 1992;48(2):361–372.
6. Nei M. Molecular evolutionary genetics. New York, New York: Columbia University Press; 1987.
7. Li CC. Population Genetics: 1st Edition. Chicago: The University of Chicago Press; 1955.
8. Hill WG. Estimation of linkage disequilibrium in randomly mating populations. Heredity 1974;33(2):229–239.
9. Hill WG. Disequilibrium among several linked neutral genes in finite population 1. mean changes in disequilibrium. Theor Popul Biol 1974;5(3):366–392.
10. Weir BS. Genetic Data Analysis II. Sunderland, Massachusetts: Sinauer Associates, Inc.; 1996.
11. Lander ES, Linton LM, Birren B, Nusbaum C, Zody MC, Baldwin J, et al. Initial sequencing and analysis of the human genome. Nature 2001;409(6822):860–921.
12. Altshuler D, Brooks LD, Chakravarti A, Collins FS, Daly MJ, Donnelly P. A haplotype map of the human genome. Nature 2005;437(7063):1299–1320.
13. Morton NE. Linkage disequilibrium maps and association mapping. J Clin Invest 2005;115(6):1425–1430.
14. Collins A, Morton NE. Mapping a disease locus by allelic association. Proc Natl Acad Sci U S A 1998;95(4):1741–1745.
15. Devlin B, Roeder K. Genomic control for association studies. Biometrics 1999;55(4):997–1004.
16. Pritchard JK, Rosenberg NA. Use of unlinked genetic markers to detect population stratification in association studies. Am J Hum Genet 1999;65(1):220–228.
17. Pritchard JK, Stephens M, Donnelly P. Inference of population structure using multilocus genotype data. Genetics 2000;155(2):945–959.
18. Stephens JC, Schneider JA, Tanguay DA, Choi J, Acharya T, Stanley SE, et al. Haplotype variation and linkage disequilibrium in 313 human genes. Science 2001;293(5529):489–493.
19. Choudhry S, Coyle NE, Tang H, Salari K, Lind D, Clark SL, et al. Population stratification confounds genetic association studies among Latinos. Hum Genet 2006;118(5):652–664.
20. Salari K, Choudhry S, Tang H, Naqvi M, Lind D, Avila PC, et al. Genetic admixture and asthma-related phenotypes in Mexican American and Puerto Rican asthmatics. Genet Epidemiol 2005;29(1):76–86.
21. Hanis CL, Chakraborty R, Ferrell RE, Schull WJ. Individual admixture estimates: disease associations and individual risk of diabetes and gallbladder disease among Mexican-Americans in Starr County, Texas. Am J Phys Anthropol 1986;70(4):433–441.
22. Shriver MD, Mei R, Parra EJ, Sonpar V, Halder I, Tishkoff SA, et al. Large-scale SNP analysis reveals clustered and continuous patterns of human genetic variation. Hum Genomics 2005;2(2):81–89.
23. Campbell CD, Ogburn EL, Lunetta KL, Lyon HN, Freedman ML, Groop LC, et al. Demonstrating stratification in a European American population. Nat Genet 2005;37(8):868–872.
24. Seldin MF, Shigeta R, Villoslada P, Selmi C, Tuomilehto J, Silva G, et al. European population substructure: clustering of northern and southern populations. PLoS Genet 2006;2(9):e143.
25. Bauchet M, McEvoy B, Pearson LN, Quillen EE, Sarkisian T, Hovhannesyan K, et al. Measuring European population stratification using microarray genotype data. American Journal of Human Genetics 2007; in press.

26. Helgason A, Yngvadottir B, Hrafnkelsson B, Gulcher J, Stefansson K. An Icelandic example of the impact of population structure on association studies. Nat Genet 2005;37(1):90–95.
27. Chakraborty R, Weiss KM. Admixture as a tool for finding linked genes and detecting that difference from allelic association between loci. Proc Natl Acad Sci U S A 1988;85(23):9119–9123.
28. Wacholder S, Rothman N, Caporaso N. Population stratification in epidemiologic studies of common genetic variants and cancer: quantification of bias. J Natl Cancer Inst 2000;92(14):1151–1158.
29. Wacholder S, Rothman N, Caporaso N. Counterpoint: bias from population stratification is not a major threat to the validity of conclusions from epidemiological studies of common polymorphisms and cancer. Cancer Epidemiol Biomarkers Prev 2002;11(6):513–520.
30. Wang Y, Localio R, Rebbeck TR. Evaluating bias due to population stratification in case-control association studies of admixed populations. Genet Epidemiol 2004;27(1):14–20.
31. Wang Y, Localio R, Rebbeck TR. Evaluating bias due to population stratification in epidemiologic studies of gene-gene or gene-environment interactions. Cancer Epidemiol Biomarkers Prev 2006;15(1):124–132.
32. Marchini J, Cardon LR, Phillips MS, Donnelly P. The effects of human population structure on large genetic association studies. Nat Genet 2004;36(5):512–517. Epub 2004 Mar 28.
33. Reich DE, Goldstein DB. Detecting association in a case-control study while correcting for population stratification. Genet Epidemiol 2001;20(1):4–16.
34. Thomas DC, Witte JS. Point: population stratification: a problem for case-control studies of candidate-gene associations? Cancer Epidemiol Biomarkers Prev 2002;11(6):505–512.
35. Pritchard JK, Donnelly P. Case-control studies of association in structured or admixed populations. Theor Popul Biol 2001;60(3):227–237.
36. Devlin B, Roeder K, Wasserman L. Genomic control, a new approach to genetic-based association studies. Theor Popul Biol 2001;60(3):155–166.
37. Dean M. Approaches to identify genes for complex human diseases: lessons from Mendelian disorders. Hum Mutat 2003;22(4):261–274.
38. Ziv E, Burchard EG. Human population structure and genetic association studies. Pharmacogenomics 2003;4(4):431–441.
39. Burnett MS, Strain KJ, Lesnick TG, de Andrade M, Rocca WA, Maraganore DM. Reliability of Self-reported Ancestry among Siblings: Implications for Genetic Association Studies. Am J Epidemiol 2006.
40. Smith MW, Lautenberger JA, Shin HD, Chretien JP, Shrestha S, Gilbert DA, et al. Markers for mapping by admixture linkage disequilibrium in African American and Hispanic populations. Am J Hum Genet 2001;69(5):1080–1094.
41. Shriver MD, Smith MW, Jin L, Marcini A, Akey JM, Deka R, et al. Ethnic-affiliation estimation by use of population-specific DNA markers. Am J Hum Genet 1997;60(4):957–964.
42. Akey JM, Zhang G, Zhang K, Jin L, Shriver MD. Interrogating a high-density SNP map for signatures of natural selection. Genome Res 2002;12(12):1805–1814.
43. Williams RC, Long JC, Hanson RL, Sievers ML, Knowler WC. Individual estimates of European genetic admixture associated with lower body-mass index, plasma glucose, and prevalence of type 2 diabetes in Pima Indians. Am J Hum Genet 2000;66(2):527–538.
44. Fernandez JR, Shriver MD, Beasley TM, Rafla-Demetrious N, Parra E, Albu J, et al. Association of African genetic admixture with resting metabolic rate and obesity among women. Obes Res 2003;11(7):904–911.
45. Gower BA, Fernandez JR, Beasley TM, Shriver MD, Goran MI. Using genetic admixture to explain racial differences in insulin-related phenotypes. Diabetes 2003;52(4):1047–1051.
46. Barnholtz-Sloan JS, Chakraborty R, Sellers TA, Schwartz AG. Examining population stratification via individual ancestry estimates versus self-reported race. Cancer Epidemiol Biomarkers Prev 2005;14(6):1545–1551.
47. Ziv E, John EM, Choudhry S, Kho J, Lorizio W, Perez-Stable EJ, et al. Genetic ancestry and risk factors for breast cancer among Latinas in the San Francisco Bay Area. Cancer Epidemiol Biomarkers Prev 2006;15(10):1878–1885.
48. Darwin C. On the Origin of Species. London: Murray; 1859.
49. Darwin C. The Descent of Man and Selection in Relation to Sex. New York: D. Appleton and Company; 1871.
50. Fisher RA. The Genetical Theory of Natural Selection. Oxford: Clarendon Press; 1930.
51. Wright S. Evolution in mendelian populations. Genetics 1931;16:97–159.
52. Haldane JBS. The Causes of Evolution. London: Longmans and Green; 1932.
53. Watson JD, Crick FH. Molecular structure of nucleic acids; a structure for deoxyribose nucleic acid. Nature 1953;171(4356):737–738.
54. Weiss KM. Genetic Variation and Human Disease: Principles and Evolutionary Approaches. Cambridge: Cambridge University Press; 1993.
55. Vogel F, Motulsky AG. Human Genetics: Problems and Approaches, Third, Completely Revised Edition. Berlin: Springer-Verlag; 1997.
56. Ayala FJ, Kiger JA. Modern Genetics, 2nd Edition. California: The Benjamin/Cummings Publishing Company, Inc.; 1984.

57. Hartl DL, Clark AG. Principles of Population Genetics: Second Edition. Sunderland, Massachusetts: Sinauer Associates, Inc.;1989.
58. Harris H, Hopkinson DA. Average heterozygosity per locus in man: an estimate based on the incidence of enzyme polymorphisms. Ann Hum Genet 1972;36(1):9–20.
59. Landsteiner K. Zur Kenntnis der antifermentativen, lytischen und agglutinierenden Wirkungen des Blutserums und der Lymphe. Zentralbl Bakteriol 1900;27:357–362.
60. Gabriel SB, Schaffner SF, Nguyen H, Moore JM, Roy J, Blumenstiel B, et al. The structure of haplotype blocks in the human genome. Science 2002;296(5576):2225–2229.
61. Daly MJ, Rioux JD, Schaffner SF, Hudson TJ, Lander ES. High-resolution haplotype structure in the human genome. Nat Genet 2001;29(2):229–232.
62. Wahlund S. Zuzammensetzung von populationen und korrelation-serscheinungen von stand pundt der vererbungslehre aus betrachtet. Hereditas 1928;11:65–106.
63. Wright S. The genetic structure of populations. Ann Eugen 1951;15:323–354.
64. Wright S. Isolation by genetic distance. Genetics 1943;28:114–138.
65. Cockerham CC. Analyses of Gene Frequencies. Genetics 1973;74(4):679–700.
66. Cockerham CC. Analyses of Gene Frequencies of Mates. Genetics 1973;74(4):701–712.
67. Nei M. Genetic distance between populations. Am Nat 1972;106:283–292.
68. Fisher RA. The use of multiple measurements in taxonomic problems. Ann Eugen 1936;7:179–188.
69. Mahalanobis PC. On the generalized distance in statistics. Proc Natl Inst Sci India 1936;2:49–55.
70. Sanghvi LD. Comparison of genetical and morphological methods for a study of biological differences. Am J Phys Anthropol 1953;11(3):385–404.


## Web Resources

http://www.hapmap.org
www.hapmap.org

# Chapter 11
# Statistical Tools for Gene Expression Analysis and Systems Biology and Related Web Resources

**Chiara Romualdi and Gerolamo Lanfranchi**

**Abstract** During the past decade advanced technologies in the field of genomics have revolutionized life sciences and medical research. Large-scale applications of these technologies are making possible the completion of the sequences of an ever growing number of genomes of a variety of organisms in animal, plant and prokaryote kingdoms. The next major task to achieve is the understanding of the function of genes and their interactions.

In this chapter we will briefly introduce the major statistical techniques proposed for gene expression data analysis and for data integration, and then we will focus on the description of the most widely used and freely accessible web tools and software dedicated to genomic data analysis (Systems Biology).

**Keywords** Boolean Networks · Asterias · Visualization · Data integration

## 11.1 Introduction

The first step in trying to understand gene function and interaction is the analysis of the transcriptional pattern of genes (transcriptome) in tissues and cells of organisms, in both physiological and pathological conditions. Novel genomic approaches such as transcriptional profiling by DNA microarrays, based on oligonucleotide and cDNA microarray platforms, allow the simultaneous analysis of tens of thousands of transcripts in a single assay. Applications of this technology are expanding exponentially. An example of the power of this approach is given by the transcriptome microarray studies of neoplastic diseases in humans that are enabling prognostic classification of cancer patients, besides standard histopathological criteria [1–2].

The peculiar structure of gene expression data characterized by the large number of features (statistical variables) and small number of experiments (statistical replicates) combined with the complex dependence structure among genes make the analysis of expression data a very challenging issue. Particularly in human studies, the small number of replicates, for example, due to paucity of biological samples and ethical restriction, leads to a decrease of statistical power in the analysis of genomics data.

In the last decade, several statistical techniques have been proposed to tackle the management of microarray expression data. Furthermore, the availability of public gene expression repositories has opened up a new realm of challenging possibilities provided by the efficient integration of gene expression data on the same issues generated by different research groups using, for example, different microarray platforms or experimental conditions. Data integration is a very useful approach in order to confirm and strengthen the results of single studies as well as to find or

C. Romualdi
CRIBI Biotechnology Centre and Department of Biology, University of Padova, via U. Bassi 58/B, 35131 Padova, Italy
e-mail: chaira.romualdi@unipd.it

S. Krawetz (ed.), *Bioinformatics for Systems Biology*, DOI 10.1007/978-1-59745-440-7_11,

complete common cellular pathways that are found altered in specific physiological or pathological processes under study. Pivotal studies of this type have been performed on cancer [3]. The analysis of multiple gene expression data-sets concerning a common biological problem, called "meta-analysis", has shown the capability of retrieving much more relevant information than single experiment data-sets.

Microarrays are just one out from the variety of "omics" techniques that are supplying high-throughput, genome-scale data. Other technologies include comparative genomic- hybridization (CGH), array CGH, chip-based high-throughput genotyping using single nucleotide polymorphism (SNP) markers, chromatin immunoprecipitation on arrays to find transcription factor binding sites (ChIP-chip), proteomics approaches including two-dimensional gel electrophoresis and mass spectrometric identification as well as large-scale yeast two-hybrid studies, and metabolomic profiling of metabolic output using mass spectrometry or nuclear magnetic resonance.

The greatest value of the 'omics' data lies in the realization that together they provide us complementary views of the biological processes we are studying. Consequently, there is a need of novel techniques and tools to manage these diverse data into a common frame, so that we can organize and facilitate data interpretation. The integration of different types of genomic data (e.g., gene sequences, transcriptional levels, functional characteristics, protein interactions) is a fundamental step in the identification of networks of molecular interaction, which will allow us to turn genomic research into accurate and robust biological hypotheses. In this context, the Systems Biology approach is a composed cycle of theory [4]. The cycle begins with proposing a computational model for specific testable hypotheses about a biological system. The next step consists of experimental validation by the acquisition of a quantitative description of the biological system is required to refine the computational model proposed in the first step of the cycle. Since the objective is a model of interactions in a system, the experimental techniques that most suit Systems Biology are those that are system-wide and attempt to be as complete as possible. Therefore, high-throughput techniques applied to transcriptomics, proteomics, and metabolomics are used to collect quantitative data for the construction and validation of models.

Individual researchers commonly report that searching or interpreting "omics" information is incredibly time consuming and sometimes discouraging since the data is scattered across the web. This common feeling is fully justified; as a matter of fact, as of this writing the NAR online Molecular Biology Database Collection, that contains a very complete survey of databases and tools dedicated to genomics and molecular biology, brings the total list of resources dedicated to "omics" data to 1078 [5].

The task of data integration and data retrieval across all these databases is not easy and needs bioinformatics, statistical, mathematical and computer science expertises. Even though several research groups are currently working on genomic, proteomic and metabolomic standards, difficulties in data retrieval still persist because of several reasons that include but not limited to the following: (i) lack of standards which effectively encapsulate genomic annotation data as a whole (annotation is available in various *ad hoc* texts and binary formats from different data sources); (ii) lack of a single common identifier for a gene or its RNA or protein products (gene/protein information cannot be interlinked syntactically); (iii) controlled vocabularies which are infrequently used, making impossible the task of semantic integration.

In addition, statistical models proposed for data analysis (normalization, identification of significant targets, meta-analysis etc.), are constantly updated and improved as described in bioinformatic papers are rarely provided in a ready-to-use fashion (e.g., software, web tools). An excellent effort in this direction is the implementation of Bioconductor [6]. It is an open source and open development software project, based on the programming language R for the analysis and comprehension of genomic data that should assist biologists and statisticians working in bioinformatics. Nevertheless, the use of Bioconductor is still difficult for researchers without a strong statistic and/or mathematical background, but it is sufficiently flexible to be implemented through a web interface.

Apart from the availability of programs and/or tools, working with "omics" data enabling data integration needs considerable computer power, as represented by the use of Web Services [7] of Systems Biology. Web services allow different applications from different sources to communicate with each other without time-consuming custom coding. Since all the communication is based on XML language, Web services are not tied to any specific operating system or programming language. For example, under XML, Java can talk with Perl, Windows applications can talk with UNIX applications, etc. An increasing number of tools and databases for genomics and bioinformatics are available as Web Services. As an example, the Taverna project discussed in this chapter is an interesting application that makes building and executing workflows accessible to bioinformaticians who are not necessarily experts in web services and programming.

## 11.2 Statistical Analysis for Gene Expression Data and "Omics" Data Integration

### *11.2.1 Gene Expression Data Normalization*

Although microarray technology has given an enormous scientific potential in the comprehension of gene regulation processes, many sources of systematic variation can affect the measured gene expression levels. In this context, performing data normalization before any other statistical analysis becomes a crucial step in data analysis. The purpose of data normalization is to minimize the effects of experimental and/or technical variations so that meaningful biological comparisons can be made and true biological changes can be found within one and among multiple experiments. Several approaches have been proposed and have been demonstrated to be effective and beneficial in the reduction of systematic errors.

#### 11.2.1.1 Normalizations for Two-channel Technology

*Global normalization* is usually directed to balance the different incorporation and signal generation efficiency of the two fluorescent molecules (commonly the cyanine Cy3 and Cy5) in two-channel technology. Global intensity normalization relies on the assumption that the quantity of RNA is the same for both the labelled samples. Furthermore, assuming a symmetrical distribution of over- and under-expressed genes for thousands of genes in a microarray platform, these changes should balance out so that the total quantity of RNA hybridizing to the array from each sample is the same. Consequently, the total integrated intensity for all spots should be the same for both the fluorescent images of microarrays obtained by the analysis of the two labelling dyes. Under this assumption, a normalization factor can be calculated and used to re-scale the intensity for each gene in the microarray.

Yang and colleagues [8] observed that the standard global median normalization can often be inadequate in cases of spatially- and intensity-dependent dye biases. Under the assumption that a significant fraction of the probes in the array are expressed at similar levels, they proposed the use of a non-linear regression technique (the widely used LOWESS, *LOcally WEighted Scatterplot Smoothing*) based on robust local regression of the log ratios Cy3/Cy5 on overall spot intensity Cy3*Cy5 (the lowess smoother for the so called MA-plots, where M is log transformation of Cy3/Cy5 and A is the log transformation of the squared root of Cy3*Cy5). According to Yang *et al.* [8], the MA plot is helpful to underline the possible spot artefacts and intensity dependent patterns in the data.

New normalization procedures have been proposed especially focussed on the problem of variance stabilization. In fact, an important characteristic of gene expression data is the dependence between mean and variance, the higher being the variance of the expression measure. A number of recent papers have addressed the importance of constant variance in the analysis of gene-expression microarray data [9–10] proposing new data transformations to stabilize variance across gene expression. Rocke and Durbin [11] presented three different families of transformations as alternatives to any other pre-processing technique and especially to log ratios, concluding through simulated results that the *generalized logarithmic transformation* (also called *glog*) performed significantly better than other types of transformations. The *glog* transformation is based on the assumption that the raw gene-expression intensities can be modelled as the sum of three components: (i) average background noise, (ii) true expression level multiplied by an exponential error term normally distributed with zero mean and constant variance, and (iii) an additive error term normally distributed with zero mean and constant variance.

#### 11.2.1.2 Normalizations for Single-channel Technology

One of the first normalization proposed for Affymetrix type of microarray data was the *quantile normalization* [12]. The goal of quantile normalization is to give the same empirical distribution of intensities to each microarray. If two data vectors have the same distribution, a quantile-quantile plot will be represented by a straight diagonal line, with slope 1 and intercept 0. Thus, if the quantiles of two data vectors are plotted against each other and each of these points are then projected onto the 45-degree diagonal line, a transformation will be obtained that gives the same distribution to both data vectors. Based upon quantiles, other normalization approaches have been proposed, [13, 14, 15].

The *cyclic loess* method is a generalization of the global loess method, [8]. When dealing with single-channel microarray data, there are pairs of arrays that are normalized to each other. The cyclic loess method normalizes intensities for a set of microarray data by working in a pair-wise manner. With only two microarrays, the algorithm is identical to that in Yang et al. (2002) [8]. With more than two arrays, only part of the adjustment is made. In this case, the procedure cycles through an all pair-wise combination of arrays, repeating the entire process until convergence is reached.

Using *Scaling/Linear method*, which was proposed by Affymetrix Inc., a baseline microarray (reference array) is chosen and all other microarrays are scaled to have the same mean intensity as this reference. This is equivalent to selecting a baseline array and then fitting a linear regression (without intercept term) between each array and the chosen array; the fitted regression line is used as the normalizing relation. A proposed modification is based on trimmed mean (the exclusion of the highest and lowest intensities when computing the mean). Affymetrix removes the highest and lowest 2% of the data.

Some variations of this last normalization approach have been proposed, such as the use of a non-linear relationship between each microarray and the baseline microarray [16], currently used in the dChip software developed [17]. Among these, cross-validated splines [16], running median lines [18] and loess smoothers [12] have also been proposed. Finally the *glog* transformation proposed by Rocke and Durbin [11] for the two-channel technology and described in the previous section can be equally applied to one-channel technology.

### *11.2.2 Inferential Statistics for the Identification of Differentially Expressed Genes*

#### 11.2.2.1 Hypothesis Testing

A common task in analyzing microarray data is the identification of genes that are differentially expressed across two or more samples obtained under two or more experimental conditions. The null hypothesis is that the gene is equally expressed in all samples compared. The alternative hypothesis (generally a two-sided hypothesis) is that the gene is either up or down regulated. Assuming the independence of genes in the platform, a statistical test is then performed for each gene. The typical result of a statistical test is the significance level (*p-value*), defined as the measure of how much evidence we have against the null hypothesis. The lower the *p-value*, lower is the probability of error in rejecting the null hypothesis. Problems related to gene expression data can be summarized in two main issues: (i) the small sample size (small number of replicates *per* sample) and (ii) the large number of statistical tests to be performed (one for each gene). Small sample size leads to reduced statistical power (defined as the probability of detecting a differentially expressed gene when it is really differentially expressed) and to a raw variance estimate of gene expression, while the large number of tests leads to an explosion of false positives (defined as the number of genes identified as differentially expressed that actually are equally expressed across samples). Recently, several statistical methods have been proposed to accomplish this goal when there are replicated samples under each condition.

#### 11.2.2.2 Statistical Tests

A straightforward method is to use the traditional two-sample t-test. There are several versions of the two-sample test, depending on the dimension of sample size and whether it is reasonable to assume that gene expression levels have an equal variance under two conditions. Because sample sizes are usually small and there is evidence to support unequal variances, the right statistical test to be used is the t-test with two independent small Normal samples with unequal variances. This t-test was proposed by Welch [19]. Two common problems with this approach are: (i) small sample size for an accurate gene specific variance estimate, and (ii) the strong parametric assumption of the null distribution of the test statistics. To handle these potential drawbacks, a number of procedures have been proposed: among others we mention Cyber-t [20], SAM [21] and Broberg [22]. Most of these methods deal with the problem of inaccurate variance estimate by inflating the denominator, a procedure that is called *moderation*. Bayesian approaches such as the empirical Bayes methods, compensate for the lack of sufficient replicates by combining information across the microarrays [23]. In addition, most of these methods use a non-parametric approach trying to estimate the null distribution directly by permutation. For a complete comparative evaluation of the statistical tests proposed for microarray data see Pan and Colleagues [24] and the more recent work of Jeffery et al. [25].

#### 11.2.2.3 Multi-comparison Problem

Statistical theory demonstrates that in the case of multi-comparison tests, the whole type I error (the event of rejection the null hypothesis) grows exponentially with the increasing number of tests. Several statistical methodologies have been proposed in the last years to overcome this problem. The most popular and definitely the easiest to apply is the Bonferroni's correction that controls the so-called *family-wise error rate* (FWER), defined as the probability of at least one false positive in the list of differentially expressed genes (null hypothesis rejection). It is a conservative method which decreases type I error of each single test by a factor equal to the number of comparisons. In this way, the whole type I error, even if increasing, will not become too large. In the case of microarray data, Bonferroni's correction is certainly too conservative due to the large amount of statistical tests to be performed. For gene expression data, the method of Benjamini and Hochberg [26] for the control of False Discovery Rate (FDR), which is defined as the expected number of false positives among the rejected hypotheses, is a valuable alternative. FDR or Q-values for each gene has been defined as:

$$Q = (p \times n)/i, \tag{11.1}$$

where $p$ is the *p-value* of the gene, $n$ the total number of genes and $i$ is the number of genes at or better than $p$.

### 11.2.3 Multivariate Analysis Applied to Genomic Data

Differential expression analysis focuses on the identification of de-regulated genes without considering similarities of expression profiles along the experimental conditions. Identifying patterns of gene expression and grouping genes into expression classes might provide greater insights into their biological functions and relevance.

Several classification techniques have been proposed for the analysis of gene expression data, including hierarchical clustering, principal component analysis (PCA) and self-organizing maps (SOMs). The first step for any classification algorithm is the choice of a mathematical description of similarity. For a series of measurements several measures of similarity of two genes can be used, such as the Euclidean distance, angle, or dot products of the two $n$-dimensional vectors representing a series of $n$-expression measurements. Anyway, standard correlation coefficient (i.e., the dot product of two normalized vectors) is informative on the intuitive biological notion of gene "co-expression". This statistic, in fact, captures similarity in "shape" without emphasis on the magnitude of the measurements.

Clustering methods can be divided into two classes: supervised and unsupervised. In supervised clustering, observations are classified with respect to known reference class. In unsupervised clustering, no predefined reference class is used. While the first is usually adopted in predictive analysis, the second is widely used for descriptive purposes. Unsupervised clustering can be further divided into hierarchical and non-hierarchical clustering techniques. The classification resulting from hierarchical clustering has an increasing number of nested classes and the result resembles a dendrogram similar to those used for phylogenetic classification. Non-hierarchical clustering techniques, such as k-means clustering, simply separate objects into different clusters, the numbers of which must be defined *a priori*, without trying to specify the relationship between the individual elements. There are various hierarchical clustering algorithms which can be applied to expression data analysis. These differ in the way by which distances are calculated between the growing clusters and the remaining members of the data set, including other clusters. The most common methods of calculating distances are described briefly below:

- Single-linkage clustering. The distance between two clusters, $i$ and $j$, is calculated as the minimum distance between a member of cluster $i$ and a member of cluster $j$. This method gives more importance to cluster cohesion with respect to cluster isolation, then, it tends to produce trees with many long, single-addition branches representing clusters that have grown by accretion.
- Complete-linkage clustering. The distance between two clusters is calculated as the greatest distance between members of the relevant clusters. This method has the tendency of producing very compact clusters of elements and clusters are often very similar in size.
- Average-linkage clustering. The most common is the unweighed pair-group method average (UPGMA). The average distance is calculated from the distances between each point in a cluster and all other points in other clusters. The two clusters with the lowest average distance are joined together to form a new cluster.
- Weighted pair-group average. This method is very similar to UPGMA, except that in computations, the size of the respective clusters (i.e., the number of objects contained) is used as a weight. This method should be used, instead of UPGMA, when the cluster sizes are suspected to be greatly uneven.

Another methodology, usually applied to multidimensional data, is the principal component analysis (PCA). PCA, also called singular value decomposition, is a mathematical technique which generates new factors (linear combination of the original features) to pull out patterns from data. It is extremely powerful in reducing the effective dimensionality of gene-expression space without significant loss of information. PCA belongs to a family of related techniques that includes factor analysis and principal coordinate analysis, providing a "projection" of complex data-sets onto a reduced, easily visualized space. In most implementations of PCA, it is difficult to define accurately the precise boundaries of distinct clusters of data or to define genes (or samples) belonging to each given cluster. However, PCA is a powerful technique for analysis of gene-expression data when used in combination with other classification techniques, such as k-means clustering or SOMs, which require the user to specify the number of clusters.

Supervised methods represent a powerful alternative which can be applied if some previous information is available concerning which genes/experiments are expected to cluster together. In the last few years several supervised statistical techniques have been proposed for analysis of gene expression data. They can be divided into parametric supervised techniques (linear and quadratic discriminant analysis), which assumes a probabilistic distribution on expression data, and non-parametric techniques (nearest neighbor, neural networks, support vector machine), which do not need any probabilistic assumptions.

Supervised techniques, also called learning machines, utilize a two step procedure. First, a training set is used to learn the system (to distinguish between members and non-members of the class on the basis of expression data). Secondly, a test set is used to predict the unknown cases according to the result obtained in the previous process. In general, procedures of cross-validation generate the train and the test sets by randomly dividing the whole gene expression data into two groups, iteratively. In this way, supervised techniques use biological information to determine expression features that are characteristic of a group and to assign genes to selected classes.

### *11.2.4 Meta-Analysis of Gene Expression Data*

Due to the huge accumulation of publicly available microarray data, it is increasingly important to develop methods to integrate the findings across studies. The combination of results from different studies should partially solve the problem of sample number (small in microarray experiments) and thus increase the power to detect differentially expressed genes. Meta-analysis has been extensively used in medical and public health applications, but only recently applied to microarray studies. The two primary methods for data integration consist of: (i) combining gene expression measures across studies, (ii) combining summary measures of expression, such as *p-values*. According to the first approach, Wang and colleagues [27] applied a weighted average procedure combining standardized mean expression differences across three independent studies. Choi et al. [28] and Stevens and Doerge [29], combined standardized gene effects into an overall mean effect using statistical modelling accounting for different sources of variation in microarray studies, including differences between studies. Hu et al. [30], extended previous techniques by incorporating a study quality index for each gene into the effect size estimate. Other investigators [31], combined "raw" gene expression data rather than gene effects, but in this case data must be comparable across studies (e.g., same gene expression technique, same experimental protocols).

Due to the difficulty in comparing cross-laboratory and cross-platform expression measures several microarray meta-analysis methods have combined the summary measures of expression rather than the expression measures themselves. Rhodes et al. [32], calculated *p-values* in individual lung cancer studies and combined the *p-values* providing an overall estimate of gene significance, while Romualdi et al. [33], applied *p-value* combination only after the construction of a meta-profile, defined as the trimmed median expression profile of all the features equally annotated. Parmigiani (2004) [34], introduced an integrative correlation approach that identified genes with consistent expression patterns across multiple platforms.

Recently, Bayesian meta-analysis models [35] analogous to the classical methods described above have been introduced. Bayesian approaches have the advantage of being well suited to the small sample sizes of microarray studies since they borrow information from all genes to estimate model parameters. They also provide a framework for incorporating information *a priori* in a systematic way, and explicitly include model and parameter variability.

### *11.2.5 Network Analysis*

The use of systematic genomic, proteomic and metabolomic technologies to construct models of complex biological systems and diseases is becoming increasingly common place. These efforts, known as Systems Biology, establish an approach to interrogate and iteratively refine our knowledge of the cell. Systems Biology integrates knowledge from diverse biological components into models of the system as a whole. At the end, this knowledge, in terms of similarity measures among diverse components, can be reported in a graph as edges, possibly weighted, connecting nodes representing genes with similar characteristics. These technologies typically depend on the knowledge of the complete sequence of the organism's genome. For instance, DNA microarrays involve spotting thousands of these gene sequences on a solid substrate to bind and detect the complementary RNAs. Global changes in RNA expression can be measured with DNA microarrays and networks can be inferred in terms of gene expression effects (e.g., *36–37*). Other technologies give us insights into molecular interactions to form a large and complex intracellular network yielding information on how the cell transmits information in response to stimuli and dynamically forms the molecular machines required for life (e.g., *38–40*). These networks are representative of different functional mechanisms of the cell. Then, the integration with weighted or unweighted procedures will be challenged in the process of regulatory network inference.

There are a large number of factors that complicate the simple mathematical modelling of gene networks, and sometimes simplifications have to be made. For example, although stochastic models show more realistic dynamics [41], models of gene networks are usually deterministic. The reason for this simplification is the difficulty to infer an underlying network if the expression patterns are the result of a stochastic process.

Weight matrices are the most established method for reconstruction of gene networks. A weight matrix consists of $n \times n$ weight values, each of which indicates the relation strength between two genes. In a certain biological system, the $n \times n$ weights composing the matrix are unknown measures at the initial step of the study. The idea is that they can be estimated and approximated from expression data during the process of network reconstruction. The estimation of the values is usually done by one of the several classical learning mechanisms, such as neural networks or genetic algorithms.

A Boolean network consists of $n$ nodes (e.g., genes), which either can be repressed or activated (the node has state 0 or 1, respectively). The dynamics of the network are determined by a list of $n$ (Boolean) functions each receiving input from $k$ specified nodes. Every node has its own specific function, which can determine its next state from the current states of all the input nodes. Compared to gene networks, Boolean networks are of course a simplification. Gene expression is not simply regulated in a on or off manner, and it is also unlikely that the number of input nodes of each gene is specifically bounded by $k$. Nevertheless, Boolean networks show comparable behavior to features of biological gene networks including global complex behavior, self-organization, redundancy [42].

Recently, linear and non-linear differential equations have been used to infer regulatory networks. These techniques search for parameters that indicate the rates of change of a gene. In this case a series of assumptions are necessary including discrete time steps for the network's next state.

## 11.3 Web Resources and Statistical Tools

In the last few years the number of databases, web tools and software for genomic and gene expression data management have proliferated. Unfortunately, few of them provide constant maintenance and updating after publication. Open source software is the right strategy for providing novel tools to the rapidly changing requirements of genomic research. In this section, we have decided to review the widely used and freely available resources, with particular attention to those that are not based on specific computer systems (platform independent). Due to the massive and still increasing amount of publications proposing novel bioinformatic resources, a selection had to be made, and therefore apologise to the Authors of tools that have been excluded from this review. Unfortunately, due to the complexity of the analysis required, there are today few tools that perform meta-analysis across diverse microarray data-sets.

### *11.3.1 Expression Data Analysis*

#### 11.3.1.1 MIDAW

MIDAW is a web interface that integrates a series of statistical algorithms for normalization, analysis and interpretation of two color microarray data [43], http://midaw.cribi.unipd.it/, Fig. 11.1A). MIDAW takes as input a tab-delimited flat file with experiment values and feature description. In the normalization phase it is able to normalize several experiments simultaneously with background correction, global and local mean and variance normalization strategies (Fig. 11.1B). The data analysis section allows for a graphical display of expression data for descriptive purposes (expression density estimation, box-plot, distribution of the mean and of the standard deviation of genes), estimation of missing values (by the k-nearest neighbors technique, row and column mean, missing data deletion), reduction of data dimension (partial least squares, principal component analysis), identification of differentially expressed genes (fold change with variation coefficient control, t-test with permutational approach and SAM test with Benjamini and Hochberg *p-value* correction), cluster analysis (hierarchical and non-hierarchical through k-means algorithm) and identification of marker genes (PAM discriminant analysis [44]) (Fig. 11.1C). The statistical results are organized in dynamic web pages and tables, where the transcript/gene probes contained in a specific microarray platform can be linked (according to user) to external databases (GeneBank, Entrez Gene, UniGene) (Fig. 11.1D). The goal of the Authors was the implementation of a web interface for some R/Bioconductor packages specifically designed for microarray data. Normalization and analysis steps can be used in a pipeline or separately according to the user's request. MIDAW is based on R/Bioconductor packages and functions, and for each analytical step the list of R commands and eventually of R error/warnings messages can be browsed. MIDAW is GPL licensed.

#### 11.3.1.2 GEPAS

This is a comprehensive web tool designed and oriented for the analysis of DNA microarray gene expression experiments ([45], Fig. 11.2A). GEPAS includes a number of interconnected tools (from the normalization step to the analysis and the interpretation of the results), implemented as individual modules (Fig. 11.2A) that can be used either independently or within the pipeline. GEPAS performs normalization either for two-channel technology, using lowess algorithm (print-tip and global, Fig. 11.2B), or for Affymetrix data, using quantile, robust quantile, loess and others. As an input file it requires GPR files for the two-channel technique and CEL files for Affymetrix experiments. Furthermore, it performs filtering (by inconsistent replicates, number of

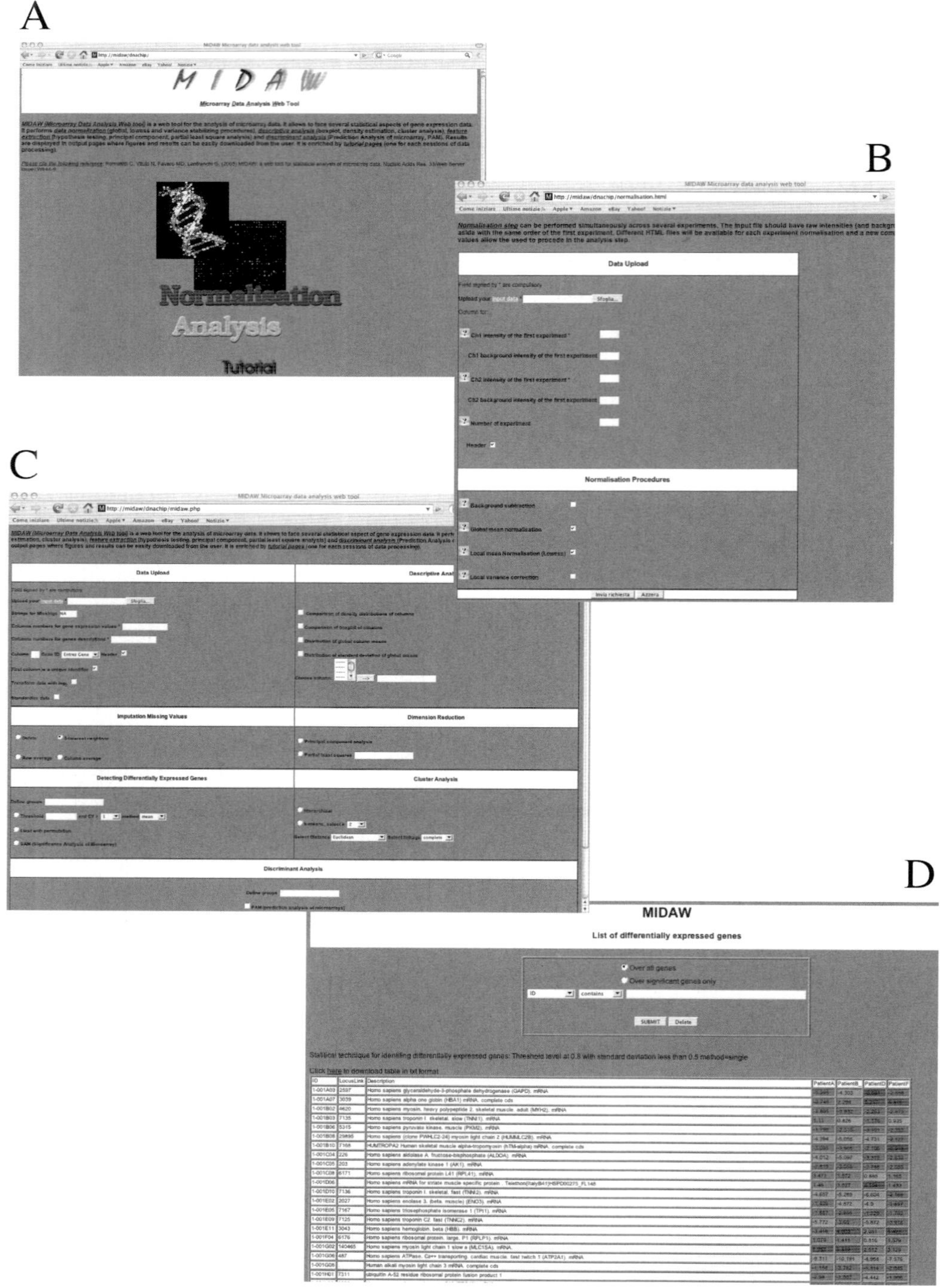

**Fig. 11.1** MIDAW web tool. In the front page (*panel A*) the user can be directed to the normalization step (*panel B*) or directly to the analyses step (*panel C*). Resulting lists of differentially expressed or of co-regulated genes will be reported in dynamic web pages linked to the external databases (according to the user choice) (*panel D*) (Copies of figures including color copies, where applicable, are available in the accompanying CD)

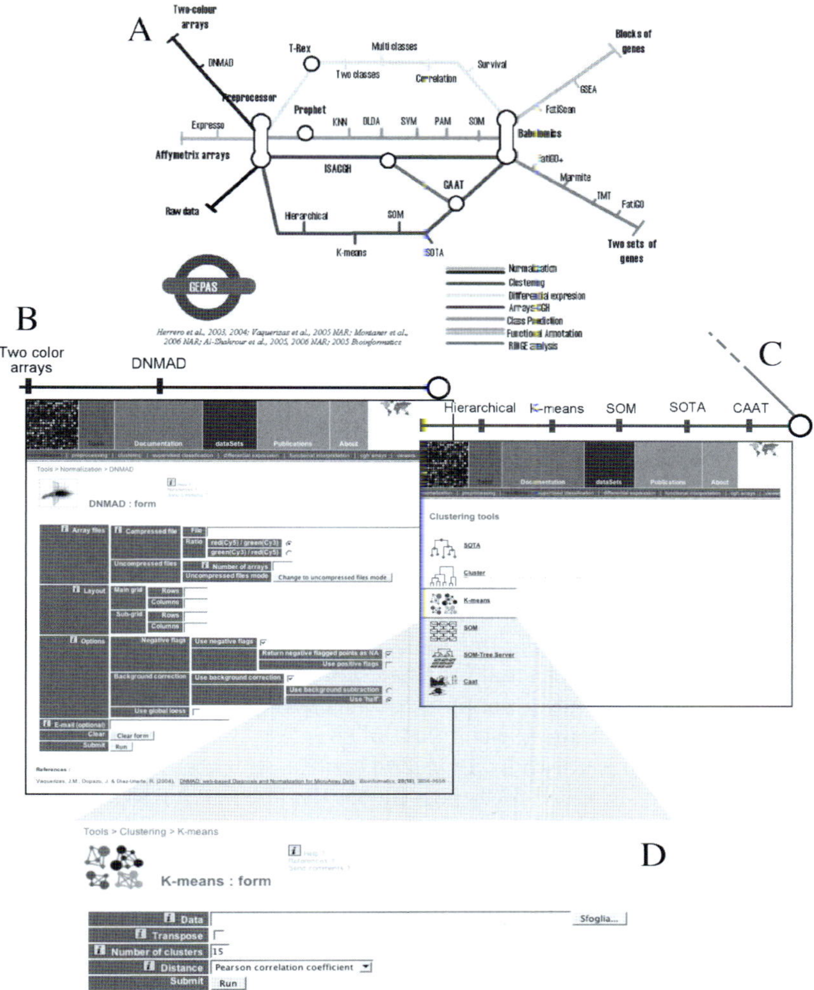


**Fig. 11.2** GEPAS web tool. The web structure of GEPAS web tool is represented by the Authors as a underground-network (*A*). Starting from the normalization steps (B DNAM packages for two-colors technology normalizations), the user can go through the networks following different strategies; e.g., cluster analysis (*B*) with *k*-means algorithm (*C*) (Copies of figures including color copies, where applicable, are available in the accompanying CD)

missing value, and standard deviation), imputation of missing values (with *knn* algorithm), clustering algorithms (SOTA, SOM, hierarchical clustering, Fig. 11.2C and D), identification of differentially expressed genes (using t-test, empirical Bayes method and SAM), discriminant analysis (like support vector machine, linear discriminant analysis, knn and PAM). Interestingly, GEPAS provides a new implementation for visualization and analysis of CGH arrays by merging DNA copy number and gene expression. GEPAS is freely available at the Centro de Investigación Príncipe Felipe, CIPF. Connected to the pipeline GEPAS, a useful tool dedicated to gene functional annotation, called BABELOMICS, is provided. BABELOMICS will be discussed in the section 11.3.2.

#### 11.3.1.3 ASTERIAS

This web tool shares a common history with GEPAS and some of the Authors who developed ASTERIAS have been also involved in GEPAS development ([46], http://asterias.bioinfo.cnio.es/, Fig. 11.3A). Differently from GEPAS, ASTERIAS perform normalizations only for two-channel technology (the same package used by GEPAS). Filtering (based on *knn*, number of missing values, gene variance estimate, Fig. 11.3B), identification of differentially expressed genes (through t-test and ANOVA models, Fig. 11.3C), discriminant analyses (the same as GEPAS with the addition of random forest algorithm) are provided as separate tools included in one single pipeline (Fig. 11.3A). ASTERIAS does not offer clustering algorithms, but includes an interesting tool, based essentially on survival models, for gene selection and signature finding in problems where the dependent variable is patient survival or, more generally, a right-censored variable. Furthermore, ASTERIAS contains the PaLs tool that filters from a list of gene/clone/protein identifiers, those that meet certain criteria related to connected references such as scientific papers indexed in PubMed, Gene Ontology terms and KEGG or Reactome Pathways (Fig. 11.3D). Apparently, ASTERIAS Authors handle most of these applications parallel computing (via MPI) using a server with 60 CPUs for computation.

#### 11.3.1.4 CARMAweb

This web resource allows the analysis of Affymetrix GeneChip, ABI microarrays and two-colors microarrays ([47], https://carmaweb.genome.tugraz.at/carma/). It is essentially based on R/Bioconductor packages. The analyses available include several normalizations and data pre-processing techniques (either for Affymetrix and for two-colors technology), detection for differentially expressed genes (t-test with permutation approach and Benjamini and Hochberg *p-value* correction, SAM, empirical Bayes test and the simple fold change), cluster analysis (hierarchical, non-hierarchical through k-means algorithms, PCA and correspondence analysis) and GO functional categories enrichment analysis. Input file type is highly flexible in CARMAweb, since Affymetrix CEL file, tab-delimited and GPR files are allowed. Users are encouraged to create an account (even a guest account is always usable) and the management guarantees password-protected access to the user's data and analysis results. After account creation, a user data directory is provided were the result files are returned after each analysis allowing a successive reloading as novel input files for further analysis. The user data directory is secured by the management that allows only the owner to access the directory. The web application is based on the Java 2 Enterprise Edition (J2EE) including Java Servlets and JSP technology, the Jakarta Struts framework and Enterprise Java Beans (EJB). Rserve package is used as interface between Java and R.

#### 11.3.1.5 RACE

This resource contains a collection of web tools designed for the analysis of DNA microarray data [48]. RACE performs probe level data pre-processing, quality checks, normalization and visualization only for Affymetrix GeneChips. In addition, RACE identifies differentially expressed genes (for all platforms type, not only for Affymetrix data) through empirical Bayes test (Bioconductor limma packages) estimating false discovery rates (FDR). A Gene Ontology term analysis assists in the biological interpretation of gene lists. The user can customize each analysis and upon submission the analysis is executed in a fully automated way. RACE is lacking any supervised and unsupervised clustering or dimension reduction algorithms.

#### 11.3.1.6 TM4

TM4 [49] is not a web application but a suite of tools, available at The Institute of Genomic Research TIGR, consisting of four major applications: Microarray Data Manager (MADAM),

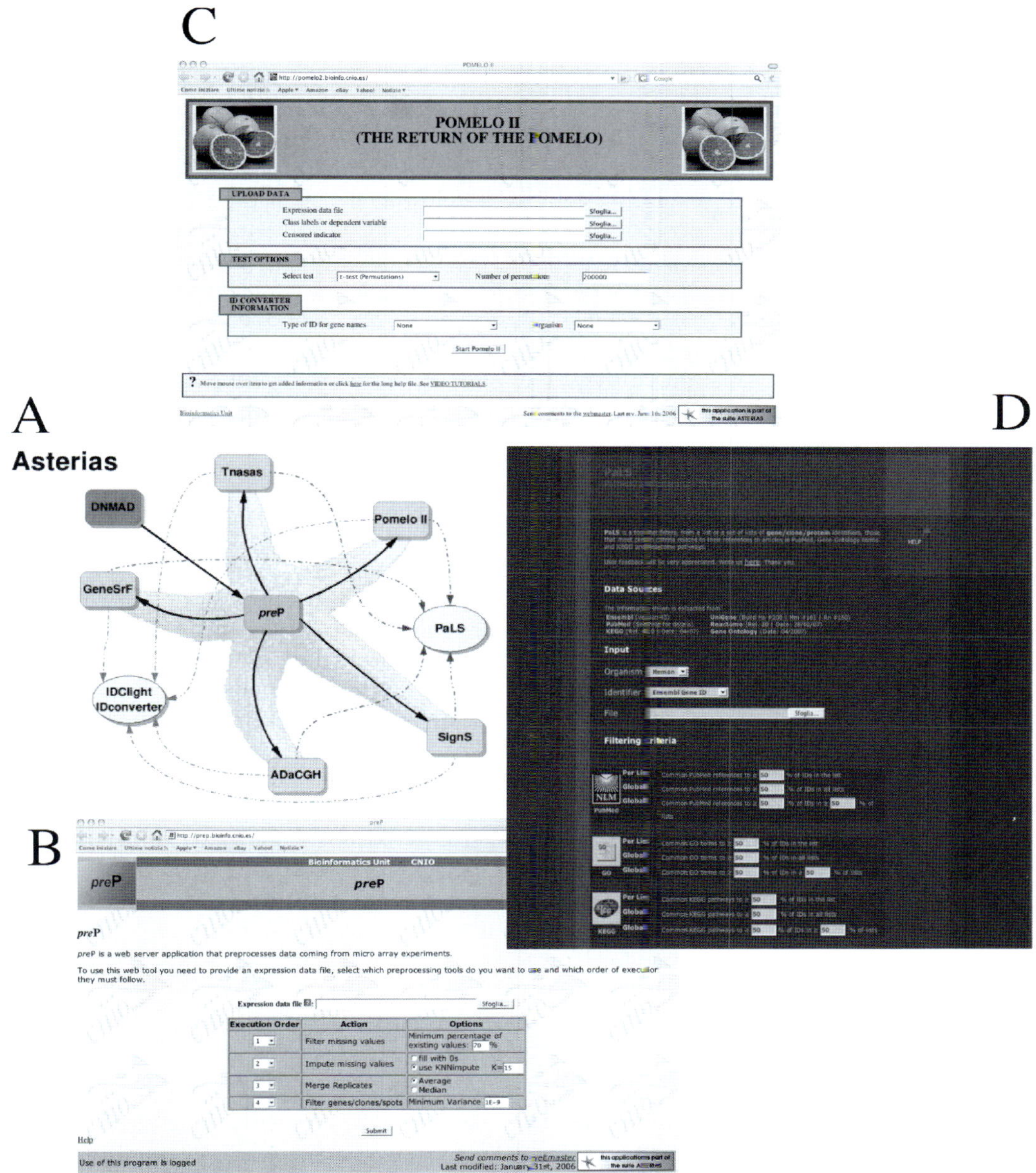


**Fig. 11.3** ASTERIAS web tool. The web structure of ASTERIAS is represented by the Authors as a see star (*panel A*). The core of the see star is represented by the starting point of data analysis, that is pre-processing (*panel B*), then the user can decide to proceed with different statistical algorithms such as testing (*panel C*) or pathways analysis (*panel D*) (Copies of figures including color copies, where applicable, are available in the accompanying CD)

TIGR_Spotfinder, Microarray Data Analysis System (MIDAS), and Multiexperiment Viewer (MeV), as well as a Minimal Information About a Microarray Experiment (MIAME)-compliant MySQL database. All four applications can be used as a pipeline of data analysis even if each single application can be used independently from the others. Most of the tools are designed for two-channel technology. Three of the TM4 applications (MADAM, MIDAS, and MeV) were developed in Java and have been tested on all platforms (Unix, Windows and Machintosh). TIGR Spotfinder was written in C/C++ and runs only on Windows systems. MADAM facilitates the entry of data into a relational MIAME compliant database and can be considered as the starting point of data management. TIGR Spotfinder was designed for the analysis of microarray images and quantification of gene expression. It reads paired 16-bit TIFF image files generated by most microarray scanners. MIDAS provides users an interface for normalization and filtering steps. Pre-processing includes normalization modules (global and lowess normalizations) and filtering (low-intensity cut-off, intensity-dependent Z-score cut-off, and replicate consistency trimming). Finally, MeV tool performs several different statistical analysis including hierarchical clustering, k-means clustering, SOMs, PCA, cluster affinity search, self-organizing trees, template matching, statistical test (t-tests, SAM), support vector machines, gene shaving, and relevance networks. Interesting features such as boot-strap and jack-knife resampling techniques that are usually missed by other are implemented here to generate consensus clusters (a measure of support for each cluster generated by clustering algorithms). TM4 and in particular MeV are constantly upgraded according to new proposed algorithms.

#### 11.3.1.7 Bioconductor

Most of the previous web tools are based on R/Bioconductor packages and therefore, we think that few words should be spent on the Bioconductor project. The Bioconductor project [6] is an initiative for the collaborative creation of extensible software for computational biology and bioinformatics. It is primarily based on the R programming language. R is a high-level interpreted language in which one can easily and quickly prototype new computational methods. The R environment includes a well-established system for packaging together related software components and documentation. The packaging system has been adopted by hundreds of developers around the world and lies at the heart of the Comprehensive R Archive Network, where hundreds of independent and interoperable packages pertaining to a wide range of statistical analysis and visualization objectives may be downloaded as open sources. Dynamicity is perhaps the most important characteristic of R; statistical computing experts and computational biologists are collaborating in the development of R receiving feedback from molecular biologists and researchers in genomics for problem solving. Results of the Bioconductor project include an extensive repository of software tools, documentation, short course materials, and biological annotation data at http://www.bioconductor.org/.

#### 11.3.1.8 Meta-Analysis Tools

The simplest strategy for comparing different gene expression studies involves the comparison of differentially expressed gene lists among related studies, visualizing overlapping genes through a Venn diagram. Automated tools that perform this type of approach are L2L [50] and LOLA [51]. This method is quick and smart but is limited by the potentially heterogeneous strategies of analysis performed by original studies from which the lists are derived.

The very recent web tool EXALT [52] enables comparative analysis of microarrays derived by different studies (also using different platforms) through a more statistically sound approach, called "signature similarity approach". It stores thousands of data-sets derived by GEO standardized database (Gene Expression Omnibus at the NCBI) and encodes them in a searchable format. The statistical approach used by the Authors is similar to that proposed by Rhodes et al. (2005) [3]. EXALT performs statistical tests and then calculates a *p-value* for each probe, separately for each study, obtaining a list of statistically de-regulated genes for each dataset. In case a probe has the same RefSeq ID, EXALT assigns to the gene ID the average *p-value* of all the probes annotated with the same ID. Then, a total identity score (TIS) is computed through the matching of the generated lists.

## 11.3.2 Annotation and Functional Class Enrichment

Iterpretation of results derived by statistical analysis of high-throughput genomic data is extremely intricate. The usual result of a microarray study is a series of lists that contain genes dysregulated in a particular experimental condition with respect to others. In this context, the researcher should find a connection among all of these genes trying to identify one or more metabolic pathways or signalling cascades. This task is extremely challenging and difficult without additional bioinformatic resources aiming at functionally annotating all these features. In the following, we will illustrate some widely used and freely accessible web tools for functional annotation of genomic data.

### 11.3.2.1 DAVID Bioinformatics Resources

The DAVID Database for Annotation, Visualization and Integration Discovery released in 2003 has been recently upgraded and expanded [53]. In the previous version, DAVID simply provided a typical batch annotation and gene-GO term enrichment analysis, highlighting the most relevant GO terms associated with a given gene list. In the current version, the enrichment analysis is extended to over 40 annotation categories, including GO terms, protein– protein interactions, protein functional domains, disease associations, bio-pathways, sequence general features, homologies, gene functional summaries, gene tissue expression, literature. (Fig. 11.4A).

The core of this resource is the DAVID Knowledgebase that consists of a collection of different and redundant databases that hold the gene's biological knowledge. DAVID Knowledgebase is built around a method developed to agglomerate gene/protein identifiers from a variety of public genomic resources, including NCBI, PIR and UniProt, into broader secondary gene clusters, called the DAVID Gene Concept. The new DAVID integrates more than 20 types of major gene/protein identifiers and more than 40 well-known functional annotation categories from dozens of public databases. In particular, it tries to address the redundant relationships among many genes-to-many terms by developing a set of novel algorithms, such as the DAVID Gene Functional Classification Tool, the Functional Annotation Clustering Tool, the Linear Searching Tool, the Fuzzy Gene-Term Heat Map Viewer. These interesting tools provide a novel way to functionally analyze a large number of genes in a high-throughput fashion by classifying them into gene groups based on their annotation term co-occurrence (Fig. 11.4B). DAVID dynamically visualizes genes from a users list within the most relevant KEGG and BioCarta pathways with the DAVID Pathway Viewer.

Enrichment analysis is based on the EASE score. EASE score is basically a modified one-tail Fisher exact test. The Fisher exact probability for over-representation is calculated using the Gaussian hyper-geometric probability distribution that describes sampling without replacement from a finite population consisting of two types of elements. In the case of microarray data, EASE defines this population of elements as the set of genes on the microarray, annotated within a given gene-classification system. For each possible classification, the two types of elements are: genes that belong to that classification; and genes that do not, (Table 11.1).

Given the number of genes of each type within the finite population, it is possible to calculate the exact probability of randomly sampling a given number of genes and observing a specific number that belongs to the classification. The one-tailed Fisher exact probability of over-representation is calculated by summing this probability with all probabilities for situations in which there is a greater number of genes within the classification. The EASE score is offered as a conservative adjustment to the Fisher exact probability that weights significance in favor of themes supported by more genes. The EASE score is calculated by penalizing (removing) one gene from the list within a given category and calculating the resulting Fisher exact probability

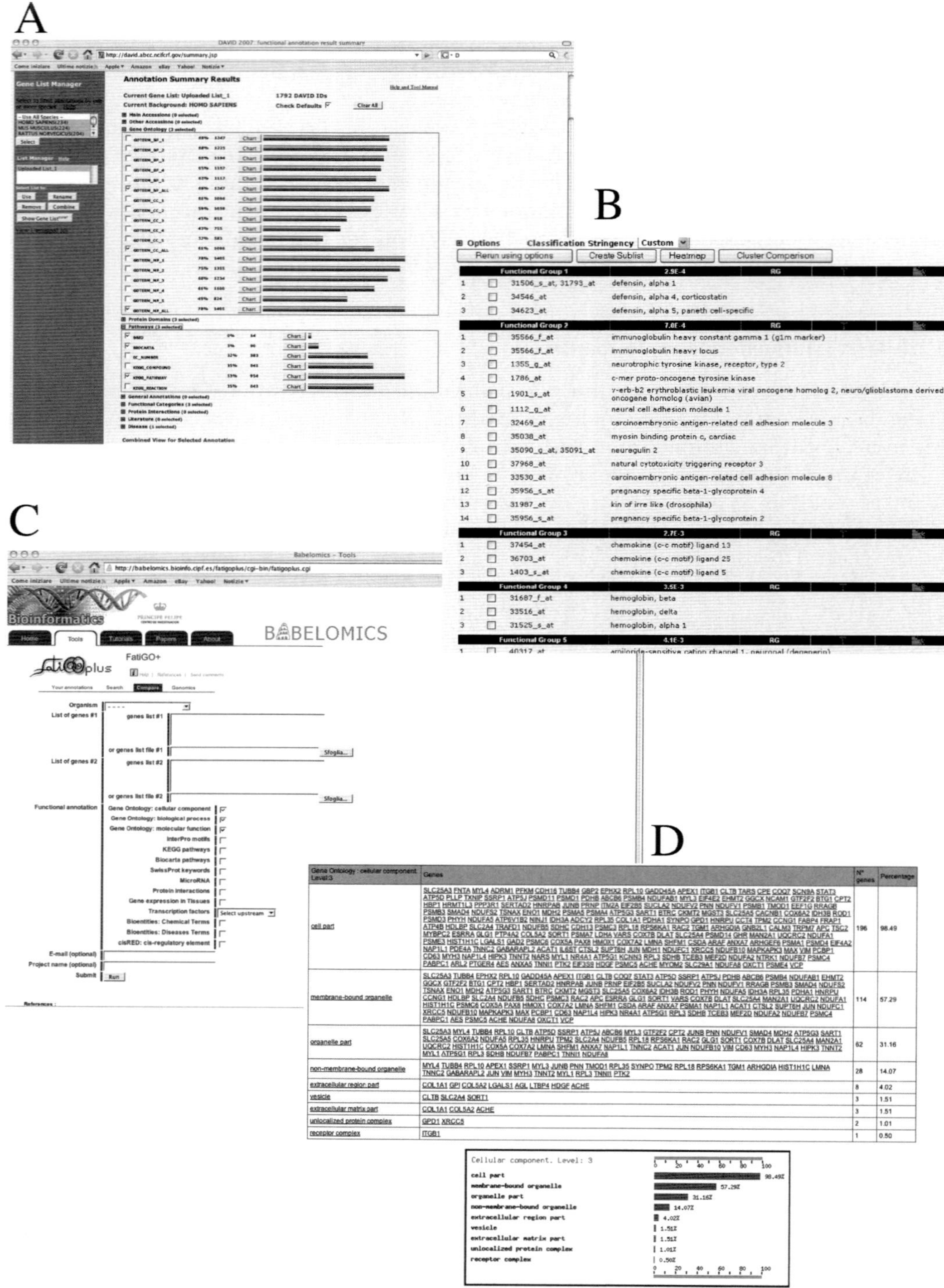


**Fig. 11.4** Annotation tools. After uploading the user gene list through the dedicated web forms (A left frame for DAVID tool; C for BABELOMICS tool), dynamic web pages display the proportion of genes belonging to annotation categories (B for DAVID and D for BABELOMICS) (Copies of figures including color copies, where applicable, are available in the accompanying CD)

**Table 11.1** Elements that belong to each classification

| Fisher Exact test 2×2 table generated for each functional category | User's genes in the submitted list | Genes in the whole genome |
|---|---|---|
| # genes belonging to category X | $n_{11}$ | $n_{12}$ |
| # genes not belonging to category X | $n_{21}$ | $n_{22}$ |

(Copies of tables are available in the accompanying CD.)

for that category. Therefore it penalizes the significance of the categories supported by few genes. Enrichment analysis in DAVID is calculated considering the entire set of genes in the genome. It should be pointed out, however, that a correct enrichment analysis should be performed on the entire set of genes represented in the microarray platform used and not in the genome. The use of with current microarray platforms are able to allocate a high number of features representative of the entire set of genes present in the most complex genomes, This calculation can be considered unbiased. In case of specific customized microarrays (such as tissue-specific array or diagnostic array), the enrichment *p-value* calculated with DAVID leads to biased results. Despite this last observation, DAVID remains a highly flexible platform, since it can accept several types of gene identifiers as input and the dynamic result pages are rapidly returned.

#### 11.3.2.2 BABELOMICS

The BABELOMICS web tool ([54], Fig. 11.4c) is the extension of the precursor FatiGO tool (Fast Assignment and Transference of Information using Gene Ontology [55]). FatiGO was developed to annotate gene lists with GO and KEGG categories or to find significant differences in the distribution of GO terms between groups of genes. Currently, BABELOMICS is composed of several modules, and includes biological information for functional annotation coming from different sources, such as GO, pathways (KEGG and BioCarta), Interpro functional motifs, tissues and chromosomal locations, miRNAs, transcription factor binding sites. Different from DAVID, BABELOMICS does not provide an enrichment *p-value* when a single gene list is provided; only the percentage of the total number of genes submitted is reported. Enrichment *p-value* is calculated when the user selects the comparison between two lists of genes (option not covered by DAVID). In this case the enrichment significance is based on the Fisher exact test with the Benjamini and Hochberg [26] correction. BABELOMICS accepts a wide range of gene accession numbers as input. Output results are stored as dynamic web pages and the result files can be mailed to the user, downloaded as zip directory with flat files or browsed on the web (Fig. 11.4D). BABELOMICS tools for gene functional annotations can be used in combination with GEPAS tools [45], to study gene expression using microarrays; in fact all the tools in BABELOMICS are connected to the proper tools of GEPAS.

### *11.3.3 Web Services Platforms: TAVERNA*

Integrating tools and databases available on the web frequently involves either parsing web pages using scripting languages like PERL or a manual copy-and-paste of data between applications. Each of these methods has its drawbacks. Parsing is a notoriously fragile method, especially in case of web page changes, while copy-and-paste of data between applications is not feasible in case of hundreds of operations and it enormously increases the execution time and the probability of mistakes by the user. Web services technology provides some solutions. Web services have several advantages: (i) tools and databases do not need to be installed locally on the user machine or the laboratory server, as they are programmatically accessible over the web, (ii) tools created using different programming languages (e.g., Python, PERL, Java, etc.) and platforms (e.g., Unix, Windows, etc.) can be accessed through the same web service interface (avoiding the need for the user to know about all the different platforms and programming languages underneath), (iii) the need for file parsing integration scripts is reduced and 'copy-and-paste' integration between web applications is eliminated, (iv) workflows, or pipelines, of web services can be custom-built to provide high-level descriptions of analyses.

Nevertheless, there are also several limitations of using web services: (i) tool and database providers should describe their application or database using the standard Web Services Description Language (WSDL), (ii) services are provided by autonomous third-parties around the world who frequently have insufficient or non-existent meta-data (invoking services relies on knowing exactly what data a service takes as input, information which is not always available), whose consequence is that many services can be difficult to find in a registry [3], (iii) joining services together into pipelines is frequently problematic, as the inputs and outputs are not directly compatible, (iv) the web services can be difficult to debug providing poor documentation by default and cryptic error messages when services fail, (iv) services accessed over the network can have unpredictable performance and reliability, when individual services fails, for whatever reason, the whole workflow can not be run.

Working with both these strengths and limitations, Taverna [56], part of the myGrid project [57], is an application that makes building and executing workflows accessible to bioinformaticians who are not necessarily experts in web services and programming (Fig. 11.5A). Taverna provides a single point of access to a range of services with programmatic interfaces, primarily web services. There are thousands of these publicly available services in molecular biology, provided by a range of third parties around the world. Currently, building workflows of these services in Taverna allows users to join these diverse resources relatively quickly. There are several services available in Taverma provided by INSDC organization (the EMBL-EBI, the NCBI Eutrez programming utilities and the DNA Databank of Japan DDBJ). Additional tools and databases are provided by the Protein Databank of Japan (PDBJ), Kyoto Encyclopedia of Genes and Genomes (KEGG), BioMART, PathPort/ToolBus tools, BioMOBY, BIND, SeqVista and Pfam from the Wellcome Trust Sanger Institute.

## *11.3.4 Networks Visualization*

Network reconstruction is a pioneering field that makes use of complex mathematical and statistical models. At the moment there are only a few freely available resources dedicated to this purpose. Here we describe two examples of interesting and well-designed software dedicated only to network visualizations (not to network construction).

### 11.3.4.1 Osprey

Osprey is a software platform for visualization and manipulation of complex interaction networks (*58*). Osprey represents genes as nodes and interactions as edges between nodes (Fig. 11.5B and C). Unlike other applications, Osprey is fully customizable and allows the user to define personal settings for the generation of interaction networks. Any interaction data-set can be loaded into Osprey using one of several standard file formats, or by uploading from an underlying interaction database. By default, Osprey uses the General Repository for Interaction Data-sets as a database (The GRID), from which the user can rapidly build out interaction networks. User-defined interactions are added or subtracted from mouse-over pop-up windows that link to the database. Networks can be saved as tab-delimited text files for future

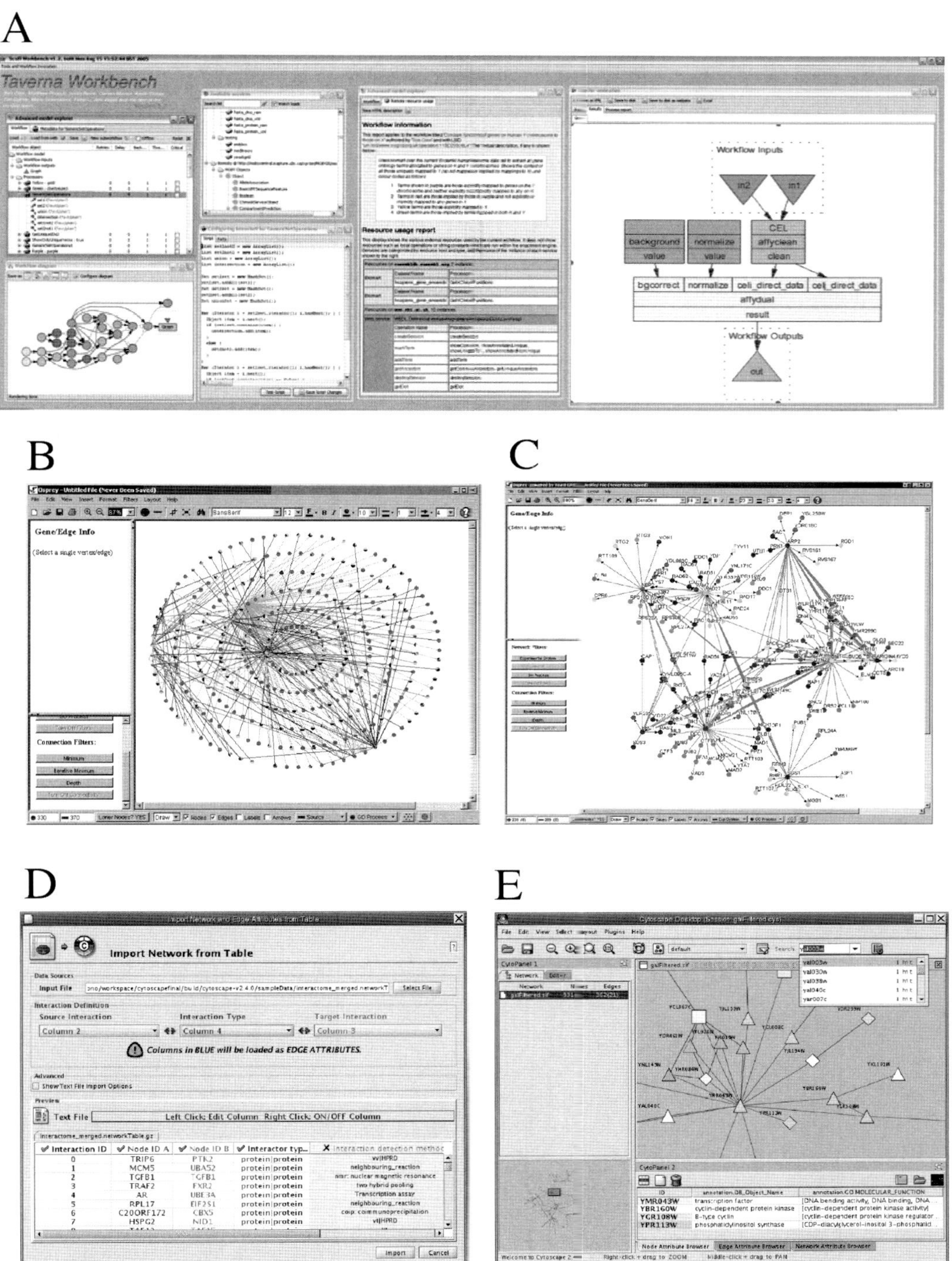


**Fig. 11.5** Example of TAVERNA work-bench and work-flow (*panel A*), Osprey and Cytoscape network visualizations (respectively panels B-C and panels D-E) (Copies of figures including color copies, where applicable, are available in the accompanying CD)

manipulation or exported as JPEG or JPG graphics, portable network graphics (PNG), and scalable vector graphics (SVG). Osprey simplifies network layouts through user-implemented node relaxation, which disperses nodes and edges according to a number of layout options. Osprey also provides several default network layouts, including circular, concentric circles, spoke and dual ring. For comparison of large-scale datasets, Osprey superimposes two or more networks in an additive manner. In conjunction with filter options, this feature allows interactions specific to any given approach to be identified. Finally, Osprey provides a series of graphical devices such as user-defined colors to indicate gene function, experimental systems and data sources. Genes can be colored according to their biological process as defined by standardized Gene Ontology annotations. Genes that have been assigned to more than one process are represented as multicolored pie charts.

#### 11.3.4.2 Cytoscape

Shannon and collaborators [59] developed the cytoscape tool as an open source software project for integrating bio-molecular interaction networks with high-throughput expression data and other molecular states. Although applicable to any system of molecular components and interactions, Cytoscape is most powerful when used in conjunction with large databases of protein-protein, protein-DNA, and genetic interactions that are increasingly available for humans and model organisms. Cytoscape software core provides basic functionality to layout and query the network, to visually integrate the network with expression profiles, phenotypes, and other molecular states; and to link the network to databases of functional annotations (Fig. 11.5D and E). Similar to Osprey, Cytoscape integrates annotations with other types of network data by transferring the desired levels of annotation onto node or edge attributes. It is possible to have many levels of annotation all active and on display at the same time, each as a different attribute on the nodes or edges of interest. Cytoscape supports a variety of automated network layout algorithms, including spring-embedded layout, hierarchical layout, and circular layout. To reduce the complexity of a large molecular interaction network, Cytoscape includes tools for the selection and display of subsets of nodes and edges. Nodes and edges may be selected according to a wide variety of criteria, including selection by name, by a list of names, or on the basis of a given attribute. Furthermore, Cytoscape has a plug-in that is able to merge different networks with the union, intersection and difference option. The Cytoscape is written in Java and has been released under an LGPL Open Source license.

## Glossary and Abbreviations

| | |
|---|---|
| LOWESS | Locally WE lighted scatterplot smoothing |
| glog | Generalized logarithmic transformation |
| FWER | Family-wise error rate |
| FDR | False Discovery rate |
| PCA | Principle Component Analysis |
| SOMs | Self-organizing maps |
| UPGMA | Unweighed pair-group method average |
| GEO | Gene Expression Omnibus |
| TIS | Total identity score |
| DAVID | Database for Annotation, visualization and Integration Discovery |
| FatiGO | Fast Assignment and Transference of Information using Gene ontology |

## Key References

Yang, Y. H., Dudoit, S., Luu, P., Lin, D. M., Peng, V., Ngai, J., and Speed, T. P. (2002) Normalization for cDNA microarray data: a robust composite method addressing single and multiple slide systematic variation, Nucleic Acids Res. 30, e15.

Jeffery, I. B., Higgins, D. G., and Cullane, A. C. (2006) Comparison and evaluation of methods for generatine differentially expressed gene lists from microarray data. BMC Bioinformatics 7, 359.

Bolstad, B. M., Irizarry, R. A., Astrand, M., and Speed, T. P. (2003). A comparison of normalization methods for high density oligonucleotide array data based on variance and bias, Bioinformatics 19, 185–193.

Ideker, T., Galitski, T., and Hood, L. (2001) A new approach to decoding life: Systems Biology, Ann. Rev. Genomics Hum. Genet. 2, 343–272.

Neerincx, P. B. and Leunissen, J. A. (2005) Evolution of web services in bioinformatics, Brief Bioinform. 6, 178–188.

Lee, T. N., Rinaldi, F., Robert, D., Odom, Z., Bar-Joseph, G., Gerber, N., Hannett, C., Harbison, C., Thompson, I., Simon, J., Zeitlinger, E., Jennings, H., Murray, D., Gordon, B., Ren, J., Wyrick, J., Tagne, T., Volkert, E., Fraenkel, D. K., Gifford, and Young, R. A. (2002) Transcriptional regulatory networks in Saccharomyces cerevisiae, Science 298, 799–804.

## Suggested Reading

### *Background*

1. Takahashi, M., Rhodes, D. R., Furge, K. A., Kanayama, H., Kagawa, S., Haab, B. B., and Teh, B. T. (2001) Gene expression profiling of clear cell renal cell carcinoma: gene identification and prognostic classification, Proc. Natl. Acad. Sci. USA 98, 9754–9759.
2. Nevins, J. R. and Potti, A. (2007) Mining gene expression profiles: expression signatures as cancer phenotypes, Nat. Rev. Genet. 8, 601–609.
3. Rhodes, D. R. and Chinnaiyan, A. M. (2005) Integrative analysis of the cancer transcriptome, Nat. Genet. 37, S31–S37.
4. Ideker, T., Galitski, T., and Hood, L. (2001) A new approach to decoding life: Systems Biology, Annu. Rev. Genomics Hum. Genet. 2, 343–372.
5. Galperin, M. Y. (2007) The Molecular Biology Database Collection: 2007 update, Nucleic Acids Res. 36, D2–D4.
6. Gentleman, R. C., Carey, V. J., Bates, D. M., Bolstad, B., Dettling, M., Dudoit, S., Ellis, B., Gautier, L., Ge, Y., Gentry, J., Hornik, K., Hothorn, T., Huber, W., Iacus, S., Irizarry, R., Leisch, F., Li, C., Maechler, M., Rossigni, A. J., Sawitzki, G., Smith, C., Smyth, G., Tierney, L. Yang, J. Y., and Zhang, J. (2004) Bioconductor: open software development for computational biology and bioinformatics, Genome Biol. 5, R80.
7. Neerincx, P. B. and Leunissen, J. A. (2005) Evolution of web services in bioinformatics, Brief. Bioinform. 6, 178–188.

### *Gene Expression Data Normalization*

8. Yang, Y. H., Dudoit, S., Luu, P., Lin, D. M., Peng, V., Ngai, J., and Speed, T. P. (2002) Normalization for cDNA microarray data: a robust composite method addressing single and multiple slide systematic variation, Nucleic Acids Res. 15, e15.
9. Durbin, B. P., Hardin, J. S., Hawkins, D. M., and Rocke, D. M. (2002) A variance-stabilizing transformation for gene-expression microarray data, Bioinformatics. 18, S105–S110.
10. Huber, W., Von Heydebreck, A., Sültmann, H., Poustka, A., and Vingron, M. (2002) Variance stabilization applied to microarray data calibration and to the quantification of differential expression, Bioinformatics. 18, S96–S104.

11. Rocke, D. M. and Durbin, B. P. (2003) Approximate variance stabilizing transformations for gene-expression microarray data, Bioinformatics. 19, 966–972.
12. Bolstad, B. M., Irizarry, R. A., Astrand, M., and Speed, T. P. (2003) A comparison of normalization methods for high density oligonucleotide array data based on variance and bias, Bioinformatics. 19, 185–193.
13. Amaratunga, D. and Cabrera, J. (2001) Analysis of data from viral DNA microchips. J. Am. Stat. Ass. 96, 1161–1170.
14. Sidorov, I. A., Hosack, D. A., Gee, D., Yang, J., Cam, M. C., Lempicki, R. A., and Dimitrov, D. S. (2002) Oligonucleotide microarray data distribution and normalization, Inf. Sci. 146, 67–73.
15. Workman, C., Jensen, L. J., Jarmer, H., Berka, R., Gautier, L., Nielser, H. B., Saxild, H.-H., Nielsen, C., Brunak, S., and Knudsen, S. (2002) A new non-linear normalization method for reducing variability in DNA microarray experiments, Genome Biol. 3, R0048.
16. Schadt, E. E., Li, C., Ellis, B., and Wong, W. H. (2001) Feature extraction and normalization algorithms for high-density oligonucleotide gene expression array data, J. Cell Biochem. Suppl. 37, 120–125.
17. Li, C. and Wong, W. H. (2001). Model-based analysis of oligonucleotide arrays: expression index computation and outlier detection. Proc. Natl. Acad. Sci. USA 98, 31–36.
18. Li, C. and Wong, W. H. (2001) Model-based analysis of oligonucleotide arrays: model validation, design issues and standard error application, Genome Biol. 2, R0032.

## *Inferential Statistics for the Identification of Differentially Expressed Genes*

19. Welch, B. L. (1947) The generalization of "student's" problem when several different population variances are involved, Biometrika. 34, 28–35.
20. Baldi, P. and Long, A. D. (2001) A Bayesian framework for the analysis of microarray expression data: regularized t-test and statistical inferences of gene changes, Bioinformatics. 17, 509–519.
21. Tusher, V. G., Tibshirani, R., and Chu, G. (2001) Significance analysis of microarrays applied to the ionizing radiation response, Proc. Natl. Acad. Sci. USA 98, 5116–5121.
22. Broberg, P. (2003) Statistical methods for ranking differentially expressed genes, Genome Biol. 4, R41.
23. Efron, B. and Tibshirani, R. (2002) Empirical bayes methods and false discovery rates for microarrays. Genet. Epidemiol. 23, 70–86.
24. Pan, W. (2002) A comparative review of statistical methods for discovering differentially expressed genes in replicated microarray experiments, Bioinformatics 18, 546–554.
25. Jeffery, I. B., Higgins, D. G., and Cullane, A. C. (2006) Comparison and evaluation of methods for generating differentially expressed gene lists from microarray data, BMC Bioinformatics 26, 7:359.
26. Benjamini, Y. and Hochberg, Y. (1995) Controlling the false discovery rate: a practical and powerful approach to multiple testing, J. R. Statist. Soc. B 57, 289–300.

## *Meta-Analysis of Gene Expression Data*

27. Wang, J., Coombes, K. R., Highsmith, W. E., Keating, M. J., and Abruzzo, L. V. (2004) Differences in gene expression between B-cell chronic lymphocytic leukemia and normal B cells: a meta-analysis of three microarray studies, Bioinformatics 20, 3166–3178.
28. Choi, J. K., Yu, U., Kim, S., and Yoo, O. J. (2003) Combining multiple microarray studies and modeling inter-study variation, Bioinformatics 19, i84–i90.
29. Stevens, J. R. and Doerge, R. W. (2005) Combining Affymetrix microarray results, BMC Bioinformatics 6, 57.
30. Hu, P., Greenwood, C. M. T., and Beyene, J. (2005) Integrative analysis of multiple gene expression profiles with quality-adjusted effect size models, BMC Bioinformatics 6, 128.
31. Park, T., Yi, S. G., Shin, Y. K., and Lee, S. (2006) Combining multiple microarrays in the presence of controlling variables, Bioinformatics 22, 1682–1689.

32. Rhodes, D. R., Yu, J., Shanker, K., Deshpande, N., Varambally, R., Ghosh, D., Barrette, T., Pandey, A., and Chinnaiyan, A. M. (2004) Large-scale meta-analysis of cancer microarray data identifies common transcriptional profiles of neoplastic transformation and progression, Proc. Natl. Acad. Sci. USA, 101, 9309–9314.
33. Romualdi, C., De Pitta, C., Tombolan, L., Bortoluzzi, S., Sartori, F., Rosolen, A., and Lanfranchi, G. (2006) Defining the gene expression signature of rhabdomyosarcoma by meta-analysis, BMC Genomics 7, 287.
34. Parmigiani, G., Garrett-Mayer, E. S., Anbazhagan, R., and Gabrielson, E. (2004) A cross-study comparison of gene expression studies for the molecular classification of lung cancer, Clin. Cancer Res. 10, 2922–2927.
35. Conlon, E. M., Song, J. J., and Liu, A. (2007) Bayesian meta-analysis models for microarray data: a comparative study, BMC Bioinformatics 7:80.


## *Network Analysis*


36. Lee, T. I., Rinaldi, N. J., Robert, F., Odom, D. T., Bar-Joseph, Z., Gerber, G. K., Hannett, N. M., Harbison, C. T., Thompson, C. M., Simon, I., Zeitlinger, J., Jennings, E. G., Murray, H. L., Gordon, D. B., Ren, B., Wyrick, J. J., Tagne, J.-B., Volkert, T. L., Fraenkel, E., Gifford, D. K., and Young, R. A. (2002) Transcriptional regulatory networks in Saccharomyces cerevisiae, Science 298, 799–804.
37. Gardner, T., diBernardo, D., Lorenz, D., and Collins, J. J. (2003) Inferring genetic networks and identifying compound mode of action via expression profiling, Science 301, 102–105.
38. Bar-Joseph, Z., Gerber, G., Lee, T. I., Rinaldi, N., Yoo, J., Robert, F., Gordon, D., Fraenkel, E., Jaakkola, T., Young, R. A., and Gifford, D. K. (2003) Computational discovery of gene modules and regulatory networks, Nat. Biotechol. 21, 1337–1342.
39. Haugen, A. C., Kelley, R., Collins, J. B., Tucker, C. J, Deng, C. Afshari, C. A., Brown, J. M., Ideker, T., and Van Houten, B. (2004) Integrating phenotypic and expression profiles to map arsenic response networks, Genome Biol. 5, R95.
40. Said, M. R., Begley, T. J., Oppenheim, A. V., Lauffenburger, D. A., and Samson, L. D. (2004) Global network analysis of phenotypic effects: protein networks and toxicity modulation in Saccharomyces cerevisiae, Proc. Natl. Acad. Sci. USA 101, 18006–18011.
41. McAdams, H. H. and Arkin, A. (1997) Stochastic mechanisms in gene expression, Proc. Natl. Acad. Sci. USA 94, 814–819.
42. Somogyi, R. and Sniegoski, C. A. (1996) Modeling the complexity of genetic networks: understanding multigenic and pleiotropic regulation, Complexity 1, 45–63.


## *Web Resources and Statistical Tools: Expression Data Analysis*


43. Romualdi, C., Vitulo, N., Del Bavero, M., Lanfranchi, G. (2005) MIDAW: a web tool for statistical analysis of microarray data, Nucleic Acids Res. 33, W644–W649.
44. Tibshirani, R., Hastie, T., Narasimhan, B., and Chu, G. (2002) Diagnosis of multiple cancer types by shrunken centroids of gene expression, Proc, Natl, Acad, Sci, USA 99, 6567–6572.
45. Vaquerizas, J. M., Conde, L., Yankilevich, P., Cabezon, A., Minguez, P., Diaz-Uriarte, R., Al-Shahrour, F., Herrero, J., and Dopazo, J. (2005) GEPAS, an experiment-oriented pipeline for the analysis of microarray gene expression data, Nucleic Acids Res. 33, W616–W620.
46. Diaz-Uriarte, R., Alibes, A., Morrissey, E. R., Canada, A., Rueda, O. M., and Neves, M. L. (2007) Asterias: integrated analysis of expression and aCGH data using an open-source, web-based, parallelized software suite, Nucleic Acids Res. 35, W75–W80.
47. Rainer, J., Sanchez-Cabo, F., Stocker, G., Sturn, A., and Trajanoski, Z. (2006) CARMAweb: comprehensive R- and bioconductor-based web service for microarray data analysis, Nucleic Acids Res. 34, W498–W503.
48. Psarros, M., Heber, S., Sick, M., Thoppae, G., Harshman, K., and Sick, B. (2005) RACE: Remote Analysis Computation for gene Expression data, Nucleic Acids Res. 33, W638–W643.

49. Saeed, A. I., Sharov, V., White, J., Li, J., Liang, W., Bhagabati, N., Braisted, J., Klapa, M., Currier, T., Thiagarajan, M., Sturn, A., Snuffin, M., Rezantsev, A., Popov, D., Ryltsov, A., Kostukovich, E., Borisovsky, I., Liu, Z., Vinsavich, A., Trush, V., and Quackenbush, J. (2003) TM4: a free, open-source system for microarray data management and analysis, Biotechniques 34, 374–378.
50. Newman, J. C. and Weiner, A. M. (2005) L2L: a simple tool for discovering the hidden significance in microarray expression data, Genome Biol. 6, R81.
51. Cahan, P., Ahmad, A. M., Burke, H., Fu, S., Lai, Y., Florea, L., Dharker, N., Kobrinski, T., Kale, P., and McCaffrey, T. A. (2005) List of lists-annotated (LOLA): a database for annotation and comparison of published microarray gene lists, Gene 360, 78–82.
52. Yi, Y., Li, C., Miller, C., and Gorge, A. L. Jr. (2007) Strategy for encoding and comparison of gene expression signatures, Genome Biol. 8, R133.

## *Web Resources and Statistical Tools: Annotation and Functional Class Enrichment*

53. Huang, D. W., Sherman, B. T., Tan, Q., Kir, J., Liu, D., Bryant, D., Guo, Y., Stephens, R., Baseler, M. W., Lane, H. C., and Lempicki, R. A. (2007) DAVID Bioinformatics Resources: expanded annotation database and novel algorithms to better extract biology from large gene lists, Nucleic Acids Res. 35, W169–W175.
54. Al-Shahrour, F., Minguez, P., Tarraga, J., Montaner, D., Alloza, E., Vaquerizas, J. M., Conde, L., Blaschke, C., Vera, J., and Dopazo, J. (2006) BABELOMICS: a Systems Biology perspective in the functional annotation of genome-scale experiments, Nucleic Acids Res. 34, W472–W476.
55. Al-Shahrour, F., Díaz-Uriarte, R., and Dopazo, J. (2004) FatiGO: a web tool for finding significant associations of Gene Ontology terms to groups of genes, Bioinformatics 20, 578–580.

## *Web Resources and Statistical Tools: Web Services and Networks Visualization*

56. Hull, D., Wolstencroft, K., Stevens, R., Goble, C., Pocock, M. R., Li, P., and Oinn, T. (2006) Taverna: a tool for building and running workflows of services, Nucleic Acids Res. 34, W729–W732.
57. Stevens, R. D., Robinson, A. J., and Goble, C. A. (2003) MyGrid: personalised bioinformatics on the information grid, Bioinformatics 19, i302–i304.
58. Breitkreutz, B.J., Stark, C., and Tyers, M. (2003) Osprey: A Network Visualization System, Genome Biol. 4, R22.
59. Shannon, P., Markiel, A., Ozier O., Baliga, N. S., Wang, J. T., Ramage, D., Amin, N., Schwikowski, B., and Ideker, T. (2003) Cytoscape: a software environment for integrated models of biomolecular interaction networks, Genome Res. 13, 2498–2504.

## Web Resources

http://www.oxfordjournals.org/nar/database/a/
http://www.gepas.org/
http://www.tm4.org/
http://www.bioconductor.org/
http://depts.washington.edu/121/
http://www.lola.gwu.edu/
http://seq.mc.vanderbilt.edu/exalt/
http://david.niaid.nih.gov
http://www.balelomics.org
http://www.fatigo.org
http://www.insdc.org
Affymentrix:www.affymentrix.com
Illumina:www.illumina.com

Agilent:www.agilent.com
www.biostat.harvard.edu/epeople/faculty/mltlee/web-front-r.html
www.cs.umd.edu/hcil/hce/power/power.html
http://genome.ucsc.edu/cgi-bin/hgLiftOver
www.nimblegen.com
www.dchip.com
http://microrna.sanger.ac.uk
www.nslij-genetics.org/microarray/soft.html
www.microarrayworld.com/SoftwarePage.html
www.chem.agilent.com
www.tigr.org/software/microarray.shtml
www.bioconductor.org
lib.stat.cmu.edu/R/CRAN/src/contrib/PACKAGES.html
www.microarraystation.com/microarray/Microarray-Bioinformatic-Tools-Databases/
www.premierbiosoft.com
www.gizinst.org/censor
www.combimatri.com
www.home.agilent.com
http://genome.ucsc.edu
www.sanger.ac.uk/Software/Artemis/
www.ncbi.nlm.nih.gov/geo/
www.mged.org/miame
http://rmaexpress.bmbolsted.com
http://quertermous.stansford.edu/heatmap.htm
http://sourceforge.net/projects/jtreeview

# Part III
# Transcriptome Analysis

## Chapter 12
# What Goes in is What Comes Out: How to Design and Implement a Successful Microarray Experiment

**Jeffrey A. Loeb and Thomas L. Beaumont**

**Abstract** Microarrays that measure thousands of gene expression changes simultaneously are powerful tools to pose far-reaching biological questions that could not have been asked previously. The trick is to understand what goes on inside the 'black box' before designing a microarray experiment. In this way, a high level of confidence will be generated that truly novel and interesting results are being generated, rather than re-affirmations of previous notions or experimental artefacts. This requires limiting the experiments to examine only the variables to be tested, designing microarray experiments with sufficient replicates to achieve high statistical power, analyzing results in an impartial way, and verifying the findings both technically and functionally. If these principles are addressed, microarray experiments can provide a rich substrate on which many exciting new discoveries can be realized.

**Keywords** Microarray · Power · Experimental design · Hybridization · Statistical analysis

## 12.1 Introduction

In a multicellular organism, what makes each cell type unique is the amount and specific pattern of genes that are expressed. Gene expression can tell us the difference between a brain cell and kidney cell, but can also reveal important clues as to what is different when that cell becomes diseased or transformed into a pathologic form, such as in cancer. It is thus not surprising that efforts to measure the expression of a large number of genes simultaneously has become a central methodology in the post-genomic era. While there are a number of other methods to measure gene expression, microarrays have taken the lead as a widely available, rapid, relatively inexpensive, and highly accurate way to measure relative gene expression that can lead to important new research directions (see review articles [1–6]). The central goal of this chapter is to take the reader, step-by-step, through the experimental design and analysis of a typical microarray experiment in a way that makes the reader comfortable with each step of the process. A key point is that each experiment must begin with a focused scientific question and should end with a useful result that answers the question(s) posed. Because of the perceived high cost and ready availability, many have designed microarray experiments using a 'black box' strategy wherein those designing, performing, and analyzing the results are not the same individuals asking the question. This leaves those asking questions often ignoring the many methodological and statistical steps required for accurate and meaningful results, and those performing and analyzing the experiment lacking an understanding of the experimental design and interpretation of the results. Unsuccessful microarray results due to

J.A. Loeb
Department of Neurology and The Center for Molecular Medicine and Genetics, Wayne State University, School of Medicine, 421 E. Canfield Ave, Detroit, MI 48201, USA
e-mail: jloeb@med.wayne.edu

S. Krawetz (ed.), *Bioinformatics for Systems Biology*, DOI 10.1007/978-1-59745-440-7_12,

poor experimental design and interpretation can therefore steer researchers into many wrong directions. Thus what goes in and what comes out of the 'box' should not overshadow what happens within it.

## 12.2 Choosing the Optimal Microarray System

Microarrays are nothing more than high density dot blots of specific nucleotide sequences that take advantage of the fundamental principle that antisense nucleotide sequences will hybridize best to their exact complementary sequence. This specific affinity thus allows the quantification of many copies of that complementary sequence that are present in a given sample or 'target', based on the amount that hybridizes to a given spot (or 'feature'). Most microarrays consist of microscopic deposits of individual coding gene sequences permanently attached to a surface so that each spot allows for the measurement of a given gene. In most cases, these surfaces are specially treated glass slides prepared in such a way to minimize non-specific association of nucleic acids. The original microarrays started out as true 'dot blots' consisting of arrays of genes spotted onto nylon membranes. Table 12.1 lists some of the more common platform microarray technologies and summarizes their differences and advantages.

One of the major differences between different microarrays that can have profound effects on the experimental results is the selection of which nucleotide or 'probe' sequences that are to be deposited. One of the major distinctions is whether smaller oligonucleotides (25–100 nucleotides) or longer and sometimes full-length cDNA probes are used. While larger cDNAs offer the advantage of increased *sensitivity*, because they cover a larger proportion of the gene, they can suffer considerably from reduced *specificity*. For example, since many genes have homologous regions to other genes, a given labeled target gene will hybridize to any probe gene sequence with sufficient homology thus giving a large number of false positives that will have to be verified in subsequent experiments. Using sophisticated algorithms, shorter oligonucleotide sequences within each gene can be identified with little to no sequence homology to other genes [7]. These can also be selected for a similar proportion of nucleotides with high affinity binding (often called the GC content) in order to minimize the biased identification of genes that have the strongest hybridization kinetics. Finally, one particular platform that uses multiple 25 nucleotide probes for each gene (Affymetrix), also includes 'mismatch' sequences that are expected to hybridize less efficiently and

**Table 12.1** Types of microarrays

| Microarray type | Probe type | Comments |
|---|---|---|
| cDNAs on filters | Large cDNA inserts | –High sensitivity, low specificity<br>–Can hybridize only one sample at a time and can strip and reprobe |
| cDNAs on glass slides | Large cDNA inserts | –High sensitivity, low specificity<br>–Can hybridize multiple samples using fluorescent targets |
| Large oligonucleotides on glass slides | Single 60-mer nucleotides (e.g., Agilent) | –High sensitivity and specificity<br>–Can hybridize multiple samples using fluorescent targets |
| Large oligonucleotides on silica beads | 50-mer nucleotides (e.g., illumina) | –High sensitivity and specificity<br>–High density, multiple arrays per slide |
| Small oligonucleotides on glass slides | Multiple 25-mer nucleotides for each gene together with mismatches (e.g., Affymetrix) | –Low sensitivity, high specificity<br>–Can hybridize only one sample at a time |

(Copies of tables are available in the accompanying CD.)

aid in identifying the false positives. While there are clear advantages to looking at multiple probes for a given gene, the use of the mismatch data from this platform has been questioned [8].

## 12.3 Defining an 'Answerable' Question by Minimizing the Variables

Figure 12.1 is a flow chart that shows a step-by-step outline, from start to finish, of a typical microarray experiment that will be referred to in subsequent sections of this chapter. The first and most important part of the experiment is defining the question and designing a highly focused experiment to answer that question, *without answering many other unintended questions,* along the way. Given that most microarray systems simultaneously measure thousands of genes, any unintended variables will produce many additional gene expression changes that cannot be differentiated from the effect of interest. For example, if one wants to determine the effect of a drug on a cell line in culture, both the treated and control cultures must be exactly the same in every way except for the drug treatment. If the cells are from a different passage or are processed to prepare RNA on a different day, your list of genes at the end of the analysis will also reflect these differences.

Similarly, when choosing what tissue to assay for gene expression, the simpler and more consistent will be better. While a tumor may seem homogeneous, some tumors may have varying degrees of vascularization (blood vessels) or may have sat at room temperature for different periods of time before being frozen or prior to RNA preparation. Unless your question is to identify genes in blood vessels or the effects of letting tumors sit at room temperature, it is crucial to make an effort in order to keep every variable, from each sample you compare, as similar as possible. This will limit the gene expression changes you observe to those associated with the effect of interest and not simply reflect the technical artefacts.

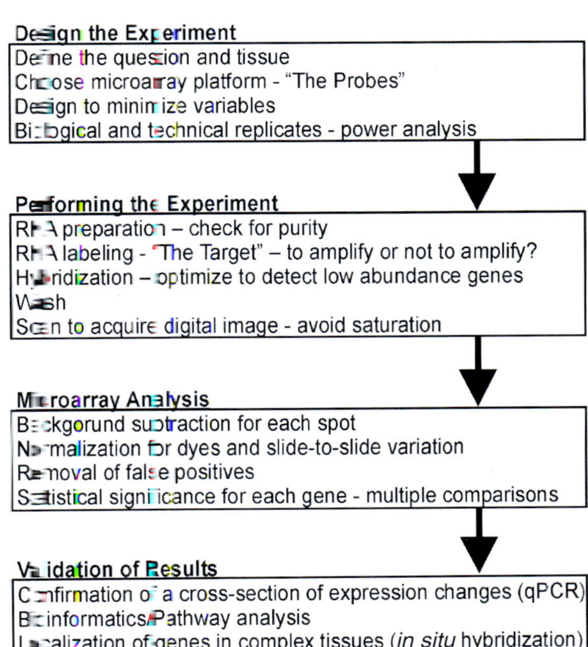


**Fig. 12.1** Flowsheet for the design and implementation of a successful microarray experiment (Copies of figures including color copies, where applicable, are available in the accompanying CD)

Keeping track of these seemingly unimportant differences is critical in biomedical applications where complex tissues are often the starting material for a microarray experiment (such as surgical or post-mortem specimens). The more complex the tissue and the experimental question, the more important it is to minimize the variables you *can* control. The bottom line is that while strictly controlled samples are always essential for all types of biological experimentation, microarrays are totally unforgiving because of the large numbers of gene measurements you are making that will reflect any and all differences between the samples.

## 12.4 Importance of Replicates

Because of their cost, many investigators have given microarrays a bad name by performing underpowered experiments with too few replicates. Error in microarray experiments can be either random (such as a piece of lint covering several spots on a given slide) or systematic. Systematic errors can be highly reproducible and due to a number of both known and unknown variables including the specific gene sequence, dye effects, or slide production defects. One thing to look for in a microarray platform is reproducibility. This includes spot morphology, signal intensity, and background from one slide to the next. This will significantly reduce the random errors. A fundamental approach to reduce both these types of errors is through a combination of biological and technical replicates that includes 'dye flipping' or control-control hybridizations for microarrays that employ two fluorescent-labeled targets hybridized together on the same slide [9].

An example of a *biological replicate* would be to set up 4 plates of cultured cells that are each treated with the same drug and processed in parallel with 4 untreated plates of cells. The gene expression ratios between the treated and untreated cells can then be determined for each pair of cultures, and then the 4 ratios can be compared to one another. Any gene expression changes due to subtle differences in experimental technique, unique to a given plate of cells, would not be seen after averaging the results of the other 3 replicates. Without these biological replicates, one cannot differentiate which gene expression differences are from the treatment and which are due to a subtle, and often unanticipated, methodological difference.

In addition to biological replicates, equally important are *technical replicates*. A technical replicate for the above experiment would be to prepare quadruplicate slides from each pair of cell cultures (treated vs. untreated). A statistical power analysis can be performed to determine the number of technical replicates necessary to measure the statistically significant expression changes based on the variance of the measurement. Power analyses [10] can be used to predict the degree of statistical confidence for a given microarray platform in a given investigator's hands by providing a statistical measure of whether an observed change in gene expression may actually be real. The example in Fig. 12.2 shows that a 1.5–fold change in expression can be measured at 92% power with quadruplicate measurements.

Ideally, combinations of both biological and technical replicates are needed to minimize errors and false positives. From the above example, 16 slides would be required to maximize both biological and technical replicates (4 biological replicates each performed in quadruplicate). Given the potential cost for this experiment, it is thus not surprising that only a few laboratories have been rigorous in their experimental design and they will either pool samples together and perform technical replicates (increasing false positives), or perform only biological replicates (with no measure of statistical significance for each experiment). There is no simple solution, but it is important to understand the inherent risks one takes with the experimental design and the consequences that it will have on the outcome.

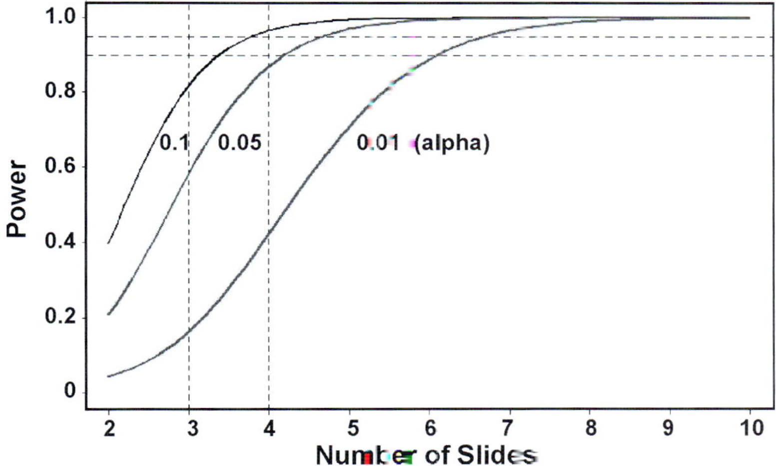


**Fig. 12.2** Power analysis to determine the number of replicates required to detect a 1.5-fold change of gene expression. This experiment was performed to find out how many replicate arrays are needed to achieve statistical significance at p-values (*alpha*) of 0.1, 0.05, and 0.01 based on the variance from the replicate Agilent 60-mer oligonucleotide microarrays hybridized with fluorescent-labeled targets. In order to reach 92% power at $\alpha = 0.05$, quadruplicate arrays are needed (shown at intersection of *dashed lines*). This type of analysis should be performed for each new microarray study, as the variance of a given measurement for each combination of platform, tissue and experimentes can be quite different. Specialized statistical software such as SAS v9, used here, or MATLAB can be used to perform such analyses (Copies of figures including color copies, where applicable, are available in the accompanying CD)

## 12.5 Designing the Experiment

While there are ways to reduce the number of slides to save money, given the above considerations, it is important to design the experiment to maximize the likelihood of finding 'true positives', through both biological and technical replicates. How the experimental design provides for these replicates is dependent on the microarray platform chosen. For example if a membrane filter is chosen, it can be re-used several times. Glass slides with cDNAs and oligonucleotides are generally used only once and then discarded. For these types of microarrays, most commonly in use today, the first consideration is whether to apply only a single sample or multiple samples with different fluorescent labels hybridized against one another on the same array. Choosing a microarray platform that uses a single dye makes both the design and subsequent analysis simpler, but requires twice the number of slides. While slides with multiple spots representing small oligonucleotides from different regions of the same gene (e.g., Affymetrix) can increase confidence in specificity, they do not truly represent the technical replicates, as each of the probes is different and cannot be equated. Similarly, they cannot account for the slide-to-slide variance. The discussion here will be limited to glass oligonucleotide slides with two different fluorescent targets (samples) applied simultaneously. Because of the two dyes, this design is more complicated than a single target per slide, but can easily be converted to a single target per slide design by doubling the number of slides. Two different experimental paradigms will be described below. The statistical treatment of each of these experimental paradigms will be discussed in a later section of this chapter.

### 12.5.1 Comparison of two or more samples to each other

If the goal of the experiment is to compare two samples with each other, such as treatment and control, 4 slides can be used with a flip-dye design to generate quadruplicate data points for each sample and reduce variances due to dye and slide effects. Fig. 12.3A shows a typical microarray experimental design for this type of comparison. This experiment generates 4 independent measurements of gene expression for each gene that can be used to compute an average fold-change together with a measure of statistical significance for each gene on the array. This is really no different than doing the experiment one gene at a time using Polymerase Chain Reaction (PCR) or other methods to measure the mRNA levels. The reason for doing half of the samples with each dye is to compensate for any systematic differences that are due solely to the differences in dye characteristics. This type of flip-dye design can also be expanded to a 'loop' design as shown for 3 or more samples in Fig. 12.3A. An advantage of the loop design is that 2 treatment effects can be interrogated without using twice the number of slides.

**A. Dye-Swap Method for Technical Qaudruplicates**

**2 Samples** **3 Samples**

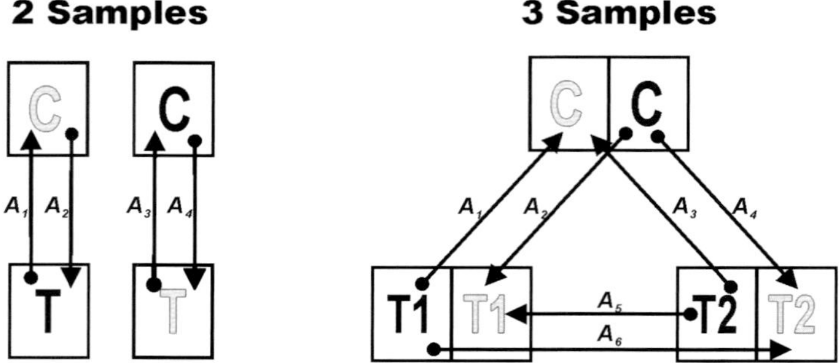


**B. Series Using a Common Reference Control**

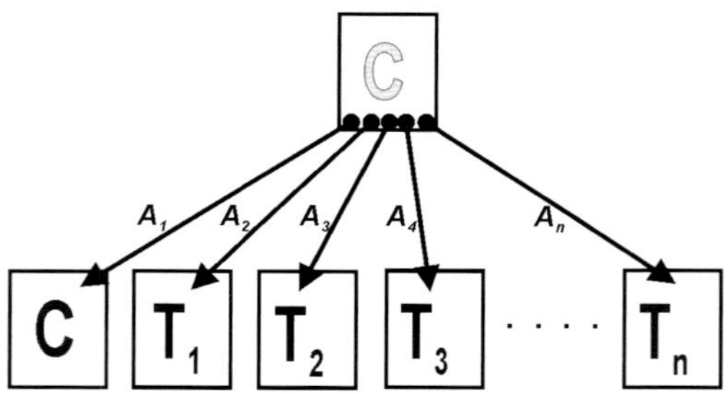


**Fig. 12.3** Designing the experiment. **(A)** Comparison of two or more samples to each other. Shown here in Kerr-Churchill notation, is the simplest type of comparison to generate quadruplicate results for each gene using two-color, dye-swap arrays. Note that each arrow labeled ($A_1$, $A_2$...) represents a single microarray. Each target is labeled with each of the two dyes. The treatment sample (T) is combined with the control sample (C) and hybridized on a single slide. Two arrays use the red (Cy5) dye for T and green (Cy3) for the C, while the other two arrays 'flip' the dye so that T is for green and C is for red. When comparing 3 or more samples a 'loop' design is used to reduce the total number of slides required. As shown on the right. For 3 samples, quadruplicate measurements can be obtained with only 6 slides. **(B)** Data series to compare sequential samples. Here a data series using an individual array for each measurement (e.g., time point or dose) and a common reference, usually the control or zero point, is used to provide a standardized reference for each point. A control-control hybridization is included in the series to identify false positives. Replicate arrays can be used for increased power but are usually not necessary with this design (Copies of figures including color copies, where applicable, are available in the accompanying CD)

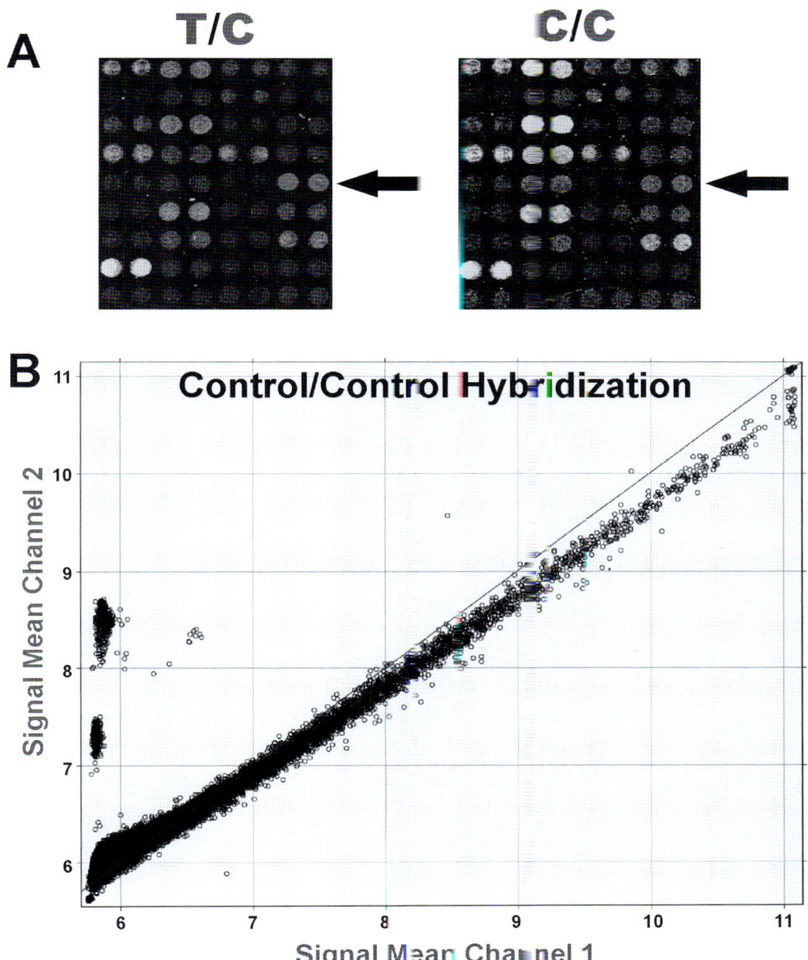


**Fig. 12.4** Control-control hybridizations to reduce false positives. **(A)** Some spotted genes artefactually fluoresce more at one wavelength than the other. This is shown for the two spots (*arrows*) that are red both in the treatment/control (T/C) experiment as well as in the control-control (C/C) experiment using a cDNA array. For the control-control experiment, mixing an equal amount of red and green should produce all yellow spots when the two images are superimposed. **(B)** When the mean signal intensities for each dye are plotted against one another, each data point should fall very near the normal, reflecting only subtle differences in dye characteristics. Those spots that deviate significantly from this line are false positives with a higher intensity of either the red or green dye. The two clusters of spots near the y-axis are control spots on the Agilent array used for this experiment (Copies of figures including color copies, where applicable, are available in the accompanying CD)

### *12.5.2 Data series to compare sequential samples*

In addition to replicates, another way to reduce extraneous variables in an experiment is to identify gene expression changes that are assumed to vary in a specific way over a series of sample points. For a given treatment effect, this could be a function of time or dosage, as might be measured in a time-course or dose-response experiment. The hypothesis is that gene expression changes will increase, decrease, or show a more complex effect (e.g. biphasic) in response to a treatment. While technical replicates add power to the comparison experiments discussed above, in a data series experiment, the main power comes from the number of time points or doses that are analyzed. The closer the data series approximates a theoretical response, such as increasing with time, the more likely the change is true. As such, instead of simple p-values, significance comes from correlation coefficients to the theoretical response. Figure 12.3B shows an example of such a series that uses duplicate measurements at each point in the series and a common reference design to reduce false positives. A flip-dye approach is not needed here as the dye effects will be constant across both the control and treatment arrays allowing the identification of gene expression changes as a function of the variable of interest such as time or dose.

This design uses the control as a common reference and also includes a control-control hybridization, which simply means that the same sample is labeled with two different dyes and applied to the same slide. This type of control is needed if absolute, rather than relative differences in gene expression are required. As shown in Fig. 12.4, a control-control hybridization should produce approximately equal intensities for both dyes, viewed as all yellow spots when using a green/red color scheme. In fact, while most spots on the microarray will be yellow, there are often many that are aberrantly more red or more green. These are systematic errors that, for unclear reasons, produce high signals for one dye or the other (for a detailed analysis of the importance of taking into account these systematic errors [9]). Even if such spurious signals represent only 1% of the spots on the array, they can lead to hundreds of false positive results, all of which could be eliminated with an experimental design that includes a control-control hybridization.

## 12.6 Target Labeling, Hybridization, and Image Analysis

### *12.6.1 RNA*

Once the experimental plan is in place, it is time to get into the laboratory and perform the actual experiment. To start with, highly purified DNA-free RNA samples from control and treatment samples are needed. Before proceeding, a critical step involved is showing that each RNA preparation is clean and not degraded. Enzymes that degrade RNA called RNases are highly stable proteins that are ubiquitous in the laboratory. Working with RNA requires extreme care in the techniques and the use of RNase free reagents and buffers. If total RNA is being used, the simplest way to verify RNA purity and integrity is by analyzing each RNA sample on a gel that resolves the

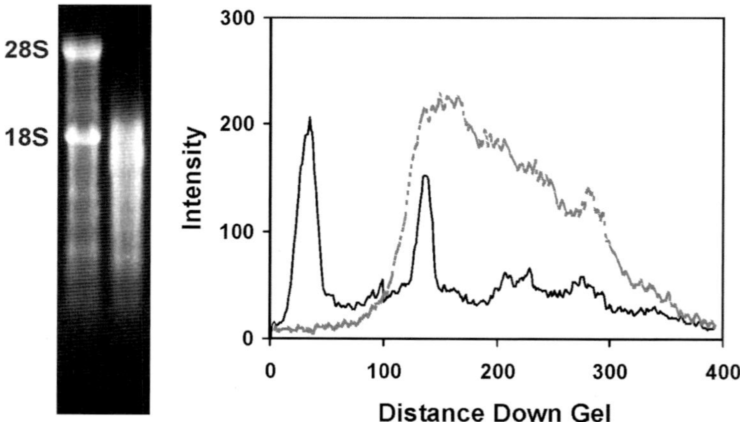


**Fig. 12.5** Measurement of RNA purity. Purified RNA that is not degraded is critical for accurate microarray results. On the left is a denaturing gel of a preparation of the total RNA made from human brain. While the sample in the first lane is intact, the sample to the right is degraded and runs as a smear. On the right a densitometric scan of this same gel shows the most abundant ribosomal RNA species at 18S and 28S (a measure of their size). For human RNA, the normal 28S/18S ratio is approximately 2.0. While the intact sample on the left shows this ratio, in the quantitation shown on the right (*dark/solid line*), the degraded sample, shows a broad distribution of sizes representing the smear seen on the gel (*lighter/dashed line*) (Copies of figures including color copies, where applicable, are available in the accompanying CD)

relatively abundant ribosomal RNA bands. A quantitative measure of the ratio of 28S/18S (2:1 for mammalian samples) is a widely used method. Figure 12.5 shows total RNA samples run on a gel and their quantitation comparing the intact and degraded samples. No matter how valuable the original sample was, it makes little sense to spend the time and money on degraded RNA samples.

### 12.6.2 Labeling

The next step is to chose the labeling method to convert the mRNA from each sample into labeled targets that can bind to the probes on your microarray. Early on, most researchers in the field used direct labeling methods that generated a 1:1 mRNA:cDNA or cRNA. This direct labeling method has the advantage of creating a direct copy of the mRNAs present in the sample without introducing non-linearity that could occur during target amplification. The major disadvantage of this method is that it requires large amounts of RNA to generate enough targets, to see a signal on a single slide (up to 40 μg of total RNA). Clearly, with the need for technical replicates, this is rarely feasible for most experiments particularly in biomedical applications where tissue is limited. More recently, this problem has been circumvented by the development of linear amplification methods that can use as little as 25 ng of the total RNA and still provide an accurate reflection of the amount of each species of mRNA in the original sample [11]. For most of the commercial microarray platforms, amplification methods that use a combination of reverse-transcription and *in vitro* transcription with RNA polymerases have been optimized by manufacturers [12,13]. Labeling of the target is achieved by incorporating labeled nucleotides into the enzymatic reactions. Most use fluorescent dyes, such as Cy3 and Cy5, when multiple samples are applied to the same slide. The amount of each labeled target that binds to each spot on the array can then be quantified using a laser scanner to measure the total amount of mRNA in each sample.

### 12.6.3 Hybridization

The labeled targets are mixed together in a hybridization solution, applied to a small volume on the slide, and allowed to hybridize at a specific temperature for 1–2 days. Small volumes are necessary to maintain the high concentrations of labeled targets required to generate a strong signal. However, working with small volumes can also result in poor coverage of the hybridization solution on the slide and hence uneven results. While the traditional method of placing a glass cover slip over the slide to distribute the hybridization solution evenly often works, it can often produce uneven hybridization or drying artefacts on the slides, thus leaving a large number of spots unreadable. One of the advantages of amplifying the target is to produce enough targets so that larger volumes of hybridization solution can be applied in a larger chamber size that will allow even spreading of the solution over the slide. The hybridization buffer, temperature, and time are the variables that can influence both the spot signal and the background signal. For cDNA arrays, the hybridization kinetics can vary for each gene spotted on the array because of the variability in the affinity of the target for each probe that depends on the GC content of nucleotides. For most oligonucleotide arrays that are designed to have similar GC contents for the probes, the conditions have been worked out by the manufacturers, and they generally perform best using the labeling and hybridization reagents and instructions provided by the manufacturer. Following hybridization, labeled nucleotides that are not specifically bound to their complementary sequences are washed away with increasingly stringent, low-salt buffers. The wash buffer and temperature can greatly influence the signal-to-noise ratio and thus, requires optimization. Slides must be dried carefully without salt precipitation and kept free of dust prior to laser scanning.

### *12.6.4 Image Analysis*

Each spot on the microarray can be viewed as an independent experiment that is performed in replicate to give information on the expression of a particular gene. As such, each spot needs to be quantified in a uniform manner. When two or more dyes are used, this has to be performed at each of the wavelengths that are optimal for the excitation/emission spectra for that dye. Differences in dyes can lead to systematic errors that have to be taken into account and, to an approximation, can be reduced by flipping the dyes as discussed in the experimental design section. Most investigators use high-resolution laser scanners that are required to resolve large numbers of small spots, up to 200,000 per glass slide. Scanners need to be set in such a way as to obtain the strongest signal possible to detect the weakest spots, but not too strong a signal that will lead to saturation of higher intensity spots. If a spot is saturated it cannot be measured. Some investigators will scan a slide twice, once at low gain and once at high gain to increase the dynamic range of the scan.

Once the slide is scanned at the optimal settings, the data is saved as an image file such as a Tagged Image File Format (TIFF) for each channel as shown in Fig. 12.6. This image is then pseudo-colored so that one dye color is shown as red and the other green. Most investigators using

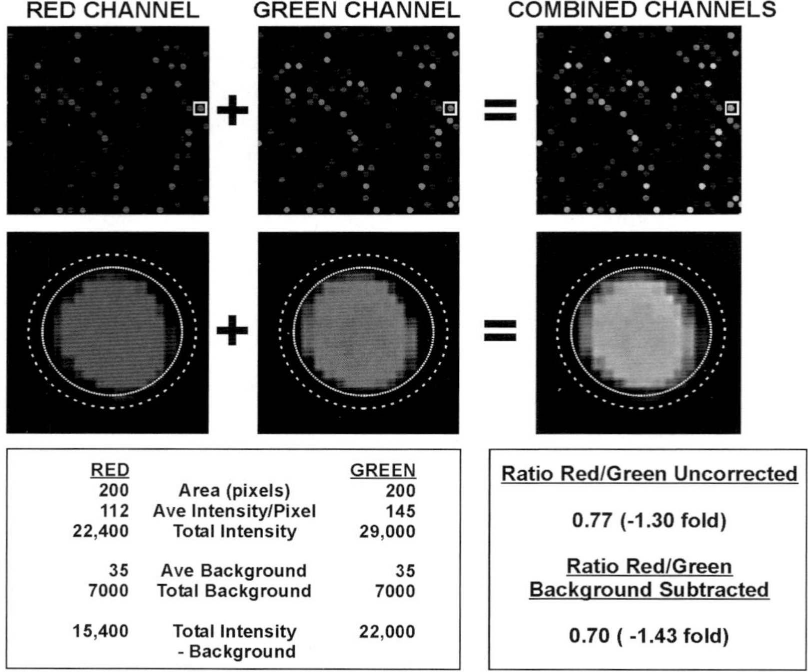


**Fig. 12.6** Microarray image analysis. For each spot on the microarray that has been hybridized with two differently colored (*wavelength*) targets, the total intensity from each color needs to be measured. The top panel shows a low-power magnification of the red and green channels separately and then combined and shows a box around a single spot that, when combined, shows a higher intensity in the red channel (see CD for color figure). The middle panels shows a higher magnification of this spot that is used for quantitation. The exact same region of interest (ROI) is drawn around the spot for each channel individually (*fine dashed line*) and the total number of pixels (*area*) as well as the average intensity per pixel and total intensity for each channel is calculated in the bottom panel. Without background correction, the ratio of the red to the green signal is 0.77 or –1.30 fold. Since the background signal can significantly contribute to this total signal, a second ROI outside the first circle (*dashed line*) is used to calculate the average background intensity between the two ROIs. This can then be multiplied by 200 pixels to give the total background signal contributing to the smaller ROI around the spot. When this background is subtracted from the total intensity, the ratio of red/green signal is reduced to 0.70 or –1.43 fold (Copies of figures including color copies, where applicable, are available in the accompanying CD)

the combination of Cy5 and Cy3 dyes, pseudo-color them red and green, respectively, even though Cy3 is in fact a lot closer to red and Cy5 is infrared and hence not visible. When combined into an Red Green Blue (RGB) image, spots where gene expression is unchanged and hence have an equal amount of red and green, appear yellow. Spots appear more red due to a higher level of gene expression in the Cy5 labeled targets and more green when the Cy3 targets are more abundant.

The next step is to quantify the intensity of each spot at each wavelength. This is achieved by a number of commercially available software programs, some of which are provided with scanners. Others that offer more flexibility can be purchased separately and used for analyzing any microarray image. How this quantitation is performed can make an enormous difference in the final results and hence is worth going through in detail. Figure 12.6 shows a step-by-step approach that illustrates this point. The first step is to define a region of interest (ROI) that is generally a circle that surrounds a circular spot. While the size of the circle can make a difference, most importantly, whatever size that is used must be the same ROI for each wavelength of the two images being compared. Within this ROI, the software can then determine the area of the spot, or total number of pixels, and the intensity of each pixel within that area. The sum of the intensity measurements represent the total intensity of the ROI. After performing these calculations at each wavelength, a ratio of the intensity of the red and green channels provides a measurement of the fold-change in gene expression.

In a perfect world, one could stop here and do this analysis for the entire slide to obtain results. However, most slides have an appreciable background signal that can be due to auto-fluorescence of the glass slide. This background signal varies for each dye/wavelength and can also vary significantly in different regions of the slide, making it often necessary to account for the *local* background signal near each spot. In fact, for spots of low intensity, the background signal can represent a significant proportion of the total signal intensity. It is therefore not surprising that a number of sophisticated image analysis methods have been developed to take into account the local effects of the background. Figure 12.6 gives a simple example of how the background is measured in a region outside the ROI, but within a larger concentric circle around the same spot. As shown in this example, the average background intensity within this local area times the total number of pixels in the ROI is then subtracted from the total intensity of the ROI. This gives a more accurate measure of the total intensity of each spot at each wavelength. This example also shows a significant difference in the ratios of gene expression calculated with and without background subtraction that will be of greatest significance for spots of low intensity.

## 12.7 Statistical Analysis

Once an accurate measure of the intensity of all the spots on the arrays is obtained for each dye, a detailed statistical analysis is required to determine the significance for each gene expression change. Performing a statistical analysis in a microarray experiment in which tens of thousands of measurements are made simultaneously is an ongoing challenge that has led to the development of a whole new area of statistical research. An important principle to keep in mind throughout this section is that statistics only provides a probability measurement that something, in this case gene expression, *might be* changed based on the magnitude of the change relative to the variance of the measurement. Having sufficient statistical grounds to reject the null hypothesis ('gene x is unchanged') does not however prove that the gene is in fact changed.

While there are a number of other chapters in this text on the statistical approaches to microarray data-sets, here we will concentrate on the initial analysis of gene expression changes based on the two types of experimental designs discussed earlier. Reducing the number of false positives and false negatives at this stage is critical for subsequent bioinformatic analyses and biological confirmation of gene expression data-sets.

### *12.7.1 Comparison of two or more samples to each other*

Fig. 12.3A shows the simplest experimental design that uses 2 samples, each labeled with a different dye and hybridized against one another on the same slide. A total of 4 slides are used that results in quadruplicate measures for each gene. By labeling each sample with each dye and then 'flipping' the dyes, systematic errors due to dye effects should be taken into account. In order to determine the difference in signal intensity between the 2 samples that is not due to differences from the array, gene, and dye variations, statisticians commonly use an ANalysis Of VAriance (ANOVA) to identify and remove these systematic and random variances. The end result attempts to determine the true differences due to differential gene expression. A sample approach used by many in the field uses a two-interconnect ANOVA model referred to as Mixed Model Analysis of Microarray Data (MANMADA) (to identify differentially expressed genes). The first step employs a normalization model for $\log_2$-transformed intensity measurements:

$$y_{ijk} = \mu + A_i + D_j + (AD)_{ij} + r_{ijkg} \qquad (12.1)$$

where $\mu$ is the sample mean, $A_i$ is the effect of $i$th array, $D_j$ is the effect of the $j$th dye (cy3 or cy5), $(AD)_{ij}$ is array-dye interaction and $r_{ijkg}$ is the residual co-variance. This residual co-variance can be taken as a "normalized" expression level and used in following gene model to obtain the treatment effect on each gene:

$$r_{ijkg} = A_{ig} + D_{jg} + T_{kg} + \varepsilon_{ijkg} \qquad (12.2)$$

where $r_{ijkg}$ is the residual of each gene from the normalization model, $A_{ig}$ and $D_{jg}$ are the array and dye effects, respectively, $T_{kg}$ is the treatment effect (control or treated) and $\varepsilon_{ijkg}$ is random error. It is important to note that if the model performed optimally, $\varepsilon_{ijkg}$ will contain only random errors and thus, be a normal distribution of values centered on 0 with a standard deviation of 1. The expression change for each gene is thus:

$$\log 2(\text{fold}) = T_{\text{treated}} - T_{\text{control}} \qquad (12.3)$$

The same model could be applied to the loop design on the right side of Fig. 12.3A. The main difference would be an additional fold-change and significance values for each treatment effect ($T_1$, $T_2$,...$T_n$) assessed. One of the main advantages of using ANOVA models is their scalability. Potential co-variance can be added and removed based on the extent to which they are found to contribute to the overall variance of the measurement.

### *12.7.2 The Multiple Comparison Problems*

At this point, we have a fold-change and statistical significance value for each gene on the microarray. Because we are performing tens of thousands of pair-wise comparisons simultaneously, it is important to adjust the significance values for multiple tests or 'comparisons.' For example, consider a microarray with 100 genes. If a statistical significance is set at 95% ($\alpha = 0.05$), we expect 5 genes to be significant by chance alone (false positives). If we expand this to a high-density microarray with 20,000 genes, we could anticipate 1000 false positives. Thus, a list of 300 gene expression changes obtained without adjustment for multiple comparisons could be entirely due to false positives. To address this problem of multiple comparisons, a number of statistical methods are used to control this family-wise error rate. The method of Bonferroni is the simplest approach, which multiplies the uncorrected p-value by the number of simultaneous tests performed. However, it is excessively restrictive for very large datasets and will likely result in very few significant gene expression changes. An alternative approach that is of much use is the False Discovery Rate (FDR) method of Benjamini, which controls for false positives with a sophisticated algorithm that adapts to the simulated proportion of true null hypotheses in the data-set [14]. As the number of genes in a microarray data-set increases, the proportion of truly unchanged genes increases and the adaptive algorithm scales down the significance adjustments to the smaller proportion of genes that are indeed changed.

### *12.7.3 Data series to compare sequential samples*

While a series measurement may not have as many replicates per point, it maintains its statistical power with the assumption that gene expression increases, decreases, or changes as a non-linear or bi-phasic function of the variable being tested. As discussed in Fig. 12.3B, two of the most simple examples would be a time course, where RNA is extracted from the same tissue or cells at increasing time intervals after a treatment; or a dose response, where increasing amounts of a substance is added to the tissue and RNA is prepared at the same time point. Often with this type of experimental design, two dyes/targets per slide are not needed, but can be used if the same 'control' sample is used as a reference for each time or dose. The advantage of using a reference sample, as similar as possible to the treatment samples, is to identify false positives more readily by having different dyes, that are similar in intensity, on all the spots except those that are changed. The first group of slides with this design would then be a control-control hybridization where any calculated changes in gene expression are, by definition, artefacts and can form a baseline measurement for any real changes observed (See [9] for details of this control correction method).

For slides in a data series, the first step is to use the ANOVA to normalize the data for each slide, for array and dye effects. This will give an absolute normalized intensity for each of the genes that will then be averaged for each point in the series. The second step is to identify genes whose expression correlates in a statistically significant way with the variable being measured, such as time or dose. A simplistic approach is to assume a linear increase (or decrease) in gene expression and calculate a statistical correlation coefficient for each gene. This can be done with a measure of correlation distance such as the Pearson metric linear regression model. Genes can then be ranked in order of their statistical significance to the variable being changed in the series. Using this scheme, it is desirable first to use an arbitrary fold-change minimum between the first and last points in the series. A more complicated way that does not have to assume a linear relationship between gene expression and the variable measured is to perform a cluster analysis to identify groups of genes with similar patterns of expression across the variable.

The end result from either of the two experimental designs and statistical analysis methods is to create lists of genes that change as a function of the variable of interest. These can then be ranked from greatest to least significant and studied further individually or group wise, based on known information about their functions and sequences. Either of these approaches can lead to exciting new research directions that would not necessarily have been predicted.

## 12.8 Using Bioinformatics to Simplify Results and Generate New Research Directions

While the goal of a sound microarray experimental design is to minimize the variables, so that only genes are changed due to the variable of interest, in many experiments the list of significant gene expression differences can be quite extensive. Often the central goal is to generate new testable hypotheses and research directions. One of the major pitfalls by investigators confronted with long lists of genes, is to simply 'pick' one's favorite genes for further studies. The experiment then becomes a self-fulfilling prophecy based on pre-existing biases of the investigator that will not lead to new research directions. If this is the case, there is really no point in doing the microarray experiment in the first place. A simple set of PCR reactions would have sufficed.

How do we allow the *data-set to tell us* what are the most important differences in gene expression from a large set of expression changes in an impartial way? The basic principle is to use bioinformatic algorithms on the genes, shown to be significantly changed, and then to divide them into smaller subgroups based on known gene functions or structural classifications. For a majority of genes in the human genome there is some knowledge of function from the scientific literature and even more information about gene structure through large-scale sequencing studies. Thus for each gene, the investigator can develop a suggestion of function that can be combined with the level of statistical significance to create a transcriptional 'fingerprint'. Browsing through the rest of this book, it should be clear to the reader that using bioinformatics to select genes and pathways for further studies can be done in many ways with many custom or commercial software programs. These include promoter analysis, ontological (functional) analysis, cluster analyses, and key word interaction maps. No matter how good one's bioinformatic algorithm is, to ascribe functions, it is hard to beat a careful literature analysis of groups of changed genes within the same biological context as the experimental system or tissue being studied.

Ultimately, the goal of any bioinformatic analysis should be to simplify groups of genes into major biological pathways or transcriptional control mechanisms. Once a pathway or transcriptional control mechanism is clearly suggested, it is still not proven until both the expression changes and biological pathways are verified. The bottom line is that microarray experiments can help form new hypotheses, each with a statistical likelihood, but other methods are required to validate those predictions. A microarray experiment without validation, while sometimes may be tantalizing, can lead to frustration and a lot of wasted time and resources.

## 12.9 Verification and Significance of Results

### *12.9.1 Other means of measuring relative mRNA levels*

The best way to assess the accuracy of a given microarray platform is to show the same gene expression change using a different method. While this cannot realistically be done for all of the genes on a microarray, it should be done on a proportion of the genes that are chosen to be studied further to rule out that the change is a false positive. False negatives, or genes that have been shown to change by another method, but do not change on the microarray, can also be discovered in a similar manner. Quantitative measurements of mRNA levels for a specific gene can be achieved in a number of ways including quantitative real-time RT-PCR (qPCR), Northern blotting, and nuclease protection.

Because of its ease of use and consistency, most microarray researchers use qPCR. This form of PCR has the key advantage of being able to verify a large number of samples quickly and inexpensively using a method that is very different from the microarray method. While it is beyond the scope of this chapter, the reader is referred to a recent review of qPCR [15]. The 'q' of qPCR is extremely important, as other non-quantitative and 'semi-quantitative' forms of PCR are generally unreliable. It is also important that, when choosing which genes to verify, genes are chosen across a wide range of intensity levels and fold changes. This provides an index of reliability across the dynamic range of the microarray platform for each experimental condition. It is not surprising that genes with low-signal intensities, relative to the background, and genes with lower fold changes are less likely to be verified. Although there are no absolute rules, the number of genes chosen to be verified should be determined based on these same parameters.

### 12.9.2 Localization of genes in complex tissues (in situ hybridization)

While in some experiments, it is possible to collect RNA from cultures of a single cell type or from specific cells within a complex tissue, such as by laser capture microdissection, in many cases this is not possible. Complex tissues containing multiple cell types are often the starting material used to generate RNA samples for microarray studies. A gene expression change seen from such a complex tissue, therefore, cannot be attributed to one or more cell types within that tissue without verification. *In situ* hybridization is a method that uses complementary hybridization of nucleotide

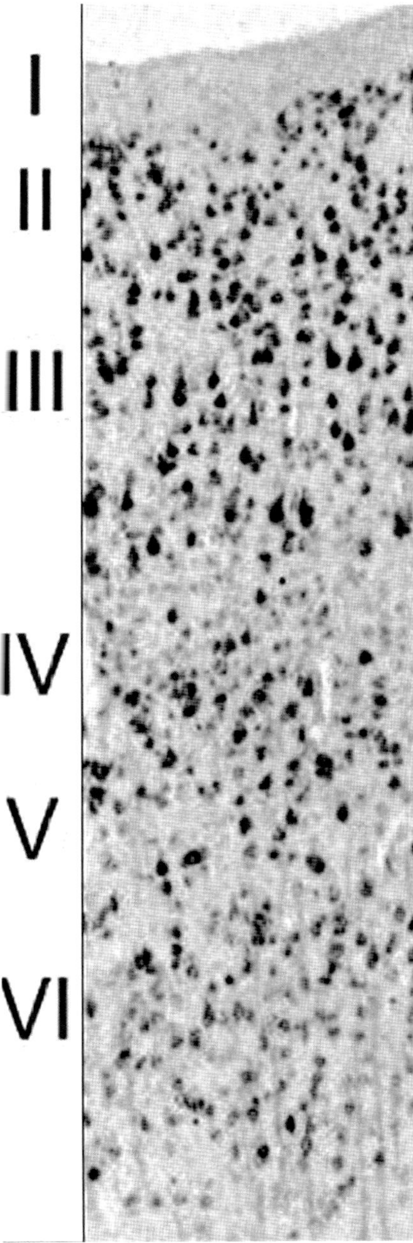


**Fig 12.7** Localization of genes expressed in complex tissues by *in situ* hybridization. Once a gene is found to be increased or decreased, it is important to show exactly in which cells this occurs, particularly in the case of complex tissues with multiple cell types. Shown here is an *in situ* hybridization of the human brain probed for a gene that is induced at sites where epileptic seizures are generated and it shows selective labeling of the neurons, mostly in the superficial cortical layers (*layers I–IV*) (Copies of figures including color copies, where applicable, are available in the accompanying CD)

sequences to identify which cells in a complex tissue express a given gene of interest. Ideally, a histological section of the same tissue that was used to prepare RNA for the microarray analysis can reveal exactly which cells express a given gene is observed to be differentially expressed. An example of an *in situ* hybridization that uses a radioactively labeled probe on human brain sections is shown in Fig. 12.7 (see [16] for an example of how *in situ* hybridization can localize differentially expressed genes in human epileptic brain tissue). This image shows that only a small proportion of cells express the gene of interest in specific neurons within the brain. This result also emphasizes the importance of having a microarray system that is as sensitive as possible. If only a small proportion of cells change their expression of a given gene, the effect could be diluted out by all of the other cells in which gene expression does not change.

### *12.9.3 Functional verification*

Ultimately, the goal of the microarray experiment is to answer an important biological question. While confirming that a given gene expression change is in fact real by qPCR and localizing that change to specific cells using *in situ* hybridization are essential steps, they cannot prove that a predicted pathway is indeed activated or inhibited. To do this one will require a biological knowledge of the pathway and further experiments with the same tissue. An example of this could be that a significant group of activated genes are known to be transcriptionally regulated by a specific transcription factor and that this transcription factor is phosphorylated when activated. Chromatin Immunoprecipitation (ChIP) for the activated transcription factor could be used to demonstrate that it is indeed bound to the proximal promoters of a group of target genes observed to be differentially expressed on the microarray. A Western blot showing this transcription factor is indeed turned on by covalent phosphorylation would further demonstrate pathway activation in an entirely independent manner from the microarray experiment.

## References


1. Armstrong, N.J. and M.A. van de Wiel, *Microarray data analysis: from hypotheses to conclusions using gene expression data*. Cell Oncol, 2004. **26**(5–6): 279–290.
2. Bueno Filho, J.S., S.G. Gilmour, and G.J. Rosa, *Design of microarray experiments for genetical genomics studies*. Genetics, 2006. **174**(2): 945–957.
3. Breitling, R., *Biological microarray interpretation: the rules of engagement*. Biochim Biophys Acta, 2006. **1759**(7): 319–327.
4. Leung, Y.F. and D. Cavalieri, *Fundamentals of cDNA microarray data analysis*. Trends Genet, 2003. **19**(11): 49–659.
5. Neal, S.J. and J.T. Westwood, *Optimizing experiment and analysis parameters for spotted microarrays*. Methods Enzymol, 2006. **410**: 203–221.
6. Simon, R., M.D. Radmacher, and K. Dobbin, *Design of studies using DNA microarrays*. Genet Epidemiol, 2002. **23**(1): 21–36.
7. Kreil, D.P., R.R. Russell, and S. Russell, *Microarray oligonucleotide probes*. Methods Enzymol, 2006. **410**: 73–98.
8. Millenaar, F.F., J. Okyere, S.T. May, M. van Zanten, L.A. Voesenek, and A.J. Peeters, *How to decide? Different methods of calculating gene expression from short oligonucleotide array data will give different results*. BMC Bioinformatics, 2006. **7**: 137.
9. Yao, B., S.N. Rakhade, Q. Li, S. Ahmed, R. Krauss, S. Draghici, and J.A. Loeb, *Accuracy of cDNA microarray methods to detect small gene expression changes induced by neuregulin on breast epithelial cells*. BMC Bioinformatics, 2004. **5**: 99.
10. Jafari, P. and F. Azuaje, *An assessment of recently published gene expression data analyses: reporting experimental design and statistical factors*. BMC Med Inform Decis Mak, 2006. **6**: 27.
11. Van Gelder, R.N., M.E. von Zastrow, A. Yool, W.C. Dement, J.D. Barchas, and J.H. Eberwine, *Amplified RNA synthesized from limited quantities of heterogeneous cDNA*. Proc Natl Acad Sci U S A, 1990. **87**(5): 1663–1667.

12. Ginsberg, S.D., *RNA amplification strategies for small sample populations*. Methods, 2005. **37**(3): 229–237.
13. Livesey, F.J., *Strategies for microarray analysis of limiting amounts of RNA*. Brief Funct Genomic Proteomic, 2003. **2**(1): 31–36.
14. Reiner, A., D. Yekutieli, and Y. Benjamini, *Identifying differentially expressed genes using false discovery rate controlling procedures*. Bioinformatics, 2003. **19**(3): 368–375.
15. Lutfalla, G. and G. Uze, *Performing quantitative reverse-transcribed polymerase chain reaction experiments*. Methods Enzymol, 2006. **410**: 86–400.
16. Rakhade, S.N., B. Yao, S. Ahmed, E. Asano, T.L. Beaumont, A.K. Shah, S. Draghici, R. Krauss, H.T. Chugani, S. Sood, and J.A. Loeb, *A common pattern of persistent gene activation in human neocortical epileptic foci*. Ann Neurol, 2005. **58**(5): 736–747.


## Key Papers in the Field


1. Breitling, R., *Biological microarray interpretation: the rules of engagement*. Biochim Biophys Acta, 2006. **1759**(7): 319–327.
2. Churchill, G.A., *Fundamentals of experimental design for cDNA microarrays*. Nat Genet, 2002. **32 Suppl**: 490–495.
3. Kerr, M.K. and G.A. Churchill, *Statistical design and the analysis of gene expression microarray data*. Genet Res, 2001. **77**(2): 123–128.
4. Leung, Y.F. and D. Cavalieri, *Fundamentals of cDNA microarray data analysis*. Trends Genet, 2003. **19**(11): 649–659.
5. Neal, S.J. and J.T. Westwood, *Optimizing experiment and analysis parameters for spotted microarrays*. Methods Enzymol, 2006. **410**: 203–221.
6. Yao, B., S.N. Rakhade, Q. Li, S. Ahmed, R. Krauss, S. Draghici, and J.A. Loeb, *Accuracy of cDNA microarray methods to detect small gene expression changes induced by neuregulin on breast epithelial cells*. BMC Bioinformatics, 2004. **5**: 99.

# Chapter 13
# Tools and Approaches for an End-to-End Expression Array Analysis

**Adrian E. Platts and Stephen A. Krawetz**

**Abstract** Microarray experiments can appear daunting because the considerations called for in their analysis cover several fields of research. To understand the data microarrays generate some knowledge of classical statistics and recent complexity theory are useful while emerging computational techniques such as XML directed workflows could aid in managing the data. These considerations are called for because as experimental tools, microarrays (arrays) exemplify the recent trend in biological research towards high dimensionality datasets. Until recently observations were made on only a few variables at a time and these were used to support or refute hypotheses, but high dimensionality datasets are generated by observing a very large number of variables (e.g. gene expression measurements) at the same time. The number of expression measurements made on arrays is not only high, but notably high when compared to the size of a typical sample population. This combination of high dimensionality and asymmetry leads to large datasets and fundamental problems when using standard approaches to interpret the data. An end-to-end approach is a general framework in which to place some useful considerations when planning an analysis. The framework described here explores the origins of signal and several sources of variance, approaches to representing high-throughput data, the statistical considerations when modeling array data and the software tools that can aid in carrying out the analysis.

**Keywords** Microarray · Statistical analysis · Data repository · False discovery rate · Promoter analysis

## 13.1 Introduction

Many reviews have presented the individual elements that separately contribute to the interpretation of array data. These include valuable treatments on quality control [1], variance [2], experimental power [3] and error rates [4]. The framework presented here is an overview and pointers into this wider literature are suggested. The approaches described are based on RNA expression arrays with a limited overview of DNA arrays. Non-nucleic acid platforms such as protein arrays, lab-on-a-chip micro-fluidics and indirect assays such as ChIP-chip fall beyond the scope of this discussion.

Expression arrays provide an alternative to established RNA sequencing techniques such as EST, 3'/5' SAGE, CAGE and the more recent pico-liter sequencing approaches [5]. Both sequencing and array approaches aim to take similar snapshots of whole-genome transcription by making large numbers of measurements on a sample. Despite the extent of this measurement, what is being quantified may vary in significance. In heterogeneous biological systems, the extent to which

A.E. Platts
Department of Obstetrics and Gynecology, Center for Molecular Medicine and Genetics, Wayne State University School of Medicine, Krawetz Lab, Mott Center, 275 East Hancock, Detroit, MI 48201, USA,
e-mail: adrian@compbio.med.wayne.edu

S. Krawetz (ed.), *Bioinformatics for Systems Biology*, DOI 10.1007/978-1-59745-440-7_13,

measurements include different cell-types or cells that are in different states is rarely negligible. Arrays nonetheless measure the total contribution of expression from all the elements that are included in the system. If only ten percent of the cells in a system express a gene at a high level, this condition would be difficult to differentiate from the state where all the cells express the gene at a much lower level. Measured systems may vary from being in synchronous to relatively incoherent states. The latter might be encountered when assessing unstable systems such as proliferating tumors. As Heng et al. have recently observed, averaging disparate states to generate tissue-level statistics may tell us little about the progression trajectories being followed at the individual cell level [6]. rtPCR has been demonstrated to be sensitive to elements in the single cell, and clinical tests usually measure parameters of the person in their environmental context. Arrays sit between, averaging the properties of cell groups or tissue systems. Before introducing an analysis framework, it is useful to connect the parameters that are measured by the array technologies.

The loss or gain of chromosomal DNA is readily visualized, but the exact property of the cell that is measured by an expression array requires more exploration. Expression studies sometimes quantify non-coding RNA species that are functional within the nucleus but are more generally designed to quantify the abundance of mRNA transcripts, and from this to infer levels of protein translation.

mRNA is most of all a highly specific intracellular messenger that is actively transported from its site of nuclear transcription to the cytoplasmic sites of protein synthesis. In the nucleus, mRNA is transcribed from potentiated genic domains [7] by RNA polymerase II. This takes place predominantly within specific regions termed transcription factories identified between the coherent spatial domains occupied by individual chromosomes [8]. Transcription factories are spatially constrained regions found to be rich in the proteins that functionally bind to gene promoter domains and that determine the rates of transcription. Within these factories, DNA that has been potentiated by decondensation [9] is bound by accessory transcription proteins and polymerase II to initiate transcription. Multiple polymerase complexes then iterate along the DNA sequence, each one assembling a complementary strand of mRNA from ambient nucleotide precursors. The origins of most mRNA lies in the nucleus, a smaller population of mtRNAs encoded by the mitochondrial DNA are transcribed in the mitochondria. Following transcription, mRNAs in the nucleus are immediately bound by splicing, capping and quality control proteins many of which remain with the transcript to form the messenger ribonuclear particles (mRNPs). In these complexes, introns are removed from the sequence and a 5' cap sequence and in most cases a 3' poly(A) sequence are added (capping and polyadenylation). The cap and poly(A) tail sequences stabilize the product allowing the mRNPs to be transported to the peripheral nuclear lamina just internal of the nuclear membrane. Here the nuclear pore complexes interact with mRNPs, in a further quality control step, that identifies poorly transcribed RNA prior to nuclear export. Exosomal complexes rapidly remove any mis-transcribed or mis-spliced RNA through nonsense mediated decay and nonstop initiated decay. Following export to the cytoplasm for translation, ribosomal complexes associate in large numbers with the individual transcripts potentially transcribing hundreds of identical proteins from each sequence in a process that gradually reduces the length of the poly(A) tail.

Injecting a pure polyadenylated mRNA into a cell's cytoplasm generates both rapid protein translation in conjunction with rapid degradation through the activity of decay bodies and ribonucleases [10]. If the cells contained only this unbound RNA, then expression array measurements would generally reflect immediate levels of protein generation. In reality, both processes of translation and degradation are moderated by the functional proteins that bind mRNA [11]. These proteins, predominantly 3' and 5' UTR binding proteins, recognize both the structures and sequences and hence vary in their transcript specificity. The availability of these stabilizing proteins is also influenced by a range of environmental conditions such as acidity, temperature and ATP availability. Since these proteins adjust transcript stability, the lifespan of the different transcripts ranges considerably, from only a few minutes for highly regulated transcripts, such as c-myc and c-fos, to several tens of hours for stable species, e g., cytochrome P450c [12]. During this lifespan, most mRNAs are actively translated into functional proteins, but examples can be found where transcripts are differentially stabilized by RNA binding proteins and maintained in a non-translating state. Both neurons and germ cells generate translin and translin-like proteins. These assist in the protracted storage of specific transcripts for transport to distant synapses (neurons) or for translation at a later time after transcription is shutdown (spermatids/oocytes) [13].

All of this has consequences for the expression array measurement much of which was quantified by Hargrove and Schmidt [14]. Their model of RNA stability and its effect on gene expression measurements and total protein turnover introduces non-transcriptional effects. The pool of cellular RNA detected by an array has a transcriptional feed and several possible drains, formed by the different RNA degradation pathways. Stabilizing RNA can increase its concentration in a cell even where the rates of transcription remain unchanged. Since the stabilizing proteins bound to the RNA are sensitive to the structure of both 3' and 5' UTRs slightly different 3' and 5' terminus iso-forms of a gene can have quite different capping complex affinities and hence be both differentially transcribed, and differentially stabilized. Microarrays generate snapshots of the populations of these transcript pools that are constantly re-populated by the processes of mRNA creation, and depleted by the complex and environmentally sensitive processes of RNA removal. Consequently, arrays measure the average properties of groups of cells that cannot be easily disaggregated from the wider systems in which they are located.

Virtually all expression arrays share common design and sample processing elements. Total RNA is first extracted from its cellular environment and purified. mRNA is then amplified by being first reverse transcribed into a cDNA before being converted back into an amplified and labeled single-stranded complementary cRNA. Thousands of somatic cells, each containing around 30 pg of RNA of which ~0.7 pg would be poly(A) mRNA, are typically sufficient as a starting template. The amplification process can take nanogram levels of RNA to produce microgram levels of the amplified product. Single- stranded oligonucleotides (short base sequences) that can bind the cRNA extracted from the sample are positioned as a series of micron sized probes on an array slide. These probes can either be printed as millions of identical oligonucleotide spots on a slide or generated *in-situ* on a gene chip through photolithography. In both cases, a precise map of where each sequence is positioned on the chip is generated. The labeled cRNA is mixed with a buffer solution and placed onto the microarray slide. Watson-Crick bonding between perfectly complementary sequences on the chip and in the labeled cRNA sample takes place in a thermally controlled environment that ensures non-specific binding is prevented by thermal disruption. The reaction takes place in a carefully controlled environment. Agitation is generally used to ensure an even hybridization. Fluorescent tags such as green Indocarbocyanine 3 (cy3, excitation $\lambda \sim 550$ nm) or red Indocarbocyanine 5 (cy5, excitation $\lambda \sim 650$ nm) are directly incorporated into the cRNA or complexed with the cRNA that will allow the RNA to be detected. All arrays gain their extreme sensitivity from the capacity to image at high resolution the binding of the fluorescently bound sequences to micron-sized features on an array thus making them sensitive to microgram levels of RNA.

While early arrays were frequently printed with lengthy cDNA sequences, the accuracy of this approach was impacted by the potential for cross-hybridization between sub-sequences and by the relatively uncontrolled thermal binding profile of the comparatively long sequence. Consequently, the trend has been to move towards shorter (oligo) reporter sequences that can be controlled for homogeneous isothermal binding, the absence of stable secondary structure and that are unique in sequence. However design trade-off exists. Shorter sequences contain less information and are more difficult to generate as uniquely specific elements. Once the length of an oligonucleotide is below ~25 bp, unique sequences proximate to the 3' end of genes are rare [15]. Hence oligonucleotides used in current arrays vary in length between platforms designed by different companies, but generally fall between 25 nucleotides on the shorter oligonucleotide platforms through around 40 bp on medium length platforms and up to 60 bp for the longer oligonucleotide platforms. Since the amplification step generally uses a poly-dT sequence that will bind the mRNAs poly(A) tail as one primer, probes tend to be located within a few hundred bases of the 3' end of a transcript. The location varies by platform, and in the case of Affymetrix a series of 25 bp probes termed probe-sets are designed that cover one or more 3' exons transcribed from the gene. The placement and design of the probes is noteworthy for the investigator and will be investigated further in the next section.

## 13.2 A Framework of Considerations

### *13.2.1 Constructing a Hypothesis and Design*

The fluorescent signal detected on an array must be interpreted with some care. In discussing what arrays measure, it was suggested that arrays could quantify elements within the transcript pool. However, arrays vary in quality as tools for the absolute quantification of transcript levels. Both the amplification stages of preparation, residual secondary structure in the hybridized sequence and the intrinsic performance metrics of the different probe sequences may impact absolute signal measurements [16]. This creates different signal levels where the same amount of RNA is hybridized to the different probe sequences. Consequently, arrays are generally used to determine differential transcript abundance between measurement and control conditions. This relies on linear sensitivity differences between the probes being cancelled in a differential analysis. An experimental approach will therefore generally use exploratory or prior evidence to develop a series of tests where the role of the arrays will be to assess differential expression.

A concise list of the question(s) being addressed with the arrays during a study design is a useful starting point. For each question, the outcome that would comprise a positive and a negative outcome are essential. These questions can then be restated as a set of testable hypotheses with respect to the conditions to be tested. Early array work was often fully data driven, but poor power considerations and the lack of a falsifiable null hypothesis led to some unreliable conclusions. Without a null hypothesis and the power to validate both this negative outcome and the alternate positive hypothesis, any results at any level of significance may be reported. A hypothesis can be multipart or nested, but the simplest and most common description is a simple binary statement that describes the null condition that has to be disproved for the alternate significant hypothesis to be suggested:

*Null* Hypothesis $H_0$: The experiment will not detect the presence of a differentially abundant set of transcripts. Reporting transcripts in this group as differentially present when in fact they are unchanged would generate a Type-I or false positive error.

*Alternate*[1] Hypothesis $H_1$: The experiment will detect the presence of a differentially abundant set of transcripts. The failure to report elements of this set when they are truly disrupted would generate a Type-II or false negative error.

The hypothesis draws its conclusion relative to a null model, that is, the anticipated distribution of signal generated in the case where the test and control conditions are the same. Since measurement variance is unavoidable, a log normal distribution of measurements is often assumed. This null distribution may be estimated post-hoc in several ways, for example by randomly comparing the distribution of signal from samples within replicate groups (biological replicates) or pooled RNA, hybridized to several different arrays (technical replicates).

---

[1] This is the one-tailed alternate hypothesis a two-tailed alternate hypothesis could have multiple conditions such as $H_1$ and $H_2$ for over and under abundance relative to null condition.

Experimental design becomes the key to implementing an approach with sufficient power in order to characterize the major sources of variance, and through a replicate structure to realistically establish the alternate hypothesis and reject the null hypothesis. The primary considerations here are sample size and the replicate structure with respect to the identified origins of spurious variance. Where biological change is small, relative to the nuisance noise in the measurement, then larger sample sizes are required to suggest that the observed changes in expression are significant, relative to the background. Large differences within population groups that have been categorized as similar can arise from the unintended influence of confounding variables such as lifestyle and environment. A large meta-study of several different cancers and the predictive power of array studies of different sizes found that, in these heavily lifestyle impacted scenarios increasing the sample size did not always lead to more predictive power.

The experiment will have sufficient power to generate a high confidence result only when the perturbation in the signal generated by the test condition can be resolved relative to other sources of variance. Since sampling variance and the extent of expression change can frequently be estimated in advance, it is often useful to conduct this assessment *a priori* in justifying a design. Design power calculations can suggest the likely experimental power, given the requirements for specificity and sensitivity. Various tools exist to establish these criteria. Lee, for example, has a web-based suite of calculators that generate design specific experimental power estimates for array experiments [17], while Seo et al., have developed a tool that uses online microarray expression archives to generate “real-world” test data-sets from which to determine experimental power [18].

When asserting $H_1$ based on array data, independent validation from other approaches can add further support to the experimental design. These may be small scale real time PCR studies or protein based. Protein based analysis such as high-resolution 2D gel electrophoresis – mass spectrometry are particularly useful, given the complexity of the transcription-translation degradation pathway.

Arrays tend to be used in a mixed mode both to test prior hypotheses and at the same time to generate data for revised hypotheses. Caution is urged in the context of validating a refined hypothesis particularly with respect to false discovery rates. Approaches such as sample splitting are generally suggested to avoid cyclical reasoning. By forming one hypothesis generating sub-group and an independent or permuted hypothesis testing sub-group, the revised hypothesis can be assessed, but at some cost to the experimental power due to the reduced testing population size.

### 13.2.2 *Selecting a Measurement Platform*

Transcript sequencing approaches make relatively few assumptions regarding the transcripts they will quantify. Arrays however can only be used to measure the levels of transcripts complementary to their probe sequences. It is therefore useful to consider the choice of sequences made by different array platforms and the ways in which these will impact the analysis. Economics of the design and reporter generation make generic species-specific arrays such as Affymetrix human U133(v2) or Mouse MG430(v2) arrays, the most commonly used in academic research. These arrays will typically have at least one reporter sequence for every well established protein-coding gene in a genome. Considerations in choosing a suitable platform to address the research hypothesis might include:

#### 13.2.2.1 Model Systems

Not all platforms cover species in equal depth and the source of the transcript sequence evidence can also vary. It can be useful to determine the provenance of the sequence data. For example, where sequence data is from a major build of the species genome, which build and how frequently are the array sequences and related bioinformatics revised. If the data is to be compared directly with archived data, it can be useful to use chips with the same design to ensure comparability, even if this is not the most current design. Conversion tools that translate between builds can be found, but they cannot fully compensate for different probe characteristics.

#### 13.2.2.2 Coverage and Redundancy

The extent to which reporters seek to cover both potential genes and the different isoforms of a gene described by the sequencing and protein literature varies. Some platforms seek to be relatively exhaustive in this coverage while others limit reporters to common somatic isoforms. It is generally useful to explore annotation files in advance for the coverage of genes of interest. The RefSeq models for a species represent the core set of gene models with strong supporting proteomic evidence. Beyond these are other model sets. The NCBI describes a set of Entrez gene models (formerly Unigene) that extend beyond the Refseq set, but are still strongly supported by at least EST evidence. Beyond these are *in-silico* gene models, generated primarily from sequence evidence that suggests a potential for open reading frames. Both Ensembl and the NCBI have sets of these models, but they require caution since much that contributes to actual transcription relates to local structure rather than sequence [19]. Recent data suggests that the transcriptome is ubiquitously transcribed, and hence the transcript evidence will be found even at low levels for many of these models and non-RefSeq genes. On some platforms the choice can be made between either a RefSeq targeted design (e.g., Illumina Human WG8) or a design that includes more extensive gene models (e.g., Illumina Human WG6).

#### 13.2.2.3 Channels

The comparison between states measured by the arrays can be indirect, where material from different conditions is hybridized to different arrays. Alternatively, multiple conditions can be compared directly by binding one fluorescent marker to the RNA from one condition and another marker that is distinct, and can be excited independently, to the RNA from another condition. In multi-color array designs the RNAs from both conditions are hybridized in conjunction to the same array and the spot's color bias is used to estimate the relative contributions from the two conditions. In order to compensate for the inherent differences between the fluorescence of the dyes in a multi-channel array, replicates are frequently run in a dye swapped design that interchanges, or flips the dyes between the measured RNA pools. This approach allows a more direct comparison, although single-color platforms are more common in recent expression work. Nonetheless multi-color arrays remain dominant in DNA array CGH approaches on platforms such as Nimblegen. Specific aspects of normalization for multi-channel platforms has been reviewed [20].

#### 13.2.2.4 Masking Options

If the study is destined to be xenobiotic or genes are to be inserted from other species to create transgenic animals, it may be useful to be able to mask from the analysis probes that can cross-hybridize between species. The potential for such a masking varies by platform. Both DChip and GCOS can mask Affymetrix platform probes. Equally valuable in this setting is the availability of comprehensive pre-computed cross-species BLAST resources that can be used to remove cross-hybridizing probes from the analysis permitting a species-specific analysis.

#### 13.2.2.5 Sequence Uniqueness

Generally, only a few complex repeat sequences such as ALUs are incorporated exonically and hence are of interest for expression analysis. Nonetheless, repeats remain of interest since their knock-out can change the distance between transcribed genes and functional upstream elements.

They are thus of interest in both a DNA and RNA array context. The potential for the signal to arise from non-specific hybridization generally forces the design of unique sequence probes. In so doing, repetitive sequences or highly similar exonic domains from homologous genes are generally excluded from the analysis. In exceptional circumstances, segments where two different repetitive regions meet to create a unique combination of sequences at their boundary can be found. Where homologous genes are the subjects of the investigation, custom designs may be necessary that place small sequence differences in the most thermally sensitive area of the probe mirroring the approach used in SNP chips.

#### 13.2.2.6 Non-Coding RNA

If the functional transcripts of interest are non-coding, small or micro-RNAs for example, hybridizing RNA to a *dense* and overlapped tiling pathway array may be considered. These arrays report the levels of RNA at a high resolution and can detect both low levels of transcription from unexpected locations as well as rare events such as run-on transcription. miRNA platforms based on databases such as the miRBase are available from, for example, Agilent.

#### 13.2.2.7 Extranuclear Expression

Most platforms will also include some reporters for mitochondrial genes, be these in the nucleus or in the mitochondrial DNA itself. Consideration again needs to be given to the separate mitochondrial ribosome pathway used to translate these genes.

#### 13.2.2.8 Bioinformatics Assets

The popularity of arrays as research tools has produced a wealth of free and commercial analysis tools. Some are highly specific and perform only one function such as clustering where as others such as GeneSpring and TIGR are suites of tools able to perform many elements of an array analysis. Many tools work across array platforms, but others such as BeadStudio and GCOS are proprietary and work only with the products of one company. Perhaps the most flexible research environment is the free Bioconductor suite of analysis tools developed for use in the R statistical development environment. In R/Bioconductor, many pre-compiled packages such as the commonly used Affy, Sma and Rma packages can be scripted to exchange their data and the results to create a complex analysis workflow. Nonetheless, some array platforms are naturally more well established than others. History has to an extent determined where bioinformatic database assets, such as probe annotations, have accumulated to provide the established platforms with more complete resources than some newer technologies. In the worst case no more than a gene ID and a reporter ID may be provided with the platform along with marginal analysis capabilities. At the other end of the spectrum systems may have highly advanced bioinformatic resources, covering numerous forms of inter-genomic and inter-chip comparison with many third party tools able to parse the probe IDs for analysis. Choosing an array with the level of support required to complete the bioinformatics tasks proposed can save considerable time.

It may be necessary to consider alternatives where no satisfactory generic platforms can be found. For example, a custom design may be warranted when no off-the-shelf array covers tissue-specific isoforms. Probe generating software such as ArrayDesigner can rapidly generate isothermal repeat-masked probes with unique sequences given a region or a set of genes of interest. Once prohibitively expensive, the gene chips currently available from Combimatrix, NimbleExpress and Agilent are printed as readily with a custom design as they are with a generic design, making the creation of custom tiling or splice arrays a relatively inexpensive option. Once created, a tiling design can be readily explored as a custom track in many genome viewers such as UCSC's genome browser or Sanger's Artemis.

Whatever array platform is selected, a standardized protocol for RNA extraction, amplification, labeling and hybridization is required. This may only be a recommended protocol and specific circumstances may require modifications. It is useful to note these modifications in the experiment description when providing the results to archives such as GEO. This can help other investigators to be aware of these changes in considering meta-studies. To this end the MIAME protocol has been developed to ensure a minimal catalog of information is available through which an array experiment can be unambiguously duplicated.

## 13.2.3 Quantifying Expression

Expression quantification is a serial process by which raw fluorescence is transformed into an expression profile, with one or more expression levels, for each gene covered by the chip. The choices made at each stage of this process will influence the experiment's outcome although many platforms now automate much of this process. Several excellent spike-in studies [21] and theoretical analyses [3] have been conducted to determine which options work well as an analysis pipeline. The choice of steps nonetheless varies by both the array platform and with the required levels of sensitivity and specificity selected. Some approaches are evidently more conservative with respect to the balance between Type-I and II error than are others, and it is useful to consider how they may contribute to the outcome of an array study.

### 13.2.3.1 Image Analysis

The first step in a microarray analysis, following hybridization, involves quantifying the level of fluorescently bound sequence that has attached to the probe. A scanner is generally used and one or more lasers excite fluorescence. The micron-sized printing of the array necessitates careful analysis to remove artefacts such as scratches and blemishes that can arise from manufacturing flaws or hybridization buffer contamination. Most manufacturers create a random or highly dispersed distribution for their probes to ensure that if multiple targets for a transcript are present, they are not physically collocated and hence are unlikely to be effected by a common surface defect. A useful step in array quality assessment can be an initial visual inspection of the hybridization to the chip. Some analysis software provides this while in other cases it may be necessary to access the raw image output from the scanner. When problems in the preparation introduce contamination into the hybridization buffer or the hybridization well has leaked, it can often be readily detected by a visual inspection of the array (Fig. 13.1).

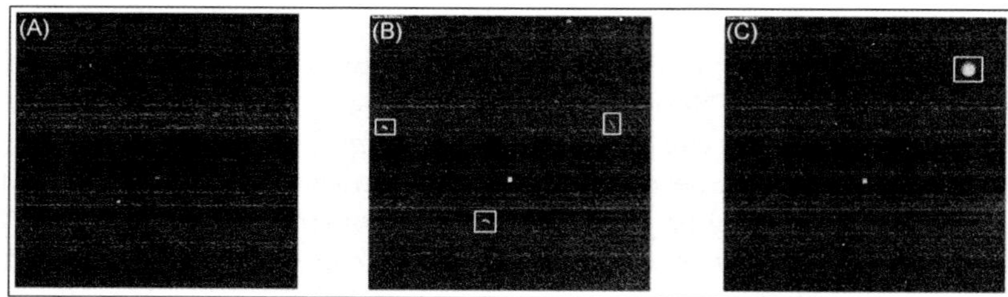


**Fig. 13.1** Slide quality control. Examples of processed Affymetrix (DChip) array images demonstrating **(A)** even hybridization **(B)** surface contamination (*white boxes*), **(C)** uneven hybridization (*white box*) (Copies of figures including color copies, where applicable, are available in the accompanying CD)

#### 13.2.3.2 Background Correction

Following hybridization of the target material to the array probes, a wash step removes sequences that have not hybridized. The effectiveness of this wash determines the threshold level below which the transcriptional signal cannot be resolved from the background. Although arrays are poor tools for determining absolute expression, the extent of the background signal will determine the confidence of a not-expressed call. Nonetheless, there will generally be some residual background signal and this may vary over the surface of the array. Various local and global adjustments can be made. The Affymetrix GCOS suite uses a set of local points from the backgrounds around each area of measurement to assess a local background level. Subtraction of this background level brings the signal as close to zero as possible for unhybridized probes. These values are frequently rounded up to the tenth percentile of the chip signal or to a small positive value to avoid problems in log analyses. Choe determined that in the Affymetrix context a localized background signal determined from a set of background points worked well [21].

#### 13.2.3.3 Array Normalization

Normalization (sometimes called standardization or scaling) is a data transformation procedure introduced to remove systematic preparation bias between arrays, allowing biologically meaningful comparisons to be made. The need for normalization arises because the preparation of the material hybridized to arrays can vary between arrays. This might lead to one array having, for example, a slightly more concentrated reagent mix or lower hybridization temperature resulting in a global shift in the expression signal. On two-channel platforms, differences in the efficiency by which the different dyes are incorporated, are removed by the normalization procedure. Due to the way arrays perform when presented with different transcript concentrations, the shift may be non-linear with some higher signal shifted more than medium or lower signal. These gross differences need to be corrected by normalization, such that two genuinely identical samples would generate virtually identical expression profiles after normalization.

In order to select the most appropriate normalization procedure it helps when the source and structure of the variance is recognized. Choosing an inappropriate normalization may over-adjust or under-adjust the data, both of which will lead to spurious results. Normalization differences are a likely sourc of a great deal of the analytical difference generated when different analysis packages quantify the same arrays [22]. The assumption made in most normalization routines is that the perturbation in the gene expression due to a treatment or condition is small, relative to the global expression of a cell and is relatively balanced with equally as much increase as decrease in expression change. Hence, the gross distribution of the signal on two chips is adjusted by these approaches to be broadly similar. Normalization approaches are numerous and range from simple median adjustments through to curve-remodeling normalizations. Five popular normalization strategies are described below alongside their software implementations:

#### 13.2.3.4 Array Median

A median normalization is a whole profile shift that centers the distributions from a number array so that their median signal is equal. Choosing the median or less often a trimmed mean as the measure of central tendency, is required to produce a robust measure that is insensitive to the high signal outliers that would otherwise bias a mean adjustment. The approach works well when the expression is shifted consistently between arrays. But it is unable to correct for the non-linear trend differences between arrays caused for example by differential saturation and non-linear sensitivity. Many expression analysis packages such as GeneSpring introduce Median normalization as a default first-stage normalization in a series of normalization steps.

#### 13.2.3.5 Invariant set

An invariant set normalization is a distribution adjustment approach. The algorithm changes the intensities of probe signals relative to each other rather than globally shifting the intensities as in the median normalization. The aim of the approach is to adjust the broad shape of multiple-signal distributions so that they match each other. While some probes may dramatically change their expression due to a treatment effect, a set of stable probes is anticipated to maintain their rank in an ordered list of expression across all arrays forming an invariant set. Even if a section of one distribution is exaggerated in a non-linear way or the global trend changes, the rank of these stable probes should not change. These rank-invariant probes are identified from all arrays and the expression of this set on an exemplar array is used to create an adjustment curve for the entire profile. The algorithm may suppress some of the signals and emphasize others in order to standardize these invariant sets of genes relative to each other. An invariant set will generally also approximate a median normalization although this cannot be fully guaranteed. The invariant set normalization is the default approach used in DChip.

#### 13.2.3.6 Quantile

The quantile normalization assumes that the global shape of the signal distribution should be similar between any two conditions with only a subset of genes changing their relative expression in a fairly symmetrical manner. Under this assumption, the median signal of the top percentile of signal expression from one chip should broadly equal that on another well normalized chip. The Quantile normalization makes this adjustment, iterating through the whole data-set on a percentile basis, adjusting the median signal for each percentile bin relative to a standard array. The quantile normalization is the default approach used in the more recent Affymetrix expression console and is available as an option in versions of DChip after 2006.

#### 13.2.3.7 Cubic-Spline

Cubic Splines are a popular example of the generic polynomial spline curves used throughout the industrial design industry to create smoothly changing surfaces. Using splines has been most generally applied to normalize the channels of two-color array platforms. The approach is optimal for systems that have a broad and low frequency non-linearity, or in simpler terms when the expression is plotted in a dot plot there is a smooth bow in the expression dot plot (see section 4). Consequently, sudden irregular hybridization at the extreme end of an array's profile may not be well corrected by the approach. Lowess (locally weighted scatter plot smoothing) is a variant of this smoothing, often also implemented with cubic polynomials due to their flexibility. The main difference between the lowess and cubic spline approaches lies in the local weighting generated in the lowess algorithm that makes the approach less sensitive to dispersed outliers. The Lowess normalization is commonly used in GeneSpring and is the default approach for many of the bioconductor analysis suites.

#### 13.2.3.8 Housekeeping/Background

Rather than using the global expression profile or elements of it determined after hybridization, another approach is to use a pre-determined set of genes. These can be spiked in, that is, added at varying levels to generate a standard by which to differentially normalize the array. Alternatively housekeeping genes can be used although these tend to be biased in expression terms and selecting good housekeeping genes can be a problem. In some cases, the background signal can be selected. These approaches are available in both GeneSpring and Illumina's BeadStudio packages.

#### 13.2.3.9 Per Gene Normalization

While the approaches listed above are intended to make one array comparable with another (per chip normalizations), it is sometimes the case where a group of genes need to be compared with each other and that their signal needs to be normalized to a baseline so that correlations between the relative changes in signal can be identified. A per gene normalization will typically set a default experiment, for example, the first time point in a time series experiment and scale the expression of all genes to a set level at this point. By applying the same per-gene scaling to subsequent arrays, the relative changes in the expression between genes can be readily displayed.

There are instances where normalization can be a source of problems notably where the necessary assumptions regarding expression profile identity between experiments and fold change symmetry are breached. In some experiments, in which the gross signal is anticipated to shift due to the experimental condition, normalization may actually introduce aberrant signal variance. For example, in X-chromosome inactivation experiments in Drosophila it would be inappropriate to normalize relative to the global signal since the X-chromosome makes up a large percentage of the genome. In this scenario, the normalizing probes might be specifically chosen from the somatic chromosomes without X-chromosome homologues. In conditions where the perturbation is even more pronounced, potentially as a result of poor data it is worth assessing whether the difference between group replicates suggests that a standard normalization might be omitted or replaced by an ANOVA normalization [23]. Where normalization is omitted, confidence limits need to be adjusted to the reproducibility evident between unnormalized biological replicates.

#### 13.2.3.10 Signal Modeling

Platforms vary in the ways in which probe replicates are designed. Some platforms have serially tiled probes and others multiple replicate probes or multiple reporter sequences per gene. If these replicates are to be utilized to infer signal stability then a modeling stage that combines multiple probe signals to generate a more reliable signal is suggested.

#### 13.2.3.11 Affymetrix

On the Affymetrix platform a tiled sequence of probes defines a probe-set. This set of oligonucleotide targets includes both a perfect match (PM) to the sequence and a central base mismatched probe (MM) to detect spurious hybridization. The tiling sequence allows a series of comparisons to be made between the background hybridization and the correctly aligned sequence signal. Interpreting the signal from these probes is the task of the modeling software. A range of options are available all of which seek to develop a stable adjusted signal level from the multiple probes that make up a probe-set. The original MAS algorithms implemented by Affymetrix in the GCOS package, used the anti-log of a robust, non-parametric, Tukey bi-weight average of the difference between the log PM and MM probe signals. Robust parametric statistics are relatively insensitive to deviations from ideal models used in parametric tests and hence are suggested for noisy probe models. Later Affymetrix approaches include Probe Logarithmic Intensity Error (PLIER) that is an option within the expression analysis suite. Independent algorithms include RMA (Robust Multi-array Analysis) available within the Bioconductor package or as a windows executable. The RMA approach is both a normalization and modeling step based solely on the PM signal. It produces a robust, that is to say a stable, $\log_2$ transformed signal value using a median polish model and quantile normalization. A variant gcRMA available as a bioconductor package (gcRMA) additionally corrects for GC to AT bias in the probe design. Li and Wong's DChip package offers several options in its signal model based expression index. The MBEI algorithm has been shown to be sensitive to signal change particularly when used in PM-MM mode. Many alternative and derivative approaches are available and discussion continues as to the merits of the respective algorithms with respect to sensitivity and stability. Irizarry presents an excellent exploration of this discussion [24]. While all of these approaches seek a stable signal value from the different measurements made along the probe-set, not all the approaches maintain the probes as elements within the probe-sets. Genomatix' ChipInspector uses a novel gene model to re-map probe level data to gene targets. Probe signals are associated with the Genomatix detailed gene models and are not re-assembled into probe-sets for analysis.

#### 13.2.3.12 General Platforms

Most manufacturers use replicates to determine the signal stability. For example, Illumina expression arrays use up to 60 randomly located beads per transcript. Signal and signal stability are then assessed from robust factors such as the median signal across beads and the variance between the replicates. DNA platforms such as Nimblegen operating in a CGH mode usually summarize signal slightly differently with similar sections of a chromosome grouped together and treated as a series of windows with confidence levels determined by the channel variance within windows. While normalization and modeling steps aim to minimize gross inter-array differences, it is useful to identify the origin of as much variance within the replicate groups as possible.

### *13.2.4 Sources of Preparation Variance and Biological Variance*

A degree of spurious variance between measurements is invariably encountered in biological experiments. Scanning and background noise tend to be relatively unbiased and on most platforms, contribute little to the variance in the final signal after modeling. Indeed technical replicates generated by re-hybridizations of the same RNA, applied independently to different arrays from the same platform, typically have correlations of around 0.995 (Illumina WG8) to 0.997 (Affymetrix U133[v2]). Other sources of extraneous variance usually contribute more and may introduce bias between the measurement sample groups where this is poorly controlled for.

#### 13.2.4.1 Preparation Variance

RNA is typically amplified either prior to and/or during the creation of labeled probes. Amplification kits typically demonstrate selectivity with respect to the transcripts that are preferentially amplified. The use of either different amplification kits or different numbers of amplification cycles when preparing the samples can consequently impact the quality of the data generated. When comparing Enzo and Ambion amplification kits on the testis total RNA, for example, the correlation ($r^2$) between preparations was 0.6 (Affymetrix U133 with DChip), while the correlation between 1 amplification cycle and 2 amplification cycles from Ambion kits was 0.8 (Illumina WG8(2) and BeadStudio). Normalization can to an extent compensate for differences. However, this is difficult to correct when transcripts have been differentially amplified relative to each other in a non-linear process. Non-parametric statistics such as rank comparisons may be necessary. To a lesser extent very different initial RNA concentrations applied to a chip can result in non-linear signal variance and although manufacturers may indicate an acceptable range of RNA hybridization quantities, most experiments benefit from using similar initial concentrations.

A hazard in RNA analysis arises from its relative instability particularly to thermal and biological degradation. If a cell sample is prepared but time elapses before either RNA stabilization with guanidinium or snap freezing, then endogenous cellular ribonuclease activity will continue to degrade the RNA in the sample. In samples with very low starting levels of RNA it is useful to optimize the DNase treatment to minimize RNase activity when removing the genomic DNA during the preparation. The general characteristics of degraded RNA can often be identified prior to hybridization. In the RNA obtained from somatic tissues the ratio of the area under the ribosomal 28S:18S peaks on an electropherogram decreases, as the longer 28S RNA degrades faster than its shorter counterpart. If the ratio drops below the optimum of 1.8, then RNA degradation may be indicated [25]. Instruments such as Agilent's bio-analyzer, use a library of total RNA electrophoretic profiles from intact and degraded RNA to develop a related RNA integrity number (RIN). This extends the 28S:18S ratio by profiling the RNA from rRNA free mRNA, as well as total RNA. Other pre-hybridization approaches include sensitive ribogreen fluorometry and the relatively less sensitive but broad 260/280 nm absorption profile. A general trend towards higher signals from probes positioned towards the 3' end of the transcript is a further indication of degradation. This ratio of 3' to 5' probe signal is reported on the Affymetrix platform as a quality variable but can also be generated from raw cel data using a bioconductor (AffyRNAdeg). If differential degradation is evident between the sample groups, then the analysis will likely be heavily compromised. DChip, however, allows the masking of a series of n 5' probes from all probe-sets where n is set by the user. This may restrict the probe signal to more intact regions of the transcript.

Freezing is a way of stabilizing the RNAs, but recognition should be given to the impact that the freeze thaw process has on the RNA quality of tissue samples. Multiple freeze-thaw events can disrupt cellular compartments and lead to RNA degradation over multiple cycles. If some samples experience more freeze-thaw cycles than others, then spurious variance may again be introduced between the samples.

#### 13.2.4.2 Biological Variance

To an extent biological variance is an inherent property of the measured system and is consequently important to characterize, rather than necessarily remove. Distinct populations can, for instance, share genetic traits that will differentiate their expression profiles [26] relative to the species average. Nonetheless, gross sources of unintended biological variance may be introduced that bias results relative to the changes being investigated. When not using cell lines this may arise from differences in the homogeneity of tissues or from tissue contamination. Contamination may arise either from the tissue collection procedure, or in tumor contexts, from the infiltration of non-tumor cells such as immune cells. The processes of fine micro-dissection, such as laser capture, required to remove these contaminating tissues may inherently stress tissues into non-physiological modes of expression. While problematic, this can be detected where suspect tissues can be identified by using marker genes that are present to a greater extent in these tissues. Where high levels of undesirable material is present it my be useful to conduct an *in-silico* study using the data from archives such as GEO, to establish the results that could be generated due to contamination. These results can then be discounted in the final analysis as potentially spurious and arising from tissue mixing. More subtle sources of biological variance can arise due to sampling of tissues at various stages of synchronization. Recent evidence suggests that many genes, in most organs, demonstrate a circadian expression rhythm such that taking measurements from biological samples at asynchronous points would introduce a degree of variance [27]. Circadian synchronization is worth noting in tissue samples, but it can be more problematic in well-synchronized cell cultures. Agonal factors such as hypoxia and acidosis have been demonstrated to significantly impact the expression profiles, and measurements taken from samples with different terminal conditions may differ significantly [28].

### *13.2.5 Visualizing Expression Distributions*

In the simplest case of two groups representing a treatment 'before' and 'after' design, the first task after having quantified expression is generally to view the expression distributions in a meaningful way. This can be challenging, as many of the standard approaches such as bar charts become overwhelmingly dense and consequently unintelligible when presented with tens of thousands of data points. When the number of conditions in the experiment is increased the optimum visualization approach will depend upon the design, specifically whether comparisons are made relative to a baseline condition or iteratively between each successive condition.

### 13.2.5.1 Histograms

Histograms are familiar tools and are readily applied to expression studies. The distribution of the signal generated by an array is viewed by placing the large number of measurements into a small set of bins. The Y-axis is generally linear while the X-axis is frequently logarithmic and divided into a series of sequential expression bins. The shape of the distribution of raw data can often be helpful in determining the most appropriate normalization procedure (Fig 13.2A). When two microarrays to be compared have distributions of similar shape, but are offset relative to each other, a weak normalization such as a median adjustment may be completely satisfactory. Where histograms

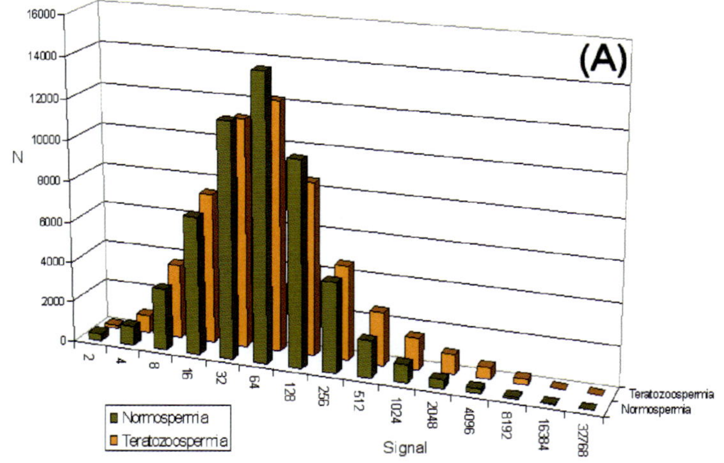


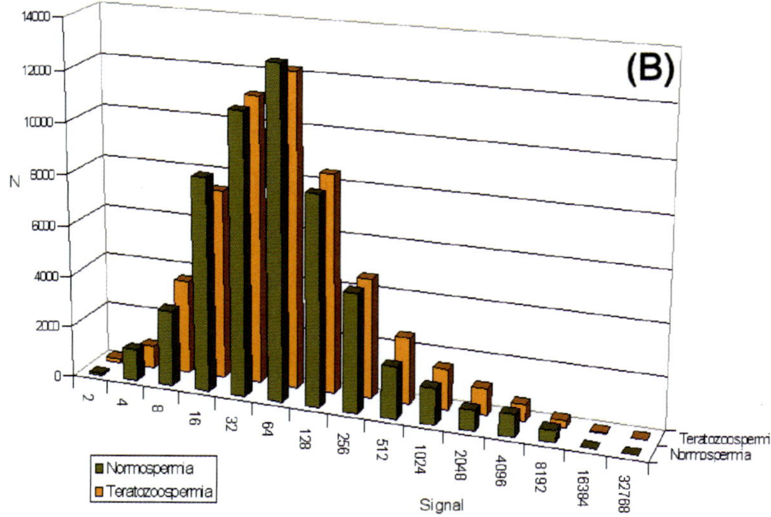


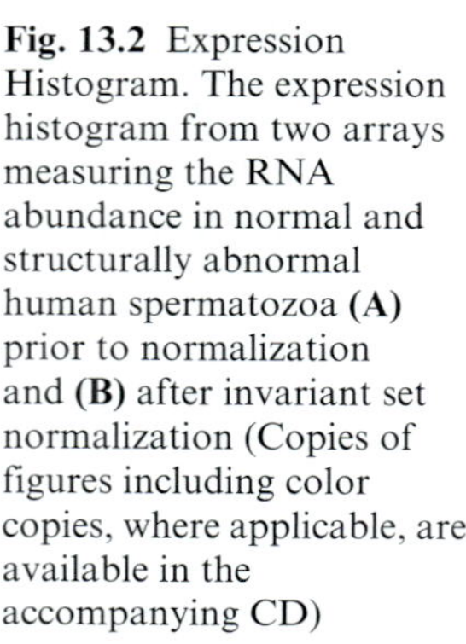
**Fig. 13.2** Expression Histogram. The expression histogram from two arrays measuring the RNA abundance in normal and structurally abnormal human spermatozoa **(A)** prior to normalization and **(B)** after invariant set normalization (Copies of figures including color copies, where applicable, are available in the accompanying CD)

indicate a more complex difference between intensity profiles a more aggressive normalization such as a quantile or an invariant set approach may be required. Following the normalization it can be useful to review the histograms to assess how effective the normalization has been and whether further or alternate approaches should be considered (Fig 13 2B). The histogram does not generate information about individual reporter probes, scatter plots are more appropriate tools for visualizing gene-level data.

### 13.2.5.2 Scatter Plots

Perhaps the simplest but also one of the most useful ways of representing high through-put data is the scatter or dot plot (Fig 13.3). This can frequently be generated in the array analysis software, suite packaged with the array platform. Caution should be exercised when using generic spread-sheet applications, since they are frequently unable to represent the extent of the data generated by an array experiment (Microsoft Office 2003: 32,000 rows, OpenOffice 1.1: 16,000 rows). A dot plot is a useful way of viewing both the level of data correlation between two experiments, the noise envelope and any skew in the broad expression profile between experiments. Given two expression experiments A and B, a dot on the chart is plotted for each reporter, based on its signal from array A plotted on X and B on the Y-axis. The result is a scatter plot that represents the gross trends in signal between the two conditions. When the signal on A is identical to the signal on B, the dot plot will lie on the central line X = Y, (Fig. 13.3A) with the points most distant from the origin indicating a high array signal on both the arrays. When the signal from A and B differs, the points plotted will move above the line where B has greater expression than A and below the line for the points where A has greater expression than B. For experiments conducted in replicate groups, the expression levels plotted will usually show the median of the group expression, potentially with error bars. While the experiments in Fig. 13.3A are normalized well, Fig. 13.3B illustrates a dot plot where there is a clear non-linearity in the data indicating a need for a normalization correction, such as cubic spline.

Like, histograms scatter plots can be useful in assessing the correlation between any two experiments. Typically, the Pearson correlation coefficient [r] is generated by attempting to fit a straight line to the data. A least squares fit seeks to fit the gradient and offset of a straight line by minimizing the sum of the distances between each experimental point and the closest point on the fitted line. The result is not only a line that optimally fits the data but that also measures correlation $r$, and the coefficient of determination $r^2$. The value of r can range from –1.0 to + 1.0 covering the

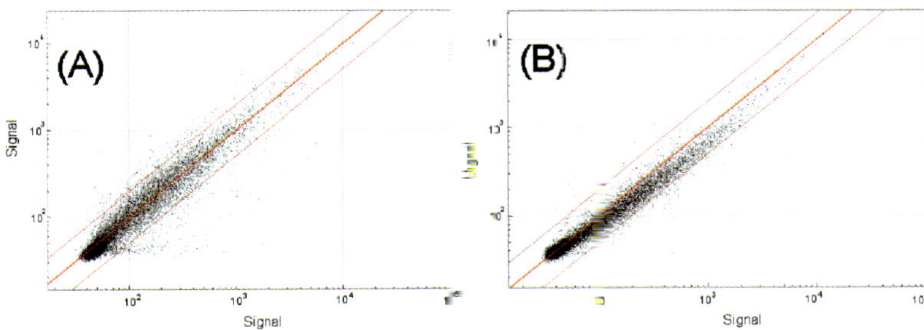


**Fig. 13.3** Scatter Plots. A scatter-plot of all chip data from two Illumina WG8 arrays is shown with logarithmic axes. The expression of each gene in the two measured conditions is shown as a dot relative to a log-log scale. The central line represents the unchanged signal between the two conditions. Upper and lower parallel lines show the limits beyond which a 3–fold change in expression is observed. Created in the BeadStudio analysis suite, plot **(A)** illustrates well-normalized signal and **(B)** data that is clearly in need of normalization (Copies of figures including color copies, where applicable, are available in the accompanying CD)

range of perfect anti-correlation to perfect correlation with the zero point indicating no relationship between two experiments. The coefficient of determination describes the percentage of the expression trend from experiment $A$ evidenced in $B$. When the $r^2$ value is 1, the two experiments perfectly co-predict each other, when 0 there is no mutual information shared.

#### 13.2.5.3 Volcano Plots

Volcano plots were devised as a way of viewing the fold change between conditions and confidence, from replicate information at the same time. Both the axes are generated on a log scale, and each dot represents one gene expression comparison between two conditions. Increasing height indicates increasing confidence in the expression change, and increasing distance from the center on the X-axis, a greater fold change (Fig. 13.4). If the plot is overly central, little expression change may have been detected between the two states, where it is low on the Y-axis but well dispersed on the X-axis, problems in reproducibility may be suggested.

#### 13.2.5.4 M-A Plots

In an M-A Plot or similar R-I (ratio-intensity) plot, the broad differences in the normalized trends between the two arrays are plotted. The log ratio of the signal between the two arrays is shown on the Y-axis (M) and the log product of the two signals on the X-axis. MA plots are generally useful in identifying the efficacy of normalization, and have been routinely used in dual-channel measurements. Following an appropriate normalization, the ratio of signal between any two arrays (Y) will not demonstrate a trend with respect to signal level (X). Figure 13.5 illustrates the condition where good normalization leads to a flat relationship relatively to the poorly normalized starting signal. Most array analysis solutions such as R/Bioconductor and DChip can generate MA plots to assess the normalization quality.

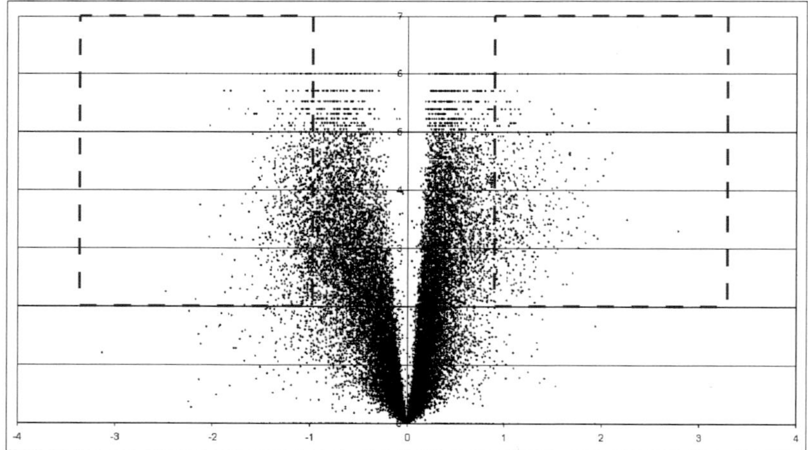


**Fig. 13.4** The Volcano Plot. The volcano plot is a form of scatter-plot used to overview chip-level data on expression change. The axes of a volcano plot represent two aspects of differential expression derived from groups of replicates. On the Y-axis is plotted $-\log_{10}(p)$ of differential expression from a t-test and on the X-axis $\log_{10}$ (*fold change*). Since transcripts with a low p-value and a high fold change are of interest, the transcripts in the domains indicated by dotted lines are consistently down regulated (*left*) and up regulated (*right*) and might be considered for further exploration (Copies of figures including color copies, where applicable, are available in the accompanying CD)

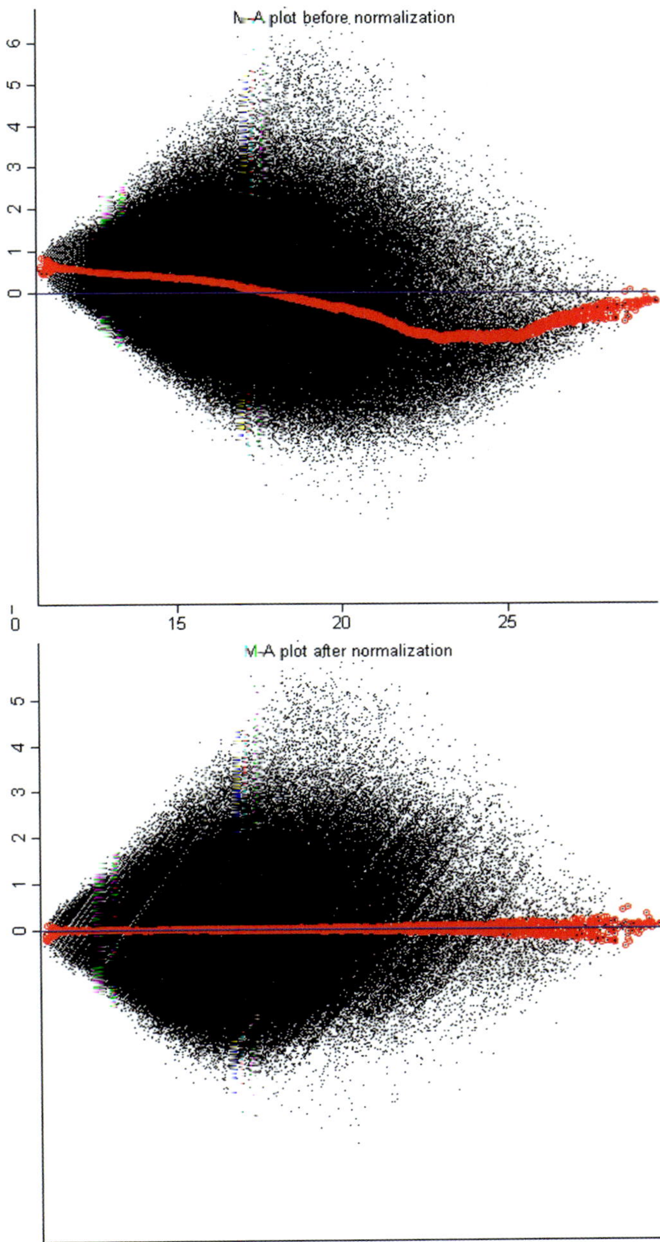


**Fig. 13.5** The M-A Plot. The M-A plot represents yet another axis variation of the scatter-plot. The purpose is to demonstrate how the fold change between two expression profiles varies as a function of expression. The Y-axis indicates the log of fold change and the X-axis the log of the expression product. Data is shown from DChip prior to normalization (*top*) and post normalization (*bottom*) by the invariant set method. The invariant probes from an Affymetrix study are highlighted in red. Clearly, a significant non-linearity with respect to the signal is corrected (Copies of figures including color copies, where applicable, are available in the accompanying CD)

#### 13.2.5.5 Series Plots

Series plots are useful ways of illustrating groups of genes that change their expression in a consensus manner between states even if their expression levels are very different. Series plots can be produced following a per gene normalization, and are typically employed to draw attention to the expression of groups of genes displaying similar trends across experiments.

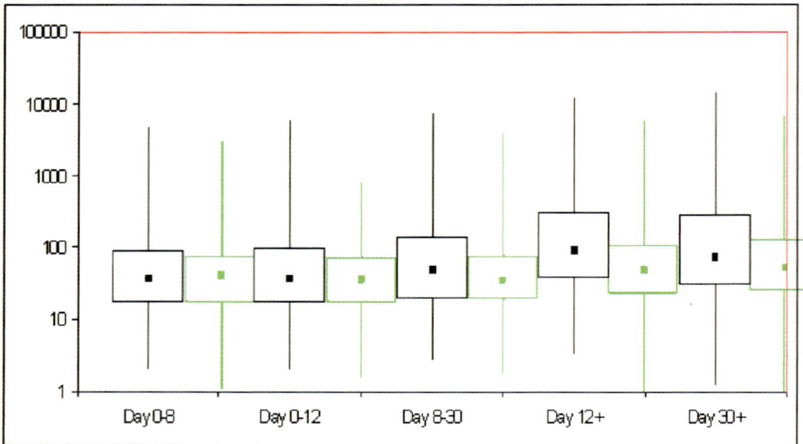


**Fig. 13.6** Box and whisker plots. Box plots (*Sigma Plot*) represent the distribution of signal sampled in a murine spermatogenesis time-course experiment. Each box and whisker summarizes expression from a single chip. The central dot represents the median signal at each interval, the box the inter-quartile range between the first and third quartiles and the lines the limits of the expression range. Dark and light boxes contrast data ranges from two different replicates prior to normalization (Copies of figures including color copies, where applicable, are available in the accompanying CD)

#### 13.2.5.6 Box (Whisker) Plots

Virtually all generic array analysis suites can use box and whisker plots to illustrate the range and dispersal of data under different conditions (Fig. 13.6). The degree of difference between the conditions can be readily visualized relative to the variance of the signal.

#### 13.2.5.7 Heat Maps

One of the most powerful tools for representing groups of genes with a similar expression pattern either across a time-course or between conditions is the heat map. Genes are typically ordered vertically and arrays in measurement groups horizontally to create a matrix. Within the matrix, color is used to represent the relative expression of genes with red frequently chosen for over expression and blue or green for relative under expression. Clustering is frequently used to order genes such that those with similar expression are next to each other (Fig. 13.7). A heat map can be another useful tool for identifying problems in array normalization. Where all the genes on an array are included and a column is constantly more red or blue than those around it normalization failure is suggested. Heat maps can be generated in most custom and generics array analysis packages such as BeadStudio, GCOS, DChip and GeneSpring. In addition, HeatMap Builder allows some visual alternative representations and JavaTreeView provides a visual interface to the output of Cluster 3.

### *13.2.6 Statistical Modeling*

Using the approaches above, trends in array data can be readily visualized. These trends can be explored in more depth within a statistical model. Below is an outline of the description of processes typically undertaken when assigning levels of statistical significance to array data. A host of further statistical approaches are available that combine or extend elements from these to make them more suitable for specific contexts of analysis. The probability value (p-value), we consider, is that of

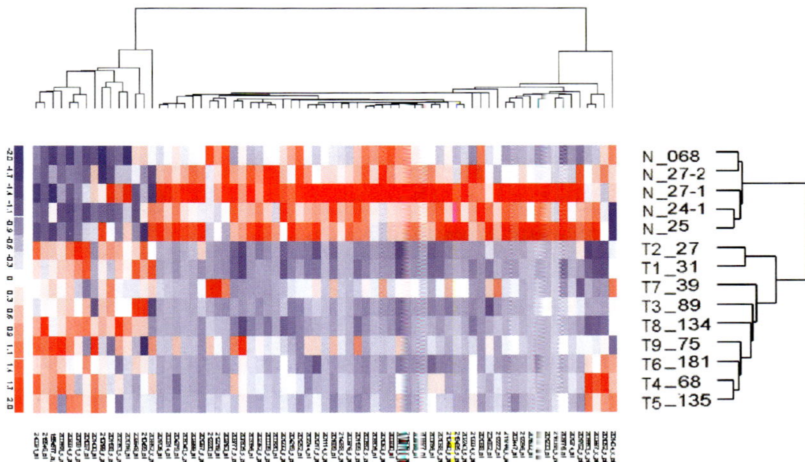


**Fig. 13.7** Heat Maps. A heat map showing the expression of proteosomal genes in which the high signal is shown darker and low is lighter. Hierarchical clustering of both samples (*right dendrogram*) and genes (*top dendrogram*) is shown. Clear evidence for clustering amongst both samples and genes can be seen (Copies of figures including color copies, where applicable, are available in the accompanying CD)

rejecting the null hypothesis. This ranges from 1, where the null hypothesis is certainly true and no real change is possible, to 0 where the null hypothesis can be completely rejected and a real change is suggested.

#### 13.2.6.1 Scale

A frequent source of confusion in interpreting an array analysis is the choice of scale. Logarithmic scales are frequently, but not exclusively, used to represent and analyze data from arrays since they hold several advantages over linear or power scales. The dynamic range of array data makes visualization over five orders of magnitude difficult to represent meaningfully on a linear scale. Additionally, the distribution of signal from arrays tends towards a normal distribution when transformed to the log scale. Since classical statistical analysis draws inference based on departures from the normal distribution, the suitability of t-tests or ANOVA for assessing significance, benefits from the log-transformed data. The log scale also aids in fold change analysis since the log-transformed fold change is symmetric around the no change condition, where log(Experiment/Control) = 0, thus readily allowing visualization of fold change in both directions. It is important not to divide two log-transformed values to generate a fold change. The log fold change can be readily arrived at for data that has already been log-transformed by subtracting log(signal) values between the two conditions. As indicated earlier, the rounding of zero and negative values in the raw normalized signal up to a small positive value avoids problems with nonsense data on the log scale. Attention in assessing array fold change data also needs to be paid to the base of the logarithm used.

#### 13.2.6.2 Fold Change

Making the assumption that arrays can report without error the fold change in the expression of a gene between two conditions, this differential expression suggests itself as the initial metric for identifying the genes of interest. Typically 2, 5 and 10-fold changes in expression, either up regulation or down regulation, from control to experimental conditions have been taken as thresholds for identifying the genes of interest. Platforms differ in the extent that exploration of a change below 2-fold can be justified. Genes selected may either be large groups with reduced confidence that they are all correctly identified (e.g., 2-fold), or smaller groups of higher confidence genes (e.g., 10-fold). As discussed in the context of Volcano plots, there are at least two dimensions to be considered when assessing genes of interest for further investigation; fold change and the p-value of this change being reproducible. It is useful to explore how the p-value value is calculated and the considerations that may drive us to look beyond classical statistics and introduce additional parameters into the confidence measure assessment.

#### 13.2.6.3 The Null Model

Typically, analytical models will generate a null distribution relative to which the measured change in the signal is assessed for the significance of this divergence. Null models vary in the extent to which sources of identifiable extraneous variance are incorporated. Parametric boot-strap models make assumptions about the distribution of the signal generally from the trends evident in large sample studies with characteristics such as Gaussian or gamma distributions. Non-parametric approaches can use the differences between samples in the biological replicate groups to achieve a model using, for example, randomization.

#### 13.2.6.4 Regression Models

In population studies, expression change is generally explored with respect to multiple possible impacting elements both genetic and environmental. If, for example, we were looking at expression in an unmatched population of diabetic and non-diabetic subjects then there would likely be age effects in combination with disease effects. Since these factors are not known in advance, we may wish to explore models where many factors are categorized for each subject and then explore the relationship between expression change and these variables. Regression models are used to explore this relationship generating a correlation strength for each variable as well as a confidence measure from either the F-statistic or t-statistic. Most statistical packages such as SAS, SPSS, Matlab and SigmaStat are well suited for the investigation of linear and non-linear regression models. For more advanced investigations, R/Bioconductor and Mathematica offer a range of enhanced approaches including, for example, penalized models that attempt to reduce over-fitting [29] or zeta theory models that can incorporate very large numbers of co-factors [30].

#### 13.2.6.5 T-Statistics

*T* and *F* statistics are examples of classical frequentist statistics generated by observations made on random subsets of very large populations. They can inform when the difference between distribution means (T) or variance (F) of two groups of observations suggest that the samples came from similar or different populations. Given two sample groups, between which there is an anticipated difference, it is important to characterize this difference relative to the within-group variance. In the case of the diabetes study, we may have identified both age and disease as influences and consequently have two sets of age matched groups. For each gene, the signal change between the tested conditions needs to be assessed relative to an appropriate variance measure such as the standard deviation within the two groups.

A t-test is well suited to a two-group unmatched design where the test is used to assess the significance of the mean difference separating two datasets relative to the variance within the two groups. The p-value of this ratio is inferred from the t-distribution and compared to a limit, $\alpha$, the acceptable risk of falsely assigning the measured variable significance. In low dimensionality experiments $\alpha$ might be set at 0.05 or 0.01 where the threshold for wrongly inferring a difference between the groups is set at 1 in 20 or 1 in 100. However, we generally need to be more rigorous in array scenarios. T-tests can be adjusted to reflect the assumptions made in the experimental design. Where an unmatched case-control design is adopted, an unpaired t-test would be appropriate. However, where a treatment effect is the design objective and each group has a matched before and after case, then a paired t-test would be applicable. Equally, one tailed or two tailed t-tests can be selected depending on whether it is change or a directional change to test and equal or unequal variance in the compared data-sets. All of these considerations will adjust the assigned significance. T-tests can be made in virtually all statistical packages and Microsoft Excel has a t-test function. Excel should be used for any array analysis with caution, however, since its inbuilt format-recognition system may irreversibly and without notification change some gene and Probe IDs into dates and other common formats.

#### 13.2.6.6 ANOVA Approaches:

T-statistics used on sets of control and experimental groups compare the dispersion of signal within the groups relative to that between groups. Analysis of Variance (ANOVA) takes this approach further. Standard ANOVA models come in 3 basic model classes:

Model 1. Fixed Effects: This most closely resembles the t-test and compares the mean difference between groups to group variance.

Model 2. Random Effects: Uses the F-test to determine whether the differences between the groups exceed what would be expected from random re-sampling of the within-group distribution.

Model 3. Mixed, Fixed and Random: This approach uses both the random effects and direct effects to suggest significance.

ANOVA can also model various levels of factor-factor interactions. A one-way approach, models the contribution to variance arising from the primary factor identified, e.g., the treatment effect. A two-way ANOVA model would take multiple contributing factors and explore the effect of each and the cross-interactions between these effects.

A one-way (model 1) ANOVA of a gene's expression in two conditions is simply a t-test. The power of ANOVA for array work lies in its extensibility beyond this simple case. ANOVA approaches employ both the t-test to characterize mean difference and the F-test to compare whether variance profiles appear similar. For array analysis, its other utility lies in the capacity to directly compare not just two, but multiple conditions. Hence, a one-way ANOVA on three conditions, can indicate where inter-group variance is significant relative to intra-group variance of the three groups (F-statistic) or where the mean difference between any is significant.

Two-way ANOVA is more generally suitable for in-depth array analysis and deals more explicitly with the sources of variance identified in an experiment with a view to isolate the signal from gene expression change. This is achieved by creating a matrix of the likely sources of variance and their (two-way) interactions with each other. Typically, an effect level for each parameter and a residual that the model was unable to incorporate will be generated in the analysis. These should be noted as indicators of whether the experiment determined the true origins of variance. Where a two-way ANOVA cannot usefully resolve the contributing variance, higher-order ANOVA models may be explored that consider three-way interacting elements.

Further development of the ANOVA approach can be found for most experimental designs. A continuous experimental variable such as time can be modeled using Continuous ANOVA (CANOVA). In designs that consider two or more treatment factors in conjunction, multivariate extensions to ANOVA are suggested. Multivariate analysis of variance (MANOVA) approaches use the off-diagonal elements in the covariance matrix to draw conclusions regarding interactions.

One limiting consideration in the standard ANOVA approach is that the effect terms are not signal dependent, and hence the effects that change strength as a function of expression can only be summarized rather than modeled. Storey and Dabney have developed an ANOVA model to incorporate intensity-dependent effects, allowing ANOVA to be used for chip level normalization after which a gene level ANOVA can be introduced to assess differential expression [23].

Most ANOVA approaches are ultimately extensions of the classical t/F-statistics and as such require a range of criteria to be met, to have confidence in their inferred significance levels. The data should be normal in distribution after appropriate adjustment (e.g., log or power transformation), the relationships between the factors investigated should be linearly co-dependent and far outliers should be avoided. Additionally, the distributions underlying the T and F statistics require a certain population size to stabilize their predictive outcomes such that for small sample sizes estimates of variance can be unreliable. Since array data is usually based on small-n populations that may not be normal in distribution and will frequently contain outliers, we have to treat the classical-parametric statistics with caution. Most array analysis packages including commercial products such as GeneSpring as well as free/open software including R/Bioconductor (multtest package), DChip and the NIA's ANOVA analysis suite, can conduct basic ANOVA. CANOVA, MANOVA and MANCOVA analyses are generally more easily conducted in specialist statistics packages or programming environments (R/Bioconductor).

#### 13.2.6.7 Non-Parametric Tests

ANOVA approaches extend t-tests but are constrained by their assumptions. Non-parametric approaches are more robust and not as limited by their prior assumptions of data distribution. The simplest and most widely used non-parametric statistic is rank. Non-parametric tests are generally regarded as being less powerful but frequently more reliable than their parametric equivalents for array data. Typically expression levels in two classes will be converted to ordinal, that is ranked, values and the difference in rank used to assess change. Specifically a Wilcoxon-signed rank test can generate a p-value for a set of genes from the comparison of ranks between two data sets. The rank value is useful because non-linear differences between the response behaviors of different experiments can be compensated. If two arrays with highly non-linear relative signal are compared, acceptable normalization may be difficult to achieve. But for a serially reducing expression pattern $A \Rightarrow C$, provided $A > B > C$ is reported by both arrays, rank order change can still be used to infer significance. Clearly this is insensitive to non-linearity and robust since outlier signal is not able to contaminate the rank model. The statistical tests that rely upon ranks include the Mann-Whitney U-test as the non-parametric equivalent of the t-test and the Kruskal-Wallis analysis for the equivalent of ANOVA/MANOVA. Statistical packages such as SPSS and Sigma-Stat as well as GeneSpring can provide non-parametric statistics.

#### 13.2.6.8 Bayesian Inference

Perhaps the most interesting development in statistical modeling arises out of a departure from frequentist approaches and adoption instead of Bayesian conditional inference. Bayesian inference differs markedly from classical- parametric approaches in that both a prior expectation of the form of a null distribution and new knowledge about the distribution are readily integrated into an analysis. This allows a more stable estimation of probability where sample sizes are small since we have prior expectation with which to weigh the evidence, but the cost, if there is one, lies in the incorporation of subjectivity into an otherwise "objective" approach. Where all prior information accords with classical log normal distributions, the Bayesian approach tends towards the classical ANOVA. Unlike the non-parametric rank tests, expression information is integrated in a fully quantitative mechanism. Baldi and Brunak present a useful introduction to the topic [31] and further specialist texts are available [32].

As with all statistical measures, the aim is to resolve the probability of the actual biological change ($\overline{H_0}$) given the data evidenced. Bayesian inference permits models with prior information about the hypothesis to be integrated with the sensitivity of the array for observing the expected evidence. This produces a posterior, or adjusted probability that can serially update the prior information. In the Bayesian approach, the evidence provided by the array is used in conjunction with our knowledge about the hypothesis likelihood within the system's context. Hence, the probability of there being a biological change given that chip evidence is calculated in two steps. The first step involves estimating from our knowledge of the system the base rate at which the tested hypothesis will be genuinely true, for example, if the hypothesis relates to fold change, from the anticipated rates of real expression change between the measured systems. This is combined with the probability of the fold change evidence, we detect for this change being observed, given the general sensitivity of our platform, for example, as a function of the signal. Secondly, the probability generated is assessed relative to the probability of observing the fold change evidence whether or not there was a differential expression. This is calculated from the sum of the first step carried out over all null and alternate hypotheses. More formally:

$$P\langle H_0 \mid Evidence\rangle = \frac{P\langle H_0\rangle \cdot P\langle Evidence|H_0\rangle}{P\langle Evidence\rangle} \tag{13.1}$$

Clearly, our result $P\langle H_0 \mid H_0 Evidence\ Evidence\rangle$ can now be used to update the prior information such as $P\langle H_0\rangle$ for our next round of analysis. This is a powerful way of moving forward, given parallel lines of evidence, since each replicate in a system can help converge our expectations of the tested hypothesis. Several applications are able to develop both t-test and Bayesian p-values from the data. Cyber-T is perhaps the most readily accessible and provides both a T-statistic and a final column with the adjusted Bayesian probability. Cyber-T also supports a modified ANOVA analysis where the distributions used are Bayesian rather than t/F-statistic based. For a **b**ayesian **a**nalysis of **g**ene **e**xpression **l**evels on two-channel experiments, the BAGEL system is a free executable, available for most platforms, while packages such as bridge and bgx are available for R/Bioconductor.

#### 13.2.6.9 Correcting for Multiple-Hypothesis Testing

While the t-test introduced by Gosset ('student') has a long tradition in both industrial and scientific statistics the application to array technologies is not without challenges. If we use arrays for generating hypotheses and estimating significance from the changes observed then we must consider the effect of multiple comparisons. This is the large $p$ (number of genes measures) relative to a small $n$ (sample size) problem, where $p>>n$ that makes over- assignment of significance a peril.

The t-statistic is based on a difference between two distributions with known form and Bayesian statistics on a modified probability distribution function. If these tests are used with p-values on a gene by gene basis between our control and experimental conditions, and if $\alpha<0.05$ is employed then 1 in 20 comparisons could trigger a false positive by chance alone. In an array analysis used in an explorative mode we are inherently testing tens, if not hundreds of thousands of gene expression change hypotheses in parallel. It is difficult then to rule out the worst case scenario, that in a comparison between samples on arrays with 50,000 probes of which 5,000 may show a fold change of interest and $\alpha=0.05$, that hundreds of the transcripts we identify as significantly different were not selected by chance alone.

Several measures have been introduced to extend the p-value for a single comparison to describe the disparity between the groups of genes selected in an analysis relative to the real population of differentially expressed genes. These include sensitivity; the power to detect real positives, and specificity; the power not to detect false positives (Fig. 13.8). The array literature is mostly

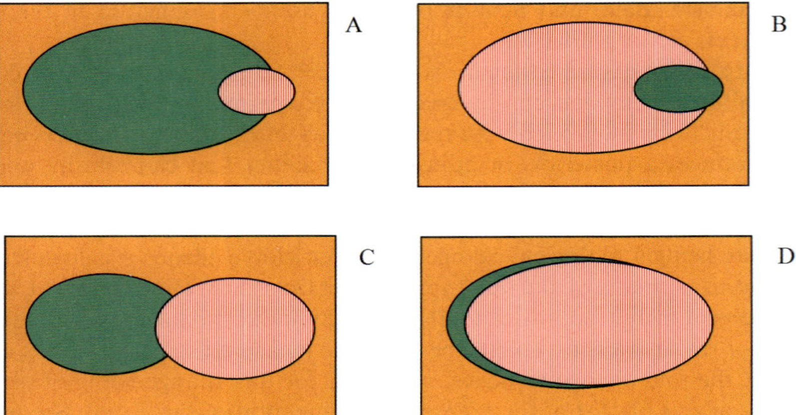


**Fig. 13.8** Specificity and sensitivity. All probes on the array (*rectangle*), probes with real biological change (*dark green*) and probes identified on the array as differentially abundant (*striped*). **(A)** High specificity low sensitivity **(B)** High sensitivity low specificity **(C)** Low specificity low sensitivity **(D)** high specificity high sensitivity (Copies of figures including color copies, where applicable, are available in the accompanying CD)

concerned with minimizing false positive (Type-I) errors without unduly prejudicing the power (sensitivity) of the analysis generally defined as 1-p (Type-II error). Strong limiting conditions have been considered such as the family-wise error rate (FWER) that describes the probability of at least 1 Type-I error appearing in the result set and the per comparison error rate (PCER). When FWER is the controlling factor in an array experiment, Type-II error would be expected to be high since the criteria to accept a transcript as genuinely differentially abundant would need to be strict. A weaker limit is the Per Family Error Rate (PFER) that can be used as a limit for the expected number of Type-I errors that should be reported. Rather than an absolute number of errors, the more commonly used weak measure adopted to control Type-I error in multiple hypothesis testing, is the proportion of results (rejected null hypotheses) that are false positives, or the False Discovery Rate (FDR) [33]. The q- value is sometimes introduced as the multiple hypothesis corrected analogue of the p-value, being the smallest FDR expected for a given p-value cutoff.

A number of approaches can be taken to adjust confidence levels relative to the multiple-hypotheses tests and thereby control these different error rates. The strongest was suggested by Bonferroni to limit the FWER. If a test is conducted with conditions that yield an error 1 in 20 times ($\alpha = 0.05$), then by repeating the test we double the probability of there being an error generated in the combined results to 1 in 10 instances. In order to maintain the 1 in 20 rate of error in the combined family of results, the stringency must be doubled. With an $\alpha = .025$ cut-off applied to both experiments, the expectation of a false result from the combined family of results would remain at the 1 in 20 level. Extending this to a microarray experiment, to maintain a FWER $\alpha$ of .05 when conducting 50,000 parallel comparisons, our cut-off should be adjusted by 0.05/10,000 = 1e-6. Clearly the Bonferroni approach will strongly control our FWER, but at the expense of grossly prejudicing our chance of ever identifying a true positive. Hence, the Bonferroni adjustment penalizes Type-I error at a prohibitive expense to Type-II error. Moreover, the Bonferroni adjustment makes the assumption of independence between the comparisons that is not the case when comparing large numbers of genes against each other. Hence there are several adjustments to the Bonferroni approach that may seek to moderate the correction, while assessing the impact this will have on the FDR, FWER and PFER. These are generally frequentist approaches to limiting error in multiple- hypotheses scenarios, but Bayesian approaches are also possible.

One of the simplest ways to assess the frequency at which a platform is producing error was suggested by Westfall and Young [34]. This approach can be used to create a non-parametric estimate of the level of Type-I error generated per chip at a given per-gene significance threshold. Taking two gene expression data-sets each of which contains biological replicates, the class assigned to each sample in the study is randomly permuted. For example if samples A-F are controls and G-L test conditions, a permuted group 1 might contain B, G, L, F, H, I and the remaining samples would form group 2. The identical tests applied to the real data, be it ANOVA or Bayesian analysis or simply a fold change with t-test, are then considered with respect to each gene in the randomly permuted groups. By generating numerous randomly permuted groups we can estimate a median chip level false-discovery rate and report this as the confidence measure. When a high FDR, relative to the stringency levels applied is observed, it may be valuable to explore whether the data distribution meets the criteria necessary for the type of statistical test being performed.

A useful extension to this approach is to examine the impact of false positives on downstream analyses such as pathways and ontology. If we anticipate by permutation that 100 genes may be included as false positives, then a pathway analysis of a set of 100 randomly selected genes can suggest the impact of a given FDR on downstream conclusions. The significance generated from the 100 random genes would become the new baseline relative to which generated p-values would need to be adjusted.

Several other false discovery rate corrections are available. These include the strong Sidak correction, Holm's step-wise adjustment formed from the ranked p-values and Storey's Q-value that uses the distribution of p-values to assess a false discovery rate. Most microarray analysis packages allow the false discovery rate to be estimated. For example, DChip permits every comparison to be accompanied by a Westfall and Young permutation. Several packages exist in the R/Bioconductor suite including OCplus and LBE. Storey's Q-value can be assessed using either the standalone Q-value executable or the R/Bioconductor package.

A final note of caution is suggested with respect to false discovery at the per hypothesis level. This can be thought of as the rate of generating a false hypothesis given a genuine fold-change. There are several million ways in which we can select any two genes from a set of 5000 potentially identified, by an array study. If three genes are selected that appear to have a potential for mechanistic linkage and it is estimated that there is only a 1e-6 chance of this linkage being spurious, then this presents itself as an attractive group for further investigation. However there are several hundred billion ways of selecting three genes from 5000 with critically a very low base-rate significance. Hence, even a 1e-6 level of significance may be suspect. It is then, not simply the misidentified genes that need concern us, but equally the potential for misconstruing the hypotheses generated from sets of genuinely differentially expressed genes.

#### 13.2.6.10 Representing Experimental Power

If spike-in data is available to quantify the levels of expression change in a data-set relative to that reported by the array, then the power of the instrument can be quantified. A useful way to represent the results of such an analysis is by a receiver operating characteristic curve (ROC) (Fig. 13.9). In an ideal situation, we would be able to obtain a false discovery rate of 0 while detecting 100% of the real transcripts that change. In the poorest circumstances even allowing the false discovery rate to approach 100% might gain us only little information on the real transcript change. In a real situation, we find that arrays as instruments lie between these two extremes. The relationship between false positive rate (Type-I error) and true positive rate (experimental power) is determined in part by the biological stability and hence the biological variance of the system and in part by the measurement error introduced. A ROC curve demonstrates this relationship. In an experiment with ideal performance, nearly 100% true positive results will be achieved with almost no false positive results.

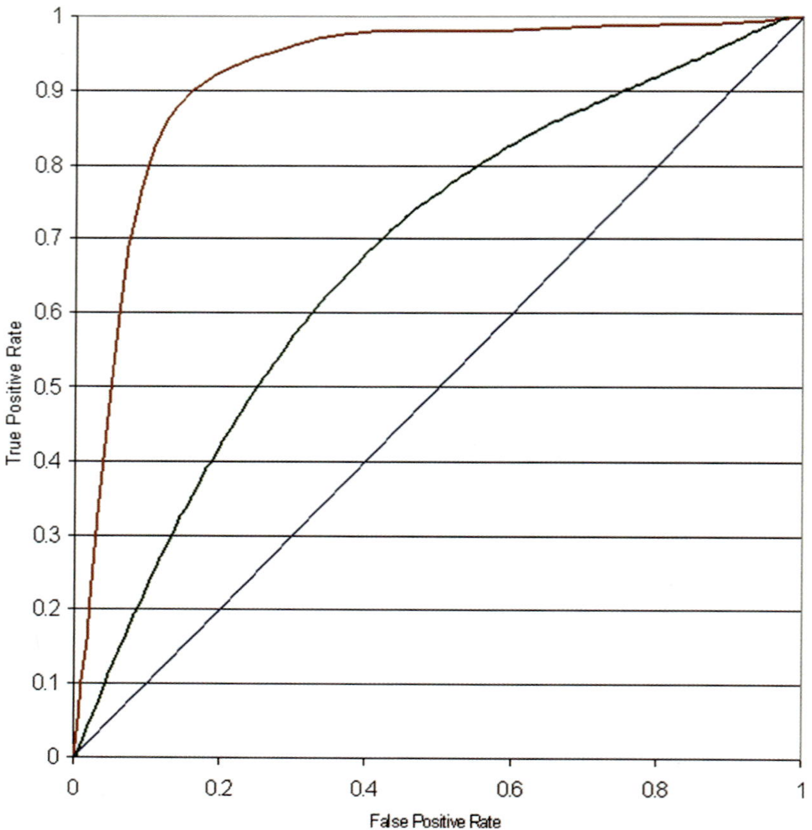


**Fig. 13.9** ROC Curves. Examples of three receiver operating characteristics illustrating experimental power. The curves shown illustrate an analysis with no power (*bottom line*), low power (*middle line*) and high power (*upper line*) (Copies of figures including color copies, where applicable, are available in the accompanying CD)

Realistically we might hope to obtain a 90% true positive rate with only a 10% false positive rate. Attempting to push past the 90% true positive rate would, in this system, introduce copious amounts of false positive data. Our worst-case scenario is the zero power experiment in which no matter what criterion is chosen we may expect to find equal false and true positive data.

#### 13.2.6.11 Concordance

The discussion above describes statistical models and confidence measures for data largely derived from a single platform. But, a question naturally arises as to how platforms concord in their results and how arrays compare with other expression measurement approaches such as sequencing. The concordance between array platforms has generated much discussion. For similar samples hybridized to the same array platform the largest difference in results was discovered between laboratories suggesting, small differences in preparation procedure may be significant. Comparisons between array platforms hybridized with the same RNA sample pools indicated that high confidence target genes accorded poorly [35]. In part, this arises from different isoform targets on different platforms but analysis pathway differences also contribute to the discordance. The differences have led the FDA to establish the MicroArray Quality Control project (MAQC), that aims to provide tools and approaches for standardizing array results. The reproducibility issue is also a core element in several software projects such as R/Bioconductor. The second part of the question is more open. Array and RT-PCR results have been shown in many instances to accord well, with an $r^2$ in excess of 0.8. Array and sequencing libraries have a much lower concordance as do array and SAGE data.

### 13.2.7 *Inferring Biological Meaning*

Clustering is one of the most flexible and visually effective ways to make classification decisions with respect to the high dimensionality data. Array data is considered to have high dimensionality, since each probe's expression measurement can be projected along its own axis in the same way that a spatial coordinate measurement generates values in three orthogonal axes. An expression study that uses a 22,000 probe platform can be considered as taking measurements in a 22,000 dimension space. We should also consider that while the array may be quantifying several tens of thousands of variables, biological systems are inherently complex and a significant amount of high dimensionality variability describing the organism, its history and environment lie beyond the array platform's measurements. Nonetheless, some work is needed to discover what may be similar between the vectors generated in this high dimensionality measurement space.

The aim of clustering is generally to infer some classification of either genes or samples into groups based on similarities evident between them. They partition the very large number of expression measurements into subgroups (sub-spaces). Within these subgroups patterns are shared between examples and between the groups large differences should be evident. Because of the many different ways in which patterns can be detected and their similarity assessed, there are many clustering approaches. Some of these are completely automated (unsupervised) and others take some training or classification information from the investigator (supervised). Many excellent reviews of the clustering approaches are available [36], and the summaries below provide only general overviews in an array context.

#### 13.2.7.1 Hierarchical

Hierarchical clustering is an unsupervised approach that is frequently performed at the gene and the sample levels. Put simply, genes or samples that behave in similar ways are placed next to each other in the output [37], the greater the similarity the greater their proximity. When differential expression is represented on a heat map using a color scale simple trends become evident. Clustering can be carried out on genes alone, for example, in a time-course experiment the genes that show similar expression trends through the study are brought together. Alternatively in a bi-clustering mode, samples and genes are used in conjunction. Groups of genes that show similar trends through the sample data are brought together on one axis, and conversely samples with a similar profile over their genes are grouped together on the other axis. In the example, clustering shown in Fig. 13.7, dendrograms are used to show the degree of difference in the expression pattern. The length of the lines represents the level of dissimilarity between the profiles. Hence, the top 5 samples are clearly similar and are linked by short lines but are very different from the lower samples, and so the line linking the upper and lower samples is relatively long. The same can be seen with respect to the proteosomal genes, where the small groups of genes on the left of the heat map clusters together and away from the larger set of genes to the right.

This clustering approach is termed hierarchical since no assumption is made in advance about the way the elements will cluster. The most similar elements are brought together to form a pair, and each pair is then organized next to its most similar pair. This process is iteratively repeated to generate a hierarchy. Several distance measures can be used that describe the profile similarity. The Pearson correlation coefficient (r) or absolute correlation ($|r|$) is popular, but several other metrics are possible. Equally, the way in which the multiple elements in sub-clusters are compared can vary. The simplest way is to use the mean profile from each cluster (centroids), but many other options are available based on, for example, the minimum or maximum distance between elements in the clusters. The choice of these options can determine how weakly or strongly similar expression profiles are brought together. Clustering algorithms seek to minimize the difference in expression within the clusters and maximize that between clusters. The exact decision as to which combination of distance metrics and clustering modes works best can, in part, be made based on this within-to-between cluster difference, but many other factors such as sensitivity to small changes in the profiles can be considered. In data that is difficult to normalize, the use of non-parametric rank correlation may prove more robust than correlation. Many investigators opt for consistency and choose to cluster with the same metrics. Indeed iterating through the clustering approaches until a desired pattern of clustering is obtained should generally be avoided.

Most array analysis suites such as DChip, BeadStudio and GCOS include clustering algorithms. Stand-alone applications with a wider range of distance metrics and clustering functions are also available such as Cluster 3.0, originally developed by Michael Eisen [37].

#### 13.2.7.2 K-Means

K-Means and K-Medians are similar to hierarchical clustering approaches, but with several notable differences. Instead of building, a hierarchy of similarity, the process takes as an input, a value [K], provided by the user and the arrays are partitioned into only these K-groups. The approach then seeks to divide the profiles into the K-groups, to maximize the similarity within the groups relative to that between groups. K-Means starts with a random selection of mean centroids, and arrays are assigned to the K-classes based on the proximity to the centroids. A new centroid is computed for each of the K-groups and the arrays are once again assigned to the closest centroid. Median clustering differs from mean clustering, in defining centroids from median values and is thus less sensitive to group outliers and hence more stable. Over thousands of iterations (N), the arrays are re-assigned between the groups in the search for an optimal segregation. If no stable segregation is found within the N optimization cycles, the clustering can generate different results because of the random initial seeding state. If there is a prior hypothesis regarding the number of clusters, for example, where samples need to be categorized into diseased or healthy groups, then K-Means/Medians can be more effective than hierarchical clustering. K-Means/Medians clustering can be performed in many applications including TM4 Cluster 3.0 and R/Bioconductor.

#### 13.2.7.3 PCA/LDA

Principal Component Analysis and its close relation the Linear Discriminant Analysis are examples of dimensionality reduction approaches. These approaches take the thousands of different probe readings (dimensions) and seek to reduce these to only two or three dimensions by plotting the expression from multiple, similarly expressed genes along the same axis. This allows samples to be positioned on a 2-dimensional X-Y plot, based on the combined expression of many genes summarized on each axis. As in hierarchical clustering, the algorithm seeks the patterns of expression that are most similar between some genes and most different between others. This forms the genes that make up the first axis. The process is repeated for patterns with less significance each time a new potential axis is generated. In some approaches, sub-groups of genes make up the compound axes while in others all genes contribute to all axes but with varying positive and negative weights accorded by their similarity to the pattern. Usually only the first two or three axes are used to plot samples (Fig. 13.10). PCA can be conducted on all genes, or on subsets of genes from different functional groups that can be used to indicate whether there is any evidence of linear separability between sample groups based on the functional group chosen. LDA can be performed in DChip while PCA is a clustering function available within Cluster 3.0 and both approaches can be explored in R/Bioconductor (pcaPP).

Many other clustering approaches can be found including self-organizing maps, support vector machines, and QT clustering. Most of these are available for R/bioconductor (kernlab) and Matlab while compiled executable of SVMLight are available for many platforms and the flexible clustering environment RapidMiner.

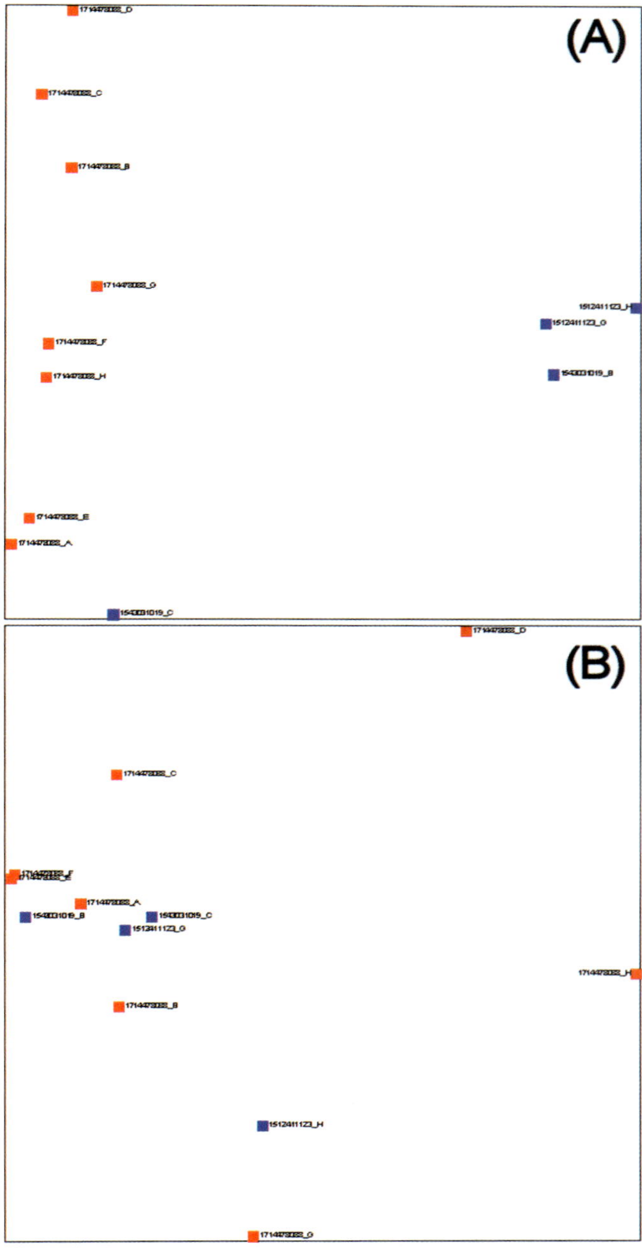


**Fig. 13.10** Principal component analysis. Compound axes are created through the linear combination of signal from multiple probes. **(A)** Signal from genes of the apoptosis pathway for two data-sets are plotted on the first two principle axes demonstrating that a linear discriminant function can be developed that separates most of sample group 1 (*dark*) from sample group 2 (*light*). **(B)** The same samples, with PCA conducted on genes involved in oxidative stress; with this group of genes no linear separability is evidenced (Copies of figures including color copies, where applicable, are available in the accompanying CD)

#### 13.2.7.4 Ontology Analysis

Ontology analysis is an example of a broader classification analysis that seeks to find commonality between the diverse sets of transcripts. The aim of an ontology analysis is to relate the changes in expression detected on an array to the genes involved in biological systems that are well established from prior research. This mapping is used to directly suggest coordinated biological systems that are undergoing disruption between the control and test conditions. In an ontology analysis, the genes that have been established as demonstrating a response to the treatment condition, for example, those identified in an ANOVA analysis, are mapped to known groups of genes (ontologies) that act together in biological mechanisms. Ontologies are currently organized by the GO consortium into three broad classes: Biological Function, Biological Process, and Cellular Compartmentalization, representing approximately orthogonal descriptions of a protein's role in the cell. Within these categories, a hierarchical relationship is formed with respect to classification. For example, a gene whose protein product locates to the nucleolus is a member of the nucleolus group with respect to cellular compartmentalization. The nucleolus, in turn, is a part of the nucleus and hence the gene is also a member of the nucleus group. Above this hierarchy lies the cellular organelle and the cell, all of which will count one gene if the transcript is present. The transcript would have similar hierarchical entries under the biological process and function.

Genes are mapped in a one-to-many manner with a single gene able to count in multiple ontology groups where it has multiple roles in the cell. Some investigators choose to investigate up regulated and down regulated genes independently. However the processes of negative feedback in biological systems suggest that up and down regulated transcripts can usefully be investigated in conjunction. Once the set of genes has been mapped to ontology groups, a p-value is generally assigned to each ontology group to demonstrate whether a disproportionate number of genes map into this group relative to a background expectation (Fig. 13.11). Again, a multiple-testing condition arises and both a t-test and FDR corrected p-value will be presented. In Fig. 13.11, where pre and post metastatic solid tumors are compared, it is noteworthy that genes interacting with the extra-cellular matrix are highlighted. These could be cataloged and further analyzed to determine if a common regulatory factor is shared in a promoter analysis. There are many other ways to classify

(A)

| Term | RT | Genes | Count | % | P-Value |
|---|---|---|---|---|---|
| development | RT | | 68 | 18.8% | 3.5E-8 |
| cell adhesion | RT | | 35 | 9.7% | 1.3E-7 |
| response to stress | RT | | 43 | 11.9% | 6.4E-6 |
| pregnancy | RT | | 8 | 2.2% | 2.9E-5 |
| physiological interaction between organisms | RT | | 8 | 2.2% | 7.9E-5 |
| phosphate transport | RT | | 10 | 2.8% | 1.4E-4 |
| morphogenesis | RT | | 26 | 7.2% | 1.4E-4 |
| reproductive organismal physiological process | RT | | 8 | 2.2% | 1.5E-4 |
| reproductive physiological process | RT | | 8 | 2.2% | 1.7E-4 |
| organismal physiological process | RT | | 70 | 19.4% | 2.2E-4 |
| interaction between organisms | RT | | 9 | 2.5% | 2.9E-4 |
| organ development | RT | | 23 | 6.4% | 4.1E-4 |
| cell differentiation | RT | | 22 | 6.1% | 5.2E-4 |
| cell death | RT | | 24 | 6.6% | 5.4E-4 |

(B)

| Term | RT | Genes | Count | % | P-Value |
|---|---|---|---|---|---|
| extracellular matrix structural constituent | RT | | 14 | 3.9% | 1.8E-8 |
| protein binding | RT | | 115 | 31.9% | 3.2E-7 |
| insulin-like growth factor binding | RT | | 6 | 1.7% | 4.0E-5 |
| growth factor binding | RT | | 8 | 2.2% | 5.5E-5 |
| iron ion binding | RT | | 16 | 4.4% | 6.9E-5 |
| calcium ion binding | RT | | 32 | 8.9% | 9.0E-5 |
| structural molecule activity | RT | | 34 | 9.4% | 9.4E-5 |
| polysaccharide binding | RT | | 9 | 2.5% | 3.2E-4 |
| cytokine binding | RT | | 8 | 2.2% | 3.5E-4 |
| pattern binding | RT | | 9 | 2.5% | 6.2E-4 |

(C)

| Term | RT | Genes | Count | % | P-Value |
|---|---|---|---|---|---|
| extracellular region | RT | | 78 | 21.6% | 1.9E-21 |
| extracellular matrix | RT | | 32 | 8.9% | 4.1E-14 |
| extracellular matrix (sensu Metazoa) | RT | | 31 | 8.6% | 1.4E-13 |
| extracellular space | RT | | 30 | 8.3% | 5.3E-9 |
| collagen | RT | | 6 | 1.7% | 6.8E-4 |

**Fig. 13.11** Ontology Analysis. An example of an ontology analysis produced by the NIH DAVID system showing **(A)** Biological Process **(B)** Molecular Function and **(C)** Cellular Compartmentalization ontologies for differentially expressed genes identified between solid tumors contrasted in their pre and post metastatic states (Copies of figures including color copies, where applicable, are available in the accompanying CD)

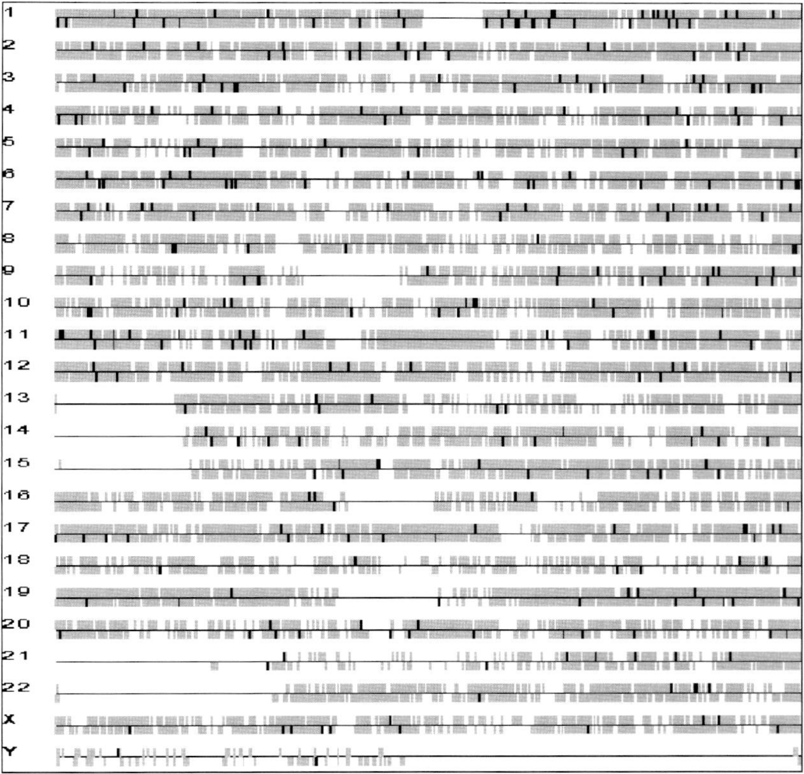


**Fig. 13.12** Genomic Clustering. Differentially expressed genes are located (DChip) on a genomic map (*gray*) and significant clustering that might suggest a co-regulated locus or aneuploidy is identified by correlation analysis (*black*) (Copies of figures including color copies, where applicable, are available in the accompanying CD)

sets of genes beyond just the basic ontologies. These can include their association with specific diseases, collocation to specific genomic loci (Fig. 13.12) or activity with respect to cohesive structures such as the proteosome.

Very many software tools exist for ontology analysis, and some of the first were OntoExpress and EASE, which was later developed as NIH DAVID. With a web interface, the NIH DAVID system offers an elegant combination of ontology, pathway and proteomics analysis. The system also offers a novel 'clustering-of-clusters', combining the different classification of clustering and pathway analyses to develop meta-clusters. The system also has valuable FDR features including the option to upload specific background lists and estimate Bonferroni, Hochberg and Westfall corrections to the t-based p-value. A plethora of ontology analysis tools now exist, many of which are linked from the GO consortium.

An extension of the ontology analysis is the Gene-Set Enrichment Analysis introduced by Subramanian [38] and recently revisited by Tibshirani [39] in exploring null models. In this approach, the significance of the individual gene expression differences between control and test conditions is combined in a non-parametric mechanism to create a score for a group of genes. This extends the standard ontology analysis, in which only the absolute number of genes above a p-value cutoff is counted to generate a rank. The genes in a gene set can be derived from standard ontology groups or groups created from other information such as *ab inito* pathway analysis. Gene sets are then ranked by this combined probability score, to identify sets of genes with probability profiles that are unusually significant. A java tool is available for GSEA analysis.

#### 13.2.7.5 Pathway Analysis

Pathway analysis is another classification system that extends the ideas employed by ontology analysis. As in ontology analysis, the aim is to map differentially expressed genes into biological pathways. The disruption to these pathways may then be indicated as linked with the test condition. Generally, pathway analysis is more concise in its indication of disruption, since there is a cause and effect process mapped by a biological pathway that can be examined for consistency with the disruption evidenced in array data. Pathway analysis then proceeds in two directions, curated and *de novo* approaches.

In the first approach, very much like ontology analysis, the differentially abundant genes, and optionally their homologues, are mapped into well-established pathways. Openly available databases of pathway maps, such as those curated at Kyoto and Biocarta, are frequently used and allow well established mechanisms to be compared with a differential expression profile. Where a pathway's disruption is suggested, it may be possible to generate downstream hypotheses concerning what else should be disrupted and to test these with respect to the data. Figure 13.13 illustrates how array data could suggest disruption to several of the pathways by which ubiquitylation is achieved and specific proteomic degradation initiated.

In the more advanced *de novo* approach, novel pathways are sought from the literature in a discovery mode. There are several ways to approach this but the most common seek to assess the abstract database of archives, such as Pubmed, for any links that have been suggested between the genes presented, either directly or by N intermediate stages where the user specifies the N. Evidently, given even a low value of N, the number of possible links between the genes and hence the possible pathways grows to unmanageable proportions. This requires careful consideration

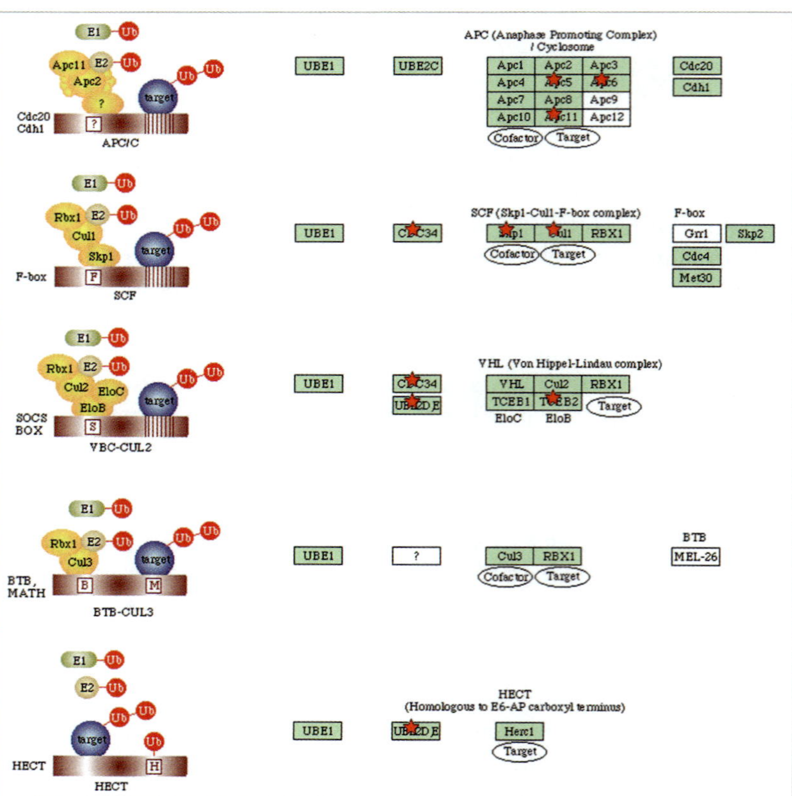


**Fig. 13.13** Curated Pathway Analysis. The ubiquitin pathway as represented in the KEGG pathway database. Genes suggested by the array analysis as differentially expressed are highlighted by stars introduced from an analysis of differentially abundant genes in the NIH DAVID system (Copies of figures including color copies, where applicable, are available in the accompanying CD)

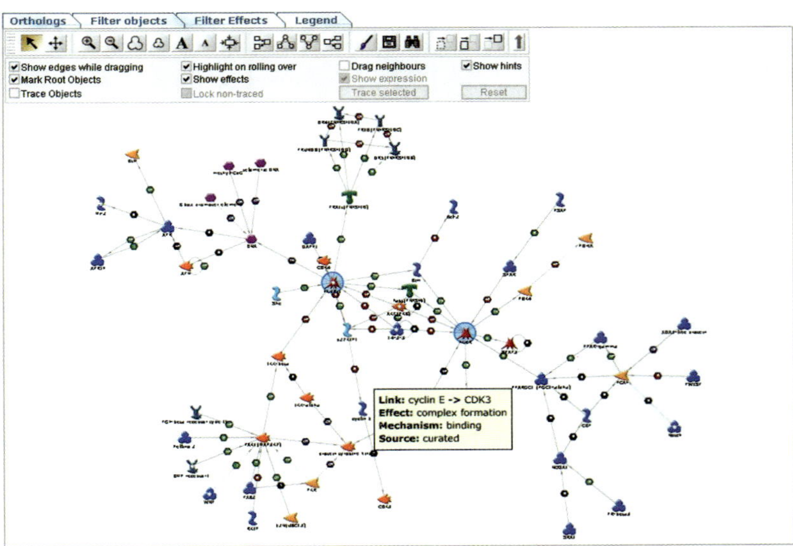


**Fig. 13.14** Hybrid pathway analysis. This network of interactions is not a fully curated biological pathway model. It was generated by combining curated pathways with data mined from the Pubmed database to locate gene-gene interactions (GeneGo). These links are used to suggest the interactions that might connect over-expressed genes (*circled*) identified in an array study of F344 Rat pituitary tumors (Copies of figures including color copies, where applicable, are available in the accompanying CD)

with respect to the discussion FDR and hypotheses. A further concern arises with some tools, as to whether computerized analysis of the literature is sufficiently powerful to generate the databases of pathway links. Several tools use teams of scientists to collate their databases where as others use natural language analysis software. Figure 13.14 illustrates a hybrid approach that extends curated pathways with *de novo* linkage.

Some tools for pathway analysis are free; the KEGG pathways can for instance be explored from within the NIH DAVID system. Other tools are available as commercial products, including the web based GeneGo, Java based Pathway Studio and the sophisticated promoter-pathway analysis suite Bibliosphere Pathway, that adds functionality to Chip-Inspector. Pathway analysis components are increasingly being incorporated within the major array analysis suites such as GeneSpring.

#### 13.2.7.6 Promoter Analysis

Any pathway or ontology analysis that suggests a set of disrupted genes within a common biological system raises the question of regulation. A promoter analysis takes small groups of genes that show evidence of co-regulation and seeks to identify the functional elements within primarily the promoter domains, 5' of transcription start sites that may help activate or moderate expression. Several systems have been developed to conduct these searches, and variation exists in the assumptions they make about promoter locations and the structure of the active elements. Some are restricted to consensus transcription start sites such as those catalogued in the DbTSS, whereas others use CAGE tags to map multiple-fuzzy promoter regions. Equally, there are those such as KSPMM, that identify functional binding with respect to simple transcription factor binding site (TFBS) motifs identified by nucleotide weight matrices, whereas others such as the Genomatix frameworker, generate models that consist of multiple regions of motif evidence discovered to have a spatial conjunction. This type of analysis can typically be propagated, in that, it will suggest transcription factors that could underlie the disruption. These can then be assessed in the expression profile for a change between the control and experimental conditions allowing evidence for a pathway disruption to be extended. More generally, promoters can be explored for regions of identity between species, while the major genome browsers such as UCSC, NCBI and Enesmbl all allow genomic features, such as CpG islands and repeat structures to be explored. For more in-depth sequence analysis, data from the ENCODE project can be visualized at UCSC although a word of caution is

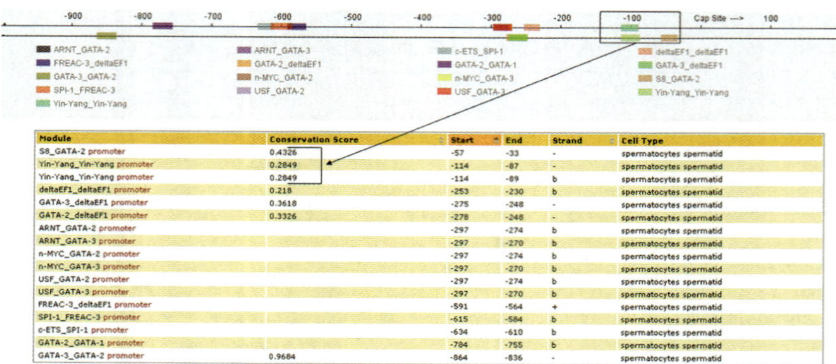


| Module | Conservation Score | Start | End | Strand | Cell Type |
|---|---|---|---|---|---|
| S8_GATA-2 promoter | 0.4326 | -57 | -33 | - | spermatocytes spermatid |
| Yin-Yang_Yin-Yang promoter | 0.2849 | -114 | -87 | - | spermatocytes spermatid |
| Yin-Yang_Yin-Yang promoter | 0.2849 | -114 | -89 | b | spermatocytes spermatid |
| deltaEF1_deltaEF1 promoter | 0.218 | -253 | -230 | b | spermatocytes spermatid |
| GATA-3_deltaEF1 promoter | 0.3618 | -275 | -248 | - | spermatocytes spermatid |
| GATA-2_deltaEF1 promoter | 0.3326 | -278 | -248 | - | spermatocytes spermatid |
| ARNT_GATA-2 promoter | | -297 | -274 | b | spermatocytes spermatid |
| ARNT_GATA-3 promoter | | -297 | -270 | b | spermatocytes spermatid |
| n-MYC_GATA-2 promoter | | -297 | -274 | b | spermatocytes spermatid |
| n-MYC_GATA-3 promoter | | -297 | -270 | b | spermatocytes spermatid |
| USF_GATA-2 promoter | | -297 | -274 | b | spermatocytes spermatid |
| USF_GATA-3 promoter | | -297 | -270 | b | spermatocytes spermatid |
| FREAC-3_deltaEF1 promoter | | -591 | -564 | + | spermatocytes spermatid |
| SPI-1_FREAC-3 promoter | | -615 | -584 | b | spermatocytes spermatid |
| c-ETS_SPI-1 promoter | | -634 | -610 | b | spermatocytes spermatid |
| GATA-2_GATA-1 promoter | | -784 | -755 | b | spermatocytes spermatid |
| GATA-3_GATA-2 promoter | 0.9684 | -864 | -836 | - | spermatocytes spermatid |

**Fig. 13.15** Promoter Analysis. Promoter analysis can be used to identify the sequences characteristic of the regulatory elements located in the promoter domains of genes showing the same expression pattern. The region shown is 1 kb upstream of the transcription start site of Protamine 1. Predictions are made based on evidence from the JASPA databases. Sites around 100 bp upstream of the transcription start site show evidence for sequence conservation (Copies of figures including color copies, where applicable, are available in the accompanying CD)

warranted for all such comparisons – ensure that the genome builds selected in the different genome browsers are the same, since promoter and gene sequences do shift coordinates between the builds. Promoter analysis can produce large data-sets of potential TFBS's, while arrays produce large data-sets of differentially expressed genes. To infer the relationship between the two, bi-clustering approaches can be used as described earlier with samples replaced by TFBS elements.

Figure 13.15 illustrates a promoter analysis using KSPMM in which multiple strands of evidence are combined. Two motif-model databases are compared alongside the evidence for functional protein binding and the conservation of non-coding promoter sequences between vertebrates. Figure 13.16 illustrates a similar analysis with the genomatix system, in which a common framework of binding sites is found in co-disrupted elements of the GABA A receptor. Many other tools

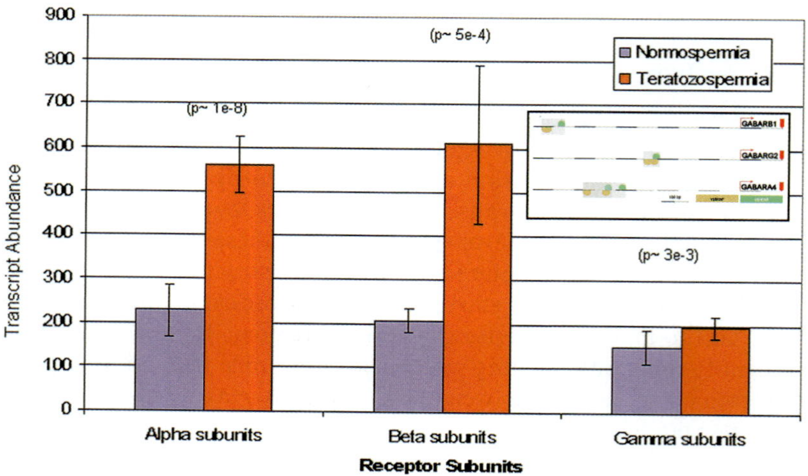


**Fig. 13.16** Expression and Promoter Analysis. Expression analysis can lead to the identification of novel functional elements. In a study cataloguing the abundance of transcripts for receptor proteins in the sperm of fertile and infertile individuals, the largest difference was evident with respect to the subunits of the GABA A receptor. Identification of the protein binding sequences (Genomatix Frameworker) common to these subunits but absent from the promoters of the unchanged subunits suggested co-regulation by a Brn-3 and HOX module (Copies of figures including color copies, where applicable, are available in the accompanying CD)

are available for promoter analysis, some of which use databases of transcription factor binding sequences such as Transfac and Jaspar, as well as others that use sequence alignment searches to identify novel patterns common to the promoter regions.

#### 13.2.7.7 Clinical Application

A few array platforms are beginning to receive the regulatory approval necessary for translation from the research lab into the clinical setting. In the clinical setting, the utility of the array may lie in taking an expression snapshot that may reveal multiple nuances of disruption that are coincident with later disease progression trajectories. Many studies have taken tumor samples and used them to try and develop a list of genes, the expression of which may be predictive of outcome. The problem with these approaches lies in the potential of the unstable tumors to undergo stochastic genomic rearrangement that can lead to almost unlimited trajectories. Nonetheless, in more stable diseases it may be possible to use arrays to predict clinical outcomes and inform treatment decision-making. Tools such as PAM have been developed to generate clinical classifiers using semi-supervised expression clustering.

#### 13.2.7.8 Data Repositories

As arrays have become more popular, the volume of expression evidence has swelled to well beyond the terabyte domain. In order to leverage the power of this extensive evidence major archives have been developed to host and aid in the exploration of this data. These include the NCBI's gene expression omnibus (GEO) archive, that archives over 160,000 high- throughput data experiments from many sources. The GEO system also permits searching and visualization within the arrays through hierarchical clustering and rank comparison. The XML interface into all of the NCBI's core databases has permitted extensions to these tools such as LaralinkGEO that re-categorizes expression evidence from across the GEO data-sets by disease and tissue. XML and web services have become powerful mechanisms for automating the identification of remote archives tools and exchanging information between them. Through interfacing databases of repetitive tasks many procedures can be co-ordinated through XML directed workflows, for example the annotation and publication of the metadata for new genome builds. Other major archives include the EBI's ArrayExpress. Smaller archives may be subject specific covering, for example, only germ cell development with the limited amount of free software available for the development of customized databases. Generally raw signal data extracted from the image data is archived, for example, Affymetrix CEL files but not DAT files, together with at least a minimal structured description of the experiment conducted (MIAME). Where the MIAME standard describes the minimal information, the MAGE object model describes XML data structures for exchanging this data between repositories.

With these archives in place it is possible to conduct a range of *in silico* experiments, in which array experiments are re-analyzed for data relating to a topic not explored in the analysis of the original study. Where multiple investigators have covered the same topic but not managed to achieve results with statistical significance, meta studies can be conducted on the archived data to extract the significance or additional insight from several independent experiments. The power of these approaches continues to increase as more evidence is generated. Should arrays be widely adopted in the clinical field, another order of magnitude of data will be generated. Whether the potential of this data to understand complex disease is realized will largely depend on surmounting the very considerable challenges associated with the sharing of anonymized clinical data between organizations.

**Acknowledgments** We would like to thank all members of the Krawetz Lab for their helpful suggestions and proof reading, particularly Claudia Lalancette, Laura Naismith and Amelia Linnemann. Some of the Affymetrix data illustrated was kindly supplied by Dr David Dix of the US EPA. The authors gratefully acknowledge the Michigan

Economic Development Corporation, the Michigan Technology Tri-corridor program (Grant 085P4001419) and the Department of Obstetrics and Gynecology, Wayne State University.

## Key References

Choe SE, Boutros M, Michelson AM, Church GM, Halfon MS. Preferred analysis methods for Affymetrix GeneChips revealed by a wholly defined control dataset. Genome biology 2005;6(2):R16.

Irizarry RA, Wu Z, Jaffee HA. Comparison of Affymetrix GeneChip expression measures. Bioinformatics (Oxford, England) 2006;22(7):789–794.

Lee ML, Whitmore GA. Power and sample size for DNA microarray studies. Statistics in medicine 2002;21(23):3543–3570.

Speed TP. Statistical analysis of gene expression microarray data. Boca Raton, FL: Chapman & Hall/CRC; 2003.

Tibshirani RJ, Efron B. On testing the significance of sets of genes. The Annals of Applied Statistics 2007;1(1):107–129.

## Suggested Reading

### *Background to Microarray Technologies*

1. Zhang W, Shmulevich I, Astola J. Microarray quality control. Hoboken, N.J.: Wiley-Liss; 2004.
2. Speed TP. Statistical analysis of gene expression microarray data. Boca Raton, FL: Chapman & Hall/CRC; 2003.
3. Lee ML, Whitmore GA. Power and sample size for DNA microarray studies. Statistics in medicine 2002;21(23):3543–3570.
4. Do K-A, Müller P, Vannucci M. Bayesian inference for gene expression and proteomics. Cambridge; New York: Cambridge University Press; 2006.
5. Bentley DR. Whole-genome re-sequencing. Current opinion in genetics & development 2006;16(6):545-52.
6. Heng HH, Stevens JB, Liu G, et al. Stochastic cancer progression driven by non-clonal chromosome aberrations. Journal of cellular physiology 2006;208(2):461–472.
7. Martins RP, Krawetz SA. Decondensing the protamine domain for transcription. Proceedings of the National Academy of Sciences of the United States of America 2007;104(20):8340–8345.
8. Martin S, Pombo A. Transcription factories: quantitative studies of nanostructures in the mammalian nucleus. Chromosome Res 2003;11(5):461–470.
9. Martins RP, Ostermeier GC, Krawetz SA. Nuclear matrix interactions at the human protamine domain: a working model of potentiation. The Journal of biological chemistry 2004;279(50):51862–51868.
10. Wilusz CJ, Wilusz J. Bringing the role of mRNA decay in the control of gene expression into focus. Trends Genet 2004;20(10):491–497.
11. Moore MJ. From birth to death: the complex lives of eukaryotic mRNAs. Science New York, NY 2005;309(5740):1514–1518.
12. Wang Y, Liu CL, Storey JD, Tibshirani RJ, Herschlag D, Brown PO. Precision and functional specificity in mRNA decay. Proceedings of the National Academy of Sciences of the United States of America 2002;99(9):5860–5865.
13. Meizel S. The sperm, a neuron with a tail: 'neuronal' receptors in mammalian sperm. Biological reviews of the Cambridge Philosophical Society 2004;79(4):713–732.

14. Hargrove JL, Schmidt FH. The role of mRNA and protein stability in gene expression. Faseb J. 1989;3(12):2360–2370.
15. Schwartz DR, Moin K, Yao B, et al. Hu/Mu ProtIn oligonucleotide microarray: dual-species array for profiling protease and protease inhibitor gene expression in tumors and their microenvironment. Mol Cancer Res 2007;5(5):443–454.
16. Dallas PB, Gottardo NG, Firth MJ, et al. Gene expression levels assessed by oligonucleotide microarray analysis and quantitative real-time RT-PCR — how well do they correlate? BMC genomics 2005;6(1):59.

## *Signal Analysis and Modeling*

17. Qiu W, Lee ML. SPCalc: A web-based calculator for sample size and power calculations in micro-array studies. Bioinformation 2006;1(7):251–252.
18. Seo J, Gordish-Dressman H, Hoffman EP. An interactive power analysis tool for microarray hypothesis testing and generation. Bioinformatics (Oxford, England) 2006;22(7):808–814.
19. ENCODE. Identification and analysis of functional elements in 1% of the human genome by the ENCODE pilot project. Nature 2007;447(7146):799–816.
20. Draghici S. Data analysis tools for DNA microarrays. Boca Raton: Chapman & Hall/CRC; 2003.
21. Choe SE, Boutros M, Michelson AM, Church GM, Halfon MS. Preferred analysis methods for Affymetrix GeneChips revealed by a wholly defined control dataset. Genome biology 2005;6(2):R16.
22. Hoffmann R, Seidl T, Dugas M. Profound effect of normalization on detection of differentially expressed genes in oligonucleotide microarray data analysis. Genome biology 2002;3(7):RESEARCH0033.
23. Dabney AR, Storey JD. A new approach to intensity-dependent normalization of two-channel microarrays. Biostatistics (Oxford, England) 2007;8(1):128–139.
24. Irizarry RA, Wu Z, Jaffee HA. Comparison of Affymetrix GeneChip expression measures. Bioinformatics (Oxford, England) 2006;22(7):789–794.
25. Online document: www.ambion.com/techlib/tn/111/8.html.
26. Williamson SH, Hubisz MJ, Clark AG, Payseur BA, Bustamante CD, Nielsen R. Localizing Recent Adaptive Evolution in the Human Genome. PLoS Genet 2007;3(6):e90.
27. Ptitsyn AA, Zvonic S, Gimble JM. Digital Signal Processing Reveals Circadian Baseline Oscillation in Majority of Mammalian Genes. PLoS Comput Biol 2007;3(6):e120.
28. Tomita H, Vawter MP, Walsh DM, et al. Effect of agonal and postmortem factors on gene expression profile: quality control in microarray analyses of postmortem human brain. Biological psychiatry 2004;55(4):346–352.

## *Statistical Approaches*

29. Wu B. Differential gene expression detection and sample classification using penalized linear regression models. Bioinformatics (Oxford, England) 2006;22(4):472–476.
30. Robson B. Clinical and pharmacogenomic data mining: 3. Zeta theory as a general tactic for clinical bioinformatics. Journal of proteome research 2005;4(2):445–455.
31. Baldi P, Brunak S. Bioinformatics : the machine learning approach. 2nd ed. Cambridge, Mass: MIT Press; 2001.
32. Carlin BP, Louis TA. Bayes and Empirical Bayes methods for data analysis. 2nd ed. Boca Raton: Chapman & Hall/CRC; 2000.

33. Benjamini, Y. Yekutieli, D. The Control of the false discovery rate in multiple testing under dependency. The Annals of Statistics 2001;29(4):1165–1188.
34. Westfall PH, Young SS. Resampling-based multiple testing: examples and methods for P-value adjustment. New York: Wiley; 1993.
35. Irizarry RA, Warren D, Spencer F, et al. Multiple-laboratory comparison of microarray platforms. Nature methods 2005;2(5):345–350.
36. Jain AK, Murty MN, Flynn PJ. Data clustering: a review. ACM Computing Surveys 1999;31(3):264–323.
37. Eisen MB, Spellman PT, Brown PO, Botstein D. Cluster analysis and display of genome-wide expression patterns. Proceedings of the National Academy of Sciences of the United States of America 1998;95(25):14863–14868.
38. Subramanian A, Tamayo P, Mootha VK, et al. Gene set enrichment analysis: a knowledge-based approach for interpreting genome-wide expression profiles. Proceedings of the National Academy of Sciences of the United States of America 2005;102(43):15545–15550.
39. Tibshirani RJ, Efron B. On testing the significance of sets of genes. The Annals of Applied Statistics 2007;1(1):107–129.

## Web Resources

www.sas.com
www.spss.com
www.mathworks.com
www.systat.com
www.wolfram.com
http://lgsun.grc.nia.nih.gov/ANOVA/
http://visitor.ics.uci.edu/genex/cybert/
http://web.uconn.edu/townsent/software.html
http://faculty.washington.edu/jstorey/qvalue/
http://www.fda:gov/nctr/science/centers/toxicoinformatics/maqu/
http://bonsai.ims.u-tokyo.ac.jp/~mdehoon/software/cluster/software.htm
www.tm4.org
http://svmlight.joachims.org
http://rapid-i.com
www.GeneOntology.org
http:vortex.cs.wayne.edu
http:david.abcc.ncifcrf.gov
http:GO.tools.shtml
www.broad.mit.edu/gsea
www.genome.ad.jp/kegg
www.biocarta.com
http://genego.com
www.ariadnegenomics.com
www.genomatrix.de
http://dbtss.hgc.jp
http://klab.med.wayne.edu/kspmm/
www.ensembl.org
http://genome.ucsc.edu/ENCODE/
www.gene-regulation.com
http://jaspar.genereg.net

http://meme.sdsc.edu
http://affymetrix.com/products/application/clinical_applications.affx
www-stat.standard.edu/~tibs/PAM/
http://klab.med.wayne.edu/cgi-bin/laralinkgeo.exe?id = PRM2
www.ebi.ac.uk/arrayexpress/
www.germonline.org
http://base.thep.lu.se
www.mged.org/Workgroups/MIAME/miame_mageom.html

# Chapter 14
# Analysis of Alternative Splicing with Microarrays

**Jingyi Hui, Shivendra Kishore, Amit Khanna, and Stefan Stamm**

**Abstract** Alternative splicing is one of the most important post-transcriptional processing steps that enhances genomic information by generating multiple RNA isoforms from a single gene. Recently, microarrays have been developed that can detect changes in splice site selection. Currently, the biggest challenge for the analysis of alternative splicing with microarrays is the bioinformatics analysis of array data and their low reproducibility by RT-PCR. Despite these problems, microarrays revealed an unexpected number of expressed RNAs, showed changes of alternative splicing in diseases and indicated that a splicing factor regulates a biologically meaningful set of genes.

**Keywords** Alternative splicing · Microarrays · RNA isoforms · Experimental validation · Exon · intron

## 14.1 Introduction

Almost all protein-coding genes in higher eukaryotes have introns, which are removed during pre-mRNA processing by RNA splicing. It is estimated that more than 88% of human genes are subject to alternative splicing, i.e., the cell decides whether to remove part of the pre-mRNA as an intron or include this part in the mature mRNA as an alternative exon [1]. Since most alternative exons encode parts of protein, the alternative splicing mechanism allows the creation of multiple proteins from a single gene, which increases the coding potential of the genome. Alternative splicing generates protein isoforms with different biological properties, such as altered protein-protein interaction, subcellular localisation or catalytic ability [2]. Bioinformatic analyses estimated that a quarter of the alternative exons introduce pre-mature stop codons that either lead to the generation of truncated isoforms or to the degradation of the mRNA by nonsense-mediated decay [3]. However, recent array analyses show that these variants are generated at a low level [4].

## 14.2 Alternative Splicing and Gene Expression

The importance of alternative splicing is evident from the large number of diseases that are associated with or caused by the wrong selection of alternative exons [5, 6]. It has been estimated that up to 50% of disease mutations localized in exons can change the splicing patterns [7]. Naturally occurring single nucleotide polymorphisms (SNP) result in different alternative splicing patterns

S. Stamm
Department of Molecular and Cellular Biochemistry, B283 Biomedical Biological Sciences Research Building, 741 South Limestone, University of Kentucky, College of Medicine, Lexington, KY 40536-0509, USA
e-mail: Stefan@stamms-lab.net

S. Krawetz (ed.), *Bioinformatics for Systems Biology*, DOI 10.1007/978-1-59745-440-7_14,

within a population that has consequences for human health. This is illustrated by CYP2D6, a member of the cytochrome P450 family of proteins. Due to an intronic SNP that changes splicing patterns, up to 10% of Caucasians express non-functional CYP2D6 and cannot metabolize certain prescription drugs [8, 9]. In addition, drugs like acetaminophen act on a specific isoform generated by alternative splicing [10], demonstrating the role of alternative splicing in drug action.

The functional differences between the isoforms created by alternative splicing are often subtle [2, 7]. The first reports of large-scale analyses in tumor tissues indicate that the combination of numerous subtle changes in isoform ratios contribute to the disease phenotpye [11], and not surprisingly, changes in alternative splicing are frequently found in cancer [12]. Finally, low abundant splicing variants are specific for individual species, where they most likely contribute to a pool of isoforms that evolution can act on [13]. This implicates that not all human-specific isoforms have been identified, an assumption that is supported by the identification of new mRNA isoforms in global array analyses [14].

The importance of alternative splicing for human gene regulation makes it necessary to create global high-throughput analysis tools. Isoform-sensitive arrays have now been developed and applied to analyze alternative splicing. We review the different systems and studies that have been performed which reveals the current experimental limitations.

## 14.3 Detection of Splicing Variants with Microarrays

### *14.3.1 Structure of Splice Variants*

Alternative splicing patterns can be classified into five basic categories: cassette exons, alternative 5' and 3' splice sites, retained introns and mutually exclusive exons (Fig. 14.1A). Combination of these events and more complicated splicing patters are possible [15]. Cassette exons are the most frequent form of alternative splicing [7]. In addition to alternative splicing, alternative mRNA isoforms can be generated by alternative promoter usage, alternative polyadenylation and RNA editing [16]. Typically a gene generates several different mRNA isoforms with often severe differences in expression levels.

### *14.3.2 Systems for Exon Detection*

cDNA arrays cannot discriminate between the splicing variants and detect the mixture of isoforms. Therefore, specific array formats have been developed that discriminate splicing events. These high-throughput analyses of splice variants were performed in two basic systems: bead-based fiber-optic arrays and oligonucleotide arrays in slide format.

#### 14.3.2.1 Fiber-optic Arrays

Bead-based fiber-optic arrays are sold by Illumina. In this technique, arrays of beads are randomly assembled onto patterned optical fiber bundles. Each bead contains an oligonucleotide probe which can detect a complementary probe. Only specified splicing variants can be detected using this system, which requires the ligation of oligonucleotides prior to detection. The system is named RASL for RNA-mediated annealing, selection, and ligation. Typically, oligonucleotides will be designed that ligate across exon-boundaries to detect differences in the isoforms. One of these oligonucleotides contains an index-sequence, which allows identification on the array. The oligonucleotides are annealed on the mRNA, which is captured on a solid phase using biotinylated oligo-dT. In a next step, these oligonucleotides are ligated and amplified by PCR. The PCR products are then hybridized to the array. In a subsequent step, the array is decoded, i.e., the beads binding to

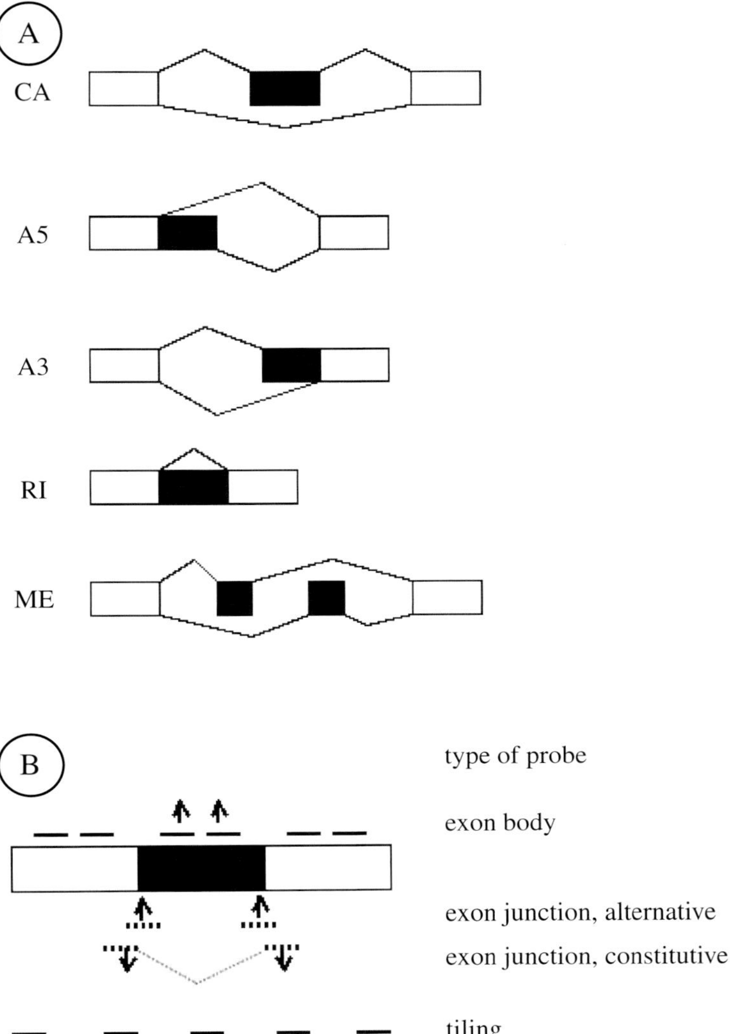


**Figure 14.1** Detection of alternative splice variants with oligonucleotide designs. **(A)** Basic types of alternative splicing. The alternative exon is indicated by a black box, constitutively used exons are indicated by a white box. The splicing patterns are indicated by lines. CA: cassette exon, A5: alternative 5' end exon, A3: alternative 3' end exon, RI: retained intron, ME: mutually exclusive exons. **(B)** Detection of alternative exons. Part of an mRNA is indicated with constitutive exons as a white and the alternative exon as a black box. Oligonucleotides are indicated as small lines, exon-junction probes as dotted lines. The exon-junction oligonucleotide is complementary to the constitutive (*white*) exons. An increase of the alternative exons would be indicated by an increase of its exon-junction and exon-body probes, and a decrease of the exon junction of the constitutive exons, as indicated by small arrows. For simplicity, only one exon body probe is shown, usually there are at least three probes, space permitting (Copies of figures including color copies, where applicable, are available in the accompanying CD)

the index-sequence are identified by hybridization to colored beads carrying the known index-sequence. The major advantage of the system is its high sensitivity and reproducibility by RT-PCR, the drawback are the limited number of events that can be studied [11, 17]. Since the detection relies on the amplification of short RNA parts, fractionated RNA resulting, for example, from storage in paraffin embedded sections can be used [11].

#### 14.3.2.2 Glass Arrays

The majority of arrays used to study alternative splicing, use oligonucleotides that are attached to glass slides. These slides can be produced by ink-jet printing (Agilent, Exonhit) or by photolithography (Affymetrix). The major advantage of ink-jet printing is that it can be easily customized, since it does not require the generation of photolithograpical masks. The drawback is the smaller number of spots per array (currently around 150,000). The major advantage of arrays generated by photolithography is their high number of spots (currently around 5,500,000). A compromise between the two systems is maskless photolithography offered by Nimblegen, which creates custom arrays with about 300,000 spots.

#### 14.3.2.3 Probe Designs

Array designs differ in the nature of probe-sets. Probes can be arranged at even spacings in tiling arrays; they can target only exon-bodies or exon-junctions (Fig. 14.1B). The use of exon-junction probes (Fig. 14.1b, dotted lines) allows the direct probing of exon-exon junctions, whereas tiling arrays indicate the relative change of exon expression. Custom-made designs often combine different types of probes. The scale of the probes is another feature that differentiate various designs. Probe-sets can be genome-wide, addressing either all exon-exon junctions [14] or all currently known exons [18]. More recently, highly focused designs were used to study a smaller number of better-characterized genes [11]. A large portion of the currently known exons is derived from EST predictions, which is reflected in a bias towards the 5' and 3' ends of genes in array designs based on these databases. This problem can be overcome by genome-wide tiling arrays. Their usage identified a large number of new transcripts that are not annotated in the current databases, which indicates that current database-generated genome-wide arrays will be incomplete [19]. In an all slide-based system, the mRNA is transcribed into cDNA. During this reversed transcription step, fluorescent dyes are incorporated into the cDNA, which is then hybridized to the array. The details of the labeling and detection procedure have been recently reviewed [20].

#### 14.3.2.4 Array Designs using Exon Junction Oligos

The detection of a splicing event with a combination of exon-junction and exon-body probes is illustrated in Fig 14.1B. Typically, several probes are made against exons (exon body probe) and a junction probe hybridizes half to the end of one exon and half to the beginning of the next exon. An increase of exon usage is indicated by a simultaneous increase of signal for its junction and body probes, whereas there is no change for the signal from the body probes from adjacent exons. In an ideal case, the signal from the junction oligonucleotide detecting the joining of both constitutive exons would decrease proportionally. However, in our laboratory we have never observed such a perfect case. The combined signal from the body probes detecting the constitutive exons indicates the general transcript level, which allows discrimination between alternative splicing and a simple change in transcript abundance. There is flexibility in the design of the exon-body probes that can be optimized for a similar hybridization temperature [21]. However, there is almost no flexibility in the design of exon-junction probes. One solution for this problem is the usage of several exon-junction probes that are offset by 1–2 nucleotides, which allows their combined analysis [14]. It is also difficult to detect small exons with oligonucleotides as the exons are too short to allow the design of body probes. Such short exons are frequently found as 3–nt long variations of alternative 3' splice sites [22].

Commercially available Affymetrix designs do not contain junction probes. Here, a change in exon usage is detected by a change in the splicing index, which is the logarithm of the ratio of the exon signal to the total signal from the gene ($\log_2$[exon/total]).

#### 14.3.2.5 Array Designs using Tiling Probes

Tiling arrays can cover the complete genome and contain 25–mer oligonucleotides that are currently spaced 5-35 bp apart, which defines the resolution of the arrays. Due to technical progress, this resolution will increase. For each oligonucleotide, a nucleotide with a mismatch serves as a control. One major advantage of tiling arrays is that they are unbiased, i.e., they do not rely on previous experiments collected in the databases. This advantage became apparent when the use of tiling arrays spaced 5-nt apart showed that more than half of the human gene expression is not yet annotated [23].

## 14.4 Analysis Tools

The most difficult part of an array experiment is the data analysis and verification. Currently, several algorithms are used without a single program emerging as a standard application, which highlights the difficulty in the analysis. Arrays detecting alternative splicing are more complicated than cDNA arrays, since they detect multiple products from one gene and have to discriminate between changes in splicing and changes in the overall gene expression.

### *14.4.1 PLIER*

The PLIER (Probe Logarithmic Error Intensity Estimate) method produces an improved signal by accounting for experimentally observed patterns in probe behavior and handling errors at low and high signal values. The PLIER algorithm was developed and released by Affymetrix in 2004. Many commercially available software packages that analyze microarray data are using PLIER (e.g., Avandis (Strand Genomic) and ArrayAssist [Stratagene]). The PLIER algorithm produces an improved gene expression value that is a summary value for a probe set, which is done by incorporating experimental observations of feature behavior. PLIER uses a probe affinity parameter, which represents the strength of a signal produced at a specific concentration for a given probe. Calculation based on the data across the arrays defines the probe affinities and the error model employed by PLIER assumes that the error is proportional to the observed intensity, rather than to background-subtracted intensity. However, the derivation of the method also assumes that the error of the mismatch probe is the reciprocal of the error of the perfect match probe.

### *14.4.2 MIDAS*

TIGR's MIDAS (Microarray Data Analysis System) is a Java based application that offers an interface to design microarray data analysis protocols combining one or more normalization and filtering steps. This assumes that the data from individual hybridizations is treated in a uniform and reproducible manner. MIDAS harbors the normalization modules that includes locally weighted linear regression (loess; [24, 25]) and total intensity normalization. These can be linked with filters, including low-intensity cutoff, intensity-dependent Z-score cutoffs, and replicate consistency trimming creating a highly customizable method for preparing expression data for subsequent comparison and analysis. Data analysis methods are constructed using a graphical scripting language and can be saved for application to other data-sets. Scatter-plots generated by the program illustrate the effects of each algorithm on the data.

### 14.4.3 ASPIRE

ASPIRE (Analysis of Splicing by Isoform Reciprocity) was designed to identify the reciprocal splicing changes between two samples i.e., it is normalized to steady-state levels. This approach allows one to identify changes in alternative splicing with high sensitivity and to discriminate them from the changes in RNA stability. Data quantification is based on the change in the fraction of exon inclusion [26].

### 14.4.4 GenASAP

GenASAP (Alternative Splicing Array Platform) predicts the level of alternate splicing for exon-skipping events detectable on custom microarray chips. It uses Bayesian learning in an unsupervised probability model to accurately predict alternate splicing levels from the microarray data. It reads the hybridization profiles of microarray data, while modeling noise processes and missing or aberrant data. It has been applied to the global discovery and analysis of AS in mammalian cells and tissues [27]. Different models have been recently reviewed [28].

### 14.4.5 Commercial Software

Algorithms have been implemented in several commercially available programs are summarized below:

*XRAY* [Biotiquesystems]: The program can run in EXCEL spreadsheets and analysis gene expression and alternative splicing events. It uses a mixed model ANOVA algorithm to discriminate between changes in alternative splicing and gene expression.

*Genomatix's ChipInspector* [Genomatix]: Carries out significance analysis on the single-exon probe level. Exon probes with significant expression ratios are annotated based on the Genomatix proprietary genome annotation and analysis system ElDorado.

*Partek Genomics Suite* [Partek]: Performs statistical analysis and allows visualization of the result. The program annotates all results and provides hyperlinks to internet databases of splicing.

Arrays from the Agilent platforms are distributed by Exonhit, which offers SpliceArray Analysis Tool (SAT). The SpliceArray Analysis Tool is an Excel application, that allows identification of changes in splicing by indicating expression values, fold changes, p-Values and Pearson correlation.

## 14.5 Experiments with Splice-site Sensitive Microarrays

Published experiments using microarrays that were designed to identify splicing variants are summarized in Table 14.1. In almost all the array experiments performed, changes in alternative splicing were validated by RT-PCR. The validation rates range from 35% [18] to 100% [26]. In the validation, only false positive events were detected. It is likely that the false negative detection rates are in the same range. Slightly higher validation rates were observed when real time-RT-PCR was used [29], which probably reflects the ability of real time PCR to detect changes over a larger range of RNA concentrations. The published data show that there is no significant difference between the available analysis programs in predicting splicing events that can be validated by RT-PCR. The highest validation rates are achieved when the result of the array experiment can be combined with other experimental data, such as binding signatures of regulatory factors [30].

**Table 14.1** Overview of array experiments performed that detect variations in alternative splicing

| Platform | Comments | Analysis | Validation | Ref |
|---|---|---|---|---|
| **Affymetrix Platton** | | | | |
| A prototype array was used to predict tissue-specific splice variants in 10 normal rat tissues. | The array represents 1600 rat genes. The expression of each gene is measured by 20 pairs of perfectly match and mismatched probes (25–mer oligonucleotide). The probes are selected from the 3' region of each gene, which biased information to the 5'end | SPLICE, NEIGHBOR-HOOD | 50%, (3/6), validated by One-Step RT-PCR | [33] |
| Prototype splice junction microarray with exon-body probes was used to examine alternative splicing in 22 adult mouse tissues. | A combination of splice junction and exon probes were designed to monitor 6216 alternative splicing events derived from alignment of mouse EST and cDNA sequences to the genome sequences. Each splice-junction probe set contains six 25–mer oligonucleotide probes tiled across the junction. | custom | 85%, validated by RT-PCR | [31] |
| Prototype splice junction microarray with exon-body probes was used to examine alternative splicing in mouse N2A cells after inhibition of NMD. | The array design was described by Sugnet *et al.*, [31]. | custom | 96%, (24/25), validated by RT-PCR | [34] |
| The Affymetrix GeneChip Human Exon 1.0 ST array was used to detect alternative splicing in 20 paired tumor-normal colon cancer samples. | The GeneChip Human Exon array contains ~5.4 million probes grouped into 1.4 million exon clusters within the known and predicted transcribed region of the entire genome. About 90% of the exons are represented by four 25–mer oligonucleotide probes. | PLIER, MIDAS, ANOVA | 35%, (15/43), validated by RT-PCR | [18] |
| An Affymetrix exon array was used to detect tissue-specific alternative splicing in a panel of 16 different adult human tissues. | The array contains more than 9.6 million probes for more than one million annotated and predicted exon clusters. ~ 90% of the exons are covered by 4 probe pairs. Each probe pair consists of a perfect match and a mismatch probe. | Simplified expression analysis (SEA), DABG, splicing index | 86%, (72/84), by RT-PCR | [35] |

**Table 14.1** (continued)

| Platform | Comments | Analysis | Validation | Ref |
|---|---|---|---|---|
| A custom Affymetrix microarray to detect alternative splicing in brains of Nova knock-out mice relative to those of wild-type mice and between different brain and immune tissues. | The microarray contains 40,443 exon-junction probe sets derived from 7,175 genes with one or more bioinformatically predicted alternative transcripts. Most probe sets consist of six 25–mer perfectly matched probes spanning the exon-exon junction. | ASPIRE | 100%, (49/49), validated by RT-PCR | [26] |
| **Agilent platform** | | | | |
| An exon-junction array was used to monitor alternative splicing in 52 tissues and cell lines. | ~125,000 different oligonucleotide probes 36 nt in length, centrally positioned with respect to every consecutive exon-exon junction are designed for all human Refseq mRNA sequences representing more than 10,000 multi-exon human genes. | | 48%, (73/153), validated by One-Step RT-PCR | [14] |
| A custom Agilent microarray was used to detect alternative splicing in 5 human tissues. | Exon-body probes (40 mer) and splice junction probes (36 mer, 5 probes per junction) were designed for 316 human genes derived from an alternative splicing database (ASAP). | | 85%, (11/13), validated by real time RT-PCR | [29] |
| Custom-spotted platform was used to study alternative splicing in several mammalian tissues and cell lines. | 3126 alternative splicing events representing 2647 distinct genes were selected based on sequence alignment of mouse cDNA and EST sequences to genome sequences. Each alternative splicing event was measured by using a probe set of 7 probes: 3 exon-body probes, 3 splice junction probes and 1 intron probe. | GenASAP | Validated by One-Step RT-PCR | [36] |
| A quantitative microarray platform (Agilent) was used to study alternative splicing during T-cell activation. | The array design was described by Pan et al., 2004 | GenASAP | 100%, (35/35), validated by One-Step RT-PCR | [37] |
| A custom exon-junction array (Agilent) was used to monitor alternative splicing events regulated by 4 splicing factors in Drosophila. | 36–mer exon-exon junction probes were designed to assay all annotated alternative-splicing events (2931 genes) in Drosophila. | | 100%, (6/6), validated by RT-PCR | [38] |

**Table 14.1** (continued)

| Platform | Comments | Analysis | Validation | Ref |
|---|---|---|---|---|
| **Other systems** | | | | |
| A custom exon-junction array was used to detect alternative splicing in mouse neuroblastoma N2A cells in which the expression of PTB, nPTB, or both proteins was knocked-down by RNAi | The arrays probed the splicing of ~1300 exons that were selected for their likely functional significance. | | >80%, validated by real time RT-PCR | [39] |
| A self-printed oligonucleotide array designed for each intron-containing yeast gene was used to detect intron retention caused by the loss of 18 different mRNA processing factors. | 40-mer oligonucleotide probes were designed to detect the splice junction, the intron, and the second exon for each yeast intron-containing gene. | Splice Junction (SJ) index, Intron Accumulation (IA) index | Validated by RT-PCR, | [40] |
| A custom alternative splicing microarray (Geniom OneR system) was used to identify different mRNA variants of neuron-specific regulators in Hodgkin Lymphoma Cells. | Expression of 86 mRNA involved in the regulation of alternative splicing was investigated. 13 pairs of 25–mer oligonucleotides were used to probe each mRNA. ~100 splicing events were detected using 13 pairs of oligonucleotides corresponding to the constitutive or alternative exons and 11 pairs of spanning alternative splice junctions. | | 90%, validated by Semi quantitative RT-PCR. | [32] |
| Maskless Array Synthesizer (MAS) technology was used to synthesize custom microarrays to generate a gene expression map for the *Drosophila melanogaster* genome. | The array consisted of 179,972 probes (36-nt) targeting exons, introns, intergenic regions, and splice junctions. | PEAB, ANOVA | Not determined | [41] |
| Cell-specific alternative splicing was detected on a fibre optic microarray platform. | 20 mer oligonucleotides were used as probes to hybridize both the exon specific sequence and the splice junctions. | | 100%, (6/6), Validated by RT-PCR, RASL(RNA mediated annealing, selection and ligation)-PCR. | [17] |

(Copies of tables are available in the accompanying CD.)

The major outcomes of the current array analyses are the identification of functionally related targets of the splicing factor NOVA–1 [30], the finding that tissue-specific exons are flanked by highly conserved intronic parts [31] and the description of widespread changes of alternative splicing in human tumors [11,32].

## 14.6 Problems with Array Analysis

The major problem with array experiments is their poor reproducibility with other methods, notably RT-PCR. The validation rate can be as low as 35%. In the majority of cases it is around 50–70% (Table 14.1). Since these numbers only address the false positive cases, the real error rates, that also include false negatives, will be much higher. One reason for the poor reproducibility could be the large amount of unknown RNAs that often overlap with known transcripts [23]. An example is shown in Fig. 14.2A,B. Here, an unknown RNA is expressed that contains the alternative exon (black box, Fig. 14.2B). Due to the up-regulation of this RNA, probes in the first alternative exon would indicate its up-regulation, which could not be confirmed by RT-PCR using primers in exons 1 and 3.

The next problem that needs to be overcome is data analysis. Currently, several programs for data analysis are available that use different algorithms. None of these programs showed consistently better prediction of splicing events. In our laboratory, we analyzed the same data-sets with different programs and surprisingly obtained different, non-overlapping predictions of changes in alternative splicing. None of the programs tested showed better reproducibility of its prediction with RT-PCR.

Another important constraint is that array experiments give no connectivity information between distant exons. In the example in Fig. 14.2C, the array cannot determine whether exon 1 is connected with its downstream exon 5. In an hypothetical experiment where up-regulation of the alternative exons 2 and 4 are observed, it cannot be determined by array analysis whether this is due to the generation of a mRNA containing both alternative exons, or due to the appearance of two mRNAs, each of which contain one alternative exon.

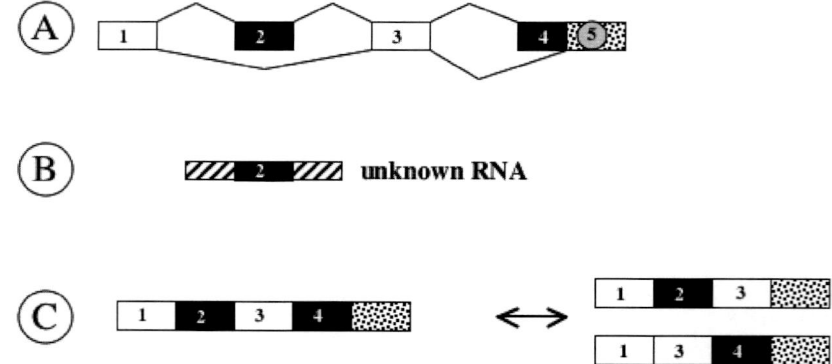


**Figure 14.2** Problems and limitations of array analysis. **(A)** Hypothetical pre-mRNA with two alternatively spliced exons, shown as black boxes. **(B)** The occurrence of new, until now unknown transcripts can simulate alternative splicing events. In this example, an unkown RNA contains the alternative exon, identical to exon 2 in Figure A (*black box*) that is flanked by novel exons (striped boxes). If this isoform is up-regulated under the experimental conditions, it is impossible to distinguish between a change in alternative splicing that up-regulates the alternative exon or an increased expression of this new RNA. **(C)** Connectivity problem. If array analysis detects changes, e.g., up-regulation of both alternative exons, it cannot be predicted whether isoform 1 is up-regulated or isoforms 2 and 3 are simultaneously up-regulated, since the array does not give information about the connection of the exons

## 14.7 Conclusion

Array analysis of alternative splicing events has emerged in the last few years as a promising high-throughput analysis tool. The use of arrays to systematically analyze transcript structures have already given significant insights into gene regulation; there are many additional transcripts expressed than annotated, splicing factors seem to regulate biological meaningful set of genes and numerous small changes are associated with tumor formation. At this stage, arrays are a detection tool and it is likely that improvements in probe design, databases and analysis tools will make arrays a reliable quantitative analysis tool.

## Glossary and Abbreviations

| | |
|---|---|
| PLIER | Probe Logarithmic Error Intensity Estimate |
| MIDAS | Microarray Data Analysis System |
| ASPIRE | Analysis of Splicing by Isoform Reciprocity |
| GenASAP | LAlternative Splicing Array Platform |
| SAT | Splice Array Analysis Tool |

## Key References

Blencowe, B.J. (2006) Alternative splicing: new insights from global analyses. *Cell,* **126,,** 37–47.

Cheng, J., Kapranov, P., Drenkow, J., Dike, S., Brubaker, S., Patel, S., Long, J., Stern, D., Tammana, H., Helt, G. *et al.* (2005) Transcriptional maps of 10 human chromosomes at 5-nucleotide resolution. *Science,* **308,,** 1149–1154.

Cuperlovic-Culf, M., Belacel, N., Culf, A.S. and Ouellette, R.J. (2006) Data analysis of alternative splicing micro arrays. *Drug Discov Today,* **11,,** 983–990.

Stamm, S., Ben-Ari, S., Rafalska, I., Tang, Y., Zhang, Z., Toiber, D., Thanaraj, T.A. and Soreq, H. (2005) Function of alternative splicing. *Gene,* **344C,,** 1–20.

## Suggested Reading

### *Alternative splicing and gene expression*

1. Kampa, D., Cheng, J., Kapranov, P., Yamanaka, M., Brubaker, S., Cawley, S., Drenkow, J., Piccolboni, A., Bekiranov, S., Helt, G. *et al.* (2004) Novel RNAs identified from an in-depth analysis of the transcriptome of human chromosomes 21 and 22. *Genome Res,* **14**, 331–342.
2. Stamm, S., Ben-Ari, S., Rafalska, I., Tang, Y., Zhang, Z., Toiber, D., Thanaraj, T.A. and Soreq, H. (2005) Function of alternative splicing. *Gene,* **344C**, 1–20.
3. Hillman, R.T., Green, R.E. and Brenner, S.E. (2004) An unappreciated role for RNA surveillance. *Genome Biol,* **5**, R8.
4. Pan, Q., Saltzman, A.L., Kim, Y.K., Misquitta, C., Shai, O., Maquat, L.E., Frey, B.J. and Blencowe, B.J. (2006) Quantitative microarray profiling provides evidence against widespread coupling of alternative splicing with nonsense-mediated mRNA decay to control gene expression. *Genes Dev,* **20**, 153–158.
5. Faustino, N.A. and Cooper, T.A. (2003) Pre-mRNA splicing and human disease. *Genes Dev,* **17**, 419–437.
6. Stoilov, P., Meshorer, E., Gencheva, M., Glick, D., Soreq, H. and Stamm, S. (2002) Defects in pre-mRNA processing as causes of and predisposition to diseases. *DNA Cell Biol,* **21**, 803–818.
7. Blencowe, B.J. (2006) Alternative splicing: new insights from global analyses. *Cell,* **126**, 37–47.
8. Bracco, L. and Kearsey, J. (2003) The relevance of alternative RNA splicing to pharmacogenomics. *Trends Biotechnol,* **21**, 346–353.
9. Hanioka, N., Kimura, S., Meyer, U.A. and Gonzalez, F.J. (1990) The human CYP2D locus associated with a common genetic defect in drug oxidation: a G1934——A base change in intron 3 of a mutant CYP2D6 allele results in an aberrant 3' splice recognition site. *Am J Hum Genet,* **47**, 994–1001.

10. Chandrasekharan, N.V., Dai, H., Roos, K.L., Evanson, N.K., Tomsik, J., Elton, T.S. and Simmons, D.L. (2002) COX-3, a cyclooxygenase-1 variant inhibited by acetaminophen and other analgesic/antipyretic drugs: cloning, structure, and expression. *Proc Natl Acad Sci U S A,* **99**, 13926–13931.
11. Li, H.R., Wang-Rodriguez, J., Nair, T.M., Yeakley, J.M., Kwon, Y.S., Bibikova, M., Zheng, C., Zhou, L., Zhang, K., Downs, T. *et al.* (2006) Two-dimensional transcriptome profiling: identification of messenger RNA isoform signatures in prostate cancer from archived paraffin-embedded cancer specimens. *Cancer Res,* **66**, 4079–4088.
12. Srebrow, A. and Kornblihtt, A.R. (2006) The connection between splicing and cancer. *J Cell Sci,* **119**, 2635–2641.
13. Xing, Y. and Lee, C. (2005) Evidence of functional selection pressure for alternative splicing events that accelerate evolution of protein subsequences. *Proc Natl Acad Sci U S A,* **102**, 13526–13531.
14. Johnson, J.M., Castle, J., Garrett-Engele, P., Kan, Z., Loerch, P.M., Armour, C.D., Santos, R., Schadt, E.E., Stoughton, R. and Shoemaker, D.D. (2003) Genome-wide survey of human alternative pre-mRNA splicing with exon junction microarrays. *Science,* **302**, 2141–2144.
15. Stamm, S., Riethoven, J.J., Le Texier, V., Gopalakrishnan, C., Kumanduri, V., Tang, Y., Barbosa-Morais, N.L. and Thanaraj, T.A. (2006) ASD: a bioinformatics resource on alternative splicing. *Nucleic Acids Res,* **34**, D46–55.

## *Detection of splicing variants with microarrays*

16. Moore, M.J. (2005) From birth to death: the complex lives of eukaryotic mRNAs. *Science,* **309**, 1514–1518.
17. Yeakley, J.M., Fan, J.B., Doucet, D., Luo, L., Wickham, E., Ye, Z., Chee, M.S. and Fu, X.D. (2002) Profiling alternative splicing on fiber-optic arrays. *Nat Biotechnol,* **20**, 353–358.
18. Gardina, P.J., Clark, T.A., Shimada, B., Staples, M.K., Yang, Q., Veitch, J., Schweitzer, A., Awad, T., Sugnet, C., Dee, S. *et al.* (2006) Alternative splicing and differential gene expression in colon cancer detected by a whole genome exon array. *BMC Genomics,* **7**, 325.
19. Carninci, P. and Kasukawa, T. and Katayama, S. and Gough, J. and Frith, M.C. and Maeda, N. and Oyama, R. and Ravasi, T. and Lenhard, B. and Wells, C. *et al.* (2005) The transcriptional landscape of the mammalian genome. *Science,* **309**, 1559–1563.
20. Hughes, T.R., Hiley, S.L., Saltzman, A.L., Babak, T. and Blencowe, B.J. (2006) Microarray analysis of RNA processing and modification. *Methods Enzymol,* **410**, 300–316.
21. Relogio, A., Schwager, C., Richter, A., Ansorge, W. and Valcarcel, J. (2002) Optimization of oligonucleotide-based DNA microarrays. *Nucleic Acids Res,* **30**, e51.
22. Hiller, M., Nikolajewa, S., Huse, K., Szafranski, K., Rosenstiel, P., Schuster, S., Backofen, R. and Platzer, M. (2007) TassDB: a database of alternative tandem splice sites. *Nucleic Acids Res,* **35**, D188192.
23. Cheng, J., Kapranov, P., Drenkow, J., Dike, S., Brubaker, S., Patel, S., Long, J., Stern, D., Tammana, H., Helt, G. *et al.* (2005) Transcriptional maps of 10 human chromosomes at 5-nucleotide resolution. *Science,* **308**, 1149–1154.

## *Analysis tools*

24. Cleveland, W.S. and Devlin, S.J. (1988) Locally Weighted Regression: An Approach to Regression Analysis by Local Fitting. *J. Ameri. Stat. Asso.,* **83**, 596–610.
25. Yang, Y.H., Dudoit, S., Luu, P., Lin, D.M., Peng, V., Ngai, J. and Speed, T.P. (2002) Normalization for cDNA microarray data: a robust composite method addressing single and multiple slide systematic variation. *Nucleic Acids Res,* **30**, e15.
26. Ule, J., Ule, A., Spencer, J., Williams, A., Hu, J.S., Cline, M., Wang, H., Clark, T., Fraser, C., Ruggiu, M. *et al.* (2005) Nova regulates brain-specific splicing to shape the synapse. *Nat Genet,* **37**, 844–852.
27. Shai, O., Morris, Q.D., Blencowe, B.J. and Frey, B.J. (2006) Inferring global levels of alternative splicing isoforms using a generative model of microarray data. *Bioinformatics,* **22**, 606–613.
28. Cuperlovic-Culf, M., Belacel, N., Culf, A.S. and Ouellette, R.J. (2006) Data analysis of alternative splicing microarrays. *Drug Discov Today,* **11**, 983–990.

## *Experiments with splice-site sensitive microarrays*

29. Le, K., Mitsouras, K., Roy, M., Wang, Q., Xu, Q., Nelson, S.F. and Lee, C. (2004) Detecting tissue-specific regulation of alternative splicing as a qualitative change in microarray data. *Nucleic Acids Res,* **32**, e180.

30. Ule, J., Stefani, G., Mele, A., Ruggiu, M., Wang, X., Taneri, B., Gaasterland, T., Blencowe, B.J. and Darnell, R.B. (2006) An RNA map predicting Nova-dependent splicing regulation. *Nature*.
31. Sugnet, C.W., Srinivasan, K., Clark, T.A., O'Brien, G., Cline, M.S., Wang, H., Williams, A., Kulp, D., Blume, J.E., Haussler, D. *et al.* (2006) Unusual intron conservation near tissue-regulated exons found by splicing microarrays. *PLoS Comput Biol,* **2**, e4.
32. Relogio, A., Ben-Dov, C., Baum, M., Ruggiu, M., Gemund, C., Benes, V., Darnell, R.B. and Valcarcel, J. (2005) Alternative splicing microarrays reveal functional expression of neuron-specific regulators in Hodgkin lymphoma cells. *J Biol Chem,* **280**, 4779–4784.
33. Hu, G.K., Madore, S.J., Moldover, B., Jatkoe, T., Balaban, D., Thomas, J. and Wang, Y. (2001) Predicting splice variant from DNA chip expression data. *Genome Res,* **11**, 1237–1245.
34. Ni, J.Z., Grate, L., Donohue, J.P., Preston, C., Nobida, N., O'Brien, G., Shiue, L., Clark, T.A., Blume, J.E. and Ares, M., Jr. (2007) Ultraconserved elements are associated with homeostatic control of splicing regulators by alternative splicing and nonsense-mediated decay. *Genes Dev,* **21,** 708–718.
35. Clark, T.A., Schweitzer, A.C., Chen, T.X., Staples, M.K., Lu, G., Wang, H., Williams, A. and Blume, J.E. (2007) Discovery of tissue-specific exons using comprehensive human exon microarrays. *Genome Biol,* **8,** R64.
36. Pan, Q., Shai, O., Misquitta, C., Zhang, W., Saltzman, A.L., Mohammad, N., Babak, T., Siu, H., Hughes, T.R., Morris, Q.D. *et al.* (2004) Revealing global regulatory features of mammalian alternative splicing using a quantitative microarray platform. *Mol Cell,* **16,** 929–941.
37. Ip, J.Y., Tong, A., Pan, Q., Topp, J.D., Blencowe, B.J. and Lynch, K.W. (2007) Global analysis of alternative splicing during T-cell activation. *Rna,* **13,** 563–572.
38. Blanchette, M., Green, R.E., Brenner, S.E. and Rio, D.C. (2005) Global analysis of positive and negative pre-mRNA splicing regulators in Drosophila. *Genes Dev,* **19,** 1306–1314.
39. Boutz, P.L., Stoilov, P., Li, Q., Lin, C.H., Chawla, G., Ostrow, K., Shiue, L., Ares, M., Jr. and Black, D.L. (2007) A post-transcriptional regulatory switch in polypyrimidine tract-binding proteins reprograms alternative splicing in developing neurons. *Genes Dev,* **21,** 1636–1652.
40. Clark, T.A., Sugnet, C.W. and Ares, M., Jr. (2002) Genomewide analysis of mRNA processing in yeast using splicing-specific microarrays. *Science,* **296,** 907–910.
41. Stolc, V., Gauhar, Z., Mason, C., Halasz, G., van Batenburg, M.F., Rifkin, S.A., Hua, S., Herreman, T., Tongprasit, W., Barbano, P.E. *et al.* (2004) A gene expression map for the euchromatic genome of Drosophila melanogaster. *Science,* **306,** 655–660.

# Part IV
# Structural and Functional Sequence Analysis

# Chapter 15
# An Introduction to Multiple Sequence Alignment — and the T-Coffee Shop. Beyond Just Aligning Sequences: How Good can you Make your Alignment, and so What?

**Steven M. Thompson**

**Abstract** I begin the chapter with a discussion of the fundamental principles of multiple-sequence alignment starting with pair-wise dynamic programming, then I move onto significance and similarity statistics, then amino acid scoring matrices, and I end the introduction with multiple-sequence alignment algorithms themselves. Reliability issues, complications, and applications of multiple-sequence alignment are discussed next. The chapter concludes with a description and tutorial about using the T-Coffee multiple-sequence alignment package.

**Keywords** Multiple sequence alignment · Dynamic programming · Significance · Reliability · T-Coffee · M-Coffee · 3DCoffee

## 15.1 Introduction

What can we know about a biological molecule, given its nucleotide or amino acid sequence? How does it fit into a particular system in some organism? What is its role in some network? We may be able to learn about it by searching for particular patterns within it that may reflect some function, such as the many motifs ascribed to catalytic activity; we can look at its overall content and composition by performing several of the gene finding algorithms; we can map its restriction enzyme or protease cut sites; and so on. However, what about comparisons with other sequences? Is it worthwhile? Yes, naturally it is. Inference through homology is fundamental to all the biological sciences. We can learn a tremendous amount by comparing and aligning our sequence against others.

Furthermore, the power and sensitivity of sequence-based computational methods dramatically increase with the addition of more data. More data yields stronger analyzes — if done carefully. Otherwise, it can confound the issue. The patterns of conservation become ever clearer by comparing the conserved portions of sequences amongst a larger and larger data-set. Those areas most resistant to change are most important to the molecule, and to the system, that molecule interacts with. The basic assumption is that those portions of sequences of crucial structural and functional value are most constrained against evolutionary change. They will not tolerate many mutations. Not that mutation does not occur in these regions, it is just that most mutations in the area are lethal, so we never see it. Other areas of the sequence are able to drift more readily, being less subject to this evolutionary pressure. Therefore, sequences end up as a mosaic of quickly and slowly changing regions over evolutionary time.

However, in order to learn anything by comparing sequences, we need to know how to compare them. We can use those constrained portions as 'anchors' to create a sequence alignment allowing

S.M. Thompson
School of Computational Science, Florida State University, Tallahassee, FL, 32306-4120, USA
e-mail: stevet@bio.fsu.edu

S. Krawetz (ed.), *Bioinformatics for Systems Biology*, DOI 10.1007/978-1-59745-440-7_15,

comparison, but this brings up the alignment problem and 'similarity.' It is easy to see that sequences are aligned when they have identical symbols at identical positions, but what happens when symbols are not identical, or the sequences are not of the same length?. How can we know when the most similar portions of our sequences are aligned, when is an alignment optimal, and does optimal mean biologically correct?

A 'brute force,' naive approach just would not work. Even without considering the introduction of gaps, the computation required to compare all possible alignments between just two sequences requires time proportional to the product of the lengths of the two sequences. Therefore, if two sequences are approximately of the same length (N), this is a $N^2$ problem. The calculation would have to repeated 2N times to examine the possibility of gaps at each possible position within the sequence, now a $N^{4N}$ problem. Waterman [1] pointed out that using this naïve approach to align two sequences, each 300 symbols long, would require$10^{88}$ comparisons, more than the number of elementary particles estimated to exist in the universe, and clearly impossible to solve. Part of the solution to this problem is the dynamic programming algorithm, as applied to sequence alignment, which will be reviewed next.

### *15.1.1 Dynamic programming*

Dynamic programming is a widely applied computer science technique, often used in many disciplines whenever optimal sub-structure solutions can provide an optimal overall solution. Let us begin with an overview of sequence-alignment dynamic programming with just two sequences. I'll use an incredibly oversimplified example first; I'll consider matching symbols to be worth one point, and will not consider gapping at all. The solution occurs in two stages. The first begins very much like the dot matrix methods; the second is totally different. Instead of calculating the 'score matrix' on the fly, as is often taught as one proceeds through the graph, I like to completely fill in an original 'match matrix' first, and then add points to those positions that produce favorable alignments next. I also like to illustrate the process working through the cells many authors prefer to work through the edges; they are equivalent. Points are added based on a "looking-back-over-your-left-shoulder" algorithm rule where the only allowable trace-back is diagonally behind and above. The illustration is given in Table 15.1.

Below are the two alignments from the path graph (f) in Table 15.1. They both have a score of three, the three matches found by the algorithm, and the highest score in the bottom row of the solved matrix:

```
 SCATS     SCA.TS
   | ||       | ||
 AC.TS     ..ACTS
```

Most software will arbitrarily (based on some internal rule) choose one of these to report as optimal. Some programs offer a HighRoad/LowRoad option to help explore this solution space.

The next example will be slightly more difficult. Unlike the previous example without gap penalties, I will now impose a very simple gap penalty function. Matching symbols will still be worth one point, non-matching symbols will still be worth zero points, but we will penalize the scoring scheme by subtracting one point for every gap inserted, unless they are at the beginning or end of the sequence. In other words, end gaps will not be penalized; therefore, both sequences do not have to begin or end at the same point in the alignment.

This zero penalty end-weighting scheme is the default for most alignment programs, but can often be changed with a program option, if desired. However, the linear gap function described here, and used in the example below, is a much simpler gap penalty function than

**Table 15.1 Dynamic programming without gap costs**

a) The completed match matrix using one point for matching and zero points for mismatching:

| | S | C | A | T | S |
|---|---|---|---|---|---|
| A | 0 | 0 | 1 | 0 | 0 |
| C | 0 | 1 | 0 | 0 | 0 |
| T | 0 | 0 | 0 | 1 | 0 |
| S | 1 | 0 | 0 | 0 | 1 |

b) Now begin to add points based on the best path through the matrix, always working diagonally, left to right and top to bottom. Keep track of those best paths. The second row is completed here:

| | S | C | A | T | S |
|---|---|---|---|---|---|
| A | 0 | 0 | 1 | 0 | 0 |
| C | 0 | 1 | 0 | 0+1 | 0+1 |
| T | 0 | 0 | 0 | 1 | 0 |
| S | 1 | 0 | 0 | 0 | 1 |

c) Continue adding the points based on the best previous path through the matrix. The third row is completed here:

| | S | C | A | T | S |
|---|---|---|---|---|---|
| A | 0 | 0 | 1 | 0 | 0 |
| C | 0 | 1 | 0 | 1 | 1 |
| T | 0 | 0 | 0+1 | 1+1 | 0+1 |
| S | 1 | 0 | 0 | 0 | 1 |

d) The score matrix is now complete:

| | S | C | A | T | S |
|---|---|---|---|---|---|
| A | 0 | 0 | 1 | 0 | 0 |
| C | 0 | 1 | 0 | 1 | 1 |
| T | 0 | 0 | 1 | 2 | 1 |
| S | 1 | 0 | 0+1 | 0+1 | 1+2 |

e) Now pick the bottom, right-most, highest score in the matrix and work your way back through it, in the opposite direction as you arrived. This is called the trace-back stage, and the matrix is now referred to as the path graph. In this case that highest score is in the right-hand corner, but it need not be:

| | S | C | A | T | S |
|---|---|---|---|---|---|
| A | 0 | 0 | 1 | 0 | 0 |
| C | 0 | 1 | 0 | 1 | 1 |
| T | 0 | 0 | 1 | 2 | 1 |
| S | 1 | 0 | 1 | 1 | 3 |

f) Only the best trace-backs are now shown in outline characters. They are both optimal alignments:

| | S | C | A | T | S |
|---|---|---|---|---|---|
| A | 0 | 0 | 1 | 0 | 0 |
| C | 0 | 1 | 0 | 1 | 1 |
| T | 0 | 0 | 1 | 2 | 1 |
| S | 1 | 0 | 1 | 1 | 3 |

(Copies of tables are available in the accompanying CD.)

that which is normally used in alignment programs. Normally [2] an 'affine' function is used, the standard '$y = mx + b$' equation for a line that does not cross the X, Y origin, where 'b,' the Y intercept, describes how much initial penalty is imposed for creating each new gap:

$$\text{total penalty} = ([\text{length of gap}] \times [\text{gap extension penalty}]) + \text{gap opening penalty}$$

To run most alignment programs with the type of simple linear DNA gap penalty used in my example below, you would have to designate a gap 'creation' or 'opening' penalty of zero, and a gap 'extension' penalty of whatever counts in that particular program as an identical base match for DNA sequences.

My example here uses two random sequences that fit the TATA promoter region consensus of eukaryotes and of bacteria. The most conserved bases within the consensus are capitalized by convention. The eukaryotic promoter sequence is along the X-axis, and the bacterial sequence is along the Y-axis in the following example. The solution illustration begins at the top of Table 15.2.

There may be more than one best path through the matrix. In this example, starting at the top and working down as we did, then tracing back, I found two optimal alignments, each with a final

**Table 15.2 Dynamic programming with a constant, linear gap cost**

a) First complete a match matrix using one point for matching and zero points for mismatching between bases, just like in the previous example:

| | c | T | A | T | A | t | A | a | g | g |
|---|---|---|---|---|---|---|---|---|---|---|
| c | 1 | 0 | 0 | 0 | 0 | 0 | 0 | 0 | 0 | 0 |
| g | 0 | 0 | 0 | 0 | 0 | 0 | 0 | 0 | 1 | 1 |
| T | 0 | 1 | 0 | 1 | 0 | 1 | 0 | 0 | 0 | 0 |
| A | 0 | 0 | 1 | 0 | 1 | 0 | 1 | 1 | 0 | 0 |
| t | 0 | 1 | 0 | 1 | 0 | 1 | 0 | 0 | 0 | 0 |
| A | 0 | 0 | 1 | 0 | 1 | 0 | 1 | 1 | 0 | 0 |
| a | 0 | 0 | 1 | 0 | 1 | 0 | 1 | 1 | 0 | 0 |
| T | 0 | 1 | 0 | 1 | 0 | 1 | 0 | 0 | 0 | 0 |

b) Now add and subtract the points based on the best path through the matrix, working diagonally, left to right and top to bottom, just as before. However, when you have to jump a box to make the path, subtract one point per box jumped, except at the beginning or end of the alignment, so that end gaps are not penalized. Fill in all the additions and subtractions, calculate the sums and differences as you go, and keep track of the best paths. The score matrix is shown with all the calculations below:

| | c | T | A | T | A | t | A | a | g | g |
|---|---|---|---|---|---|---|---|---|---|---|
| c | 1 | 0 | 0 | 0 | 0 | 0 | 0 | 0 | 0 | 0 |
| g | 0 | 0+1=1 | 0+0–0 =0 | 0+0–0 =0 | 0+0–0 =0 | 0+0–0 =0 | 0+0–0 =0 | 0+0–0 =0 | 1+0–0 =1 | 1+0=1 |
| T | 0 | 1+1–1 =1 | 0+1=1 | 1+0 or+1–1 =1 | 0+0–0 =0 | 1+0–0 =1 | 0+0–0 =0 | 0+0–0 =0 | 0+0–0 =0 | 0+1=1 |
| A | 0 | 0+0–0 =0 | 1+1=2 | 0+1=1 | 1+1=2 | 0+1–1 =0 | 1+1=2 | 1+1–1 =1 | 0+0–0 =0 | 0+0–0 =0 |
| t | 0 | 1+0–0 =1 | 0+1–1 =0 | 1+2=3 | 0+1=1 | 1+2=3 | 0+2–1 =1 | 0+2=2 | 0+1=1 | 0+0–0 =0 |
| A | 0 | 0+0–0 =0 | 1+1=2 | 0+2–1 =1 | 1+3=4 | 0+3–1 =2 | 1+3=4 | 1+3–1 =3 | 0+2=2 | 0+1=1 |
| a | 0 | 0+0–0 =0 | 1+0–0 =1 | 0+2=2 | 1+3–1 =3 | 0+4=4 | 1+4–1 =4 | 1+4=5 | 0+3=3 | 0+2=2 |
| T | 0 | 1+0–0 =1 | 0+0–0 =0 | 1+1=2 | 0+2=2 | 1+3=4 | 0+4=4 | 0+4=4 | 0+5=5 | 0+5–1 =4 |

c) Clean up the score matrix next. I'll only show the totals in each cell in the matrix shown below. All the best paths are highlighted:

| | c | T | A | T | A | t | A | a | g | g |
|---|---|---|---|---|---|---|---|---|---|---|
| c | 1 | 0 | 0 | 0 | 0 | 0 | 0 | 0 | 0 | 0 |
| g | 0 | 1 | 0 | 0 | 0 | 0 | 0 | 0 | 1 | 1 |
| T | 0 | 1 | 1 | 1 | 0 | 1 | 0 | 0 | 0 | 1 |
| A | 0 | 0 | 2 | 1 | 2 | 0 | 2 | 1 | 0 | 0 |
| t | 0 | 1 | 0 | 3 | 1 | 3 | 1 | 2 | 1 | 0 |
| A | 0 | 0 | 2 | 1 | 4 | 2 | 4 | 3 | 2 | 1 |
| a | 0 | 0 | 1 | 2 | 3 | 4 | 4 | 5 | 3 | 2 |
| T | 0 | 1 | 0 | 2 | 2 | 4 | 4 | 4 | 5 | 4 |

d) Finally, convert the score matrix into a trace-back path graph by picking the bottom-most, furthest right and highest scoring coordinate. Then choose the trace-back route that got you there, to connect the cells all the way back to the beginning using the same 'over-your-left-shoulder' rule. The two best trace-back routes are now highlighted with outline font in the trace-back matrix below:

| | c | T | A | T | A | t | A | a | g | g |
|---|---|---|---|---|---|---|---|---|---|---|
| c | 1 | 0 | 0 | 0 | 0 | 0 | 0 | 0 | 0 | 0 |
| g | 0 | 1 | 0 | 0 | 0 | 0 | 0 | 0 | 1 | 1 |
| T | 0 | 1 | 1 | 1 | 0 | 1 | 0 | 0 | 0 | 1 |
| A | 0 | 0 | 2 | 1 | 2 | 0 | 2 | 1 | 0 | 0 |
| t | 0 | 1 | 0 | 3 | 1 | 3 | 1 | 2 | 1 | 0 |
| A | 0 | 0 | 2 | 1 | 4 | 2 | 4 | 3 | 2 | 1 |
| a | 0 | 0 | 1 | 2 | 3 | 4 | 4 | 5 | 3 | 2 |
| T | 0 | 1 | 0 | 2 | 2 | 4 | 4 | 4 | 5 | 4 |

(Copies of tables are available in the accompanying CD.)

score of 5, using our example's zero/one scoring scheme. This score is the highest, bottom-right value in the trace-back path graph, the sum of six matches minus one interior gap in one path, and the sum of five matches minus no interior gaps in the other. This score is the number optimized by the algorithm, not any type of a similarity or identity percentage. This first path is the GCG Wisconsin Package [3] Gap program High Road alignment found with this example's parameter settings (note that GCG uses a score of 10 for a base match here, not 1):

GAP of: Euk_Tata.Seq to: Bact_Tata.Seq

Euk_Tata: A random Eukaryotic promoter TATA Box
Preferred region: center between −36 and −20.
Bact_Tata: A random sequence that fits the consensus from the standard E. coli RNA polymerase promoter 'Pribnow' box −10 region.

| | |
|---|---|
| Gap Weight: **0** | Average Match: 10.000 |
| Length Weight: **10** | Average Mismatch: 0.000 |

| **HighRoad option** | | **LowRoad option** | |
|---|---|---|---|
| **Quality:** | **50** | **Quality:** | **50** |
| Ratio: | 6.250 | Ratio: | 6.250 |
| Percent Similarity: | 75.000 | Percent Similarity: | 62.500 |
| Length: | 10 | Length: | 10 |
| Gaps: | 2 | Gaps: | 80 |
| Percent Identity: | 75.000 | Percent Identity: | 62.500 |

HighRoad option:

```
1 cTATAtAagg 10
  |   |||||
1 cg.TAtAaT. 8
```

LowRoad option:

```
1 cTATAtAagg 10
     |||||
1 .cgTAtAaT. 8
```

The GCG Low Road alignment is my second, equivalent path. Notice that even though it has 62.5% identity as opposed to 75% identity in the HighRoad alignment, it has exactly the same score! Another way to explore the dynamic programming solution space is to reverse the entire process. This can often discover other alignments; therefore, see alternative by reversing the sequences.

To recap, and for people who like mathematics, an optimal pair wise alignment is defined as an arrangement of two sequences, 1 of length $i$ and 2 of length $j$, such that:

1) you maximize the number of matching symbols between 1 and 2;
2) you minimize the number of gaps within 1 and 2; and
3) you minimize the number of mismatched symbols between 1 and 2.

Therefore, the actual solution can be represented by the following recursion:

$$S_{ij} = s_{ij} + \max \left\{ \begin{array}{ll} S_{i-1\quad j-1} & \text{or} \\ \max\limits_{2 < x < i} \; S_{i-x\quad j-1} + w_{x-1} & \text{or} \\ \max\limits_{2 < y < i} \; S_{i-1\quad j-y} + w_{y-1} & \end{array} \right\}$$

where $Sij$ is the score for the alignment ending at i in sequence 1 and $j$ in sequence 2,

$sij$ is the score for aligning $i$ with $j$,
$wx$ is the score for making a $x$ long gap in sequence 1,
$wy$ is the score for making a $y$ long gap in sequence 2; and
allowing gaps to be any length in either sequence.

However, just because dynamic programming guarantees an optimal alignment, it is not necessarily the only optimal alignment. Furthermore, the optimal alignment is not necessarily the 'right' or biologically relevant alignment! Significance estimators, such as Expectation values and Monte Carlo simulations can give you some handle on this, but you must always question the results of any computerized solution, based on what you know about the biology of the system. The above example illustrates the Needleman and Wunsch [4] global solution. Later refinements [5] demonstrated how dynamic programming could also be used to find optimal local alignments. To solve dynamic programming using local alignment (without going into all the gory details), the following two tricks can be used:

1) Mismatches are penalized by using a match function that assigns negative numbers to them. Therefore, bad paths quickly become very bad. This leads to a trace-back path matrix with many alternative paths, most of which do not extend the full length of the graph.
2) The best trace-back within the overall graph is chosen. This does not have to begin or end at the edges of the matrix — it is the best segment of alignment.

### *15.1.2 Significance*

The concept of homology versus similarity is particularly misunderstood: there is a huge difference! Similarity is merely a statistical parameter that describes how much two sequences, or portions of them, are alike according to some set scoring criteria. It can be normalized to ascertain statistical significance as in database searching methods, but it is still just a number. Homology, in contrast and by definition, implies an evolutionary relationship — more than just the fact that all life evolved from the same primordial 'slime.' You need to be able to demonstrate some type of evolutionary lineage between the organisms or genes of interest in order to claim homology. Better yet, one can demonstrate experimental evidence, structural, morphological, genetic, or fossils, that corroborates your assertion. There really is no such thing as percent homology; something is either homologous or it's not. Walter Fitch (personal communication) explained with the joke, "homology is like pregnancy — you can't be 45% pregnant, just like something can't be 45% homologous. You either are or you are not." Do not make the mistake of calling any old sequence similarity homology. Highly significant similarity can argue for homology, not the other way around.

So, how do you tell if a similarity, in other words, an alignment discovered by some program, means anything? Is it statistically significant, is it truly homologous, and even more importantly, does it have anything to do with real biology? Many programs generate percent similarity scores; however, as seen above, these really do not mean a whole lot. Don't use percent similarities or identities to compare sequences except in the roughest way. They are not optimized or normalized in any manner. Quality scores mean a lot more but are difficult to interpret. At least they take the length of similarity, all of the necessary gaps introduced, and the matching of symbols all into account, but quality scores are only relevant within the context of a particular comparison or search. The quality ratio is the metric optimized by dynamic programming divided by the length of the shorter sequence. As such, it represents a fairer comparison metric, but it also is relative to the particular scoring matrix and gap penalties used in the procedure.

A traditional way of deciding alignment significance relies on an old statistical trick — Monte Carlo simulations. This type of significance estimation has implicit statistical problems, however, few practical alternatives exist for just comparing two sequences, and they are fast and easy. Monte Carlo randomization options in dynamic programming alignment algorithms compare an actual score, in this case the quality score of an alignment, against the distribution of scores of alignments of a randomized sequence. These options randomize your sequence at least 100 times after the initial alignment and then generate the jumbled alignment scores and a standard deviation based on their distribution. Comparing the mean of the randomized sequence alignment score to the original score, using a 'Z-score' calculation. can help you decide the significance. An old 'rule-of-thumb' is, if the actual score is much more than three standard deviations above the mean of the randomized scores, the analysis may be significant; if it is much more than five, than it probably is significant; and if it is above nine, then it is definitely significant. Many Z-scores measure this distance from the mean using a simplistic Monte Carlo model assuming a normal Gaussian distribution, in spite of the fact that 'sequence-space' actually follows an 'extreme value distribution'; however, this simplistic approximation estimates significance quite well:

$$\text{Z-score} = \frac{[(\text{actual score}) - (\text{mean of randomized scores})]}{(\text{standard deviation of randomized score distribution})}$$

When the two TATA sequences from the previous dynamic programming example are compared to one another using the same scoring parameters as before, but incorporating a Monte Carlo Z-score calculation, their similarity is not found to be significant. The mean score based on 100 randomizations was 41.8, +/− a standard deviation of 7.4. Plugged into the formula: (50 – 41.8) / 7.4 = 1.11, i.e., there is no significance to the match in spite of 75% identity. Composition can make a huge difference — the similarity is merely a reflection of the relative abundance of A's and T's in the sequences

The FastA [6, 7], BLAST [8, 9], Profile [10], and HMMer [11] search algorithms, all use a similar approach, but base their statistics on the distance of the query matches from the actual, or on a simulated extreme value distribution from the rest of the 'insignificantly similar' members of the database being searched. For alignments without gaps, the math generalizes such that the Expectation value $E$ relates to a particular score $S$ through the function $E = Kmne^{-\lambda s}$ [12]. In a database search, $m$ is the length of the query and $n$ is the size of the database in residues. $K$ and $\lambda$ are supplied by statistical theory, dependent on the scoring system and the background amino acid frequencies, and calculated from the actual or simulated database alignment distributions. Expectation values are printed in scientific notation and the smaller the number, i.e., the closer it is to 0, the more significant the match. Expectation values show us how often we should expect a particular alignment to occur merely by chance alone in a search of database particular size. In other words, in order to assess whether a given alignment constitutes evidence for homology, it will help to know how strong an alignment can be expected from chance alone. Rough, conservative guidelines to Z-scores and Expectation values from a typical protein search are given in Table 15.3.

**Table 15.3 Rough, conservative guidelines to Z-scores and Expectation values from a typical protein search**

| ~Z-score | ~E value | Inference |
|---|---|---|
| $\leq 3$ | $\geq 0.1$ | little, if any, evidence for homology, but impossible to disprove! |
| $\cong 5$ | $\cong 10^{-2}$ | probably homologous, but may be due to convergent evolution |
| $\geq 10$ | $\leq 10^{-3}$ | definitely homologous |

(Copies of tables are available in the accompanying CD.)

Be very careful with any guidelines such as these, because they are entirely dependent on both the size and content of the database being searched as well as on how often you perform the search. Think about it: the odds are way different for a rolling dice depending on how many dice you roll, whether they are 'loaded' or not, and how often you try.

Another very powerful empirical method of determining significance is to repeat a database search with the entry in question. If that entry finds more significant 'hits' with the same sort of sequences as the original search, then the entry in question is undoubtedly homologous to the original entry, i.e.,the homology is transitive. If it finds entirely different types of sequences, then it probably is not. Modular proteins with distinctly separate domains confuse issues considerably, but the principles remain the same, and can be explained through domain swapping and other examples of non-vertical transmission. And, finally, the 'gold-standard' of homology is shared structural folds — if you can demonstrate that two proteins have the same structural fold, then, regardless of similarity, at least that particular domain is homologous between the two.

### *15.1.3 Scoring Matrices*

However, what about protein sequences — conservative replacements and similarities, as opposed to identities? This will certainly be an additional complication. Particular amino acids are very much alike, structurally, chemically, and genetically. How can we take advantage of the amino acid similarity in our alignments? People have been struggling with this problem since the late 1960's. Dayhoff [13] unambiguously aligned closely related protein data-sets (no more than 15% different, and in particular cytochrome), available at that point in time and noticed that certain residues, if they mutate at all, are prone to change into certain other residues. As it works out, these propensities for change fell into the same categories that chemists had known for years — those same chemical and structural classes mentioned above — conserved through the evolutionary constraints of natural selection. Dayhoff's empirical observation quantified these changes. Based on the multiple-sequence alignments that Dayhoff created and the empirical amino acid frequencies within those alignments, the assumption that estimated mutation rates in closely related proteins can be extrapolated to more distant relationships, and with matrix and logarithmic mathematics, Dayhoff was able to empirically specify the relative probabilities at which different residues mutated into other residues through evolutionary history, as appropriate within some level of divergence between the sequences considered. This is the basis of the famous PAM (corrupted acronym of 'accepted point mutation') 250 (meaning that the matrix has been multiplied by itself 250 times) log odds matrix.

Since Dayhoff's time, other bio-mathematicians [esp. see ref- 14], BLOSUM series of matrices, and for a somewhat controversial matrix see [15], have created matrices that are regarded more accurate than Dayhoff's original, but the concept remains the same. Furthermore, Dayhoff's original PAM 250 matrix remains a classic, as historically the most widely used amino acid substitution matrix. Collectively, these types of matrices are known as symbol comparison tables, log odds matrices, and substitution or scoring matrices, and they are fundamental to all sequence comparison techniques.

The default amino acid scoring matrix for most protein similarity comparison programs is now the BLOSUM62 table [14]. The "62" refers to the minimum level of identity within the ungapped sequence blocks that went into the creation of the matrix. Lower BLOSUM numbers are more appropriate for more divergent data-sets. The BLOSUM62 matrix is given below in Table 15.4; values of magnitude $\geq \pm 4$ are drawn in shadowed characters to make them easily recognizable.

Notice that positive identity values range from 4 to 11, and negative values, for rare substitutions, go as low as $-4$. The most conserved residue is tryptophan with an identity score of 11;

**Table 15.4 The BLOSUM62– amino acid scoring matrix**

| | A | B | C | D | E | F | G | H | I | K | L | M | N | P | Q | R | S | T | V | W | X | Y | Z |
|---|---|---|---|---|---|---|---|---|---|---|---|---|---|---|---|---|---|---|---|---|---|---|---|
| A | 4 | −2 | 0 | −2 | −1 | −2 | 0 | −2 | −1 | −1 | −1 | −1 | −2 | −1 | −1 | −1 | 1 | 0 | 0 | −3 | −1 | −2 | −1 |
| B | −2 | 6 | −3 | 6 | 2 | −3 | −1 | −1 | −3 | −1 | −4 | −3 | 1 | −1 | 0 | −2 | 0 | −1 | −3 | −4 | −1 | −3 | 2 |
| C | 0 | −3 | 9 | −3 | −4 | −2 | −3 | −3 | −1 | −3 | −1 | −1 | −3 | −3 | −3 | −3 | −1 | −1 | −1 | −2 | −1 | −2 | −4 |
| D | −2 | 6 | −3 | 6 | 2 | −3 | −1 | −1 | −3 | −1 | −4 | −3 | 1 | −1 | 0 | −2 | 0 | −1 | −3 | −4 | −1 | −3 | 2 |
| E | −1 | 2 | −4 | 2 | 5 | −3 | −2 | 0 | −3 | 1 | −3 | −2 | 0 | −1 | 2 | 0 | 0 | −1 | −2 | −3 | −1 | −2 | 5 |
| F | −2 | −3 | −2 | −3 | −3 | 6 | −3 | −1 | 0 | −3 | 0 | 0 | −3 | −4 | −3 | −3 | −2 | −2 | −1 | 1 | −1 | 3 | −3 |
| G | 0 | −1 | −3 | −1 | −2 | −3 | 6 | −2 | −4 | −2 | −4 | −3 | 0 | −2 | −2 | −2 | 0 | −2 | −3 | −2 | −1 | −3 | −2 |
| H | −2 | −1 | −3 | −1 | 0 | −1 | −2 | 8 | −3 | −1 | −3 | −2 | 1 | −2 | 0 | 0 | −1 | −2 | −3 | −2 | −1 | 2 | 0 |
| I | −1 | −3 | −1 | −3 | −3 | 0 | −4 | −3 | 4 | −3 | 2 | 1 | −3 | −3 | −3 | −3 | −2 | −1 | 3 | −3 | −1 | −1 | −3 |
| K | −1 | −1 | −3 | −1 | 1 | −3 | −2 | −1 | −3 | 5 | −2 | −1 | 0 | −1 | 1 | 2 | 0 | −1 | −2 | −3 | −1 | −2 | 1 |
| L | −1 | −4 | 1 | −4 | −3 | 0 | −4 | −3 | 2 | −2 | 4 | 2 | −3 | −3 | −2 | −2 | −2 | −1 | 1 | −2 | −1 | −1 | −3 |
| M | −1 | −3 | −1 | −3 | −2 | 0 | −3 | −2 | 1 | −1 | 2 | 5 | −2 | −2 | 0 | −1 | −1 | −1 | 1 | −1 | 1 | −1 | −2 |
| N | −2 | 1 | −3 | 1 | 0 | −3 | 0 | 1 | −3 | 0 | −3 | −2 | 6 | −2 | 0 | 0 | 1 | 0 | −3 | −4 | −1 | −2 | 0 |
| P | −1 | −1 | −3 | −1 | −1 | −4 | −2 | −2 | −3 | −1 | −3 | −2 | −2 | 7 | −1 | −2 | −1 | −1 | −2 | −4 | −1 | −3 | −1 |
| Q | −1 | 0 | −3 | 0 | 2 | −3 | −2 | 0 | −3 | 1 | −2 | 0 | 0 | −1 | 5 | 1 | 0 | −1 | −2 | −2 | −1 | −1 | 2 |
| R | −1 | −2 | −3 | −2 | 0 | −3 | −2 | 0 | −3 | 2 | −2 | −1 | 0 | −2 | 1 | 5 | −1 | −1 | −3 | −3 | −1 | −2 | 0 |
| S | 1 | 0 | −1 | 0 | 0 | −2 | 0 | −1 | −2 | 0 | −2 | −1 | 1 | −1 | 0 | −1 | 4 | 1 | −2 | −3 | −1 | −2 | 0 |
| T | 0 | −1 | −1 | −1 | −1 | −2 | −2 | −2 | −1 | −1 | −1 | −1 | 0 | −1 | −1 | −1 | 1 | 5 | 0 | −2 | −1 | −2 | −1 |
| V | 0 | −3 | −1 | −3 | −2 | −1 | −3 | −3 | 3 | −2 | 1 | 1 | −3 | −2 | −2 | −3 | −2 | 0 | 4 | −3 | −1 | −1 | −2 |
| W | −3 | −4 | −2 | −4 | −3 | 1 | −2 | −2 | −3 | −3 | −2 | −1 | −4 | −4 | −2 | −3 | −3 | −2 | −3 | 11 | −1 | 2 | −3 |
| X | −1 | −1 | −1 | −1 | −1 | −1 | −1 | −1 | −1 | −1 | −1 | −1 | −1 | −1 | −1 | −1 | −1 | −1 | −1 | −1 | −1 | −1 | −1 |
| Y | −2 | −3 | −2 | −3 | −2 | 3 | −3 | 2 | −1 | −2 | −1 | −1 | −2 | −3 | −1 | −2 | −2 | −2 | −1 | 2 | −1 | 7 | −2 |
| Z | −1 | 2 | −4 | 2 | 5 | −3 | −2 | 0 | −3 | 1 | −3 | −2 | 0 | −1 | 2 | 0 | 0 | −1 | −2 | −3 | −1 | −2 | 5 |

(Copies of tables are available in the accompanying CD.)

cysteine is next with a score of 9; histidine gets 8; both proline and tyrosine get scores of 7. These residues get the highest scores because of two biological factors: they are very important to the structure and function of proteins, and they are the rarest amino acids found in nature. Also check out the hydrophobic substitution triumvirate — isoleucine, leucine, valine, and to a lesser extent methionine — all easily swap places. So, rather than using the zero/one match function that we used in the previous dynamic programming examples, protein sequence alignments use the match function provided by an amino acid scoring matrix. The concept of similarity becomes very important with some amino acids being 'more similar' than others.

### *15.1.4 Multiple sequence dynamic programming*

Dynamic programming reduces the pair wise alignment problem's complexity to order $N^2$. But how do you work with more than just two sequences at a time? It becomes a much harder problem. You could manually align your sequence data with an editor, but some type of an automated solution is desirable, at least as a starting point for manual alignment. However, solving the dynamic programming algorithm for more than just two sequences rapidly becomes intractable. Dynamic programming's complexity, and hence its computational requirements, increases exponentially with the number of sequences in the data-set being compared (complexity = $[\text{sequence length}]^{\text{number of sequences}}$). Mathematically this is an N-dimensional matrix — quite complex, pair wise dynamic programming solves a two-dimensional matrix and the complexity of the solution is equal to the length of the longest sequence squared. A three-member standard dynamic programming sequence comparison would be a matrix with three axes, the length of the longest sequence cubed, and so forth. You can at least draw a three-dimensional matrix, but more than that becomes impossible to even visualize. It quickly boggles the mind!

Several different heuristics have been employed over the years to simplify the complexity of the problem. One program, MSA [16], attempts to simultaneously solve the N-dimensional matrix recursion using a bounding box trick. However, the algorithm's complexity precludes its use in most situations, except with very small data-sets. One way to simultaneously solve the algorithm and yet reduce its complexity is to restrict the search space to only the most conserved 'local' portions of all the sequences involved. This approach is used by the program PIMA [17]. MSA and PIMA are both available through the Internet at several bioinformatics servers (in particular see 18], or they can be installed on your own machine.

### *15.1.5 How the Algorithms Work*

Most implementations of automated multiple alignment modify dynamic programming by establishing a pair wise order in which to build the alignment. This heuristic modification is known as pair wise, progressive dynamic programming. Originally attributed to Feng and Doolittle [19], this variation of the dynamic programming algorithm generates a global alignment, but restricts its search space, at any one time, to a local neighborhood of the full length of only two sequences. Consider a group of sequences. First, all are compared to each other, pair wise, using some quick variation of standard dynamic programming. This establishes an order for the set, most to least similar, like a 'guide-tree' if you will. The sub-groups are clustered together similarly. The algorithm then takes the top two, most similar sequences, and aligns them. It then creates a quasi-consensus of those two and aligns that to the third sequence. Next create the same sort of quasi-consensus of the first three sequences and align that to the forth most similar. The way the program makes and uses this 'consensus' sequence is one of the big differences between the various implementations. This process, all using standard, pair wise dynamic programming, continues until it has worked its way through all of the sequences and/or sets of clusters, to complete the full multiple-sequence alignment.

The pair wise, progressive solution is implemented in several programs. Perhaps ClustalW [20] and its multi-platform, graphical user interface ClustalX [21] are the most popular. ClustalW made the first major advances, beyond the basic Feng and Doolittle algorithm, by incorporating variable sequence weighting, dynamically varying gap penalties and substitution matrices, and a neighbor-joining (NJ) [22] guide-tree. The programs can be downloaded from the ClustalX homesite, ftp://ftp-igbmc.u-strasbg.fr/pub/ClustalX/, to install on your own machine, or they can be run through the World Wide Web (WWW) at several sites. ClustalX is available for most windowing Operating Systems: most UNIX flavors, Microsoft Windows, and Macintosh. Complete documentation comes with the program and is accessed through a "Help" menu. The GCG program, PileUp implements a similar method, but without the later innovations, and ClustalW is included in the GCG package as well.

Several variations on the theme have come along in recent years. T-Coffee (Tree-based Consistency Objective Function For alignment Evaluation, [23]) was one of the first after ClustalW, and it has gained much favor. It will be presented in further detail in this chapter's final section. Its biggest innovation is the use of a pre-processed, weighted library of all the pair wise global alignments between the sequences in the data-set, plus the ten best local alignments associated with each pair of sequences. This helps build both the NJ guide-tree and the progressive alignment. Furthermore, the library is used to assure consistency and help prevent errors, by allowing 'forward-thinking' to see whether the overall alignment will be better one way or another after particular segments are aligned one way or another. Notredame [24] makes the apt analogy of school schedules: everybody, students, teachers and administrators, with some folk being more important than others, i.e., the weighting factor, puts the schedule they desire in a big pile, i.e., T-Coffee's library, with the trick being to best fit all the schedules to one academic calendar, so that everybody is happiest, i.e., T-Coffee's final multiple sequence alignment. T-Coffee is one of the most accurate multiple sequence alignment methods available because of this rationale, but it is not the fastest.

Muscle [25] is another relatively new multiple-sequence alignment program. It is incredibly fast, yet nearly as accurate as T-Coffee. Muscle is an iterative method that uses weighted log-expectation profile scoring along with a slew of optimizations. It proceeds in three stages: draft progressive using k-mer counting, improved progressive using a revised tree from the previous iteration, and refinement by sequential deletion of each tree edge with subsequent profile realignment. Another fairly new program, MAFFT [26], can be run either in the fast, approximate mode, using a Fast Fourier Transformation, where its capability to handle large data-sets and its speed is similar to Muscle, or in a slow, iteratively refined, optimized mode, where its results and capabilities are similar to T-Coffee. Perhaps the most accurate new multiple-sequence alignment program is ProbCons [27]. It uses the Hidden Markov Model (HMM) techniques and posterior probability matrices that compare random pair wise alignments to expected pair wise alignments. Probability consistency transformation is used to re-estimate the scores, and a guide-tree is then constructed, which is used to compute the alignment, which is then iteratively refined. These methods and more are all tied into T-Coffee as external modules, as long as they are all installed on your system.

### *15.1.6 Coding DNA Issues*

All of these alignment algorithms, pair wise, multiple, and database similarity searching, are far more sensitive at the amino acid level than at the DNA level. Twenty match symbols are just much easier to align then only four; the signal to noise ratio is so much better. Plus, the concept of similarity applies to amino acids, but generally not to nucleotides. Furthermore, many DNA base changes (especially third position changes) do not change the encoded protein. All of these factors drastically increase the 'noise' level of the DNA; typically giving the protein searches a much greater 'look-back' time, at least doubling it. Therefore, database searching and sequence alignment should always be done on a protein level, unless one is dealing with noncoding DNA, or if the sequences are so similar as to not cause any problems. Therefore, if dealing with coding sequences, translate the DNA to its protein counterpart, before performing multiple-sequence alignment.

Even if one is dealing with very similar coding sequences, where the DNA can be directly aligned, it is often best to align the DNA along with its corresponding proteins. In addition to the much more easily achieved alignment, this also ensures that the alignment gaps are not placed within codons. Phylogenetic analysis can then be performed on the DNA rather than on the proteins. This is especially important when dealing with data-sets that are quite similar, since the proteins may not reflect many differences hidden in the DNA. Furthermore, many people prefer to run phylogenetic analyzes on DNA rather than protein regardless of how similar they are — the multiple substitution models have a long and well-accepted history, and yet are far simpler. In fact, some phylogenetic inference algorithms do not even take advantage of amino acid similarity when dealing with protein sequences; they only count the identities, though many others can use PAM style models. However, the more diverged a data-set becomes, the more random the third and eventually the first codon positions become, which introduces noise (error) into the analysis. Therefore, often the third positions and sometimes first positions are masked out of the data-sets. Just like in most of computational molecular biology, one is always balancing signal against noise. Too much noise or too little signal, both degrade the analysis to the point of nonsense.

Several scripts and programs, as well as some Web servers, can perform this sort of codon-based alignment, but they can be a bit tricky to run. Examples include mrtrans [28] (also available in EMBOSS [29] as tranalign and in BioPerl [30] as aa_to_dna_aln), transAlign [31], RevTrans [32], protal2dna [33], and PAL2NAL [34]. Dedicated sequence analysis editors, such as GCG's SeqLab, based on Smith's Genetic Data Environment (GDE), [35] can also be used for this in a manual process. The logic to this manual paired protein and DNA codon alignment approach is as follows:

1) The easy case is where you can align the DNA directly. If the DNA sequences are directly alignable because they are quite similar, then one can use whatever automated tool one wants to create one's DNA alignment and load it into the multiple sequence editor. Next, the editor's align translation function is used to create aligned corresponding protein sequences. Select the region to translate based on the CDS reference in each DNA sequence's annotation. Be careful of CDS entries that do not begin at position 1 — the GenBank CDS feature annotation "/codon_start = " identifies which position the translation begins within the first codon listed for each gene. One may also have to trim the sequences down to just the relevant gene or exons, especially if they're genomic. Group each protein to its corresponding DNA sequence, if the option is available, so that subsequent manipulations will keep them together.
2) The way more difficult case is where one will need to use the protein sequences to create the alignment because the DNA is not directly alignable. In this case, one needs to create the protein sequence alignment first, and then load their corresponding DNA sequences into the editor. The DNA sequence accession codes can be found in the annotation of the protein sequence entries. Next translate the unaligned DNA sequences into new protein sequences with the align translation function and group these to their corresponding DNA sequences, just as above. However, this time the DNA along with their translated sequences are not aligned as a set, just the other protein set is aligned. Also, group all of the aligned protein

data-set together, separately from the DNA/aligned translation set. Now comes the manual part; slide the original aligned protein sequence set apart to match the codons of the DNA along with its aligned translation, inserting gaps in whichever set needs them to recreate the alignment. Merge the newly aligned sequences into the existing alignment group as you go, and then start on the next one. It sounds difficult, but since one is matching up two identical protein sequences, the DNA translation and the original aligned protein, it's really not too bad.

Multiple sequence alignment is much more difficult if one is forced to align nucleotides, because the region does not code for a protein. Automated methods may be able to help as a starting point, but they are certainly not guaranteed to come up with a biologically correct alignment. The resulting alignment will probably have to be extensively edited, if it works at all. Success will largely depend on the similarity of the nucleotide data-set.

## 15.2 Reliability?

One liability of most global progressive, pair wise methods is that they are entirely dependent on the order in which the sequences are aligned. Fortunately, ordering them from most similar to least similar usually makes biological sense and works quite well. However, the techniques are very sensitive to the substitution matrix and gap penalties specified. Some programs allow 'fine-tuning' of areas of an alignment by realignment with different scoring matrices and/or gap penalties; this can be extremely helpful. However, any automated multiple sequence alignment program should be thought of as only a tool to offer a starting alignment that can be improved upon, not the 'end-all-to-meet-all' solution, guaranteed to provide the 'one-true' answer. Although, in this post-genomics era, when one is dealing with giga bases of data, it does make sense to start with the 'best' solution possible. This is the premise of using a very accurate multiple-sequence alignment package, such as T-Coffee [23].

Regardless of the program used to create an alignment, always use comparative approaches to help assure its reliability. After the program has offered its best guess, try to improve it further. Think about it — a sequence alignment is a statement of positional homology — it is a hypothesis of evolutionary history. It establishes the explicit homologous correspondence of each individual sequence position, in each column in the alignment. Therefore, ensure that you have prepared a good one — be sure that it makes sense — devote considerable time and energy towards developing the best alignment possible.

Editing alignments to ensure that all columns are truly homologous should be encouraged. Dedicated sequence alignment editing software such as GCG's SeqLab [3], Jalview [36], Se-Al [37], and SeaView [38] are great for this, but any editor will do, as long as the sequences end up properly formatted afterwards. Use your understanding of the system to help guide your judgment. Look for conserved functional sites and other motifs — they should all line up. Searches of the *PROSITE Database of protein families and domains* [39] for catalogued structural, regulatory, and enzymatic consensus patterns or 'signatures' in one's data-set can help, as can *de novo* motif discovery tools like the MEME [40] and MotifSearch [41] program pair.

Make subjective decisions. Is it good enough: do things line up the way that they should? Assure that the known enzymatic, regulatory, and structural elements all align. Look for columns of strongly conserved residues such as tryptophans, cysteines, and histidines; important structural amino acids such as prolines, tyrosines and phenylanines; and the conserved isoleucine, leucine, valine substitutions. If, after all else, you decide that you just can't align some region, or an entire sequence, then get rid of it. Another alternative is to use the mask function available in some programs. Cutting an entire sequence out of an alignment may leave columns of gaps across the entire alignment that will have to be removed. The extreme amino- and carboxy-termini (5' and 3' in DNA) seldom align nicely; they are often jagged and uncertain, and should probably be excluded. The results of subsequent analyses are absolutely dependent upon the quality of your alignment.

Researchers have successfully used the conservation of co-varying sites in ribosomal and other structural RNA alignments to assist in alignment refinement. That is, as one base in a stem structure changes the corresponding Watson-Crick paired base will change in a corresponding manner. This principle has guided the assembly of rRNA structural alignments at the Ribosomal Database Project at Michigan State University [42] and at the University of Gent, Belgium, at the European Ribosomal RNA database [43].

Be sure an alignment makes biological sense — align things that make sense to align. Beware of comparing 'apples and oranges.' Be particularly suspectful of sequence data-sets found through text-based database searches such as Entrez [44] and GCG's LookUp (based on the Sequence Retrieval System [SRS] of Etzold and Argos [45]). For example, don't try to align receptors and/or activators with their namesake proteins. Be wary of trying to align genomic sequences with cDNA, when working with DNA: the introns will cause all sorts of headaches. Similarly, aligning mature and precursor proteins, or alternate splicing forms, from the same organism and locus, doesn't make evolutionary sense, as one is not evolved from the other, rather one is the other. Watch for redundant sequences; there are many in the databases. If creating alignments for phylogenetic inference, either make paralogous comparisons (i.e., evolution via gene duplication) to ascertain gene phylogenies within one organism, or orthologous (within one ancestral loci) comparisons to ascertain gene phylogenies between organisms (which should imply organismal phylogenies). Try not to mix them up without complete data representation. Otherwise, confusion can mislead the interpretation, especially if the nomenclature of the sequences is inconsistent. These are all easy mistakes to make; try your best to avoid them.

Remember the old adage "garbage in — garbage out!" Some general guidelines to remember include the following [46]:

- If the homology of a region is in doubt, then throw it out.
- Avoid the most diverged parts of molecules; they are the greatest source of systematic error.
- Do not include sequences that are more diverged than necessary for the analysis at hand.

## 15.3 Complications

Sequence data format is a huge problem in computational molecular biology. The major databases all have their own distinct format, plus many of the different programs and packages require their own. Clustal [47] has a specific format associated with it. The FastA database similarity-searching package [6] uses a very basic sequence format. The National Center for Biotechnology Information (NCBI) uses a library standard called ASN.1 (Abstract Syntax Notation One), plus it provides the GenBank flat file format for all sequence data. GCG uses three sequence formats: Single Sequence Format (SSF), Multiple Sequence Format (MSF), and SeqLab's Rich Sequence Format (RSF) that contains both sequence data and annotation. PAUP* (Phylogenetic Analysis Using Parsimony [and other methods, pronounced "pop star"] 48), MrBayes [49], and many other phylogenetic analysis packages, have a required format called the NEXUS file. The PAUP* interface in the GCG Package, PAUPSearch, generates the NEXUS format directly from GCG alignments. Even PHYLIP (PHYLogeny Inference Package, 50) has its own unique input data format. Format standards have been argued over for years, such as using XML for everything, but until everybody agrees, which is not likely to happen, it just won't happen. Fortunately, several freeware programs are available to convert formats back and forth between the required standards, but it all can get quite confusing. BioPerl's SeqIO system [30] and ReadSeq [51] are two of the best. T-Coffee [23] comes with one built in named "seq_reformat."

Alignment gaps are another problem. Different program suites may use different symbols to represent them. Most programs use hyphens, "–"; the GCG Package uses periods, ".", for interior gaps, and tildes, "~", for placeholder gaps. Furthermore, not all gaps in the sequences should be interpreted as deletions. Interior gaps are probably okay to be represented this way, as, regardless of whether a deletion, insertion or a duplication event created the gap, logically they are treated the same by the algorithm. These are known as 'indels.' However, end gaps should not be represented as indels, because a lack of information before or beyond the length of any given sequence may not be due to a deletion or an insertion event. It may have nothing at all to do with the particular stretch of sequence being analyzed. It just may not have been sequenced! These gaps are just placeholders for the sequence. Therefore, it is safest to manually edit an alignment, to change the leading and trailing gap symbols to "x"'s which mean "unknown amino acid," or "n"'s which mean "unknown base," or "?"'s which is supported by many programs, but not all, and means "unknown residue or indel." This will assure that the programs don't make incorrect assumptions about your sequences.

## 15.4 Applicability?

Now that we understand some of the principles and problems of multiple sequence alignment, what's so great about doing it anyway; why would anyone want to bother? Multiple sequence alignments are:

- very useful in the development of PCR primers and hybridization probes;
- great for producing annotated, publication quality, graphics and illustrations;
- invaluable in structure/function studies through homology inference;
- essential for building HMM profiles for remote homology similarity searching and alignment; and
- required for molecular evolutionary phylogenetic inference programs.

A multiple sequence alignment is useful for probe and primer design by allowing one to visualize the most conserved regions of an alignment. This technique is invaluable for designing phylogenetic specific probes as it clearly localizes the areas of high conservation and high variability in an alignment. Depending on the data-set that you analyze, any level of phylogenetic specificity can be achieved. Pick out the areas of high variability in the overall data-set that correspond to areas of high conversation in the phylogenetic category subset data-sets, to differentiate between universal and phylo-specific potential probe sequences. After localizing general target areas on the sequence, one can then use any of several primer discovery programs, such as GCG's Prime, or MIT's Primer3 [52], or the commercial Oligo program (National Biosciences, Inc.), to find the best primers within those regions, and to test those potential probes for common PCR conditions and problems. (See my workshop tutorial illustrating this technique using GCG and SeqLab at bio.fsu.edu/~stevet/PrimerDesign.pdf if you are interested.) The technique is illustrated in Fig. 15.1, below where potential primer locations are identified that should differentiate between the major capsid protein genes (L1), of the highly carcinogenic Human Papillomavirus (HPV) Type 16 strains from the rest of the Type 16 relatives.

Graphics prepared from multiple sequence alignments can dramatically illustrate functional and structural conservation. These can take many forms of all or portions of an alignment — shaded or colored boxes or letters for each residue (e.g., BoxShade [53], and the various PostScript output options in GCG's SeqLab), cartoon representations (e.g., WebLogos [54] and GCG's SeqLab graphical feature representation), running line graphs of overall similarity (as seen above with

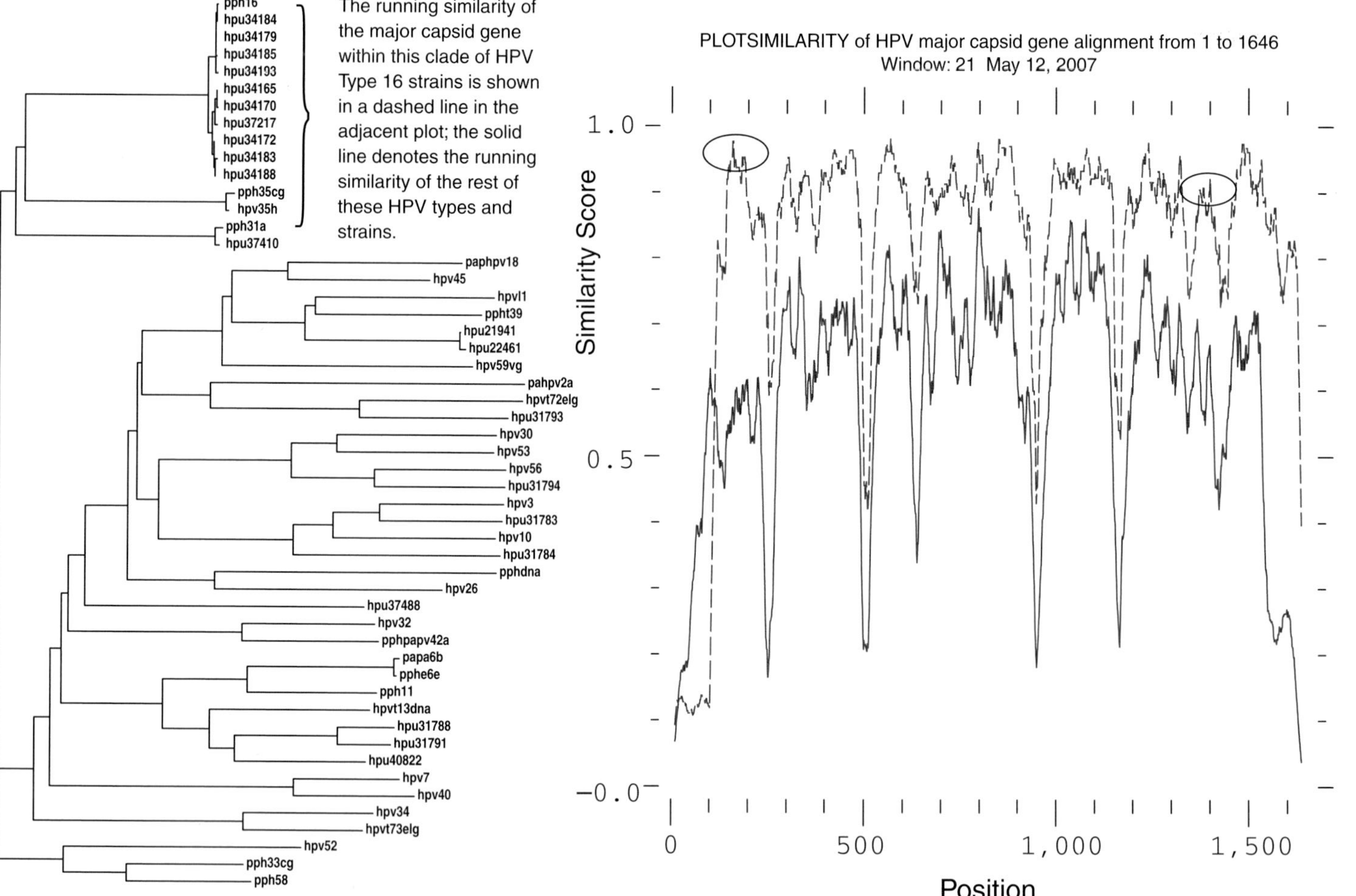


**Fig. 15.1** A phylogram of the HPV type assemblage, most closely related to Type16 based on the L1 major capsid protein, and the corresponding GCG PlotSimilarity traces. The ellipses denote potential areas in which to localize the PCR primers within the gene that would differentiate the Type 16 clade from it's closest relatives. These are areas of high L1 conservation in the Type 16 clade (the dashed upper line) that correspond to areas of much weaker conservation in the other clades (the solid lower line) (Copies of figures including color copies, where applicable, are available in the accompanying CD)

**Fig. 15.2** A GCG SeqLab Post-Script graphic of the most conserved portion of the HMG-box DNA binding domain from a collection of paralogous human HMG-box protein sequences (Copies of figures including color copies, where applicable, are available in the accompanying CD)

GCG's PlotSimilarity and as displayed by ClustalX), overlays of attributes, various consensus representations, etc. — all can be printed at high-resolution, usually in color or gray tones. These can make a big difference in a poster or manuscript presentation. Fig. 15.2 shows a multiple sequence alignment of the most conserved portion of the HMG DNA-binding domain from several paralogous members of the human HMG-box super family.

Conserved regions of an alignment are important. In addition to the conservation of the primary sequence, structure and function is also conserved in these crucial regions. In fact, recognizable structural conservation between true homologues extends way beyond statistically significant sequence similarity. An often cited example is in the serine protease super family. *S. griseus* protease A, demonstrates remarkably little sequence similarity when compared to the rest of the super family (Expectation values $E \geq 10^{1.8}$ in a typical search), yet its three-dimensional structure clearly shows its allegiance to the serine proteases (RMSD of less than 3 Å with most of the family) (Pearson, W.R., personal communication). These principles are the premise of 'homology modeling' and it works remarkably well. An automated homology-modeling tool is even available on the ExPASy server in Switzerland. Supported by the Swiss Institute of Bioinformatics (SIB) and GlaxoSmithKline, the Swiss-Model [55] has dramatically changed the homology modeling process. It is a relatively painless way to get a theoretical model of a protein structure. While not always successful, the minimal amount of effort involved to make the attempt makes it an excellent time investment. It will not always generate a homology model for your sequence, depending on how similar the closest sequence with an experimentally solved structure is to it; however, it is a very reasonable first approach and will often lead to remarkably accurate representations. I submitted a *Giardia lamblia* Elongation Factor 1α sequence to Swiss-Model in "First Approach mode." The results were e-mailed back to me in less than five minutes. Figure 15.3 displays a RasMac [56] "Strands" graphic of the *Giardia* EF–1α structural model from the Swiss-Model superimposed over the eight most similar solved structural templates.

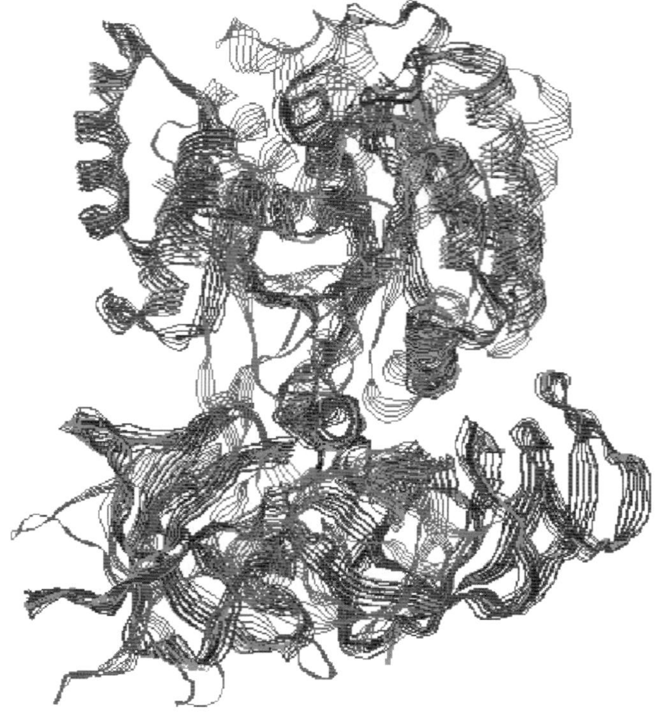

**Fig. 15.3** A RasMac representation of the Swiss-Model *Giardia* lamblia EF-1α structure superimposed over the eight most similar solved structures (Copies of figures including color copies, where applicable, are available in the accompanying CD)

Profiles are a position specific scoring matrix (PSSM) description of an alignment or a portion of an alignment. Gap insertion is penalized more heavily in conserved areas of the alignment than it is in variable regions, and the more highly conserved a residue is, the more important it becomes. Profiles are created from an existing alignment of related sequences, and then they are used to search for remote sequence similarities and/or to build larger multiple sequence alignments. Originally described by Gribskov [10], and then automated by NCBI's PSI-BLAST [9], later refinements have added more statistical rigor (see e.g., Eddy's Hidden Markov Model profiles [11]). The original Gribskov style profiles require a lot of time and skill to prepare and validate, and they are heuristics based. An excess of subjectivity, and a lack of formal statistical rigor are also drawbacks. Eddy's HMMer (pronounced "hammer") package uses Hidden Markov modeling, with a formal probabilistic basis and consistent gap insertion theory, to overcome these limitations. The HMMer package can build and manipulate HMMer profiles and profile databases, search sequences against HMMer profile databases and vice versa, and can easily create multiple sequence alignments using HMMer profiles as a 'seed.' This ability to easily create larger and larger multiple sequence alignments is incredibly powerful and faster than starting all over each time you want to add another sequence to an alignment. The 'take-home' message is that HMMer profiles are much easier to build than traditional profiles, and they do not need to have as many sequences in their alignments in order to be effective. Furthermore, they offer a statistical rigor not available in Gribskov profiles, plus they have all the sensitivity of any profile technique. In effect, they are like the 'old-fashioned' profiles pumped up on steroids! One big difference between HMMer profiles and others is when the profile is built, one has to specify the type of eventual alignment it will be used with, rather than when the alignment is built. The HMMer profile will either be used for global or local alignment, and it will occur in multiples or singly on a given sequence. All profile techniques are tremendously powerful; they can provide the most sensitive, albeit, extremely computationally intensive, database similarity search possible.

Finally, we can use multiple sequence alignments to infer phylogeny. Based on the assertion of homologous positions in an alignment, many, different methods can estimate the most reasonable evolutionary tree for that alignment. A few of the packages that incorporate these methods were mentioned earlier in the complications section with regard to format issues, e.g., PAUP* [48], MrBayes [49], and PHYLIP [50]. This is a very huge, complicated, and highly contentious field of study. (See the Woods Hole Marine Biological Laboratory's excellent summer short course, the *Workshop on Molecular Evolution*). However, always remember that regardless of the algorithm used, any form of parsimony, all of the distance methods, all maximum likelihood techniques, and even all types of Bayesian phylogenetic inference, all make the absolute validity of your input alignment matrix their first and most critical assumption refer to [57].

Therefore, the accuracy of your multiple sequence alignment is the most important factor in inferring reliable phylogenies; your interpretations are utterly dependent on its quality. Structural alignments are the 'gold-standard,' but the luxury of having homologous solved structures is not always available. In fact, many experts recommend not using any questionable portions of sequence data at all. These highly saturated regions have the property known as 'homoplasy.' This is a region of a sequence alignment where so many multiple substitutions have occurred at homologous sites that it is impossible to know if those sites are properly aligned, and thus, impossible to ascertain the relationships based on those sites. Phylogenetic inference algorithm's primary assumption is most violated in these regions, and this phenomenon increasingly confounds evolutionary reconstruction, as divergence between the members of a data-set increases. Because of this, only analyze those sequences and those portions of your alignment that assuredly align. If any sequences or portions are in doubt, exclude them. This usually means trimming down or masking the alignment's terminal ends and may require internal trimming or masking as well. These decisions are somewhat subjective by nature, experience helps, and some software, such as ASaturA [58] and T-Coffee [23], have the ability to evaluate the quality of particular regions of your alignment as well. Bio-computing is always a delicate balance — signal against noise — and sometimes it can be quite a balancing act!

## 15.5 The T-Coffee Shop

I call this section the T-Coffee shop because T-Coffee [23] is much more than merely a program for doing multiple sequence alignment. Much like a Starbucks® coffee shop that offers many different flavors and types of coffee drinks, the T-Coffee command line offers an entire suite of multiple sequence alignment tools. Notredame [24] has done a very good job of providing documentation with the package's distribution. In particular, be sure to read the entire *Tutorial and FAQ*. It's quite good, and I cannot do justice to it in my description here. Therefore, I will attempt to distil the most vital portions of the documentation, and illustrate a subset of T-Coffee's potential in a 'bare-bones' manner, just enough to get a novice user started in exploring the package.

As mentioned in the algorithm section, T-Coffee is one of the most accurate multiple sequence alignment tools around, and it does this in its default mode. It achieves its accuracy by producing the multiple sequence alignment that has the highest consistency level with a library of pre-processed, global, and local pair wise alignments. However, it can do much more than that. In addition to merely aligning a sequence data-set, it can combine pre-existing alignments, evaluate the consistency of alignments, extract a series of motifs to create a local alignment, perform all sorts of data manipulation, format operations, and with SAP (Structure Alignment Program, [59]) installed it can even use structural information to make the most accurate protein alignment possible.

T-Coffee's "seq_reformat" tool can perform standard data reformatting operations and change the appearance of your alignment, but it is incredible, as well, for extracting or combining subsets of your data based on sequence names, patterns, coordinates in the alignment, and/or level of consensus. It even has the ability to translate DNA sequences into their corresponding protein sequences, or to generate DNA alignments based on the corresponding protein alignment, either using the actual DNA sequences (*ala* mrtrans and relatives) or by using a random back-translation procedure. Furthermore, "seq_reformat" can read phylogenetic trees in Newick format to compare two trees or to prune tips off of a tree. Another practical utility in T-Coffee is "extract_from_pdb;" it allows you to download either the three-dimensional coordinates or the FastA format sequence of structures held at the PDB (Protein Data Bank, [60]), using UNIX's "wget" command. There is little that cannot be done with the T-Coffee utility tools when it comes to data organization and manipulation. One of my favorites is using it to remove sequences that are, or are nearly, redundant. Pages 11 through 25 of Notredame's tutorial [24] cover all these operations very well, and I encourage you to work through these examples; I will not take the space to review them here.

Okay, how do we begin? I'm making the presumption that you already have T-Coffee installed on either your own computer or on a server that you have access to. If this is not the case, then refer to Notredame's [24] *Technical Documentation*, and either install it yourself, or get a local systems administrator to do it for you. I'm using version 5.05 here. I'll be consistent in my command syntax in these examples, but realize that Notredame's tutorial mixes up command syntax a bit, freely replacing equal signs and commas with spaces in some examples and not in others, and T-Coffee doesn't mind. Also realize that all commands are typed on a single line that will wrap in your terminal window. They are just shown here on multiple lines to simplify reading them. Let's first look at T-Coffee's default mode. Issue the following command to see all of T-Coffee's default parameter settings:

```
Prompt% t_coffee -help
```

The list is huge; scroll back to skim over the entire thing. Notice the "–seq" parameter usage: "List of sequences in any acceptable format." T-Coffee will accept several input formats, but it works most reliably if you have your input files in FastA format. If I have a file containing unaligned sequences in FastA format named "unaligned.fa," then the following command will run T-Coffee without any options or parameter specifications:

```
Prompt% t_coffee unaligned.fa
```

This will produce a screen trace of the program's progress; an output alignment named "unaligned.aln" in Clustal Aln format; another alignment file named "unaligned.html" in HTML format for Web browsers, with columns color-coded based on reliability; and a file named "unaligned.dnd," that contains the Newick format tree used to guide the alignment. If you have your data-set spread around in more than one file, then you can use the "–seq" option followed by a comma-delimited list of input files. This option will also strip the gaps out of any input file that might already be an alignment. And if you don't want the Clustal Aln format, use the "–output" option. Here I'll use it to generate a FastA format output alignment file named "unaligned1.fasta_aln" from two FastA input files (T-Coffee uses the first file's primary name to identify the output file). The only output alignment file produced this way is in FastA format:

```
Prompt% t_coffee -seq=unaligned1.fa,unaligned2.fa -output=fasta_aln
```

The T-Coffee "–output" option supports alignment formats with the following identifiers: "msf_aln" (for GCG MSF), "pir_aln" (for PIR), "fasta_aln" (for FastA), "phylip" (for PHYLIP), as well as its default "clustalw_aln" and "html". You can produce output files in more than one format by comma separating the identifiers. So, you cannot get directly to NEXUS format, but PAUP* has the ability to import GCG MSF, PIR, or PHYLIP format with the ToNEXUS command. Plus, if you do not like T-Coffee's default output file naming convention, you can use the "–outfile" option to specify any name you might want.

### *15.5.1 Alignment Parameters*

As in nearly all sequence alignment programs, the substitution matrix and the gap penalties are very important run parameters. In most cases, the T-Coffee matrix defaults, the BLOSUM62 and 50 matrices for its global and local pair wise alignment steps, respectively, will work just fine. And, in fact, T-Coffee only uses these matrices in its first pass through your data-set, when it builds its consistency library. It replaces the usual BLOSUM style matrix when building its final multiple sequence alignment with the optimal position specific scores of all the potential pair wise matings in its library. Regardless, using the optional "–matrix = blosum30mt" flag (or blosum40mt or blosum45mt depending on your data's level of divergence) is a great idea, when dealing with sequences that are quite dissimilar. Furthermore, gap penalties can be changed, if you really want to, but the default gap opening penalty of –50 and gap extension penalty of zero, changed through the "–gapopen" and "–gapext" options are, as Notredame [24] says, only "cosmetic," changing the final alignment's appearance by changing how residues slide around in 'unalignable' regions, since all the 'alignable' regions are found from the previously built library. It's much more complicated if you want to change how the library alignments are built, and I don't suggest messing with it. If you insist, the parameters are specified through the "–method" option, and two methods build the pair wise alignment library by default (several others are available by option for special cases): a global, "slow_pair," one, and a local, "lalign_id_pair," one. To change their respective default behaviors a combination of "MATRIX" specification and "GOP" and "GEP" parameters are used. The defaults for the global library alignments are a "GOP" of –10 with a "GEP" of –1, and for the local alignment library a "GOP" of –10, with a "GEP" of –4. For instance, if I have a real difficult data-set, with barely discernible homology, and with no structural homologues, then perhaps using a combination of parameters such as the following would produce a more accurate and pleasing looking multiple sequence alignment:

```
Prompt% t_coffee -seq=lousydata.fa -matrix=blosum30mt
  -gapopen=-100 -gapext=0 -output=fasta_aln,clustalw_aln,html
  -method=slow_pair@EP@GOP@-5@GEP@-1,lalign_id_pair@EP@GOP@-5@GEP@-4
```

Notice the bizarre syntax: the at sign, "@," is used as a method parameter separator, and "EP" stands for "Extra Parameter." This command would run both the global and local library builds with the BLOSUM30 matrix, will double the 'cosmetic' gap opening penalty, would cut the penalties in half for opening a gap in both the global and local library alignments, and would keep all the extension penalties at their default levels. Additionally, it would produce output alignments in FastA, ClustalW, and HTML formats. If you want to use different substitution matrices for the different methods, then you would add e.g., "MATRIX@blosum45mt" after "@EP@" and before "@GOP@ for the appropriate method. But, will this actually produce a 'better' alignment?

### *15.5.2 Quality*

This brings up the heart of T-Coffee: consistency. T-Coffee's method relies on reconciling its internal pair wise alignment library as best as it can with its eventual multiple sequence alignment; the more these agree, the more consistent the alignment and, we assume, the more accurate. This premise allows us to use T-Coffee to evaluate and compare alignments. The easiest way to see how accurate T-Coffee 'thinks' its alignment is, is to look at the "SCORE" it receives in its ClustalW or HTML format output. The higher this score value the more closely the alignment overall agrees with the internal pair wise alignment library. Notredame [24] says that values above 40 are "usually pretty good." Every T-Coffee ClustalW and HTML format output alignment has this value associated with it. However, what if you don't have the right output format, or you want to see how the output from some other alignment program ranks, or you are interested in how different alignments compare to each other, or you are interested in what portions of an alignment are good and what portions are bad? T-Coffee can do all of this.

I'll discuss methods that do not rely on structure first. Structural methods will follow when I discuss T-Coffee's ability to integrate sequence and structure. T-Coffee's CORE index is the local consistency level of each position within your alignment. All the T-Coffee HTML format output alignments represent this index with color-coding, plus there are specific score output file options. Position colors range across the spectrum from blue, to green, to yellow, to orange, and finally red, corresponding to an increase in the consistency level from none to absolute. These colors correspond to local consistency values of 0 through 9. To test a pre-existing alignment with the CORE index use the "–infile" specification for your alignment, the "–evaluate" (replaces the deprecated "–score" flag, and in default mode equivalent to "–special_mode = evaluate") option, and minimally specify HTML output format:

```
Prompt% t_coffee -infile=lousydata.fasta_aln -evaluate -output=html
```

Notice we need to use "–infile" rather than "–seq", in order to run T-Coffee in this manner. There are even ways to automatically filter unreliable columns from your alignment, based on the CORE index; however, the various commands' syntax are quite complicated, and I refer you to page 45 of Notredame's [24] tutorial.

### *15.5.3 Comparing Alignments*

T-Coffee has several ways to compare existing alignments of the same sequences beyond just looking at their consistency scores. The "aln_compare" module is one of the more powerful, and can tell you how different two alignments are. It needs to be launched with the "–other_pg" option, which tells T-Coffee that you want to use an external module. This option must be the first parameter on the command line after "t_coffee." "aln_compare" supports several further options that can help with the visualization. Here, the "aln_compare" module is used to analyze the difference between two existing alignments with the "–al1" and "–al2" options to produce an output screen trace of the first alignment where all residues with less than 50% of their pairing partners in the other alignment are represented as an "x:"

```
Prompt% t_coffee -other_pg=aln_compare -al1=trial1.fasta_aln
  -al2=trial2.fasta_aln -output_aln -output_aln_threshold 50
  -output_aln_modif x
```

The same command without the "–output_align" parameters will produce summary statistics of the percentage of similarity between the two alignments by counting the sum of all pairs of residues in those alignments. Type the command without any parameters to see all that "aln_compare" offers:

```
Prompt% t_coffee -other_pg=aln_compare
```

All of T-Coffee's built in external modules support this help syntax, versus its standard "–help" option.

Another way to compare alignments is to turn one into T-Coffee's library and leave the other as an alignment. You need to use the "–aln" option to do this. This option tells T-Coffee to use the specified input alignment file to build its library. The following command will show how well the alignment "somedata1.fasta_aln" agrees with the library produced from the alignment "somedata2.fasta_aln:"

```
Prompt% t_coffee -infile=somedata1.fasta_aln
  -aln=somedata2.fasta_aln -evaluate -output=html
```

The HTML alignment output will highlight those residues that are either in agreement or not between the two input alignments, using T-Coffee's standard CORE index-coloring scheme.

### 15.5.4 Combining Alignments

Again, there's a slew of ways to combine alignments with T-Coffee. One neat way is to not really worry how alignments compare and just turn them all into a T-Coffee library so that they will combine together yielding one optimal alignment that best agrees with all the input alignments. They do not even need to all have the same sequences to do this. Turn the three specified alignments into a library and produce an output alignment in Clustal Aln, FastA, and HTML format with the command below:

```
Prompt% t_coffee -aln=one.fasta_aln,two.fasta_aln,three.fasta_aln
    -output=clustal_aln,fasta_aln,html
```

And, of course, you could easily add some unaligned input sequences to the mix with the "–seq" option as well.

As discussed in this chapter under multiple sequence alignment applications, profiles are a very powerful technique for building larger and larger alignments. T-Coffee can deal with profiles in several ways, though they are not quite the same sort of profile as, for instance, Gribskov [10] or Eddy [11] envisioned. T-Coffee defines profiles as multiple sequence alignment matrices that will not have their gaps removed, rather than a true PSSM where residues receive higher weights in more conserved regions. Regardless, T-Coffee can take as many different profiles and sequences as you want to specify, and combine all of them into one alignment (given that it is biologically correct to attempt to align them):

```
Prompt% t_coffee
    -profile=one.fasta_aln,two.fasta_aln,three.fasta_aln
    -seq=lousydata.fa,evenworsedata.fa
    -output=clustal_aln,fasta_aln,html
```

This command will feed three alignments to T-Coffee, such that, the sequences within them will not have their gaps removed, it will add more gaps to reconcile those three alignments, and it will add two more sequences to the resulting alignment in the most consistent manner.

And to get the most accurate profile alignment add "–profile_comparison = full," which runs the profile alignment in a slower, more exact mode, "on a library that includes every possible pair of sequences between the two profiles," as opposed to the above command, which "vectorizes" the multiple sequence alignments designated as profiles [24].

### 15.5.5 Combining Methods

T-Coffee has a special 'Meta' mode named M-Coffee [61]. This gives T-Coffee an incredible amount of power. M-Coffee is amazing for those situations where you just do not know what alignment tools to trust, and you don't want to have to build and test many alternatives.

It automatically runs up to eight different multiple-alignment programs (by default, more external methods can be added) on your data, and combines the best parts of each, to come up with one, most consistent, consensus alignment. Your system needs to have ClustalW [20], POA [62], Muscle [25], ProbCons [27], MAFFT [26], Dialign-T [63], PCMA [64], and T-Coffee [23] all installed for this to work. If I want to see how M-Coffee handles that difficult data FastA format file I have, then I would issue the following command to run M-Coffee in its default mode:

```
Prompt% t_coffee -seq=lousydata.fa -special_mode=mcoffee
  -output=clustal_aln,fasta_aln,html
```

The output Clustal and HTML format files will list the alignment's overall score as a percentage of consistency between all the methods. The HTML format will additionally provide T-Coffee's usual color-coded position consistencies. Or, If you prefer some methods to others, you can select the particular methods to combine with the "–method" option, with syntax like the following:

```
Prompt% t_coffee -seq=lousydata.fa
    -method=t_coffee_msa,mafft_msa,muscle_msa -output=fasta_aln,html
```

Here's some general guidelines as to which of T-Coffee's integrated external multiple sequence alignment methods are best in which situations (based on [24 and 65]):

| | |
|---|---|
| clustalw_msa | neither the fastest nor the most accurate, but a reasonable 'industry-standard.' |
| probcons_msa | uses consistency and Bayesian inference to provide ultra-accurate, but runs very slow. |
| muscle_msa | very, very fast for large data-sets, especially; uses weighted log-expectation scoring. |
| mafft_msa | in fast mode (FFT-NS-i) screamingly quick on large data-sets, but not incredibly accurate; in slow mode (L-INS-i) very accurate but quite slow, especially with large data-sets. |
| pcma_msa | combines ClustalW and T-Coffee strategies. |
| poa_msa | very accurate local alignments using partial order graphs. |
| dialignt_msa | accurate local, segment-based, progressive alignment. |

Pick and choose among the most appropriate methods and let M-Coffee combine the best aspects of each.

### *15.5.6 Local Multiple Sequence Alignments*

If you know the coordinates of some pre-defined sequence pattern in one sequence, you can use T-Coffee's mocca routine (Multiple OCCurrences Analysis, [66]) to find all the occurrences of similar patterns in other sequences and assemble a local multiple sequence alignment of them. Mocca is a perl script that launches T-Coffee, computes a T-Coffee library from the input sequences, and then prompts you with an interactive menu to extract the homologous motifs and assemble the alignment. It is designed to find and align motifs of 30% and greater identity that are at least 30 amino acids long. The interactive menu can be confusing, so I recommend that you place the sequence with your identified motif first in the data-set, and specify the beginning and the length of your motif on the command line, rather than in the menu. I'll use mocca here to prepare a local sequence alignment of the motifs in a data-set named "motifdata.fa," where I know the first occurrence of the motif is at absolute position 35 and runs for 65 residues in my first sequence of the data-set:

```
Prompt% t_coffee -other_pg=mocca motifdata.fa -start=35 -len=65
```

You cannot use the "–seq" option with mocca to specify your input file in this command. Specifying your motif coordinates is also a bit tricky, since all the input sequences have their gaps removed (if it was an alignment) and are then concatenated together. That's why it's easiest if you put your known motif sequence first. After mocca computes the optimal local alignment with your motif, it pauses and displays its menu. Type a capital "X" to exit the program and write the default Clustal Aln and HTML alignment files.

### *15.5.7 The 'Gold-standard:' Creating Structure based Alignments*

As mentioned earlier, you need to minimally have SAP [59] installed on your system for T-Coffee's structure based alignment mode to actually use structural information. And, even better yet, have the FUGUE package [67] installed as well. Additionally you need "wget" on your system, to access PDB files over the Internet, but mostly all UNIX/Linux installations should include this utility. T-Coffee uses a special mode named 3DCoffee [68] to create structure-based alignments. By default, 3DCoffee uses four methods to create the T-Coffee consistency library, if you specify an input sequence data-set: the standard T-Coffee global "slow_pair" and local "lalign_id_pair," SAP's "sap_pair," and FUGUE's "fugue_pair." If I have a data-set that includes some PDB structures, and those sequences are named using the PDB's identifier with a chain name (e.g., 1EFTA for chain A of the *Thermus aquaticus* elongation factor Tu structure [69]), then the following command will produce the most consistent alignment of them, based on all the available structural pairs and T-Coffee's usual pairs:

```
Prompt% t_coffee -seq=elongation.fa -special_mode=3dcoffee
```

You'll get three output files by default with this command: "elongation.aln," "elongation.html," and "elongation.dnd". The alignment files will have T-Coffee's standard reliability index. If you specify that your input data-set is already in alignment, then the local "lalign_id_pair" will not be used, and the alignment will be turned directly into T-Coffee's library along with the SAP and FUGUE pairs:

```
Prompt% t_coffee -aln=elongation.fasta_aln -special_mode=3dcoffee
```

These 3DCoffee analyses will even produce output alignments if you don't have SAP or FUGUE installed, but they will report error warnings for every pair, and, naturally, no structural information will be used in the production of the alignment. If you don't like the warning messages, just specify the particular methods you have available, for e.g.:

```
Prompt% t_coffee -aln=elongation.fasta_aln
  -method=sap_pair.slow_pair
```

A nice trick is to combine two existing, related, but only distantly so, alignments with 3DCoffee, if they both have at least one sequence where the structure has been solved, and they follow proper naming conventions. Suppose I have one alignment of elongation factor Tu sequences containing the sequence for the solved structure for *Thermus aquaticus*, and another alignment of elongation factor 1α sequences containing the sequence of the solved human structure. Do this analysis with the "–profile" input specification:

```
Prompt% t_coffee
  -profile=elongationla.fasta_aln.elongationtu.fasta_aln
  -special_mode=3dcoffee
```

The output alignment will combine the two existing alignments in the most consistent manner with the structural alignment of the human and *Thermus aquaticus* sequences.

### 15.5.8 Using Structure to Evaluate Alignments

Your existing alignment needs to have at least two members with solved structures in order to evaluate it using T-Coffee's structural method, and, as above, those sequences need to be named according to their PDB identifiers as well as their chain. T-Coffee uses a special version of Root Mean Square Distance Deviation analysis, not dependent on specific α carbon backbone super positioning called iRMSD (the "i" stands for intra-catener [70]), for structural alignment evaluation. It also reports a normalized NiRMSD not dependent on the alignment's length, and it reports an older measure, APDB, not as powerful as iRSMD, based on the fraction of residue pairs with 'correct' structural alignments. The smaller the iRMSD Å numbers are, the bigger the APDB percent will be, and the better will the alignment correspond to structural 'reality.' As with the other integrated external methods in the T-Coffee package, iRMSD is launched with the "–other_pg" option. Specify your alignment's file name with the "–aln" option, and specify an output file with the "–apdb_outfile" option, otherwise the output will just scroll to screen. Optionally generate an HTML alignment output as well with "–outfile":

```
Prompt% t_coffee -other_pg=irmsd -aln=elongation.fasta_aln
  -apdb_outfile=irmsd.out=outfile=irmsd
```

The summary statistics at the end of the output file are the most telling:

```
    # TOTAL for the Full MSA
            TOTAL EVALUATED: 81.58 %
            TOTAL APDB:      78.25 %
            TOTAL iRMSD:      0.75 Angs
            TOTAL NiRMSD:     0.93 Angs
# EVALUATED: Fraction of Pair wise Columns Evaluated
# APDB:      Fraction of Correct Columns according to APDB
# iRMDS:     Average iRMSD over all evaluated columns
# NiRMDS:    iRMSD*MIN(L1,L2)/Number Evaluated Columns
# Main Parameter: -maximum_distance 10.00 Angstrom
# Undefined values are set to -1 and indicate LOW Alignment Quality
# Results Produced with T-COFFEE (Version_5.05)
# T-COFFEE is available from http://www.tcoffee.org
```

My elongation factor 1α/Tu alignment is pretty good with an overall NiRSMD of less than 1 Å and an APDB statistic of 78%. The old T-Coffee external method specification "–other_pg = apdb" produces the same output. The optional HTML output shows which residues in the solved structures (only) are in agreement with the structural alignment, using the APDB coloring scheme, where blue corresponds to 0% and red corresponds to 100% (get iRSMD coloring with "–color_mode = irmsd").

### 15.5.9 Special Situations

Even though T-Coffee is great for so many things, its default run parameters are far from ideal for some special situations. It does not work very well for larger data-sets, i.e., anything with over 100 sequences, it does not particularly like DNA/RNA alignments, and it has some problems with jumping over huge gaps that one would see when aligning splicing variants or cDNA to genomic DNA containing introns. Fortunately, there are ways around each one of these scenarios, and most of all multiple sequence alignment programs have trouble with the same situations as well.

Let's start with real large alignments. Both Muscle and MAFFT (in fast mode) are more appropriate for data-sets with more than around 100 sequences; however, T-Coffee does have a special mode that will at least allow it to estimate an approximate alignment with such data-sets. This should work when your data-set has an overall identity of 40% or more:

```
Prompt% t_coffee -seq=hugedata.fa -special_mode=quickaln
```

The resulting alignment should be about as accurate as that built with ClustalW. For data-sets between 50 and 100 sequences, T-Coffee automatically switches to another heuristic mode named DPA (double progressive alignment).

Naturally, DNA and RNA alignments are harder for T-Coffee to perform. The rationale for this phenomenon is explained in the introduction, and it confounds every multiple sequence alignment program. DNA is just really hard to align, unless the sequences are 90% or so identical. One way to help T-Coffee with an alignment that must be built using DNA or RNA, especially if the locus does not code for a protein, is to specify that the sequences are DNA with the "–type" option:

```
Prompt% t_coffee -seq=DNadata.fa -type=dna
```

T-Coffee should detect this sequence type automatically, but it can't hurt to declare it up front. When T-Coffee realizes that it is working with DNA, it uses specific DNA optimized methods to build its library, "slow_pair4dna" and "lalign_id_pair4dna." These methods have lower built-in gap penalties and use a DNA specific scoring matrix. And if your DNA is a particularly noisy coding locus, but you just can't figure out the translation, because it is so noisy, then T-Coffee's special "cdna_fast_pair" method takes potential amino acid similarity considering frame shifts into account, and may help:

```
Prompt% t_coffee -seq=noisy_cDNAdata.fa -method=cdna_fast_pair
```

You may want to try this when aligning cDNAs to the genomic DNA as well, in order to jump over the introns. If the results look good, you can even use T-Coffee's external "seq_reformat" module to translate it to the appropriate protein translation in spite of how noisy the original DNA sequences were:

```
Prompt% t_coffee -other_pg=seq_reformat

  -in=noisy_cDNAdata.fasta_aln

  -action=+ clean_cdna, + translate > noisy_pep.fasta_aln
```

Notice that "seq_reformat" does not support "–outfile;" you need to use UNIX ">" output redirection. " + clean_cdna" is a small HMM that tries to maximize the appropriate frame choice at every point in the sequences to best match the alignment of all the other sequences, and " + translate," obviously, does the translation.

T-Coffee is actually pretty good at jumping over big gaps in protein sequence alignments, that can be present when trying to align various protein splicing variants. It achieves this by relying heavily on the local, pair wise knowledge gained in its internal "lalgn_id_pair" method. If you can't find the alignment using default parameters, try to restrict your method to the local pair library only:

```
Prompt% t_coffee -seq=splicingvariant.fa -method=lalgn_id_pair
```

Another trick that can work well with EST sequences is to increase the default ktuple size from 2 to 5, along with specifying the "cfasta_pair_wise" "dp_mode" option:

```
Prompt% t_coffee -seq=EST.fa -dp_mode=cfasta_pair_wise -ktuple=5
```

This should use a "checked" [24] version of the FastA algorithm with a word size of five to create T-Coffee's consistency library. A DNA alignment can be produced much faster this way than with other methods given sufficient similarity, but difficult regions will be less accurate. Try mixing and matching the various methods most appropriate for your data to come up with the 'most satisfying' multiple sequence alignment you can.

### *15.5.10 T-Coffee Servers and Expresso*

Finally, if all this command line stuff just bewilders you, there are several T-Coffee Web servers around, e.g., the primary one at http://www.tcoffee.org/; they just can't do all the things that can be done in the package from the command line. T-Coffee Web servers do, however, offer "Regular" and "Advanced" modes. Furthermore, at the time of this writing, T-Coffee Web servers are the only way to run Expresso [71]. Expresso, is T-Coffee's latest and greatest mode. It's the triple-shot espresso, extra whip cream, Irish whiskey enriched, grandé cappuccino, of modes! It's a logical pipeline. The server runs an all against all BLAST search of your data-set against the sequences in PDB, finds all the templates with greater than 60% identity to any of your sequences (if they exist), and then uses T-Coffee/3DCoffee to align your data-set using that structural information to build the consistency library, as well as T-Coffee's usual library methods, all in an automated fashion. Give it a try. It's very slick, and impressively accurate.

## 15.6 Conclusion

The comparative method is a cornerstone of the biological sciences, and key to understanding systems biology in so many ways. Multiple sequence alignment is the comparative method on a molecular scale, and is a vital prerequisite to some of the most powerful bio-computing analyzes available, such as structure/function prediction and phylogenetic inference. Understanding something about the algorithms and the program parameters of multiple sequence alignment is the only way to rationally know what is appropriate. Knowing and staying well within the limitations of any particular method will avert a lot of frustration. Furthermore, realize that program defaults may not always be appropriate. Think about what these default values imply and adjust them accordingly, especially if the results seem inappropriate after running through a first pass with the default parameters intact. Consistency based approaches, such as those implemented in T-Coffee, can help with these decisions.

Oftentimes you will need to deal with quite complicated data-sets — distantly related local domains, perhaps not even in syntenic order between sequences; or widely divergent paralogous systems resulting from large gene expansions; or extremely large sequence collections with mega bases of genomic data; often you'll even need to resort to manual alignment, at least in some regions — these are the situations that will present vexing alignment problems and difficult editing decisions. These are the times that you'll really have to think. A comprehensive multiple sequence editor such as GCG's SeqLab, or alternative freeware and public-domain editors, can be a lifesaver in these situations. As can the powerful evaluation and comparison modes built into the T-Coffee multiple sequence alignment package.

Gunnar von Heijne in his quite readable but dated treatise, *Sequence Analysis in Molecular Biology; Treasure Trove or Trivial Pursuit* [72], provides an appropriate conclusion:

> "Think about what you're doing; use your knowledge of the molecular system involved to guide both your interpretation of results and your direction of inquiry; use as much information as possible; and do not blindly accept everything the computer offers you . . . . Don't expect your computer to tell you the truth . . . if any lesson is to be drawn . . . it surely is that to be able to make a useful contribution one must first and foremost be a biologist, and only second a theoretician . . . . We have to develop better algorithms, we have to find ways to cope with the massive amounts of data, and above all we have to become better biologists. But that's all it takes."

## References

1. Waterman MS. Sequence alignments. In: Waterman MS, ed. Mathematical Methods for DNA Sequences, Boca Raton: CRC Press, 1989.
2. Gotoh O. An improved algorithm for matching biological sequences. J Mol Biol 1982;162:705–708.
3. Genetics Computer Group (GCG®) Program Manual for the Wisconsin Package®, version 11, San Diego: Accelrys, Inc., ©1982–2007.
4. Needleman SB, Wunsch CD. A general method applicable to the search for similarities in the amino acid sequence of two proteins. J Mol Biol 1970;48:443–453.
5. Smith TF, Waterman MS. Comparison of bio-sequences. Adv Appl Math 1981;2:482–489.
6. Pearson WR, Lipman DJ. Improved tools for biological sequence analysis. Proc Natl Acad Sci U S A 1988;85:2444–2448.
7. Pearson WR. Empirical statistical estimates for sequence similarity searches. J Mol Biol 1998;276:71–84. FastA package available at http://fasta.bioch.virginia.edu/fasta_www2/fasta_down.shtml
8. Altschul SF, Gish W, Miller W, Myers EW, Lipman D.J. Basic Local Alignment Tool. J Mol Biol 1990;215:403–410.
9. Altschul SF, Madden TL, Schaffer AA, Zhang J, Zhang Z, Miller W, Lipman DJ. Gapped BLAST and PSI-BLAST: a new generation of protein database search programs. Nucleic Acids Res 1997;25:3389–3402. Server at http://www.ncbi.nlm.nih.gov/BLAST/ and source code at ftp.ncbi.nih.gov/blast/
10. Gribskov M, McLachlan M, Eisenberg D. Profile analysis: detection of distantly related proteins. Proc Natl Acad Sci U S A 1987;84:4355–4358.
11. Eddy SR. Profile hidden Markov models. Bioinformatics 1998;14:755–763.
12. Karlin S, Altschul SF. Methods for assessing the statistical significance of molecular sequence features by using general scoring schemes. Proc Natl Acad Sci U S A 1990;87:2264–2268.
13. Schwartz RM, Dayhoff MO. Matrices for detecting distant relationships. In Dayhoff MO, ed. Atlas of Protein Sequences and Structure, vol. 5. Washington D.C: National Biomedical Research Foundation, 1979:353–358.
14. Henikoff S, Henikoff JG. Amino acid substitution matrices from protein blocks. Proc Natl Acad Sci U S A 1992;89:10915–10919.
15. Gonnet GH, Cohen MA, Benner SA. Exhaustive matching of the entire protein sequence database. Science 1992;256:1443–1145.
16. Gupta SK, Kececioglu JD, Schaffer AA. Improving the practical space and time efficiency of the shortest-paths approach to sum-of-pairs multiple sequence alignment. J Comput Biol 1995;2:459–472. MSA available at www.ncbi.nlm.nih.gov/CBBresearch/Schaffer/msa.html
17. Smith RF, Smith TF. Pattern-induced multi-sequence alignment (PIMA) algorithm employing secondary structure-dependent gap penalties for comparative protein modeling. Protein Eng 1992;5:35–41. Available at genamics.com/software/downloads/pima-1.40.tar.gz
18. Smith RF, Wiese BA, Wojzynski MK, Davison DB, Worley KC. BCM Search Launcher — an integrated interface to molecular biology data base search and analysis services available on the World Wide Web. Genome Research 1996;6:454–462. See the Baylor College of Medicine's Search Launcher at http://searchlauncher.bcm.tmc.edu/
19. Feng DF, Doolittle RF. Progressive sequence alignment as a prerequisite to correct phylogenetic trees. J Mol Evol 1987;25:351–360.
20. Thompson JD, Higgins DG, Gibson TJ. CLUSTALW: improving the sensitivity of progressive multiple sequence alignment through sequence weighting, positions-specific gap penalties and weight matrix choice. Nucleic Acids Res 1994;22:4673–4680. Available at http://www.ebi.ac.uk/clustalw/
21. Thompson JD, Gibson TJ, Plewniak F, Jeanmougin F, Higgins DG. The ClustalX windows interface: flexible strategies for multiple sequence alignment aided by quality analysis tools. Nucleic Acids Res 1997;24:4876–4882. Available at http://ftp-igbmc.u-strasbg.fr/pub/ClustalX/
22. Saitou N, Nei M. The neighbor-joining method: A new method of constructing phylogenetic trees. Mol Biol Evol 1987;4:1406–1425.

23. Notredame C, Higgins DG, Heringa J. T-Coffee: a novel method for multiple sequence alignments. J Mol Biol 2000;302:205–217.
24. Notredame C. T-Coffee: Tutorial and FAQ and Technical Documentation. 2006. Included with the distribution through http://www.tcoffee.org/Projects_home_page/t_coffee_home_page.html
25. Edgar RC. MUSCLE: multiple sequence alignment with high accuracy and high throughput. Nucleic Acids Res 2004;32:1792–1797. Available through http://www.drive5.com/muscle/
26. Katoh K, Kuma K, Toh H, Miyata T. MAFFT version 5: improvement in accuracy of multiple sequence alignment. Nucleic Acids Res 2005;33:511–518. Available at http://align.bmr.kyushu-u.ac.jp/mafft/software/
27. Do CB, Mahabhashyam MSP, Brudno M, Batzoglou S. ProbCons: Probabilistic Consistency-based multiple sequence alignment. Genome Res 2005;15:330–340. Available at http://probcons.stanford.edu/download.html
28. Pearson WR. Rapid and sensitive sequence comparison with FASTP and FASTA. Methods Enzymol 1990;183:63–98. MrTrans included with the FastA package at http://fasta.bioch.virginia.edu/fasta_www2/fasta_down.shtml
29. Rice P, Longden I, Bleasby A. EMBOSS: The European Molecular Biology Open Software Suite. Trends Genet 2000;16:276–277. Available at http://emboss.sourceforge.net/
30. Stajich JE, Block D, Boulez K, Brenner SE, Chervitz SA, Dagdigian C, Fuellen G, Gilbert JG, Korf I, Lapp H, Lehvaslaiho H, Matsalla C, Mungall CJ, Osborne BI, Pocock MR, Schattner P, Senger M, Stein LD, Stupka E, Wilkinson MD, Birney E. The Bioperl toolkit: Perl modules for the life sciences. Genome Res 2002;12:1611–1618. Available at http://www.bioperl.org/
31. Bininda-Emonds ORP. transAlign: using amino acids to facilitate the multiple alignment of protein-coding DNA sequences. BMC Bioinformatics 2005;6:156. Available through http://www.personal.uni-jena.de/~b6biol2/ProgramsMain.html
32. Wernersson R, Pedersen AG. RevTrans — Constructing alignments of coding DNA from aligned amino acid sequences. Nucleic Acids Res 2003;31:3537–3539. Available at http://www.cbs.dtu.dk/services/RevTrans/download.php
33. Letondal C, Schuerer K. Pasteur Institute, Paris, France, www.pasteur.fr/english.html, 2003. ProtAl2DNA available at http://ftp.pasteur.fr/pub/GenSoft/unix/alignment/protal2dna and in BioPerl
34. Suyama M, Torrents D, Bork P. PAL2NAL: robust conversion of protein sequence alignments into the corresponding codon alignments. Nucleic Acids Res 2006;34:609–612. See http://coot.embl.de/pal2nal/
35. Smith SW, Overbeek R, Woese CR, Gilbert W, Gillevet PM. The Genetic Data Environment, an expandable GUI for multiple sequence analysis. Comput Appl Biosci 1994;10:671–675. The original Sun OS version is at megasun.bch.umontreal.ca/pub/gde/. See Linux and Mac OS X GDE ports at http://www.bioafrica.net/GDE-linux/index.html and http://www.msu.edu/~lintone/macgde/
36. Clamp M, Cuff J, Searle SM, Barton G.J. The Jalview Java Alignment Editor, Bioinformatics 2004;20:426–427. Available at http://www.jalview.org/
37. Rambaut A. Se-Al: Sequence Alignment editor. 1996. Available at http://evolve.zoo.ox.ac.uk/software.html?id = seal
38. Galtier N, Gouy M, Gautier C. SeaView and Phylo_win, two graphic tools for sequence alignment and molecular phylogeny. Comput Appl Biosci 1996;12:543–548. Available through http://pbil.univ-lyon1.fr/software/
39. Bairoch A. PROSITE: A dictionary of sites and patterns in proteins. Nucleic Acids Res 1992;20:2013–2018
40. Bailey TL, Elkan C. Fitting a mixture model by expectation maximization to discover motifs in biopolymers. Proc Int Conf Intel Syst Mol Biol 1994;2:28–36
41. Bailey TL, Gribskov M. Combining evidence using p-values: Application to sequence homology searches. Bioinformatics 1998;14:48–44
42. Cole JR, Chai B, Farris RJ, Wang Q, Kulam-Syed-Mohideen AS, McGarrell DM, Bandela AM, Cardenas E, Garrity GM, Tiedje JM. The ribosomal database project (RDP-II): Introducing myRDP space and quality controlled public data. Nucleic Acids Res 2007;35:D169–D172. See http://rdp.cme.msu.edu/
43. Wuyts J, Perriere G, Van de Peer Y. The European ribosomal RNA database. Nucleic Acids Res 2004;32:D101–D103. See http://www.psb.ugent.be/rRNA/
44. National Center for Biotechnology Information (NCBI) Entrez, public domain software distributed by the authors. National Library of Medicine, National Institutes of Health, Bethesda. See http://www.ncbi.nlm.nih.gov/Entrez/
45. Etzold T, Argos P. SRS — an indexing and retrieval tool for flat file data libraries. Comput Appl Biosci 1993;9:49–57
46. Olsen G. Inference of Molecular Phylogenies, University of Illinois at Urbana-Champaign; lecture, September 3, 1992.
47. Higgins DG, Bleasby AJ, Fuchs R. CLUSTALV: improved software for multiple sequence alignment. Comput Appl Biosci 1992;8:189–191.
48. Swofford DL. PAUP* (Phylogenetic Analysis Using Parsimony, and other methods) version 4.0+. ©1989–2007. Home page at paup.scs.fsu.edu/, distributed through Sunderland: Sinaeur Associates, Inc. at http://www.sinauer.com/

49. Ronquist F, Huelsenbeck JP. MRBAYES 3: Bayesian phylogenetic inference under mixed models. Bioinformatics 2003;19:1572–1574. See http://mrbayes.scs.fsu.edu/.
50. Felsenstein J. PHYLIP (Phylogeny Inference Package). Distributed by the author. Department of Genome Sciences, University of Washington, Seattle. 1980–2007. Available at http://evolution.genetics.washington.edu/phylip.html
51. Gilbert DG. ReadSeq. Distributed by the author. Biology Department, Indiana University, Bloomington, 1990–2006. See http://iubio.bio.indiana.edu/soft/molbio/readseq/
52. Rozen S, Skaletsky H. Primer3 on the WWW for general users and for biologist programmers. In: Krawetz S, Misener S, eds. Bioinformatics Methods and Protocols: Methods in Molecular Biology, Totowa: Humana Press, 2000:365–386. Available at http://primer3.sourceforge.net/
53. Hofmann K, Baron M. 1999. BOXSHADE server at www.ch.embnet.org/software/BOX_form.html and software available at http://www.isrec.isb-sib.ch/pub/boxshade
54. Schneider TD, Stephens RM. Sequence logos: A new way to display consensus sequences. Nucleic Acids Res 1990;18:6097–6100. See http://www-lmmb.ncifcrf.gov/~toms/sequencelogo.html
55. Guex N, Diemand A, Peitsch MC. Protein modelling for all. Trends Biochem Sci 1999;24:364–367. See http://swissmodel.expasy.org//SWISS-MODEL.html
56. Sayle RA, Milner-White EJ. RasMol: Biomolecular graphics for all. Trends Biochem Sci 1995;20:374–376. See http://www.umass.edu/microbio/rasmol/ and openrasmol.org/
57. Lunter G, Miklos I, Drummond A, Jensen JL, Hein J. Bayesian coestimation of phylogeny and sequence alignment. BMC Bioinformatics 2005;6:83.
58. Van de Peer Y, Frickey T, Taylor JS, Meyer A. Dealing with saturation at the amino acid level: A case study based on anciently duplicated zebrafish genes. Gene 2002;295:205–211. ASaturA available at http://bioinformatics.psb.ugent.be/software_details.php?id = 6
59. Taylor WR. Protein structure comparison using iterated double dynamic programming. Protein Sci 1999;8:654–665. SAP available for non-profit academic work at http://mathbio.nimr.mrc.ac.uk/ftp/wtaylor/sap/
60. Berman, HM, Henrick K, Nakamura H. Announcing the worldwide Protein Data Bank. Nat Struct Biol 2003;10:980. See http://www.rcsb.org/pdb/
61. Wallace IM, O'Sullivan O, Higgins DG, Notredame C. M-Coffee: Combining multiple sequence alignment methods with T-Coffee. Nucleic Acids Res 2006;34:1692–1699. Included in T-Coffee distribution.
62. Lee C, Grasso C, Sharlow M. Multiple sequence alignment using partial order graphs. Bioinformatics 2002;18:452–464. POA available at http://bioinfo.mbi.ucla.edu/poa/
63. Subramanian AR, Weyer-Menkhoff J, Kaufmann M, Morgenstern B. DIALIGN-T: An improved algorithm for segment-based multiple sequence alignment. BMC Bioinformatics 2005;6:66. Available at http://dialign-t.gobics.de/
64. Pei J, Sadreyev R, Grishin NV. PCMA: fast and accurate multiple sequence alignment based on profile consistency. Bioinformatics 2003;19:427–428. Available at http://iole.swmed.edu/pub/PCMA/
65. Edgar RC, Batzoglou S. Multiple sequence alignment. Curr Opin Struct Biol 2006;16:368–373.
66. Notredame C. Mocca: semi-automatic method for domain hunting. Bioinformatics 2001;17:373–374. Included in T-Coffee distribution.
67. Shi J, Blundell TL, Mizuguchi K. FUGUE: Sequence-structure homology recognition using environment-specific substitution tables and structure-dependent gap penalties. Journal of Mol Biol 2001;310:243–257. See http://www-cryst.bioc.cam.ac.uk/fugue/
68. O'Sullivan O, Suhre K, Abergel C, Higgins DG, Notredame C. 3DCoffee: Combining protein sequences and structures within multiple sequence alignment. J Mol Biol 2004;340:385–395. Included in T-Coffee distribution.
69. Kjeldgaard M, Nissen P, Thirup S, Nyborg J. The crystal structure of elongation factor EF-Tu from Thermus aquaticus in the GTP conformation. Structure 1993;1:35–50.
70. Armougom F, Moretti S, Keduas V, Notredame C. The iRMSD: a local measure of sequence alignment accuracy using structural information. Bioinformatics 2006;22:35–39. Included in T-Coffee distribution.
71. Armougom F, Moretti S, Poirot O, Audic S, Dumas P, Schaeli B, Keduas V, Notredame C. Expresso: automatic incorporation of structural information in multiple sequence alignment using 3D-Coffee. Nucleic Acids Res 2006;34:604–608. See http://www.tcoffee.org/
72. von Heijne G. Sequence Analysis in Molecular Biology; Treasure Trove or Trivial Pursuit. San Diego: Academic Press, 1987.


## Web Resources

http://ftp-igbmc.u-strasbg.fr/pub/ClustalX/
http://bio.fus.edu/nstevet/PrimerDesign.pdf
http://workshop.molecularevolution.org/

# Chapter 16
# A Spectrum of Phylogenetic-Based Approaches for Predicting Protein Functional Sites

**Dukka Bahadur K. C. and Dennis R. Livesay**

**Abstract** This chapter introduces four distinct, yet aligned, computational approaches to predict protein functional sites. The methods range from simple alignment conservation scores to more sophisticated phylogenetic-based approaches (i.e., evolutionary trace, ConSurf, and phylogenetic motifs). Each approach and its theoretical underpinnings are discussed in detail. Key differences within later revisions of each method (when appropriate) are also provided. Applications of the method and how it has been assessed are also discussed for several methods. Finally, information on availability of each method is provided.

**Keywords** Multiple sequence alignment · Phylogeny · Shannon entropy · Sum of pairs · Evolutionary trace · Consurf · Phylogenetic motif

## 16.1 Introduction

High-throughput genome sequencing projects have resulted in an explosion of known protein sequences. Assigning functions to these proteins is a key bioinformatics challenge. There are various bioinformatics approaches for the analysis of protein function [1]. However, while global descriptions of function are clearly important, only interrogation at specific functional residues can provide mechanistic details. Traditionally, these interrogations have been carried out through experimental site-directed mutagenesis. The collective knowledgebase gained from these efforts has dramatically advanced our understanding of protein function. Unfortunately, the time and expense associated with site-directed mutagenesis limits its wide scale utility in the post-genomic era. In response, there are currently dozens of computational protein functional-site prediction techniques described in the literature. The theoretical underpinnings of these methods are highly variable, ranging from methods that base predictions on conservation across a Multiple- Sequence Alignment (MSA) to sophisticated biophysical calculations.

In this chapter, we introduce the most commonly applied protein functional-site predictions methods, namely those based on the concepts of phylogeny. In fact, we discuss four broad classes of phylogenetic-based methods; each varies significantly in its approach and predictions. Nevertheless, all of the methods are based on the straightforward premise that conservation of function is the ultimate evolutionary driving force. The first set of methods are based on the simple premise that alignment positions that are more conserved are more likely to be functional than those that are not conserved. The second set of methods, called *evolutionary trace*, goes one step further and includes alignment positions that are not strictly conserved, but whose variations are consistent with known

D.R. Livesay
Department of Computer Science and Bioinformatics Research Center, University of North Carolina at Charlotte, 9201 University City Blvd., Charlotte, NC, 28223, USA
e-mail: drlivesa@uncc.edu

S. Krawetz (ed.), *Bioinformatics for Systems Biology*, DOI 10.1007/978-1-59745-440-7_16,

functional variability. Meaning, these positions are likely responsible for the subfamily functional differences. The third set of methods, which are based on the *ConSurf* algorithm, is ostensibly a hybrid of the first two; however, in actuality it is closer to the first class of methods than the second. ConSurf uses sophisticated Bayesian (or maximum likelihood) statistics and phylogenetic tree topology to improve the accuracy of its predictions over simple conservation scores. Finally, the fourth class of methods discussed is based on *phylogenetic motifs*. Phylogenetic motifs are alignment fragments that reproduce the overall familial phylogeny, which have been shown to represent good functional site predictions.

It should be pointed out that all of the methods discussed here are fundamentally sequence-based: none explicitly requires structural information. Yet, common usage of most of the methods discussed (especially evolutionary trace and ConSurf), use structures to improve their predictions. By structurally clustering their predictions, better accuracy is achieved when predicting protein functional sites. This approach is well justified due to the amount of functional details encoded within the structure. However, the expansion of sequence-space is increasing much faster than that of structure-space. As such, accurate methods that do not rely on structure are also necessary. For example, the phylogenetic motif method has been explicitly designed to fill this need. Usage of evolutionary trace and ConSurf in the absence of structures is also discussed.

## 16.2 A (Very) Brief Overview of Phylogenetic Reconstruction

Before delving into the various methods of predicting protein functional sites, a brief introduction of phylogeny is in order. A *phylogeny* describes an inferred evolutionary history across a sample set based on their relatedness. For example, the field of *systematics* attempts to infer these relationships at the species level. In molecular biology, phylogenetic trees are generally constructed for sets of evolutionarily related proteins and/or genes. The tree describes the evolutionary history of a set of related proteins based on their sequence similarity. Each protein sequence (sometimes called a *leaf*) is clustered based upon its similarity/dissimilarity to the others within the data-set. In an alignment of $N$ sequences, there will be $N$ leaves within the phylogenetic tree. In addition, phylogenetic reconstruction attempts to describe the common ancestor at each *node* (branch point) within a tree. For example, the following two sequences (AICGS and AVCGT) could have evolved from AVCGS. In each case, a single mutation to the common ancestor leads to the observed pair. The earliest common ancestor is called the *root*.

There are a wide number of computational algorithms to infer evolutionary relationships (called *phylogenetic reconstruction*) from molecular sequence data. The simplest of these methods are the *distance-based* approaches. Distant-based methods use overall sequence similarity to approximate phylogenetic trees. While computationally fast, these methods are generally the least accurate approaches. More advanced, and more computationally costly methods include parsimony, maximum-likelihood, and Bayesian inference. A complete discussion of each is outside the scope of the present chapter. Interested readers are encouraged to refer to the excellent text by Felsenstein [2] for a deeper discussion of phylogenetic reconstruction.

## 16.3 Conservation Methods

Using conservation within an alignment to predict evolutionarily important positions is a fundamental tenet of biology and bioinformatics. Strong evolutionary pressures must be present to counter-balance random divergence. The standard interpretation of these pressures is that they are present to conserve the structure and/or function of the protein. For example, the catalytic triad of serine proteases is exactly conserved across an alignment of the family. There is a wide variety of conservation scores described within the literature; here we describe a few of the most common. Note that, strictly speaking, conservation scores are not based on "phylogenetics" since the evolutionary relationships between each protein sequence are ignored. However, since conservation is a similar concept, they are frequently discussed together.

### 16.3.1 Methods that Ignore Residue–Residue Relationships

The simplest way to measure conservation is to simply evaluate the *percent conservation* of each alignment position. The most abundant residue occurring in each position is identified and counted. The final conservation score is simply the count divided by the number of sequence present. While commonly used, this method suffers from two key theoretical shortcomings. The first is that the method is sufficient for very conserved positions, but what about the pathological problem of a column that is 51% Ala and 49% Val? Its conservation score is exactly the same as a column that was 51% Ala, 19% Val, 10% Ile, 10% Phe, and 10% Met. Clearly, the first column is more "conserved" than the second, even though the conservation score for each is 51%. Similarly, the second limitation is based on the assumption that each amino acid identity is an independent state, meaning that chemical and/or statistical relationships between the various amino acids are ignored. For example, in these methods, a column that is 70% Lys and 30% Arg is treated exactly the same as a column with 70% Val and 30% Cys. However, the amount of *chemical conservation* is clearly different since Lys and Arg are chemically similar (they are both basic amino acids), whereas Val and Cys are vastly different. Nevertheless, in spite of these two gratuitous limitations, these methods can be quite powerful. Additionally, their popularity is further supported by their ease of use and implementation.

The *Shannon entropy* resolves the first incongruity of percent conservation. In bioinformatics, the Shannon entropy is a useful extension of Shannon's early work in information theory [3]. The Shannon entropy of column $i$ (designated $H_i$) is calculated by:

$$H_i = -\sum_{i=1}^{20} P_i \log_2 P_i \tag{16.1}$$

where the count is over all the twenty amino acids and $P_i$ is the probability of each residue occurring within the alignment column. The values of such an expression range from zero (strictly conserved) to 4.32 (which is the value if each of the twenty amino acids have uniform probability). Going back to the pathological problem, the Shannon entropy of the first column is 1.0, whereas the entropy of the second (less conserved) is 1.9. Note that the Shannon entropy is unit less.

### 16.3.2 Sum of Pairs

In bioinformatics, a scoring matrix describes the likelihood of one amino acid being changed (mutated) to another. As such, a method to resolve the second incongruity of the percent conservation score is available. The most common scoring matrices are those based on the work of Margaret Dayhoff (called PAM matrices) [4]. While a complete discussion of scoring matrices is outside the scope of this chapter, they reflect chemical similarity of the amino acids through observed substitutions. For example, the substitution score of the residue pair $a$,$b$ (designated $s[a,b]$) is larger when the substitution is more likely to occur, which can be assumed to indicate greater chemical similarity. The self-substitution scores, $s(a, a)$, are generally the largest values within the matrix. The sum of pairs (SP) score uses scoring matrix substitution values to determine the amount of conservation within an alignment column. The SP score of column $m_i$ (designated $S[m_i]$) is calculated as:

$$S(m_i) = \sum_{k<l} s(m_i^k, m_i^l) \qquad (16.2)$$

where the sum is overall all possible pairs within a single alignment column. Since larger $s(a, b)$ values indicate greater similarity, larger SP scores indicate greater conservation. Meaning that we now have a means to resolve the conundrum in the first paragraph of Section 16.3.1(namely, which is more "conserved" — a position with 7 Lys and 3 Arg or a position with 7 Val and 3 Cys?). Using the PAM250 scoring matrix, the SP score of the first example is 138.0 (indicating strong conservation), whereas the SP score of the second is only 43.5. While not present in this simple example, alignment positions with gaps are not defined in $s(a, b)$. The standard application of the method is to simply evaluate gap characters the same way as the twenty amino acids (meaning, gaps are considered the 21st residue type). As such $s(a,-)$ and $s(-,a)$ are simply equal to the gap penalties.

### 16.3.3 Availability

It is very simple to implement any of the above methods into computer code. Doing so would be a good exercise and may be the best way of computing their values when dealing with large data-sets that require specialized scripts and/or programs. Nevertheless, when one needs only to compute the values on a small number of data-sets, re-implementation of the methods is impractical. Currently, there are a large number of web-servers and freely available code to compute the various conservation scores. One of the best is the *Scorecons* server [5] from the Janet Thornton's group at the European Bioinformatics Institute. Scorecons includes online implementations of a wide variety of conservation scores and allows the user to define a variety of input parameters (i.e., scoring matrix, gap parameters, etc.). Scorecons is on the web at http://www.ebi.ac.uk/thornton-srv/databases/cgi-bin/valdar/scorecons_server.pl

## 16.4 Evolutionary Trace

Computational prediction of functionally important residues in proteins from a MSA is a long-standing paradigm. The method relies on the two basic tenets of protein family evolution, namely functionally important residues have fewer mutations than less critical positions, and protein structures descending from a common ancestor maintain a conserved structure. As such, strictly conserved positions that group together in structure-space make very good functional site predictions. However, it is also well known that alignment positions that are overall variable, but whose variability is limited to boundaries between subfamilies can be functionally important as well. It is assumed that these subfamily specific residues mediate functional differences (i.e., substrate specificity, differences in catalytic efficiency, etc.) between the various groups. *Evolutionary Trace* (ET), which is one of the first phylogenetically based method of prediction of protein functional sites [6], uses this concept to improve functional site predictions. The method is widely used (as of June 2007, the original ET paper has been cited over 380 times) and its predictions have been experimentally verified [7]. Oliver Lichtarge, who was then in Fred Cohen's laboratory at the University of California at San Francisco, originally developed the ET approach. Since then, Lichtarge has started his own laboratory at the Baylor College of Medicine. Most of the ET methodologies discussed here were developed by Lichtarge or by the members of his laboratory.

As the name suggests, ET attempts to trace the evolutionary history of the protein family in order to identify functionally important residues. ET uses information from phylogenetic analyses, sequence conservation and 3D structure. The fundamental steps in Evolutionary Trace are:

1. Generation of MSA from homologues of the query sequence. It is important that at least one member of the homologues has a solved 3D structure.
2. Construction of a phylogenetic tree from the MSA and division of sequences into subfamilies. When functional interrelationships are known, the partitions are chosen based on expert curation. Otherwise, the partitions are chosen with reference to a phylogenetic tree. Various partition identity cutoffs (PICs) are chosen to define partitions. Grouping together sequences that branch off from a node to the right of a PIC line generates a partition of the entire sequence. At low PIC, partitions consists of a few large clusters and at high PIC, partitions consists of many smaller clusters.
3. ET is constructed for each partition, which is done in two steps: (i) A consensus sequence for each group is calculated where positions varying within the group are left blank and are termed as 'neutral sites' and the invariant position in each group is assigned the conserved residue type and is termed 'class-specific'. (ii) All consensus sequences are aligned to obtain the evolutionary trace for the entire partition. In the alignment, any position is called *neutral* if it is variable within any of the consensus sequences. Otherwise, the position is termed *conserved* if all the consensus sequences are invariant, or *class-specific* if the residue type varies between consensus sequences. Class-specific residues are commonly called '*Trace Residues*' (TRs). However, since the conserved (invariant) residues are class-specific at rank $i = 1$, TRs generally refer to the union of class-specific and conserved residues.
4. TRs are mapped onto the known 3D-structure. Residues that form clusters over the 3D-structure are considered to have some important information, and residues that are dispersed on the 3D-structure are considered as noise. As such, structural information is used to filter out ET results that are less likely to be functional. Nevertheless, the underlying TR identification is based solely on sequence-based information (namely, an alignment and a phylogeny).

One of the most important steps in the above procedure is step [2], which defines the evolutionary partitions. For example, let us consider the phylogenetic tree in Fig. 16.1A. If it is known *a priori* that the sequences in the tree can be functionally grouped, then the known functional discriminations defines the considered phylogenetic partitions. For example, in the above example, suppose sequences in Group A have function A', Group B has function B' and Group C has function C', meaning the tree is partitioned into three sub-families as shown. The consensus sequences for the three groups are (shown in Fig. 16.1B) –L–A–D–, ML–A–Q– and –L–A–Q–, respectively. The corresponding ET sequence is –L–A–X–, where L and A are conserved residues and X is class-specific, meaning that these three alignment positions constitute the TRs that are projected onto the structure. However, a key challenge is to define the partitions when functional annotation is lacking. While it is ideal to use known functional relationships to define the number of phylogenetic partitions, unfortunately, complete knowledge of the functional relationships is rare.

In such scenarios, PIC is introduced to identify the phylogenetic partitions from tree topology. The vertical lines X, Y, Z in Fig. 16.1A, represent different PICs that define a series of phylogenetic partitions to be considered. For each PIC, a partition of sequences is generated by grouping together sequences that branch-off from a node to the right of the PIC line, which corresponds to grouping sequences together such that the lowest level of sequence identity inside the group is greater than the value of PIC for that line. At low PIC partitions, partitions consist of a few large clusters, whereas partitions obtained at higher PICs achieve higher functional resolution as large clusters are fragmented into smaller ones. The ET sequence corresponding to the three shown PICs are provided in Fig. 16.1C. The partition corresponding to line X corresponds to a PIC value of $x\%$,

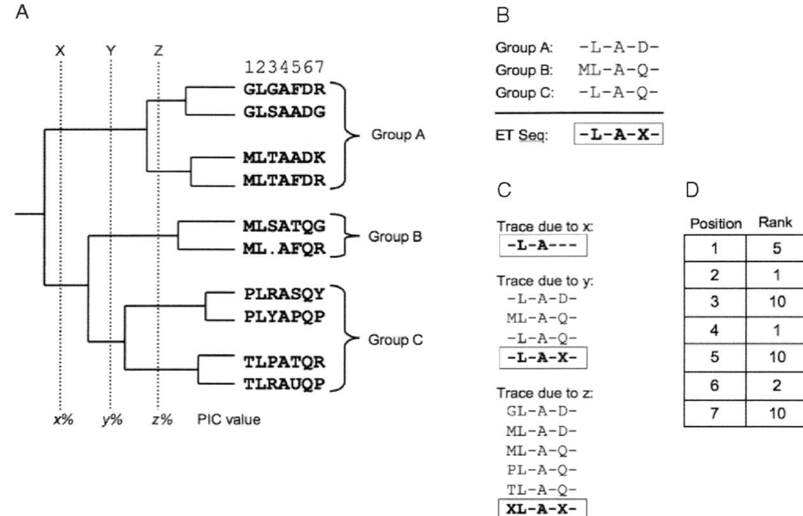


**Fig. 16.1** A schematic diagram of Evolutionary Trace (ET). **(A)** A hypothetical phylogenetic tree and its underlying multiple-sequence alignment. Each sequence is associated to a group based on outside functional annotation. Dashed vertical lines represent different partition identity cutoffs (PIC). **(B)** For each functional group, a class consensus sequence is created and compared to others from which, the ET sequence is generated. Note that the ET sequence is within the box, and that class-specific residues are signified by X; conserved (invariant) residues are simply labeled with their one-letter amino acid code. Class-specific and conserved residues constitute the group of 'trace residues'. When outside functional annotation is unavailable, PIC can be used. The class consensus and ET sequences at each PIC are shown in **(C)**. The resultant ET rankings for the example are provided in **(D)** (Copies of figures including color copies, where applicable, are available in the accompanying CD)

line Y to a PIC value of $y\%$ and line Z to a PIC value of $z\%$, and $x < y < z$. The trace produced by line X (PIC value of $x\%$), partitions the sequences into one group and the evolutionary trace sequence corresponding to this partition is –L–A——. Conversely, the trace produced by line Z (PIC value of $z\%$), partitions the sequences into five groups whose class consensus sequences are: GL–A–D–, ML–A–D–, ML–A–Q–, PL–A–Q–, TL–A–Q–, meaning the ET sequence is XL–A–X–.

A key point of this example in Fig. 16.1 is that the set of TRs varies for the various different traces. Moreover, as PIC increases, the number of TRs must also increase. Initial implementations of the ET method relied on the visual inspection of mapped residues on the structure for the determination of PIC [6]. In particular, PIC was chosen such that the number of TRs scattered or dispersed while mapped on the structure was the minimum.

The above process can be repeated such that each residue position is assigned an evolutionary rank. The evolutionary rank of a residue is the minimum number of branches into which the tree must be divided for it to become class-specific. For example, the first position of the above example does not become a TR till the PIC value of $z\%$ (corresponding to vertical line Z), which partitions the sequences into five groups — consequently, its rank is 5. The third position of the alignment does not become a TR till some PIC value which partitions the tree into ten groups and hence its rank is 10. However, the 2nd and 4th positions of the alignment become trace residues at a PIC value of $x\%$, which partitions the sequences into one group and hence for each of these position, meaning their rank is 1. Top ranks (1, 2, 3, ...) are assumed to be intimately related to function, whereas lower ranks are assumed to be of limited importance. In the above example (Fig. 16.1d), the corresponding ET rankings for each alignment position are provided.

After initial development of the ET algorithm in 1996 [6], it has been tested in SH2 and SH3 domains, type II zinc-fingers from nuclear-hormone receptors, heterotrimeric G proteins, RGS proteins, and G protein-coupled receptors. Moreover, ET has been also successfully applied in PBTI, heregulin, TGF-beta and related growth factors, and in PHD zinc-fingers. Although, ET was tested on a large variety of proteins for functional site prediction, two key issues that limited the applicability of the method on a large scale were ignored until Madabushi's 2002 paper [8].

### *16.4.1 Dealing with Gaps and Automation*

Two key limitations of the original ET implementation were resolved in the 2002 paper by Madabushi et al. [8]. The first is that the basic algorithm completely ignores gapped alignment positions. Hence, the selection of sequences is drastically limited due to the number of gaps within the MSA from even moderately distant homologues. On this basis, the method assumes that any position with a gap cannot be functional. For example, in the example within Fig. 16.1a, position 3 is treated as a neutral position and hence can never be a TR. In newer implementations of the method, gaps are treated as 21st amino acid, meaning the above restriction has been removed. For a simple example, let us consider the following three class consensus sequences: AE–TFT, VERT–T and ADR.YT, meaning the ET sequence is XX—–T. However, when allowing for gaps, the fourth position changes from neutral (no consensus) to a TR (XX–X–T). Gaps often occur in blocks and these blocks indicate some functional importance at the location of these gaps. Hence, in practice the ability to rank gapped positions eliminates "holes" from ET analyses. The comparison of the original, gap-intolerant method, to the new gap-tolerant method was performed in order to evaluate the approach, and it was observed that the newer method achieved more significant clusters.

The second limitation is that the original implementation of the method required visual interpretation of the results to identify clusters of top-ranked residues in structure space. A few large clusters would be considered as a true signal, whereas small-scattered clusters would be considered as noise. However, near the signal-to-noise threshold this approach becomes too subjective. For the automation of the PIC, objective statistics to assess the significance were introduced [8]. Moreover, it was shown that the top-ranked residues cluster spatially within the protein structure. The statistical significance measures are based on the size of the largest cluster and the number of identified clusters. Clusters of random residues will be scattered throughout the structure and thus will tend to give rise to many small clusters. By comparing the number of clusters and the size of the largest cluster to the random expectation, it was demonstrated that these quantities could be used as an objective assessment of the cluster with the greatest statistical significance.

### *16.4.2 Weighted ET*

Originally, ET [6] only considered those residues that are strictly conserved within the cluster for subsequent trace analysis and neglected positions that are not conserved within the cluster. However, when dealing with a large numbers of sequences it becomes important to be able to consider those residue positions where there are limited numbers of mutations. In such cases, it is better to employ some sort of quantitative measure of residue variation at each position. In this regard, Landgraf et al. [9] introduced a quantitative measure of residue variation at each position within the ET protocol, which they termed 'weighted evolutionary trace' (WET). The logic behind the WET method is that if one is interested in the specificity of a set of sequences within a fraction of the alignment (subfamily), a given position could be conserved within the fraction, but otherwise variable. Meaning, one can infer that the particular position contributes specifically to subfamily functional specificity.

In order to quantify the above considerations, a measure of variability was calculated in each position based on an amino acid substitution matrix. Each amino acid sequence is weighted according to its uniqueness and the variability in each position. Conserved residues in one subfamily that are variable in another are considered possible specificity-conferring residues. The variability measure $V_p(M)$ in each position $p$ of the MSA $M$ is calculated as:

$$V_p(M) = \sum_{i=1}^{n} \frac{V_{i,p}}{n} \tag{16.3}$$

where $n$ is the number of sequences in the multiple alignment and $V_{i,p}$ is the variability measure at position $p$ using sequence $i$ as a reference. The value of $V_p(M)$ ranges from 0 to 1 where zero means no variability. The individual variability score $V_{i,p}$ is calculated as:

$$V_{i,p} = \sum_{j \neq i, j=1}^{n} \frac{S_{M(i,p),M(i,p)} - S_{M(i,p),M(j,p)}}{S_{M(i,p)M(i,p)}} \times W_i \tag{16.4}$$

where $S_{M(i,p),M(i,p)}$ is the value of substituting the amino acid at position $p$ in sequence $i$ with the amino acid in the same position, but of sequence $j$, as given by Gonnet matrix [9]. Gaps are also allowed and given maximum substitution penalty. Weights are assigned to each sequence according to its level of similarity in order to avoid over-representation of identical or nearly identical sequences [9]. The sum of all weights equals 1.0. Finally, a cluster formed by $m$ sequences that has variability $V_p(m)$, the variability ratio at position $p$, is given by:

$$C_m(p) = \frac{V_M(p) + 1}{V_m(p) + 1} - 1 \tag{16.5}$$

$C_m(p)$ is designed to assign high scores to positions that display variability across the entire alignment position, yet are conserved within the cluster of $m$ sequences.

The method was tested on a set of sequences that represent the heregulin family of proteins, which is a subset of the family of EGF-like growth and differentiation factors. The application of weighted evolutionary trace method identified two distinct clusters of residues, which are thought to represent binding sites that are specific to the heregulin family and "...reflect differences between hrg (heregulin) ligands and the EGF-like ligands as a whole. Besides the preference for a different subset of receptors (HER2, 3 and 4 versus EGFR), hrg also shows a strong preference for the interaction with receptor heterodimers versus the homodimeric interaction seen between EGF and EGFR" [9]. Using this idea, heregulins were conferred a distinct specificity within the family of EGF-like growth factors.

As suggested above, the WET method is likely to be of use in the search for clusters that identify functional sites, specific to a particular subfamily and contributing to the specificity of the functional site. However, even though the WET method proved useful in this particular study, it should be noted that it is not general enough to be applicable to any type of sequence data and thus cannot be applied on a large scale. As noted by the authors, the method is useful mostly when one is interested in a subfamily with a high level of sequence similarity between the sequences used in the alignment.

### 16.4.3 Modified ET with a Focus on Invariant Polar Residues

As stated above, the original algorithm used visual inspection to identify ET clusters. Aloy et al. developed an automated approach to identify the cluster on the 3D-structure focusing on invariant polar residues [10]. The approach improved the basic ET in the sense that no manual intervention was required to determine the ideal cluster. Due to their functional importance, this method focused on the analysis of invariant polar residues (D, E, H, K, N, Q, R, S, T). In essence, the slightly modified version of the algorithm for determination of clusters for invariant residues consists of the following steps:

1. An alignment obtained by using the master protein sequence (of known structure) as a query sequence is generated. A phylogenetic tree is constructed from the MSA. Subsequently, the query sequence is traced along the tree and every time that a new sequence or a group of sequences, is added to its own cluster, a consensus sequence is constructed. At each alignment position, residues that are totally conserved are taken as invariant and the number of invariant residues in the MSA divided by the length of the query expressed as a percentage is called 'attained identity' (AI).
2. Starting at a low AI (say 10%), the *invariant polar residues* corresponding to the phylogenetic tree are then mapped to the structure. Spatial clustering is applied to identify those residues that are proximal in space. Residues are clustered by a single linkage, meaning a residue is added to the cluster if it is less than 8 Å from at least one residue already within the cluster. If a spatial cluster is formed, go to step 4; otherwise, go to step 3.
3. The tree is examined and the alignment is modified, and sequences that are far in the phylogenetic tree are removed in order to achieve a higher level of attained identity and then spatial clustering is performed as described in step 2. If a cluster is formed, go to step 4; otherwise, repeat step 3.
4. A sphere containing all the clustered residues is built using the geometric centroid of the cluster as the center and the largest distance from the center to any C-beta atom of the clustered residues as the radius. The sphere and all the residues inside form the predicted functional site.

A key result of the Aloy et al. paper was their unbiased assessment of the method [10]. By comparing their predicted functional site sphere to a sphere defined by known functional sites, they were able to quantify how well the method performed. If the overlap of the two spheres was greater than 50% of the volume of the predicted functional site, the predicted site was considered *correct*. If the spheres overlapped to a lower degree, the method was deemed *useful*. Otherwise, the prediction was wrong. Impressively, 79% of the predictions were correct, a further 15% were useful and only 6% were *wrong*.

## 16.4.4 Ranking of Residue Importance using ET

Fundamentally, ET combines information obtained from phylogenetics and structure to predict the functional sites in a protein. However, there are cases when the structural information is not available. Considering such scenarios, various versions of ET has been developed such that it can be useful for the prediction of functional sites of a protein. Essentially, these types of methods rank the residues of a protein in the order of their functional importance. Here, we describe two types of ranking scores developed using ET analysis.

### 16.4.4.1 Integer-Valued ET

As discussed above, it is useful to rank each position within the input alignment on the basis of its relative importance. In Fig. 16.1, the original ET algorithm is repeated for each PIC, meaning, the positions can be ranked based on their evolutionary importance. However, the definition of rank in the original ET paper was based on tree topology and lacked mathematical formulation. In this regard, Mihalek et al. [11] developed the integer-valued ET (ivET) procedure. The mathematical formulation precisely defines the tree partitions using the following iterative process:

1. Identify the nearest node, which corresponds to the most recent common ancestor on the tree. This node is labeled $n$ = N – 1, where N is equal to the number of leaves in the tree. The node defines a partition (just like PIC did above), meaning that leaves to the right of the node are grouped.
2. The next nearest node is identified, which is labeled $n$ = N – 2. This process is iterated throughout the tree, with every node label decreasing by 1. This process ensures that the label of the root is 1 (see Fig. 16.2A).

Additionally, at each step in the iteration corresponding to the assignment of node label $n$, the leaf set is divided into $n$ groups, which corresponds to a partition occurring there. For example, Fig. 16.2a highlights the two groups corresponding to node $n$ = 2. Subsequently, the ivET rank for each alignment position $i$ is defined as:

$$r_i = 1 + \sum_{n=1}^{N-1} \begin{cases} 0 \text{ if position } i \text{ is conserved within each group } g \\ 1 \text{ otherwise} \end{cases} \tag{16.6}$$

where $g$ simply represents each of the groups defined by node $n$. In other words, the summation stops at a division into groups such that the position $i$ is conserved within the group. The 1 added before the sum is simply provided to ensure that the ivET and original ET results are consistent.

It is to be noted here that equation [6] simply constitutes a mathematical description of the ranking of the original ET algorithm. In fact, the rankings of each position calculated using the original ET and ivET are exactly the same. Nevertheless, the importance of ivET is that it now places the ET rankings into a rigorous mathematical equation and, perhaps more importantly, lays the foundation for a hybrid ET/entropy method (see next section).

In Fig. 16.2B, the ivET rank is shown for each residue calculated using equation 16.6. For example, position 1 is not conserved within groups corresponding to nodes 1, 2, 3 and 4 and is

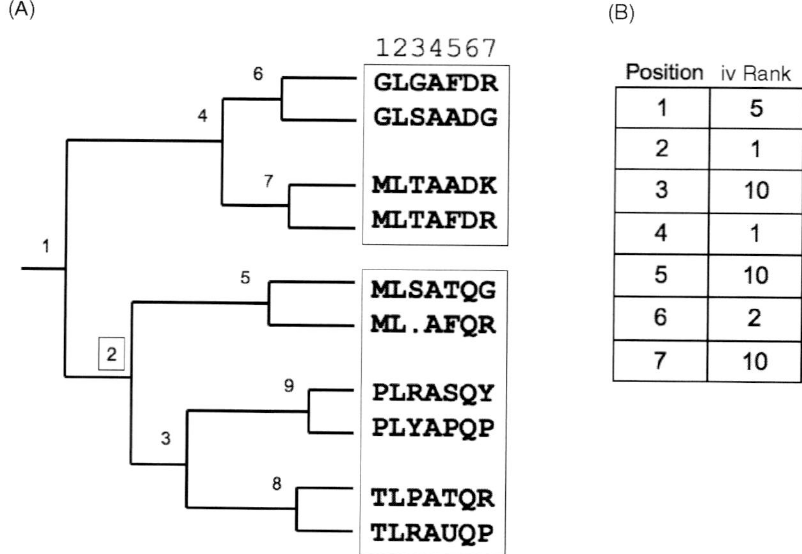


| Position | iv Rank |
|---|---|
| 1 | 5 |
| 2 | 1 |
| 3 | 10 |
| 4 | 1 |
| 5 | 10 |
| 6 | 2 |
| 7 | 10 |

**Fig. 16.2** Calculation of integer-valued ET (ivET) rankings. **(A)** The method labels each node $n = 1$ (which corresponds to the root) to $N$–1 (which corresponds to the nearest node); $N$ is equal to the number of leaves within the tree. The node labeled $n$, divides the tree into $n$ groups. For example, at node $n$ = 2, the tree is split into two groups as shown. Subsequently, the ivET rank is evaluated for each position using equation 16.6. Integer-valued ET ranks are provided in **(B)**, which are exactly the same as the ranking using original ET as in Fig. 16.1 (Copies of figures including color copies, where applicable, are available in the accompanying CD)

conserved within groups corresponding to nodes 5, 6, 7, 8 and 9. Hence, its ivET rank is $1+(1+1+1+1+0+0+0+0+0) = 5$. Similarly, position 2 is conserved throughout, meaning its ivET rank is $1+(0+0+0+0+0+0+0+0+0) = 1$. Conversely, position 3 is never conserved, thus its ivET rank is $1+(1+1+1+1+1+1+1+1+1) = 10$.

#### 16.4.4.2 Real-Valued ET

Mihalek et al. [11] combined ivET and a conservation-based approach into an expression for a hybrid score called real-valued ET (rvET) (sometimes called hybrid ET). This method is a marriage between two different functional site prediction classes, namely methods based on sequence conservation (Shannon entropy) and methods based on phylogenetic analysis (all-or-none approaches). Consequently, it maintains the added power of phylogenetic-based methods, yet is more robust against deviations from the ideal family-tree picture. Meaning, conservation scores, unlike ET, are sensitive to cases where alignment positions are nearly conserved, whereas ET is sensitive to positions that can be subdivided into two or more out-groups. Conservation scores fail in the latter as they only see global conservation.

Real-valued ET combines both into an expression giving the rank of a residue belonging to position $i$ by:

$$\rho_i = 1 + \sum_{n=1}^{N-1} \frac{1}{n} \sum_{g=1}^{n} \left( - \sum_{a=1}^{20} f_{ia}^{g} \ln f_{ia}^{g} \right) \tag{16.7}$$

where$f_{ia}^{g}$ is the frequency of the amino acid of type $a$ within a sub-alignment corresponding to group $g$, and index $n$ refers to the number of groups. The first part of the equation (1 + the first sum) coincides with the ivET ranking for position $i$, whereas the second sum is simply the Shannon entropy (see equation 16.1). This expression can be viewed as an extension of the ET method; a group $g$ contributes 0 to the score if the residue at position $i$ is conserved and contributes any value between 0 to ln 20 if the position is not conserved. Residue-importance ranking is based on its rvET score $\rho_\iota$.

It has been verified using a number of protein sets that rvET performs better than the ivET [11]. Real-valued ET is able to handle raw sequence input and its robustness makes it potentially applicable on the proteomic scale. In essence, rvET is more robust to sequence errors and sequence fragments. Real-valued ET [11] is particularly robust and suitable for making blind predictions about the evolutionary behavior of residues. Recent version of ET server [12], at Lichtarge's group also uses rvET for the calculation of ET. Moreover, there is another variant of the hybrid method for ranking of residues called '*zoom*' described in [11] and interested readers are advised to refer to this paper.

### *16.4.5 Availability*

There are various web-based tools available for performing ET analysis. Here, we will discuss a few of those freely available to academic users.

#### 16.4.5.1 ET Server at Baylor College of Medicine

The ET server at Lichtarge's group is available as two front-end programs for running and viewing evolutionary trace results:

(i) *Evolutionary trace report maker* [13]: This takes a PDB identifier as its input and uses different sources for protein sequence, structure and elementary annotation, and infers the evolutionary behavior at each site using real-valued evolutionary trace method and is available for academic use at http://mammoth.bcm.tmc.edu/report_maker.

(ii) *Evolutionary trace viewer* [12]: ET viewer is a set of integrated modules that make such predictions of functional sites and specificity determinants widely available. ETV allows a user to launch traces, and then interactively view the alignment, the related phylogenetic tree, and a molecular graphics display of top-ranked residues mapped on the structure for any user-specified adjustable rank threshold. This also requires only the PDB id as the input. ETV is implemented using Java to run across different operating systems using Java Web Start technology and is available for download from http://mammoth.bcm.tmc.edu/traceview/index.html. An example of the ET viewer output on the glycolytic enzyme triose-phosphate isomerase (TIM) is provided in Fig. 16.3.

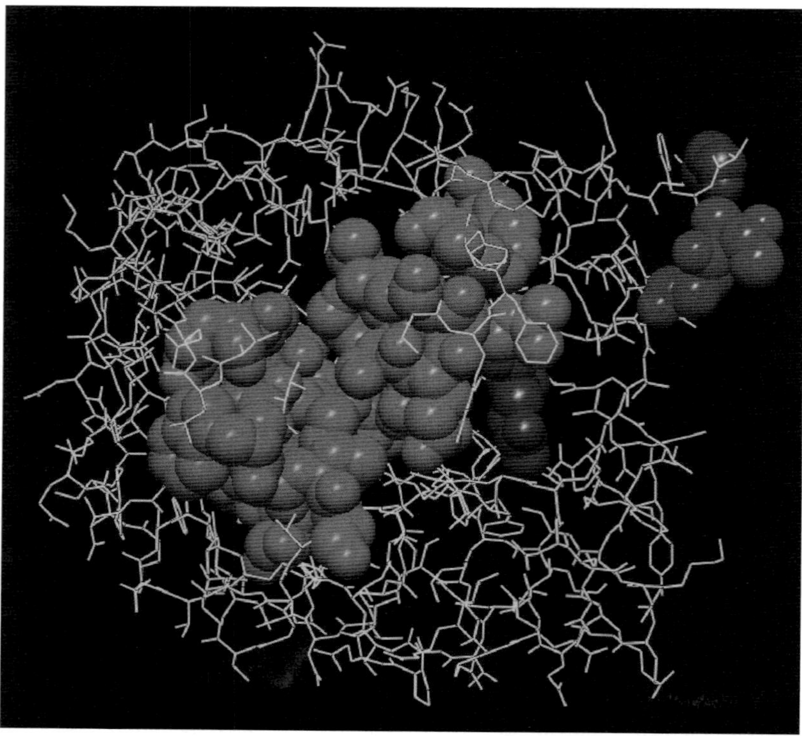

**Fig. 16.3** Example of the evolutionary trace (ET) output on the glycolytic enzyme triose-phosphate isomerase, using the ET Server at Baylor College of Medicine. Color-coding corresponds to the distinct trace residue clusters; clusters of just one residue are colored dark grey. The largest cluster (*colored red*) coincides with the enzyme's active site (Copies of figures including color copies, where applicable, are available in the accompanying CD)

#### 16.4.5.2 TraceSuite II

TraceSuite II [14] is a fully automated evolutionary trace server developed at Blundell's group and is available at http://www-cryst.bioc.cam.ac.uk/~jiye/evoltrace/evoltrace.html. The server takes a sequence alignment as the input and performs ET analysis. Any sequence with a PDB code as the sequence name will be regarded as a known structure and the results of the ET will be mapped onto this structure.

#### 16.4.5.3 JEvTrace

JEvTrace [15] is a JAVA graphical interface of the Evolutionary Trace method at Fred Cohen's group in UCSF and is available for download at http://www.cmpharm.ucsf.edu/~marcinj/JEv-Trace/.

#### 16.4.5.4 MASH

MASH is a tool for extraction of functional sites by evolutionary trace and related methods. This tool consists of different variants of ET methods and various conservation-based methods. The service is developed and maintained at the Toh lab at Kyushu University. The service can be used as a web server and is also available for download at http://timpani.bmr.kyushu-u.ac.jp/~mash/.

## 16.5 ConSurf

A second set of phylogenetic-based approachs to protein functional site prediction is based on the *ConSurf* algorithm. ConSurf also relies on the assumption that key residues that are important for functions should be conserved throughout evolution; hence the algorithm is closely related to the simple sequence conservation scores (i.e., Shannon entropy or sum of pairs). While conservation scores can be easily estimated from an input alignment, a key problem is that sequence-space is generally sampled unevenly: the value of the conservation score is only as good as the alignment it is calculated from. Furthermore, all-or-none consensus sequence-based treatments in ET [6], treat all columns with variable amino acid residues as non-conserved regardless of the physio-chemical similarity between them. For example, even when only one sequence differs in a certain position, the position is considered to be non-conserved. The research group of Nir Ben-Tal at the Tel Aviv University has developed ConSurf (Conservation Surface Mapping) in direct response to this limitation [16]. The process used in ConSurf consists of the following steps:

1. Search for homologous sequences and generate MSA. Subsequently, a phylogenetic tree is reconstructed. In the original implementation of ConSurf, a parsimony reconstruction method (as implemented in PHYLIP [17]) was used. Parsimony methods generate a large number of possible trees and presents the simplest tree as the correct one — parsimony is often compared to the principle of *Occam's razor*, which states that the simplest explanation is generally the best. One caveat of the parsimony reconstruction is that it frequently identifies multiple equally good trees.
2. Calculate the ConSurf conservation score for each site as defined by the phylogenetic tree. Each exchange between amino acids $i$ and $j$ is multiplied by a weight factor $M_{ij,}$ calculated by Miyata et al., determined the physio-chemical similarity between the $ij$ pair [18]. Meaning, $M_{ij}$ is a substitution matrix score; however, it is based on chemistry-derived rules instead of observed substitutions (i.e., the PAM matrices). The conservation grade $P_k$ at position $k$ in the alignment is calculated as:

$$P_k = \sum_{m=1}^{N} (A_{ij}^{m}(k) M_{ij}) \tag{16.8}$$

   where $A_{ij}(k)$ is 0 if there is no substitution and 1 when there is substitution; $N$ denotes the number of sequences within the alignment. In cases where multiple good trees are identified, ConSurf simply averages over each.
3. After calculating the conservation score for each position in the alignment, the score is normalized such that the average score for all positions is zero and the standard deviation is one.
4. Finally, the conservation scores are projected onto the 3D-structure. The lowest scores represent more conserved positions, whereas higher scores indicate plasticity. To aid in the visualization of the coloring scheme, the score is coarse-grained to an integer between 1 and 9. Variable sites (color grade = 1) are colored blue, whereas conserved sites (color grade = 9) are colored Bordeaux (see Fig. 16.4).

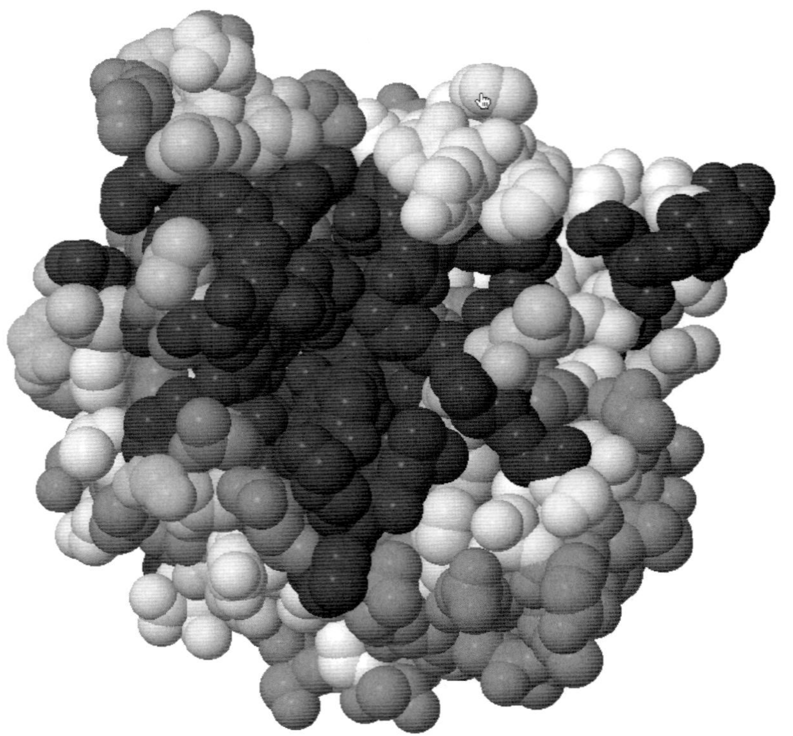

**Fig. 16.4** Example of the ConSurf output on the glycolytic enzyme triose-phosphate isomerase (the orientation is the same as in Fig. 16.3). The most conserved residues (*colored Bordeaux*) coincide with the enzyme's active site (Copies of figures including color copies, where applicable, are available in the accompanying CD)

A careful consideration of equation 16.8 reveals that it is very similar to the SP score defined in equation 16.2. However, there is a key difference. ConSurf uses inferred phylogenetic information to correct for unevenly sampled alignments. For example, consider an alignment that is enriched in mammalian sequences with the conserved position $k$. Is position $k$ conserved due to evolutionary pressures, or is its high conservation score simply a fact of poor sampling of non-mammalian sequences? Incorporating evolutionary relationships in the form of phylogenetic trees helps to resolve such issues. A very good description of how the process works is provided by Pupko et al. [19]. However, in practice, our (unpublished) results indicate that ConSurf is surprisingly insensitive to differences in tree topology or phylogenetic reconstruction accuracy.

ConSurf was originally applied on SH2 and PTB domains, and it was shown that the patches of conserved residues correlate well with the known functional regions of the domains [16]. Moreover, comparison to ET shows similar mapping of the conservation of the peptide binding face. Moreover, ConSurf detected a contact area of the SH2 domain with the other domains of the Src protein that was not identified by ET.

Since the original implementation of the approach, newer versions of ConSurf have been developed with the intent of improving the position-specific conservation score calculation. Recently, two new conservation calculations have been implemented into ConSurf; one is based on maximum-likelihood (ML), whereas the other is based on Bayesian inference methods [19]. Both provide results that are substantially better than the original method described in [16]. Testing of ConSurf indicates that the Bayesian conservation score provides better results than the ML method [20]. Hence, the current version of the Consurf server uses it as a default [21]. However, note that the improved performance comes at the cost of additional computational complexity.

### 16.5.1 ConSeq

For functional site prediction in families of unknown structure, a variant of ConSurf, called ConSeq [22], has been developed. ConSeq performs the same conservation analysis as ConSurf; however, the mapping of the results to the structure is, of course, omitted. ConSeq does provide some additional features to better inform the functional site prediction from sequence. For example, the solvent accessibility of each residue is predicted from an Artificial Neural Network algorithm. Based on whether a conserved position is exposed or buried, sites are classified as functionally or structurally important, respectively. The output of the ConSeq also maps the conservation scores, using the same color scheme from above, onto the sequence alignment. Predicted solvent accessibility is also indicated (see Fig. 16.5).

### 16.5.2 Rate4Site

The conservation algorithm used by ConSurf and ConSeq is called Rate4Site [19]. Rate4Site has been implemented into a stand alone program that can be used locally in large-scale analyses.

### 16.5.3 Availability

The ConSurf server is available at http://consurf.tau.ac.il/ and the ConSeq server is available at http://conseq.bioinfo.tau.ac.il/. Rate4Site is available for free download as either a Windows executable or UNIX/LINUX (C++) source code at http://www.tau.ac.il/~itaymay/cp/rate4site.html.

**Legend:**

**The conservation scale:**

1 2 3 4 5 6 7 8 9

**Variable** **Average** **Conserved**

e - **An exposed residue according to the neural-network algorithm.**

b - **A buried residue according to the neural-network algorithm.**

f - **A predicted functional residue (highly conserved and exposed).**

s - **A predicted structural residue (highly conserved and buried).**

X - **Insufficient data - the calculation for this site was performed on less than 10% of the sequences.**

**Fig. 16.5** Example of the ConSeq output on the glycolytic enzyme triose-phosphate isomerase. The sequence provided is that of the query. Color-coding on the sequence is the same as used by ConSurf (Copies of figures including color copies, where applicable, are available in the accompanying CD)

## 16.6 Phylogenetic Motifs

As pointed out in the excellent review by Jones et al. [23], protein functional site prediction algorithms can generally be classified into one of the two groups: methods based on sequence conservation or methods based on "feature" (or phylogenetic) conservation. ET is clearly a feature conservation method, whereas ConSurf is closer to a sequence conservation approach (even though it does use some phylogenetic information). The *phylogenetic motif* [24] approach, which was developed in our laboratory, is unique in that it overlaps the two classifications. Moreover, PMs are unique from the rest of the methods discussed herein since they identify stretches of residues versus single alignment positions. This difference comes at the cost of specificity; however, it improves sensitivity.

The PM method was first described by La et al. [24]. The approach is based on a previous observation that motifs taken from regions known to be functionally important *a priori* conserve the overall phylogeny of the family. Specifically, using the ubiquitous enzyme copper-zinc superoxide dismutase (CuZnSOD), we have shown that a contiguous subsequence taken from three functionally annotated motifs conserve the overall phylogeny of the family [25]. The functional subsequence represented ~10% of the overall alignment. Randomly chosen subsequences of similar size never reproduced the overall phylogeny as well as how the functional one did. This process was later repeated on a small number of additional protein families. The PM method simply reverses this scenario. Meaning, the algorithm uses alignment fragments that parallel the overall phylogeny as its functional site predictions.

### 16.6.1 PM Identification

The PM algorithm is based on a sliding sequence window (SSW) approach. The SSW generates all possible alignment fragments of fixed width (generally five alignment positions). Subsequently, a "phylogenetic" tree is constructed on each fragment. Finally, the method compares each of the fragment trees to the overall phylogenetic tree. Fragments with tree topologies closest to that of the overall familial tree are deemed as PMs. Above, phylogenetic is highlighted in quotes because using this term is not exactly correct. A phylogeny describes the overall evolution of a group of evolutionarily related proteins. Here, the sequences within the alignment fragment are not complete, so the term phylogeny is technically wrong, even though we use exactly the same reconstruction algorithms for building the window trees.

Tree topology comparisons are done using the ubiquitous partition-metric algorithm [26]. The partition metric algorithm, which is also called the symmetric difference or the Robinson-Foulds distance, quantifies how *different* two trees are. The partition metric simply counts the number of partitions, which are defined by the tree branch points that vary between the pair. (Note that the partition metric used here is slightly modified from the true symmetric difference. The modification and its rationale are beyond the scope of this discussion, but are described completely in [27]) The raw partition-metric counts for each alignment fragment are converted into z-scores, which are then plotted as a *phylogenetic similarity spectrum* (Fig. 16.6). The smaller the phylogenetic similarity z-score (PSZ), the more similar the fragment tree is to the overall tree. All the overlapping windows (in sequence) that score past the user-defined PSZ threshold are defined as a PM. In the example in Fig. 16.6, there are seven PMs, which are composed of 38 alignment fragments. The shortest PM is equal to two fragments, whereas the largest is composed of ten fragments. Comprehensive testing of window widths between 2 and 20 on a structurally heterogeneous data-set indicated that a window size of 5 and a PSZ threshold between −1.5 and −2.0 are best for identifying functional regions [24].

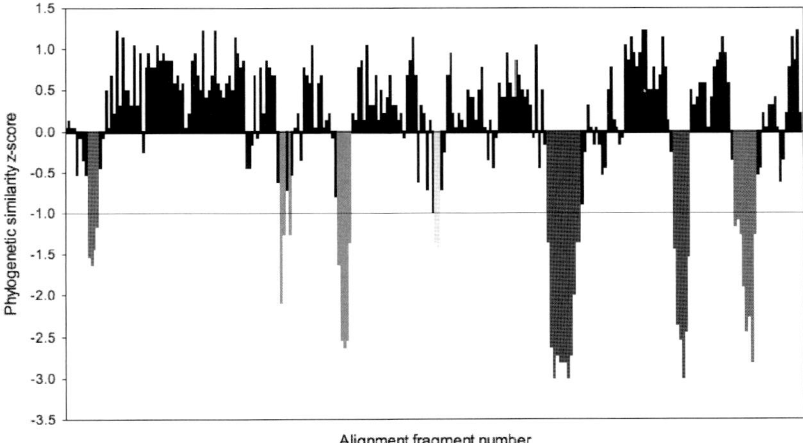


**Fig. 16.6** The phylogenetic similarity spectrum of triose-phosphate isomerase. The tree of each alignment fragment (*plotted on the abscissa*) is compared to the overall tree using the partition metric algorithm, which is plotted on the ordinate as a phylogenetic similarity z-score (PSZ). When the PSZ values are below a user-defined threshold (here equal to −1), all overlapping windows are grouped into a single phylogenetic motif. In this example, there are seven PMs identified (Copies of figures including color copies, where applicable, are available in the accompanying CD)

### *16.6.2 Prediction Assessment*

The PM example shown in Fig. 16.6 is again from TIM. In Fig. 16.7A, the PMs have been mapped to structure. It can clearly be seen that despite little proximity in sequence, PMs structurally cluster near the TIM active site. In fact, the PMs cover the two catalytic residues of TIM and all residues that define substrate specificity (Fig. 16.7B). Moreover, application of the method to a set of 14 other structurally and functionally diverse protein families indicates that annotating the identified PMs as functional is consistent with structural and biochemical data.

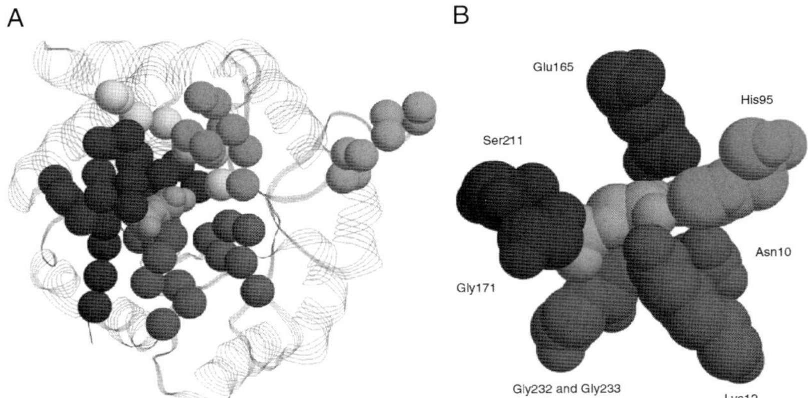


**Fig. 16.7** All phylogenetic motifs (PMs) are plotted (*left*) (A) onto the triose-phosphate isomerase (TIM) structure. The color-coding is the same as in Fig. 16.6; orange indicates the substrate analog bound within the enzyme's active site. The orientation is the same as in Figs. 16.3 and 16.4. The identified PMs include all TIM-substrate electrostatic contacts, as shown to the right (B). In fact, Glu165 and His95 are the enzyme's two catalytic residues, meaning they directly partake in the enzyme-mediated reaction mechanism (Copies of figures including color copies, where applicable, are available in the accompanying CD)

### 16.6.3 Comparisons to Traditional Motifs and ET

It was observed early on that PMs are strongly conserved in sequence, despite being solely identified from tree topology. Hence, PM scores have been directly compared to traditional motif scores that quantify alignment fragment conservation. In order to quantify the relationship, a False Positive Expectation (FPE) is calculated for each alignment fragment. FPE provides a probability of randomly encountering a sequence described by the motif within the SwissProt [28] database. Fragments with FPEs close to zero are conserved, whereas values above ~0.15 are indicative of plasticity. Fig. 16.8 plots PSZ vs. FPE for each TIM alignment fragment. The shape of the plot is very typical, indicating that PMs are rarely unconserved; however, there are several conserved windows that are not PMs. A cursory analysis of these windows indicates that they are less likely to be functional. PM results have also been shown to be consistent with MEME [29] results, which is a very common motif identification procedure.

Comparison of all identified PMs to ET predictions indicates that PMs are significantly populated by TRs. This similarity is to be expected since the methods are based on similar evolutionary arguments — PM windows generally contain two to four TRs, whereas the remaining positions approach strict conservation. However, a complete analysis of PM and ET predictions (without structure) concluded that PM predictions structurally cluster better around known functional sites, especially around substrate-binding epitopes. In essence, it appears that the sequence clusters of TRs that allow PMs to be identified focus the predictions to the active site region. While it remains to be tested, it is possible that an algorithm that simply identifies sequence clusters of ET residues could have the same effect.

### 16.6.4 Automated Similarity Threshold Selection

As stated above, the original implementation of the PM method [24] required manual selection of the PSZ threshold. This requirement made large-scale application of the method intractable. As such, we have automated the signal to noise problem inherent in PSZ threshold selection. There are many ways of dealing with signal to noise problems; here, we use a simple clustering algorithm to detect the optimal threshold. In addition, some data manipulation is performed which has the effect of "sharpening" the phylogenetic similarity spectrum. The automated algorithm [30] for determining the PSZ thresholds consists of following three steps:

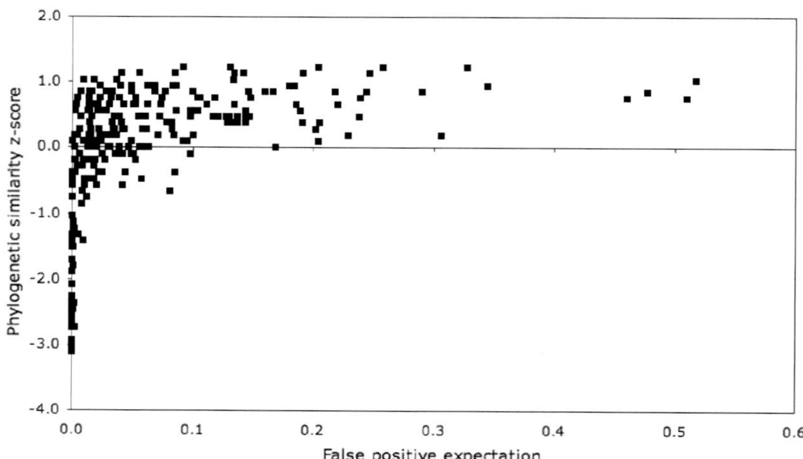


**Fig. 16.8** An example, in this case triose-phosphate isomerase, of a plot of phylogenetic similarity z-score (PSZ) vs. false positive expectation (FPE). Note that windows with large FPEs have very large PSZs, meaning they are not identified as phylogenetic motifs (Copies of figures including color copies, where applicable, are available in the accompanying CD)

1. Raw data preprocessing: Based on our earlier investigations [24], we know that any alignment fragment with a PSZ $< -2.0$ should definitely be included within a PM. Further, any fragment with a PSZ $> -1.0$ should definitely not be included within a PM. The remaining fragments, with $-1.0 <$ PSZ $> -2.0$, constitute the set of fragments whose annotation is unclear. In keeping with the PM algorithm, namely that overlapping fragments are grouped into a single PM, the data within the gap is "sharpened" to accentuate differences within the PSZ distribution. Within a set of local values inside the gap, the smallest (most negative) value is used to represent the entire set; while the other values are eliminated (see Fig. 16.9).
2. Subsequently, the sharpened values are clustered using Partition Around Medoids Clustering (PAMC) [31] into two groups — the cluster of larger values represents noise, whereas the cluster of smaller (more negative) values represent PM signals. The number of data points in a signal cluster is counted. If the signal cluster contains five or less data points, the threshold is set to the least negative value in that cluster.
3. An algorithmic override is invoked if any of the following three situations occur: (i) the cluster contains more than five data points, (ii) if less than three sharpened points occur in the gap — it does not make sense to cluster so few data points into two groups, or (iii) no PSZs $< -2.0$ are present within the distribution. In each case, the override defines the PSZ threshold as the first (rank ordered) PSZ value is greater than $-2.0$. This process normalizes the number of putative functional sites, preventing both too few and too many predictions.

Using a quantitative assessment similar to that as discussed by Aloy et al., above for ET, [10], we have benchmarked the accuracy of the automated PM method. On a small data-set of 32 protein families, the method had 69% *correct* predictions and 23% *useful* predictions. Only 11% were deemed *wrong*. The ability of our sequence-based predictions to get close to the performance of the structurally informed ET predictions is exciting. Interestingly, in the case of arginyl-tRNA synthetase, PMs correspond to regions surrounding both the amino acid/tRNA receptor stem and enzyme-anticodon interactions (see Fig. 16.10). Presumably, the PMs occurring at the anticodon stem help orient properly the tRNA onto the surface of the synthetase enzyme.

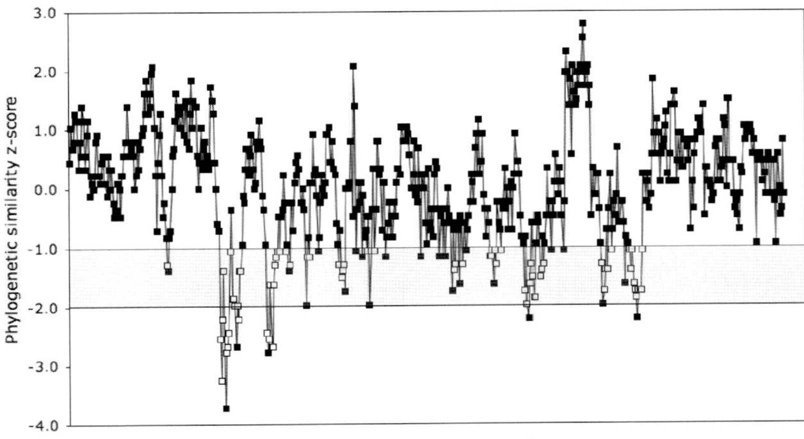


**Fig. 16.9** The phylogenetic similarity spectrum of arginyl-tRNA synthetase is shown. In the traditional implementation of the phylogenetic motif identification algorithm, the user must determine the appropriate phylogenetic similarity z-score (PSZ) threshold. The automated threshold identification algorithm attempts to classify (signal or not) values between PSZ = –1.0 and –2.0 (termed the gap). The algorithm begins by "sharpening" the phylogenetic similarity spectrum, meaning, values colored yellow are reset to PSZ = 0. Subsequently, the scores remaining within the gap are clustered using Partitioning Around Medoids Clustering (PAMC) into two groups, representing the signal and noise clusters (Copies of figures including color copies, where applicable, are available in the accompanying CD)

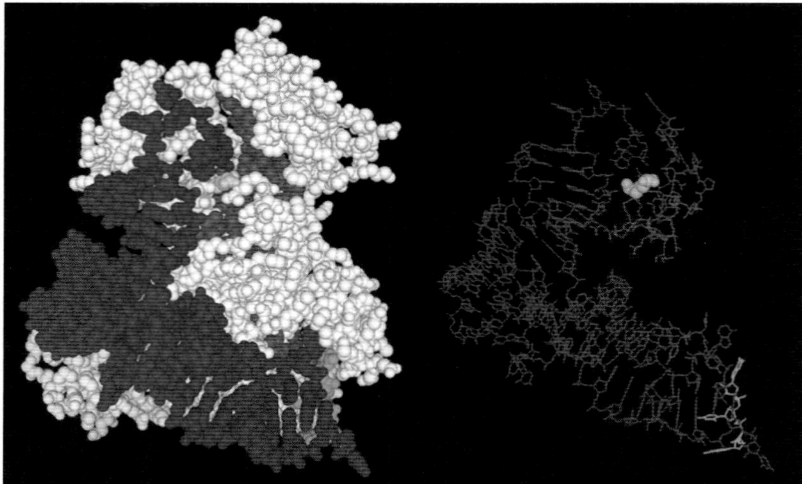

**Fig. 16.10** Arginyl-tRNA synthetase phylogenetic motifs (PMs) are highlighted. Residues colored red indicate PMs identified using the automated PSZ threshold identification algorithm. The arginine substrate is colored yellow; the tRNA is colored blue; and the anticodon is colored cyan. Two structurally unique phylogenetic motif clusters are identified. The larger cluster corresponds to the enzymes active site, whereas the smaller coincides with three H-bonds between the enzyme and the anticodon arm of tRNA. It is believed that these interactions help orient the tRNA properly onto the enzyme surface (38). The orientation of the figure on the right is the same as the left, but non-PM regions of the enzyme have been removed for clarity (Copies of figures including color copies, where applicable, are available in the accompanying CD)

### *16.6.5 Improved PM Detection using Parsimonious Trees*

Heretofore, all phylogenetic trees have been computed using the Neighbor Joining (NJ) method as implemented within ClustalW [32]. NJ, which is a distance-based approach, is used for computational tractability. For example, in the case of TIM, over 250 trees must be computed. Nevertheless, it is well known that parsimony phylogenetic reconstruction methods are generally superior to distance-based approaches. As such, a parsimony implementation of the method has been developed. Our results indicate that parsimony-PMs do a better job in accurately predicting functional sites; this is especially true in divergent data-sets.

### *16.6.6 Availability*

MINER [33], which is our implementation of the PM identification algorithm, is freely available from the web at http://coit-apple01.uncc.edu/MINER/. The input to MINER is any MSA or a set of unaligned sequences, which MINER will align using ClustalW. Note that the easiest way to improve PM prediction accuracy is to simply improve the alignment. As such, we strongly recommend alignments from MUSCLE [34,35], ProbCons [36], or Probalign [37]. Optionally, a PDB structure may be submitted to better highlight the PM regions, however, this is only used for assessment of the method. PM identification currently does not consider any structural information. The MINER output is a framed HTML file that provides Phylogenetic similarity versus window number plots, an annotated structure, and an annotated MSA. Moreover, interactive structural visualization of the identified PMs is achieved via Jmol. Upon request, a stand alone version of MINER is available (see the MINER website for details).

## 16.7 Which Method Is Best?

A perfectly legitimate question for the novice to ask is, "*Which of the discussed methods is best?*" Unfortunately, this question cannot be answered uniquely. In a recent analysis of the relative predictive ability of five methods (sequence conservation, ET, ConSurf, PMs, and FPE) against the neurotransmitter/sodium symporter (NSS) family, we found that PMs had the best sensitivity, whereas ConSurf had the best specificity (unpublished results). Surprisingly, a simple percent conservation metric had the best overall balance between the two. However, if one wanted to identify positions that define subfamily functional divergence, then ET would be the most appropriate technique. Note that the above analysis was done in the absence of structure. Undoubtedly, the results would change had structure considerations been included.

A second surprising result from this work was the observation that the various methods tend to provide results that are orthogonal to each other. We developed an orthogonality scale from 0 to 1, where 0 = perfectly symmetrical predictions and 1 = perfectly orthogonal results. Across all method pairs, the average orthogonality score was 0.62, meaning that the methods provide somewhat distinct prediction sets. However, in spite of this result, accuracy is substantially improved when residues are predicted by multiple methods. In fact, when any position was identified by any three methods, both sensitivity and specificity are greater than 0.5.

This brief discussion is not meant to establish a clear ranking of the various methods. In fact, applying the methods to other families result in somewhat different assessments. Rather, a discussion meant to demonstrate that simultaneous application of various methods is the most prudent approach. And then, after applying the various methods, the collective knowledge can be used to inform one's conclusions and guide downstream experimental studies.

**Acknowledgments** The authors thank Jim Mottonen and Andrei Istomin (both of the University of North Carolina at Charlotte) for proofreading the manuscript and providing several suggestions. Oliver Lichtarge (Baylor College of Medicine) and Nir Ben-Tal (Tel Aviv University) are thanked for clarifications about ET and ConSurf, respectively. David La (Purdue University) is acknowledged for developing and maintaining the MINER webserver.

## References

1. Toh, H. (2004) Bioinformatics for the Analysis of Protein Function (In Japanese). 1st ed. Okawa Publication.
2. Felsenstein, J. (2004) Inferring Phylogenies. 1st ed. Sinauer Associates, Sunderland, Massachusetts.
3. Shannon, C.E. (1948) The mathematical theory of communication. The Bell System Technical J, 27, 379–423.
4. Dayhoff, M.O., Schwartz, R.M. and Orcutt, B.C. (1978) In Dayhoff, M.O. (ed.), Atlas of Protein Sequence and Structure, Vol. 5, 345–352.
5. Valdar, W.S. (2002) Scoring residue conservation. Proteins, 48, 227–241.
6. Lichtarge, O., Bourne, H.R. and Cohen, F.E. (1996) An evolutionary trace method defines binding surfaces common to protein families, J. Mol. Biol. 257, 342–358.
7. Sowa, M.E., He, W., Slep, K.C., Kercher, M.A., Lichtarge, O. and Wensel, T.G. (2001) Prediction and confirmation of a site critical for effector regulation of RGS domain activity, Nat. Struct. Biol. 8, 234–237.
8. Madabushi, S., Yao, H., Marsh, M., Kristensen, D.M., Philippi, A., Sowa, M.E. and Lichtarge, O. (2002) Structural clusters of evolutionary trace residues are statistically significant and common in proteins, J. Mol. Biol. 316, 139–154.
9. Landgraf, R., Fischer, D. and Eisenberg, D. (1999) Analysis of heregulin symmetry by weighted evolutionary tracing, Protein Eng. 12, 943–951.
10. Aloy, P., Querol, E., Aviles, F.X. and Sternberg, M.J. (2001) Automated structure-based prediction of functional sites in proteins: applications to assessing the validity of inheriting protein function from homology in genome annotation and to protein docking, J. Mol. Biol. 311, 395–408.
11. Mihalek, I., Res, I. and Lichtarge, O. (2004) A family of evolution-entropy hybrid methods for ranking protein residues by importance, J. Mol. Biol. 336, 1265–1282.
12. Morgan, D.H., Kristensen, D.M., Mittelman, D. and Lichtarge, O. (2006) ET viewer: an application for predicting and visualizing functional sites in protein structure, Bioinformatics 22, 2049–2050.

13. Mihalek, I., Res, I. and Lichtarge, O. (2006) Evolutionary trace report-maker: a new type of service for comparative analysis of proteins, Bioinformatics 22, 1656–7.
14. Innis, C.A., Shi, J. and Blundell, T.L. (2000) Evolutionary trace analysis of TGF-beta and related growth factors: implications for site-directed mutagenesis, Protein Eng. 13, 839–847.
15. Joachimiak, M.P. and Cohen, F.E. (2002) JEvTrace: refinement and variations of the evolutionary trace in JAVA, Genome Biol. 3, RESEARCH0077.
16. Armon, A., Graur, D. and Ben-Tal, N. (2001) ConSurf: an algorithmic tool for the identification of functional regions in proteins by surface mapping of phylogenetic information, J. Mol. Biol. 307, 447–63.
17. Felsenstein, J. (1989) PHYLIP - Phylogeny Inference Package (Version 3.2), Cladistics 5, 164–166.
18. Miyata, T., Miyazawa, S. and Yasunaga, T. (1979) Two types of amino acid substitutions in protein evolution, J. Mol. Evol. 12, 219–236.
19. Pupko, T., Bell, R.E., Mayrose, I., Glaser, F. and Ben-Tal, N. (2002) Rate4Site: an algorithmic tool for the identification of functional regions in proteins by surface mapping of evolutionary determinants within their homologues, Bioinformatics 18 Suppl 1, S71–S77.
20. Mayrose, I., Graur, D., Ben-Tal, N. and Pupko, T. (2004) Comparison of site-specific rate-inference methods for protein sequences: empirical Bayesian methods are superior, Mol. Biol. Evol. 21, 1781–1791.
21. Landau, M., Mayrose, I., Rosenberg, Y., Glaser, F., Martz, E., Pupko, T. and Ben-Tal, N. (2005) ConSurf 2005: the projection of evolutionary conservation scores of residues on protein structures, Nucleic Acids Res 33, W299–W302.
22. Berezin, C., Glaser, F., Rosenberg, J., Paz, I., Pupko, T., Fariselli, P., Casadio, R. and Ben-Tal, N. (2004) ConSeq: the identification of functionally and structurally important residues in protein sequences, Bioinformatics 20, 1322–1324.
23. Jones, S. and Thornton, J.M. (2004) Searching for functional sites in protein structures, Curr. Opin. Chem. Biol. 8, 3–7.
24. La, D., Sutch, B. and Livesay, D.R. (2005) Predicting protein functional sites with phylogenetic motifs, Proteins 58, 309–320.
25. Livesay, D.R., Jambeck, P., Rojnuckarin, A. and Subramaniam, S. (2003) Conservation of electrostatic properties within enzyme families and superfamilies, Biochemistry 42, 3464–2473.
26. D. Penny, M.H. (1985) The use of tree comparison metrics, Syst. Zool. 34(1), 75–82.
27. Roshan U., L.D., La D. (2005) Improved Phylogenetic motif identification using parsimony. The IEEE 5th Symposium on Bioinformatics and Bioengineering BIBE05, 19–26.
28. Boeckmann, B., Bairoch, A., Apweiler, R., Blatter, M.C., Estreicher, A., Gasteiger, E., Martin, M.J., Michoud, K., O'Donovan, C., Phan, I. et al. (2003) The SWISS-PROT protein knowledgebase and its supplement TrEMBL in 2003, Nucleic Acids Res. 31, 365–370.
29. Bailey, T.L. and Gribskov, M. (1998) Methods and statistics for combining motif match scores, J. Comput. Biol. 5, 211–221.
30. La, D. and Livesay, D.R. (2005) Predicting functional sites with an automated algorithm suitable for heterogeneous data-sets, BMC Bioinformatics, 6, 116.
31. Kaufman L., R.P. (1990) Finding Groups in Data: An introduction to Cluster Analysis, Wiley.
32. Thompson, J.D., Higgins, D.G. and Gibson, T.J. (1994) CLUSTAL W: improving the sensitivity of progressive multiple sequence alignment through sequence weighting, position-specific gap penalties and weight matrix choice, Nucleic Acids Res. 22, 4673–4680.
33. La, D. and Livesay, D.R. (2005) MINER: software for phylogenetic motif identification, Nucleic Acids Res. 33, W267–W270.
34. Edgar, R.C. (2004) MUSCLE: a multiple sequence alignment method with reduced time and space complexity, BMC Bioinformatics 5, 113.
35. Edgar, R.C. (2004) MUSCLE: multiple sequence alignment with high accuracy and high throughput, Nucleic Acids Res. 32, 1792–1797.
36. Do, C.B., Mahabhashyam, M.S., Brudno, M. and Batzoglou, S. (2005) ProbCons: Probabilistic consistency-based multiple sequence alignment, Genome Res. 15, 330–340.
37. Roshan, U. and Livesay, D.R. (2006) Probalign: multiple sequence alignment using partition function posterior probabilities, Bioinformatics 22, 2715–2721.

## Key References

Armon, A., Graur, D. and Ben-Tal, N. (2001) ConSurf: an algorithmic tool for the identification of functional regions in proteins by surface mapping of phylogenetic information, J. Mol. Biol. 307, 447–463.

La, D., Sutch, B. and Livesay, D.R. (2005) Predicting protein functional sites with phylogenetic motifs, Proteins 58, 309–320.

Lichtarge, O., Bourne, H.R. and Cohen, F.E. (1996) An evolutionary trace method defines binding surfaces common to protein families, J. Mol. Biol. 257, 342–358.

Madabushi, S., Yao, H., Marsh, M., Kristensen, D.M., Philippi, A., Sowa, M.E. and Lichtarge, O. (2002) Structural clusters of evolutionary trace residues are statistically significant and common in proteins, J. Mol. Biol. 316, 139–154.

Valdar, W.S. (2002) Scoring residue conservation, Proteins 48, 227–241.

## Web Resources

http://www.ebi.ac.uk/thornton-srv/databases/cgi-bin/valdar/scorecons_server.pc
http:/mammoth.bcm.tmc.edu/report_maker
http://mammoth.bcm.tmc.edu/traceview/index.html
http://www-cryst.bioc.cam.ac.uk/vjiye/evoltrace/evoltrace.html
http://www.cmpharm.ucsf.edu/vmarcinj/JEv_Trace/
http://timpani.bmr.lcyshu-u.ac.jp/vmash/
http://consurf.tau.ac.il/
http://conseq.bioinfo.tau.ac.il/
http://www.tau.ac.il/~itaymay/cp/rate4site.html
http://coit-apple01.uncc.edu/MINER/

# Chapter 17
# The Role of Transcription Factor Binding Sites in Promoters and Their *In Silico* Detection

**Thomas Werner**

**Abstract** As detailed in this chapter TFBSs are among the most important elements of transcription control in promoters, enhancers, locus control regions, and Scaffold/Matrix Attachment regions, to name only the best known. So far, the best way to identify TFBSs in genomic sequences is by sequence similarity searches with whatever method is suitable for the task. As detailed, nucleotide weight matrices are the most popular and developed tools for this purpose. However, all of these methods locate TFBSs one by one and independent of each other, yielding what is called physical binding sites. The only answer such results can provide is the physical binding probabilities of a whole group of different TFs. Neither specific binding of a particular TF can be deduced from such data nor can any functional properties of the stretch of DNA where the TFBS was found be determined. TFBSs do not act as isolated individual binding sites but always as part of a larger context.

**Keywords** Transcription factors · Frameworks · ModelInspector · Bindingsite · Promoter

## 17.1 Introduction

Transcriptional regulation is the key link between the static genomic sequence and all the variable events that can be summarized as life. Therefore, regulatory regions in the genome influencing the level of transcription at various levels represent a natural focus for research, linking genomics and transcriptomics or as defined in a previous article "regulomics" [1]. Regulatory regions share several features despite their obvious divergence in sequence and function such as promoters, enhancers, locus control regions, or scaffold/matrix attachment regions. Most of these features are not fixed nucleotide sequences but are variable in sequences restrained by functional requirements. Therefore, understanding of the major components and events during the formation of regulatory DNA-protein complexes is crucial for the design and evaluation of algorithms for the analysis of regulatory regions. One of the most important classes of proteins acting in this context is represented by transcription factors (TFs), many of which interact directly with the genomic DNA.

Algorithms for the analysis and recognition of transcription factor binding sites (TFBSs), necessarily rely on the underlying biological principles in order to generate suitable computational models. Therefore, a brief overview over the biological properties and mechanisms is required to understand why an individual algorithm has been developed in a particular way. The choice of the parameters and implementation of the algorithms largely control the sensitivity and speed of a program. The specificity of software recognizing TBFBs in DNA is determined, to a large extent, by how closely the algorithm follows what will be called the biological model from

T. Werner
Genomatix Software GmbH, Bayerstr. 85A, D-80335, München, Germany
e-mail: werner@genomatix.de

S. Krawetz (ed.), *Bioinformatics for Systems Biology*, DOI 10.1007/978-1-59745-440-7_17,

here on. However, every biological model of individual TFBSs is necessarily incomplete as it became quite evident that a focus on individual TFBSs is too narrow to capture biological function. Extension of the models to consider the relevant context in particular, combinations of several TFBSs, are required to match biology. Nevertheless, even such sophisticated models ultimately have to rely on recognition of the individual TFBSs as a first step.

## 17.2 Transcription Factors

Transcription factors (TFs) represent a special class of proteins that comes in two flavors: Those TFs that bind directly to DNA and exhibit at least some sequence specificity or better selectivity in their binding and those who form protein-protein networks important in transcription control but do not bind to DNA directly. This latter group are also known as mediators and they will be largely ignored in this chapter due to the fact that they remain invisible to any TFBSs detection. The focus is on the TFs that do bind to DNA directly and more specifically on the corresponding TFBSs of these factors.

### *17.2.1 Expression, Modifications, and Localization*

The extraordinary importance of TFs in the general pool of genes and proteins is clearly visible from the greater variability of expression exhibited by transcription factor genes as compared to other genes. Transcription factor genes produce more alternative transcripts resulting in alternative proteins (62%), than other genes (29%), and also produce more tissue-specific isoforms than the average gene [2]. Often, alternative transcripts of TF genes are also associated with alternative promoters adding another level of complexity to the regulation of TF gene expression. However, the presence and functionality of TFs is controlled on many levels far beyond the initial transcription and translation of all their isoforms.

TFs quite naturally have to enter the nucleus in order to exert any effect on the genomic DNA. Since they are translated, like most other proteins, in the cytoplasm they must cross the nuclear membrane to reach the chromatin, the site of their action. TFs may be regulated in several ways by prevention or facilitation of this cytoplasmic-nuclear transfer, as is the case for NfkB, for example. The classical regulator is IkB, which in itself comes in three isoforms, a, b, c [3], hindering the nuclear transfer of NFkB. It is destructed proteolytically, releasing NFkB triggered by signaling pathways, which is known as the canonical pathway of NFkB activation. However, a variant form of a NFkB precursor protein, P100 can act as a fourth IkB protein and prevent the NFkB complex from nuclear transfer as well. This mechanism of cytoplasmic NFkB retention and its release by other signaling pathways is known as the non-canonical pathway of NFkB activation [4]. Adding the sequestration of NFkB components by other factors, e.g., glucocorticoid receptors illustrates the enormous complexity of even such a simple mechanisms as sequestration. This is by no means restricted to NFkB, as other TFs are also subject to sequestration such as the glucocorticoid receptor, which is guided by heat shock proteins throughout the activation process [5].

## 17.3 Transcription Factor Binding Sites (TFBSs)

The counterparts of the TFs on the genomic DNA are their binding sites (TFBSs), which attract TFs to the appropriate sites of genomic DNA. Despite the enormous variability of such binding sites and the different selectivity of such TFBSs with respect to the TFs that will bind to them, they do exhibit a few common features, which will be summarized below as physical properties of TFBSs.

### *17.3.1 Physical Properties*

TFBSs generally consist of about 10–30 nucleotides, only a few of which are crucial for specific protein binding. Therefore, individual TFBSs can vary in sequence considerably, even if they bind to the same protein. Nucleotides contacted by the protein in a sequence-specific manner are usually the best-conserved parts of a binding site. Other nucleotides involved in the DNA backbone contacts, i.e., contacting the sugar-phosphate framework of the DNA helix (not sequence specific as they do not involve the bases A, G, C, or T) are much less conserved. The least conserved regions are the internal "spacers" that are not contacted by the protein at all. In general, protein-binding sites exhibit enough sequence conservation to permit the detection of candidates by a variety of sequence similarity-based approaches. However, potential binding sites can be found almost all over the genome and are by no means restricted to (known) regulatory regions. Quite a number of binding sites outside the regulatory regions are also known to bind their respective binding proteins [6], indicating that the abundance of predicted TFBSs is not just a shortcoming of the detection algorithms but at least in some cases reflects biological reality.

Another important feature of TFBSs is their actual lack of specificity. They do bind selectively to TFs but by no means are restricted to a single TF. In many cases, several distinct TFs can bind to the same DNA sequence complicating the identification of the actually binding protein (e.g., STAT1 and STAT3 competing in the IL10 promoter for the same binding site [7]).

It is important to keep in mind that binding of the TF to its cognate binding site is only half of the story. Timely dissociation is at least as important, as transcriptional complexes must disassemble in time to allow for the release of the DNA at the end of the action. As a consequence, nature has not only optimized TFBSs for binding of the TFs but also for releasing them, which is probably one of the reasons why we always have to deal with suboptimal binding sequences in all analyses of real DNA sequences. This has some very important consequences for the development and performance of TFBSs detection algorithms.

### *17.3.2 Functional Properties–Functional Context*

Often it is not possible to identify individual binding proteins as they might bind as part of multi-protein complexes [8]. This illustrates another important point already raised in the introduction: TF-binding *in vivo* is usually context-dependent. The isolated TF will bind to a cognate site quite differently if brought together in a reaction tube as a naked protein and DNA probe (e.g., in a gel shift assay) than *in vivo* where the adaptive DNA structure, chromatin, other TFBSs, and a host of other proteins are around. As it became evident from several chromatin immunoprecipitation (ChIP) studies even *in vivo* binding of a TF does not automatically imply a function in transcription control as was found in a genome-wide study, which identified many more CREB binding sites than CREB regulated genes [9].

#### 17.3.2.1 Epigenetic Context

There is a very simple method to prevent a TF from exerting any effect on a corresponding binding site: Hiding that DNA stretch from the protein by any means will do the trick. There are several mechanisms that will effectively sequester TFBSs efficiently summarized as epigenetic events. The simplest way to prevent many TFs from binding to their cognate sites is by DNA methylation which changes the structure sufficiently to inhibit TF binding [10, 11]. Another mechanism is to inhibit protein access to the DNA by packing the DNA more tightly as is achieved by deacetylation of histones resulting in a denser chromatin structure [12]. All of these modifications cannot be read from the DNA sequence directly and are thus necessarily ignored in any sequence-based analysis method such as computational TFBSs detection, which is another reason why potential and real binding sites do not necessarily correspond.

#### 17.3.2.2 Promoters

The context-dependency of TFBSs can be best illustrated by the example of eukaryotic polymerase II (pol II) promoters. The TFBSs within the promoters (and most likely in other regulatory sequences as well) do not show any general patterns with respect to location and orientation within the promoter sequences although particular functionality may be associated with a specific location or association within the promoter [13]. However, even functionally important binding sites for a specific transcription factor may occur almost anywhere within a promoter if a large number of promoters are analyzed statistically. However, different locations of TFBSs in individual promoters very often are correlated with specific and distinct functions of such TFBSs. For example, functional AP-1 (Activating protein 1, a complex of two TFs, usually one from the fos and one from the Jun family) binding sites can be located far upstream, as in the rat bone sialoprotein gene where an AP-1 site located about 900 nucleotides upstream of the transcription start site (TSS) inhibits expression [14]. An AP-1 site located close to the TSS is important for the expression of Moloney Murine Leukemia Virus [15]. Moreover, functional AP-1 sites have also been found inside exon 1 (downstream of the TSS) of the proopiomelanocortin gene [16] as well as within the first intron of the fra-1 gene [17], both located outside the promoter. AP1 is only one example, the principles outlined here also apply to other TFBSs, illustrating why the overall statistical correlation of TFBSs within promoters is not meaningful with respect to the biological function of the TFBSs. The context of a TFBS is one of the major determinants of its role in transcription control and the context in this case has to be defined functionally. Physical vicinity of two TFBSs may or may not be functionally relevant; the particular interactions of the binding TFs determines which TFBSs are just nearby or essential for transcriptional function. I will refer to a set of functionally interacting TFBSs in any regulatory region (not restricted to promoters) as *transcriptional modules*. The same TFBSs can be part of distinct transcriptional modules depending on the condition or cell type as has been shown experimentally numerous times (e.g., [18]). A subset of overlapping transcriptional modules from the RANTES promoter is shown in Fig. 17.1. The original definition of transcriptional modules by Arnone and Davidson was more general than the one used here [19].

#### 17.3.2.3 FrameWork Concept

The context of a TF-site is one of the major determinants of its role in transcription control [18]. As a consequence of context requirements, TF sites are usually grouped together in functional groups (frameworks), which have been described in many cases and conveying a specific promoter function will require more than one site (e.g., [20]). When the mutual dependency of TFBSs

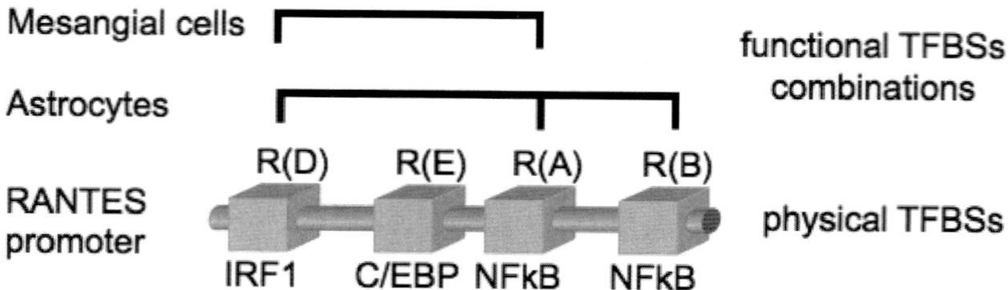


**Fig. 17.1** Proximal transcriptional modules in the human RANTES promoter R(A) – R(D) identify experimentally verified promoter regions in the RANTES promoter [18]. Below the boxes, the identifiers of the corresponding TFBSs are given (Copies of figures including color copies, where applicable, are available in the accompanying CD)

in a framework has been experimentally verified, such frameworks are also called transcriptional modules (Fig. 17.1). The organization of the binding sites (and probably also of other elements) of a transcriptional module appears to be much more restricted than what the apparent variety of TFBSs and their distribution in the whole promoter suggests. Within a transcriptional module, both sequential order and distance can be crucial for function indicating that these modules may be the critical determinants of a regulatory region rather than the individual binding sites. Promoter modules are always constituted of more than one binding site. Since regulatory regions, such as promoters, can contain several modules that may use overlapping sets of binding sites, the conserved context of a particular binding site cannot be determined from the primary sequence (Fig. 17.1). This is also the reason why analysis of a DNA sequence solely for individual TFBSs will miss the functional context and thus can only serve as the first step in elucidating transcriptional functions in a DNA sequence. The corresponding modules must be either known *a priori*, determined by comparative sequence analysis or experimentally in suitable expression assays.

## 17.4 How Transcription Factors Bind to DNA

Transcription factors bind to DNA via a multitude of atomic interactions that are either van der Waals hydrophobic contacts or supported by juxtaposition of oppositely charged amino acids and DNA components. Generally, two basic modes of molecular interactions can be distinguished:

1. The first involves nonspecific contacts between the protein side chains and the so-called backbone of the DNA, which consists of the sugar-phosphate structure linking the bases together (Fig. 17.2). Such contacts can form anywhere on a DNA (double)strand and is responsible for the general tendency of TFs to associate with the DNA. No sequence specific effects are involved in this interaction.
2. The second mode is the sequence-specific recognition achieved by direct contact of amino acid side chains with particular bases of the DNA. Therefore, these contacts can only be formed where there is a suitable succession of bases, i.e., a specific nucleotide sequence (Fig. 17.2). However, as the protein mainly recognizes the DNA structure this allows for some sequence variation as long as the binding structure is maintained.

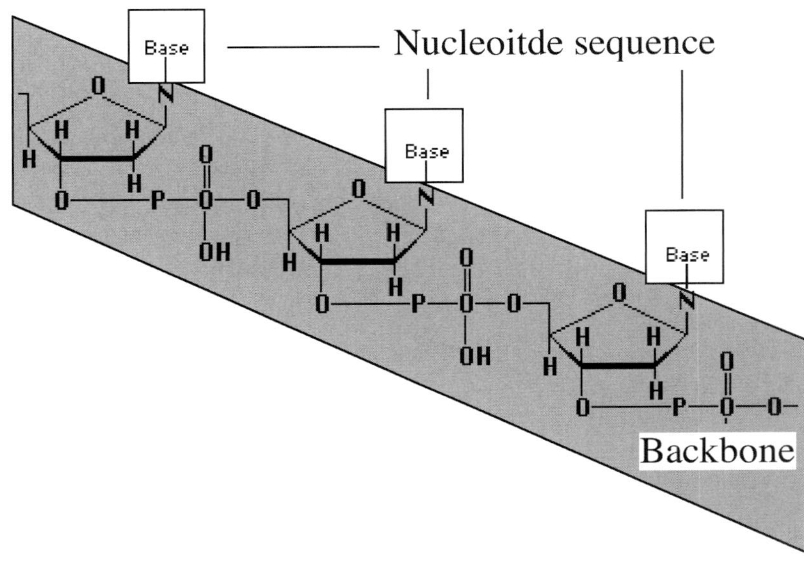


**Fig. 17.2** Basic structure of a single DNA strand. The gray area represents the sugar-phosphate backbone involved in the nonspecific protein contacts. The white boxes represent the bases, which are responsible for the sequence specific contacts with proteins (Copies of figures including color copies, where applicable, are available in the accompanying CD)

## 17.5 Transcription Factor Binding Site Detection *In Silico*

TFBSs need to be differentiated from all the other possible sequences by any detection algorithm. Algorithms used to analyze and detect TFBSs are necessarily based on some kind of a usually simplified model of what a particular TFBS should look like. All models used are inevitably compromising between accuracy with respect to the biological model (the standard of truth) and the computational feasibility of the model. For example, a computational model based on *a priori* three-dimensional structure prediction derived from molecular dynamics using sophisticated force fields may be the most accurate model for a TFBS but cannot be used for the analysis of real data due to excessive demand on computational resources as well as the limited knowledge about structure-sequence relations. Nevertheless, it would represent the ideal biological model, as proteins can only interact with structures and not with letters, we call sequences.

### *17.5.1 Models of TFBSs*

Over time, various different models have been introduced to describe TFBSs. There are several approaches to take the inherent variability of TFBSs into account. The basic models will be discussed in the order of rising complexity.

#### 17.5.1.1 Direct Sequences

The easiest way of course is to put all real sequences of TFBSs into a database and then locate the exact matches only. Figure 17.3 shows the binding sites for glucocorticoid receptor – binding sites as collected in MatBase (Genomatix Software GmbH, Munich) as an example. The biggest disadvantage of such an approach is that as the level of abstraction of this "model" is zero, sequences are taken as they are, and there is no way of inferring onto anything similar. While

| Name | Alignment | Matrix similarity | References |
|---|---|---|---|
| remmo1 | TGAGCTCTTAGT**GTTC**TAT | (0.906) | 1995608 |
| ocotglo5 | GCGTTCCAAGCT**GTTC**TCC | (0.905) | 3453115 |
| HUMMET2A | CCGGTACACTGT**GTCC**TCC | (0.932) | 2881624 |
| MMTV2 | TTGGTATCAAAT**GTTC**TGA | (0.932) | 2881624 |
| HIVBRUCG_2 | GGCTAACTATAT**GTCC**TAA | (0.867) | 7684876 |
| RATANFA | CTGCCTGTTTGT**GTTC**TGA | (0.843) | 1835978 |
| ratatrc | CAGGACTTGTTT**GTTC**TAG | (0.893) | 2881624 |
| RATTATGRE3 | GGGGTACAGGTT**GTTC**TGA | (0.974) | 2881624 |
| RATTOG5A | TATGCACAGCGA**GTTC**TAG | (0.864) | 2881624 |
| ratTAT | GCTGTACAGGAT**GTTC**TAG | (0.962) | 2881624 |
| RATTOG5B | CCCTTTCATGAT**GTCC**TGG | (0.881) | 2881624 |
| HUMGH1 | TGGGCACAATGT**GTCC**TGA | (0.943) | 2881624 |
| rnafpg | GAAGTGGTCTTT**GTCC**TTG | (0.861) | 2454390 |
| MSV2 | GCTGTTCCATCT**GTTC**TTG | (0.952) | 2881624 |
| MSV1 | TGGGGACCATCT**GTTC**TTG | (0.953) | 2881624 |
| gglyshre | CCAGTTTGTACA**GTTC**TGG | (0.839) | 3416833 |
| MMTV1 | TGGTTACAAACT**GTTC**TTA | (0.943) | 2881624 |
| HUMBGPG | AGGGTATAAACA**GTGC**TGG | (0.841) | 2038339 |

**Fig. 17.3** Binding sites for glucocorticoid receptors collected from various genes and mammalian organisms. Name: gene symbol, Alignment: actual sequence, Matrix Similarity: MatInspector score, Reference: PubMedID (Copies of figures including color copies, where applicable, are available in the accompanying CD)

**Fig. 17.4** The 15 letter IUPAC ambiguity code (Copies of figures including color copies, where applicable, are available in the accompanying CD)

| | | | |
|---|---|---|---|
| **A** = A | **C** = C | **G** = G | **T** = T (U) |
| **W** = A or T | **S** = C or G | **K** = G or T | |
| **M** = A or C | **Y** = C or T | | |
| **R** = A or G | | | |
| **B** = C, G, T = **(non A)** | **D** = A, G, T = **(non C)** | **H** = A, C, T = **(non G)** | **V** = A, C, G = **(non T)** |
| **N** = A, C, G, | | | |

this will yield only proven binding sites, this approach will miss all variant binding sites and becomes unfeasible if huge numbers of TFBS sequences are to be considered. Naturally, direct sequence matching is not used in any computational approach.

However, those collections of real binding sites are a prerequisite for all modeling approaches and thus databases containing collected TFBSs are valuable resources providing the training sequences for all algorithms used so far.

#### 17.5.1.2 IUPAC Consensi

The most rudimentary improvement to matching direct sequences is to allow for mismatches but this allows indiscriminative matching of all sequences differing by the number of allowed mismatches, taking none of the TF-specific restrictions into account. The simplest model introducing such restrictions is based on simple sequence similarities detected by a IUPAC consensus (using additional letters to describe ambiguities such as R = A or G). Such IUPAC-sequences can be easily used on a computer. Figure 17.4 shows the complete IUPAC ambiguity code.

The IUPAC-sequences can deal with variant sequences but a major drawback of the method is that the definition of the consensus sequence is highly arbitrary depending on which rules are being used to determine the "prevalent" nucleotides as well as the number of sequences considered (e.g., [21]). IUPAC consensi are also very promiscuous in allowing patterns that do not occur anywhere in reality by odd combinations of substitutions. They treat all positions as equally important, which is in stark contrast to the biological reality (base-specific, backbone-, non-binding), i.e., a mismatch at a specific contact site is scored the same as a mismatch in a spacer region. Below is the IUPAC representation of the GRE sequence alignment from Fig. 17.3:

GRE-IUPAC: N N G G T W C W N N N T G T T C T N R

It is easy to see that a lot of sequences not matching any of the founding TFBS sequences will be accepted as well and no particular scoring is possible.

The next model in terms of complexity would be the positional weight matrices. They have been proven to be the most widely used models as of today and will be discussed after the more complex Hidden Markov Models.

#### 17.5.1.3 Hidden Markov Models (HMM)

This is a sophisticated statistical model that has been successfully applied to describe protein-sequence alignments [22]. In brief, HMMs are fully probabilistic models, allowing manipulation and optimizing parameters using Bayesian probability theory. To explain this in simpler terms, consider flipping coins where there are just two possibilities for the outcome. Either the digit is visible or the picture (an eagle if you throw US quarters) is visible. You may attempt to find out what is the probability of either side to show up next just given the outcome of the last event. This would constitute a simple Markov chain. Now consider, there is somebody behind a curtain (so you cannot observe the actual action) throwing coins and telling you the outcome. Again, you attempt to predict the outcome from the previous result. However, this time things are complicated by the fact that the person behind the curtain has several coins to choose from and would not tell you which one was thrown. So, what you are told is the final result while you do not know the start condition (i.e., which coin was thrown). The selection of the coins is supposed to be a stochastic background process, which cannot be observed directly (hidden). Calculating probabilities for outcome scenarios from the previous outcome involving such a hidden background process constitutes a Hidden Markov Model (HMM).

There are strong as well as the weak points with respect to TFBSs descriptions. In order to train a probabilistic model sufficiently well, a huge training set of instances (here sequences) is required. In terms of binding sites this is about 1,000 or more sequences per model, i.e., binding sites, while realistic numbers for verified sequences is typically in the range of 5–50. Although HMMs represent by far the best mathematical model to describe the full range of parameters influencing the TFBSs (Fig. 17.5), they require far more sequences than those available at present. For this sole reason, HMMs are still not the method of choice for TFBS descriptions. Another problem is that even sophisticated HMMs still do not account for mutual dependencies of nucleotide positions within the TFBSs, which are known to occur in a number of real TFBSs.

New high-throughput methods such as ChIP-on-chip provide us with more and more experimentally verified TFBS-sequences, so one could expect that HMMs may be well- posed to replace the weight matrix based methods one day. However, simultaneously we are learning that subtle differences in such large collections of TFBSs may actually have important functional (i.e., context-dependent) consequences, arguing against throwing them together for HMM training. It is too early to come up with a definitive answer as to whether the final nod goes to HMMs or weight matrices, or something else.

#### 17.5.1.4 Positional Weight Matrices (PMW)

This is a simple and robust model that has been around for more than a decade and is still the prevalent method in every day application [23–26]. In brief, the method starts with an alignment of pre selected sequences (TFBSs for a given factor) and derives a nucleotide distribution matrix from that alignment (counting the number of occurrences for each nucleotide at each position of the alignment). Then weighting is applied, which depends on the particular algorithm used. The resulting positional weight matrix (PWM) is then used to score sample sequences for how well they fit into the matrix. Figure 17.6 shows the nucleotide distribution matrix for the GRE alignment shown in Fig. 17.3.

At the bottom of Fig. 17.6 the Consensus Index (Ci) is shown, which is used to weight the individual positions in the scoring of any sequence. This weighting is based on the concept of

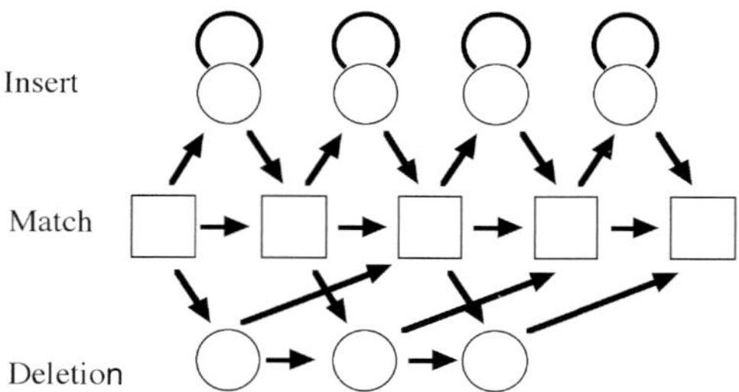


**Fig. 17.5** Schematic architecture of a HMM describing the alignment of sequences with a length of five nucleotides. The square boxes indicate matching positions. The circles above the boxes indicate insertions and the circles below indicate deletions. Note that each sequence can feed through a different path in this HMM. The bold arrows represent the transitions and carry the parameters determining the relative probabilities for these transitions (Copies of figures including color copies, where applicable, are available in the accompanying CD)

| Pos. | 1 | 2 | 3 | 4 | 5 | 6 | 7 | 8 | 9 | 10 | 11 | 12 | 13 | 14 | 15 | 16 | 17 | 18 | 19 |
|---|---|---|---|---|---|---|---|---|---|---|---|---|---|---|---|---|---|---|---|
| A | 1 | 3 | 3 | 0 | 2 | 10 | 0 | 9 | 8 | 5 | 4 | 3 | 0 | 0 | 0 | 0 | 0 | 5 | 6 |
| C | 5 | 6 | 2 | 1 | 4 | 2 | 12 | 3 | 2 | 1 | 6 | 0 | 0 | 0 | 5 | 18 | 0 | 2 | 2 |
| G | 6 | 7 | 10 | 13 | 1 | 1 | 2 | 1 | 4 | 4 | 5 | 0 | 18 | 0 | 1 | 0 | 0 | 7 | 9 |
| T | 6 | 2 | 3 | 4 | 11 | 5 | 4 | 5 | 4 | 8 | 3 | 15 | 0 | 18 | 12 | 0 | 18 | 4 | 1 |
| IUPAC | N | N | G | G | T | W | C | W | N | N | N | T | G | T | T | C | T | N | R |
| Ci | 22.4 | 20.7 | 27.4 | 54.7 | 35.4 | 32.5 | 47.3 | 27.8 | 20.9 | 24.8 | 15.8 | 72.0 | 100.0 | 100.0 | 51.1 | 100.0 | 100.0 | 19.1 | 30.6 |

**Fig. 17.6** Nucleotide distribution at the 19 positions of the GRE alignment from Fig. 17.3. IUPAC: IUPAC consensus sequence, Ci: Consensus Index as calculated by the MatInspector algorithm (Copies of figures including color copies, where applicable, are available in the accompanying CD)

mutual information content (Shannon entropy), which reflects the varying importance of the different positions of a TFBS due to different contact restrictions as discussed. The algorithm was originally described by [26] and further improvements and extensions are described by [23].

The graphical representation of the Ci in comparison with biological evidence from Fig. 17.6 is shown in Fig. 17.7, illustrating the enhanced sensitivity of the weighted matrix towards biological features (sequence specific binding, backbone contacts and spacer region).

So, how can all of that be dealt with in a practical application? There are many different approaches to detect TFBSs and the reader is directed to a recent review for an overview [27]. This chapter will use the MatInspector approach for illustration as this is the only program directly connected to further analyses that take functional context into account and thus at least allows one to go beyond the mere physical results.

## *17.5.2 TFBS Variability and Multiple TF Binding*

All search programs attempting to locate potential TF binding sites in the genomic DNA basically face similar challenges. The most notorious one is incomplete data, preventing the generation of high

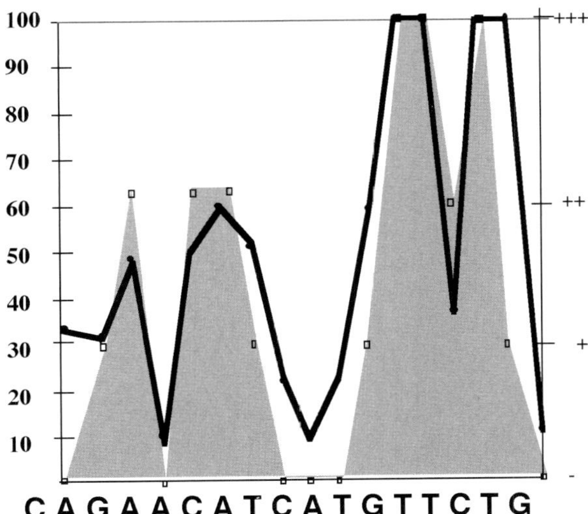


**Fig. 17.7** Comparison of calculated importance (consensus index based on Shannon entropy) and experimental evidence for the Glucocorticoid Receptor Binding site. The scale on the left shows the normalized consensus index [28] derived from the nucleotide weight matrix of the GRE [26]. The experimental evidence is given as follows: – = no DNA contact by the protein, + = backbone contact, + + = unspecific base contact, + + + = sequence specific base contact (Copies of figures including color copies, where applicable, are available in the accompanying CD)

quality descriptions by any means as discussed in the context of the HMMs. If only two examples for a specific binding site are available, it is meaningless (but not impossible) to construct a weight matrix and even a simpler IUPAC string will not be very useful. The situation is very close to that of a non-existent description, which is the next most frequent problem. Just because there is no IUPAC or matrix available for a binding site does not mean there is no specific binding site for that protein.

Now, let us consider the case where there is a sufficiently well defined weight matrix and this is used to scan sequences. There are two additional features of TFBSs to be taken into consideration. The first is the inherent suboptimal binding affinity (due to dissociation requirements), which renders strict scoring optimization unsuitable for obtaining biologically meaningful results. The second problem is that very often, similar sequences are known as binding sites for different but related TFs. As already discussed, such TFs (e.g., within the STAT or NFkB families) can actually bind to their mutual binding sites quite well. One solution to take these biological facts into account is the concept of matrix families uniting closely similar TFBS matrices into groups that are then used to score DNA sequences, as has been successfully implemented in the MatInspector algorithm based on grouping of TFBS matrices by self organizing maps [23].

### *17.5.3 Threshold Issues*

Regardless of the model used to describe TFBSs there is one common factor to all of them– the use of one or more thresholds to determine whether a given sequence should be considered a match or not. Choosing the right threshold for matrix detection is always a choice between Scylla and Carybdis. A very sensitive approach will minimize the amount of false negative TFBS predictions and thus is oriented towards a complete annotation. However, this inevitably requires accepting large numbers of additional TFBS hits which might be false positives and easily outnumber the true positive predictions (i.e., verified ones) by an order of magnitude. Just to complicate things even more, all the TFBS methods are designed to detect physical TFBS, i.e., TFBS that will be able to bind to the respective TF, e.g., in a gel shift assay. However, most of the time researchers are not interested in such physical binding sites, but want to know about binding sites that are functional with respect to the transcriptional control of a particular gene. Due to the context-dependency of the TFBS functions, a rather weak match may be the biologically functional one, while a very strong match elsewhere is void of biological function in a particular context.

For these reasons every threshold used in finding TFBS matches in sequences is necessarily a compromise between sensitivity, selectivity, likelihood of TF binding and some kind of 'black box' called biological functionality. What became more than clear over the past decade is that there is definitely no single threshold of the one-size-fits-all type. Matrices are quite different in length and relative sequence conservation, which mandates a more differentiated treatment of thresholds. This is the reason why MatInspector uses individually optimized thresholds for all matrices.

Again, the dilemma between finding TFBS individually (the physical TFBSs) and functional requirements (such as in frameworks and transcriptional modules) has an influence on threshold selection. While a relatively low threshold does not make any sense in scanning large sequences, it is perfectly fit – even required – in a close functional context such as a transcriptional module. The reason is that cooperative binding often substitutes for the requirement of a strong TFBS, as a weaker TFBS might be sufficient in connection with the protein-protein interaction between the two binding TFs. However, such information cannot be retrieved or used by any means if the method focuses on the detection of single TFBSs. Therefore, TFBS-model thresholds remain a thorny issue and whatever choice is made it remains suboptimal in other circumstances.

Just to complicate things a bit further it is also becoming increasingly difficult to determine true positive and true negative matches of TFBS models due to the context dependency of functional "truth". A TFBS void of function in one tissue may very well be functionally required within the same promoter in another context (Fig. 17.1).

## 17.6 How to Define Unknown Transcription Factor Binding Sites?

Especially in the light of an ever-growing number of full-scale genomic sequencing projects, it is very important to go after new, unknown binding sites. However, since the generation of a weight matrix requires a set of known binding sites, how is this accomplished for unknown binding sites?

There are several ways to get out of this dilemma, at least partially. So far, there is no way to go for (TFBSs) from a large totally anonymous sequence. Some previous knowledge is required in order to use pattern definition algorithms to produce new patterns that can be turned into IUPAC consensus sequences or nucleotide weight matrices. A very effective and relatively simple approach is the experimental determination of the binding site spectrum for a given protein. This is called SELEX, and is actually an *in vitro* selection of binding sequences from a large collection of random oligonucleotides. Only the sequences with sufficient affinity to the protein will be bound, the rest can be washed away and then sequencing of the bound oligonucleotides reveals individual binding sequences that can be used to derive a nucleotide weight matrix. The catch is that before that can be done an extensive purification of the protein has to be carried out.

Another experimentally oriented approach is the evaluation of expression array data. Here, large amounts of gene probes are arrayed onto a filter or glass chip. This array is then hybridized to the RNA (via complementary DNA) isolated from cells that underwent some treatment. The amount of signal over each spot indicates the approximate level of RNA present for this gene. By comparing such values with other experiments with untreated cells, it is possible to detect which genes changed their RNA levels under treatment. Then it is possible to cluster the genes according to their expression patterns. Analyzing the promoter regions from genes with a very similar expression pattern and at least within a common biologically functional context can be used to identify common patterns from scratch, many of which are transcription factor binding sites. Brazma et al. have demonstrated this approach successfully in the yeast system [29]. Fortunately, promoters in higher eukaryotes, like mammals, are now as readily available as in yeast, due to improved promoter prediction (e.g., [30]) as well as large-scale experimental efforts to locate TSS [31].

Regardless of the way the promoter sequences are acquired, the sequences need to be analyzed for unknown motifs hidden in the set. There are many programs available to go after pattern definition in a set of sequences. The most popular methods are the Gibb's sampler [32], expectation maximization algorithms (e.g., [33], and a variety of other approaches (e.g., [34]).

Discussing all of these algorithms is beyond the scope of this article but all of these methods have specific strong and weak points. There is no single program that will do in all the cases, so a considerable effort in trying out different methods with one set of sequences is still mandatory. The reader is again referred to the excellent review of Tompa et al. [27].

## 17.7 How to Proceed from Physical TFBSs to Functional Context?

It is quite clear that a single sequence will yield few clues to what the functional context might be, especially as the same sequence may contain organizational structures that are linked to many different functional scenarios. Comparative sequence analysis is one of the most powerful methods to deduce regulatory features and organization, because a selection of sequences to be compared can be based on functional similarities of the sequences (e.g., same pathway or expression behavior). Basically, two types of comparative analysis can be distinguished. The first approach compares the regulatory regions, e.g., promoters within one species such as promoters co-expressed under particular conditions (e.g., from microarray studies). The second approach compares only orthologous regulatory sequences across several species (again promoters are the most prominent representatives) in order to elucidate which features and elements remained conserved in evolution. Such features should be closely associated with functional conservation of the corresponding regulatory regions. While comparative analysis within species not necessarily allows distinguishing between pure statistical findings and functional conservation, phylogenetic analysis of orthologous regulatory sequences should indicate the predominantly functionally conserved features. However, intra-genomic comparison may differentiate between individual functions especially when based on selection methods such as microarrays [35], while phylogenetic analysis will always yield a summary over all the conserved functions. Thus, very often a combination of both approaches is the best way to go [36, 37].

Most of this approach has been implemented into the program package GEMS Launcher® (Genomatix Software GmbH), which contains MatInspector®, to locate individual TFBSs and the program FrameWorker to carry out automatic comparative sequence analysis for TFBS-frameworks as well as the program ModelInspector® to locate other regulatory sequences in whole genomes containing the same organizational structures as described in a recent publication [23]. This way, the crucial step from physical to functional sequence analysis becomes possible and the principle has been applied successfully in numerous projects and publications (e.g., [36, 38, 20, 35, 7]).

## 17.8 Summary

Therefore, elucidation of the role of TFBSs in promoters requires taking the functional context (frameworks) into account, which none of the individual detection programs is capable of. Therefore, in the light of all accumulated knowledge about the mechanisms of transcriptional regulation, the title of the chapter should actually read: "The role of transcription factor binding sites in promoters and *in silico* detection of their functional context." Still, finding the individual TFBSs is and remains to be an important task, but stopping there is like collecting all words of a book without looking at the sentences. You will have all the data and almost none of the information contained within.

Frameworking as explained in that chapter is not the only way to elucidate functional context of TFBSs but the only way to do so is based on comparative sequence analysis alone. All other approaches require a lot more experimental evidence such as ChIP data or functional assays (deletion or mutation studies, etc.). Therefore, framework analysis appears to be the natural first step, not excluding or replacing any of the other methods. It is also a very efficient way to take advantage of the enormous wealth of information hidden in the vast amount of genomic sequences available today.

**Acknowledgments** Part of this work was supported by Biochance-PLUS-3 grant 0313724A (Germany).

## References

Werner, T. (2004). Proteomics and regulomics: the yin and yang of functional genomics. *Mass Spectrom Rev* **23**(1), 25–33.

Taneri, B., Snyder, B., Novoradovsky, A., and Gaasterland, T. (2004). Alternative splicing of mouse transcription factors affects their DNA-binding domain architecture and is tissue specific. *Genome Biol* **5**(10), R75.

Scheinman, R. I., Gualberto, A., Jewell, C. M., Cidlowski, J. A., and Baldwin, A. S., Jr. (1995). Characterization of mechanisms involved in transrepression of NF-kappa B by activated glucocorticoid receptors. *Mol Cell Biol* **15**(2), 943–953.

Basak, S., Kim, H., Kearns, J. D., Tergaonkar, V., O'Dea, E., Werner, S. L., Benedict, C. A., Ware, C. F., Ghosh, G., Verma, I. M., and Hoffmann, A. (2007). A fourth IkappaB protein within the NF-kappaB signaling module. *Cell* **128**(2), 369–381.

Pratt, W. B., Morishima, Y., Murphy, M., and Harrell, M. (2006). Chaperoning of glucocorticoid receptors. *Handb Exp Pharmacol*(172), 111–138.

Kodadek, T. (1998). Mechanistic parallels between DNA replication, recombination and transcription. *Trends Biochem Sci* **23**(2), 79–83.

Ziegler-Heitbrock, L., Lotzerich, M., Schaefer, A., Werner, T., Frankenberger, M., and Benkhart, E. (2003). IFN-alpha induces the human IL-10 gene by recruiting both IFN regulatory factor 1 and Stat3. *J Immunol* **171**(1), 285–290.

Panne, D., Maniatis, T., and Harrison, S. C. (2004). Crystal structure of ATF-2/c-Jun and IRF-3 bound to the interferon-beta enhancer. *Embo J* **23**(22), 4384–4393.

Impey, S., McCorkle, S. R., Cha-Molstad, H., Dwyer, J. M., Yochum, G. S., Boss, J. M., McWeeney, S., Dunn, J. J., Mandel, G., and Goodman, R. H. (2004). Defining the CREB regulon: a genome-wide analysis of transcription factor regulatory regions. *Cell* **119**(7), 1041–54.

Douet, V., Heller, M. B., and Le Saux, O. (2007). DNA methylation and Sp1 binding determine the tissue-specific transcriptional activity of the mouse Abcc6 promoter. *Biochem Biophys Res Commun* **354**(1), 66–71.

Kim, J., Kollhoff, A., Bergmann, A., and Stubbs, L. (2003). Methylation-sensitive binding of transcription factor YY1 to an insulator sequence within the paternally expressed imprinted gene, Peg3. *Hum Mol Genet* **12**(3), 233–245.

Ling, G., Wei, Y., and Ding, X. (2006). Transcriptional Regulation of Human CYP2A13 Expression in the Respiratory Tract by C/EBP and Epigenetic Modulation. *Mol Pharmacol.*

Tronche, F., Ringeisen, F., Blumenfeld, M., Yaniv, M., and Pontoglio, M. (1997). Analysis of the distribution of binding sites for a tissue-specific transcription factor in the vertebrate genome. *J Mol Biol* **266**(2), 231–245.

Yamauchi, M., Ogata, Y., Kim, R. H., Li, J. J., Freedman, L. P., and Sodek, J. (1996). AP-1 regulation of the rat bone sialoprotein gene transcription is mediated through a TPA response element within a glucocorticoid response unit in the gene promoter. *Matrix Biol* **15**(2), 119–130.

Sap, J., Munoz, A., Schmitt, J., Stunnenberg, H., and Vennstrom, B. (1989). Repression of transcription mediated at a thyroid hormone response element by the v-erb-A oncogene product. *Nature* **340**(6230), 242–244.

Boutillier, A. L., Monnier, D., Lorang, D., Lundblad, J. R., Roberts, J. L., and Loeffler, J. P. (1995). Corticotropin-releasing hormone stimulates proopiomelanocortin transcription by cFos-dependent and -independent pathways: characterization of an AP1 site in exon 1. *Mol Endocrinol* **9**(6), 745–755.

Bergers, G., Graninger, P., Braselmann, S., Wrighton, C., and Busslinger, M. (1995). Transcriptional activation of the fra-1 gene by AP-1 is mediated by regulatory sequences in the first intron. *Mol Cell Biol* **15**(7), 3748–3758.

Fessele, S., Maier, H., Zischek, C., Nelson, P. J., and Werner, T. (2002). Regulatory context is a crucial part of gene function. *Trends Genet* **18**(2), 60–63.

Arnone, M. I., and Davidson, E. H. (1997). The hardwiring of development: organization and function of genomic regulatory systems. *Development* **124**(10), 1851–1864.

Naschberger, E., Werner, T., Vicente, A. B., Guenzi, E., Topolt, K., Leubert, R., Lubeseder-Martellato, C., Nelson, P. J., and Sturzl, M. (2004). A NF-kappaB motif and ISRE cooperate in the activation of guanylate binding protein-1 expression by inflammatory cytokines in endothelial cells. *Biochem J* **Pt**.

Cavener, D. R. (1987). Comparison of the consensus sequence flanking translational start sites in Drosophila and vertebrates. *Nucleic Acids Res* **15**(4), 1353–1361.

Eddy, S. R. (2004). What is a hidden Markov model? *Nat Biotechnol* **22**(10), 1315–1316.

Cartharius, K., Frech, K., Grote, K., Klocke, B., Haltmeier, M., Klingenhoff, A., Frisch, M., Bayerlein, M., and Werner, T. (2005). MatInspector and beyond: promoter analysis based on transcription factor binding sites. *Bioinformatics*. **21**(13), 2933–2942.

Chen, Q. K., Hertz, G. Z., and Stormo, G. D. (1995). MATRIX SEARCH 1.0: a computer program that scans DNA sequences for transcriptional elements using a database of weight matrices. *Comput Appl Biosci* **11**(5), 563–566.

Kel, A. E., Gossling, E., Reuter, I., Cheremushkin, E., Kel-Margoulis, O. V., and Wingender, E. (2003). MATCH: A tool for searching transcription factor binding sites in DNA sequences. *Nucleic Acids Res* **31**(13), 3576–3579.

Quandt, K., Frech, K., Karas, H., Wingender, E., and Werner, T. (1995). MatInd and MatInspector: new fast and versatile tools for detection of consensus matches in nucleotide sequence data. *Nucleic Acids Res* **23**(23), 4878–4884.

Tompa, M., Li, N., Bailey, T. L., Church, G. M., De Moor, B., Eskin, E., Favorov, A. V., Frith, M. C., Fu, Y., Kent, W. J., Makeev, V. J., Mironov, A. A., Noble, W. S., Pavesi, G., Pesole, G., Regnier, M., Simonis, N., Sinha, S., Thijs, G., van Helden, J., Vandenbogaert, M., Weng, Z., Workman, C., Ye, C., and Zhu, Z. (2005). Assessing computational tools for the discovery of transcription factor binding sites. *Nat Biotechnol* **23**(1), 137–144.

Frech, K., Herrmann, G., and Werner, T, (1993). Computer-assisted prediction, classfication, and delimitation of protein binding sites in nucleic acids. *Nucleic Acids Res* **21**(7), 1655–1664.

Brazma, A., Jonassen, I., Vilo, J., and Ukkonen, E. (1998). Predicting gene regulatory elements in silico on a genomic scale. *Genome Res* **8**(11), 1202–1215.

Scherf, M., Klingenhoff, A., and Werner, T. (2000). Highly specific localization of promoter regions in large enomic sequences by promoterInspector: a novel context analysis approach. *J Mol Biol* **297**(3), 599–606.

Kodzius, R., Kojima, M., Nishiyori, H., Nakamura, M., Fukuda, S., Tagami, M., Sasaki, D., Imamura, K., Kai, C., Harbers, M., Hayashizaki, Y., and Carninci, P. (2006). CAGE: cap analysis of gene expression. *Nat Methods* **3**(3), 211–222.

Neuwald, A. F., Liu, J. S., and Lawrence, C. E. (1995). Gibbs motif sampling: detection of bacterial outer membrance protein repeats. *Protein Sci* **4**(8), 1618–1632.

Cardon, L. R., and Stormo, G. D. (1992). Expectation maximization algorithm for identifying protein-binding sites with variable lengths from unaligned DNA fragments. *J Mol Biol* **223**(1), 159–170.

Wolfertstetter, F., Frech, K., Herrmann, G., and Werner, T. (1996). Identification of functional elements in unaligned nucleic acid sequences by a novel tuple search algorithm. *Comput Appl Biosci* **12**(1), 71–80.

Seifert, M., Scherf, M., Epple, A., and Werner, T. (2005). Multievidence microarray mining. *Trends Genet* **21**(10), 553–558.

Cohen, C. D., Klingenhoff, A., Boucherot, A., Nitsche, A., Henger, A., Brunner, B., Schmid, H., Merkle, M., Saleem, M. A., Koller, K. P., Werner, T., Grone, H. J., Nelson, P. J., and Kretzler, M. (2006). Comparative promoter analysis allows de novo identification of specialized cell junction-associated proteins. *Proc Natl Acad Sci U S A* **103**(15), 5682–5687.

Doehr, S., Klingenhoff, A., Maier, H., Hrabe de Angelis, M., Werner, T., and Schneider, R. (2005). Linking disease-associated genes to regulatory networks via promoter organization. *Nucleic Acids Res* **33**(3), 864–872.

Masuda, K., Werner, T., Maheshwari, S., Frisch, M., Oh, S., Petrovics, G., May, K., Srikantan, V., Srivastava, S., and Dobi, A. (2005). Androgen Receptor Binding Sites Identified by a GREF_GATA Model. *J Mol Biol* **353**(4), 763–771.

# Chapter 18
# *In Silico* Discovery of DNA Regulatory Sites and Modules

**Panayiotis V. Benos**

**Abstract** In this chapter we describe methods commonly applied for pattern representation and some analogies to the physical world of protein–DNA interactions. Next we present the general methodology for *de novo* DNA pattern discovery in a set of promoter sequences with and without prior information and we discuss some well-established algorithms. The common problems of pattern matching, i.e., the prediction of a site based on prior information are discuss along with the contribution of the evolutionary information to the pattern discovery and pattern matching algorithms. Finally, in we introduce the topic of *cis-regulatory modules* (CRMs) and some of the algorithms designed to find them.

**Keywords** Bioinformatics · Genetics · Genomics · Transcription · DNA regulatory regions

## 18.1 Introduction

### *18.1.1 Role of TFs in Regulation of Gene Expression*

Systems Biology aims to the understanding of the interactions between the various molecular components in the cell and between the cell and its extra-cellular environment. Cells respond continuously to a constantly changing environment by adjusting the expression levels of their genes. This applies to responses to extracellular stimuli, developmental needs, and differentiation. A fundamental control of the expression of every gene takes place at *transcription*, during which the genomic DNA is copied ("transcribed") into RNA. The *rate of transcription* is influenced by various factors. On a large scale, the status of the chromatin ("opened" or "closed" form) determines whether the genomic DNA is accessible or not to interacting proteins, which has been associated with activation or silencing of large parts of the genome. On a gene-by-gene basis, transcription regulation is achieved for the most part by the presence or absence of *transcription factor* (*TF*) proteins in the nucleus. TFs recognize short DNA "signals" (typically 6–15 bp long) in the vicinity of the genes' *transcription start site* (TSS) and they have the potential to initiate (*activators*) or repress (*repressors*) the transcription of nearby genes. These "DNA signals" are commonly referred to as *transcription factor binding sites* (TFBSs) or more generally as *cis-regulatory elements*.

P.V. Benos
Department of Computational Biology, School of Medicine, 3501 Fifth Avenue, Suite 3064 BST3, Pittsburgh, PA, 15260, USA
e-mail: benos@pitt.edu

S. Krawetz (ed.), *Bioinformatics for Systems Biology*, DOI 10.1007/978-1-59745-440-7_18, 

#### 18.1.1.1 Organization of the Promoter in Prokaryotes and Eukaryotes

The genomic region with the TFBSs that control the expression of a gene is called the *promoter*. The size and the organization of a promoter differ depending on the organism and the gene under study. Protein coding genes have a set of *core promoter elements*, which are characteristic of the domain the organism belongs. Prokaryotic organisms (like bacteria) have two distinct regulatory elements in their core promoters: the *Pribnow box*, which is located about 10 bp *upstream* of the TSS (typical sequence: TATAAT) and a second element located around position −35 (typical sequence: TTGACA). Eukaryotic core promoters are more variable. However, many of them contain a *TATA box* at position −30 from the TSS (typical sequence: TATAAA) and a *CAAT-box* at around 75–100 bases upstream of the TSS (typical sequence: CCAAT).

Besides these core elements, a number of gene-specific DNA elements regulate the appropriate transcription rate of a gene. These are the target sites of the sequence-specific transcription factor proteins. They are located either upstream of the gene's TSS or downstream, in its exons or introns (for the eukaryotic genes). The distance at which these elements can be found varies from some hundreds of bases in bacteria or single cell eukaryotes (e.g., yeast) to 10–20,000 bases in complex eukaryotes (e.g., in the fruitfly *Drosophila melanogaster*). Sometimes the more distant regulatory elements, also known as *enhancers*, are not sufficient to drive transcription but they rather play an assisting role. Mutations on TFBSs have been associated with diseases, like β-thalassemia in humans. Schematic views of typical prokaryotic and eukaryotic core promoters are presented in Fig. 18.1.

#### 18.1.1.2 Protein-DNA interactions

The DNA *cis*-regulatory sites are recognized by the transcription factor proteins *via* the formation of chemical bonds between the amino acids and the DNA. Hydrogen bonds as well as van der Waals interactions and water-mediated bonds have been observed in crystal structures. Most of these interactions, are not sequence-specific, i.e., they are formed between the DNA backbone and the protein. As such they do not contribute much to the binding specificity except through *indirect reading*, which is associated with the "bendability" potential of the DNA and other DNA

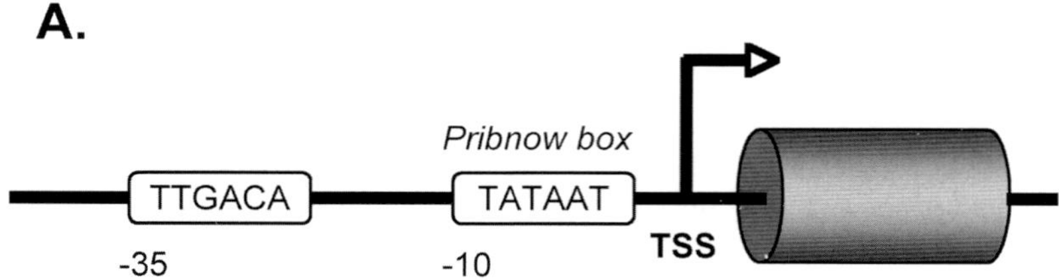


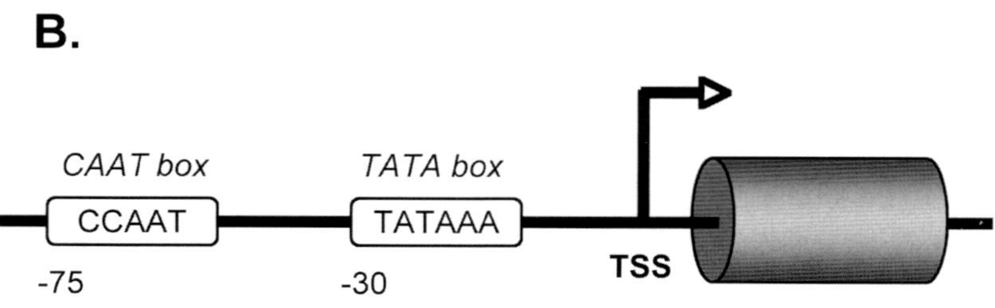


**Fig. 18.1** Schematic view of prokaryotic **(A)** and eukaryotic **(B)** core promoters. The elements of the general transcription factors as well as examples of gene-specific binding sites are depicted (Copies of figures including color copies, where applicable, are available in the accompanying CD)

properties. Nevertheless, most of the sequence specificity is obtained *via* hydrogen bonds between the amino acids and bases.

In most cases, the TF proteins use α-helices to attack the DNA major groove. Each helix can "read" only 3–4 bases, which is an insufficient length for unique target recognition in a genome of thousands or millions of bases. Therefore, TFs utilize multiple α-helices to build composite DNA patterns (e.g., proteins of the $C_2H_2$ zinc-finger family), or a combination of multiple helices and amino acid tails that contact the bases in both the major and the minor grooves (e.g., some homeodomain proteins, like Drosophila *engrailed*). In other cases, proteins with short DNA targets (4–6 bases long) can form homodimers, hence expanding their total target length (e.g., bZIP and bHLH proteins). In addition, eukaryotic organisms are able to inactivate large parts of the genome, by making them inaccessible to TFs through chromatin packaging or *insulator* molecules (like CTCF protein).

In terms of computational searching for DNA patterns in the promoter regions of the genes, two questions are frequently asked: (i) given a set of co-regulated genes (e.g., from microarray experiments), what are the putative DNA binding sites of the (unknown) TFs that are regulated by them? (ii) given a single promoter, does TF *X* regulate the associated gene? A number of methods have been developed to address the first question. The underlying hypothesis of these methods is that all or most of the promoters contain binding sites for the same TF(s). In Section 18.3, we will describe the most important of these strategies and algorithms. For the second question, the straight-forward approach is to scan the promoter of the gene of interest with the binding site pattern of the TFs of interest. In this way, putative sites are predicted based on some user-defined threshold. Typically, this method results in a high number of false positive predictions, which can be reduced by utilizing evolutionary information. The underlying hypothesis here, is that biologically important parts of the promoter (e.g., those containing binding sites of TFs important for that gene's expression) will evolve at a slower rate and thus will tend to be more conserved between species.

However, the hypotheses underlying both these methodologies have been known to be true only to a certain extent. Noise in microarray data and uncertainties associated with the data analyses may result in the inclusion of irrelevant genes in the set of co-regulated genes. This will increase the noise of the method, especially when analyzing promoters of complex eukaryotic organisms, where usually large parts of the genomic DNA upstream and downstream of the TSS are considered. For the second methodology, it has been recently shown that in some organisms TFBSs have a high turnover. In other words, sites that are functional in one organism may become non-functional in another organism and may get replaced by sites in other parts of the promoter region. Despite those known limitations, these general methodologies are used to address these two very important questions.

## 18.2 DNA Pattern Representation

Typically, each TF recognizes multiple DNA target sites in a sequence-specific manner. The target sites are frequently viewed as variants of a "preferred" (*consensus*) site. Given a set of known binding sites of a TF, the first question is: what is the best way to organize the information so that it becomes useful in the search of other, yet unknown, sites? When the number of known TFBSs is small or their variation is limited, they can be used directly in simple string searches to identify new occurrences. This is the case of the *c-myc* TF. The vast majority of the known *c-myc* targets contain the sequence CACGTG, although another target CATGTG or its reverse-complement (CACATG) has also been reported. With such limited repertoire, one can directly scan a genomic DNA sequence and seamlessly identify all potential *c-myc* sites. Given that most of the known *c-myc* sites are CACGTG, biologists will naturally have more "trust" in predictions with this hexanucleotide than in the other two forms when it comes to prioritizing their tests.

### *18.2.1 Representation by Consensus*

Most of the TFs have more than three known target forms, in which case a straightforward method to summarize and present the binding preferences is done by calculating their *consensus sequence*. The targets are aligned on the top of each other and for each position the IUPAC code is used to denote bases and base ambiguities (Table 18.1). A regular expression search can then be employed to predict new sites of this TF. Consensus target representation is used frequently in the literature to effortlessly present positional base preferences, but it has limited value when it comes to searching for new sites. In order to better understand why, let's consider the *c-myc* example above. The consensus representation of the three *c-myc* sites would be CAYRTG. This representation will capture all three known sites, but it will also yield sequence CATATG as putative *c-myc* target, which is not known to be true. For TFs with longer and more variable targets, the consensus may quickly deteriorate into a useless pattern as more variable functional sites are discovered and added to the consensus. Consider, for example, the aligned target sequences of Fig. 18.2A. If we knew only the top eight targets, then the consensus pattern would have been GGRHKTYCCC, which would have detected 2.3 putative "hits" on an average on a random DNA sequence about 100,000 bases long (assuming equal background probabilities for all bases). These are the sites generated under a background model, so they represent the amount of "noise" or false positive predictions one might expect. When all known sites of Fig. 18.2A are considered, the consensus pattern becomes RGRNDNYYMH, which will detect 4.4 sites for every 1,000 bases on average. So, the latter pattern will generate ~200 times more false positive sites than the former, although the former will miss more of the (total) known sites of Fig. 18.2A.

Furthermore, the consensus representation provides no insights on the quantitative nature of the binding. The example of the three *c-myc* sites, it neglects the fact that CACGTG is a much more frequent *c-myc* target than all the rest. Clearly, a probabilistic way to represent these patterns should be more powerful.

**Table 18.1** IUPAC codes for DNA bases and their combinations

| Symbol | Combination | Name |
|---|---|---|
| A | | Adenosine |
| C | | Cytidine |
| G | | Guanine |
| T | | Thymidine |
| M | A C | *Amino* |
| R | A G | *Purine* |
| W | A T | *Weak* |
| S | C G | *Strong* |
| Y | C T | *Pyrimidine* |
| K | G T | *Keto* |
| V | A C G | |
| H | A C T | |
| D | A G T | |
| B | C G T | |
| N | A C G T | |

(Copies of tables are available in the accompanying CD.)

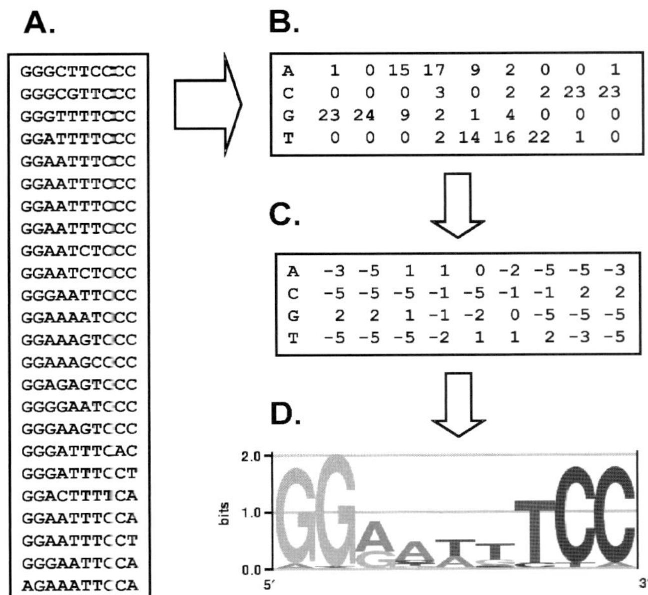


**Fig. 18.2** PSSM model representation. A set of aligned target sites **(A)** can be used to generate a position-specific *count matrix* **(B)** The count matrix is later transformed into a log-frequency matrix **(C)** (with or without background correction). Any of the three forms of information can be graphically represented as a LOGO of symbols **(D)** using the appropriate software (Copies of figures including color copies, where applicable, are available in the accompanying CD)

## *18.2.2 Representation by Weight Matrices*

The *Position-Specific Scoring Matrices* (*PSSMs*; also known as *Position Weight Matrices* or *PWMs*) is the most widely used way to capture and represent binding site information. Like consensus sequences, PSSM models are generated from a set of aligned known sites, but each position is represented by a set of four weights that correspond to the likelihood of each base appearing in this position of the target sequence. The construction method of a PSSM model is illustrated in Fig. 18.2. For each position, $I$, of the alignment, the four-dimensional vector constitutes the log-likelihood of the observed frequency of each base at this position over an expected background frequency (Fig. 18.2C). These weights provide a quantitative measure of how frequently a particular base is observed in the set of known sites as opposed to the background (e.g., in the genome). Positions that are more important for the TF binding tend to have higher log-likelihood ratios. The composite model of all $L$ vectors ($L$ is the length of the pattern) is the $4 \times L$ PSSM model. The average log-likelihood for each position is the *relative entropy*, which is formally defined as:

$$RH(I) = \sum_{b=A}^{T} f(b, I) \cdot \ln \frac{f(b, I)}{P_{ref}(b)} \tag{18.1}$$

where $f(b,I)$ is the estimated frequency of base $b$ at position $I$ of the pattern and $P_{ref}(b)$ is the background frequency of base $b$ (e.g., in the genome). Averaging all the $L$ positions, we obtain the average relative entropy of the motif. A number of motif finding algorithms identify patterns that maximize either the overall log-likelihood of the motif or its relative entropy.

Although they are derived from Shannon information theory, PSSM models have an interesting explanation that is based on the thermodynamic properties of the proteins and DNA [1]. Assuming that a TF, $P$, interacts freely with its genomic DNA targets (i.e., no protein-protein synergistic or competing actions take place and no change in the protein concentration occurs during the interaction), then under equilibrium one would expect that the probability that a given target, $D$, is bound by the TF would be:

$$P(D \cdot P) = \frac{P_{ref}(D) \cdot e^{-H(D,P)/RT}}{\sum_{D_i} P_{ref}(D_i) \cdot e^{-H(D_i,P)/RT}} \tag{18.2}$$

where $P_{ref}(D)$ is the background frequency of $D$ and $H(D, P)$ is the energy of the interaction between the target and the protein. The denominator is the *partition function*, which ensures that the sum of the probabilities over all targets, $D_i$, will be 1. Based on that, the log-likelihood PSSM score (see above) is equal to the negative energy of the interaction in RT units (plus some constant).

PSSM models, like the consensus patterns, assume that the positions in the DNA target are *independent* in their contribution to the overall TF-DNA specificity. In other words, the observed base frequencies at position $I$ are independent of the frequencies in any other position. In equation 18.2, this translates into: the energy function being further partitioned on individual base-amino acid contacts to simplify their calculation. This is known as the *additivity assumption* and has been highly debated over the years. Studies on individual TFs show that although this assumption is not completely accurate, it is still useful in many cases, in the sense that it produces reliable predictions [2]. From a biologist point of view, this is what matters most. Few models have been proposed that consider higher order of position dependencies [3, 4], but the limited data availability for most TFs makes these models inefficient due to the problem of *overfitting*.

### *18.2.3 Scoring Sites with a PSSM Model*

There are many ways to interpret a PSSM model. The most useful one is perhaps as a probabilistic model of a set of known sites (for a very good review, see [5]). According to that, a frequency matrix is a multinomial distribution that can generate target sites. Hence, we can measure the *probability* that a given site, $D_k$, was generated by this frequency matrix by multiplying the probabilities of the $D_k$ bases at the corresponding positions. However, multiplying the probabilities can quickly lead to a memory underflow problem. It is usually more convenient to calculate the sum of log-likelihoods instead. When the PSSM model, $W$, is the log-likelihood ratio of the frequency matrix over the background frequencies, the following score function can be used:

$$S(\vec{D}_k|W) = \sum_{i=1}^{L} \sum_{b=A}^{T} w(b,i) \cdot \delta(D_k(i), b) \tag{18.3}$$

where $w(b, i)$ is the PSSM weight of base $b$ at position $I$ in $D_k$, and $\delta$ is the *Kronecker's delta* function, which is 1 if the $i$-th base of the $D_k$ is $b$ and zero otherwise. Assuming that the PSSM scores have taken into consideration the background frequencies, a positive composite score will generally mean that the sequence, *seq_x,* is more likely to have been produced by the model that generated the real target sites than by the background model. For example, the highest scoring sequence in the model of Fig. 18.2, is GGAATTTCC (score = 14), whereas one of the lower scoring sequences is TTTCCAAAT (score = −38).

### *18.2.4 Visual Pattern Representation with LOGOs*

We, humans, can much easily comprehend and appreciate biological patterns when they are presented in a graphical rather than in a numeric form. In 1986, Schneider and colleagues developed a method to graphically represent the DNA and amino acid patterns [6, 7]. The method represents each position of the multiple-sequence alignment as a stack of symbols (bases, amino acids). For DNA sequences, the total height of the stack at position *I* is equal to:

$$H(I) = 2 + \sum_{b=A}^{T} f(b, I) \cdot \log_2 f(b, I) \tag{18.4}$$

where $f(b, I)$ is the observed frequency of base $b$ at position $I$. This formula calculates the decrease in *entropy* (from a maximum value of two for DNA sequences in $\log_2$ scale) due to uncertainty. For a column with maximum uncertainty (equally probable bases), $H(I) = 0$. For a column with maximum conservation or zero uncertainty [when one base has $f(b,I) = 1$], $H(I)$ takes its maximum value, i.e.,$H(I) = 2$. Once the stack's height, $H(I)$, has been determined, the height of the individual symbols within the stack is calculated proportionally to its relative frequency, $f(b,I)$. The LOGO of the frequency matrix of the alignment in Fig. 18.2A is presented in Fig. 18.2D as an example. Note that the *relative entropy* in equation 18.1 is the *entropy* in equation 18.4 normalized for the background frequencies.

## 18.3 De Novo Pattern Discovery

Let's now focus on the most common problem related to the identification of regulatory sequences (for a very good review of various methods, see [5]). Suppose we have a list of genes that are found to be co-regulated. These, for example, can be genes resulting from microarray or other gene expression data analysis. A reasonable assumption is that a large proportion of these co-regulated genes share a set of TFs that regulate them by binding to conserved DNA elements in their corresponding promoter regions. In most cases the identity of these TFs and their DNA binding preferences are unknown so the problem can be formalized as: *given a set of (unaligned) promoter regions identify common DNA elements that are likely to be targets of some TF.*

Several methods have been developed to address this problem. Typically, these methods search for DNA patterns over-represented in these sequences compared to some "background" or "expected" frequency. The methods differ on the objective function they try to optimize, the computational algorithm they use for this optimization and the background model they consider. Below we describe some general principles of these methods.

### *18.3.1 Finding Patterns with a Greedy Search*

One of the first algorithms for finding patterns in unaligned DNA sequences was *Consensus*, developed by Hertz and Stormo in 1990 [8, 9]. Program *Consensus* (not to be confused with the *consensus representation* of binding patterns described before) uses a greedy algorithm in order to identify the set of sequences of a given length that maximize the pattern's *information content*, formally defined as:

$$IC(pattern) = \sum_{I=1}^{L} \sum_{b=A}^{T} f(b, I) \cdot \log_2 f(b, I) \tag{18.5}$$

where *f(b,I)* is the frequency of base *b* at position *I*. For a given pattern length *L*, the algorithm starts by creating one sequence matrix for each sequence *L*-mer "word". In the subsequent cycles, this matrix is compared against all the other sequences and ranks the resulting pairwise alignments according to their information content. In order to reduce the search space and time, in each cycle, only a percent of the examined patterns is retained (e.g., 10% matrices with the highest information content). In its present form the algorithm manages to keep the calculation cost very low [$O(N^2)$ in time and $O(N)$ in space for $N$ sequences].

### *18.3.2 Finding Patterns with Iterative Optimization*

One general algorithm for the maximum likelihood parameter estimation in probability mixture models, is the *expectation-maximization* or *EM*. In 1990, Lawrence and Reilly applied it for the first time in biological pattern finding [10], although perhaps the most famous implementation as of today is program *MEME*, developed in 1993 by Bailey and Elkan [11]. Assuming a set of observed quantities (e.g., *binding sites*) determined from a set of *hidden* or *missing data* (e.g., PSSM model of TF-DNA binding), the EM algorithm seeks to identify the PSSM model that best explains the observed quantities. However, since the binding sites are also unknown, the EM algorithm iterates between predicted sites and PSSM models. For each iteration the current PSSM model is used to calculate the *expected likelihood* of all subsequences of length *L* in the unaligned promoter data-set. The set that *maximizes* this expectation is the new (predicted) set of binding sites, which is used to calculate the new PSSM model to be used in the next cycle. In other words, in each cycle the algorithm selects the parameters (PSSM model) that maximize the expected likelihood of the observed data calculated over the values of the missing data and model parameters. When the PSSM model does not change significantly or when a maximum number of iterations have been reached, the algorithm terminates. Initializations of the PSSM model can be done by performing a one step EM with all possible motif-starting points and then evaluating the results.

One of the disadvantages of the EM algorithms is that they can be trapped in local maxima. Another very popular algorithm, *Gibbs sampling*, is designed to overcome this problem. Gibbs sampling is very similar to EM, but in each cycle the candidate sites are selected randomly with their probability of selection calculated by their match to the current PSSM model. In this way, suboptimal patterns (with respect to the current PSSM model) can enter the next cycle hence help escaping local optima. Initialization of the Gibbs sampling algorithm is typically done by randomly selecting one "word" from every input sequence and by using them to calculate the first PSSM model. The first Gibbs sampling algorithm implementation for DNA and protein sequences was performed in 1993 by Charles Lawrence and colleagues [12], since then many other algorithms have implemented it [13, 14].

### *18.3.3 Other Pattern Finding Methods*

A number of other strategies have been explored in the search for better methods for DNA pattern identification. These include artificial neural networks like perceptrons and self-organizing maps (e.g., programs *ANN-SPEC* [15] and *SOMBRERO* [16]). Also, they include the so called *dictionary-based* or *word enumerating* methods. The basic principle of the dictionary-based methods is that the biologically important patterns will show relatively small variability and there will be significantly more frequent in the test set (e.g., unaligned promoter sequences) than in some background. Thus, enumeration of all "words" and comparison of their frequencies to the expected background can help in the identification of the most statistically significant ones. In this context, the concept of the "word" can be extended to include patterns with a predetermined variability (e.g., $k$ of $L$ bases to be conserved) or regular expression patterns (e.g., GG[AG]A[TA][TG][TC]CC for the pattern in Fig. 18.2D).

## 18.4 Pattern Matching

Now we switch to a different kind of problem. Assuming we know the binding preferences of a TF in a form of a weight matrix of length $L$, how can we predict if and where it binds in a given promoter? One straightforward way is to use the PSSM model to "scan" the promoter region and score each subsequence of length $L$ (both strands) using equation 18.3. Note that since the weights, $w(b,i)$, in equation 18.3 are a function of the logarithms of the base probabilities from the model of the *observed* (known) sites, the above sum is the log-likelihood that the target comes from that set (corrected for the background base frequencies). Having calculated a score for each subsequence of the promoter region, the initial problem now changes into a problem of determining an appropriate threshold that discriminates between true positive and true negative sites. Unfortunately, this is not a trivial problem. The prediction of TF binding sites is notoriously difficult mainly due to a relatively low signal-to-noise ratio. Part of the problem is the incomplete modeling of the binding preferences. Higher order models have been proposed in the past, but with limited usefulness, due to the small number of known sites for most TFs. However, the high false positive rate may also be an inherent property of the TFs themselves. After all, a TF needs to specifically recognize its targets, but the protein-DNA association also needs to be easily reversible.

Having noted this, there are two main methodologies one can follow to set up a threshold. According to the first method, given a set of known target sites and their resulting PSSM model, a score can be calculated for each of these sites based on the model. Then, a score threshold can be defined as the score that would detect the top $x$ percent of the known targets ($x$ is a user-defined cutoff, e.g., 90%, 95% or 100%). This cutoff will correspond to the expected *false negative rate*.

The second method calculates a score according to a user-defined expected *false positive rate* or *false discovery rate* (*FDR*). In order to do so, we first need a model for the background DNA. For the purposes of this test, *background DNA sequence* is a sequence that is not expected to contain binding sites for the TF of interest. On a first approximation, the background can be modeled by the single nucleotide or higher order frequencies of the genome (di-nucleotides, tri-nucleotides, etc.). A number of random background sequences can be generated based on these frequencies. Alternatively, as background, we can consider the coding nucleotide regions or the intronic regions, since these are not expected to contain TF binding sites (although some times the introns –especially the first intron- may contain TF binding sites). The background sequences will then be scanned and each subsequence will be scored according to the PSSM model. Since the background is not expected to have *any* binding sites, all predictions can be considered "false positives". The score of the top $x$th percentile subsequence is the threshold with $x$ percent FDR. For example, if we have 10 background sequences, each of length 1,000 bp, then the PSSM score of the 10th top scoring subsequence (20th if we scan both strands) will constitute the threshold for FDR = 1%. *MatInspector*, a popular scanning program, uses FDR $= 3\times10^{-4}$ as its default value. We note that the FDR is sometimes referred to as the *p-value* by some researchers.

These two methods for determining the threshold are based on different expectations (false negative *vs.* false positive predictions). So, it will be useful to compare the two. Fig. 18.3 presents graphically the performance of these two methods for four TFs with good PSSM model site representation. For the background, we have used the first order Markov model of 1,000 randomly selected genomic pieces, 5,000 bases long.

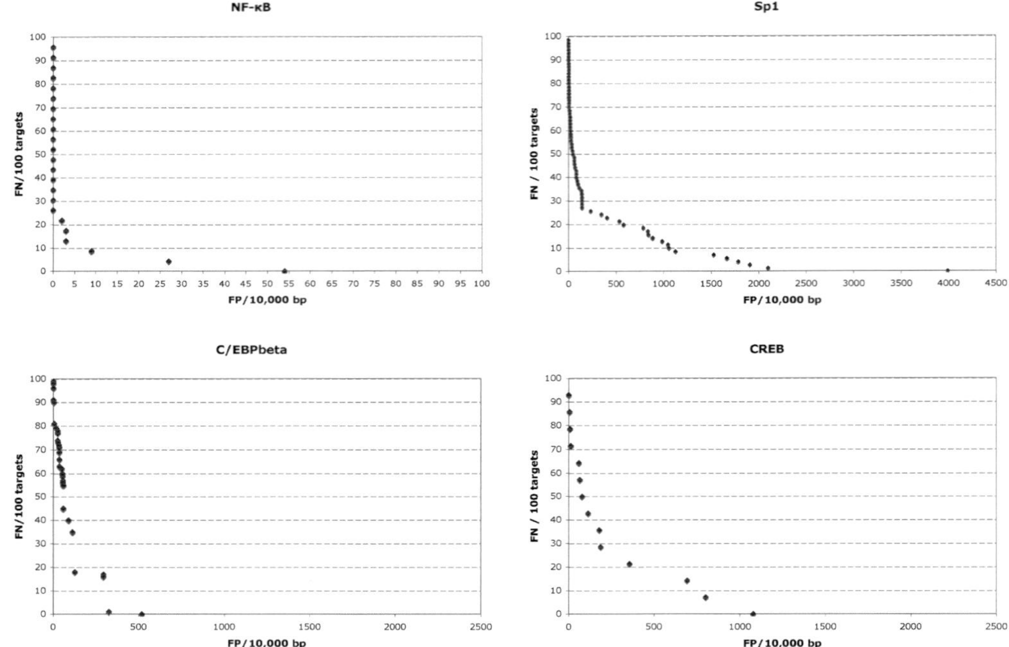


**Fig. 18.3** False positive (FP) and false negative (FN) rate comparison for various TFs. The FP to FN rate is very different for the four well-studied mammalian TFs. These plots illustrate the difficulty one has to establish an objective PSSM threshold for predicting new binding sites (Copies of figures including color copies, where applicable, are available in the accompanying CD)

## 18.5 The Light of Evolution

In order to overcome the problem of the low signal-to-noise ratio on pattern matching algorithms, evolutionary information can be taken into account. The idea is simple–"true" (i.e., biologically relevant) TF binding sites should be conserved between related species. Of course, sites can be lost or replaced and new sites can be generated and the mechanism of this process is still unclear. It has been shown, however, that the functional sites tend to congregate in evolutionary conserved regions. In a recent study, we analyzed the promoters of 513 human genes in 8 other vertebrate species. We compared the promoter percent conservation with the number of known sites found in conserved regions. Table 18.2 shows the results of this analysis (all species compared to human, the

**Table 18.2** Promoter and TF binding site conservation

| Human vs. | No. ortholog. genes | Detectable sites | Promoter conservation | *Sites in conserved regions* |
|---|---|---|---|---|
| Chimp | 512 | 1157 | 94.06% | 94.81% |
| Mouse | 506 | 1146 | 24.20% | 72.34% |
| Opossum | 389 | 912 | 6.72% | 41.23% |
| Chicken | 189 | 451 | 3.21% | 21.73% |
| Tetraodon | 166 | 363 | 2.50% | 12.12% |

Average percent conservation of the 5 kb regions upstream of the transcription start site of 1,162 protein coding genes in six vertebrate species. Percent of 513 known human TF binding sites located in the corresponding conserved regions. Conserved regions are defined as a window of at least 50 bp in length with >65% nucleotide identity. All species conservation is measured with respect to the human genes. Data from [17]
(Copies of tables are available in the accompanying CD.)

reference genome in this study). We see that although both promoter sequences and site conservation rates decrease with the evolutionary distance, the sequence conservation does so much faster.

This idea of evolutionary conservation has been explored by various pattern-matching algorithms, in order to reduce the large number of false positive predictions that are almost inevitable in such searches. This concept is known as *phylogenetic footprinting*. *rVista* [18], a popular phylogenetic footprinting method, scans one of two homologous promoters with PSSM models of known TFs and then evaluates the putative sites based on the degree of conservation of the site themselves and the interval in which they are located between the two species under comparison. Sites should be "globally aligned" in the homologous promoters in order to be reported, meaning, they should be located within a specific window length (typically: 21 bp). *ConSite* [19] is another phylogenetic footprinting algorithm. Unlike *rVista*, *ConSite* scans *both* promoter regions and reports those predicted sites that are located in equivalent positions in the conserved regions of the two homologous promoters. Like in *rVista*, conserved intervals are also calculated using a sliding window approach. *FOOTER* [20] is the newest of these methods. Unlike the other two, *FOOTER* uses both the location and the PSSM score to statistically evaluate the potential of a pair of sites to be functional. Also, predicted sites on both the conserved and non-conserved regions are examined.

## 18.6 *Cis*-Regulatory Modules

Sydney Brenner once said: "complex organisms evolve from simpler ones not by constantly inventing new genes, but by fine-tuning the regulation of existing ones".[1] With the maximum number of genes set to a moderate number of less than 25,000, complex eukaryotic organisms, such as flies and mammals, are expected to have developed complicated mechanisms for gene regulation. This has been shown to be true in many cases. Multiple TFs are co-operating or competing in order for the gene to be properly expressed under a given condition. At the DNA level, the binding sites of competing TFs are frequently overlapping or located in close proximity on the promoter. In this way, when one of the proteins is bound to its target, it inhibits the binding of its competitor. Similarly, the binding sites of co-operating TFs are expected to be located close to each other, so that the proteins can facilitate binding through cross-talk *via* protein-protein interactions on a given promoter. Sometimes, proteins that bind DNA targets, thousands of bases from each other, can "communicate" *via DNA looping* (see Fig. 18.4).

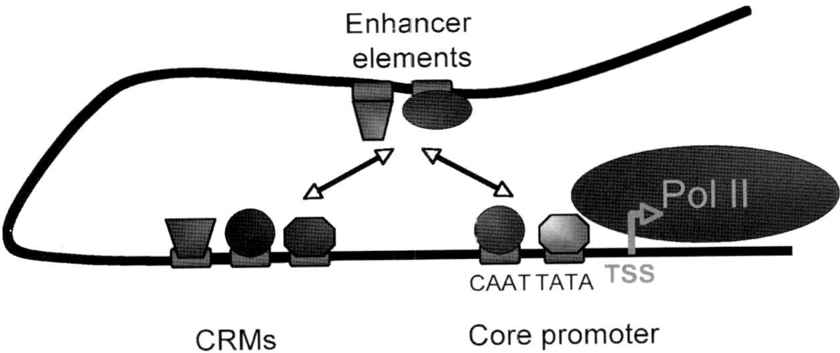


**Fig. 18.4** Schematic view of a typical eukaryotic promoter. The elements of the general transcription factors as well as examples of gene-specific binding sites (CRMs) and enhancer elements are presented. Arrows indicate possible communication between the enhancer binding proteins and the CRMs on the core transcription complex. The interactions can be direct or *via* other proteins (Copies of figures including color copies, where applicable, are available in the accompanying CD)

[1] From Sydney Brenner's seminar "*Return of the human genome*", Washington University in St. Louis, St. Louis, Missouri, 2000.

A number of algorithms have been developed to identify co-occurrences of TF binding sites, also known as *cis-regulatory modules* or *CRMs*. Some of these algorithms incorporate a *de novo* motif finding method with CRM identification. *Co-Bind* [21], for example, applies a Gibbs sampling method to identify concurrently *two* motifs located within a window of a user-defined length. Other methods attempt to assign significance on the observed motif co-occurrences. The significance can be assigned by various methods, including *p*-value calculation (e.g., method *MSCAN*), and Monte Carlo simulations (e.g., method *SCORE*).

Despite the success of such algorithms in identifying some of the known CRMs, it is still unclear what constitutes a CRM or how general their observed properties may be. For example, in some cases the order and spacing of the individual TF binding sites is important. This is the case of the well-known SRY module in the promoters of the *major histocompatibility complex* (*MHC*) genes. In this case, the distance between the individual *cis*-regulatory elements as well as the distance of the whole SRY module from the transcription start site of the downstream gene is important. In other cases, the order and/or the spacing of the individual sites as well as the exact location of the CRM does not seem to be very important. This is the case of many enhancers that have been found to work equally well from tenths of thousands of bases away from the transcription start sites and when placed near it. This variability in the CRM properties makes the design of efficient algorithms for CRM detection more difficult.

## 18.7 On-Line Resources

Although links to on-line resource pages tend to quickly become outdated, we mention here some that we believe will be useful to the readers.

### *DNA Motif and* cis-*regulatory site databases*

- JASPAR (open access database): http://jaspar.genereg.net/
- PLACE (plant): http://www.dna.affrc.go.jp/htdocs/PLACE/
- TRANSFAC (public version): http://www.gene-regulation.com/pub/databases.html

### *Promoter retrieval*

- EPD: http://www.epd.isb-sib.ch/
- dbTSS: http://dbtss.hgc.jp/index.html
- SCPD (yeast): http://rulai.cshl.edu/SCPD/
- TRED promoter database: http://rulai.cshl.edu/TRED/

### *Motif finders*

- AlignACE: http://atlas.med.harvard.edu/
- ANN-SPEC: http://www.cbs.dtu.dk/services/DNAarray/ann-spec.php
- BioProspector: http://robotics.stanford.edu/~xsliu/BioProspector/
- Consensus: http://bifrost.wustl.edu/consensus/html/Html/interface.html
- Gibbs motif sampler: http://bayesweb.wadsworth.org/gibbs/gibbs.html
- MEME: http://meme.nbcr.net/meme/intro.html
- MotifSampler: http://homes.esat.kuleuven.be/~thijs/Work/MotifSampler.html

### *Phylogenetic footprinting*

- ConSite: http://www.phylofoot.org/consite
- FOOTER: http://www.benoslab.pitt.edu/Footer
- rVista: http://rvista.dcode.org/

**Acknowledgments** The author would like to thank Patrick Shea, who kindly provided the data for Fig. 18.3. This work was supported by NSF grant MCB0316255, NIH grants RR014214 and NO1 AI-50018 and by a tobacco settlement grant from the Pennsylvania Department of Health. PVB was also supported by NIH grant 1R01LM007994-01, TATRC/DoD USAMRAA Prime Award W81XWH-05-2-0066 and by intramural funds from the Department of Computational Biology, University of Pittsburgh and the University of Pittsburgh Cancer Institute (UPCI).

## Glossary and Abbreviations

| | |
|---|---|
| PSSM | Position-Specific Scoring Matrix |
| SNP | Single Nucleotide Polymorphism |
| TF | Transcription Factor |
| TFBS | Transcription Factor Binding Sites |
| TSS | Transcription Start Site |

## References

### *Protein-DNA interactions*

1. Benos PV, Lapedes AS, Stormo GD: Probabilistic code for DNA recognition by proteins of the EGR family. *J Mol Biol* 2002, **323**(4):701–727.
2. Benos PV, Bulyk ML, Stormo GD: Additivity in protein-DNA interactions: how good an approximation is it? *Nucleic Acids Res* 2002, **30**(20):4442–4451.
3. Barash Y, Elidan G, Friedman N, Kaplan T: Modeling Dependencies in Protein-DNA Binding Sites. In: *Seventh Annual International Conference on Computational Molecular Biology (RECOMB): 2003*; 2003.
4. Zhou Q, Liu JS: Modeling within-motif dependence for transcription factor binding site predictions. *Bioinformatics* 2004, **20**(6):909–916.

### *DNA pattern representation*

5. Stormo GD: DNA binding sites: representation and discovery. *Bioinformatics* 2000, **16**(1):16–23.
6. Schneider TD, Stephens RM: Sequence logos: a new way to display consensus sequences. *Nucleic Acids Res* 1990, **18**(20):6097–6100.
7. Schneider TD, Stormo GD, Gold L, Ehrenfeucht A: Information content of binding sites on nucleotide sequences. *J Mol Biol* 1986, **188**(3):415–431.

### *De novo motif finding*

8. Hertz GZ, Hartzell GW, 3rd, Stormo GD: Identification of consensus patterns in unaligned DNA sequences known to be functionally related. *Comput Appl Biosci* 1990, **6**(2):81–92.
9. Hertz GZ, Stormo GD: Identifying DNA and protein patterns with statistically significant alignments of multiple sequences. *Bioinformatics* 1999, **15**(7–8):563–577.
10. Lawrence CE, Reilly AA: An expectation maximization (EM) algorithm for the identification and characterization of common sites in unaligned biopolymer sequences. *Proteins* 1990, **7**(1):41–51.
11. Bailey TL, Baker ME, Elkan CP: An artificial intelligence approach to motif discovery in protein sequences: application to steriod dehydrogenases. *J Steroid Biochem Mol Biol* 1997, **62**(1):29–44.

12. Lawrence CE, Altschul SF, Boguski MS, Liu JS, Neuwald AF, Wootton JC: Detecting subtle sequence signals: a Gibbs sampling strategy for multiple alignment. *Science* 1993, **262**(5131):208–214.
13. Liu X, Brutlag DL, Liu JS: BioProspector: discovering conserved DNA motifs in upstream regulatory regions of co-expressed genes. *Pac Symp Biocomput* 2001:127–138.
14. Hughes JD, Estep PW, Tavazoie S, Church GM: Computational identification of cis-regulatory elements associated with groups of functionally related genes in Saccharomyces cerevisiae. *J Mol Biol* 2000, **296**(5):1205–1214.
15. Workman CT, Stormo GD: ANN-Spec: a method for discovering transcription factor binding sites with improved specificity. *Pac Symp Biocomput* 2000:467–478.
16. Mahony S, Hendrix D, Golden A, Smith TJ, Rokhsar DS: Transcription factor binding site identification using the self-organizing map. *Bioinformatics* 2005, **21**(9):1807–1814.
17. Mahony, Corcoran, Benos (2007) Genome Biol **8**:R84.

## *Phylogenetic footprinting in motif detection*

18. Loots GG, Ovcharenko I, Pachter L, Dubchak I, Rubin EM: rVista for comparative sequence-based discovery of functional transcription factor binding sites. *Genome Res* 2002, **12**(5):832–839.
19. Lenhard B, Sandelin A, Mendoza L, Engstrom P, Jareborg N, Wasserman WW: Identification of conserved regulatory elements by comparative genome analysis. *J Biol* 2003, **2**(2):13.
20. Corcoran DL, Feingold E, Dominick J, Wright M, Harnaha J, Trucco M, Giannoukakis N, Benos PV: Footer: a quantitative comparative genomics method for efficient recognition of cis-regulatory elements. *Genome Res* 2005, **15**(6):840–847.
21. Guha Thakurta D, Stormo GD: Identifying target sites for cooperatively binding factors. *Bioinformatics* 2001, **17**(7):608–621.

## Key References

Stormo GD: DNA binding sites: representation and discovery. *Bioinformatics* 2000, **16**(1):16–23.

Bailey TL, Baker ME, Elkan CP: An artificial intelligence approach to motif discovery in protein sequences: application to steriod dehydrogenases. *J Steroid Biochem Mol Biol* 1997, **62**(1):29–44.

Lawrence CE, Altschul SF, Boguski MS, Liu JS, Neuwald AF, Wootton JC: Detecting subtle sequence signals: a Gibbs sampling strategy for multiple alignment. *Science* 1993, **262**(5131):208–214.

Tompa M, Li N, Bailey TL, Church GM, De Moor B, Eskin E, Favorov AV, Frith MC, Fu Y, Kent WJ, Makeev VJ, Mironov AA, Noble WS, Pavesi G, Pesole G, Regnier M, Simonis N, Sinha S, Thijs G, van Helden J, Vandenbogaert M, Weng Z, Workman C, Ye C, Zhu Z: Assessing computational tools for the discovery of transcription factor binding sites. *Nat Biotechnol* 2005, **23**(1):137–144.

# Part V
# Literature Mining for Association and Meaning

# Chapter 19
# Mining the Research Literature in Systems Biology

**Keir T. Reavie**

**Abstract** The research literature in Systems Biology is growing exponentially, making it difficult for the researchers to efficiently search, retrieve, analyze, and apply the literature to new research in a timely fashion. This chapter helps researchers to efficiently search the major biomedical literature databases *PubMed* and the *Web of Science*, stay current with the literature using alert services, and manage personal article citation databases using bibliographic management software like *EndNote*. Tricks for getting the most out of database search interfaces and using them in innovative ways are presented. The chapter also discusses new bioinformatics tools that go beyond the standard database search interfaces and present the published literature in ways that elucidate hidden relationships within the Systems Biology research, which can lead to new information discovery and more productive research.

**Keywords** Databases, bibliographic · Database management systems · Information storage and retrieval · Libraries, digital · PubMed · MEDLINE · Web of science · Science citation index

## 19.1 Introduction

Any researcher who tries to stay current of the literature in Systems Biology knows that it proliferates at an exponential rate and ensuring that the literature can be retrieved, analyzed, and applied to new research in a timely fashion is an increasingly complex task. The *MEDLINE* database, which constitutes the largest part of *PubMed,* contains over 15 million records, with 2,000 to 4,000 new article citations added every week [1], and *PubMed* is not the only venue indexing the extensive Systems Biology literature. This chapter discusses how to efficiently access, stay current with, and manage the published research literature. In addition, we briefly discuss new bioinformatics tools specifically designed to mine the literature and efficiently extract relevant published research using methods that can lead to new information discovery and more productive research.

## 19.2 Searching the Research Literature

### *19.2.1 Overview*

Access to the published literature in Systems Biology is available from a handful of key resources and a large number of new and growing resources that provide access to the literature in different contexts. The most common resource used to search biomedical research literature is *PubMed.*

---

K.T. Reavie
Peter J. Shields Library, University of California, Davis, 100 North West Quad, Davis, CA, 95616, USA
e-mail: ktreavie@ucdavis.edu

S. Krawetz (ed.), *Bioinformatics for Systems Biology*, DOI 10.1007/978-1-59745-440-7_19,

*PubMed* embodies the *MEDLINE* database and is a vital tool in enabling access to the research literature in Systems Biology. *PubMed* indexes articles from the journal literature. The database is freely available to the general public, is well constructed and easy to use.

Another important literature index for searching the Systems Biology literature, although not openly available to the public, is the *Web of Science* (*WoS*), which includes the *Science Citation Index* (*SCI*). The *WoS* is designed to track the research literature via analysis of citations within journal articles. *PubMed* and *WoS* duplicate much of the research literature, but each index has unique information and search features that when used judiciously, enable researchers to complete comprehensive searches and stay current with the literature.

Many researchers will also be familiar with search systems like *Google Scholar*, mainly for their ease of use and ability to retrieve large comprehensive sets of research publications. We will not discuss *Google Scholar* in this chapter, as the data from which it extracts information is not as comprehensive as that found in databases like *PubMed* and *WoS*, and it does not allow for very precise subject searching. *Google Scholar* is useful for locating specific articles, or articles by specific authors and institutions, but it is not recommend for efficient subject searching and staying current with the literature.

## *19.2.2 PubMed*

### 19.2.2.1 MEDLINE

*MEDLINE* is a database produced by the National Library of Medicine (NLM), that indexes clinical and biomedical research literature. The database originally enabled access to literature back to 1966. *PREMEDLINE*, which included earlier citations to the literature, has been absorbed into *MEDLINE*, enabling access to literature dating back to approximately 1950. Older material is being added to *MEDLINE* all the time, pushing this date back on a continuing basis. *MEDLINE* now indexes articles from approximately 5,000 journals. The database currently holds approximately 15 million references [1].

The subject coverage in *MEDLINE* is extensive. It is easily the most relevant database for searching biomedical research literature, especially in the area of Systems Biology. *MEDLINE* is available through a number of third party vendors that provide a variety of search interfaces, both free and commercial, to search the database. *PubMed*, produced by the National Center for Biotechnology Information (NCBI) and NLM, is one of the best search interfaces for access to *MEDLINE*, particularly when searching the literature in Systems Biology, as it provides numerous links for easy access to information in the NCBI suite of databases.

### 19.2.2.2 Medical Subject Headings

All citations entered into the *MEDLINE* database are indexed using the *Medical Subject Headings* (*MeSH*). *MeSH* is the NLM indexing vocabulary. It is updated on a regular basis [2] and it enables the NLM indexers to control the terminology used to index articles in *MEDLINE*. It is also invaluable to the researcher, who can use this vocabulary to extract information from the database. By using *MeSH* in the search process, researchers can be both precise and comprehensive in retrieving citation records from *MEDLINE*.

Citation records are added to *PubMed* everyday. However, the indexing of each citation can vary. Important journals are indexed quickly (perhaps within days) and more obscure titles may not be indexed for several months. Because of the time lag in the indexing process, recent citations in *PubMed* cannot be retrieved by searching the *MeSH* field of the database, as the most recent citations do not yet have the *MeSH* terms added. Fortunately, *PubMed* provides an algorithm to enable easy retrieval of both indexed and un-indexed citations (see next Section of this chapter).

#### 19.2.2.3 PubMed

*PubMed* provides an easy to use interface for basic searching but also provides sophisticated search features that enable researchers to quickly access the literature, perform precise searches, and retrieve a comprehensive list of citations for journal articles. The main features of the *PubMed* interface that facilitate searching are the mapping algorithm and the **MeSH Database**.

The *PubMed* screens are laid out with links, to search services in a menu, to the left. Tabs below the main search box provide easy access to the variety of search features. There is also a context-sensitive **Help** link in the left-hand menu that can be used at any time to get assistance with a service or feature.

Any search phrase entered into the main search line of *PubMed* is processed using a mapping algorithm that attempts to identify and then use *MeSH* terms to complete the search. The mapping algorithm can also recognize and extract author names and journal titles from the author and journal title indexes, and then formulate and complete a search based on these terms. The mapping algorithm is designed to improve your search, to make it specific, as well as comprehensive and to ensure that no important references are missed. Be aware, however, that the algorithm is not infallible. Depending on the search terms entered, *PubMed* can easily mistake author names and journal titles for *MeSH* terms and retrieve irrelevant search results. To ensure relevant retrieval it is a good idea to think carefully about your search terms and search strategy before approaching *PubMed*. Searching may also require the use of synonymous terms and trial and error to locate the relevant literature. One cannot always expect *PubMed* to provide relevant retrieval the first time a search is performed. It is an iterative process that often requires many steps.

Basic PubMed Searching

To see how *PubMed* operates let's try a simple search to retrieve articles on the mapping of protein interactions in yeast. The search may look something like this:

```
protein interaction map AND yeast
```

Protein interaction map and yeast are combined using the Boolean operator **AND**, to retrieve citations that discuss both of these topics together. The two other Boolean operators used when searching *PubMed* are **OR** and **NOT**. **OR** will retrieve articles on one topic or the other, or both, and **NOT** will remove any articles on a selected topic from the search set. Boolean operators should be entered in the upper case when searching *PubMed*. When PubMed sees a Boolean operator in the upper case, it knows to treat it as a Boolean operator. If it is in the lower case, it will attempt to treat the term like any other in the search phrase. This is particularly important when combining search sets on the **History** screen, which we will discuss in a moment.

When searching *PubMed* from the main search box, the mapping algorithm breaks the phrases into components to identify relevant *MeSH* terms for the search strategy. When the algorithm fails to find *MeSH* terminology, it will break the search phrase into smaller parts and continue the mapping process. If Boolean operators are not present in the search phrase, *PubMed* will automatically assume an **AND** operator and place it between the phrases and *MeSH* terms. To avoid any confusion in the search strategy, it is a good practice to always include relevant Boolean operators in the search phrase.

Enter the search into the *PubMed* search box and hit **Enter** or click the **Go** button to the right of the search box. *PubMed* will respond with a list of citations. To understand how *PubMed* retrieved the citations, click the **Details** tab to view how the mapping algorithm identified the *MeSH* terms and then performed the search. For the above search, the mapping algorithm creates the following search strategy:

```
(("protein interaction mapping" [TIAB] NOT Medline[SB]) OR "protein interaction mapping" [MeSH Terms]
OR protein interaction map[Text Word]) AND (("yeasts" [TIAB] NOT Medline[SB]) OR "yeasts" [MeSH
```

Terms] OR ("saccharomyces cerevisiae" [TIAB] NOT Medline [SB]) OR ""saccharomyces cerevisiae" [MeSH Terms] OR yeast [Text Word])

Some of the information in this search strategy is a little complicated and incomprehensible, but it is important to note that *PubMed* was able to identify the *MeSH* terms protein interaction mapping and saccharomyces cerevisiae as the most appropriate for this search strategy. *PubMed* also searched the *MeSH* and originally entered terms as text words in citation titles and abstracts. In this way, *PubMed* is attempting to make the search more comprehensive, so that you do not miss any relevant articles. It is important to search the title and abstract fields, since the most recent literature added to *PubMed* may not yet be indexed with *MeSH* terms. If we perform *PubMed* searches on the *MeSH* field only, these recent articles will not be retrieved in our search results.

You will notice that *PubMed* uses parentheses to provide an order in the search strategy. When the *PubMed* interface sees parentheses, it will process the Boolean operators within the parentheses first. If you do not use parentheses in a *PubMed* search, *PubMed* processes the Boolean operators from left to right, which can lead to irrelevant search results. Most search strategies will require that all **OR** operators be executed first, so when using **OR** to combine terms, parentheses should be placed around the **OR** phrase. You can see how the mapping algorithm has done this in the above search strategy from the **Details** page.

Note that the **Details** screen allows the searcher to edit the search and run it again against the database by clicking on the **Search** button. This enables quick and easy search modification, to retrieve additional and more relevant citations.

### Using the MeSH Database

The **MeSH Database** is an excellent tool to help increase the accuracy of searches in *PubMed*. The database enables a researcher to identify *MeSH* terms, build a search strategy, and then run the strategy against the *PubMed* database. The **MeSH Database** link is in the **PubMed Services** list in the left-hand menu.

If we enter yeast into the **MeSH Database** search box, it will respond with a list of *MeSH* terms from which you can select the appropriate term to search. You will see that saccharomyces cerevisiae appears at the top of the list. The **MeSH Database** uses its cross-referencing features to identify appropriate *MeSH* terms. It identifies terms differently than the mapping algorithm in the *PubMed* search interface, so you may find that the **MeSH Database** does not always find appropriate *MeSH* terms in the way the mapping algorithm is capable.

Select the appropriate *MeSH* term by clicking on its link. This will open the **Detailed MeSH** screen, which includes the term, a definition of how it is defined for use as an index term in *PubMed*, the year it was added to *MeSH* as an index term, and a list of **Subheadings** that enable the searcher to focus the search to more specific aspects of a subject. If we were specifically interested in locating research on yeast genetics, we could select genetics from the **Subheading** list by selecting the box to the left of the term. This will focus the search on that specific topic. Information on how each **Subheading** is defined can be found in the **PubMed Help**. Any combination of the **Subheadings** can be selected and included in the search strategy. If two or more **Subheadings** are selected, the **MeSH Database** will place an **OR** operator between the search terms. For example, if we are interested in the genetics and cytology of saccharomyces cerevisiae, we would select those two **Subheadings** and the **MeSH Database** would create the following search:

"Saccharomyces cerevisiae/cytology" [Mesh] OR "Saccharomyces cerevisiae/genetics" [Mesh]

The **MeSH Details** screen also enables the researcher to limit a search to only those articles where the topic is defined as a major subject. To do this, select the **Restrict search to Major Topic headings only**. All citations in *PubMed* are indexed with major and minor topics. This will reduce the size of your search retrieval set. It is recommended that you do not use this feature unless necessary, for example, if your search retrieval set is very large. If you limit to the major topics only, there is a chance that you could miss some relevant articles where the subject being searched was considered minor by the indexer.

In addition, it is possible to turn off the **Explode** function when searching *MeSH* terms. All *MeSH* terms are ordered in a hierarchy of terminology from general to specific subjects. At the bottom of the screen, you will see where the *MeSH* term selected, is located in the **MeSH Tree** hierarchies. When using *MeSH* terms to search *PubMed* the **Explode** function is turned on all the time, unless you request it be turned off by selecting the **Do not Explode this term** option. The **Explode** function in *PubMed* searches all subordinate terms to the *MeSH* term you have selected, enabling the retrieval of citations indexed with more specific *MeSH* terms. This is important in *PubMed*, since the rule for indexing articles in the database is to use the most specific *MeSH* term available to the indexer. For example, if we search sequence analysis in the **MeSH Database**, we can see in the **MeSH Tree** that there are a number of more specific *MeSH* terms listed as subordinate to sequence analysis, such as sequence analysis, protein. By searching sequence analysis we can retrieve all articles indexed under sequence analysis and all the more specific terminology subordinate to it. In this way we are able to make our search more comprehensive, and not miss important citations. It is recommended that you leave the **Explode** feature turned on at all times when searching *PubMed*. The **Explode** function is also in play when searching from the main *PubMed* search box. Once the mapping algorithm identifies *MeSH* terms for the search strategy, it will **Explode** those terms when it performs the search.

Once you have selected a *MeSH* term or **Subheading**, it can be added to a search box by using the **Send to** pull down menu near the top of the screen. With the first term you can select any of the **Search Box** options in the menu. A search box containing the selected search term will appear. It is now possible to look up additional *MeSH* terms and add them to the box as well, in order to build more complex searches, usually using the **Search Box with AND** option in the **Send to** menu, to combine terms using the **AND** Boolean operator. You will see that the **OR** or **NOT** operators are also available to help you build a search. The search constructed in the search box can be run against the *PubMed* database at any time by clicking the **PubMed Search** button below the box.

### Search Limits

After completing the initial subject search in *PubMed* the search results can be furthered filtered using the **Limits** options. **Limits** are accessible through the **Limits** tab below the search box. The most common search limits for filtering searches include: date limits, both publication date and date entered into *PubMed*; **English** language; and **Human** or **Animal** studies. A quick browse of the **Limits** page will show that searches can also be limited by: **Type of Article**, such as **Review** articles; studies on a particular **Gender**; studies on specific **Age** groups; **Subsets** of journal collections or subject areas in *PubMed*; and to articles that are available in **Full Text** online. You will notice that there is also a limit to provide access to articles from journals that are available in **Free Full Text**. This will limit your search citations for articles that are freely available online. It is recommended that the researcher use limits sparingly when they are trying to complete comprehensive searches. Any one limit has the potential to filter out an important article.

### Display Options

When the *PubMed* search (Fig 19.1) has been completed, *PubMed* will respond with a list of articles in a **Summary** format, which includes authors, article title, journal title, journal volume, issue, page numbers, and the **PMID**, or **PubMed ID**. All citations in the *PubMed* database have a unique **PMID**. This ID can be used to quickly retrieve the article in future by simply typing it into the *PubMed* search box and clicking on **Go**. This ID is also used to track the article and obtain a copy if

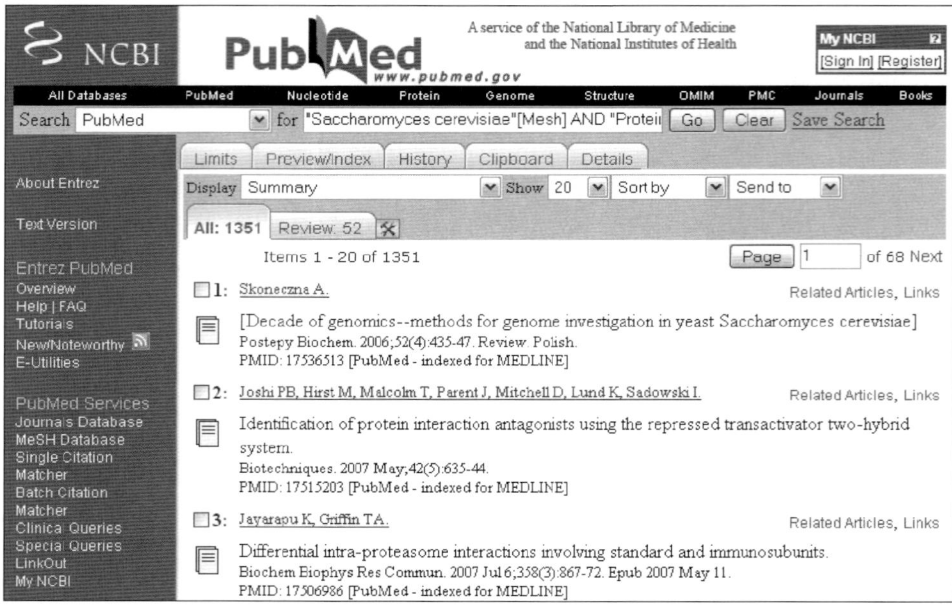


**Fig. 19.1** The PubMed search retrieval screen (Copies of figures including color copies, where applicable, are available in the accompanying CD)

it is not readily accessible online or from a local library. You should consult with your institution's library to get help for retrieving articles not readily available online or in the library.

If you pull down the **Display** menu at the top of the *PubMed* retrieval screen you will see numerous display formats. Many of these are formats enable linking to other NCBI databases, including links to gene and protein information. It is recommended that you try these links to get an idea of the ease the researcher can link from the literature to gene and protein sequence data.

The basic options for displaying citations in *PubMed* are **Abstract**, **AbstractPlus**, **Citation,** and **MEDLINE**. The **Abstract** format displays the **Summary** information, as well as the first author's institutional affiliation, the abstract and full text links. **AbstractPlus** displays the same information as the **Abstract** display, but also includes links to what *PubMed* considers **Related Articles**. These links can lead researchers to other similar articles in the database. **Related Articles** are derived from an algorithm that looks at similarities in *MeSH* terminology, and words and phrases in titles and abstracts of citations. They are ordered from most relevant to least relevant, so they are not in chronological order like references on the main retrieval screen. If you click on an author's name on the main *PubMed* retrieval page, the single citation is displayed in the **AbstractPlus** format. The **Citation** display is the same as **Abstract**, but also includes the *MeSH* index terms for the citation. The **MEDLINE** format displays all fields in the *PubMed* records with a two or four letter tag. The **MEDLINE** format is read by bibliographic management software like *EndNote*, *Reference Manager*, *ProCite* and *RefWorks* (see Section 19.2.3.5 below). You should use this format when downloading *PubMed* references to be loaded into these software packages.

By default *PubMed* displays 20 references on the retrieval screen. You can display more references by clicking the **Show** menu and selecting a larger number, up to 500.

### Print and Download Options

*PubMed* citations can be printed or saved in a variety of ways using the **Send to** menu at the top of the citation list. Before saving or downloading records, select the citations you wish to keep by clicking the box to the left of the citation. If you do not select any citations, *PubMed* will automatically print or save the entire retrieval set. Next, select the display method in which you want to print or download the references. Remember that saving references to be read into bibliographic management software like *EndNote* requires the records to be in the **MEDLINE** format. After selecting the display format, select the appropriate option from the **Send to** menu. The **Printer** option creates a page without all the *PubMed* interface graphics, so the references can be easily printed. The **Download** option will invoke the download operation in your operating system. The **Send to** menu also enables you to **Email** references to yourself or another researcher, and set up an **RSS** feed for your *PubMed* search. RSS enables quick retrieval of new citations in *PubMed* from within a web browser's bookmark menu. We will discuss this more in **Section 19.3.2**.

Another option you will notice in the **Send to** menu is **Clipboard**. The **Clipboard** provides a temporary storage area into which you can save citations. This is a useful feature when performing multiple searches in *PubMed*, as it enables you to save citations as you go. You can open the **Clipboard** by selecting the **Clipboard** tab, and then print or download all the citations saved in the **Clipboard** as a group.

#### Access to Full Text Articles Online

Access to full text articles online is available from the **Abstract** or **AbstractPlus** displays in *PubMed*. To the top right of each abstract there is a publisher's link to the full article. Note that the publisher's link will only appear in *PubMed* if the publisher has made an agreement with *PubMed* to allow the link. Not all abstracts will have a full text link, and citations without a link may still be available online. Alternatively, your institution may be using one of *PubMed's* linking services, which allow institutions to set up links to online journals for which it has subscriptions. If this is the case, you will see another button next to the publisher's, which links to your institutional subscriptions. This linking service will vary from institution to institution, so it is important to check with your library to obtain information on how it works.

#### Search History

Another useful tool in *PubMed* is the **History**, which can be accessed by selecting the **History** tab. The **History** screen displays all recent *PubMed* searches. For researchers performing more complex searches, **History** enables the creation of numerous searches that can then be combined in different ways using Boolean operators, to obtain different search results.

### *19.2.3 Web of Science*

The *Web of Science (WoS)* is comprised of three databases: the *Science Citation Index* (*SCI*), *Social Science Citation Index*, and the *Arts and Humanities Citation Index*. In this section we are concerned only with *SCI*, since it is the most relevant database for Systems Biology. *WoS* is produced by the *Institute for Scientific Information* (*ISI*), who developed and maintained a number of resources related to journal citation analysis. *WoS* is the search interface for the citation indexes. The citation indexes as a whole contain information from approximately 9,300 journals, with well over 6,000 of these titles related to scientific research. The entire *WoS* database has over 30 million references [3]. The great strength of *WoS* is that it tracks journal article citations. *WoS* is designed to enable quick access to the articles that have been cited by others, making it possible to easily follow research fronts in the literature and locate new articles on related research. Note that the *WoS* is a commercial resource and is not freely available like *PubMed*. You will need to be located at an institution that subscribes to this resource to obtain access.

#### 19.2.3.1 Subject Searching the SCI in WoS

Subject searching on the *WoS* is more difficult than *PubMed*, mainly because the *WoS* does not provide a controlled vocabulary like *MeSH* to assist in the search process. We are left to search *WoS* with the vocabulary that we, the researcher, thinks is most appropriate. We need to think carefully about search terms and any synonymous vocabulary that may be relevant in retrieving records from the *WoS*, so that we do not miss important citations. *WoS* does provide for the searching of author provided keywords, which are extracted from the journal publication, and something called **Keywords Plus**. **Keywords Plus** terms are derived by scanning the titles of references in the indexed articles and extracting commonly occurring terms and phrases. Author keywords and **Keywords Plus** do provide some capacity to make searches more precise and comprehensive, but we need to remember that they are derived from diverse and uncontrolled sources, unlike *MeSH*, which is maintained by an editorial board. In summary, subject searching in the *WoS* can be problematic, especially given its enormous size.

Subject searches can be performed from the main search page. It should be noted that the *WoS* is a great database to search for publications by authors from specific institutions. *ISI* maintains a comprehensive database of author addresses, and normally provides the addresses of all authors on an article in the citation record. Addresses can be searched in the **Address** line.

Subject searches can be entered into the **Topic** search box. Terms are searched in the keywords, titles, and abstracts of citations. Because there is no controlled vocabulary, if we want to search for articles on yeast, we should use the search phrase yeast OR saccharomyces cerevisiae to retrieve all citations on yeast, since the authors may have used either yeast or saccharomyces cerevisiae as a keyword. In *WoS,* the Boolean operators do not need to be in upper case, like in *PubMed.* However, it is good practice to search this way, to be consistent in your approach to searching the literature in all databases. The other search term from our example search, protein interaction map, should be truncated using an asterisk (*). Again, when using a controlled vocabulary like *MeSH*, truncation is not as important, but in this case, searching for protein interaction map* will retrieve all citations where authors have used the terms map, maps, and mapping, etc. as keywords. The full search strategy would look like the following:

(yeast OR saccharomyces cerevisiae) AND protein interaction map*

Since we are using two different Boolean operators, **AND** and **OR**, in our search strategy, it is important to use parentheses like we would in *PubMed*, to ensure the search terms are processed in the correct order. Search phrases in parentheses are processed first by the *WoS* interface. The *WoS* search interface processes **AND** operators first, followed by **OR** operators, unless it sees parentheses. Without parentheses on this search strategy, we would retrieve all articles using the keywords saccharomyces cerevisiae and protein interaction map*, and then all articles using the keyword yeast, whether or not the articles have anything to do with protein interaction maps.

#### 19.2.3.2 Citation Searching

The strength of the citation indexes is that they index and provide access to the article references for all citations in the database. This enables cross-referencing of articles and discovery of new research through citations and cited research. The **Cited Reference Search** option in *WoS* allows us to track who has cited a particular article or author. In other words, we can take a published research article that is 5 or 10 years old and enter the citation information into *WoS* to see who has cited it, and thus obtain access to more recent research.

Select **Cited Reference Search** button on the top left of the *WoS* screen to do a cited reference search. Simply enter an author's name to retrieve a list of all articles that have cited that particular author, or be more specific and add a journal title (in the search box called **Cited Source**) and year, to see who has cited a specific article. Note that when entering the **Cited Source** information, *WoS* requires that you use the journal title abbreviation from the *ISI* journal title index. Click the search link and enter the beginning of a journal title into the search box. This is a little cumbersome if you have no idea what the abbreviation might be, so this process can take a bit of browsing. Once the title abbreviation is found, click the **ADD** button to the left of the title. The title will appear in a box at the bottom of the screen. Click the **OK** button to add the title to the **Cited** box.

If we enter only an author's name into the **Cited Reference Search**, *WoS* will provide a list of all articles by that author. On the far right are links that enable you to view the actual citation for an article in *WoS*. Some articles in the list may not have a link to a citation in the database, either because they are too old or come from a publication that is not indexed in the citation indexes. On the retrieval screen select all the citations for which you wish to view the citing articles, and then click the **Finish Search** button at the top or bottom of the screen. *WoS* will provide a list of citations for all articles that have cited the selected citations. Be careful that you select all citations in which you are interested. The *WoS* contains many typos that transfer directly from typos in the reference lists of journal articles. The same publication may appear in a variety of ways on your cited reference retrieval list.

To trace even more relevant research using the cited reference search feature, use two or three journal articles on the topic of interest. Look up each article separately in the **Cited Reference Search** and then use the *WoS* **Search History** feature to identify recent articles that have cited all of these publications together. Once you have completed each **Cited Reference Search**, click the **Search History** button on the top of the screen. You will see a list of all your searches. Select the articles you wish to combine in the history by checking the boxes to the left of each search. Select the **AND** option in the **Combine Sets** area and then click the **Combine** button. The *WoS* will provide you with a set of articles that cite all the articles in your searches. The citations retrieved in this combined set should be more relevant, since they cite more than one related article.

#### 19.2.3.3 Limiting Search Results

The *WoS* search retrieval screen has a number of options that enable the researcher to filter a search retrieval set. Select any option in the **Refine results** links to the left of the retrieval list. A window will open that provides options to limit the search to **Subject Categories**, **Source Titles**, **Document Types**, **Authors,** and **Publication Years**. For Example, selecting **Source Titles** will display a list of journal titles that appear in the search retrieval set, indicating how many articles in the set appear within that journal. Search retrieval can be limited to specific journals by selecting the box to the left of the title, and then clicking on the **View Records** button.

The *WoS* also allows researchers to limit a search by searching for additional keywords within the current retrieval set. Use the **Search within these results** search box to perform searches on the retrieval set.

#### 19.2.3.4 Display Options

*WoS* displays 10 records per page. Up to 50 records can be displayed per page by changing the **Show 10** menu to either 25 or 50. The records are displayed with the basic citation information: author, title, journal, volume, issue, and page numbers. WoS does not provide additional display options to view the abstracts. Abstracts must be viewed one at a time by clicking on the citation's title. The abstract display includes the abstract, keywords and author addresses, as well as links to **Cited References** and **Times Cited**. These links provide access to the article references and those articles that cite the one being displayed.

#### 19.2.3.5 Print and Download Options

To the left of the abstract display in *WoS* there is a box with the heading **Output This Record**, which provides numerous options for printing, downloading, or transferring the record to bibliographic management software. You can also send records to the *WoS EndNote Web* (see Section 19.5). This box may also contain your institution's link to the full text article online, assuming they have this option set up and subscribe to the online version of the journal. This link also appears on the main search retrieval screen under each citation. You should check with your institution's library to obtain more information about access to full text journal articles online, as the process will vary from institution to institution.

While you can print and download records individually in *WoS*, it is best to browse and select the records you wish to save by clicking the box to the left of each citation. After selecting the records you want to save, click the **Add to Marked List** button to the right of the screen. The **Marked List** is a temporary storage area for citations. The **Marked List** button on the top of *WoS* pages opens the marked or saved list of citations, so that it can be printed, downloaded or transferred as a group.

#### 19.2.3.6 Analyzing WoS Search Retrieval

Recent versions of the *WoS*, enable analysis of search results to identify authors, research institutions, and journals that appear frequently in search retrieval sets (Fig. 19.2). To analyze search results, click the **Analyze** button in the left hand menu of the search retrieval screen. The **Analyze** tool allows ranking of the citations by **Author**, **Country/Territory**, **Document Type**, **Language**, **Institution Name**, **Publication Year**, **Source Title** (Journal), and **Subject Category**. The subject categories are defined by *ISI*. These are broad subject areas used to order journals in the *ISI Journal Citation Reports*, which ranks journals in the subject categories based on how often their articles are cited. *WoS* is able to analyze up to 100,000 records, so you could do very broad subject

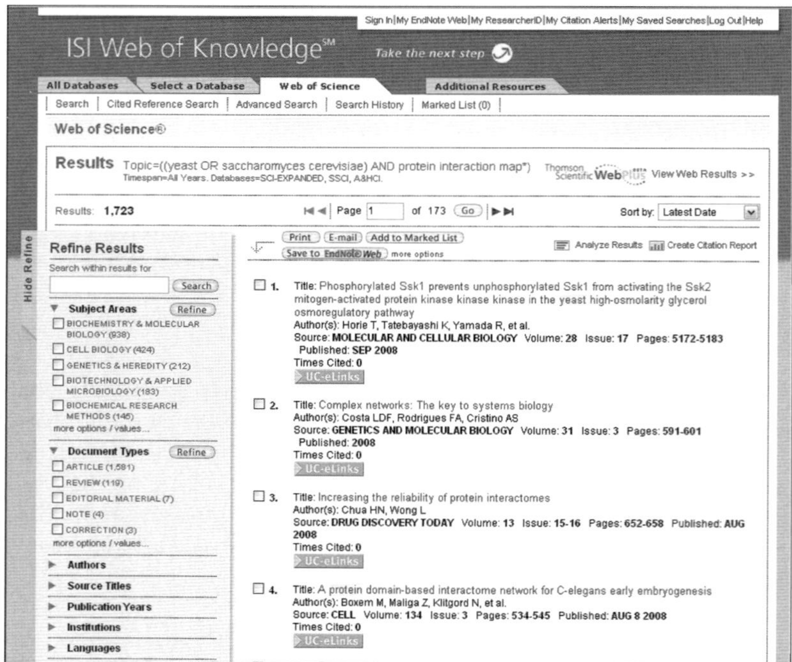


**Fig. 19.2** The WoS search retrieval screen (Copies of figures including color copies, where applicable, are available in the accompanying CD)

searches and view for example what journals or institutions are publishing the most in a particular research area.

### *19.2.4 BIOSIS Previews*

There are a number of other databases that may be useful for retrieving literature in Systems Biology, although *PubMed* and *WoS* will cover all major scientific journals. Of note is the *BIOSIS Previews* database. *BIOSIS Previews* is a large comprehensive database that indexes the biological literature, and is useful in providing access to the literature of proceedings, books and patents, which *PubMed* and *WoS* do not. The database indexes 5,500 sources from around the world, including 1,500 regularly published proceedings for conferences. The full database contains over 18 million records going back to 1926, so *BIOSIS Previews* is also a good source for older literature, if needed [4].

Depending on the topic being searched, it is advisable to consult with a science or medical librarian, familiar with the breadth of available literature databases, so they can direct you to other resources that may have unique content relevant for the search.

## 19.3 Alert Services

Alert services enable researchers to stay current with new literature as it is being published. When doing thorough research, one cannot do it in one day. Good research requires constant searching of the literature and following citations to ensure nothing important is missed. Alert services are an efficient way to stay on top of the research as it is being published. Both *PubMed* and *WoS* provide alert services, but it is important to note that almost all, if not all databases that index the scientific literature provide some form of alert service. Many electronic journals also provide table-of-contents alert services, so you can scan the contents of your favorite journals in your email as new issues are published.

### *19.3.1 My NCBI*

The **My NCBI** service in *PubMed* is simple to use. **My NCBI** links are found at the top right hand corner of the *PubMed* search screen. Registration is quick and only requires that you provide a login and email address which will be used to send the alerts. Once signed into **My NCBI**, you can enter the **My NCBI** services by clicking the **My NCBI** link in the left hand menu of the *PubMed* screens. At sign in, **My NCBI** gives you the option of remaining signed in all the time. Selecting this option ensures that you will always be signed into the service when searching *PubMed*. **My NCBI** offers a number of services, including the ability to set up alerts in other NCBI databases.

To save a *PubMed* search as an alert, click the **Save Search** link to the far right of the *PubMed* search box. This will invoke **My NCBI**. You can simply save the search, in which case **My NCBI** allows you to re-run the search at any time and retrieve only the references that have been added to *PubMed* since the search was last run. To save the search as an alert, choose **Yes** for the **Would you like to receive e-mail updates** option. This opens a larger window that provides alert options, like: the day to run the search, how often to run it, the format in which you wish to view the citations, and whether the search results should be emailed to you in a text or HTML format.

**HTML** is the preferred format, as this enables direct linking from the email alert to *PubMed* abstracts.

### *19.3.2 PubMed RSS Feeds*

*PubMed* searches can also be saved as an **RSS** (Really Simple Syndication) feed. To use the RSS feed you will need an RSS reader. There are a variety of different RSS readers available on the web, like *Bloglines* or *My Yahoo*. However, the easiest way to set up the feed is by a bookmark in your web browser. The bookmark displays a list of titles for *PubMed* citations, with the most recent citations added to *PubMed* at the top of the list.

If you are using the *Firefox*, complete your *PubMed* search, and then click the orange RSS feed link to the right of the web address in your browser's address box. You will be asked to add the feed to the bookmarks. The bookmark is then available at any time to view the titles of new citations in *PubMed*.

If you are using *Internet Explorer* or *Safari*, or a web-based RSS reader, after completing the *PubMed* search select **RSS Feed** from the **Send to** menu. Choose the number of references you would like to view in the feed at one time, up to 100, from the **Limit items if more than** menu, give the feed a **Name**, and then click the **Create Feed** button. To complete this process, click the orange **XML** button. *PubMed* will give you the choice of adding the feed as a **Live Bookmark** in your browser, or in another RSS feed service.

### *19.3.3 WoS Alerts*

One of the salient features of creating an alert in the *WoS* is the ability to do a cited reference search on an author or paper and have the *WoS* email you new citations for articles that have cited the author or paper.

To create alerts for new citations in *WoS* after completing a search, click the **Search History** button at the top of the search retrieval screen. The **Search History** displays your completed searches. Choose the search strategy you wish to create an alert for, by selecting the box to the left of the search, and then click the **Save History** button at the top right of the **Search History**. You will be asked to sign in. If you do not have a sign in for *WoS* it can be easily set up by clicking the **Register** link. When saving the search be sure to check the **Send Me E-mail Alerts** option. You will see similar options to those in **My NCBI** for saving an alert. The search can be run weekly or monthly, and delivered in a text or HTML format. Again, the email alerts should be delivered in an **HTML** format, so you can link directly from references in the email to the abstracts in *WoS*.

After saving a search in *WoS* you are also given the option to save it as an RSS feed. After clicking the **Save** button, you will see a screen that contains an orange **XML** button. Click this button to save an RSS feed to your bookmarks, like in *PubMed*. You can also do this by accessing your saved search history. The **Saved Search History** page provides the **XML** option for each saved search. Simply click the orange **XML** button for any saved search to set up the RSS feed in your bookmarks.

## 19.4 Text Mining Tools

*PubMed* provides open protocols for searching and extracting data from the database. This has enabled development of search interfaces by third parties that can target specific user groups, more easily navigate the literature, and mine data not easily extracted using the standard *PubMed* interface. The current state and future development of literature mining resources available to the systems biologist has been discussed in the literature [5,6]. The basic operations of these search interfaces, which for the most part extract records from *PubMed*, use a combination of gene, protein, disease and drug terminology to scan article titles, abstracts and indexes using algorithms that identify relationships between terminology that in turn elucidate for example, protein-protein or protein-disease interactions not evident by simply browsing abstracts. Interfaces are also being designed to help researchers visualize relationships in the literature, and in turn save the countless hours normally required to browse citations and manually identify links between disparate research and publications. This can hopefully accelerate new discovery and research directions.

Algorithms used for mining the literature and extracting protein-protein interactions use a combination of weighting and proximity tools that derive search terminology from *MeSH* and the *Gene Ontology* (*GO*). Different tools have developed various methods of mapping and extracting term relationships between *MeSH* and *GO*, and that information has been used to mine data from *PubMed* records and highlight term relationships that enable researchers to visualize the relationships between different publications that are not apparent in the traditional *PubMed* search interface. Explanations on the design of numerous text-mining algorithms, tools and studies on their success in navigating and mining the published Systems Biology research can be found in the bioinformatics literature [7–12], including *iHOP* (*Information Hyperlinked Over Proteins*), which is discussed at length in chapter 22 of this book. The number of tools available to navigate the literature is too great to be discussed here. Readers are free to explore the variety of text mining tools in the cited references.

In addition to mining tools, there are tools being developed to increase the capabilities of search interfaces like *iHOP*. One such tool is the *iHoperator*, which enables direct access to specific *PubMed* records based on term relationships and the ability to visualize links between *PubMed* references [13]. It is easy to see that in time we will have access to a variety of tools to make searching more efficient and help researchers navigate the vast published research literature, leading them to the most important and relevant articles quickly.

It should be noted that the extensive use of *PubMed* as the source of mined data from the literature, due to its open accessibility, does present some limitations to linking together all available research publications. *PubMed* for the most part will retrieve all literature from the major biological and biomedical research journals, but may miss articles published in more obscure titles indexed in the *WoS*, and conference proceedings indexed in databases like *BIOSIS Previews*. Over time we can expect to see the development of more mining tools that extract information from databases other than *PubMed*.

## 19.5 Bibliographic Management Software

In addition to staying current with the literature, it is important to manage the literature for future reference and publication purposes. There are several bibliographic management software (BMS) packages available at a variety of prices that can manage references to the published literature in personal databases. The most popular packages are *EndNote*, *Reference Manager*, *ProCite* and *RefWorks*. *RefWorks* differs from the other packages in that it is web-based and is available on a subscription basis. The other packages require a one-time purchase of the software, although software like *EndNote* releases a new version every year, requiring an annual purchase to maintain the latest version. *EndNote* also has a stripped down web-based version available in the *WoS*, so researchers can do basic management of citations while searching this database. However, to use the software's more powerful features, like formatting papers for publication, the records need to be transferred into the full *EndNote*, or another BMS package.

All of these packages enable the researcher to import citations from the literature databases, in some case directly manage a database of citations, and format publications with references and endnotes. All the software mentioned above provide formatting tools designed for specific journals. For example, if you are sending a publication to *Nature*, you simply have a software like *EndNote* format the publication for submission to *Nature*.

There are also a number of other BMS tools freely available on the web, or that can be attached to your web browser. One tool available for *Firefox* is *Zotero*. *Zotero* runs in conjunction with the browser and allows you to quickly grab citation information for articles viewed online or web sites. *Zotero* can export references to the more powerful BMS tools like *EndNote*.

### *19.5.1 Communal Citation Management and Alert Services*

The web now provides numerous resources that enable scientists to collaborate with other scientists around the world and share information about the research literature. Some resources, like *Connotea* available through Nature Publishing Group, enable researchers to work together and develop a library of publications relevant to the researchers' interests, and to also discuss the significance of specific publications. *Connotea* allows researchers to freely set up their own collaborative groups, to which anyone can join and discuss their specific research interests together.

Other resources like *Faculty of 1000*, a commercial resource available through *BioMed Central*, use leading scientific researchers (the Faculty) to identify, review and rate specific journal articles as they are published, to help direct researchers to important and breakthrough research as it becomes available. *Faculty of 1000* has a Biology and Medicine resource, each of which are divided into specific subject areas that are easily browsed and searched.

## Suggested Reading

### *MEDLINE*

1. MEDLINE Fact Sheet. United States National Library of Medicine, 2006. (Accessed July 21, 2007, at http://www.nlm.nih.gov/pubs/factsheets/medline.html)

### *Medical Subject Headings*

2. Medical Subject Headings (MeSH) Fact Sheet. 2005. (Accessed July 21, 2007, at http://www.nlm.nih.gov/pubs/factsheets/mesh.html)

### *Web of Science*

3. Web of Science. Thompson Scientific, 2007. (Accessed July 21, 2007, at http://scientific.thomson.com/products/wos/)

### *BIOSIS Previews*

4. BIOSIS Previews. Thompson Scientific, 2007. (Accessed July 21, 2007, at http://scientific.thomson.com/products/bp/)

## *Text Mining Tools*

5. Jensen LJ, Saric J, Bork P. Literature mining for the biologist: from information retrieval to biological discovery. Nat Rev Genet 2006;7(2):119–129.
6. Kersey P, Apweiler R. Linking publication, gene and protein data. Nat Cell Biol 2006;8(11):1183–11389.
7. Ananiadou S, Kell DB, Tsujii J. Text mining and its potential applications in Systems Biology. Trends Biotechnol 2006;24(12):571–579.
8. Chun HW, Tsuruoka Y, Kim JD, et al. Extraction of gene-disease relations from Medline using domain dictionaries and machine learning. Pac Symp Biocomput 2006:4–15.
9. Ling X, Jiang J, He X, Mei Q, Zhai C, Schatz B. Automatically generating gene summaries from biomedical literature. Pac Symp Biocomput 2006:40–51.
10. Pan H, Zuo L, Kanagasabai R, et al. Extracting information for meaningful function inference through text-mining. In: Discovering Biomolecular Mechanisms with Computational Biology; 2006:57–73.
11. Perez-Iratxeta C, Bork P, Andrade MA. Literature and genome data mining for prioritizing disease-associated genes. In: Discovering Biomolecular Mechanisms with Computational Biology; 2006:74–81.
12. Roberts PM. Mining literature for Systems Biology. Brief Bioinform 2006;7(4):399–406.
13. Good BM, Kawas EA, Kuo BY, Wilkinson MD. iHOPerator: User-scripting a personalized bioinformatics Web, starting with the iHOP website. BMC Bioinformatics 2006;7(1):534.

## Web Resources

http://pubmed.gov/
http://isiknowledge.com/wos/
http://www.GeneOntology.org/
http://www.ihop-net-org/UniPub/iHOP/
http://www.endnote.com/
http://www.refman.com/
http://www.procite.com/
http://www.refworks.com/
http://www.zotero.org/
http://www.connotea.org/
http://www.f1000biology.com/browse/

# Chapter 20
# GoPubMed: Exploring PubMed with Ontological Background Knowledge

**Heiko Dietze, Dimitra Alexopoulou, Michael R. Alvers, Liliana Barrio-Alvers, Bill Andreopoulos, Andreas Doms, Jörg Hakenberg, Jan Mönnich, Conrad Plake, Andreas Reischuck, Loïc Royer, Thomas Wächter, Matthias Zschunke, and Michael Schroeder**

**Abstract** With the ever increasing size of scientific literature, finding relevant documents and answering questions has become even more of a challenge. Recently, ontologies—hierarchical, controlled vocabularies—have been introduced to annotate genomic data. They can also improve the question and answering and the selection of relevant documents in the literature search. Search engines such as GoPubMed.org use ontological background knowledge to give an overview over large query results and to answer questions. We review the problems and solutions underlying these next-generation intelligent search engines and give examples of the power of this new search paradigm.

**Keywords** PubMed · Literature search · Ontology · Intelligent search

## 20.1 Introduction

*Which techniques use the Prominin-1 (CD133) marker? Which proteins are related to Alzheimer's disease? Which hormone is Autistic Disorder associated with? Is apoptosis a hot topic? Which are leading centers and scientists for liver transplantation? Where is the main research done for dengue and leprosy? What treatments does the web discuss for Alzheimer disease?*

The scientific literature and the web hold answers to all of these questions, but it is difficult to obtain them with the classical search engines, as they merely present, possibly, a long lists of search results. In contrast, ontology-based search engines can use their hierarchical background knowledge to provide an intelligent filing system, which categorizes the results. The categorization gives an overview over large result sets and can be used to answer questions. For example to find the techniques associated with CD133, a query for CD133 will return many documents as a long list in a classical search engine. In contrast, a search engine with ontological background knowledge will identify flow cytometry as a technique and categorize the documents accordingly. The user can then use this hierarchical filing system to select the few articles mentioning the techniques and even fewer mentioning flow cytometry.

Key to this new search paradigm is the background knowledge, which is used to categorize documents. With efforts such as the Gene Ontology [1] and MeSH, the needed knowledge is readily available. MeSH contains, for instance, the fact that flow cytometry is a technique and Gene Ontology contains, that, apoptosis is also known as programmed cell death and that caspases are part of the apoptotic program.

The central problem of ontology-based search is the mapping of ontology terms to the text. This task, known as term extraction, is difficult, as authors do not write their abstracts with an ontology in mind. For instance the mapping must be flexible and map the ontology term "transcription factor binding" to the text "... a transcription that binds ...", although it does not appear literally.

M. Schroeder
Biotec, TU Dresden, Dresden, Germany
e-mail: ms@biotc.tu-dresden.de

S. Krawetz (ed.), *Bioinformatics for Systems Biology*, DOI 10.1007/978-1-59745-440-7_20,

In the remainder of this chapter, we give a brief introduction to ontologies, finding ontology terms and entity recognition in text. We show how GoPubMed.org, a search engine which uses the Gene Ontology and MeSH to index PubMed, can answer the introductory questions and more. Furthermore, we present GoWeb, an application using the GoPubMed features to introduce an ontological knowledge base web search. We conclude by comparing several other search engines, including other PubMed search engines and ontology-based search engines.

## 20.2 Ontology-based Text Mining

### *20.2.1 Ontologies*

A fundamental aspect for the work of researchers is the need to share knowledge. In the beginning, this was often done without the help of a controlled vocabulary or nomenclature. This is in particular applicable for the biomedical area and life sciences. There are many genes and proteins that have multiple names or identifiers. An example is Hnrpa1 which is also known as Tis, Fli-2, heterogeneous nuclear ribonucleoprotein A1, helix-destabilizing protein, single-strand-binding protein, hnRNP core protein A1, HDP-1, and topoisomerase-inhibitor suppressed. But there are also names such as Cleopatra, Ariadne, Groucho, Lost in Space, Brokenheart, Hairy, Superman and many more. Of course, there have also been efforts to standardize names or at least to reach a consensus for naming. For instance in the context of yeast research and for human genes there are widely used standards, even if they are not always adhered to in literature.

Similar issues arise, if the task is to annotate genes and their function within the categories biomedical process, molecular function, and cellular components. You can find that

- Cellulose 1,4-beta-cellobiosidase is also known as exoglucanase,
- superoxide-generating NADPH oxidase as cytochrome B-245,
- thiamin as vitamin B1,
- pyrexia as fever,
- heme as haem, and
- Apoptosis as cell death.

The aim of ontologies is to reduce this problem. They include concepts, synonyms, and their relationships.

One prominent example for a widely used ontology is Gene Ontology [1]. In the beginning, it was developed for the annotation of the fruit fly genome. Later the Gene Ontology was adapted and expanded for mouse and other genomes and it now covers biomedical processes, molecular functions, and cellular components. It uses two kinds of relationships to model the dependencies between the concepts: is-a and part-of. Today the Gene Ontology is part of the Open Biomedical Ontology (OBO) effort, which houses over 60 ontologies covering many areas of interests. This includes anatomy, chemical compounds, development, experimental conditions, phenotype, taxonomy and more.

The second example is the Medical Subject Headings (MeSH). The MeSH thesaurus is developed by the U.S. National Library of Medicine (NLM). Its main purpose is to provide an index for articles, books, and other media in the National Library of Medicine. It tries to cover all the relevant topics for the medical area and this includes disease, anatomy but also others like geographic locations and experimental techniques.

There are other medical ontologies, e.g., GALEN, SNOMED, and UMLS [2]. An overview of all presented Ontologies is available in Table 20.1. The Unified Medical Language System (UMLS) has a different approach. It tries to integrate as much relevant ontology as possible. The UMLS consists of three parts: a meta-thesaurus, a semantic network, and the specialist lexicon. Whereas

**Table 20.1** URLs for ontologies, literature search engines and ontology-based literature search engine

| | |
|---|---|
| Ontologies | |
| Gene Ontology.org | Ontology with ≥20.000 terms on biomedical processes, molecular functions and cellular component |
| nlm.nih.gov/mesh | Medical Subject Headings created by the U.S. National Library of Medicine, taxonomy with ≥150.000 terms |
| opengalen.org | formal medical ontology, with ≥70.000 terms |
| snomed.org | commercial medical ontology, which contains ≥350.000 terms |
| nlm.nih.gov/research/umls/ | Unified Medical Language System created by the U.S. National Library of Medicine, contains ≥1.000.000 terms |
| obofoundry.org | Open Biomedical Ontology, collection of over 60 specialized biomedical ontologies |
| Search engines | |
| pubmed.org | NIH's literature search engine |
| hubmed.org | "PubMed rewired" |
| invention.swmed.edu/etblast/ | Text similarity: an alternative way to search Medline |
| glycosciences.de/tools/PubFinder/ | PubFinder |
| www-tsujii.is.s.u-tokyo.ac.jp/medie | MEDIE answering questions |
| ihop-net.org | iHOP, gene network for navigating the literature |
| scholar.google.com | Google's literature search engine |
| academic.live.com | Microsoft's literature search engine |
| scopus.com | Elsevier's literature search engine |
| clustermed.info | Document clustering on the fly with Vivisimo |
| Ontology-based literature search engines | |
| gopubmed.org | Explore PubMed with ontological background knowledge |
| textpresso.org | Wormbase full texts with many ontologies |
| xploremed.org | Classification with high-level MESH headings and word co-occurrences |

(Copies of tables are available in the accompanying CD.)

the meta-thesaurus represents the concepts including the synonyms, the semantic network corresponds to categories and the specialist lexicon acts as a kind of index.

A non-trivial aspect is the design and evolution of ontologies. With many thousands of concepts and definitions how does one keep it all including the relations consistent. Although this starts with the question: How is consistency defined in the first place? Gene Ontology follows an informal approach. The transitive closure still has to hold. This means, if a concept A is-a B and B is-a C then A is-a C has to be true. These inferred redundant relationships are not kept directly in the ontology. This helps to ease the maintenance of the ontology, as corrections, modifications, and additions only need to check if their direct relations are still valid.

Even though this consistency definition is a pragmatic solution there are more formal approaches. One such idea is the usage of description logics to formally define concepts and their relations. This was used for instance in the GALEN and SNOWMED ontologies. The advantage of the formal definitions is the chance to automatically check for inconsistencies in the ontology. Imagine that one adds the new fact that heparin is-a glycosaminoglycan, but it was not yet stated that heparin biosynthesis is-a glycosaminoglycan biosynthesis. Because of the formally defined relations and concepts, this additional relation can be inferred with this new fact in the knowledge base.

### *20.2.2 Finding Ontology Terms in Text*

The ontologies presented above have been designed to annotate data or to be used as classification schemes. But they were not designed for the purpose of building novel search engines. Therefore, the identification of ontology entities in free text remains a challenging task. For instance, a recent assessment for extracting Gene Ontology terms revealed performance around 20% success rate [3]. The difficulties of automating manual annotation is evident from the fact that as few as 15% of manually annotated terms appear literally in the associated abstracts. Biomedical text mining uses various techniques and algorithms, e.g., natural language processing, information retrieval, and machine learning, to identify the relevant entities [8] and consider groups of problems.

#### 20.2.2.1 Ad-hoc Variations of Names

To begin with, terms in vocabularies and labels of concepts in ontologies appear in many, slight or severe, variations in natural language texts.

- orthographic: IFN gamma, Ifn-$\gamma$
- morphological: Fas ligand, Fas ligands
- lexical: hepatic leukaemia, liver leukemia
- structural: cancer in humans, human cancers
- acronyms/abbreviations: MS, Nf2
- synonyms: neoplasm, tumor, cancer, carcinoma
- paragrammatical phenomena/typographical errors: cerevisae, nucleotid

Some of the terms encountered in texts are rather ad-hoc creations, which cannot be found in any term list.

#### 20.2.2.2 Synonymity of Ontological Terms

As mentioned before, terms in a vocabulary or ontology might not appear literally in a text, but authors rather use synonyms for the same concept. First of all, this complicates proper searches: When searching for "digestive vacuole", results should also contain texts that mention "phagolysosome"; mentioning of "ligand" should refer to the concept "binding"; an "entry into host" might occur as an "invasion of host". In the Plant ontology for example, many synonyms exist for the same structure in different species. "Inflorescence" is referred to as "panicle" in rice, and as "cob" in sorghum, and "spike" in wheat, for instance. We note that there are also intra-ontology synonymities: "eye" in AnoBase can refer to the eye spot or the adult compound eye.

#### 20.2.2.3 Ambiguity of Ontological Terms

Terms can have a very specific meaning in biomedical research, but may mean other things in other contexts. Examples are "development", "envelope", "spindle", "transport", and "host". Protein names such as "Ken and Barbie", "multiple sclerosis" or "the" that resemble common names, diseases, or common English words are especially hard to disambiguate. The same problems arise from drug names like "Trial" or "Act".

#### 20.2.2.4 Stemming and Missing Words

Some aspects for finding terms in text refer to the actual processing of natural language and appear rather technical. Very often, words will appear in different forms, such as "binding" and "binds". These refer to the same concept, which can be solved by resolving words to their stem ("bind"). However, the analogous reduction of "dimerization" to "dimer" is more questionable. The former talks about the process, the latter about the result. A similar example is "organization", where a transformation into "organ" is invalid.

Texts contain additional words that are missing in the ontological term. This happens, for instance, when a text contains further explanations that describe the findings in more detail. An example is "tyrosine phosphorylation of a recently identified STAT family member" that should match the ontology term "tyrosine phosphorylation of STAT protein." In general, matching is allowed to ignore words such as "of", "a", "that", "activity", but obviously not "STAT".

Additional background information on term variations is needed to know that a "family member" can refer to a protein. Formatting of terms represents another source for potential matching errors. Terms in ontologies contain commas, dashes, brackets, etc., which require special treatment. For "thioredoxin-disulfide" the dash can be dropped, for "hydrolase activity, acting on ester bonds" the clause after the comma is important, but unlikely to appear as such in text. Terms containing additions such as "(sensu Insecta)" may have important contextual information, but are also less likely to appear in text.

#### 20.2.2.5 Ontology Specific Issues

*Term overlaps*—some concepts can overlap in their labels or synonyms: in many cases there is a difference between what the authors write and what they actually mean to express. Unfortunately, researchers do not have strict and formal ontologies or nomenclatures in their minds when composing a scientific article; in most of the cases they might use parent terms to refer to a child term, or vice versa. For example, many people are treating the MeSH terms 'cardiovascular disease' and 'coronary artery disease (CHD, CAD)' the same, although the latter is a child of the first.

*Descriptive labels*—in most of the cases, the labels in annotation ontology cannot be used directly for text mining, often due to their explanatory nature. For example, it is unlikely that the Gene Ontology term "cell wall (sensu Gram-negative bacteria)" will appear as such in text. Terms like "positive regulation of nucleobase, nucleoside, nucleotide and nucleic acid metabolism" and "dosage compensation, by inactivation of X chromosome" are almost complete sentences and are also unlikely to be found as such in text.

*Ambiguity*—results either from identical abbreviations for different terms, or, in general, tokens that can refer to terms that may or may not be of our interest. An example of an ambiguous *abbreviation* is "CAM" that can stand for "constitutively active mutants", "cell adhesion molecule", or "complementary alternative medicine". The second category of ambiguities—and the most difficult to handle—is that of terms that (in the context of anatomy) can refer to different species. An example of such ambiguities is "embryo", which can be a chicken, mouse, human, or even zebrafish embryo.

### *20.2.3 Entity Recognition*

Although finding ontological concepts in free text is important, there are many more relevant things to find in the text for instance:

- proteins,
- genes,
- species, or
- mutations.

The task to find these entities is called entity recognition. The identification of ontology terms can be seen as a sub species of the more general task of entity recognition. As a consequence, many of the techniques and problems described above are also valid for entity recognition. But, for instance, for protein and gene name identification there are some other difficulties [12].

One challenge is the increased ambiguity and synonymity of names. Often, the gene name and the protein are used by the authors as synonyms or a gene has the same name in different organisms. Another task is to deal with the number of entities one can find. As an example, the UniProtKB/TrEMBL protein database contains over 4,500,000 entries. The real number to match is even higher as one has to integrate all the synonyms and variants of the protein names and genes. A recurring problem is for the case of identification of which species an article describes that the species is sometimes never mentioned in the text. For mentioning point mutations, one has to recognize the mutations and the related proteins to have a useable result [13, 14]. Moreover many cases of ambiguities and missing concepts can only be resolved if one tries to use any information available from the text. For example with additional information on species or point mutations, ambiguous gene or protein names can be filtered. If a mentioned protein candidate does not match the species from text, it's less likely to be correct. Furthermore if the protein doesn't have the corresponding residue for the point mutation, it can be ignored.

The importance of entity recognition and their relations has been acknowledged by the scientific community. There have been efforts to establish benchmarks and competitions to advance the research. Examples of this are the "bio-entity recognition task at JNLPBA" [10] or the "Critical Assessment of Information Extraction in Molecular Biology" (BioCreAtIvE) [9]. In the BioCreAtIvE II, gene mention task, the best systems [11] could achieve a precision of 78.9% and a recall of 83.3%. An example for the current state of the art.

All of the above problems mean that extracting entities from literature will not be error-free. However, despite all of these problems, ontology-based literature with text mining can answer questions as posed in the introduction. Next, we introduce GoPubMed and illustrate how they help to answer questions.

## 20.3 Question and Answering with GoPubMed

Traditional keyword based searching gives a possible very long list of results. But finding the relevant documents is only the start; the user has to check if the results are relevant to him. Often, there is a question behind a query. GoPubMed can answer all the introductory questions, as it uses the ontological background knowledge, namely the Gene Ontology and MeSH to index search results. This allows GoPubMed to categorize the search results, identify relevant terms in the result set and summarize trends for a topic. This topic can either be a term and its children or the result set of a query. For an ontology-enhanced web search, the GoWeb system is available. Figure 20.1 shows a screen shot of GoPubMed. The main panel contains the search results and the panel on the left the relevant categories from the ontologies in a summary and as a tree.

Now let us consider the questions and more importantly the answers in detail. Please note that these questions were answered with GoPubMed in July 2007, and due to the increasing number of publications the results may vary in the future.

**Question:** Which techniques use the Prominin-1 (CD133) marker?
**Answer:** Search in GoPubMed for "CD133" and open Techniques and Equipment in "top five & more" on the left. Listed as first is "Flow Cytometry". If you hover with the mouse above this term, you will see the description in a tool tip. The listed articles for flow cytometry contain statements like:

> "CD133+ and CD34+ cells were analyzed by flow cytometry to assess expression of cell division antigens" (Denner L. et al., Cell Prolif., 2007).

Other interesting terms are "Cell Separation" and "Immunohistochemistry" There you can find a statement like:

> "Microarray screening, single and dual-label immunocytochemistry and RT-PCR were performed to detect embryonic and neuronal stem cell markers, such as Oct3/4, Nanog, CD133, and Musashi-1." (Seigel GM et al., Mol. Vis., 2007).

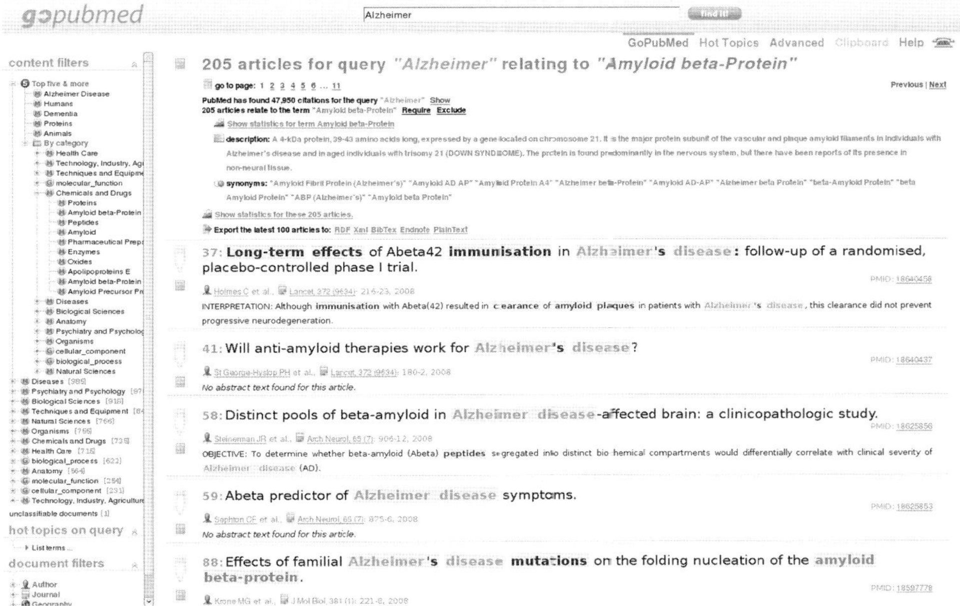


**Fig. 20.1** Which proteins are related to Alzheimer's disease? GoPubMed uses its ontological background knowledge to index search results according to the Gene Ontology and MeSH. The interface consists of three parts. The top most part contains the input field for the query, in this example it is "Alzheimer". You can submit a query by using the "find it!" button. The panel below, comprises the results for the query and is split into to a left and a right part. The left panel contains the ontological background knowledge relevant to your query. A summary over all identified terms in your result is presented in "top five & more". If you open the category "Chemicals and Drugs" you can find also proteins. In the "hierarchy of content" the complete induced ontology tree is available for browsing all the concepts found.
On the right side, you can browse the found articles. The articles are shown with title, authors, journal, abstract and affiliation, also Wikipedia links and links to proteins identified from the text are offered if available. The picture shown here is the abbreviated version for articles for faster browsing. You may switch between the full and short variant with the provided buttons. On top of the articles there is a summary with details for the query. This may include a link to dynamic Hot Topics and if your query matched an ontological concept a link to the corresponding term Hot Topics. There are also links to export the results to citation mangers.
After selecting a term from the left side, here "Amyloid beta-Protein", the result view is updated. It now shows the articles containing the selected concept. This also includes all the child terms of the selected term. Please note that the initial result set size of 1000 articles was reduced down to 50 relevant articles in two clicks. In the summary field, the term description and term synonyms are listed. In case of "Amyloid beta-Protein" there are currently 10 synonyms listed. To select an interesting article into the built-in clipboard, use the paper clip icon provided directly next to the each article. To export a single article you can use the export icon. To view the content of your clipboard select Clipboard link in the top bar. There you can also find the link to Hot Topics, Advanced Search, Help and a contact form (Copies of figures including color copies, where applicable, are available in the accompanying CD)

A follow up question might be "Which types of cells are often targeted with these techniques?" The answer is already present "Stem Cells"; it is the top term for the query.

**Question:** Which proteins are related to Alzheimer's disease?
**Answer:** Type in Alzheimer and open chemicals and drugs in "top five & more" on the left. Among others, there are "Amyloid", "Amyloid beta-Protein" and "Cholinesterase Inhibitors" listed as related proteins. By clicking on Amyloid beta-Protein, we can reduce from 1000 to 60 relevant articles and get the following definition:

> "A 4-kDa protein, 39–43 amino acids long, expressed by a gene located on chromosome 21. It is the major protein subunit of the vascular and plaque amyloid filaments in individuals with Alzheimer's disease and in aged individuals with trisomy 21 (DOWN SYNDROME). The protein is found predominantly in the nervous system, but there have been reports of its presence in non-neural tissue."

The article from Ohyagi Y et al. in 2007 mentions, e.g., "Inhibition of aggregation of amyloid p-protein (AP) ... are known as potent therapeutic tools for Alzheimer's disease (AD)." Another article (Chiarini A. et al., Ital J. Anat. Embryol., 2006), states "Reportedly, beta-amyloid peptides (Abeta40 and Abeta42) induce the neurodegenerative changes of Alzheimer's disease (AD) ...".

**Question:** Was Abeta42 already used in a clinical setting?
**Answer:** Enter "Abeta42 drug" into the GoPubMed system and go on the result page to hierarchy of content on the lower left. Open first the category "Chemicals and Drugs" and than "Organic Chemicals". By clicking on "hydrocarbons" you reduce the result set to only 41 articles. A quick skimming over of the abstracts reveals statements like

- "... minocycline treatment did not alter the cerebral deposition of Abeta ..." (Fan R et al., J Neurosci, 2007), or
- "... naproxen that do not lower Abeta42 ..." (Cole GM. et al., Ann. NY Acad. Sci., 2004).

Select the next category "Carboxylic Acids", this will display 36 articles. On the top of the results you can find again the definition of the term but also a link, the Wikipedia article about carboxylic acids. The list of articles includes statements such as

- "... ibuprofen possess preferential Abeta42-lowering activity ..." (Leuchtenberger S. et al., Curr. Pharm. Des., 2006).

The GoPubMed system also provides links for protein names mentioned in an article, e.g., APP. This links opens the EBI-Swissprot database showing a list of all proteins related to APP.

**Question:** Which hormone is Autistic Disorder associated with?
**Answer**: Submit Autistic Disorder as query in GoPubMed. In "hierarchy of content" open "Chemicals and Drugs", then "Hormones, Hormone Substitutes and Hormone Antagonists", and then select "Hormones", which reduces the number of relevant articles to 49. For more details on special hormones you can browse in the "Gonadal Hormones" category which also has the term "Testosterone". Selecting testosterone, the result now shows 5 articles, with sentences like

- "... that high fetal testosterone levels could play a role in the aetiology of autism." (de Bruin EI. et al., Dev. Med. Child Neurol., 2006),
- "Fetal testosterone and sex differences in typical social development and in autism" (Knickmeyer RC. et al., J Child Neurol, 21 (10): 825–845, 2006), or
- "... high levels of testosterone influences some autistic traits and that hormonal factors may be involved in vulnerability to autism" (Knickmeyer R. et al., Horm. Behav., 2006).

For more examples of questions and answers have a look at Table 20.2.

### *20.3.1 Hot Topics*

Despite the overall growth of literature, some topics are hot and take-off while others are stagnant or are in a cool down phase. Bibliometric analyses aim to shed light on such developments and help to identify emerging trends. Such analyses date back to the 1960s [4] and are typically focused on research topics [5], specific journals [6], or on the researchers themselves [4,7]. The Hot topic feature of GoPubMed, features views on ontology terms from the knowledge base. It considers a term and all its children as one topic. For each topic a bibliometric analysis is provided.

**Table 20.2** More example questions answered with GoPubMed

Which diseases are associated with HIV?

**Answer:** Type "HIV" and wait for the tree on the left to appear. Go to "top five & more" and click on "disease". Among others hepatitis and tuberculosis are mentioned. Clicking on tuberculosis retrieves the relevant articles including statements such as "HIV and parasitic co-infections in tuberculosis patients".

Which anatomical structure is affected by the bacterium helicobacter pylori?

**Answer:** Type "helicobacter pylori", go to "top five & more" and open "anatomy" Among the terms listed is "gastric mucosa". Hovering the mouse over the term reveals an explanation, which mentions that gastric mucosa is the lining of the stomach.

Which biological process is the protein Rab5 involved in and where is located in the cell?

**Answer:** Type "rab5" and wait for the tree on the left to appear. Go to "top five & more". Click on biological process shows "endocytosis" and clicking on "cellular component" shows "endosomes". Hovering over the terms displays brief explanations of what endocytosis and endosomes are.

In which organisms is toluene degradation studied?

**Answer:** Type "toluene degradation" and wait for the tree on the left to appear. Go to "top five & more" and open "organisms". The bacteria pseudomonas is listed first. A click retrieves the relevant articles.

Which enzymes are inhibited by aspirin?

**Answer:** Type "aspirin" and wait for the tree on the left to appear. Go to "hierarchy of content" and then "chemicals and drugs" and "enzymes and co-enzymes". From there always click the top child until you reach "cyclooxygenase 1" and "cyclooxygenase 2". Clicking reduces the articles to a few which mention that aspirin inhibits cyclooxygenases.

Which enzymes are important for congenital muscular dystrophy?

**Answer:** Type "congenital muscular dystrophy" and wait for the tree on the left to appear. Go to "hierarchy of content" and then "chemicals and drugs", "enzymes and co-enzymes", "enzymes", "transferases". There are a number of articles with statements such as "glycosyltransferases has revealed a novel mechanism for congenital muscular dystrophy."

Which techniques are frequently used to study zebrafish development?

**Answer**: Search for "zebrafish development". Under "top five & more" open "techniques and equipment". In situ hybridization is listed first. Clicking the term retrieves relevant articles.

Which process are osteoclasts involved in?

**Answer:** Search for "osteoclast". Under "top five & more" open "biological process". The first entry is "bone resorption".

What are common histone modifications?

**Answer:** Search for "histone modification". Under "top five & more" open "biological sciences" and find methylation and acetylation.

Which diseases are associated with wnt signalling?

**Answer:** Search for "wnt signalling". Under "top five & more" open "disease" and find "carcinoma" and many other cancer terms.

Were there clinical trails focusing on Abeta42 and were any side effects observed?

**Answer:** Search for "Abeta42 clinical trail". In "top five & more" open Diseases and click on "Meningoencephalitis". The result now shows 4 articles, with titles like "Subacute meningoencephalitis in a subset of patients with AD after Abeta42 immunization". So, yes there where clinical trials, but there were also severe side effects like brain inflammation.

Which molecular function is Autistic Disorder associated with?

**Answer:** Search for Autistic Disorder. Under "top five & more" open "Molecular Function" and find "neurexin binding".

Which disease is Autistic Disorder associated with?

**Answer:** Search for Autistic Disorder. Under "top five & more" open "Diseases" and find for instance "Fragile X Syndrome" as a related disease.

(Copies of tables are available in the accompanying CD.)

The hot topic page for an ontology term includes two graphs (Fig. 20.2) showing the absolute number of publications per year for a topic. The second graph shows the relative share compared to the total number of publications per year in PubMed. An increase in the share indicates that the topic is growing faster than the overall number of publications. Both graphs can be used to check whether the publication activity in a topic is decreasing, stagnant, or growing. In addition, to the publication count you can find a list of the most active authors, the list of journals with the most publications for this topic and a list of cities and countries with the most publications. To visualize co-authorship, which author publishes together with which other authors, we provide a co-author network image. Publications between authors are denoted as edges between the author nodes. If no edge exists then the authors did not yet publish together, according to the publications listed in PubMed for this topic. The last feature is a world map where red dots indicate where all the publications are located for the current topic. All these features of the hot topics page are pre-calculated using the list of authors and affiliation of an article and the annotations from the GoPubMed system for all 16 Million PubMed articles.

To check the hot topics in GoPubMed for a term there are two options. The first way is to just search for the term in the normal search field and select the link from the list after "Show statistics for term:". Or by the second option, one can directly use the "Hot Topics" mode by selecting it in top bar. There you could also choose to use the advanced search, use the help page, a contact form, or see the content of your clipboard.

**Question:** Is apoptosis a hot topic?
**Answer:** Use the hot topics to search for apoptosis. There are two apoptosis entries available, one from the Gene Ontology and the other from MeSH. Select one of them by clicking on it. To answer the questions about trends, have a look at the two graphs in Publications over time. They both reveal that the topic has been growing since the early 1990s. This is in line with Garfield and Melino's [5] investigation of the field. But the second graph with the relative research interest shows that in the last 3 years the growth was not faster than the average growth of the whole PubMed literature.

**Question:** Which are leading centers and scientists for liver transplantation?
**Answer:** Query GoPubMed for "liver transplantation" and open the hot topics statistics for this term (see also Fig. 20.2). Among the top authors is "Neuhaus P" and among the top cities is "Berlin". Prof. Peter Neuhaus works at the Charité Hospital Berlin, Germany. He is a leading specialist in the field. A look in the co-author graph reveals with whom Peter Neuhaus has worked and published with.

**Question:** Where is the main research done for dengue and leprosy?
**Answer:** Retrieve the term statistics for Dengue. You will find that in the list of top cities there are Bangkok and Rio de Janeiro as the two top cities. In the top countries Brazil, Thailand and India are in the top 4.

For the term Leprosy you will find in the countries section India is the top country. This is also reflected in the list of important cities, where one can find several cities located in India. Both terms show that the local occurrence of diseases can be shown in GoPubMed.

All the examples for the usage of hot topics were based on the pre-computed statistics using the ontology terms from the knowledge base as topics. But the result set of a given query may also be seen as a topic. This dynamic hot topics feature of GoPubMed offers you a bibliometric analysis of any result set of a query. This analysis contains the graphs about the publications over time, the lists of top authors, journals, cities and countries. It also includes the world map for the visualization of the geographic locations.

The dynamic hot topics can of course also be used to answer questions, for instance:

**Question:** Who are the top authors for Abeta42 Protein?
**Answer:** Use the GoPubMed site to search for "Abeta42". This query finds currently 767 articles. In the query summary field, above the articles, there is a link saying "Show statistics for these 767 articles". Clicking on this link will lead to the dynamically created hot topics. After the two graphs for the publications over time, there is the list of top authors. Listed there you can find for instance "Bennow K" as top author. The number of shown authors can be increased by clicking on the "more" link below the table. The publications for the author can be retrieved by using the provided link with the author name from the table.

**Question:** Who publishes most at the Max Planck Institute of Molecular Cell Biology and Genetics, Dresden (MPI-CBG)?

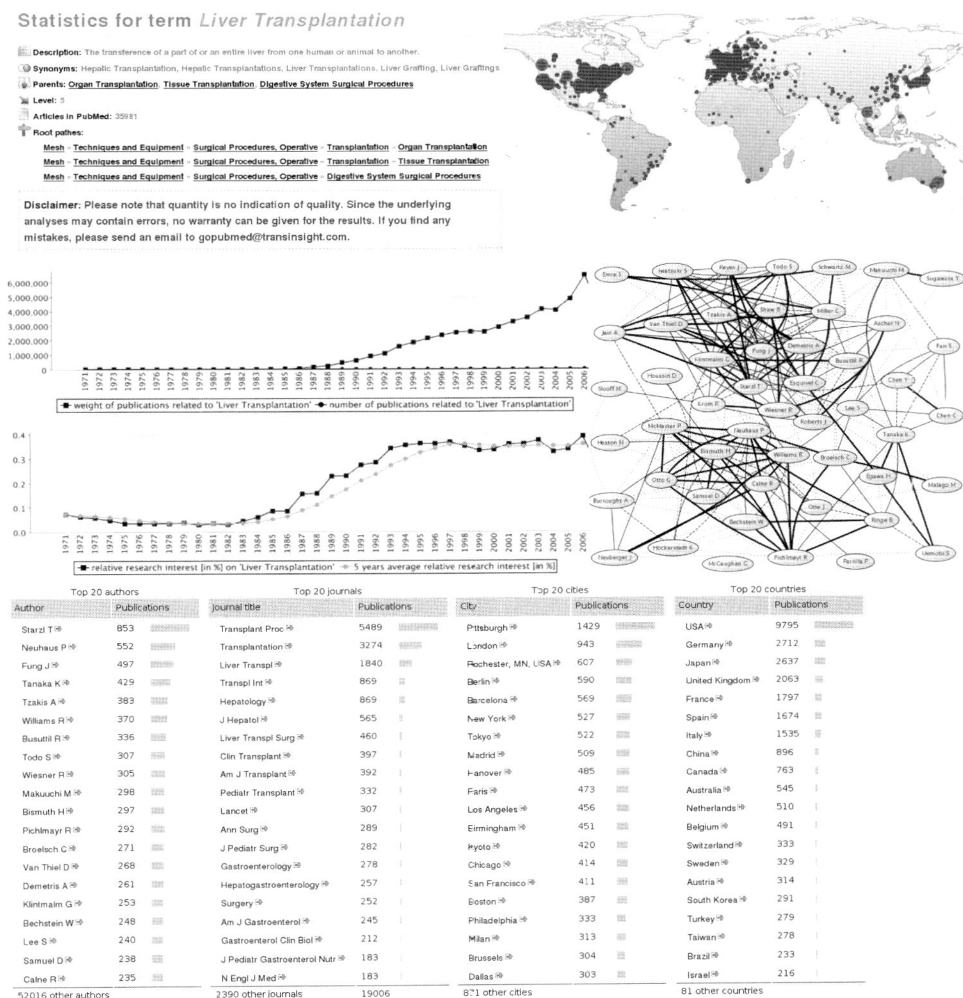


Top 20 authors

| Author | Publications |
|---|---|
| Starzl T | 853 |
| Neuhaus P | 552 |
| Fung J | 497 |
| Tanaka K | 429 |
| Tzakis A | 383 |
| Williams R | 370 |
| Busuttil R | 336 |
| Todo S | 307 |
| Wiesner R | 305 |
| Makuuchi M | 298 |
| Bismuth H | 297 |
| Pichlmayr R | 292 |
| Broelsch C | 271 |
| Van Thiel D | 268 |
| Demetris A | 261 |
| Klintmalm G | 253 |
| Bechstein W | 248 |
| Lee S | 240 |
| Samuel D | 238 |
| Calne R | 235 |
| 52016 other authors | |

Top 20 journals

| Journal title | Publications |
|---|---|
| Transplant Proc | 5489 |
| Transplantation | 3274 |
| Liver Transpl | 1840 |
| Transpl Int | 869 |
| Hepatology | 869 |
| J Hepatol | 565 |
| Liver Transpl Surg | 460 |
| Clin Transplant | 397 |
| Am J Transplant | 392 |
| Pediatr Transplant | 332 |
| Lancet | 307 |
| Ann Surg | 289 |
| J Pediatr Surg | 282 |
| Gastroenterology | 278 |
| Hepatogastroenterology | 257 |
| Surgery | 252 |
| Am J Gastroenterol | 245 |
| Gastroenterol Clin Biol | 212 |
| J Pediatr Gastroenterol Nutr | 183 |
| N Engl J Med | 183 |
| 2390 other journals | 19006 |

Top 20 cities

| City | Publications |
|---|---|
| Pittsburgh | 1429 |
| London | 943 |
| Rochester, MN, USA | 607 |
| Berlin | 590 |
| Barcelona | 569 |
| New York | 527 |
| Tokyo | 522 |
| Madrid | 509 |
| Hanover | 485 |
| Paris | 473 |
| Los Angeles | 456 |
| Birmingham | 451 |
| Kyoto | 420 |
| Chicago | 414 |
| San Francisco | 411 |
| Boston | 387 |
| Philadelphia | 333 |
| Milan | 313 |
| Brussels | 304 |
| Dallas | 303 |
| 871 other cities | |

Top 20 countries

| Country | Publications |
|---|---|
| USA | 9795 |
| Germany | 2712 |
| Japan | 2637 |
| United Kingdom | 2063 |
| France | 1797 |
| Spain | 1674 |
| Italy | 1535 |
| China | 896 |
| Canada | 763 |
| Australia | 545 |
| Netherlands | 510 |
| Belgium | 491 |
| Switzerland | 333 |
| Sweden | 329 |
| Austria | 314 |
| South Korea | 291 |
| Turkey | 279 |
| Taiwan | 278 |
| Brazil | 233 |
| Israel | 216 |
| 81 other countries | |

**Fig. 20.2** Hot topics for "Liver Transplantation"
The result page for a query to Hot Topics starts with a summary for the selected concept including a description, synonyms or the number of all publications in PubMed. Under Publication over time you can find two graphs. The first graph displays the number of publications related to this term per year. The second graph visualizes the fraction of publications on the topic over the total number of publications in that year. For "Liver Transplantation" the first graph displays the growing number of publications, but the second graph denotes over the last years stagnation in comparison to the overall publication growth in PubMed. The top authors, journals, cities, and countries are presented as tables. All table entries are links that retrieve the related articles. If you would click on "Neuhaus P" you can retrieve all the publications which have an author with this name. The co-author graph shows which author published together with whom. The more thick a line, the more articles contain their names as co-authors. The world map shows the regional distribution of the articles (Copies of figures including color copies, where applicable, are available in the accompanying CD)

**Answer:** Search for "Dresden[AD] Planck[AD] Genetics[AD]" and click on the link "show statistics for these 305 articles". Currently, the top author is Kai Simons with 41 publications, but this will probably change when new articles are published. In addition to the people, one can also easily retrieve all publications in the Science journal from the MPI-CBG by clicking on the provided link "Science" in the top journals list.

This example can be extended to be used with any institution mentioned in the affiliations of PubMed articles. One might also consider to use date ranges (e.g., years), to check for changes in the publication profiles over time.

### *20.3.2 GoWeb*

Sometimes the search with PubMed is not enough and the user wants to use a general purpose search engine like Google or Yahoo. With GoWeb we offer an internet search with ontological background knowledge. Some of the resources you can search with are, for instance, full text articles not included in PubMed, non-scientific sources like wikipedia or web based patent databases, commercial sites and vendors for equipment, special interest sites like the alzforum.org, or even news sites.

GoWeb uses standard web search engines and categorizes the results with its annotation algorithms. Normally web searches return the url and the title as well as a short text snippet from the result page containing your searched keywords. These texts are text mined, and the resulting terms are used in the same way as in GoPubMed to present the results of your search. You can use the ontological background knowledge to answer questions and reduce the result in a fast an efficient way without the need to read all the presented results. It includes, if available, also wikipedia links and protein names. Some example questions and answers are:

**Question:** Are there antibodies for ADDL?
**Answer:** GoPubMed can also search the web. Go to gopubmed.org/goweb and type ADDL antibody. Open "Chemicals and Drugs" and click on "Antibodies, monoclonal". The search results are now reduced from 100 to 8. Besides many pages of the Alzforum, there is the news that "Acumen and Merck Enter Into Alzheimer's Collaboration" which talks about: "... exclusive rights Acumen's ADDL technology monoclonal antibodies ... million development approval milestones first antibody product is commercialized. ..."

**Question:** What treatments does the web discuss for Alzheimer?
**Answer:** Go to gopubmed.org/goweb and type "Alzheimer treatment". Go to "Chemicals and Drugs", there you can find the term Memantine and also the term Vitamins. For more information on Vitamins click on the term. This will reduce the result set from 100 to 2 documents. In the result snippets you can find a statement like: "... vitamin may also be an ideal natural treatment for Alzheimer's disease too. ... Over the course of a small study, researchers at the University of Wisconsin ..."

## 20.4 Comparison and Conclusion

Currently, there is a lot of interest in literature searching as evidenced by the recent search engines such as Google Scholar or Microsoft's Windows Live Academic (see Table 20.1). This also includes publishers like Elsevier with Scopus. These Engines offer a more comprehensive or different document base, than the classical PubMed, but they currently do not include intelligence to answer questions.

GoPubMed [15] indexes PubMed search results with ontological background knowledge, such as Gene Ontology and MeSH. As shown above, this novel approach to search can help answer questions. In particular, the summary of important terms in "top five & more" is a most helpful feature for answering questions or reducing the big initial result to a smaller set of relevant articles in one click. With GoWeb, the ontological background knowledge can also be applied to normal web search and be able to answer questions from non-PubMed sources to answer questions.

GoPubMed's hot topics feature, additionally allows users to get an overview of research trends, relevant journals, key authors, and regional research interests. This feature is not provided by any of the other engines so far. GoPubMed is scalable and the system currently handles a user's search result of up to 10.000 documents. It also provides additional useful features like links to wikipedia pages and mentioned proteins in SwissProt.

There are a number of related tools (see also Table 20.1):

- *HubMed* [16] is the direct front end to PubMed It offers tools for the citation management of PubMed articles. It also provides options for expanding the query or clusters the results in categories. This is all based on the MeSH terms directly provided by PubMed. If there are no MeSH concepts available for an article, then this features does not work, because no term matching is done by HubMed itself. This is usually the case for the more recent articles. As an alternative they offer a tagging system where you can add your own tags to an article.
- *iHOP*[21] uses genes and proteins as hyperlinks between sentences and abstracts. It converts the information in PubMed into one navigatable resource.The navigation along the gene network allows for a stepwise and controlled exploration of the information space. Each step through the network produces information about one single gene and its interactions.
- *eTBLAST* [18] is quite a different approach to search the PubMed articles. It is based on text similarity and allows you to search for related articles using a relevancy, ranking different from PubMed. Input a paragraph/abstract which is relevant for your search and eTBLAST returns a list of articles. For a search result one can list relevant authors, journals and a timeline.
- *PubFinder* [17] Similar to eTBLAST, it can find related articles from a set of abstracts. It derives a list of discriminating words, which is subsequently used for scoring all defined PubMed abstracts for their probability of belonging to the defined scientific topic.
- *Textpresso* for C. elegans [19] has been developed as part of the Wormbase effort. It currently offers about 100 concepts such allele, anatomy, association, characterization, clone, comparison, consort, developmental stage, disease, drugs, effect, entity feature, gene, involvement, life stages, mutants, nucleic acid, organism, pathway, phenotype, purpose, regulation, reporter gene, restriction enzyme, sex, spatial relation, strain, time relation, transgene, transposon, vector and including also a subset of Gene Ontology concepts. It searches only abstracts and full text articles relevant for *C. elegans*. Textpresso does not offer an ontology tree for the exploration of a result set.
- *Vivisimo ClusterMed* does not use existing ontologies, but clusters documents hierarchically, although it distinguishes between categories like title and abstract, authors, affiliation, or publication date. From the document clusters, it derives representative terms. This automated hierarchy generation inevitably merges concepts of different nature, as the algorithm is only guided by the given documents, thus missing a lot of background knowledge a human uses in the creation of an ontology. Since Vivisimo clusters documents on the fly, there is a limit to its scalability.
- *XploreMed* [20] filters the PubMed results by the eight main MeSH categories and then extracts topic keywords and their co-occurrences. Abstracts can be retrieved for co-occurring keywords. The topic keywords are single words, usually occurring with a high frequency. Thus, multi-word concepts such as "Stem Cell" are not proposed as keyword. Currently XploreMed has a limited scalability and searches are restricted to 500 documents.

The combination of text mining and ontology-based background knowledge holds the possibility for intelligent search either in the literature or in the web. With a new generation of emerging search engines, biomedical researchers can answer questions and get an overview of a topic.

**Acknowledgments** We kindly acknowledge funding by the projects Sealife.

## References

### The Ontology

1. Ashburner, M., Ball, CA., Blake, JA., Botstein, D., Butler, H., Cherry, JM., Davis, AP., Dolinski, K., Dwight, SS., Eppig, JT., Harris, MA., Hill, DP., Issel-Tarver, L., Kasarskis, A., Lewis, S., Matese, JC., Richardson JE., Ringwald, M., Rubin, GM., and Sherlock, G. (2000) Gene ontology: tool for the unification of biology. The Gene Ontology Consortium, Nat. Genet. 25(1):25–29.
2. Bodenreider, O. (2004) The Unified Medical Language System (UMLS): integrating biomedical terminology, Nucleic Acids Res., 32(Database issue):D267–270.
3. Ehrler, F., Geissbhler, A., Jimeno, A. and Ruch P. (2005) Data-poor categorization and passage retrieval for gene ontology annotation in Swiss-Prot, BMC Bioinformatics, 6 Suppl 1:S23.

### Text and Literature Mining

4. Price, DJ (1965) Network of scientific papers, Science, 149(3683):510–515.
5. Garfield, E. and Melino, G. (1997) The growth of the cell death field: an analysis from the ISI-science citation index, Cell Death Differ, 4(5):352–361.
6. Boyack, KW. (2004) Mapping knowledge domains: characterizing PNAS, Proc. Natl. Acad. Sci. USA, 101 Suppl. 1, 5192–5129.
7. Newman, M. (2004) Coauthorship networks and patterns of scientific collaboration, Proc. Natl. Acad. Sci. USA, 101 Suppl. 1, 5200–5205.
8. Jensen, LJ., Saric, J. and Bork P. (2006) Literature mining for the biologist: from information retrieval to biological discovery, Nat. Rev. Genet., 7(2):119–129.
9. Hirschman, L., Yeh, A., Blaschke, C. and Valencia, A. (2005) Overview of BioCreAtIvE: critical assessment of information extraction for biology, BMC Bioinformatics, 6 Suppl. 1, S1.
10. Kim, J., Ohta, T., Tsuruoka, Y. and Tateisi, Y. (2004) Introduction to the bioentity recognition task at JNLPBA, Proceedings of the Joint Workshop on Natural Language Processing in Biomedicine and its Applications Geneva, 70–76.
11. Hakenberg, J., Royer, L., Plake, C., Strobelt, H. and Schroeder, M. (2007) Me and my friends: gene mention normalization with background knowledge, Proc. 2nd BioCreAtIvE Challenge Evaluation Workshop:141–144.
12. Finkel, J., Dingare, S., Manning, C., Nissim, M., Alex, B. and Grover, C. (2005) Exploring the boundaries: gene and protein identification in biomedical text, BMC Bioinformatics, 6 Suppl. 1, S5.
13. Lee, L., Horn, F. and Cohen, F. (2007) Automatic Extraction of Protein Point Mutations Using a Graph Bigram Association. PLoS Comput. Biol., 3, e16.
14. Baker, CJ. and Witte, R. (2006) Mutation Mining—A Prospector's Tale Information Systems Frontiers, Kluwer Academic Publishers, 8, 47–57.

### Searching

15. Doms, A. and Schroeder, M. (2005) GoPubMed: exploring PubMed with the Gene Ontology, Nucleic Acids Res., 33, W783–6.
16. Eaton, A. (2006) HubMed: a web-based biomedical literature search interface. Nucleic Acids Res., 34, W745–747.
17. Goetz, T. and von der Lieth, CW. (2005) PubFinder: a tool for improving retrieval rate of relevant PubMed abstracts. Nucleic Acids Res., 33, W774–778.
18. Lewis, J., Ossowski, S., Hicks, J., Errami, M. and Garner, H. (2006) Text similarity: an alternative way to search MEDLINE, Bioinformatics, 22, 2298–2304.
19. Müller, H., Kenny, E. and Sternberg, P. (2004) Textpresso: an ontology-based information retrieval and extraction system for biological literature, PLoS Biol., 2, e309.
20. Perez-Iratxeta, C., Pérez, A., Bork, P. and Andrade, M. (2003) Update on XplorMed: A web server for exploring scientific literature, Nucleic Acids Res., 31, 3866–3868.
21. Hoffmann, R. and Valencia, A. (2004) A gene network for navigating the literature, Nature Genetics, 36, 664.

## Key References

Doms, A. and Schroeder, M. (2005) GoPubMed: exploring PubMed with the Gene Ontology, Nucleic Acids Res., 33, W783–786.

Müller, H., Kenny, E. and Sternberg, P. (2004) Textpresso: an ontology-based information retrieval and extraction system for biological literature, PLoS Biol., 2, e309.

Jensen, LJ., Saric, J. and Bork P. (2006) Literature mining for the biologist: from information retrieval to biological discovery, Nat. Rev. Genet., 7(2):119–129.

Eaton, A. (2006) HubMed: a web-based biomedical literature search interface. Nucleic Acids Res., 34, W745–747.

Hoffmann, R. and Valencia, A. (2004) A gene network for navigating the literature, Nature Genetics, 36, 664.

## Web Resources

http://GoPubMed.org
http://gopubmed.org/goweb

# Chapter 21
# BiblioSphere — Hypothesis Generation in Regulatory Network Analysis

**Anton Epple and Matthias Scherf**

**Abstract** With its microarray analysis strategy BiblioSphere introduced a new analysis concept for the interpretation of microarray results. The first level of analysis integrates multiple complementary data sources, to compile a network that covers the entire knowledge and data available for the biological entities in the input data set to enable the interactive analysis of the data. A broad range of filters together with ranking techniques are available as tools to enable the sub grouping of entities and relations to focus the analysis to the user's interest.

To complement this highly dynamic and interactive process, BiblioSphere introduces a second level of analysis by modelling Genomatix expert knowledge in an integrated expert system for the rule-based analysis of input data. Strategies that have proven successful in network analysis for the assessment of regulatory relations over the years have been formalized to an expert system for the automated generation of hypotheses.

While the first level enables the interactive analysis of microarrays by experts, the second level of analysis makes network analysis available for high-throughput microarray data evaluation.

**Keywords** Textmining · Microarray · Network · Statistical analysis

## 21.1 Introduction

In recent years, the availability of high-throughput methods like microarray analysis has dramatically changed biological sciences. Before the availability of microarrays, the applicability of expression analysis was limited to testing a given hypotheses. As only a few transcripts could be tested in parallel, an initial hypothesis had to be developed based on the data available, before the experimental design could decide which transcripts should be tested under experimental conditions. The change from being able to conduct only a small number of experiments at the same time to monitoring the whole transcriptome led to a change in experimental design. Previously the limited technical possibilities forced scientists to carefully choose the biological entities they would monitor in an experiment, to verify or falsify a given hypothesis. The new techniques make expression analysis available for data collection as the first step of the scientific method. The process of hypothesis generation shifted from the experimental design to the interpretation of the results. Instead of being the starting point, the hypothesis is now the desired result of most expression analysis experiments.

While high throughput techniques enable a new experimental design, they also pose new challenges for the analysis and interpretation of these data and they require new tools that help with this process. The aim of such tools is to guide the user in generating hypotheses to explain the observed expression patterns. In an ideal setting, the user would be presented with a ranked list of hypotheses to

A. Epple
Genomatix Software GmbH, Bayerstr. 85A, D–80335 Müenchen, Germany
e-mail: epple@genomatix.de

S. Krawetz (ed.), *Bioinformatics for Systems Biology*, DOI 10.1007/978-1-59745-440-7_21,

choose from as a starting point for further analysis. Statistical methods proved helpful in identifying biological entities that show significant differences under experimental conditions, and clustering algorithms help to group entities by common features. However, biological interpretation of results remains the most challenging part of the analysis. This article focuses on a two level analysis approach to provide a workflow for biological interpretation. The first level of analysis classifies input sets by the biological topics and compiles the network of potential relations from independent lines of evidence. The second level of analysis uses a rule based approach for the semantic interpretation of these relations in a biological context. These techniques for classification, graph analysis and rule based hypothesis generation have been combined into a workflow for automated hypothesis generation that is available in the BiblioSphere program package.

## 21.2 Data and Data Sources

To enable the integration of multiple complementary lines of evidence, BiblioSphere provides an ontology of biological entities, each of which can be assigned to catalogs of pre-defined classes, as well as a formal semantic describing their relationships in regulatory networks. Entities, entity relations and classifications are mapped to common standard identifiers to allow cross-referencing.

### *21.2.1 Biological Entities*

The catalog of molecular components, that BiblioSphere supports for modelling regulatory networks, consists of a set of base entities that are further classified and sub grouped by additional annotations. Most pathway and protein interaction databases as well as the major part of the relevant literature, focus on genes or proteins. Therefore, most available tools also focus on these elements as the base units of regulation. As BiblioSphere also includes *in-silico* sequence analysis techniques, its data model is more fine-grained and based on alternative transcripts and their respective promoters from the ElDorado database (Genomatix Software GmbH), as base elements of gene regulation.

Nevertheless, the replication units are mapped to EntrezGene loci [1] and their respective protein products, to integrate available databases or data derived from literature analysis. In addition, BiblioSphere makes use of an expert curated database of regulatory relevant protein complexes with a focus on transcription factors, kinases and receptor complexes. ChemIdPlus identifiers are used for referencing biologically relevant small molecules. Table 21.1 shows the biological entities referenced in BiblioSphere together with their referenced datasource.

**Table 21.1** Biological entities

| Biological entity | Source |
|---|---|
| Small molecules | PubChem (http://pubchem.ncbi.nlm.nih.gov/) |
| Transcript | Genomatix |
| Gene | EntrezGene |
| Protein | EntrezGene |
| Protein complex | Genomatix / NetPro |
| Transcription factor | MatBase (Genomatix Software GmbH) |
| Gene/protein | EntrezGene |

(Copies of tables are available in the accompanying CD.)

**Table 21.2** Classes of biological entities currently supported

| Biological entity | Class catalog | Annotation source | Evidence type |
|---|---|---|---|
| transcript | Genomatix tissue catalog | GNF | Experimental data |
| transcript | Genomatix tissue catalog | dbEST | Experimental data |
| Gene/protein | Genomatix Pathway catalog | EntrezGene | Expert curated/ electronically inferred |
| Gene/protein | GO biological process | EntrezGene | Expert curated/ electronically inferred |
| Gene/protein | GO cellular component | EntrezGene | Expert curated/ electronically inferred |
| Gene/protein | GO molecular function | EntrezGene | Expert curated/ electronically inferred |
| Gene/protein | Genomatix MatBase | EntrezGene | Expert curated/ electronically inferred |

(Copies of tables are available in the accompanying CD.)

## 21.2.2 *Entity Classes*

Experiments indicate correlations within groups of biological entitites (usually RNAs or proteins). The initial step to associate groups with a pertinent biological context is biological classification. Statistical methods can be applied to deduce significantly over represented classes in standardized catalogs or ontologies from the class assignments of individual constituents of a set. BiblioSphere makes extensive use of this approach integrating publicly available and proprietary catalogs and data sources. Expert curated or electronically inferred annotations are added as attributes to the entities. Table 21.2 gives an overview of the available annotation sets for biological entities in BiblioSphere.

## 21.2.3 *Entity Relations*

While classification of biological entities identifies the biological processes affected by the experimental settings, they do not supply information on the role of the individual biological entities involved. This can be achieved by mapping the entities to a network compiled from pair wise relations between biological entities. Supported relations in BiblioSphere range from protein-protein interactions, predicted transcription factor binding sites in promoters to interactions of small molecules and proteins. BiblioSphere is, at the time of this writing, the only application that integrates relations derived from such different data sources like natural language processing, expert curated manual annotations, *in silico* promoter analysis and expression profiles.

### 21.2.3.1 Promoter Analysis

The key mechanism underlying results from microarray experiments is transcriptional regulation. *In-silico* promoter analysis can be applied to predict regulatory relations between transcript and potential regulators. To enable this, the BiblioSphere database includes a genome-wide *in-silico* promoter analysis data. Transcription factor binding site matrices from MatBase (Genomatix Software GmbH, Munich) are used to scan all known promoters from the Genomatix promoter resources for potential binding sites using MatInspector® [2]. Background information on the analysis of promoters for transcription factor binding sites and more details on the MatInspector program is given in another chapter of this book (Chapter 17 by Thomas Werner).

### 21.2.3.2 Natural Language Processing

BiblioSphere analyzes the PubMed database for co-occurrence of protein names using a large expert curated synonym database (Fig. 21.1).

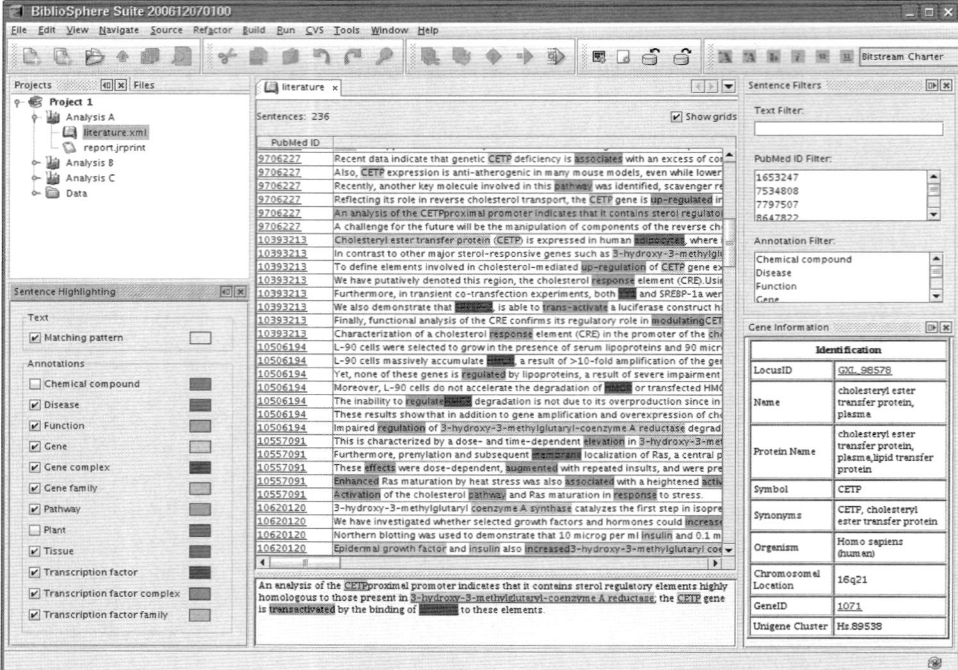


**Fig. 21.1** BiblioSphere literature viewer with syntax highlighting for identified entities (Copies of figures including color copies, where applicable, are available in the accompanying CD)

Techniques in information extraction range widely from the mere co-citation analysis to the highly sophisticated semantic analysis. While simple techniques suffer from a large number of false positive matches and therefore show only low specificity, they usually have a high recall (depending on the lexicon); vice versa, more sophisticated techniques tend to show a lower recall factor while providing a better specificity (Fig. 21.2).

BiblioSphere enables users to dynamically set the level of specificity. Specificity levels range from simple co-citation analysis, over template based and context aware information extraction techniques to entity relations curated by domain experts:

- Level 0: Two biological entities are co-cited somewhere within an abstract of a publication.
- Level 1: Two biological entities are co-cited within the same sentence.
- Level 2: Two biological entities are co-cited in the same sentence and the sentence also contains a "function word".
- Level 3: Sentence matches a pre-defined information extraction template.
- Level 4: Manually confirmed connection of two biological entities.

The decision whether it is sensible to go for a higher recall or for a higher specificity depends on the application. If the information extracted from literature constitutes the single source of relational

$$precision = \frac{|\{relevant\} \cap \{retrieved\}|}{retrieved} \qquad recall = \frac{|\{relevant\} \cap \{retrieved\}|}{relevant}$$

**Fig. 21.2** Precision is defined as the proportion of retrieved *and* correct facts to all the facts retrieved, while recall defines the proportion of relevant facts that are retrieved, out of all the correct facts available (Copies of figures including color copies, where applicable, are available in the accompanying CD)

information, or if the data base is very large, it is advisable to go for higher specificity, unless a broad statistical base is needed, whereas in cases where literature independent information is available (other lines of evidence), it is favorable to go for higher recall and to combine the literature derived information with these other lines of evidences to rule out false positives. BiblioSphere already systematically integrates several independent lines of evidence to leverage this approach.

#### 21.2.3.3 Pathway Databases and Expert Curated Knowledge

In addition to relations extracted from the scientific literature by Genomatix domain experts, BiblioSphere also integrates external databases. Molecular Connections NetPro database is an expert curated database, covering the whole range of relations relevant to network analysis including protein-DNA binding, protein-protein interactions and small molecules. More than 115,000 individual expert curated relations for biological entities have been imported into the BiblioSphere knowledgebase.

#### 21.2.3.4 Expression Profiles

Besides relations manually or automatically extracted from literature or inferred by *in-silico* sequence analysis, BiblioSphere also includes experimental expression data. A series of microarray analysis experiments in different tissues has been performed to create tissue profiles for individual transcripts. Entity relations are based on the similarity of these profiles as an indicator for co-regulation.

#### 21.2.3.5 User Data

In addition to these sources of relations, user-defined data can be used to complement the data sources of BiblioSphere. Thus, BiblioSphere can be enhanced to reflect new types of connections such as entity relations based on e.g. microRNAs

### *21.2.4 Entity Relation Classes*

Like the biological entities themselves, relations between them can be used for the classification of entity sets. The same statistical methods that are used to find significantly over represented classes for a set of biological entities can be applied when annotations are available for a relation. If the relation is derived from Genomatix proprietary pathway annotation, the respective pathway catalog can be utilized to classify the set of relations and thereby indirectly the set of related entities. Relations derived from PubMed can be evaluated using their classification with the hierarchical MeSH catalog. Table 21.3 shows the different entity relation classes included in the BiblioSphere knowledge base.

**Table 21.3** Entity relation classes in BiblioSphere

| Entity Relation | Class catalog | Annotation source | Evidence type |
|---|---|---|---|
| Protein-Protein | MeSH: Disease | PubMed | Expert curated |
| Protein-Protein | MeSH: Chemicals and Drugs | PubMed | Expert curated |
| Protein-Protein | MeSH: Anatomy | PubMed | Expert curated |
| Protein-Protein | MeSH: Biological Sciences | PubMed | Expert curated |
| Protein-Protein | MeSH: Analytical, Diagnostic and Therapeutic Techniques and Equipment | PubMed | Expert curated |
| Protein-Protein | MeSH: Disease | PubMed | Expert curated |

(Copies of tables are available in the accompanying CD.)

### 21.2.5 Same Information – Different Sources: Multiple Lines of Evidence

As shown before, BiblioSphere integrates a wide range of independent lines of evidence for individual entity relations. The idea behind this concept is that there are many indicators pointing to a certain interaction, which are too weak to build a useful hypothesis when evaluated in isolation. Therefore, BiblioSphere enables the user to filter interactions that are supported by independent lines of evidence indicating the same relationship. While this approach is not capable of verifying or falsifying a hypothetical relationship, it helps in ranking relationships and identifying the most promising candidates for further evaluation. As an example, co-expression taken as a single line of evidence, is only a weak indicator of co-regulation as is the occurrence of the same transcription factor module in the promoter of two transcripts; taken together they form a stronger indication. If the transcription factors, known to bind this module, are found to be co-cited in literature with one of the gene products of these transcripts, this further strengthens the hypothesis. The gene products might also be annotated to belong to the same biological process, show a similar tissue profile, and so on. All of these individual findings are weak when viewed in isolation, while taken together they provide strong evidence for certain type of relationships based on the consistency of the different lines of evidence. While small data sets may be manually evaluated by presenting these integrated lines of evidence to the user, this is not possible for data-sets derived from high-throughput experiments with millions of potential relations. Here, it is necessary to have a tool that helps identify the most likely candidates. To do so, BiblioSphere allows the users to rank relationships by evidence and to dynamically filter for relations that are supported by many lines of evidence.

## 21.3 First Level Analysis

A widely accepted technique for biological interpretation is to classify groups of biological entities for identifying processes and pathways affected by the experimental settings usually called GO-ranking. A variety of methods and tools are available for this task including DAVID [3], GoSurfer [4] or FuncAssociate [5]. BiblioSphere similarly analyzes annotations of biological entities and relations, with controlled vocabularies, to find categories that are overrepresented for the input set but not limited to GO-terms. Another common method of many bioinformatics software tools for the analysis of microarray experiments is to project the experimental data onto precompiled pathways, to facilitate the interpretation of the results. While this method helps to bring experimental data into the context of common knowledge from textbooks, pathway databases and scientific articles, it limits the analysis to the small subset of the analyzed data for which such information is available. BiblioSphere also allows such projections of expression data but it is not restricted to this limited set of expert curated information. Due to its multiple lines of evidence approach as described above it enables a much larger portion of the data to be analyzed.

### 21.3.1 Statistical Analysis of Entity-Sets

All entity class and entity relation class catalogs can be used to subgroup the entities and relations between them by their respective annotations. In addition, these controlled vocabularies or ontologies can be used to detect classes that are overrepresented in the set. BiblioSphere uses a method called p-value ranking as a statistical test for the overrepresented groups. P-value ranking corrects for wrong positives that result from performing thousands of null hypothesis tests at once, a problem known as "multiple-hypothesis testing". Figure 21.3 shows an example report for a set of genes that have been found to be upregulated on stimulation with Platelet-derived growth factor (PDGF) in a time series experiment. Original microarray data from [6], is publicly available from Gene Expression Omnibus (GEO, accession number GSE1484).

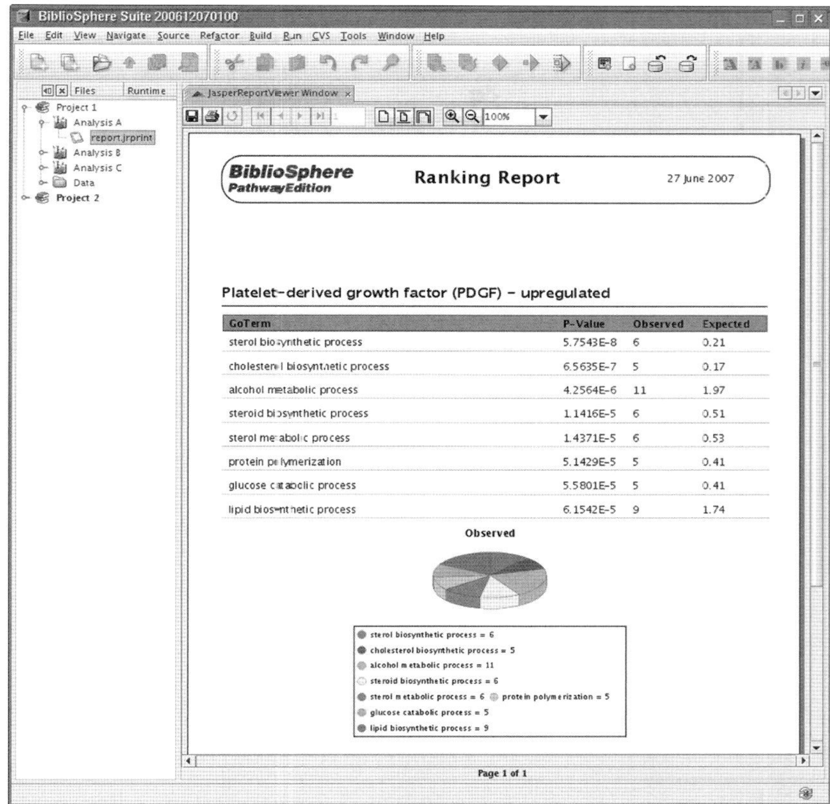


**Fig. 21.3** Example report for overrepresented groups in an input set of 105 genes (Copies of figures including color copies, where applicable, are available in the accompanying CD)

The most overrepresented groups are consistent with the findings of the original study, where Sterol biosynthesis was identified as the key process activated by PDGF stimulation. It is also possible to compare two results to test how the over represented categories differ for two different experimental conditions or time points (Fig. 21.4).

## *21.3.2 Network Analysis*

Networks are compiled from all relationships that exist between the biological entities in an input set. BiblioSphere networks can be mathematically interpreted as graphs consisting of vertices (the biological entities) and edges (the entity relations). Standard graph algorithms may therefore be applied. The main purpose of network visualization in BiblioSphere is to give an overview on how entities interact, to identify higher-order functional modules and to understand the flow of information in and between these functional modules.

Due to the large knowledgebase behind BiblioSphere, networks tend to be too large to be comprehensible when displayed unprocessed. For a list of 1000 transcripts, which is the average input size for the analysis of microarray results, the number of relations in BiblioSphere database can easily exceed 100,000. It is a general theory that the size of networks should not exceed 100 nodes. Consequently, the graph needs to be reprocessed before being presented to the user. This coincides with the typical size of known functional modules or pathways, as contained in metabolic pathway databases like KEGG [7] or signal transduction pathways like STKE which typically have a size of 5 to 60 entities.

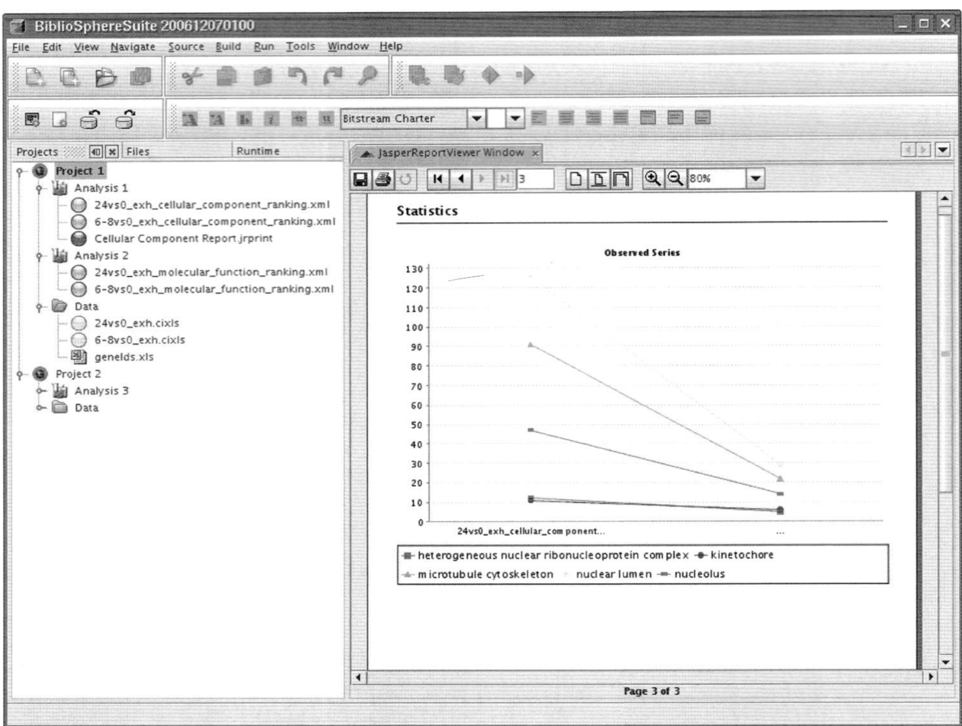


**Fig 21.4** A report to compare the ranking results for different rankings (Copies of figures including color copies, where applicable, are available in the accompanying CD)

The two main strategies to reduce graph complexity in BiblioSphere are filtering and clustering by connectivity. Bibliosphere offers a potent system of filters to customize the analysis output according to your needs. This includes filters based on the content of the literature itself, as well as on functional analysis using hierarchical annotation terms. Table 21.4 gives an overview of the available filters.

A second strategy for the reduction of complexity makes use of dense sub-graphs or clusters in the network. Obviously, clusters that have a lot of relations between their constituents have a greater likelihood of belonging to a common process than loosely coupled components. The connectivity of a cluster in biological networks has been shown to be correlated with biological function [8].

It has already been shown that functional grouping provides a very simple way to identify subgroups that belong to a common category, but it has a major drawback when used to analyze groups of genes. Grouping nodes or edges by their attributes in functional clustering makes a clear separation between nodes having a certain attribute or not, while connectivity is less restrictive. Functional clustering represents the current knowledge about a process, given that all known entities that belong to a group are correctly annotated, and eventually reveal new relations between the constituent members of a cluster. Connectivity clustering, however, can identify new members of a group, and it can even identify groups that represent a new category.

While these methods are able to reduce the number of nodes in a graph, they do not reduce the number of relations between them. Even a mathematically scarce graph can be “dense” in terms of visualization. The upper graph (A) in Fig. 21.5, shows a sub-network for genes upregulated in PDGF-stimulated fibroblasts showing only the relations that have been verified by domain experts. It is still hard to comprehend the individual relations between pairs of entities or even identify a coherent information flow.

**Table 21.4** Filters available in BiblioSphere

| Filter Name | Source | Filtered Entitiy | Type |
|---|---|---|---|
| GO: Molecular Function | Gene Ontology / EntrezGene | Gene | Hierarchical |
| GO: Biological Process | Gene Ontology / EntrezGene | Gene | Hierarchical |
| GO: Cellular Component | Gene Ontology / EntrezGene | Gene | Hierarchical |
| MeSH: Disease | MeSH / PubMed | Abstract | Hierarchical |
| MeSH: Chemicals and Drugs | MeSH / PubMed | Abstract | Hierarchical |
| MeSH: Anatomy | MeSH / PubMed | Abstract | Hierarchical |
| MeSH: Biological Sciences | MeSH / PubMed | Abstract | Hierarchical |
| MeSH: Analytical, Diagnostic and Therapeutic Techniques and Equipment | MeSH / PubMed | Abstract | Hierarchical |
| Co-Citation: frequency | Natural Language Processing | Gene | Plain |
| Co-Citation: Specificity level | Natural Language Processing | Gene | Plain |
| Co-Citation: Connectivity | Natural Language Processing | Gene | Plain |
| Free text | Natural Language Processing | Abstract | Plain |

(Copies of tables are available in the accompanying CD.)

A simple strategy to reduce the complexity here is based on the observation that users usually start exploring the network from a single biological entity of interest.

To focus the graph on one entity, only the shortest path of this node to all other nodes in the set is displayed. This converts the graph to a tree structure and hides all the other edges. The lower graph (B) in Fig 21.5, illustrates how this reduces the complexity and enhances readability of a network. This strategy allows users to explore the network in a stepwise manner by the exploration of focussed sections. It is important to note, that all filtering operations still work on the full set of edges, the reduction is only affecting the display.

## 21.4 Second Level Analysis

The major shortfall of simply projecting the data onto pathways or regulatory networks is that the results are only visualized, while the process of semantic interpretation of the results is left to the user. Given the thousands of relations between biological entities this can be a challenging, if not impossible, task. To help users with the interpretation of these regulatory networks, BiblioSphere introduces a rule-based generation of hypotheses based on experimental data as a second level of data analysis.

### *21.4.1 Rule-Based Hypothesis Generation*

It has already been shown how multiple lines of evidence can be used to rank relations between individual entities by simply counting the number of distinct classes of complementary relations identified. While this method is useful, it does not allow for a presumption on the semantic nature of the connection.

BiblioSphere uses a production rule system to cope with this problem. A schematic view of this system is shown in Fig. 21.6. The core component of a production rule system is an inference engine that matches facts and data against production rules, to infer conclusions and trigger actions.

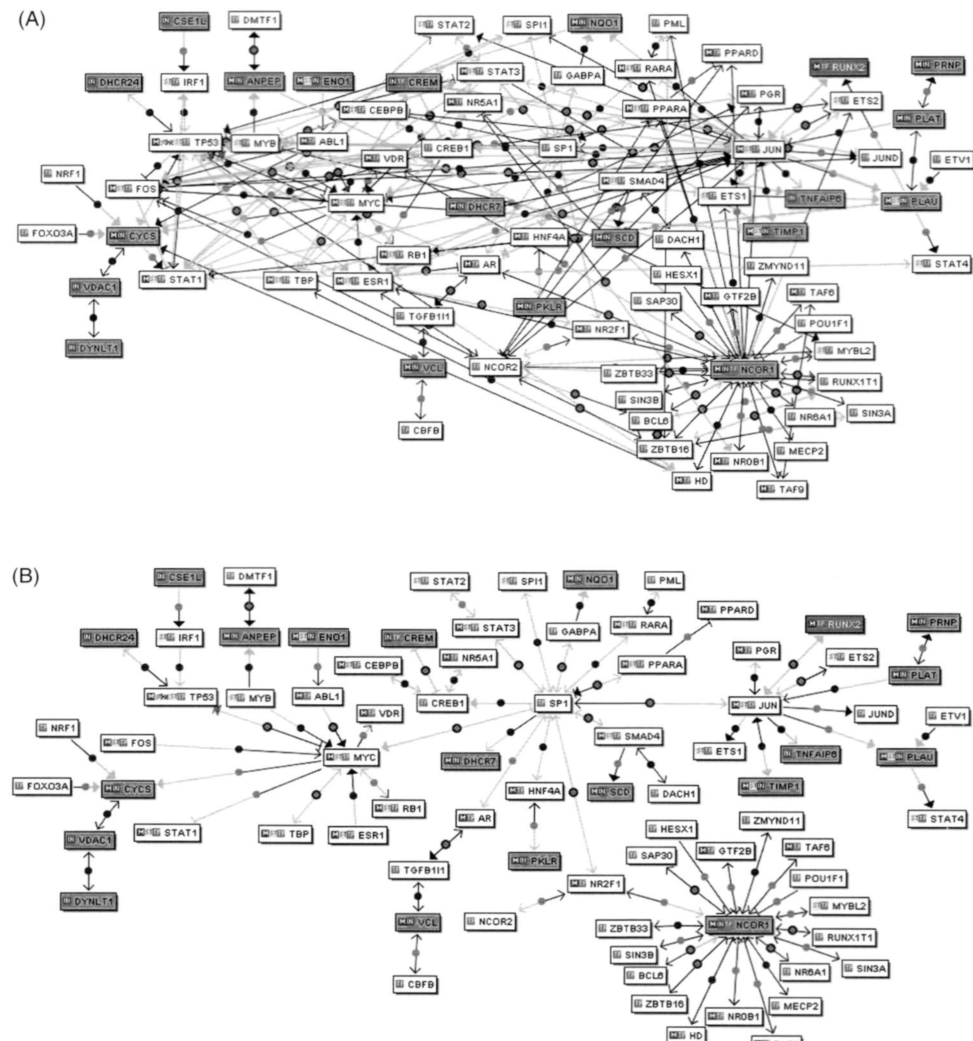


**Fig. 21.5** Both networks show a view of the same graph. The network created from an input set of 105 genes upregulated in PDGF-stimulated fibroblasts has been filtered for relations supported by expert curated annotations. **(A)** An unrestricted view of the graph; **(B)** Only the shortest path for node Jun is displayed (Copies of figures including color copies, where applicable, are available in the accompanying CD)

A Production Rule is a two-part structure using the first order logic for knowledge representation. The left hand side of the rule is a condition, while the right hand side defines the action invoked in case the condition is true.

One of the major advantages of using a rule-based solution is the separation of the logic rules from the data model. Rules can thus be written in a domain specific language by users without programming experience. Genomatix experts have used their domain knowledge to build a knowledge base with rule sets, for various strategies of hypothesis generation. These rules can be adjusted to reflect new semantic knowledge or extended to cover other types of entities or relations. All data about entities, relations as well as their class attributes including data supplied by the user (such as expression values), build the input to the system. Figure 21.7 shows an example of rule set for the

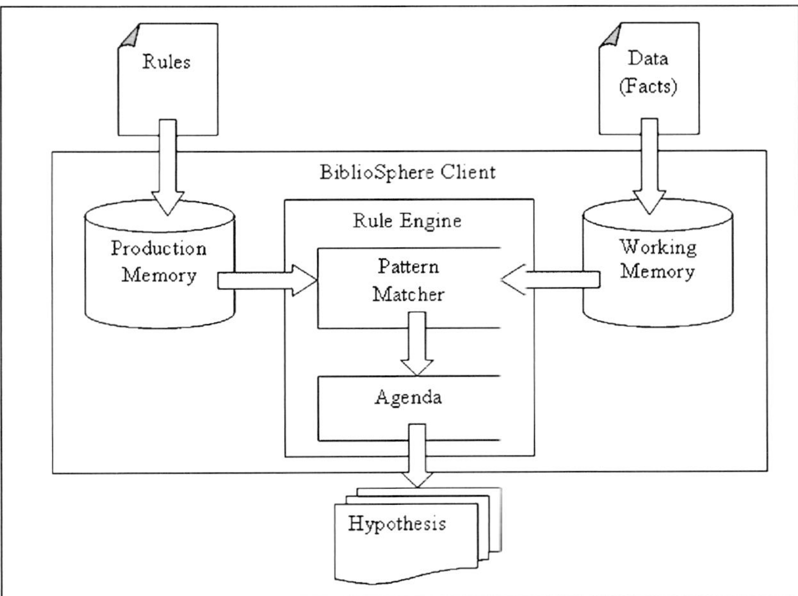


**Fig. 21.6** A schematic overview of the hypothesis generation process in BiblioSphere. Users provided or pre-defined sets of rules are loaded into the production memory. Input data and relational data from the BiblioSphere database are uploaded to the working memory. The rule engines pattern matching mechanism determines all matching rules and creates an agenda for conflicting rules. The generation of a new hypothesis is triggered by matching its respective rule (Copies of figures including color copies, where applicable, are available in the accompanying CD)

```
Rule "Transcription factor activates Transcript"
when
   $t1: Transcript( upregulated == true )
   $t2: Transcript( upregulated == true )
   $t2 ( $tf: transcriptionfactor == true )
   $tf  ( matches contains $t1.Promoter )
   $t1 ( cocited contains $t2 )
then
   hypotheses.add (new Hypothesis ("activates",$t1, $t2, 2)) ;

Rule "transcript coregulated transcript"
when
   $t1: Transcript( regulated == true )
   $t2: Transcript( regulated == true )
   $m: Module ( matches contains $t1)
   $m ( matches contains $t2)
then
   hypotheses.add(new Hypothesis("coregulated",$t1, $t2, 4);
```

**Fig. 21.7** Two simplified rules for hypothesis generation. The first rule "Transcription factor activates Transcript", tests for transcriptionally upregulated transcription factor-transcript pairs based on the uploaded expression array data. If the transcription factor has a binding site in the promoter of a transcript and both have been found co-cited, a new hypothesis is added to the data model. The second rule "transcript co-regulated transcript" tests for upregulated transcripts that share the same promoter module (Copies of figures including color copies, where applicable, are available in the accompanying CD)

generation of hypotheses on gene regulation. Pre-defined rule sets cover transcriptional regulation and protein-protein interactions. Every generated hypothesis can be manually evaluated since the rule engine allows to keep track of the decisions and the evidence for these decisions.

## References


1. Maglott D., Ostell J., Pruitt, K.D., and Tatusova T. (2005). Entrez Gene: gene-centered information at NCBI. *Nucleic Acids Res.* 33(Database Issue): D54–D58.
2. Cartharius, K., Frech, K., Grote, K., Klocke, B., Haltmeier, M., Klingenhoff, A., Frisch, M., Bayerlein, M., and Werner, T. (2005). MatInspector and beyond: promoter analysis based on transcription factor binding sites. *Bioinformatics* **21**, 2933–2942.
3. Dennis, G. Jr., Sherman, B.T., Hosack, D.A., Yang, J., Gao, W., Lane, H.C., and Lempicki, R.A. (2003), DAVID: Database for Annotation, Visualization and integrated Discovery. *Genome Biol.* **4**(5):P3.
4. Zhong S., Storch F., Lipan O., Kao M.J., Weitz C., and Wong W.H. (2004). GoSurfer: a graphical interactive tool for comparative analysis of large gene sets in Gene Ontology space. *Appl. Bioinformatics*, **3**(4),261–264.
5. Berriz, G.F., King, O.D., Bryant, B., Sander, C., and Roth, F.P. (2003) Characterizing gene sets with FuncAssociate. *Bioinformatics*. **18**,2502–2504.
6. Demoulin, J.B., Ericson J., Kallin A., Rorsmann C., Ronnstrand L., and Heldin, C.H. (2004) Platelet-derived growth factor stimulates membrane lipid synthesis through activation of phosphatidylinositol 3-kinase and sterol regulatory element binding-proteins. *J. Biol Chem.* 279(34):35392–35402.
7. Kanehisa, M., Goto, S., Hattori, M., Aoki-Kinoshita, K.F., Itoh, M., Kawashima, S., Katayama, T., Araki, M., and Hirakawa, M. (2006). From genomics to chemical genomics: new developments in KEGG. *Nucleic Acids Res.* 34, D354–357.
8. Samuel Lattimore, B., van Dongen S., and Crabbe, M.J. (2005). GeneMCL in microarray analysis. *Comput Biol Chem.* 5,354–359.


## Web Resources

http://www.molecluarconnections.com
http://www.gnf.org/

# Chapter 22
# Biological Knowledge Extraction

## A Case Study of iHOP and Other Language Processing Systems

**Florian Leitner, Robert Hoffmann, and Alfonso Valencia**

*Knowledge is of two kinds. We know a subject ourselves, or we know where we can find information on it.*

—Samuel Johnson (1709–1784)

**Abstract** Text Mining is the process of extracting [novel] interesting and non-trivial information and knowledge from unstructured text (Google$^{TM}$ search result for "define: text mining"). Information retrieval, natural language processing, information extraction, and text mining provide methodologies to shift the burden of tracing and relating data contained in text from the human user to the computer. The emergence of high-throughput techniques has allowed biosciences to switch its research focus on Systems Biology, increasing the demands on text mining and extraction of information from heterogeneous sources. This chapter will introduce the most fundamental uses of language processing methods in biology and present the basic resources openly available in the field. The search for information about a common disease, chronic myeloid leukemia, is used to exemplify the capabilities. Tools such as PubMed, eTBLAST, METIS, EBIMed, MEDIE, MarkerInfoFinder, HCAD, iHOP, Chilibot, and G2D – selected from a comprehensive list of currently available systems – provide users with a basic platform for performing complex operations on information accumulated in text.

**Keyword** iHOP · CML · Text mininig · Language processing

## 22.1 Introduction

During the turn of the millennium, the emergence of high-throughput methods in Molecular Biology posed a significant change in the experimental approach. After the field of genomics had completed the most ambitious biological project of the last century, the sequencing of the human genome, and established the gene microarray expression profiling, other areas have profited from these large-scale experimental approaches. In proteomics, the chip technology has been expanded to protein arrays for analyzing a multitude of protein interactions with nucleic acids, antigens, and other proteins. Metabolomics is exploring the interaction of the genome and proteome with the chemical environment of the cell. Building on data contributed by all these "Omics", Systems

A. Valencia
Spanish National Cancer Research Centre, CNIO, Structural and Computational Biology Group, C/ Melchor Fernandez Almagro, 3, E-28029, Madrid, Spain
e-mail: valencia@cnio.es

S. Krawetz (ed.), *Bioinformatics for Systems Biology*, DOI 10.1007/978-1-59745-440-7_22,

Biology has emerged as a completely new discipline of research utilizing this information generation and a broad range of biological facts.

The wealth of information produced by these large-scale experimental approaches is being analyzed by various data mining methods to produce new insights and potential research targets, e.g., assembling list of genes potentially associated to cancer. These data are stored in dedicated databases, which index and distribute the content in a well-structured form. Yet, database projects cataloging the generated facts are struggling to keep up with this information flood, annotating publications and depositing the contained data in tabulated units. Because of this bottleneck, by far the largest repositories of biomedical data still are the publications. In 2006, the **MEDLINE**© PubMed© collection of biomedical abstracts surpassed the 16 million mark[1] – accordingly, every day it is becoming increasingly more difficult to keep track of publications and new information. MEDLINE is the "meta-database" of all biomedical publications, containing title, abstract, author names and several annotations for each published article, while the full texts of articles are stored with each journal separately.

This imposes some limitations on the usability of the data: the full text is not freely accessible and the publications are often distributed in PDF format, which is a proprietary format and very difficult to disassemble. Because of the restricted access, searches over the full text of documents is only possible on a single repository at a time, for e.g., while reading a publication on Nature.com, it is not possible to search the Science magazine in the same interface. To counter these limitations, several **Open Access** initiatives have been created, e.g., the Public Library of Science (PLoS), BioMed Central (BMC), the National Institute of Health (NIH, with the PubMed Central (PMC) repository), or the Wellcome Trust, to solve this issue. One of the advantages of repositories such as PLoS, BMC, or PMC is that they provide free access to the full text, which is distributed in both human (PDF) and machine (XML) readable format.

To provide consistent descriptors for documents, in 1954 the Medical Subject Heading (**MeSH**) thesaurus was created. **Thesauri** (or *subject headings*) are used to annotate objects (e.g., MeSH is used to annotate MEDLINE documents), and can be understood as indexed collections of terms agreed upon by the community that uses them. Although this supplies additional information for each publication, they do not make direct connections to database objects. The respective entities in the text, like the gene names, although annotated to the MEDLINE record as a MeSH term, are not directly linked to their database entries, e.g., the gene sequence.

We Imagine a world where scientific content is freely accessible and exhaustively annotated. Searches across all (full text) repositories would become available, making it possible to locate specific information at a full level of detail, e.g., searching for publications describing a specific diseases in conjunction with a special compound and a unique mutation. While reading publications, the respective mentions of the entities in the text are hyperlinked with their database objects. These objects in turn link back to all the publications relevant to them, making it easy to find former publications for a specific gene, extracting a set of RNA sequences for a disease, or locating chemical compounds interacting with a given protein. Carried to the extreme, complete signaling pathways and interaction networks could be browsed and extracted in such an environment.

Apart from the bureaucratic, political, and commercial restrictions mentioned, we will attempt to describe the technical limitations that scientific research is aiming to overcome to fulfill this vision. We will be looking at increasingly more complex problems on this path, and then will

[1] As of beginning 2007, the MEDLINE database of biomedical abstracts contains 16,120,074 entries, of which 686,406 were added just in 2006.

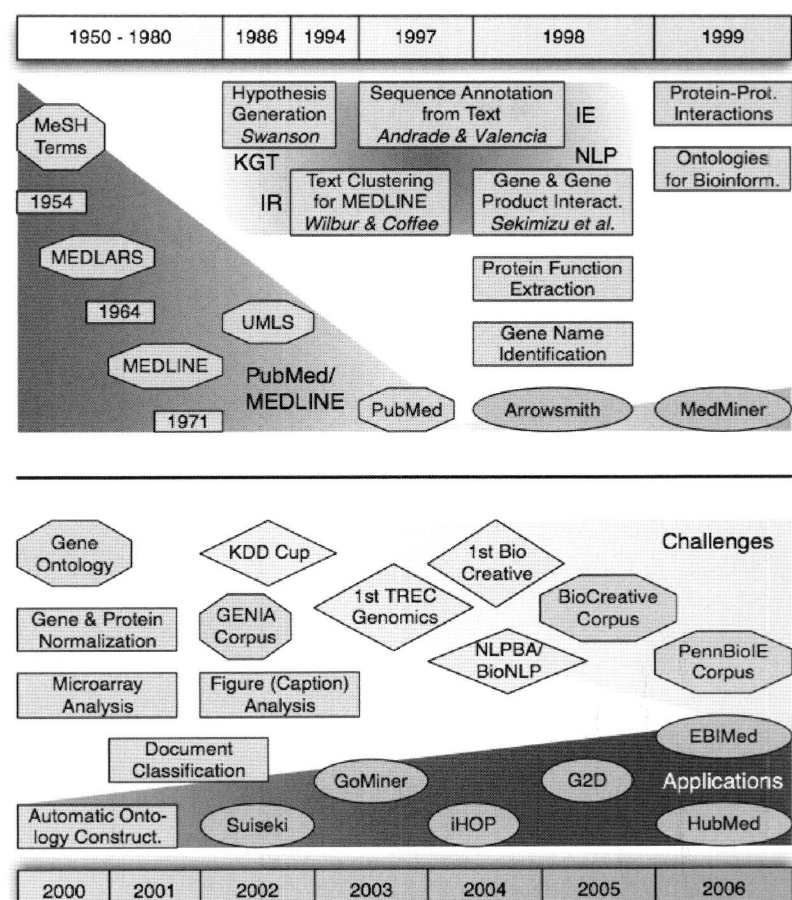


**Fig. 22.1** Timeline explaining the evolution of biological information extraction. After making the MEDLINE resource publicly available, the development of algorithms and applications began to flourish. Through the introduction of cups and challenges, standards to compare and assess these tools were established. For explanations see text **Legend:** octagons – resources (thesauri, corpora, ontologies); diamonds – cups and challenges; squares – breakthroughs and novel techniques in BioNLP/TM; circles – major applications. Abbreviations: NLM: [U.S.] National Library of Medicine. IE: Information Extraction, IR: Information Retrival, KGT: Knowledge Generation from Text, NLP: Natural Language Processing. (Copies of figures including color copies, where applicable, are available in the accompanying CD)

introduce existing applications at least partially closing the gap to such an ideal world of open and interconnected information. Beginning with tools that allow locating and retrieving publications with functionalities going beyond simple search engine queries, we take the reader all the way to the applications that allow one to view and explore networks.

Although **biomedical** text mining *per se* might be relatively new, with the first approaches published in the 80s of the last century, its building blocks come from several different long-running areas (for an historic overview, see Fig. 22.1).

## 22.2 Applications

There is a broad range of text mining tools available by now. Here, we will focus on a small subset of tools useful to the biologists for retrieving and discovering information. Apart from the tools presented in this section, an extensive list of tools can be found in Table 22.4. To exhibit the utilization and abilities of each tool, we will use a well-known cancer, Chronic Myeloid Leukemia (CML), as example topic throughout this section.

### 22.2.1 Chronic Myeloid Leukemia

CML (OMIM[2] ID 608232) is characterized by the increased and unregulated clonal proliferation of myeloid cells from the pluripotent bone marrow stem cells. It is mostly [3] caused by a chromosomal translocation forming an abnormally small chromosome, the Philadelphia chromosome (Ph), as a result of a reciprocal recombination between chromosome 9 and 22 (Ch 9 and Ch 22). This translocation (t[22;9][q34;q11]) commonly creates a fusion gene on the Ph from the two genes breakpoint cluster region (bcr, Ch 22) and Abelson murine leukemia viral oncogene homolog 1 (c-abl, also known as abl1, on Ch 9) called bcr-abl.

The fusion protein BCR-ABL is exclusively found in the cytoplasm and can cross-phosphorylate its own Tyr residues. It has been shown to have transforming potential:

- Enhancing proliferation by preventing CDK-activity and cell cycle arrest, by gradually inducing myc transcription, and by altering the cells response to various growth factor signals;
- Reducing bone marrow matrix adhesion to the stromal layers and fibronectin and phosphorylation of focal adhesion proteins; and
- Inhibiting apoptosis via activating the STAT5, RAS and PI3K/AKT pathways and reducing apoptosis induced by TNF$\alpha$ signaling.

Most of the interactions originate from the uncontrolled activity of the SRC-homology 1 (SH1) Tyr kinase domain contributed by the abl gene.

Until the 80's of the last century, CML was considered incurable until allogenic stem cell transplantations became available. Recent treatment strategies mainly have focused on silencing the ABL SH1 domain using Tyr kinase inhibitors (Imatinib mesylate – also known as Gleevec), while some approaches also attempt to target downstream effectors of BCR-ABL. The latest review on CML and the pathways involved is in [2], describing a downstream effector model for CML. Another very complete overview can be found in [3].

### 22.2.2 Overview

Table 22.1 is an overview of the applications treated in this chapter. The applications will be grouped by the results they produce. The first part will treat systems that allow the user to retrieve abstracts based on their query. This section will begin with results presented as lists of articles, and then advance to more sophisticated ways of presenting the results based on evidence passages or biological entities (genes, proteins). The second part will focus on applications that allow constructing interaction networks from the queries in graphical form. Finally, a short introduction to hypothesis generation systems will be presented, together with a tool to mine genes related to inherited diseases.

### 22.2.3 Retrieving Abstracts

#### 22.2.3.1 Search Engines

The first step in locating information, usually, is the retrieval of documents or data objects relevant to a certain topic and is called Information Retrieval (IR). Typical examples of such IR systems are

[2] Online Mendelian Inheritance in Man, a catalog of human genes and genetic disorders.

[3] see Ref. [1]

**Table 22.1** Overview of applications in this section (without search engines)

| Application | Query Data | Result Description |
|---|---|---|
| eTBLAST | Text paragraph | Similar abstracts |
| MedBlast | Sequence or BLAST result | Related abstracts |
| METIS | Sequence or SwissProt ID | Related abstracts |
| EBIMed | Keywords | Abstracts indexed by biological concepts and binary relations |
| MEDIE | Semantic keywords | Related sentences |
| MIF | Genomic markers | Related abstracts |
| HCAD | Genes or genomic bands | Genes, terms and related sentences |
| iHOP | Gene/protein symbols and chemical compounds | Related sentences, entities and network graph |
| Chilibot | Gene/protein symbols and keywords | Related sentences and network graph |
| G2D | Genomic location or gene ID | Candidate genes |

(Copies of tables are available in the accompanying CD.)

search engines like Google® or PubMed. They commonly employ an index of terms mapping the association of each term with the documents that contain them and storing the frequency of each term (term frequency, TF – the number of occurrences of a term in a document) [4]. Documents containing a query term can then be retrieved in ranked order by dividing the TF by the number of documents in which the term appears (document frequency, DF). It can be assumed that the document with the highest TFIDF (Term Frequency times Inverted Document Frequency) score is the most specific document for that term. Such a system was used for example to cluster MEDLINE documents [5].

All three major scientific search engines – Google Scholar™, Scirus®, and PubMed – have some functionality in common: Boolean queries (using the "AND" and "OR" operators between terms – where the default joining of words and phrases is "AND"), and searching for phrases by enclosing multiple words in double quotes – a search using *cell cycle* will return a larger result than searching for *"cell cycle"* (i.e., the former retrieves all documents containing the words cell AND cycle anywhere in the document). They do not just index the document but assign the content to specific **fields**, and PubMed even adds manually annotated terms (medical subject headings, MeSH) to the documents. Yet, each engine has a different set of fields available, which are summarized in Table 22.2. All the query fields available in Scholar and Scirus are also available with PubMed, but instead of the 'field:term' syntax, PubMed (and all of the NCBI query forms) uses a slightly

**Table 22.2** Search engine specific query fields

| Field | Scholar | Scirus | PubMed |
|---|---|---|---|
| Author | author: | au: | [AU] |
| Author affiliation | - | af: | [AD] |
| Domain | n/a | Dom: | n/a |
| EC Number | - | - | [RN] |
| First author | - | - | [1AU] |
| Journal | * | jo: | [TA] |
| Keywords | - | ke: | [MH] (MeSH) |
| Publication date | * | * | [DP] |
| Publication type | * | - | [PT] |
| PubMed ID | - | - | [PMID] |
| Title | intitle: | ti: | [TI] |
| Topic | - | - | [SB] |
| URL | site: | url: | n/a |

*available in advanced search mode

(Copies of tables are available in the accompanying CD.)

different form –'term[field]'. Another functionality offered by all three engines is the retrieval of similar (related) articles.

### 22.2.3.2 Text-based Searches

Keyword searches are obviously the domain of search engines. Yet, instead of just listing results by some ranking system that then has to be scanned manually, more sophisticated approaches for result presentation would be desirable. The systems described in the remaining part of this section do not directly provide the user with lists of abstracts, but first present sentences that were relevant to the query. By highlighting important concepts, the user can quickly locate interesting articles by scanning through these passages, instead of having to deduce the relevance of the article from the title (or worse, the complete abstract).

Just like, the "related article" functions of the search engines, abstracts can be retrieved based on the content of a given example document. Documents which align best to the query string (i.e., a linear set of characters) are retrieved [6]. This is analogous to the BLAST sequence alignment, and basically uses the same dynamic programming approach to find the most similar string. The eTBLAST search engine provides such a text similarity search interface [7]. The eTBLAST system fetches the first 400 high-scoring MEDLINE abstracts that have the most similar distribution of words as the input example. These 400 documents are then re-ranked by similarity to the input sample, aligning the characters of each input sentence to the sentences in the documents.

An interesting example would be to retrieve highly specific articles for a given topic. Using the abstract from "BCR-ABL-induced oncogenesis is mediated by direct interaction with the SH2 domain of the GRB-2 adaptor protein", Pendergast et al., 1993, PMID 8402896, describing the binding of Grb2 to Tyr177 phosphorylated BCR in the fusion protein (this interaction is subsequently required for the RAS pathway activation), we run another eTBLAST. Although the system returns some unspecific results, most top hits are either describing GRB2s role in the RAS pathway and/or BCR-ABLs role as a tyrosine kinase. While it might be as effective to search PubMed using the query *grb2 bcr abl*, this system returns documents which do not necessarily have to match all the query terms and can be useful for retrieving less specific articles, but still well related to the input (i.e., at better recall).

Apart from retrieving documents based on a query string, a classifier can be trained for discerning the records. The currently best approach of such **Document Classification** is the use of Support Vector Machines (SVMs) [8]. As with all supervised data mining methods, this requires a training and a test set with samples tagged as positive or negative, and is the most labor-intensive part of this approach. The sample texts are tokenized into their single words, creating a word frequency dictionary for each document (a **Bag-of-Words**). Each word in the complete test set then represents a dimension (or *feature*) and the frequency of a word is the coordinate value on that dimension for a document. The SVM models a hyperplane in this multi-dimensional space, which best separates the positive from negative documents. Using statistical analysis, the most significant words in the positive set can be extracted.

### 22.2.3.3 Sequence-Based Searches

A more abstract way of retrieving articles is by using a biological sequence as the starting point. Such sequence-based search facilities are provided by MedBlast [6] and METIS [9]. Using one or several protein (or genomic) sequences as input, these systems retrieve articles related to the submitted sequences. A two-step process achieves this: first, the sequences are submitted to BLAST to find the most similar known sequences. For these known sequences then the database annotations and linked MEDLINE articles are fetched. This is especially useful to help explore completely new or unknown sequences. MedBlast retrieves all the articles from a BLAST result up

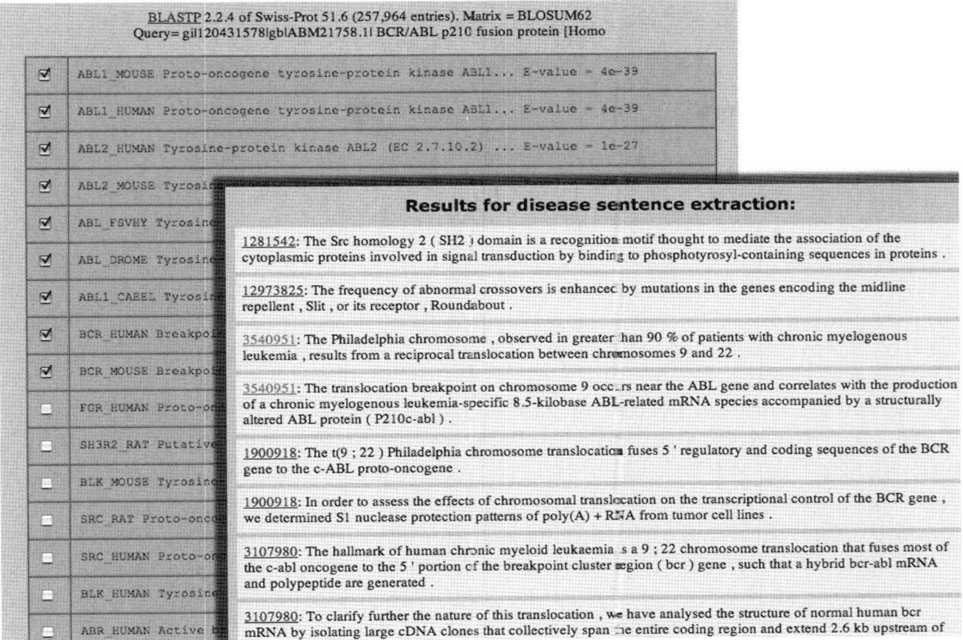


**Fig. 22.2** METIS search – BLAST result selection and evidence sentences for diseases related to the BCR-ABL fusion gene (Copies of figures including color copies, where applicable, are available in the accompanying CD)

to a user-defined e-value cutoff, while METIS allows selecting specific BLAST hit pairs used to further extract functional, structural or disease-related evidence sentences for the selected BLAST result. These sentences are extracted using machine learning (SVM) and pattern-based methods.

Assuming we would not know what sequence we had, submitting the BCR-ABL p210 fusion protein (NCBI Entrez GI 120431578) sequence to METIS and then selecting all significant BLAST results (the top 9 hit pairs up to BCR MOUSE, e-value 4e-6), we could use this system to make some hypothesis on the sequence. The disease extraction results clearly identify CML and the Ph chromosome for this sequence. The structure results return many evidence sentences on the involved SH2 and SH3 domains, while the functional annotations hint on the kinase activity of ABL1 and the fusion protein. Each evidence sentence is marked up for keywords and contains links to the PubMed abstract (see Fig. 22.2). The system can help to quickly highlight important aspects when examining a new sequence.

#### 22.2.3.4 Enhanced Keyword Searches

The next step is to analyze the retrieved documents, extracting factual knowledge from the text – this is the concept of **Information Extraction** (IE). The most obvious single pieces of information in biological research papers are the gene names. Finding such gene names, as well as other terms, like proteins, diseases, chemical compounds, etc. is called **Named Entity Recognition** (NER).

The most successful algorithms for locating entities in texts are Conditional Random Fields (CRFs), a machine learning technique very similar to Hidden Markov Models (HMMs) [3] [10].

[3] The main difference between CRFs and HMMs is that in CRFs states are not independent of each other, while in an HMM, the transition probabilities only depend on the current state. That is, for the elements in an HMM, the independence assumption has to be true, for a CRF not to be true.

Recognized gene names are then further *normalized*, which means to identify the correct sequence database ID corresponding to a gene name (therefore, called **Gene and Protein Normalization**) [11]. This is commonly done using pattern matching to dictionaries of known gene names (e.g., extracted from the NCBI Entrez database), retrieving a unique database identifier for the gene name. The main handicap is *disambiguating* genes with multiple names (synonyms) and names assigned to multiple genes (homonyms) to their correct entries, taking into account the necessity to discern the correct species in the process. Although all this might seem trivial at a glance, it can get very tricky. For example, the word 'arm' can be a valid gene name (see SwissProt accession O46082) and 'arm' is also assigned to Q9XZT1. The correct organism mapping is not always straightforward, e.g., for human and mouse publications, the respective organism is often not explicitly quoted. Beyond these obvious limitations, the name usage of genes is constantly evolving [12], making it hard to keep track of literature on a specific gene, and it is not uncommon that the official gene name nomenclature is not used in the literature [13].

Using recognized entities, another possible improvement on search engines is to present the search results based on biomedical concepts. This is precisely what the EBIMed service provides: systematic retrieval of MEDLINE abstracts based on co-occurrence of biomedical entities (genes/proteins, drugs and diseases) and concepts [14]. It indexes all the MEDLINE documents using Lucene[4]. The Lucene query systems allows one to query using Boolean and wildcard operators as provided by other search engines, but adds some interesting functionality. After submitting a query, the system processes the results, indexing protein, gene, drug, and species name matches, as well as matching terms from the Gene Ontology (see next section) to the query result. Instead of providing a list of articles, it returns a summary table to navigate the concepts found. From this summary, a detailed investigation of a specific biomedical concept related to the query is possible, and EBIMed focuses on the retrieval of HitPairs, i.e., co-occurrence of two terms in a sentence.

Using our CML example, lets look for sentences providing evidence and useful abstracts about the fusion protein. Bcr-abl might also be written as bcr/abl, BCR-ABL1, Bcr-Abl, bcr/c-abl, etc., so we can use the special Lucene proximity search syntax *"bcr abl"<1* to state we want abstracts containing bcr and abl within one word distance, which will commonly be the fusion protein mention. Additionally, we add the keywords *fusion* and *kinase* to the search, making the query more specific for abstracts about the fusion protein (or gene) and the tyrosine kinase activity. Following the number links, we are presented with the sentences providing evidence for the fusion protein, its tyrosine kinase activity, the interaction with the Imatinib (Gleevec) kinase inhibitor, and the chromosome translocation process required for CML development. Following other Protein/Gene links for ABL in the summary table, we can easily recover sentences for other interactions, such as the activation of the RAS/RAF, AKT, or STAT5 pathways by the oncogene. Using the various sorting functions of the table, we can find associations of our query with other biological concepts: if we sort by the *Cellular component* column, we can use the HitPair "chromosome" and "pathogenesis" to discover abstracts explaining the translocation leading to the fusion protein; Sorting by "Drug", we can quickly find publications about Imatinib (Gleevec).

Going beyond searches for evidence passages based on keywords and fields, MEDIE provides an interface to create semantic queries [15]. This "Semantic Search", which consists of up to three terms entered in the fields for subject, verb and/or object, retrieves documents based on the semantic similarity of the query to sentences in MEDLINE articles, which have all been parsed and grammatically analyzed by the system beforehand. MEDIE then presents sentences containing the query terms in the given context (i.e., used as subject/verb/object), plus links to the MEDLINE abstracts the sentences that were extracted. For example, to find BCR-ABL

[4] http://lucene.apache.org.

inhibitors, we would enter "inhibit" in the field verb and "bcr-abl" in object. We get a large list of sentences describing the inhibition of BCR-ABL, and we notice the system has some built-in intelligence: e.g., it knows that words such as "suppress" or "block" are also used to describe inhibition and includes them in the results. Certain keywords are linked to databases, such as gene mentions to their most probable EntrezGene entry, or disease names to the OMIM database. Using this query logic, we run a more interesting search: what are the targets of BCR-ABL? We know the oncogene works via its SH1 kinase domain, so we search using *bcr-abl* in the subject field and *phosphorylate* as the verb. We can immediately extract a list of targets from the 9 sentences matching our query.

#### 22.2.3.5 Entity-Based Searches

After having located the correct entity, annotating the functional properties describing the entity in the text can be attempted. The first approach was to extract keywords for protein families that are used significantly more frequently with those proteins only [16]. Using such **statistical inference**, the text can be analyzed for these keywords and then accordingly assigned to a protein family. After investigating these methods and because of the growth of biological databases, it soon became obvious that such annotations require a strict naming convention if they are to be used across different databases and applications [17].

Out of this requirement, the **Gene Ontology** (GO) was born. Ontologies are, just like thesauri introduced in the first section, a form of **controlled vocabularies**. The main difference between a thesaurus and an ontology is that the latter additionally provides *qualified* relationships between its entities (in the case of GO, those are either "part of" or "is a" relations). Actually, GO consists of three distinct ontologies for biological processes, cellular components, and molecular functions. For example, the Abelson murine leukemia viral oncogene homolog 1 (c-abl) gene is annotated with the following GO terms: it is associated with the term *nucleus* (GO:0005634, cellular component), has *protein-tyrosine kinase activity* (GO:0004713, molecular function), and plays a role in the *regulation of progression through cell cycle* (GO:0000074, biological process). Therefore, functional annotations extracted from text now maps the functional description in the text to a unique identifier of the corresponding ontology [18]. Using the GO, reveals some problems related to consistency and coherence. It is unified for all organisms, not all processes and functions in the ontology that exists in every organism. The limitation to two relations types can be insufficient (especially the broad use of "part of" relations); and the strict separation of the three different hierarchies makes it impossible to relate terms in separate categories. The main obstacle to overcome in language processing is the tentativeness of finding the correct ontology term for a given textual mention, as the jargon used in scientific publications usually does not coincide with the descriptive expressions used in ontologies.

Often we will already have a list of IDs (such as gene IDs, OMIM IDs, etc.) about which we want to know more, or we will know other well-characterized facts, that are represented in an ontology such as the Gene Ontology. Instead of using keywords, text or sequence, it is possible to use such unique identifiers to extract information.

One such attempt is the tool MarkerInfoFinder (MIF), which allows the direct retrieval of publications relevant to a genomic marker, such as cytobands, STSs, SNPs, or genes and proteins [to be published]. Although the main focus of this system is to extract literature a given NCBI SNP ID, we will not reproduce such an example directly here. Extracting SNP IDs, e.g., bcr and then submitting these to the system would be more like an exercise to validate the system. Yet, for genome-wide SNP association studies or for just identifying the significance of a single SNP this tool relieves the researcher from manually extracting genomic markers around the given SNPs and then searching MEDLINE with these marker names for abstracts.

Instead of using the SNP approach, we will search for abstracts for our two genes bcr (Entrez ID 613, UniGene ID Hs.517461) and abl (Entrez ID 25, UniGene ID Hs.431048). The resulting page presents us with the top most frequent MeSH headers (see the PubMed search engine description and subsection 22.1.2.2) in the related documents. Using a limit of 100 MeSH descriptors, this produces a list of the 100 most frequent MeSH terms associated to the articles found for the given gene IDs. As compared to searching PubMed with the query *abl bcr human*, which returns more than 4,300 articles, these sets are rather small, from 355 papers tagged for "Philadelphia Chromosome" to as little as 3 articles for "Drug Resistance, Neoplasm" (which must be attributed to the very shallow annotation of MeSH terms, not the MIF system).

Another literature mining-based tool is the Human Chromosome Aberration Database (HCAD) [19]. It allows the user to retrieve evidence sentences about chromosomal aberrations for a given cytoband or gene name. For example, searching for the cytoband 9q34 (ABL1) in the Breakpoint Browser, we will find CML as the first entry in the list of MeSH terms associated with this breakpoint region. Additionally, genes associated with this breakpoint are listed, including ABL1 and BCR, for which direct links to UniProt and OMIM (among others) are provided. Selecting that term, we are presented with sentences mentioning the chromosomal region in conjunction with CML. From this view we can either investigate a given MeSH term or gene, or we can read the abstract the sentence is associated to. This leads us to the second possibility of accessing HCAD, via genes. Searching for ABL1 in the Gene Browser, we are again directed to the 9q34 breakpoint. Thus, HCAD provides a very rapid way to find literature associated to a chromosomal breakpoint, which can be either based on the location of the breakpoint itself or on a gene closely associated to the breakpoint. Further, it allows identifying other genes also associated to the same breakpoint and because of the association with MeSH terms, breakpoint-to-disease associations can be browsed readily.

### *22.2.4 Reconstructing Networks*

All applications discussed so far only retrieve or extract single pieces of information, making the data accessible as lists or tables. Yet, biological data can be provided by modeling their relationships and presenting the user with networks to view and navigate the data associated to the nodes and connectors of these networks [20]. Going beyond single entities, **Relationship Extraction** attempts to combine two or more objects based on some correlation. A typical example would be mining for protein-protein interactions (PPIs) in text. The first attempt at this task was to extract PPIs based on *linguistic patterns* of *interaction verbs*, a technique used in **Natural Language Processing** (NLP) [21]. That is, sentences describing PPIs are analyzed for typical verbs joining the two proteins, such as "ABL1 *phosphorylates* GRB-2". Patterns of the form "*proteinA .... interaction-verb . . . proteinB*" are then used to mine documents for new PPIs. Although such pattern approaches produce high quality results (high **precision**), because of the versatile nature of the language, many interaction mentions are missed (low **recall**). Alternatively, statistical approaches have been applied, which achieve better recall values in exchange for precision. Therefore, current approaches commonly apply a harmonic mixture[5] from both the linguistic and statistical domain.

This section will present two network exploration tools, Chilibot and iHOP. According to the Alexa web traffic statistics, iHOP probably is the most used online text mining application for biology if ranked by Internet traffic and excluding search engines (i.e., PubMed, Google Schoolar, Scirus, etc.), other highly used tools with slightly lower usage statistics include HubMed and EBIMed.

[5] The combined value from recall and precision is called **F-Score**.

#### 22.2.4.1 iHOP information resource

The iHOP information resource takes an interactive approach to exploring the biomedical literature [22]. The basic idea behind iHOP is to make the network of concurring biological entities accessible for navigation. In iHOP, with genes and proteins acting as hyperlinks between sentences and abstracts, a large part of the information in PubMed becomes a giant navigable information network exhibiting all the advantages of the Web Hence the original name of iHOP: Information Hyperlinked over Proteins.

Under the hood, iHOP employs state-of-the-art text-mining methods to screen, index and annotate information specific on genes and proteins, their physical interactions and regulatory relationships, their relevance in pathologies, and their interactions with chemical compounds. Every day, thousands of novel PubMed documents are mined and information specific to thousands of different biological molecules is collected. As of May 2007, iHOP contains information pertaining to 104,000 genes, 41,000 chemical compounds and 13,000 MeSH terms gathered from more than twelve million sentences. This wealth of information makes iHOP the most comprehensive and up-to-date resource for literature-derived gene information on the Web.

However, iHOP users should always be aware that information in iHOP is generated with automatic text mining methods and can thus exhibit certain systematic errors. The details and estimated efficiency of the employed text mining technologies are published [23]. For instance, a major problem for text-mining methods in the biomedical domain is the correct detection and identification of entities (e.g. genes and proteins) in the natural text, because of the high semantic overloading of abbreviations and synonyms. Even human experts can have difficulties to identify the correct gene or chemical compound for very ambiguous synonyms. However, many text-mining problems can be mitigated, because of the interactive nature of iHOP: iHOP users navigate between sentences taken directly from their source abstracts and thus retain control over the reliability of the information they obtain (Table 22.3). In particular, three features assist the user and maximize transparency of the iHOP system: First, biomedical entities (e.g., genes, proteins, chemical compounds, pathologies, etc.) are highlighted within the original sentences based on a consistent color scheme and mouse-over information is always available. Second, the iHOP text-mining algorithms use context information to assign confidence scores to every gene reference in the text. Third, users are always informed if there are reasonable doubts concerning the correct identification of a gene. This notion of maximal transparency provides iHOP users with all the benefits of automatic text mining, while minimizing the downsides.

Recent advances in high-throughput methods such as microarrays and protein interaction screens, enable the systematic exploration of functional gene groups, but present a formidable challenge to data analysis. In iHOP, experimental network data (e.g., from protein interaction screens) can be

**Table 22.3** The different views on genes and proteins in iHOP

| View | Description |
|---|---|
| Minimal information view | Contains basic information about the gene, like the name and links to external resources. It is available for most genes; all other views are only available for genes that have been found in the scientific literature. |
| Defining information view | Lists the most informative sentences that mention the gene together with chemical compounds or MeSH terms. Sentences in this view contain no other genes beside the selected gene to increase specificity. The most frequent associations will always rank higher in this view, so that the user can quickly obtain an idea of the function of a protein. |
| Interaction view | Lists the most informative sentences about the association between the selected gene and other genes. |
| Most recent information view | Contains the most recent informative sentences on a gene from the literature of the past two years. This view shows associations between the gene and any other entity in a mixed fashion. |

(Copies of tables are available in the accompanying CD.)

superimposed upon the literature network to support simultaneous analysis of novel and established knowledge. Technically, this is achieved by highlighting sentences in the iHOP network that contain protein associations for which external experimental evidence exists (Fig. 22.4).

Another important feature to make navigation in iHOP efficient is a history function, which allows users to keep track of their movements through the network. In the course of navigation, interesting sentences can be added to the Gene Model, which also serves as a "log book" and represents the results of an investigation. The log book also includes a dynamically generated figure to provide a graphical representation of all associations between the genes in collected sentences (Fig. 22.5). This graph represents the condensed result of a literature search, but also remains hyperlinked to the corresponding sentences. In this way, users can familiarize themselves with the newly acquired information in an interactive manner and further extend the model.

In the context of chronic myeloid leukemia, iHOP can be used to explore the detailed functions and relationships of the genes in the current model, but also to extend the model. Before diving into the sentences, it is a good idea to explore the summary page of a gene, for instance the defining information view on ABL1 (Fig. 22.3). This view shows the chemical compounds, and MeSH terms and the frequencies at which these concur with ABL1. The most frequently concurring chemical is L-tyrosine, which is expected since ABL1 is a protein tyrosine kinase. The second chemical in the ranked list is the drug Gleevec. By clicking on the term Gleevec, one may retrieve the chemical structure from the PubChem database or zoom-in on the sentences that describe the relationship between Gleevec and ABL1. From these sentences, it is immediately obvious that Gleevec has been studied as an ABL tyrosine kinase inhibitor and thus in the treatment of chronic myeloid leukemia. Similarly, it is possible to zoom-in on the relationship with other frequent MeSH terms, i.e., Leukemia, Myeloid, Chronic or Translocation, to access the source literature on these relationships.

**Fig. 22.3** This screenshot illustrates the iHOP summary page of the *defining information view* on ABL1. According to the current filter settings, it lists all MeSH terms, chemical compounds and the number of concurrences with ABL1 in the literature (Copies of figures including color copies, where applicable, are available in the accompanying CD)

Sentences in this view contain interactions of ABL1 - Interaction Information is available whenever you see this symbol - Read more.
For a summary overview of the information in this page click here. new

High impact journals
Order by relevance

BCR [BCR]-ABL [ABL1]-**induced** inhibition of TRAIL transcription in hematopoietic [HCKAP1L] cells may provide a novel mechanism for tumorigenicity in chronic myeloid leukemia. [2003]

BCR [BCR]-ABL [ABL1]-**induced** oncogenesis is **mediated** by direct interaction with the SH2 domain of the GRB-2 [GRB2] adaptor protein. [1993]

These findings implicate activation of Ras function as an important component in BCR [BCR]-ABL [ABL1]-**mediated** transformation and demonstrate that GRB-2 [GRB2] not only functions in normal development and mitogenesis but also plays a role in oncogenesis. [1993]

Tumorigenic activity of the BCR [BCR]-ABL [ABL1] oncogenes is **mediated** by BCL2 [BCL2]. [1995]

BCR [BCR]/ABL [ABL1] significantly **enhances** the expression of RAD51 [RAD51] and several RAD51-Paralogs. [2001]

BCR [BCR]-ABL [ABL1] **suppresses** C/EBPalpha expression through inhibitory action of hnRNP E2 [PCBP2]. [2002]

BCR [BCR]-ABL1 [ABL1] kinase activity is **linked** to defective pre-B cell receptor signaling and the expression of a truncated isoform of the pre-B cell receptor-**associated** linker molecule SLP65 [BLNK]. [2004]

BCR [BCR]/ABL [ABL1]-**mediated** leukemogenesis requires the activity of the small GTP-binding protein [rab1c] Rac. [1998]

We show that the tumor suppressor p53 [TP53] is selectively **activated** by imatinib in BCR [BCR]-ABL [ABL1]-**expressing** cells as a result of BCR [BCR]-ABL [ABL1] kinase inhibition. [2006]

Signal transducer and activator of transcription 5 (STAT5 [STAT5A]) is constitutively **activated** by BCR [BCR]/ABL [ABL1], the oncogenic tyrosine kinase responsible for chronic myelogenous leukemia. [2002]

Mutation of Y177 to phenylalanine (Y177F) abolishes GRB-2 [GRB2] **binding** and abrogates BCR [BCR]-ABL [ABL1]-induced Ras activation. [1993]

Upon transfection with a vector carrying the dominant-negative N17Rac, BCR [BCR]/ABL [ABL1]-**expressing** myeloid precursor 32Dcl3 cells retained the resistance to growth factor deprivation-**induced** apoptosis but showed a decrease in proliferative potential in the absence of interleukin-3 [Il3] (IL-3 [Il3 / IL3]) and markedly reduced invasive properties. [1998]

Together, these results suggest that FUS [FUS] might **function** as a regulator of BCR [BCR]-ABL [ABL1] leukemogenesis, promoting growth factor independence and preventing differentiation via modulation of cytokine receptor expression. [1998]

These results show that BCR [BCR]-ABL [ABL1]-**expressing** leukemic stem cells depend to a greater extent on CD44 [CD44] for homing and engraftment than do normal HSCs, and argue that CD44 [CD44] blockade may be beneficial in autologous transplantation in CML. [2006]

Here we demonstrate that RAD51 [RAD51] is important for resistance to cisplatin and mitomycin C in cells **expressing** the BCR [BCR]/ABL [ABL1] oncogenic tyrosine kinase. [2001]

EPO [EPO] treatment also resulted in biological **effects** that were remarkably similar to those of BCR [BCR]/ABL [ABL1], including improved viability, altered integrin function, and a weak mitogenic signal. [1997]

We show here that the small GTP-binding protein [rab1c] Rac is **activated** by BCR [BCR]/ABL [ABL1] in a tyrosine kinase-dependent manner. [1998]

Grb2 [GRB2 / Grb2 / Grb2] forms a **complex** with BCR [BCR]-ABL [ABL1] and the nucleotide exchange factor Sos that leads to the activation of the Ras protooncogene [APPBP1]. [1995]

**Fig. 22.4** This is a screenshot of the iHOP *interaction view* on ABL1, which contains all the gene-gene associations of ABL1 found in PubMed. Sentences are ranked according to frequency, impact, date of publication and optional parameters. Flasks at the right indicate experimental evidence on specific interactions (Copies of figures including color copies, where applicable, are available in the accompanying CD)

The other important view in iHOP lists the interactions and associations of ABL1 with other genes (Fig. 22.4). As expected, the association with BCR is one of the most frequent observations, but also the disease relevant interactions with IFNA1, CRKL, GRB2 and STAT5A can be found among the top sentences. By clicking on a gene in one of the sentences one can zoom-in on the specific interaction of ABL1 and this gene. For example, clicking on CRKL makes it immediately clear that CRKL is studied as a target for phosphorylation by the BCR-ABL fusion protein. At any step, the user can collect relevant relationships or observations into the Gene model to create a summary of the explored genes (Fig. 22.5).

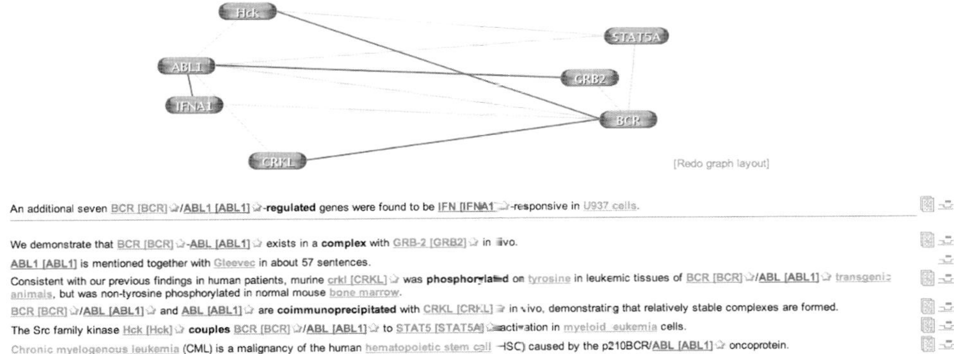


**Fig. 22.5** This illustrates an iHOP gene model or log book created by the user. Sentences can be added to the gene model over the course of literature exploration. Associations in the collected sentences are illustrated in an interactive graph (Copies of figures including color copies, where applicable, are available in the accompanying CD)

Besides the interactive web interface for individual researchers, much of the information in iHOP is accessible through a programmatic web service API [24]. This allows integrating iHOP information in bioinformatic programs and workflows, and to facilitate large-scale data-set analysis. iHOP is currently hosted by the cBio/MSKCC and updated on a daily basis.

#### 22.2.4.2 Chilibot Relationship Mining

Another such system to identify and browse networks constructed from MEDLINE abstracts is Chilibot [25]. In contrast to iHOP, Chilibot creates networks not exclusively based on gene and protein names, but also includes the use of other keywords, such as diseases or drugs, but requires an input of all components of the network beforehand. The search strategy of Chilibot is always based on binary relations; i.e., a query consists of two or more terms. Searches can be made for the relation between the two terms (binary, 1 to 1), all binary relations in a list of terms (pair wise, n to m) or all binary relations from one list of terms to another list of terms (1 to n). Additionally, a context for the search can be added, for example a disease name that has to co-occur with all terms. Chilibot makes a linguistic analysis of the relationships and integrates this information into the graph by color-coding. Using 1 to 1 binary searches it presents the user with a list of sentences in which these terms were co-occurring and their links to the originating MEDLINE abstracts. When using pair wise search for a list of terms, the user is presented with a graph representing the relational strengths between each term: the vertices are the search terms, while the edges have numbers representing the number of sentences found which contain the two terms, color coded for the relationship type, the edge represents stimulative, inhibitory, both, neutral, and parallel relationships as well as abstract relations (i.e., co-occurrence only)). Clicking on such a graph edge presents the user with the sentences from which this relationship was extracted, facilitating the evaluation of the relationship type and giving access to the PubMed display of the MEDLINE abstract.

It is easier to understand what Chilibot can do by using the leukemia example. First let us examine the concept space of some of the more prominent proteins involved in CML. We will try to find cancer-related functional properties of prominent proteins in the disease development. Therefore, our first list will consist of the terms apoptosis, proliferation, and adhesion, plus our two proteins BCR and ABL1, while in the second list we will put a list of proteins[6], which are affected by the fusion protein. On the advanced page we restrict our context to CML[7] and remove the line containing "!ABL" from the ABL1 synonyms (as we want to allow ABL as a synonym for ABL1) and add "CD56" to the synonyms of N-CAM. Because of the list of terms we use, the query will take a while – returning later through the *Saved Sessions*, the results are ready to view (Fig. 22.6). For example, lets explore the relations of MYC to the list of terms: MYC enhances proliferation as we can see from the texts extracted, although the link is red, not green, because most sentences describe the relation using negations – which means that relationship types are to be verified by the user before trusting them, and should only be used as quick guides. MYC also attenuates apoptosis, correctly displayed as a red link and has a neutral relation with adhesion, a result from unspecific co-occurrences of the two terms MYC and adhesion (to the knowledge of the author, there is no direct relation between myc and cell adhesion). To summarize, although the graph alone might have not been informative by itself, maybe even misleading, a short examination of the evidence sentences will tell us the correct correlation rapidly and directly leads us to the publications relevant for the given relation.

Microarray analysis augmented with text mining tools is a type of application belonging in this section. Here, evaluating microarray experiments has been shown to be possible together with the

---

[6] MYC, RAS, STAT5, AKT, BCL-X, BCL-2, ICSBP, LFA-3, N-CAM.

[7] Using *chronic myeloid leukemia OR CML OR chronic myelogenous leukemia OR chronic granulocytic leukemia* as context keywords.

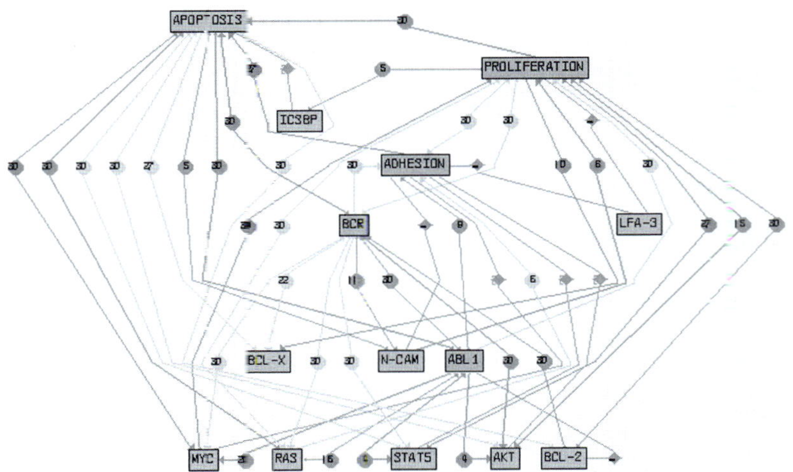


**Fig. 22.6** Resulting Chilibot network graph for the CML example query (see text). Sentences providing evidence for the relation (edges) can be browsed, and the quality of the relations is coded in the color of the edges (not shown) (Copies of figures including color copies, where applicable, are available in the accompanying CD)

**Table 22.4** List of tools and applications for biological information extraction

| Name | Type | URL<br>Description |
|---|---|---|
| Anni | MA | http://biosemantics.org/applications/main.html<br>analyze microarray datasets to cluster genes based on functional relations mined from the literature |
| Arrowsmith | IM | http://arrowsmith.psych.uic.edu/arrowsmith_uic/index.html<br>identify meaningful links between two sets of MEDLINE search results |
| BioIE | AR | http://www.bioinf.manchester.ac.uk/dbbrowser/bioie/<br>extract informative sentences from MEDLINE based on a keyword search |
| BioRAT | AR | http://bioinf.cs.ucl.ac.uk/biorat/<br>NLP-based automated update service |
| BITOLA | IM | http://www.mf.uni-lj.si/bitola/<br>discover potentially new relations between biomedical concepts |
| Chilibot | NW | http://www.chilibot.net/<br>investigate relations between proteins, genes, or keywords from MEDLINE in graph form |
| ConceptLink | NW | http://project.cis.drexel.edu/conceptlink/<br>graphically explore medical concept relations in MEDLINE |
| Dragon TF AM | IM | http://research.i2r.a-star.edu.sg/DRAGON/<br>mining potential associations between transcription factors, biomedical terms, and diseases from MEDLINE |
| EBIMed | AR | http://www.ebi.ac.uk/Rebholz-srv/ebimed/index.jsp<br>identify MEDLINE records for a concept or binary relation retrieved from a keyword search |
| eTBLAST | AR | http://invention.swmed.edu/etblast/etblast.shtml<br>retrieve MEDLINE records similar to a given input paragraph |
| FABLE | AR | http://fable.chop.edu/<br>find MEDLINE records for a specific gene or list genes associated to a keyword |
| G2D | IM | http://www.ogic.ca/projects/g2d_2/<br>scan a human genomic region for genes related to an inherited disease |
| GeneScene | NW | http://genescene.arizona.edu/index.html<br>navigate genetic regulatory pathway relations extracted from biomedical literature |

**Table 22.4** (continued)

| Name | Type | URL / Description |
|---|---|---|
| GoMiner | MA | http://discover.nci.nih.gov/gominer/<br>interpret microarray data by relating genes to processes, functions and components of GO |
| GOPubMed | AR | http://www.gopubmed.org/<br>ontology-based keyword search of MEDLINE records |
| HAPI | MA | http://array.ucsd.edu/hapi/<br>interpret conceptual similarities of a cluster of genes based on MEDLINE MeSH terms |
| HCAD | AR | http://www.pdg.cnb.uam.es/UniPub/HCAD/<br>analyze chromosome breakpoints and their related genes |
| HubMed | AR | http://www.hubmed.org/<br>an alternative interface to PubMed for searching MEDLINE |
| iHOP | NW | http://www.ihop-net.org/<br>navigate the network of genes and proteins in MEDLINE to construct graphical Gene Models |
| KinasePathway | NW | http://kinasedb.ontology.ims.u-tokyo.ac.jp<br>database of protein kinases and their interaction pathways mined from the literature |
| LitMiner | AR | http://andromeda.gsf.de/litminer<br>find frequently co-cited genes, molecules, phenotypes and tissues for a given key term |
| MIF | AR | http://brainarray.mhri.med.umich.edu/brainarray/<br>search, identify and retrieve MEDLINE records for different types of genetic markers or genetic variations |
| MedBlast | AR | http://medblast.sibsnet.org/<br>retrieve MEDLINE records based on a input sequence |
| MedGene | IM | http://medgene.med.harvard.edu/MEDGENE/main.do<br>database of diseases and their associated genes mined from MEDLINE |
| MEDIE | AR | http://www-tsujii.is.s.u-tokyo.ac.jp/medie/<br>retrieve biomedical correlations from MEDLINE records based on semantic queries |
| MedMiner | AR | http://discover.nci.nih.gov/textmining/main.jsp<br>extract relevant sentences in the literature based on gene, gene-gene, or gene-drug queries |
| MedMOLE | MA | http://medmole.cineca.it/<br>group genes co-regulated on the basis of MEDLINE co-occurrence mining |
| METIS | AR | http://www.bioinf.manchester.ac.uk/<br>retrieve evidence sentences from MEDLINE based on a input sequence |
| microGENIE | MA | http://www.cs.vu.nl/microgenie/<br>batch query PubMed using thousands of GenBank or UniGene Ids |
| PubGene | NW | http://www.pubgene.org/<br>visualize relationships between proteins, genes, and terms from MEDLINE and sequence data |
| PubMatrix | NW | http://pubmatrix.grc.nia.nih.gov/<br>retrieve pair-wise co-occurrence frequencies from two lists of terms in MEDLINE |
| ReleMed | AR | http://www.relemed.com/<br>interface to PubMed for searching MEDLINE, ranking results be relevance |
| SGO | MA | http://shad.cs.utk.edu/sgo/sgo.html<br>cluster genes based on conceptual relationships derived from MEDLINE abstracts |
| Textpresso | AR | http://www.textpresso.org/<br>*C. elegans*-specific literature information retrieval and extraction tool |
| XplorMed | AR | http://www.ogic.ca/projects/xplormed/<br>explore associations between MEDLINE abstracts from a keyword query |

AR: abstract retrieval, IM: inference methods, MA: microarray analysis, NW: network retrieval
(Copies of tables are available in the accompanying CD.)

iHOP tool. Yet, there are also tools specifically designed for this task only, which are discussed in other chapters of this book. For reference, these tools have also been included in Table 22.4.

### 22.2.5 *Inferring Relations*

Only beyond extracting relevant information, the field of **text mining** begins, attempting to uncover patterns and associations in and across documents. Although the term "text mining" is often used interchangeably with IE, it is distinct from IE in that it is concerned with generating a new hypothesis based on the mined information. It is a subset of knowledge discovery; therefore, text mining is also called Knowledge Discovery in Text (KDT). The classical reference for KDT is Swanson's relation between fish oil and Raynaud's disease [26]: Raynaud patients have high blood viscosity and platelet aggregation, and they suffer from vasoconstriction. Fish oil, or rather its active ingredient eicosapentaenoic acid, lowers viscosity and aggregation, and causes vasodilation. Mining these facts from different sources in the literature, Swanson inferred that fish oil might aid patients suffering from Raynaud's disease. This type of application emphasizes that discovering associations not directly extractable from a single text alone, require multiple text sources, even using sequence data with other biological knowledge. Ultimately, these tools are used to infer results based on connections between the data, which are not readily extractible from their single sources.

Probably the most straightforward tool of this kind is Genes to Disease (G2D) [27]. This system aims at extracting candidate genes related to a genetic disease and a genomic region established by genetic linkage mapping. After determining a linkage mapping for a disease, the last step is the identification of the gene(s) contributing to this phenotype. To improve this bottleneck of finding candidate genes, the system uses a combination of data and text mining (phenotype data) and gene sequence analysis (genotype data). As an input, the system requires a disease definition in the form of an OMIM ID and a genomic region that should be searched for potentially associated genes. The inference then made by G2D is the assumption that for phenotypically similar diseases, genes known to be associated to the diseases can be used to discover genes currently not reported to have a relation. If a gene with no known association to the disease has functions in common with some known gene, the unknown gene is a candidate for being associated to the same disease. Although functional similarity here is based on sequence similarity and might be a very unspecific method, it has a great advantage that genes lacking functional annotations will be detected as well. Also, as this method first searches MEDLINE abstracts annotated with the input disease, it can use the phenotype to mine known genes for the disease instead of requiring a fixed set of input genes.

As an example lets attempt to detect one of the two rare fusion proteins also found in a few CML patients with slightly different cytogenetic abnormalities: ZNF198-FGFR1 (t[8;13][p11;q12]) and TEL-PDGFRB (t[5;12][q33;p13]). Both fibroblast growth factor receptor 1 (FGFR1) and platelet-derived growth factor receptor β (PDGFRB) have cytoplasmatic tyrosine kinase domains, consequently they should be detectible by the system. Checking for the human FGFR1 on NCBI, we find that it is in the genomic position around 38,400,000 on chromosome 8. Using the *PHENOTYPES* query box of G2D, we enter *189980* as the OMIM ID (can be selected also from the list available at the interface) and in the location box we put *36000000 41000000* as our search region, selecting *Position* and chromosome *8* (Fig. 22.7). On the result page, the first hit is the FGFR1. We found the functional homolog of ABL1 using only a phenotype and the region around the breakpoint as input. Compared to searching for the corresponding gene using fluorescence *in-situ* hybridization (FISH) and cloning of the potential regions to vectors to test for cytogenetic abnormalities, a significantly shorter procedure, although in this case the gene (FGFR1) is a known alternative to the bcr-abl translocation in CML, therefore this is slightly "cheating."

The more generalized KDT tools require much background knowledge of the subject of interest and a firm grasp of the inference model underlying the system. Although these systems might seem

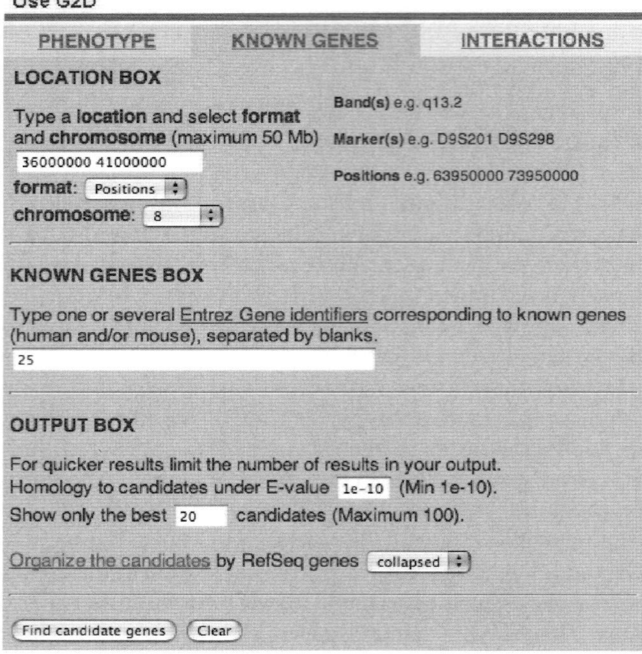


**Fig. 22.7** Screenshot of the G2D input for finding the FGFR1 Tyr kinase on chromosome 8 (a rare form of CML with the znf198-fgfr1 fusion gene) (Copies of figures including color copies, where applicable, are available in the accompanying CD)

straightforward and fairly simple to understand, without firm background knowledge of the field investigated and the assumptions being made it is rather unlikely to derive any discovery using these tools. They do allow making conclusions and hypothesis that would have been impossible without them, but using them for successful knowledge generation is far from trivial. Therefore, we have not included the more complex systems like Arrowsmith [28] and BITOLA [29]. In the final section of this chapter, some introductory literature to the topic of hypothesis and knowledge generation can be found.

## 22.3 Evaluation

To evaluate a literature-mining method, either an expert needs to judge a reasonable sample of the results, or the performance of the system is measured against a gold standard. Community-wide assessments of biological information extraction are held at several international workshops and challenges. The first effort of this kind was the annual KDD (Knowledge Discovery and Data mining) Cup 2002, focusing on bioinformatics and text mining in that year. Currently, there are two recurring events, the Text Retrieval Conference (TREC) Genomics track[8] and BioCreative (Critical Assessment of Information Extraction systems in Biology). While TREC traditionally focuses on IR – retrieving passages from large document sets for a given topic – the BioCreative tasks are modeled after typical IE needs of biologists and biological database annotators.

The two BioCreative (BC) challenges so far were split into two task groups, the first dealing with finding the gene and protein names and mapping these to database identifiers. The second task group focused on assigning Gene Ontology (see Entity-based Searches) term annotations to protein mentions in full text articles (BC I), and the extraction of protein-protein interactions, including the experimental evidences for these interactions (BC II). From these challenges it is becoming evident

[8] the last track was held in 2008 – TREC objectives change every few years.

that some tasks such as gene name normalization are reaching mature levels, while the more complex problems, such as correct protein-protein interaction extraction, are fields still requiring research and development [30]. These challenges are the basis for creating standards in the scientific community and they strongly promote the development of knowledge extraction applications.

## 22.4 Outlook

Providing a more in-depth coverage of the techniques in information extraction and text mining is far beyond the scope of this chapter. For interested readers, some starting points in the form of reviews and books are assembled here. Although the theoretical part covered some of the technical aspects of language processing, the reader might be interested in a more systematic introduction on the topic of NLP and text mining. The complex and concise nature of the biomedical literature renders most generic natural language processing tools, as used, e.g., on news wire text, inapplicable. Instead, a wide range of applications adapted to the domains of biology and medicine have been developed and are described in a review on text mining for bioinformatics applications. Finally, KDT systems are only briefly covered in this chapter, as they are not easily applicable, require very complex background, and except for the presented tools have no straightforward user interfaces. The interested reader will find a good review on knowledge discovery. For all the three topics (theory, applications, and KDT) complete books are available (see Key References).

A comprehensive list of applications related to the purposes of information retrieval, information extraction and text mining in the field of Molecular Biology is supplied in Table 22.4. It contains links to all the tools discussed in this chapter and many others, which cannot all be presented in the context of a single chapter. A short description will help the reader identify tools of potential interest.

Apart from these applications dedicated to end users, one of the main topics on natural language processing in biology is extending and enhancing annotations of entities in databases [31]. For this purpose, journal content has to be openly accessible to databases and information extraction tools. Unrestricted access to full text is also necessary to extract data not contained in abstracts, such as experimental evidence, kinetic rates, or sequence motifs.

The Web can be considered as the main presentation layer of scientific information; Web services and grid computing based on standardized protocols and data objects enable information exchange to present compound data originating from many different sources, while research interests currently are about developing ontologies (e.g., the Open Biomedical Ontologies (OBO) Foundry) and standardized annotations (e.g., the Systems Biology Markup Language, SBML). The Semantic Web is the most prominent exponent of this intent. It is an extension to the WWW to annotate web content in both human and machine readable form, thus permitting to share, integrate and relate information more easily. Currently, journals and database are separate, discreet entities. A vision of tomorrow's information technology is the semantic publishing of scientific texts [32], i.e., adding markup to publications, like the Semantic Web markup (Web Ontology Language, OWL). This will lead to simplified information extraction from text and will allow to link descriptions in journal articles to their corresponding database entries (and vice versa). Going even beyond this data assimilation, semantically annotated documents can be processed using deductive reasoning and inference, bringing text mining and hypothesis generation to an all-new scale. For this scenario to become a reality, the journals need to enforce annotation standards for authors; NLP tools could simplify the annotation process, e.g., automatically suggesting keywords and database identifiers to the author.

A researcher's interest will be in finding new functional and structural properties of the genes or proteins being investigated, spotting pathway connections and regulatory influences that are unknown to the researcher him, elucidating possible phenotypes generated by the given genotypes, discover unrelated work significant to the researchers line or research, etc. – just to name a few objectives. We hope this chapter has provided the reader with a good overview of tools utilizing natural language documents as data sources. It should help to significantly decrease the time spent searching and browsing for such information, providing substantial aid in pursuing the goals above.

## References


1. Clarkson, B., Strife, A., Wioniewski, D., Lambek, C.L. and Liu, C. Chronic myelogenous leukemia as a paradigm of early cancer and possible curative strategies. *Leukemia* 2003; 17: 1211–1262.
2. Van Etten, R.A. Oncogenic signaling: new insights and controversies from chronic myeloid leukemia. *J Exp Med* 2007;204(3):461–465.
3. Goldman, J.M., and Melo, J.V. Chronic myeloid leukemia—advances in biology and new approaches to treatment. *N Engl J Med* 2003;349(15):1451–1464.
4. Salton, G., Wong, A., and Yang, C.A vector space model for automatic indexing. *Communications of the ACM* 1975;18(11):613–620.
5. Wilbur, J.W., and Coffee, L. The effectiveness of document neighboring in search enhancement. Inf *Process Manage* 1994;30(2):253–266.
6. Tu, Q., Tang, H., and Ding, D. MedBlast: searching articles related to a biological sequence. *Bioinformatics* 2004;20(1):75–77.
7. Lewis, J., Ossowski, S., Hicks, J., Errami, M., and Garner, H.R. Text similarity: an alternative way to search MEDLINE. *Bioinformatics* 2006;22(18):2298–2304.
8. Joachims, T. Text categorization with support vector machines: Learning with many relevant features. *Proceedings of ECML-98, 10th European Conference on Machine Learning* 1998; 1398: 137–142.
9. Mitchell, A.L., Divoli, A., Kim, J.H., Hilario, M., Selimas, I., and Attwood, T.K. METIS: multiple extraction techniques for informative sentences. *Bioinformatics* 2005;21(22):4196–4197.
10. Lafferty, J., McCallum, A., and Pereira, F, Conditional random fields: Probabilistic models for segmenting and labeling sequence data. In: Proceedings of the International Conference on Machine Learning (ICML-2001): Morgan Kaufmann, San Francisco, CA; 2001:282–289.
11. Krauthammer, M., Rzhetsky, A., Morozov, P., and Friedman, C. Using BLAST for identifying gene and protein names in journal articles. *Gene* 2000;259(1-2):245–252.
12. Hoffmann, R., and Valencia, A. Life cycles of successful genes. *Trends Genet* 2003;19(2):79–81.
13. Tamames, J., and Valencia, A. The success (or not) of HUGO nomenclature. *Genome Biol* 2006;7(5):402.
14. Rebholz-Schuhmann, D., Kirsch, H., Arregui, M., Gaudan, S., Riethoven, M., and Stoehr, P. EBIMed—text crunching to gather facts for proteins from Medline. *Bioinformatics* 2007;23(2):e237–244.
15. Ohta, T., Tsuruoka, Y., Takeuchi, J., et al. An intelligent search engine and GUI-based efficient MEDLINE search tool based on deep syntactic parsing. Proceedings of the COLING/ACL on Interactive presentation sessions 2006:17–20.
16. Andrade, M.A., and Valencia, A. Automatic annotation for biological sequences by extraction of keywords from MEDLINE abstracts. Development of a prototype system. Proc Int Conf Intell Syst Mol Biol 1997;5(1553-0833 [Print]):25–32.
17. Bard, J.B.L., and Rhee, S.Y. Ontologies in biology: design, applications and future challenges. *Nat Rev Genet* 2004;5(3):213–222.
18. Blaschke, C., Leon, E.A., Krallinger, M., and Valencia, A. Evaluation of BioCreAtIvE assessment of task 2. BMC *Bioinformatics* 2005;6 Suppl 1(1471–2105 [Electronic]):S16.
19. Hoffmann, R., Dopazo, J., Cigudosa, J.C., and Valencia, A. HCAD, closing the gap between breakpoints and genes. *Nucleic Acids Res* 2005;33(Database issue).
20. Rzhetsky, A., Zheng, T., and Weinreb C. Self-correcting maps of molecular pathways. PLoS ONE 2006;1.
21. Blaschke, C., Andrade, M.A., Ouzounis, C., and Valencia, A. Automatic extraction of biological information from scientific text: protein-protein interactions. In: Proc Int Conf Intell Syst Mol Biol. Protein Design Group, CNB-CSIC, Madrid, Spain.; 1999:60–67.
22. Hoffmann, R., and Valencia, A. A gene network for navigating the literature. *Nat Genet* 2004;36(7).
23. Hoffmann, R., and Valencia, A. Implementing the iHOP concept for navigation of biomedical literature. *Bioinformatics* 2005;21 Suppl 2.
24. Fernández, J.M.M., Hoffmann, R., and Valencia, A. iHOP web services. Nucleic Acids Res 2007 May.

25. Chen, H., and Sharp, B.M. Content-rich biological network constructed by mining PubMed abstracts. BMC *Bioinformatics* 2004;5:147.
26. Swanson, D.R. Fish oil, Raynaud's syndrome, and undiscovered public knowledge. *Perspect Biol Med* 1986;30(1):7–18.
27. Perez-Iratxeta, C., Wjst, M., Bork, P., and Andrade M.A. G2D: A Tool for Mining Genes Associated with Disease. BMC *Genet* 2005;6(1).
28. Smalheiser, N.R., Swanson, D.R. Using ARROWSMITH: a computer-assisted approach to formulating and assessing scientific hypotheses. Computer Methods and Programs in Biomedicine 1998;57(3):149–153
29. Hristovski, D., Peterlin, B., Mitchell, J.A, and Humphrey, S.M. Using literature-based discovery to identify disease candidate genes. *Int J Med Inform* 2005;74(2–4):289–298.
30. Hirschman, L., Yeh, A., Blaschke, C., and Valencia, A. Overview of BioCreAtIvE: critical assessment of information extraction for biology. BMC *Bioinformatics* 2005;6 Suppl 1(1471–2105 [Electronic]):S1.
31. Valencia, A. Search and retrieve. Large-scale data generation is becoming increasingly important in biological research. But how good are the tools to make sense of the data? *EMBO Rep* 2002;3(5):396–400.
32. Bourne, P. Will a biological database be different from a biological journal? *PLoS Comput Biol* 2005;1(3):179–181.

# Key References

## *Books*

- C.D. Manning and H. Schutze, *Foundations of Statistical Natural Language Processing*. The MIT Press, June 1999.
- S. Ananiadou, ed., *Text Mining for Biology and Biomedicine*. Artech House, December 2005.
- C.J. Baker and K.-H. Cheung, eds., *Semantic Web Revolutionizing Knowledge Discovery in the Life Sciences*. Springer, 2007.

## *Reviews*

- **Systematic introduction:** A. M. Cohen and W. R. Hersh, "A survey of current work in biomedical text mining", *Brief Bioinform*, vol. 6, no. 1, pp. 57–71, 2005.
- **Text mining applications:** M. Krallinger and A. Valencia, "Text-mining and information-retrieval services for molecular biology", *Genome Biol*, 2005; 6(7):224.
- **Knowledge discovery:** L. J. Jensen, J. Saric, and P. Bork, "Literature mining for the biologist: from information retrieval to biological discovery", *Nat Rev Genet*, vol. 7, no. 2, pp. 119–129, 2006.

## *iHOP*

- R. Hoffmann and A. Valencia, "A gene network for navigating the literature", *Nat Genet*, vol. 36, July 2004.
- R. Hoffmann and A. Valencia, "Implementing the iHOP concept for navigation of biomedical literature", *Bioinformatics*, vol. 21 Suppl. 2, September 2005.
- J.M.M. Fernández, R. Hoffmann, and A. Valencia, "iHOP web services", *Nucleic Acids Res*, May 2007.
- R. Hoffmann, M. Krallinger, E. Andres, J. Tamames, C. Blaschke, A. Valencia, "Text mining for metabolic pathways, signaling cascades, and protein networks", *Sci STKE* 2005, 2005(283):pe21.

# Web Resources

http://lucene.apache.org

# Part VI
# Genomic Databases

# Chapter 23
# Using KEGG in the Transition from Genomics to Chemical Genomics

**Kiyoko F. Aoki-Kinoshita and Minoru Kanehisa**

**Abstract** KEGG is well known as a useful pathway reference database, containing all of the major metabolic and signaling pathways such as carbohydrate, energy, lipid and amino acid metabolism, membrane transport and signal transduction. The latest addition to KEGG, the KEGG BRITE database, is a resource of hierarchical classifications of biological data, including pathway-based gene ortholog information, which, as a result, provides genetic information, computed in the biological context within which genes are expressed. Moreover, BRITE contains chemical compound data derived from the KEGG COMPOUND database, which has been classified based on compound structure similarity such that they may be analyzed as ligands via hierarchical classifications. Thus, KEGG provides a valuable resource for genomic analysis in terms of its wealth of data in the PATHWAY, GENES and ENZYME knowledgebases, while at the same time providing a unique but important resource for understanding the chemical environment in which these biological processes occur. The concept of chemical compound similarity is increasingly being utilized in the latest research of ligand prediction and drug design. Such research will be able to make use of the data in KEGG BRITE as well as the new KEGG DRUG database containing maps of drug development and drug classifications. In this chapter, we will provide an introduction to the KEGG databases as well as the latest research in chemical genomics using KEGG. We will also describe some of the available practical means of accessing KEGG, such as directly via a computer program using the KEGG API.

**Keywords** KEGG · Pathways · Orthologs · Chemical genomics

## 23.1 Introduction

KEGG, a well known useful pathway reference database, contains all of the major metabolic and signaling pathways such as carbohydrate, energy, lipid and amino acid metabolism, membrane transport and signal transduction. All of the pathway maps in KEGG PATHWAY are drawn manually based on textbook and literature references. Feedback is quickly incorporated and the maps are continuously and manually updated with information from the latest resources. The pathways are all linked with genetic, enzymatic and chemical information, which are available in KEGG GENES, KEGG ENZYME and KEGG LIGAND, respectively. Moreover, the latest addition, KEGG BRITE, includes pathway-based ortholog information, which, as a result, provides genetic information computed in the biological context within which the corresponding gene is expressed. Inversely, the chemical compounds in the KEGG COMPOUND database have been

---

K.F. Aoki-Kinoshita
Department of Bioinformatics, Faculty of Engineering, Soka University, 1-236 Tangi-machi, Hachioji, Tokyo, 192-8577, Japan
e-mail: kkiyoko@t.soka.ac.jp

S. Krawetz (ed.), *Bioinformatics for Systems Biology*, DOI 10.1007/978-1-59745-440-7_23,

classified based on compound structure similarity such that they may be analyzed via hierarchical classifications as ligands. These classifications are also contained in BRITE. Thus, KEGG provides a valuable resource for genomic analysis in terms of its wealth of knowledge in the PATHWAY, GENES and ENZYME databases, while at the same time providing a unique but important resource for understanding the chemical environment in which these biological processes occur [Aoki and Kanehisa, 2005]. The concept of chemical compound similarity is increasingly being utilized in the latest research in ligand prediction and drug design. In contrast to traditional drug discovery techniques where potential chemical compounds as drugs are scanned against a particular drug target for binding affinity, new research is finding that the proteins acting as ligands for a particular compound may be scanned, in not only predicting drug targets but also in surveying possible ligands causing side-effects [1]. Such research will be able to make use of the data in KEGG BRITE as well as the new KEGG DRUG database containing maps of drug development and drug classifications.

## 23.2 KEGG for Genomics

KEGG is a bioinformatics resource for understanding the functions and utilities of cells and organisms from both high-level and genomic perspectives [2]. It consists of four major data sources called PATHWAY, GENES, BRITE, and LIGAND.

### *23.2.1 BRITE*

The KEGG BRITE database serves as the foundation for the genomic and chemical classifications on top of which the PATHWAY maps are constructed. KEGG BRITE is a collection of hierarchical classifications of biological systems, including genes, their interactions and chemical compounds. In contrast to KEGG PATHWAY, which is limited to molecular interactions and reactions, KEGG BRITE incorporates information on a variety of inter-relationships. Thus, the mapping of genomic and molecular data to KEGG BRITE (by the KO system described below) supplements the KEGG PATHWAY mapping for inferring higher-order functions.

At the time of this writing, KEGG BRITE consists of hierarchies and classifications for (1) Genes and Proteins, (2) Compounds and Reactions, (3) Drugs and Diseases, and (4) Cells and Organisms. This information is available from the main BRITE page at www.genome.jp/kegg/brite.html. One of the most ubiquitous of these classifications is the KEGG Orthology, or KO system. The KO hierarchy contains classifications of orthologous genes (determined by best-best hit relationships) based on pathway information. The KO groups are identified by the prefix "K," and each group contains those orthologous genes that are found to participate in the same reaction of the same pathway. In this way, the biological context is maintained in these KO groups that are generated in the genomic context. That is, the genes in the same KO groups are ensured to be those that are not only orthologous, but also functionally equivalent, due to the fact that they are known to participate in the same pathway performing the same functions.

### *23.2.2 PATHWAY*

PATHWAY contains the latest knowledge on metabolisms, signal transduction, etc., for various biological processes. They consist of manually drawn pathway maps that can be visualized in a number of ways. The default pathway is called the Reference pathway, which provides a birds-eye

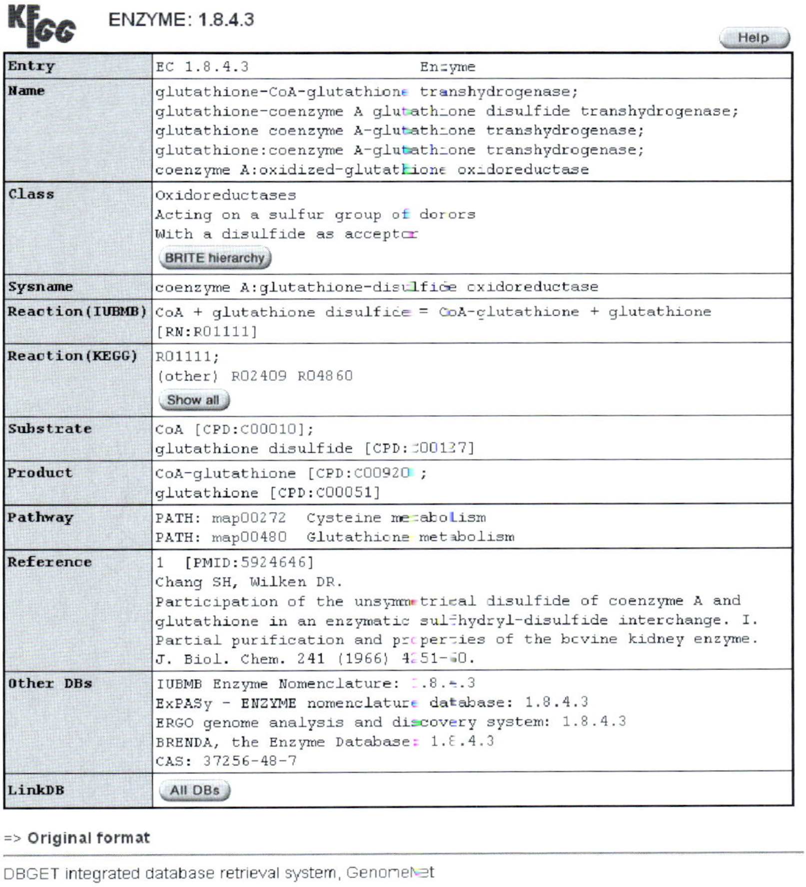

KEGG ENZYME: 1.8.4.3 Help

| Entry | EC 1.8.4.3 Enzyme |
|---|---|
| Name | glutathione-CoA-glutathione transhydrogenase;<br>glutathione-coenzyme A glutathione disulfide transhydrogenase;<br>glutathione coenzyme A-glutathione transhydrogenase;<br>glutathione:coenzyme A-glutathione transhydrogenase;<br>coenzyme A:oxidized-glutathione oxidoreductase |
| Class | Oxidoreductases<br>Acting on a sulfur group of donors<br>With a disulfide as acceptor<br>BRITE hierarchy |
| Sysname | coenzyme A:glutathione-disulfide oxidoreductase |
| Reaction(IUBMB) | CoA + glutathione disulfide = CoA-glutathione + glutathione<br>[RN:R01111] |
| Reaction(KEGG) | R01111;<br>(other) R02409 R04860<br>Show all |
| Substrate | CoA [CPD:C00010];<br>glutathione disulfide [CPD:C00127] |
| Product | CoA-glutathione [CPD:C00920];<br>glutathione [CPD:C00051] |
| Pathway | PATH: map00272 Cysteine metabolism<br>PATH: map00480 Glutathione metabolism |
| Reference | 1 [PMID:5924646]<br>Chang SH, Wilken DR.<br>Participation of the unsymmetrical disulfide of coenzyme A and glutathione in an enzymatic sulfhydryl-disulfide interchange. I. Partial purification and properties of the bovine kidney enzyme.<br>J. Biol. Chem. 241 (1966) 4251-60. |
| Other DBs | IUBMB Enzyme Nomenclature: 1.8.4.3<br>ExPASy - ENZYME nomenclature database: 1.8.4.3<br>ERGO genome analysis and discovery system: 1.8.4.3<br>BRENDA, the Enzyme Database: 1.8.4.3<br>CAS: 37256-48-7 |
| LinkDB | All DBs |

=> **Original format**

DBGET integrated database retrieval system, GenomeNet

Fig. 23.1 An enzyme entry page containing a variety of information corresponding to this enzyme (Copies of figures including color copies, where applicable, are available in the accompanying CD)

view of all known genes among all known organisms containing this pathway. Genes are represented by rectangles with their EC (Enzyme Commission) numbers, which are written over the arrows for the enzymatic reaction that they catalyze. Circles are the chemical compounds that are being catalyzed. Large ovals are links to other pathways. The genes are hyperlinked to their corresponding ENZYME entry page (Fig. 23.1), and compounds are linked to their COMPOUND entry page (Fig. 23.2). A species-specific pathway can be viewed by selecting an organism name from the pull-down menu at the top. Fig. 23.3 is a snapshot of this pathway for *Homo sapiens*. These pathway maps are colored to indicate those genes that are known to be found in the given species. Therefore, clicking on these green colored boxes will display their corresponding KEGG GENES entry pages, as illustrated in Fig. 23.4.

The pull-down menu at the top of each pathway contains the list of species that can be viewed, and the options for other types of pathways, including Reference pathways for Reactions, those for KOs, and "all organisms in KEGG." In the Reaction Reference Pathway, the genes are hyperlinked to their respective REACTION Entry (Fig. 23.5). The KO Reference Pathway is colored with purple-colored boxes hyperlinked to the KO Entry for the corresponding group of orthologous genes in the pathway (Figs. 23.6 and 23.7).

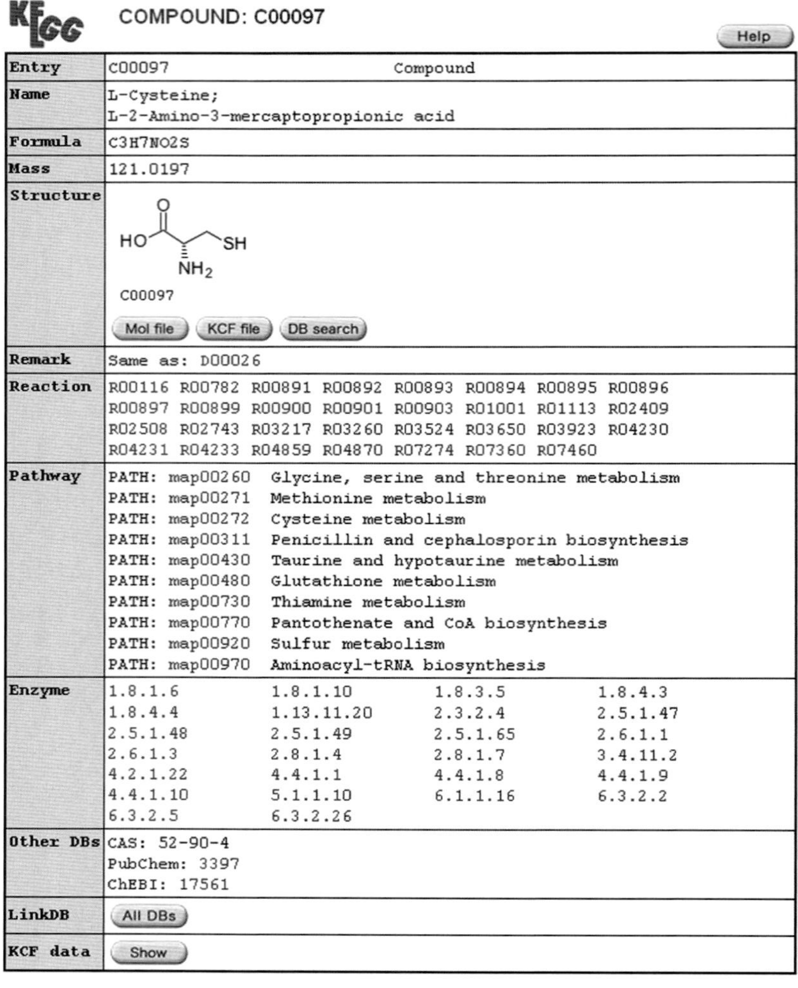

KEGG COMPOUND: C00097

Help

| | |
|---|---|
| Entry | C00097 Compound |
| Name | L-Cysteine;<br>L-2-Amino-3-mercaptopropionic acid |
| Formula | C3H7NO2S |
| Mass | 121.0197 |
| Structure | HO SH NH2<br>C00097<br>Mol file KCF file DB search |
| Remark | Same as: D00026 |
| Reaction | R00116 R00782 R00891 R00892 R00893 R00894 R00895 R00896<br>R00897 R00899 R00900 R00901 R00903 R01001 R01113 R02409<br>R02508 R02743 R03217 R03260 R03524 R03650 R03923 R04230<br>R04231 R04233 R04859 R04870 R07274 R07360 R07460 |
| Pathway | PATH: map00260 Glycine, serine and threonine metabolism<br>PATH: map00271 Methionine metabolism<br>PATH: map00272 Cysteine metabolism<br>PATH: map00311 Penicillin and cephalosporin biosynthesis<br>PATH: map00430 Taurine and hypotaurine metabolism<br>PATH: map00480 Glutathione metabolism<br>PATH: map00730 Thiamine metabolism<br>PATH: map00770 Pantothenate and CoA biosynthesis<br>PATH: map00920 Sulfur metabolism<br>PATH: map00970 Aminoacyl-tRNA biosynthesis |
| Enzyme | 1.8.1.6 1.8.1.10 1.8.3.5 1.8.4.3<br>1.8.4.4 1.13.11.20 2.3.2.4 2.5.1.47<br>2.5.1.48 2.5.1.49 2.5.1.65 2.6.1.1<br>2.6.1.3 2.8.1.4 2.8.1.7 3.4.11.2<br>4.2.1.22 4.4.1.1 4.4.1.8 4.4.1.9<br>4.4.1.10 5.1.1.10 6.1.1.16 6.3.2.2<br>6.3.2.5 6.3.2.26 |
| Other DBs | CAS: 52-90-4<br>PubChem: 3397<br>ChEBI: 17561 |
| LinkDB | All DBs |
| KCF data | Show |

=> **Original format**

DBGET integrated database retrieval system, GenomeNet

**Fig. 23.2** A COMPOUND entry page, containing the structure, reactions, and pathways, among others, corresponding to this chemical compound (Copies of figures including color copies, where applicable, are available in the accompanying CD)

## *23.2.3 GENES*

The GENES database contains all the genes known to be found in all of the organisms in the KEGG taxonomy. The list of organisms and the data source from which they have been retrieved can be viewed at http://www.genome.jp/kegg/catalog/org_list.html. This table also lists the KEGG Organism ID, which is a three-letter code that prefixes each gene ID in the KEGG GENES database. Each GENE entry lists the given gene's definition including its EC number, its KO number, the pathways in which it is involved, links to entries in motif databases such as Pfam and PROSITE, links to other sequence databases such as NCBI, its chromosomal position and its amino acid and nucleotide sequences (See Fig. 23.4). Links to other tools such as BLAST are also available, indicated by grey oval buttons. For example, the "DB Search" button in Fig. 23.4, next to

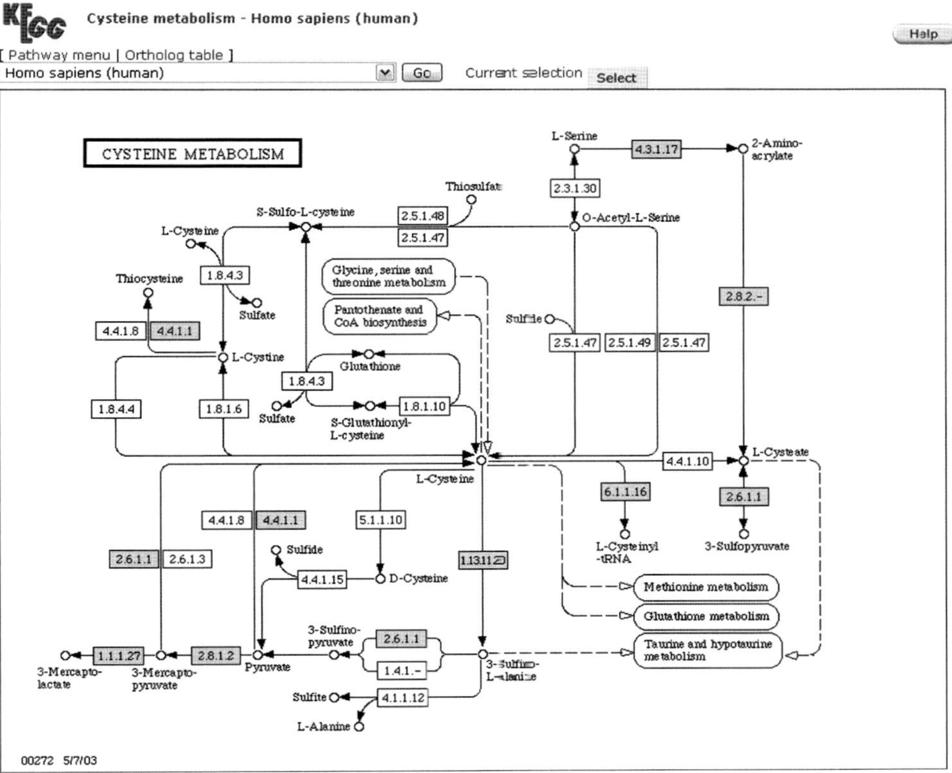


**Fig. 23.3** The cysteine metabolism pathway for human, where colored genes represent those found in human (Copies of figures including color copies, where applicable, are available in the accompanying CD)

the amino acid sequence, is linked to the BLAST search form for the given sequence, as in Fig. 23.8. On this form, the given sequence can be compared against (a) all of KEGG GENES, (b) a selected list of organisms, (c) all of KEGG GENES and genes from draft genomes in DGENES, (d) genes from viral genomes, or (e) other outside databases such as NCBI, Swiss-Prot, UniProt, RefSeq and PDB. The standard BLAST options are also available at the bottom. The results of this BLAST search will appear as in Fig. 23.9. The results list the most similar genes by KEGG Gene ID. From here, several tools can be run by selecting the desired list of similar genes (using the checkboxes) and clicking on the pull-down menu as in the figure. Multiple-sequence alignment by ClustalW, MAFFT and PRRN are available. The alignments can be visualized using the "Draw alignment" options, and the sequence motifs in Pfam or PROSITE that are common to the selected sequences can be found using the respective "Search common motifs" options.

## 23.3 KEGG for Chemical Genomics

In order to analyze the chemical compounds involved in a particular pathway, the computational model representing a pathway needs to be defined. Pathways viewed as networks can be turned around to focus on the chemical compounds underlying the network. In computer science terms, a graph, which describes the sets of objects (called nodes) and the relationships (called edges) between pairs of these objects, is usually used to represent a network. Then, a line-graph $L(G)$ of a graph $G$ is a graph whose nodes are the edges in $G$ and whose edges are the nodes in $G$. Thus, the inverse $G$ is

KEGG Homo sapiens (human): 10993

Help

| | |
|---|---|
| **Entry** | 10993 CDS H.sapiens |
| **Gene name** | SDS |
| **Definition** | serine dehydratase [EC:4.3.1.17] |
| **Orthology** | KO: K01752 L-serine dehydratase |
| **Pathway** | PATH: hsa00260 Glycine, serine and threonine metabolism<br>PATH: hsa00272 Cysteine metabolism |
| **Class** | BRITE hierarchy |
| **SSDB** | Ortholog Paralog Gene cluster |
| **Motif** | Pfam: PALP DUF1805<br>PROSITE: DEHYDRATASE_SER_THR<br>Motif |
| **Other DBs** | OMIM: 182128<br>NCBI-GI: 33469958<br>NCBI-GeneID: 10993<br>UniProt: P20132 |
| **LinkDB** | PDB All DBs |
| **Position** | 12q24.13 |
| **AA seq** | 328 aa AA seq DB search<br>MMSGEPLHVKTPIRDSMALSKMAGTSVYLKMDSAQPSGSFKIRGIGHFCKRWAKQGCAHF<br>VCSSAGNAGMAAAYAARQLGVPATIVVPSTTPALTIERLKNEGATVKVVGELLDEAFELA<br>KALAKNNPGWVYIPPFDDPLIWEGHASIVKELKETLWEKPGAIALSVGGGGLLCGVVQGL<br>QEVGWGDVPVIAMETFGAHSFHAATTAGKLVSLPKITSVAKALGVKTVGAQALKLFQEHP<br>IFSEVISDQEAVAAIEKFVDDEKILVEPACGAALAAVYSHVIQKLQLEGNLRTPLPSLVV<br>IVCGGSNISLAQLRALKEQLGMTNRLPK |
| **NT seq** | 987 nt NT seq +upstream 0 nt +downstream 0 nt<br>atgatgtctggagaacccctgcacgtgaagacccccatccgtgacagcatggccctgtcc<br>aaaatggccggcaccagcgtctacctcaagatggacagtgcccagccctccggctccttc<br>aagatccggggcattgggcacttctgcaagaggtgggccaagcaaggctgtgcacatttt<br>gtctgctcctcggcgggcaacgcaggcatggcggctgcatatgcggccaggcaactcggc<br>gtccccgccaccatcgtggtgcccagcaccacacctgctctcaccattgagcgcctcaag<br>aatgaaggtgccacagtcaaggtggtgggtgagttattggatgaagccttcgagctggcc<br>aaggccctagcgaagaacaacccgggttgggtctacattcccccctttgatgacccctc<br>atctgggaaggccacgcttccatcgtgaaagagctgaaggagacactgtgggaaaagccg<br>ggggccatcgcgctgtcagtgggcggcgggggcctgctgtgtggagtggtccaggggctg<br>caggaggtgggctgggggacgtgcctgtcatcgccatggagacttttggtgcccacagc<br>ttccacgctgccaccaccgcaggcaaacttgtctccctgcccaagatcaccagtgttgcc<br>aaggccctgggcgtgaagactgtgggggctcaggccctgaagctgtttcaggaacacccc<br>attttctctgaagttatctcggaccaggaggctgtggccgccattgagaagttcgtggat<br>gatgagaagatcctggtggagcccgcctgcggggcagccctggccgctgtctatagccac<br>gtgatccagaagctccaactggaggggaatctccgaaccccgctgccatccctcgtggtc<br>atcgtctgcgggggcagcaacatcagcctggcccagctgcgggcgctcaaggaacagctg<br>ggcatgacaaataggttgcccaagtga |

=> **Original format**

DBGET integrated database retrieval system, GenomeNet

**Fig. 23.4** The GENE entry for the human gene serine dehydratase (Copies of figures including color copies, where applicable, are available in the accompanying CD)

*L(G)*. Considering a pathway, where nodes represent genes and edges represent the compounds involved in the catalysis, the graph can be transformed such that nodes represent the compounds and the edges represent the genes catalyzing the reaction. A pathway can thus be described as a dual-network. It has been shown that the dual graph (or dual-network) has some interesting properties, including the fact that it still follows a power-law distribution, thus maintaining the scale-free property. A scale-free network, in general, is described as one containing hubs of nodes; i.e., a few of the nodes in the network are heavily connected with other nodes, while others are sparsely connected. In the case of dual-networks, we may consider these hubs as those chemical compounds that serve key roles in the network [3].

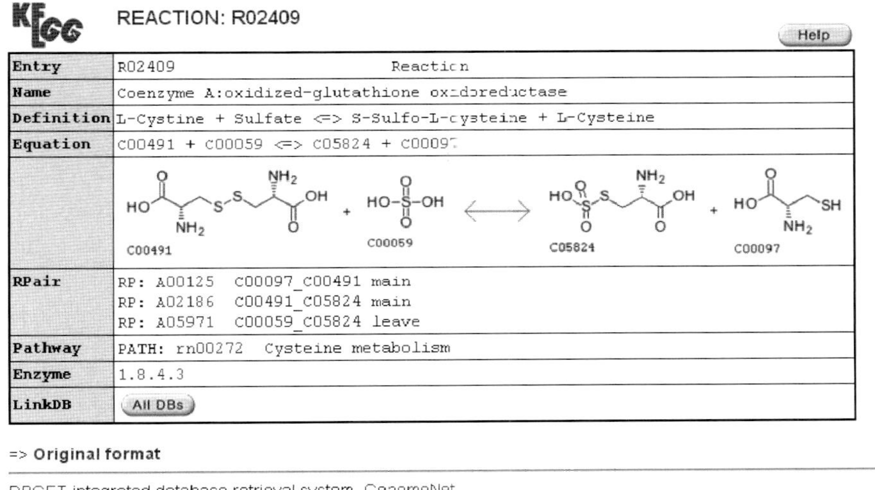

REACTION: R02409

Help

| | |
|---|---|
| Entry | R02409 Reaction |
| Name | Coenzyme A:oxidized-glutathione oxidoreductase |
| Definition | L-Cystine + Sulfate <=> S-Sulfo-L-cysteine + L-Cysteine |
| Equation | C00491 + C00059 <=> C05824 + C00097 |
| | C00491 + C00059 ⟷ C05824 + C00097 |
| RPair | RP: A00125 C00097_C00491 main<br>RP: A02186 C00491_C05824 main<br>RP: A05971 C00059_C05824 leave |
| Pathway | PATH: rn00272 Cysteine metabolism |
| Enzyme | 1.8.4.3 |
| LinkDB | All DBs |

=> Original format

DBGET integrated database retrieval system, GenomeNet

**Fig. 23.5** The REACTION entry for coenzyme A: oxidized-glutathione oxidoreductase (Copies of figures including color copies, where applicable, are available in the accompanying CD)

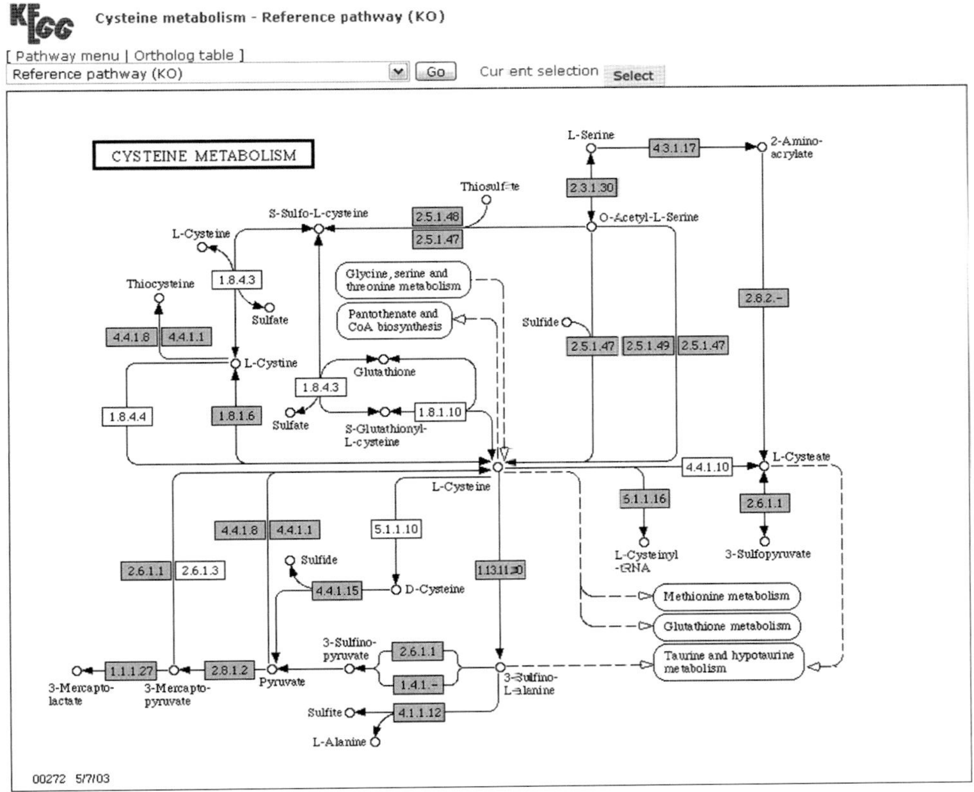


**Fig. 23.6** The KO Reference pathway for cysteine metabolism, where colored genes are linked to their corresponding KO entries (Copies of figures including color copies, where applicable, are available in the accompanying CD)

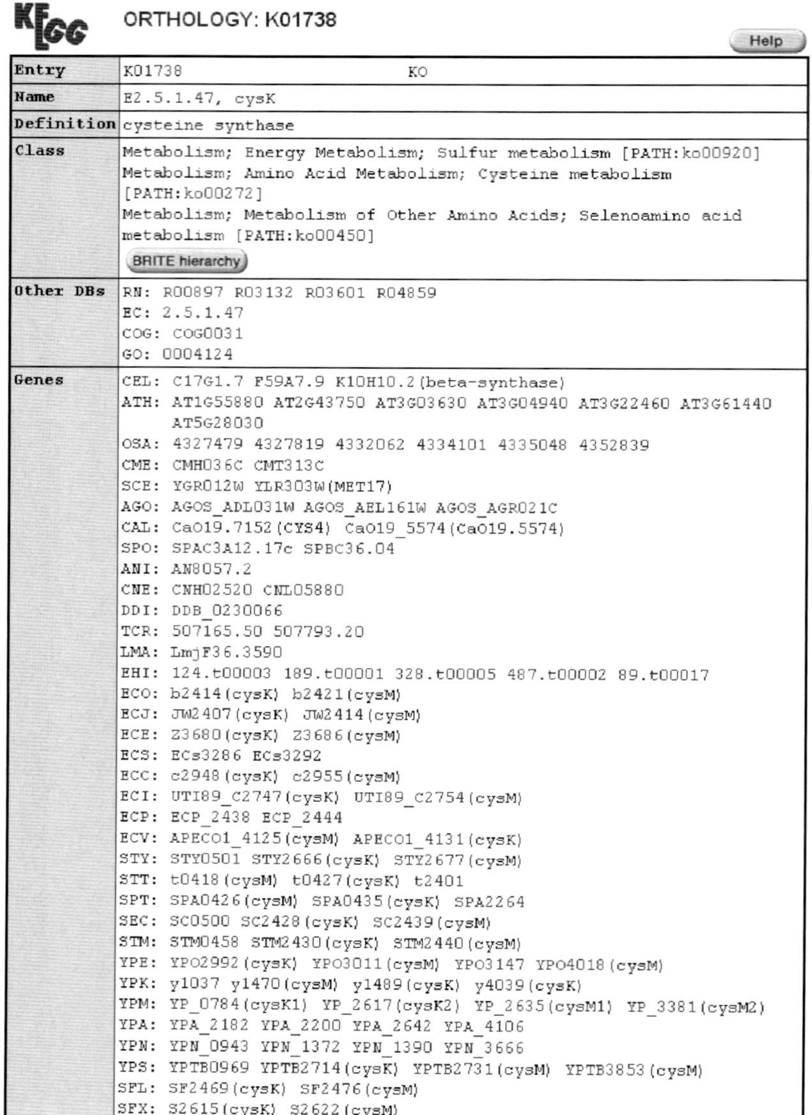

ORTHOLOGY: K01738

Help

| | |
|---|---|
| Entry | K01738 KO |
| Name | E2.5.1.47, cysK |
| Definition | cysteine synthase |
| Class | Metabolism; Energy Metabolism; Sulfur metabolism [PATH:ko00920]<br>Metabolism; Amino Acid Metabolism; Cysteine metabolism [PATH:ko00272]<br>Metabolism; Metabolism of Other Amino Acids; Selenoamino acid metabolism [PATH:ko00450]<br>BRITE hierarchy |
| Other DBs | RN: R00897 R03132 R03601 R04859<br>EC: 2.5.1.47<br>COG: COG0031<br>GO: 0004124 |
| Genes | CEL: C17G1.7 F59A7.9 K10H10.2(beta-synthase)<br>ATH: AT1G55880 AT2G43750 AT3G03630 AT3G04940 AT3G22460 AT3G61440 AT5G28030<br>OSA: 4327479 4327819 4332062 4334101 4335048 4352839<br>CME: CMH036C CMT313C<br>SCE: YGR012W YLR303W(MET17)<br>AGO: AGOS_ADL031W AGOS_AEL161W AGOS_AGR021C<br>CAL: CaO19.7152(CYS4) CaO19_5574(CaO19.5574)<br>SPO: SPAC3A12.17c SPBC36.04<br>ANI: AN8057.2<br>CNE: CNH02520 CNL05880<br>DDI: DDB_0230066<br>TCR: 507165.50 507793.20<br>LMA: LmjF36.3590<br>EHI: 124.t00003 189.t00001 328.t00005 487.t00002 89.t00017<br>ECO: b2414(cysK) b2421(cysM)<br>ECJ: JW2407(cysK) JW2414(cysM)<br>ECE: Z3680(cysK) Z3686(cysM)<br>ECS: ECs3286 ECs3292<br>ECC: c2948(cysK) c2955(cysM)<br>ECI: UTI89_C2747(cysK) UTI89_C2754(cysM)<br>ECP: ECP_2438 ECP_2444<br>ECV: APECO1_4125(cysM) APECO1_4131(cysK)<br>STY: STY0501 STY2666(cysK) STY2677(cysM)<br>STT: t0418(cysM) t0427(cysK) t2401<br>SPT: SPA0426(cysM) SPA0435(cysK) SPA2264<br>SEC: SC0500 SC2428(cysK) SC2439(cysM)<br>STM: STM0458 STM2430(cysK) STM2440(cysM)<br>YPE: YPO2992(cysK) YPO3011(cysM) YPO3147 YPO4018(cysM)<br>YPK: y1037 y1470(cysM) y1489(cysK) y4039(cysK)<br>YPM: YP_0784(cysK1) YP_2617(cysK2) YP_2635(cysM1) YP_3381(cysM2)<br>YPA: YPA_2182 YPA_2200 YPA_2642 YPA_4106<br>YPN: YPN_0943 YPN_1372 YPN_1390 YPN_3666<br>YPS: YPTB0969 YPTB2714(cysK) YPTB2731(cysM) YPTB3853(cysM)<br>SFL: SF2469(cysK) SF2476(cysM)<br>SFX: S2615(cysK) S2622(cysM) |

**Fig. 23.7** The KO entry for cysteine synthase, containing all the genes for this ortholog group (Copies of figures including color copies, where applicable, are available in the accompanying CD)

It has been shown that chemical compound similarity can be effectively used as a measure to group related proteins based on their ligands [1]. Proteins with similar ligands that are different in sequence may actually be related by the common ligand shared by them, serving as a clue to their function. In this work, targets that would have been considered unrelated were found to be closely related according to their ligands. Chemically similar drugs were found to bind to proteins that were different in both sequence and structure. These proteins may explain why specific drugs cause unexpected side effects due to seemingly unrelated proteins binding to them. As a result, proteins were classified based on the drugs that they were found to interact, consequently grouping proteins together which were previously believed to be unrelated.

**Fig. 23.8** BLAST search page from KEGG using the selected sequence (Copies of figures including color copies, where applicable, are available in the accompanying CD)

## *23.3.1 LIGAND*

KEGG LIGAND consists of six major components: COMPOUND, DRUG, GLYCAN, REACTION, RPAIR and ENZYME. These correspond to data on chemical compounds, compounds used as therapeutic drugs, glycans, biochemical reactions, reactant-pair alignments and enzyme nomenclature. The structures are represented in KEGG Chemical Function, or KCF, format so that they can be conveniently stored and re-used. KCF represents compound structures also as graphs. In the case of chemical structures, nodes represent atoms and edges represent their bonds. KCF is also used to represent polysaccharides, or glycans, with nodes representing monosaccharides, and edges representing their glycosidic linkages. In either case, the graphic representation allows one to use graph-based computations for comparing chemical compounds (or glycans) with one

**BLASTP Search Result**

**Database: genes**

**Protein sequence database entries related to hsa:10993** - 499 hits

Top 10 | Clear | Select operation | Exec

Select operation
CLUSTALW
MAFFT
PRRN
Draw alignment
Draw alignment(SSDB)
Search common motifs(pfam)
Search common motifs(prosite)
List definitions

| Entry | bits | E-val |
|---|---|---|
| ☑ hsa:10993 SDS; ser ... 01752 L-serine... | 578 | e-164 |
| ☑ ptr:452274 SDS; se | 573 | e-162 |
| ☑ cfa:486281 LOC4862 ... [EC:4.3.1.17... | 511 | e-144 |
| ☑ bta:514346 MGC1396 | 473 | e-132 |
| ☑ mmu:231691 Sds; se ... K01752 L-serin... | 473 | e-132 |
| ☑ rno:25044 Sds; ser ... 01752 L-serine... | 452 | e-126 |
| ☑ bta:790340 LOC7903 ... 40 | 370 | e-101 |
| ☑ hsa:113675 SDSL; serine dehydratase-like | 332 | 8e-90 |
| ☑ gga:417030 SDSL; serine dehydratase-like | 331 | 1e-89 |
| ☑ bta:516783 LOC516783; similar to Serine dehydratase-like | 328 | 9e-89 |
| ☐ cfa:486282 LOC486282; similar to serine dehydratase-like | 326 | 4e-88 |
| ☐ mmu:257635 Sdsl; serine dehydratase-like | 324 | 2e-87 |
| ☐ rno:360816 Sdsl_predicted; serine dehydratase-like (predicted) | 320 | 4e-86 |
| ☐ xtr:496756 sdsl; serine dehydratase-like [EC:4.3.1.19]; K01754 t... | 316 | 6e-85 |
| ☐ xla:398771 MGC68790; hypothetical protein MGC68790 [EC:4.3.1.19]... | 315 | 1e-84 |
| ☐ spu:594261 LOC594261; similar to serine dehydratase-like | 270 | 4e-71 |
| ☐ ptr:452275 SDSL; serine dehydratase-like | 256 | 6e-67 |
| ☐ ppu:PP_2930 L-serine dehydratase, putative | 234 | 2e-60 |
| ☐ pen:PSEEN2008 L-serine dehydratase [EC:4.3.1.17]; K01752 L-serin... | 228 | 2e-58 |
| ☐ hch:HCH_03183 threonine dehydratase | 223 | 7e-57 |
| ☐ nmu:Nmul_A0340 pyridoxal-5'-phosphate-dependent enzyme, beta sub... | 218 | 1e-55 |
| ☐ ddi:DDB_0230212 L-serine ammonia-lyase [EC:4.3.1.17]; K01752 L-s... | 210 | 4e-53 |
| ☐ rle:RL4706 ilvA; putative threonine dehydratase biosynthetic [EC... | 210 | 5e-53 |
| ☐ dre:563642 LOC563642; similar to serine dehydratase-like | 210 | 5e-53 |
| ☐ ret:RHE_CH04092 putative L-serine dehydratase protein [EC:4.2.3.... | 207 | 3e-52 |
| ☐ reh:H16_B0620 L-serine dehydratase, putative | 206 | 5e-52 |
| ☐ bam:Bamb_1136 pyridoxal-5'-phosphate-dependent enzyme, beta subunit | 201 | 2e-50 |
| ☐ cme:CMR455C similar to serine dehydratase | 197 | 4e-49 |
| ☐ bvi:Bcep1808_1213 pyridoxal-5'-phosphate-dependent enzyme, beta ... | 193 | 4e-48 |
| ☐ cne:CNE02390 serine family amino acid catabolism-related protein... | 190 | 4e-47 |
| ☐ bur:Bcep18194_A4400 L-serine ammonia-lyase [EC:4.3.1.17]; K01752... | 189 | 6e-47 |
| ☐ mxa:MXAN_6166 pyridoxal phosphate-dependent enzyme | 188 | 1e-46 |
| ☐ bch:Bcen2424_1257 L-serine ammonia-lyase [EC:4.3.1.17]; K01752 L... | 183 | 6e-45 |
| ☐ bcn:Bcen_0776 L-serine ammonia-lyase [EC:4.3.1.17]; K01752 L-ser... | 183 | 6e-45 |
| ☐ pfo:Pfl_3526 pyridoxal-5'-phosphate-dependent enzyme, beta subunit | 164 | 3e-39 |
| ☐ cne:CNK00180 L-serine ammonia-lyase | 159 | 1e-37 |
| ☐ ago:AGOS_AFR747W AFR747Wp | 141 | 3e-32 |
| ☐ sce:YCL064C CHA1; catabolism of hydroxy amino acids [EC:4.3.1.17... | 134 | 3e-30 |

**Fig. 23.9** BLAST search result page, when the selected (checked) sequences can be used as input to analysis such as Clustal W and PRRN (Copies of figures including color copies, where applicable, are available in the accompanying CD)

another. Subsequently, algorithms for such computations have been developed and implemented in KEGG as tools. SIMCOMP is available for comparing compounds [4], SUBCOMP computes the most similar substructures, and KCaM computes glycan similarity [5]. These tools are used in practice to perform structure-based searches on the COMPOUND and GLYCAN databases.

The data in KEGG LIGAND are all linked with the pathway maps where applicable. That is, those compounds that are involved in any pathway in KEGG are hyperlinked to their corresponding entry in LIGAND. To browse all of LIGAND, a portal interface to LIGAND is available at http://www.genome.jp/kegg/ligand.html. This page lists all ligand-related information, including tools for chemical compound analysis. For example, e-zyme is a tool to automatically assign EC numbers to reactant pairs of compounds, and PathComp is a tool to generate all possible reaction paths given a list of enzymes. This latter tool takes as input a list of enzymes and uses the reactions that they catalyze to predict the consecutive reactions (paths) that could possibly take place between them.

### *23.3.2 KEGG BRITE*

The development of the BRITE classification of chemical compounds, reactions, drugs and diseases is intended to organize chemical data in order to understand its interactions with other biomolecules. By looking at the genome from the perspective of chemical compounds as the main players, new insights may be gained into the function of seemingly unrelated proteins. In KEGG, compounds are grouped by structural similarity and organized hierarchically. The portal page for BRITE is at http://www.genome.jp/kegg/brite.html. Here, classifications for genes/proteins, compounds/reactions, drugs/diseases and cells/organisms are available. As such, the annotations between groups at various levels of hierarchy from genes/proteins and chemical compounds may be compared with new classifications or with one another such that the current knowledge of such groups of data can be applied for new analyses.

The BRITE hierarchies of compounds and reactions provide classifications of compounds, reactions, and compound interactions. These classifications are further classified according to various types of interactions and biological functions. Table 23.1 lists the classifications currently available at the time of this writing. For example, under Compound Interactions, there is the hierarchy of GPCR Ligands, as in Fig. 23.10. The down-arrows across the top of the hierarchy can be used to open or close the hierarchy at various levels. Clicking on the right-most arrow will open all the levels. At the lowest level for GPCR Ligands, the specific entries are listed as compound or drug entries, depending on the database in which they are found (COMPOUND or DRUG, respectively). These entries are grouped in terms of KO annotation at the third level of the hierarchy. Each KO group can be viewed by clicking on the given K-number, whose entry provides the list of pathways or BRITE hierarchies in which the given ortholog group can be found. Alternatively, the hierarchy of Receptors and Channels under Protein families in BRITE lists as one group GPCRs, which groups KO orthologs by receptor.

From the chemical compound perspective, the hierarchies under Compounds represent classifications of compounds based on structural similarity. Furthermore, the Reactions classifications are hierarchies of reactions based on similarity. These sorts of classifications can be used to determine their correspondence to the ligands with which they bind [Izrailev and Farnum 2004]. New classifications based on such binding affinity data will likely produce interesting results

**Table 23.1** BRITE classifications of compounds and reactions

| Class | Subclass |
|---|---|
| Compounds | Compounds with biological roles |
| | Lipids |
| | Phytochemical compounds |
| | Polyketides and nonribosomal peptides |
| | Bioactive peptides |
| Reactions | IUBMB reactions |
| | IUBMB reaction hierarchy |
| | Glycosyltransferase reactions |
| Compound interactions | Enzyme ligands |
| | Transporter substrates |
| | Ion channel ligands |
| | GPCR ligands |
| | Nuclear receptor ligands |
| | Cytochrome P450 substrates |

(Copies of tables are available in the accompanying CD.)

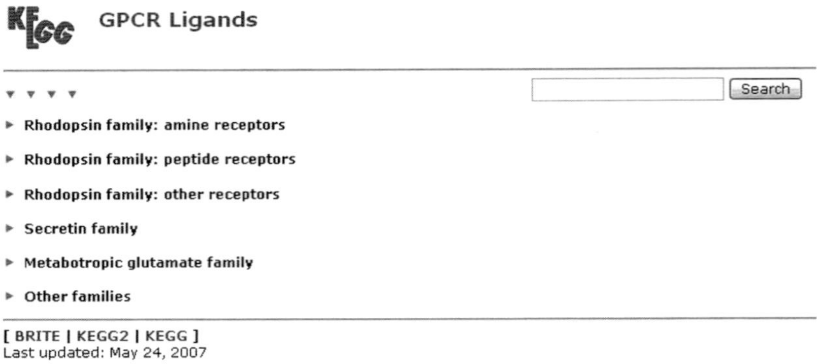

KEGG GPCR Ligands

Search

- Rhodopsin family: amine receptors
- Rhodopsin family: peptide receptors
- Rhodopsin family: other receptors
- Secretin family
- Metabotropic glutamate family
- Other families

[ BRITE | KEGG2 | KEGG ]
Last updated: May 24, 2007

**Fig. 23.10** The hierarchy of GPCR Ligands in BRITE (Copies of figures including color copies, where applicable, are available in the accompanying CD)

whereby drug development may benefit by producing drugs that are both effective and produce fewer side-effects.

A recent component of KEGG BRITE is the Pathway Module hierarchy, whose entries are indicated as M-numbers. It is useful to look at pathways as a whole to get a bird's-eye view of various biological processes. However, sometimes the details may be overwhelming, in which case one attempts to break down the whole into meaningful components [Paolini et al. 2006]. The Pathway Module hierarchy organizes the pathways into meaningful modules of sub-networks, from which one can analyze the KO groups as well as chemical reaction and compound information involved in the module. Thus from a single module, the chemistry behind the modules can be analyzed for further exploration.

Another addition to KEGG BRITE includes the hierarchies for Drugs and Diseases. The Drugs hierarchies include those for therapeutic categories of drugs and the classification of drugs as defined in [6]. The therapeutic categories of drugs can be viewed in two ways: with target information or pathway map number. There are currently two hierarchies for Diseases: "Infectious diseases" and "Metabolic disorders." Table 23.2 gives an overview of these hierarchies. Currently, the Infectious Diseases category mainly consists of bacterial and viral infections, along with Protozoal diseases and Mycoses. Metabolic disorders mainly cover those related to the existing pathways in KEGG, including disorders associated with carbohydrate, lipid, glycan and amino acid metabolism.

### *23.3.3 Linking Genomes to Endogenous Chemical Substances*

With the chemical-based resources in KEGG, it is possible to make predictions regarding the chemical repertoire of a biological system from genomic information. For example, glycosyltransferases are enzymes that synthesize glycan (carbohydrate sugar chain) structures. With microarray technology, it is possible to measure the expression of glycosyltransferases and other carbohydrate enzymes. Since KEGG contains the GLYCAN database of carbohydrate structures, predictions can be made as to which structures correspond to a given glycosyltransferase expression profile [7,8].

This process involves various pieces of information including the glycosidic bonds synthesized by a glycosyltransferase gene, the glycosidic bonds comprising a glycan structure, and a method of computing the likelihood of a given glycan structure given a microarray expression profile. First, a glycosidic bond profile is generated for each glycan structure, where the

**Table 23.2** BRITE hierarchies of drugs and diseases

| Hierarchy | Contents |
|---|---|
| Therapeutic category of drugs | Agents affecting the nervous system and sensory organs |
| | Agents affecting individual organs |
| | Agents affecting metabolism |
| | Agents affecting cellular function |
| | Crude drugs and Chinese medicine formulations |
| | Agents against pathologic organisms and parasites |
| | Agents not mainly for therapeutic purposes |
| | Narcotics |
| Drug classification | Autonomic drugs |
| | Cardiovascular-renal drugs |
| | Drugs that act on smooth muscle |
| | Drugs that act on the central nervous system |
| | Drugs used to treat diseases pertaining to blood inflammation and gout |
| | Endocrine drugs |
| | Chemotherapeutic drugs |
| | Others |
| Infectious diseases | Bacterial infections – gamma proteobacteria |
| | Bacterial infections – beta proteobacteria |
| | Bacterial infections – delta/epsilon proteobacteria |
| | Bacterial infections – alpha proteobacteria |
| | Bacterial infections – gram-positive bacteria |
| | Bacterial infections – actinobacteria |
| | Bacterial infections – Chlamydia |
| | Bacterial infections – spirochetes |
| | Bacterial infections – bacteroides |
| | Protozoal diseases |
| | Mycoses |
| | Viral infections – dsDNA viruses |
| | Viral infections – retroviruses |
| | Viral infections – ssRNA viruses |
| | Viral infections – ssDNA viruses |
| Metabolic disorders | Disorders of Carbohydrate Metabolism |
| | Disorders of Lipid Metabolism |
| | Disorders of Glycan Metabolism |
| | Disorders of Amino Acid Metabolism |

(Copies of tables are available in the accompanying CD.)

distribution of bonds is computed. Thus, a library of these profiles is accumulated. Next, a coefficient is calculated for every pair of co-occurring bonds, such that the co-occurrence of two glycosidic bonds amongst all known glycan structures is obtained. In this manner, the glycosyltransferases that co-express can be associated with the corresponding co-occurrence scores of glycosidic bonds to predict which glycan structure is most likely being synthesized in the given microarray experiment.

### 23.3.4 Linking Exogenous Chemical Substances to Genomes

Conversely, it is also possible to make predictions at the genomic level when given information regarding chemical substances. Just as similarity scores for sequences are calculated using the Smith-Waterman algorithm or BLAST, similarity scores for small molecules can also be calculated. This is done using atom types. The list of atom types used in KEGG is provided at http://www.genome.jp/kegg/reaction/KCF.html. These atom types are the basis by which the SIMCOMP and SUBCOMP calculations are made. Using these atom types, then, chemical reactions can also be compared. That is, enzymatic reactions can be viewed as a set of atom type modifications between pairs of compounds, called reactant pairs. Thus, in addition to the reaction and chemical compound information in the KEGG LIGAND database, a novel database of reactant pair alignments is available as the KEGG RPAIR database [9]. This database is a collection of compound structure transformation patterns involved in all the reactions in KEGG REACTION. These patterns are defined as RDM patterns for the changes between the substrate and product molecules involved in the reactions. RDM patterns involve the KEGG atom type changes at the reaction center (R), the difference region (D) and the matched region (M) for each reactant pair. Recent work has shown that these patterns are useful for characterizing enzymatic reactions by comparing them against the pathways in which characteristic RDM patterns are found [10].

All of the pathways in KEGG were analyzed by looking at all the reactions involved and listing the RDM patterns that were found. These are summarized in Table 23.3 by listing the pathway categories analyzed along with the number of reactions, number of chemical compounds, number of RDMs and the number and percentage of unique RDMs. From this table, one can see that the pathway category with the largest percentage of unique RDM patterns is the Xenobiotics pathway category. With this knowledge, we may assume that the reactions involved in xenobiotics degradation are distinct from those in other pathways, so a prediction method can be applied to predict the bio-degradation pathways given a query compound.

This method basically works as follows. The given query compound is first searched globally against the COMPOUND database using SIMCOMP. The resulting matching compounds are then locally searched for, against the RDM pattern library, to produce a list of possible RDM patterns possibly involving the query compound. These RDM patterns are then applied against the query compound to generate a list of product compounds that may be produced from the original compound. As a result, we now have a list of possible pathways containing one reaction. These product compounds are then used as query compounds to search for the next step in the predicted pathways.

Further analysis showed that there were cases where consecutive RDM patterns of three to five were conserved amongst distinct pathways. Such "RDM pattern profiles" may be considered as "pathway modules" which may also aid in new prediction schemes, including the prediction of missing enzymes and even new pathways.

**Table 23.3** Unique RDM patterns in KEGG pathway categories

| Metabolic pathway category | # Reactions | # Compounds | # RDMs | # Unique RDMs | (%) |
|---|---|---|---|---|---|
| 1. Carbohydrate | 727 | 452 | 444 | 303 | 68.2 |
| 2. Energy | 171 | 114 | 128 | 62 | 48.4 |
| 3. Lipid | 560 | 450 | 261 | 169 | 64.8 |
| 4. Nucleotide | 256 | 143 | 102 | 66 | 64.7 |
| 5. Amino acid | 723 | 534 | 430 | 254 | 59.7 |
| 6. Other amino acids | 170 | 147 | 119 | 55 | 46.2 |
| 7. Glycan | 243 | 96 | 42 | 14 | 33.3 |
| 8. PK/NRP | 210 | 226 | 97 | 66 | 68.0 |
| 9. Cofactor/vitamin | 336 | 279 | 224 | 151 | 67.4 |
| 10. Other secondary metabolites | 533 | 514 | 245 | 161 | 65.7 |
| 11. Xenobiotics | 580 | 522 | 347 | 275 | 79.3 |
| Total | 4,256 | 3,057 | 1,901 | 1,576 | |

(Copies of tables are available in the accompanying CD.)

## 23.4 Using the KEGG Resources

There are a number of ways to access the KEGG resources other than via the web at http://www.genome.jp/. Originally, the DBGET and LinkDB systems were developed such that related data, including those outside of KEGG, but related to KEGG, could be retrieved together. Because these tools are being developed under a separate project, the usage of these tools is beyond the scope of this manuscript, and the interested reader is referred to the online documentation at http://www.genome.jp/en/about_dbget.html.

The KEGG API was later developed such that users could program the code to directly access relevant KEGG data on-the-fly. The classic approach for obtaining KEGG data using FTP is also available at ftp://ftp.genome.jp/pub/kegg/, which can be downloaded by academic users. Furthermore, KEGG data can be downloaded in a specialized XML format called KGML.

### *23.4.1 KEGG API*

The most direct approach is to use the KEGG API, which is a SOAP/WSDL interface to the data in KEGG. Currently, libraries are available for Perl, Ruby, Python and Java. Documentation is available for setting up one's programming environment to be able to retrieve data from the available KEGG resources.

### *23.4.2 KGML*

KGML, which stands for KEGG Markup Language, supports BioPAX Level 1 format, such that all the metabolic pathways can be easily transferred between systems supporting BioPAX. (Further information on BioPAX, a standard exchange format for pathway data) In KGML, pathways are specified as graph objects, with the entry elements as its nodes and the relation and reaction elements as its edges. The relation and reaction elements indicate the connection patterns of rectangles (gene products) and the connection patterns of circles (chemical compounds), respectively, in the KEGG pathways. The two types of graph objects, those consisting of entry and relation elements and those consisting of entry and reaction elements, are called the protein network and the chemical network, respectively. Since the metabolic pathway can be viewed both as a network of proteins (enzymes) and as a network of chemical compounds, metabolic pathways in KEGG are viewed as both protein networks and chemical networks whereas regulatory pathways in KEGG are viewed as protein networks only. All of the metabolic and regulatory pathways in KEGG are available for download as KGML files at ftp.genome.jp/pub/kegg/xml/.

## 23.5 Summary

In the post-genomic era, it is no longer sufficient to analyze sequence similarity for understanding protein function. To this end, many databases attempt to organize protein-related information, such as those based on protein-protein interactions and pathways. In terms of the latter, many factors are involved, the major one being the substituents involved in an enzyme's catalytic reaction. KEGG's utility in providing a number of generalized and comprehensive resources for pathways, protein and ligand information is acknowledged worldwide. As the era of chemical genomics begins, KEGG will continue to provide the latest and most comprehensive information towards the understanding of human diseases and the development of pertinent approaches for drug therapy.

## References

1. Keiser, M.J., Roth, B.L., Armbruster, B.N., Ernsberger, P., Irwin, J.J., and Shoichet, B.K. Relating protein pharmacology by ligand chemistry. Nat. Biotech. 2007;25(2):197–206.
2. Kanehisa, M., Goto, S., Hattori, M., Aoki-Kinoshita, K.F., Itoh, M., Kawashima, S., Katayama, T., Araki, M., and Hirakawa, M., From genomics to chemical genomics: new developments in KEGG. Nucl Acids Res. 2006;34:D 354–357.
3. Schwartz, J.M., Gaugain, C., Nacher, J.C., de Daruvar A., and Kanehisa M. Observing metabolic functions at the genome scale. Genome Biol. 2007;8, R123.
4. Hattori, M., Okuno, Y., Goto, S., and Kanehisa, M. Development of a chemical structure comparison method for integrated analysis of chemical and genomic information in the metabolic pathways. J. Am. Chem. Soc. 2003:125;11853–11865.
5. Aoki, K.F., Yamaguchi, A., Okuno, Y., Akutsu, T., Ueda, N., Kanehisa, M. and Mamitsuka, H. Efficient Tree-Matching Methods for Accurate Carbohydrate Database Queries. Genome Informatics, 2003;14:134–143.
6. Katzung, B.G. Basic & Clinical Pharmacology. McGraw-Hill Medical. 2006, VK.
7. Kawano, S., Hashimoto, K., Miyama, T., Goto, S., and Kanchisa, M. Prediction of glycan structures from gene expression data based on glycosyltransferase reactions. Bioinformatics. 2005; 21:3976–3982.
8. Suga, A., Yamanishi, Y., Hashimoto, K., Goto, S., and Kanehisa, M. An improved scoring scheme for predicting glycan structures from gene expression data. Genome Informatics. 2007; 18:237–246.
9. Kotera, M., Okuno, Y., Hattori, M., Goto, S., and Kanehisa, M. Computational assignment of the EC numbers for genomic-scale analysis of enzymatic reactions. J. Am. Chem. Soc. 2004;126:16487–16498.
10. Oh. M., Yamada, T., Hattori, M., Goto, S., and Kanehisa, M. Systematic analysis of enzyme-catalyzed reaction patterns and prediction of microbial biodegradation pathways. J. Chem. Inf. Model 2007;47:1702–1712.

## Key References

- Izrailev, S., and Farnum, M.A. Enzyme Classification by Ligand Binding. PROTEINS: Struct., Funct., Bioinf. 2004;57:711–724.
- Paolini, G.V., Shapland, R.H.B., van Hoorn, W.P., Mason, J.S., and Hopkins, A.L. Global mapping of pharmacological space. Nat. Biotech. 2006;24(7):805–815.

## Suggested Reading

### *KEGG Overview*

1. Aoki, K.F., Kanehisa, M., Using the KEGG database resource, chapter 1.12 in "Current Protocols in Bioinformatics," Current Protocols, (series ed. Liz Miranker), John Wiley & Sons, Ltd., August, 2005.

## Web Resources

www.genome.jp/kegg/brite.html
http://www.genome.jp/kegg/catalog/org-list.html
http://www.genome.jp/kegg/ligand.htmt
http://www.genome.jp/kegg/reaction/KCF.html
http://www.genome.jp/
http://www.genome.jp/en/about_dbget.html
http://ftp.genome.jp/pub/kegg/
http://www.biopax.org/
ftp.genome.jp/pub/kegg/xml

# Chapter 24
# Ensembl

## Open-Source Software for Large-Scale Genome Analysis

**Anne Parker and Fiona Cunningham**

**Abstract** Ensembl is a Perl-based application designed for the easy retrieval and analysis of genomic data. Both a command-line API and a web interface are available, allowing the manipulation of sequences and associated genomic features. Additionally, the web interface displays this information in a variety of graphical formats. All software and associated data are freely downloadable and are released under an open-source licence

**Keywords** Gene · Prediction · Analysis · Annotation · Perl · Web · Interface · Display

## 24.1 Introduction

Biological sequence analysis presents computational challenges as great as those faced by other "big science" projects such as particle physics and climate simulation. Since the completion of the human genome project (HGP) in 2003, research and computational biologists have had access to genomic data on an unprecedented scale. The reference sequence alone consists of approximately three billion base pairs of DNA, enough to fill 134 complete sets of the Encyclopaedia Britannica, according to some estimates.

Data on this scale presents enormous computational problems, in storage, memory management and delivery. Conventional methods of genome data curation and annotation cannot keep up with the rate at which data are accumulating. And with the move from the study of a single reference genome to the comparison of sequences from many individuals or populations, this gap will continue to widen.

Ensembl [1], a joint project between the Wellcome Trust Sanger Institute (WTSI) and the European Bioinformatics Institute (EBI), was designed to meet the need for an automated annotation system capable of handling large volumes of finished and unfinished sequence data. It provides researchers with access to state-of-the-art automated annotation of genomic data, based on sequences from projects around the globe. Initial development was naturally concentrated on human genomic data, but the model is now applied to a wide variety of eukaryote organisms, including the classic experimental species such as mouse, fly and worm, as well as a selection of representative chordates.

## 24.2 Ensembl: An Open-Source Tool

One of the major successes of the HGP was to make the human genome sequence freely available to all. However, at that time, the computational resources necessary to analyze the data were not widely available outside of large private companies. This could have had the effect of confining the

A. Parker
Wellcome Trust Sanger Institute,
Cambridge, UK
e-mail: ap5@sanger.ac.uk

S. Krawetz (ed.), *Bioinformatics for Systems Biology*, DOI 10.1007/978-1-59745-440-7_24,

usefulness of the raw genome data to a small number of well-funded groups. One of Ensembl's principal aims has been to bring these data to the widest possible audience.

A central tenet of the Ensembl philosophy is openness: all data, software and associated information is freely available and without restriction. The project has enthusiastically embraced the open-source ethic, with all the code released under an open-source licence, and contributors around the globe supplying expertise and time free of charge [2].

## 24.3 The Ensembl Code Base

The Ensembl project generates a substantial body of Perl code [3], all of which can be downloaded from the website. The code base consists of three main parts:

- the analysis pipeline, which adds new data and analyses to the databases
- the API (application programming interface), which gives structured access to these data
- the web interface, which uses the API and graphics code to generate user-friendly views of the data

Other implementations building on the elements of the code base, such as Java clients, stand-alone applications and so forth, are certainly possible and indeed encouraged.

### *24.3.1 The Analysis Pipeline*

The Ensembl analysis pipeline runs a series of *in silico* analyses on an individual genome, integrating data from protein, cDNA and EST evidence to predict and refine gene models. Gene predictions may be labelled as 'known' or 'novel'. 'Known' indicates that the prediction is based on evidence from the same species as the assembly and with a high matching score; the label 'novel' is applied to predictions based on proteins from other species, or matches which had a lower score. The finished gene set is then compared with manual annotations from the Havana project [4], and genes that match are assigned to a 'gold' merged set. The final non-redundant set of gene models, and the evidence used to build these models, are presented in a single database.

Additional data from external sources, such as microarray probesets, single nucleotide polymorphisms (SNPs), and gene ontologies, may be included in the species data-set where available. Comparative genomics analyses provide sequence alignments and homology predictions based on multi-species comparison.

### *24.3.2 The API*

The Ensembl API provides a representation of the data in an Ensembl database in terms of model biological objects: genes, clones, contigs, etc. This abstraction shields programmers from the underlying complexity of the data, making it easier to retrieve information in a meaningful form. This makes the API an extremely powerful tool for biologists. It is simple, for example, to extract a specific clone object from the database and query it for biological properties such as contigs, length, sequence, and any features added by the analysis pipeline such as genes, repeats and CpG islands. The API also enables access to the additional databases, so that sequence objects can be queried for variations, orthologues and other external data.

### *24.3.3 The Web Interface*

While the Perl API offers a rapid and efficient way to access the Ensembl databases, the website enables visual representations of often complex data, making it more human-readable. Perl modules auto-generate images from the genomic data, allowing the display not just of sequences and their associated annotations, but of karyotypes, gene trees, and other genomic data.

While most Ensembl data can be accessed through either the website or the Perl API, a few features are web-only. These include LD (linkage disequilibrium) values, which are calculated on the fly (using code written in C) then displayed via Ensembl's Perl drawing modules, and search tools such as BLAST (see Section 23.5), using the Ensembl website.

Another web-only feature is the system of user accounts, which enables quick access to your favorite Ensembl pages from any computer, as well as facilitating collaboration and data sharing. More on adding your own data to an Ensembl display and sharing the results can be found below.

## 24.4 Ensembl Data Concepts

The Ensembl data model revolves around two central ideas: an *assembly* and a *slice*. Genomic sequence data is accumulated as a set of overlapping clones, each containing one or more sequence fragments called *contigs*. In the *unfinished* data, the order and orientation of these contigs is often unknown. *Finished* data has been fully assembled, such that each clone comprises only a single contig, i.e., the clone has been fully sequenced. The clone overlap information, combined with data from genetic maps, can be used to assemble the clones into a continuous pathway along a chromosome. This pathway through the data, containing the non-redundant sequence, is sometimes referred to as the *golden path*.

It is often necessary to work with regions of an assembly that do not map exactly to a single clone or contig. For example, one might wish to examine the first megabase of a chromosome. The Ensembl API allows this manipulation of arbitrary regions by providing an abstraction of the underlying assembly, called a *slice*. A slice behaves as if it were a real contig, but may actually be constructed "behind the scenes" from multiple contigs and fragments of contigs.

## 24.5 Using the Ensembl Website

The Ensembl website displays genomic data in a variety of graphical forms, including sequence alignments, wiggle plots and tree diagrams. It offers a hierarchical interface to the various genomes, so that the user can *drill down* into the data by selecting successively more detailed views. For example, clicking on a chromosome takes you to a high-level display of the region, and from there to detailed displays of individual clones, genes, transcripts, proteins and other sequence features.

As well as drilling down into the data, the user can search for particular identifiers, or find sequence locations using alignment methods such as BLAST [5]. The website also provides extensive facilities for exporting data in a variety of forms, including FASTA, EMBL, tab-delimited lists, and several image formats, and incorporates the BioMart data-mining tool [6] which allows extraction of information from combinations of disparate data-sets. A wide variety of learning materials are now available to help you use Ensembl, both on the website itself (under 'Help & Information') and in print [7].

### *24.5.1 Displaying Other Data in Ensembl*

In addition to the annotations provided by the Ensembl pipeline, you can view your own data alongside the genome. The simplest method is to upload a file in one of the commonly used formats such GTF, BED or PSL; data can either be uploaded directly from your computer, or via a URL anywhere on the Internet. Alternatively, you can display data in DAS (Distributed Annotation System) format, by attaching any source from a public DAS server to Ensembl. DAS [8] is a standard used by institutions around the world to share genomic resources without having to copy vast amounts of data back and forth.

The Ensembl website includes a DAS Registry, which is a selection of publicly accessible DAS sources. These include data from projects based at the Wellcome Trust Sanger Institute, such as Decipher (Database of Chromosomal Imbalance and Phenotype in Humans using Ensembl Resources), plus a selection of other public sources such as NCBI RefSeq. Many of these DAS sources are maintained within the WTSI or the EBI, either on the Ensembl DAS server or within individual research projects. Other sources are located on servers around the world and may occasionally be unavailable from within Ensembl. For more information on using these data, see the examples, below.

## 24.6 Website Tour

As an introduction to the Ensembl website, we will investigate ***SGCB***. This is the gene for beta-sarcoglycan, a glycoprotein associated with dystrophin and hence potentially involved in muscular dystrophy.

*Note*: Screenshots for these examples were taken from the Ensembl release 44 (April 2007) – see the relevant archive at http://apr2007.archive.ensembl.org. A revised and reorganised version of the site will be released in the second half of 2008, but the basic features remain unaltered.

### *24.6.1 Searching for a Feature*

The website uses the Exalead search engine to provide a full index of all the Ensembl databases, enabling fast and comprehensive searches. A search box can be found in the top right corner of most pages, or you can start from the main search on the Ensembl **home page** (Fig. 24.1).

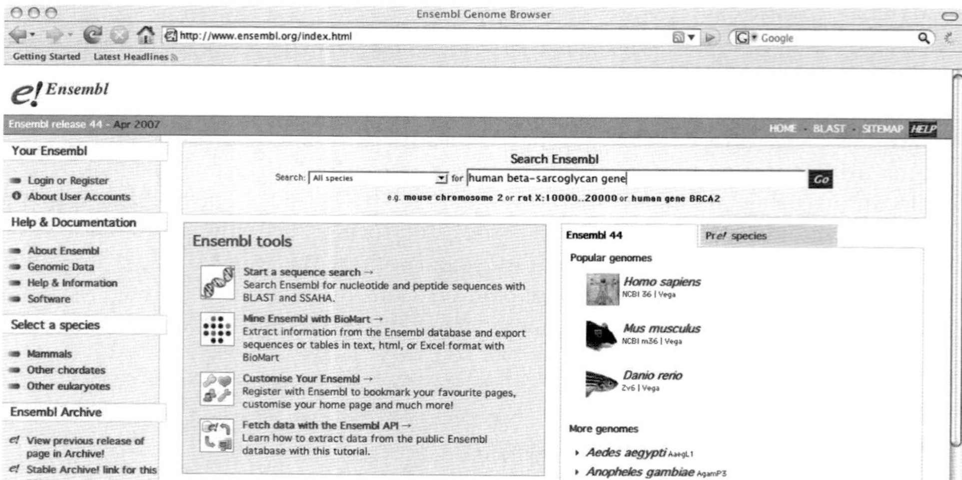


Fig. 24.1 The Ensembl website home page (Copies of figures including color copies, where applicable, are available in the accompanying CD)

*SGCB* is an HGNC Symbol, i.e., a short, unique name for the gene which has been assigned by the HUGO Gene Nomenclature Committee. A gene symbol can map to more than one Ensembl prediction; however in this case, entering 'human beta-sarcoglycan gene' in the text field gives us only one hit – Ensembl gene ENSG00000163069. All Ensembl genes, transcripts, exons and proteins have an ID beginning with 'ENS' – this *stable id* can be tracked from release to release, so you can always be sure you are looking at the same information. Non-human genes also have a three letter code identifying the species (e.g., 'MUS' for *Mus musculus*). Each identifier then has a single letter to identify the object type – G for gene, T for transcript, P for protein and so on – followed by a numeric ID.

Clicking on the name of the gene in the SearchView takes us to **GeneView**, where information about the gene is displayed.

### 24.6.2 GeneView

The GeneView page (Fig. 24.2) contains much detail and information about *SGCB* including its Ensembl identifier, description, genomic location, and transcripts. Links in the left-hand menu bar lead to further information about the gene, including regulatory regions, variation displays and phylogenetic trees across Ensembl species.

*SGCB* lies on the reverse strand, so in Ensembl it is shown below the broad blue line representing the contig (labelled 'AC093858' in the figure); genes aligned to the forward strand are shown above the contig.

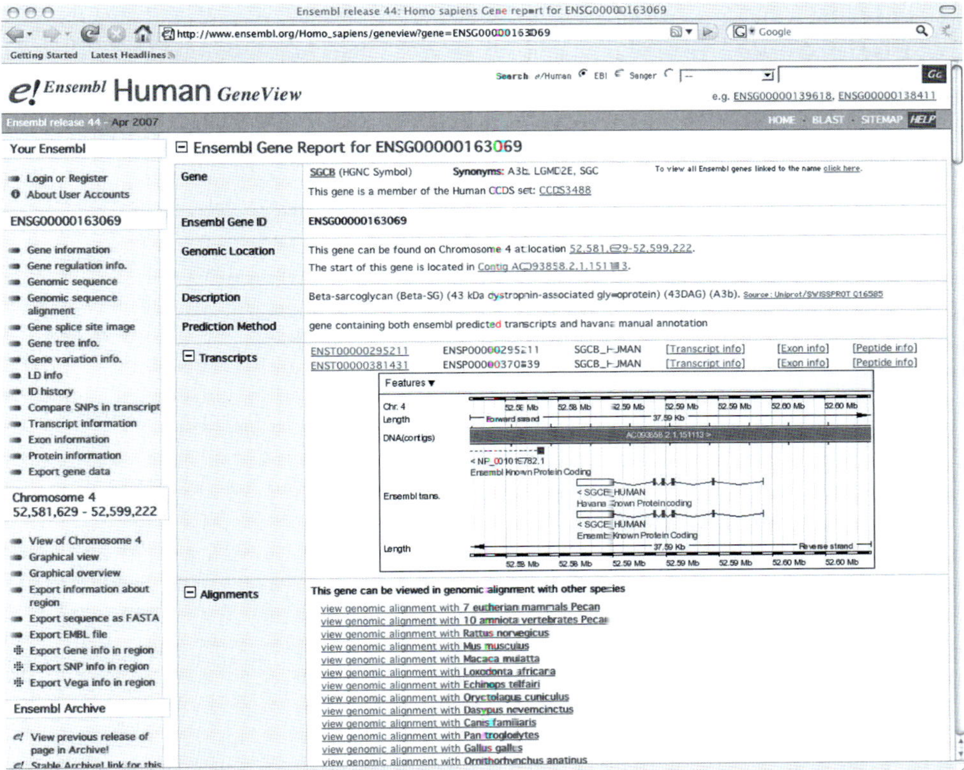


**Fig. 24.2** The Gene Report for *SGCB* (Copies of figures including color copies, where applicable, are available in the accompanying CD)

Also, *SGCB* is a member of the Consensus CDS set (a subset of human genes agreed upon between Ensembl, UCSC and NCBI) and has been manually annotated by the Havana project, so the transcript diagram shows both the Ensembl prediction (the lower transcript, in dark red on the web) and the Havana annotation (the upper transcript, in blue). Exons are shown as boxes, introns as angled lines; the empty box at the left-hand end of each transcript represents the non-coding UTR (untranslated region) portion of the gene.

Additional tracks such as EST genes, SNPs and regulatory regions can be displayed alongside the transcript by clicking on the "Features" menu button in the yellow bar at the top of the image. Further down the page are sections on individual transcripts, including orthologues in other species, associated GO (gene ontology) terms, and Interpro protein domain matching. To find out more about the protein translation for this gene, you can click on the link 'Protein Information' in the left-hand menu, or there are links above the transcript neighborhood diagram.

### 24.6.3 ProtView

ProtView displays more detailed information about the protein produced from a transcript, including domains, protein families and the amino acid sequence. The sequence display can be customized to show the introns and exons in different colors and also to highlight SNPs (Fig. 24.3).

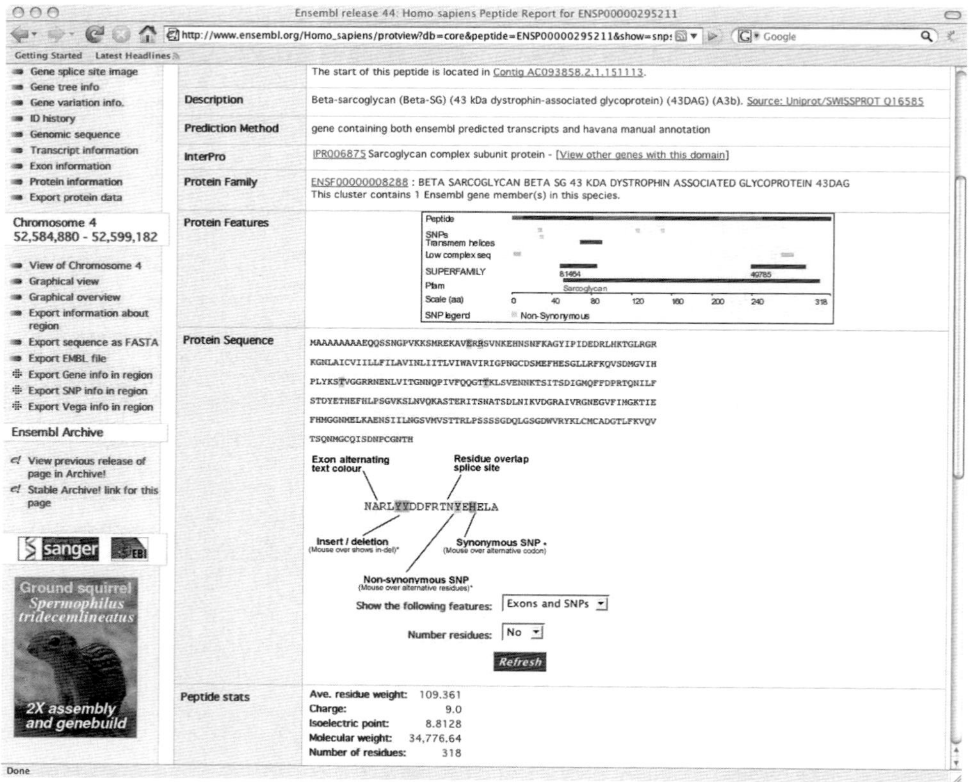


**Fig. 24.3** ProtView, showing the amino acid sequence of *SGCB* (Copies of figures including color copies, where applicable, are available in the accompanying CD)

From ProtView (or GeneView), we can go to the chromosomal region for this gene, by clicking on either the 'Graphical View' link in the left-hand menu or on the coordinates link in the Genomic Location section of the main table.

## 24.6.4 ContigView

ContigView is the main genome browser display for the Ensembl website. It allows the user to walk freely up and down any chromosome, and displays all the sequence features appropriate to the current region. There is a variety of navigation tools provided to enable the user to move to any location, zoom in and out, and focus on a specific area.

The ContigView display is customizable. The user can turn various feature tracks on and off, including gene sets, markers, aligned proteins, oligo probes, and repeats. A number of standard DAS tracks are provided, or you can attach a DAS source from any external server (see below). Finally, the size of the image can be configured and legends, rulers, pop-up menus, etc., can be turned on-or-off. All of these customizations can be saved in your Ensembl user account, so that you can access exactly the same display from any computer you use, without having to reset all the options.

Clicking on any of the features will display a pop-up menu with more feature information and links to alternative views such as GeneView and ExportView. As can be seen in Fig. 24.4, ContigView shows a hierarchy of views. The topmost view shows the chromosome that we are looking at.

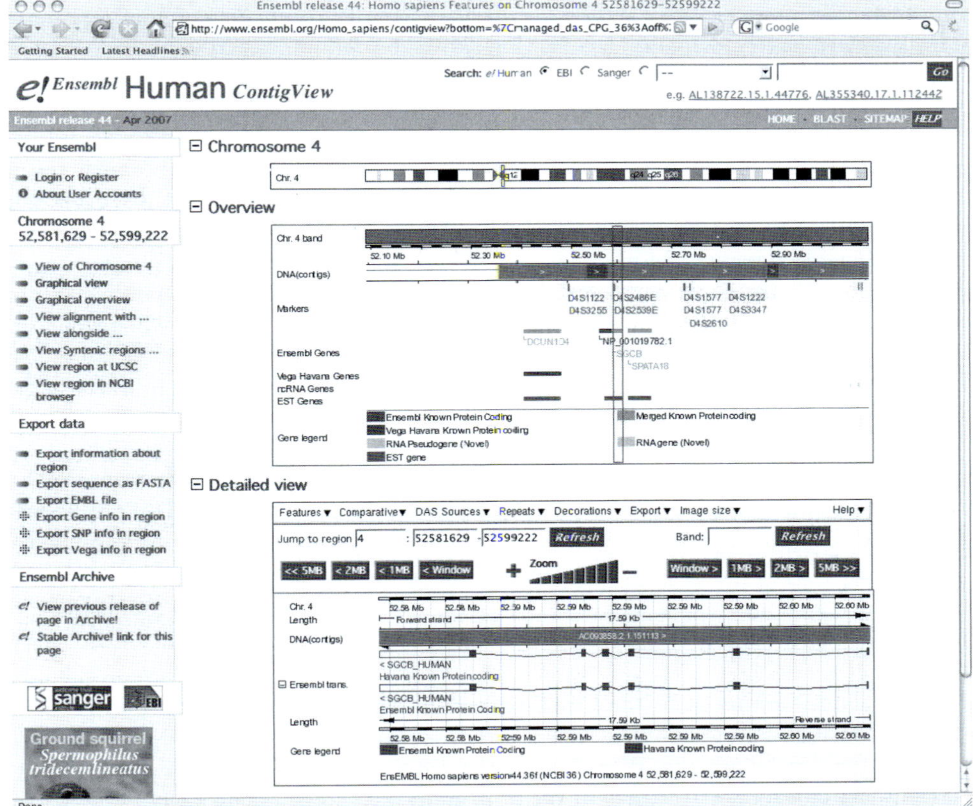


Fig. 24.4 ContigView (Copies of figures including color copies, where applicable, are available in the accompanying CD)

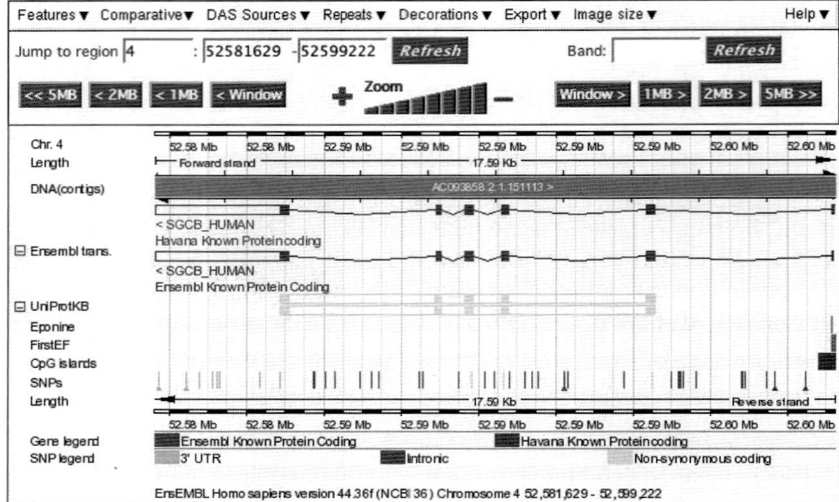


**Fig. 24.5** Additional tracks in ContigView (Copies of figures including color copies, where applicable, are available in the accompanying CD)

A box marks the region (in this case, close to the centromere of chromosome 4) that is represented in the next image- the overview display. The overview shows, by default, one megabase of DNA, and is a high-level view of the region. For this reason, it shows only landmark features such as markers and genes. Clicking anywhere on the overview brings up a generic zoom menu, or you can click and drag the mouse to outline the area you are interested in and bring up a context-sensitive zoom menu.

Below the overview is the detailed display (see Fig. 24.5). It shows a relatively short region of the genome, but can display all the features present in that area. Only a few tracks are turned on in Fig 24.4, but you can see the Ensembl and Havana transcripts below the contig, as in the transcript image on the GeneView page.

One final display, basepair view, is available below the detailed view, but is collapsed by default and is not visible on the screenshot. It shows a 100 bp window whose location will be marked in the detailed view when the basepair view is open. Just click on the plus sign next to its name to display the individual base pairs and possible amino acid translations in all six reading frames (forward and reverse), along with the restriction enzymes.

Figure 24.5 shows a close-up of the detailed view, with additional tracks turned on from the 'Features' menu. The Eponine and FirstEF tracks are indicators of possible transcription start sites, while CpG islands are associated with the promoter regions. Also turned on is the SNP track, showing single nucleotide polymorphisms in and around the gene.

### *24.6.5 Sharing your Data with DAS*

As well as all the data produced by the Ensembl pipeline and associated projects, you can view any other annotation data via DAS sources. A selection of useful sources is pre-configured, and available via the **DAS sources** menu in ContigView.

In Fig. 24.6, we can see a DAS track at the bottom of the detailed view, displaying RefSeq data selected from the basic list. Ensembl also includes a registry of additional DAS sources, or you can provide the URL from another server. In either case you can configure the color, layout and style of these extra tracks (see Fig 24.7).

⊟ Detailed view

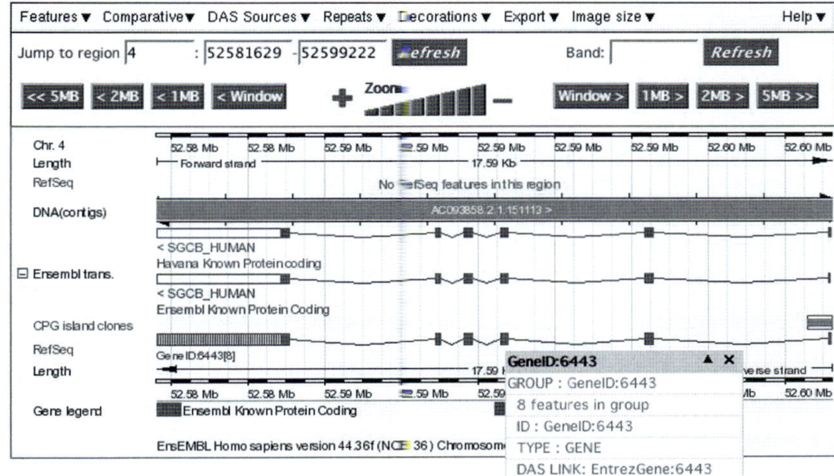


**Fig 24.6** DAS track in ContigView (Copies of figures including color copies, where applicable, are available in the accompanying CD)

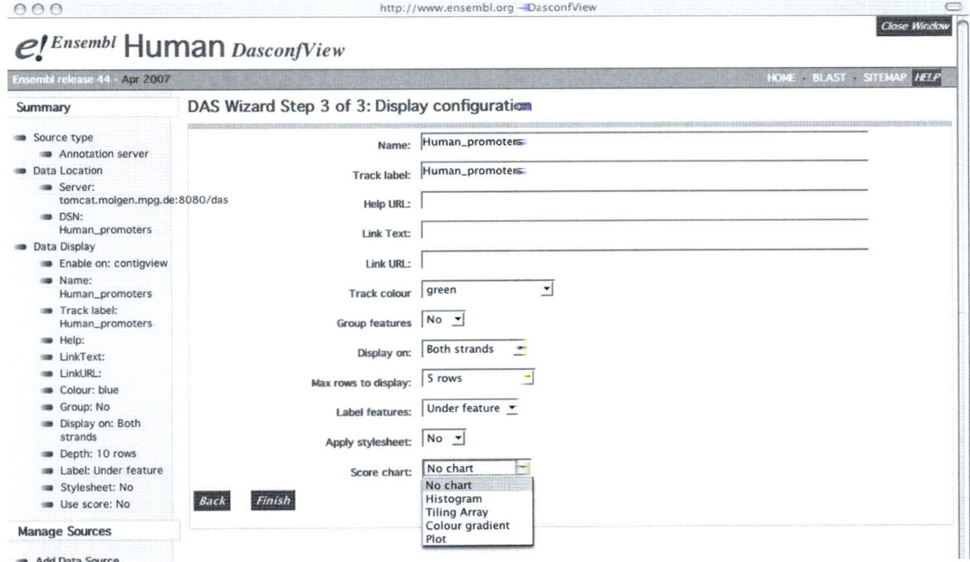


**Fig. 24.7** Configuring your DAS track (Copies of figures including color copies, where applicable, are available in the accompanying CD)

## 24.7 Using the Ensembl API

### *24.7.1 Installation*

Before using the Ensembl API, you must first obtain and install the Ensembl software modules. The components you will require for a basic installation are:

- Bioperl – provides some foundation objects used by the Ensembl API [9]
- Ensembl – the core functionality and database objects

These components, along with instructions for retrieving and installing them, can be found on the Ensembl website, www.ensembl.org. You will also need a local installation of the Perl programming language – we recommend upgrading to the latest version of Perl and associated packages, especially standard modules such as DBI and CGI, to avoid bugs.

You will also need an Ensembl database to run your code. While it is possible to download all the Ensembl MySQL databases and make a local installation, at the time of writing the compressed downloads total around 500 GB, so in most cases it will be quicker and easier to use the public database servers at the Wellcome Trust Sanger Institute. These are the servers we will be using in this example. Note that this is a separate installation from the databases used by the website; we strongly encourage you to use the public databases for your scripts, as it avoids degrading the service for other users!

*Important note*: the database schema typically changes from release to release, so you will need to make sure that your API version matches the version of the database you are using. The CVS checkout command (see the online installation instructions) will tell you which release of the code you are getting; the database names include the version number as well. Names of all the current databases can be found via the MySQL links on http://www.ensembl.org/info/data/ft-links/html or by logging onto ensembldb.ensembl.org using a MySQL client and user 'anonymous'.

### 24.7.2 Setup of the Ensembl Software Environment

Once you have installed the necessary software, you will need to configure your environment so that your Perl script will know where to find the modules. The easiest way to do this is to add the location of the modules to the PERL5LIB environment variable.

For example, if you installed Ensembl and Bioperl into a directory called */home/bob*, then set the environment as follows:

In UNIX:

```
csh:  setenv PERL5LIB /home/bob/ensembl/modules:/home/bob/bioperl
bash: export PERL5LIB = /home/bob/ensembl/modules:/home/bob/bioperl
```

On Windows:

```
set PERL5LIB c:\home\bob\ensembl\modules;c:\home\bob\bioperl
```

Note that the /ensembl/modules directory contains only the **core API** for accessing sequences, genes and other transcription features in the core Ensembl databases. If you wish to use the comparative genomics or variation API calls to access features from other databases (as in the example below), these are in separate directories and they will also need to be added to your environment.

The code shown below only scratches the surface; the API has many powerful features for retrieving genomic data, including a database registry which makes setting up database connections even quicker and easier than in the example. Additional details are available at the website, which always contains the latest API documentation.

### 24.7.3 Worked Example

In the following example, we will use the API to find all the SNPs on a particular gene in the mouse, *Mus musculus*. The gene we have chosen is *Tyr,* responsible for the production of the enzyme tyrosinase which catalyses several steps in the conversion of tyrosine to melanin. Mutations in this gene can thus cause albinism in mice (and other mammals) by disrupting pigment production.

Firstly, we need to load the database adaptor modules, which will allow us to connect to the database. Genes are stored in the core database, but SNPs are stored in the variation database for a species, so we will need to load both the adaptors. The modules are:

- Bio::EnsEMBL::DBSQL::DBAdaptor
- Bio::EnsEMBL::Variation::DBSQL::DBAdaptor

To make the connection, we need to specify the host machine that is serving the database (ensembldb.ensembl.org), the name of the database to connect to (e.g., mus_musculus_core_44_36e), and the username to connect as (ensembldb accepts the user *anonymous*).

The code to make the connection is:

```
my $dbCore = Bio::EnsEMBL::DBSQL::DBAdaptor->new
       (-host => 'ensembldb.ensembl.org',
       -dbname => 'mus_musculus_core_44_36e',
       -species => 'mouse',
       -group => 'core',
       -user => 'anonymous');
```

We can repeat this process for the variation database, using the same connection details and the database name mus_musculus_variation_44_36e. If either of the connections fail, you will get an error message. Otherwise you are ready to start working with the data in the databases.

As described in the opening section on the Ensembl code base, the API offers an abstract representation of all the various genomic "objects" stored in the database. There is an adaptor for each type of object, which gives you easy access to a number of standard method calls, such as 'fetch_all_by_description'. For this example we will need adaptors for three types of objects: gene, SNP and slice.

The gene and SNP objects both contain their location in the form of a region name (usually a chromosome), start and end coordinates, and strand; however, rather than having dozens of specific calls such as 'fetch_SNPs_by_gene', 'fetch_SNPs_by_exon' and so on, the API uses the generic slice object (see Section 4, Ensembl Data Concepts) to map features onto one another. The slice object is thus one of the most often-used objects in the Ensembl API.

We create our object adaptors by making calls on the appropriate database adaptor:

```
my $gene_adaptor = $dbCore->get_GeneAdaptor();
my $slice_adaptor = $dbCore->get_SliceAdaptor();
my $vf_adaptor = $dbVariation->get_VariationFeatureAdaptor();
```

Note that SNPs are not the only kind of data that can be stored in the variation database (insertions and deletions are also types of variation, although there are presently none in Ensembl), so we can use the generic Variation Feature Adaptor to get the SNPs rather than a specialized SNP adaptor.

Once we have our adaptors, we can start creating the genomic objects and extracting data from them. The gene name *Tyr* is an MGI symbol (a short unique name assigned by the Mouse Genome Informatics project); in other words, it has been assigned outside Ensembl. Therefore, the first thing we need to do is to find out which gene predictions within Ensembl have been identified as corresponding to this gene.

It is possible for an external name to map to more than one Ensembl stable ID, though as it happens there is only one mapping for *Tyr* in mouse, namely ENSMUSG00000004651.

The command we need is straightforward:

```
my @genes = @{$gene_adaptor->fetch_all_by_external_name('Tyr') };
```

Note that, the method call `fetch_all_by_external_name` always returns a reference to an array, so we need to de-reference it to get our gene objects. We can then loop through the array to get information on each Ensembl gene. Of course if we knew in advance that we would definitely only get one Ensembl gene returned for *Tyr*, we could simple take the first element of the array, but the script will be more reliable and reusable if we assume the possibility of multiple gene objects.

Within our foreach loop, we first need to get the stable ID from the gene object. Unsurprisingly, the method name is stable_id:

```
my $id = $gene->stable_id;
```

Once we have that, we can find the DNA sequence underlying the gene:

```
my $slice = $slice_adaptor->fetch_by_gene_stable_id($id, 0);
```

The fetch_by_gene_stable_id takes two parameters: the stable ID, and an integer representing the length of the flanking sequence required. In this script, we will confine our search to the gene itself, but if we were interested in, for example, regulatory mutations, we could extend the search area by any number of base pairs.

Finally, we can search our slice for SNPs, using a call on the VariationFeature adaptor:

```
my $vfs = $vf_adaptor->fetch_all_by_Slice($slice);
```

Again, the method returns an array reference, which we can de-reference and loop through to get information about each SNP:

```
print "Variation: ", $vf->variation_name, " with alleles ",
    $vf->allele_string, " in chromosome ", $slice->seq_region_name,
        "and position ", $vf->start,"-",$vf->end,"\n";
```

A screenshot of the resultant printout can be seen in Fig. 24.8.

#### 24.7.3.1 Finished Script

```
#!/usr/bin/perl

use strict;
use warnings;
# load required API modules
use Bio::EnsEMBL::DBSQL::DBAdaptor;
use Bio::EnsEMBL::Variation::DBSQL::DBAdaptor;
```

```
File Edit View Terminal Tabs Help
[ap5]24: setenv PERL5LIB /ensemblweb/shared/bioperl/bioperl-release-1-2-3:/ensemblweb/ap5/www/branch_44/ensembl/modules:/ensemblweb/ap5/www/branch_44/ensembl-variation/modules
[ap5]25: perl snps_on_a_gene.pl

SNPS in gene ENSMUSG00000004651 (Tyr)
Variation: rs31840595 with alleles T/C in chromosome 7 and position 19-19
Variation: rs36971348 with alleles A/G in chromosome 7 and position 75-75
Variation: rs39092187 with alleles A/G in chromosome 7 and position 110-110
Variation: rs36300750 with alleles T/G in chromosome 7 and position 194-194
Variation: rs36925470 with alleles C/A in chromosome 7 and position 392-392
Variation: rs36433386 with alleles A/C in chromosome 7 and position 442-442
Variation: rs31391415 with alleles A/C in chromosome 7 and position 471-471
Variation: rs31047739 with alleles A/C in chromosome 7 and position 476-476
Variation: rs31748517 with alleles A/C in chromosome 7 and position 783-783
Variation: rs36259722 with alleles A/T in chromosome 7 and position 793-793
Variation: rs37647005 with alleles T/C in chromosome 7 and position 837-837
Variation: rs38005625 with alleles C/T in chromosome 7 and position 856-856
Variation: rs37763927 with alleles T/C in chromosome 7 and position 907-907
Variation: rs36390126 with alleles T/A in chromosome 7 and position 1026-1026
Variation: rs38112931 with alleles C/T in chromosome 7 and position 1029-1029
Variation: rs31821576 with alleles C/T in chromosome 7 and position 1030-1030
Variation: rs32205624 with alleles A/C in chromosome 7 and position 1059-1059
Variation: rs31691095 with alleles A/T in chromosome 7 and position 1065-1065
Variation: rs32475702 with alleles G/A in chromosome 7 and position 1119-1119
```

**Fig. 24.8** Sample printout from the API script (Copies of figures including color copies, where applicable, are available in the accompanying CD)

```
# connect to Core database
my $dbCore = Bio::EnsEMBL::DBSQL::DBAdaptor->new
       (-host =>'ensembldb.ensembl.org',
       -dbname =>'mus_musculus_core_44_36e',
       -species =>'mouse',
       -group =>'core',
       -user =>'anonymous');
# connect to Variation database
my $dbVariation = Bio::EnsEMBL::Variation::DBSQL::DBAdaptor->new
       (-host =>'ensembldb.ensembl.org',
       -dbname =>'mus_musculus_variation_44_36e',
       -species =>'mouse',
       -group =>'variation',
       -user =>'anonymous');

# get all the adaptors we'll need
# (more efficient to get them outside of any loops!)
my $gene_adaptor = $dbCore->get_GeneAdaptor();
my $slice_adaptor = $dbCore->get_SliceAdaptor();
my $vf_adaptor = $dbVariation->get_VariationFeatureAdaptor();

# find gene stable ids corresponding to the MGI symbol 'Tyr'
my @genes = @{ $gene_adaptor->fetch_all_by_external_name('Tyr') };

# loop through array of genes
foreach my $gene (@genes) {

  my $id = $gene->stable_id;
  print "SNPs in gene $id \n";

  my $slice = $slice_adaptor->fetch_by_gene_stable_id($id, 0);
  # return all variations defined in $slice
  my $vfs = $vf_adaptor->fetch_all_by_Slice($slice);

  # print out the results
  foreach my $vf (@{$vfs}){
  print "Variation: ", $vf->variation_name, " with alleles ",
    $vf->allele_string, " in chromosome ", $slice->seq_region_name,

        "and position ", $vf->start,"-",$vf->end,"\n";

  }
}
```

## 24.8 Future Developments

Ensembl is under active development. The already rich API is growing constantly to provide convenient access to a wider range of genomic data resources. Each release of the database sees the inclusion of additional data of ever-increasing quality and scope, including more species and new data types. The website is also being adapted to represent these underlying improvements in the API and data, and to meet the increasingly complex demands of the research community. Future developments currently being considered include:

- further extending the data beyond single reference genomes, particularly in the areas of functional genomics and variation
- additional development of web collaboration and customization tools

## 24.9 Contact List

The Ensembl project has a very active development mailing list called *ensembl-dev*. All of the Ensembl team participate in the list, so if you have a question or comment about Ensembl you can be assured it will be answered there. There is also a low-traffic mailing list, *ensembl-announce*, which is used to announce major updates or a new release. To subscribe to these lists, send an email to majordomo@ebi.ac.uk with either *subscribe ensembl-announce* or *subscribe ensembl-dev* in the body of the mail. These mailing lists are archived on the Ensembl website.

Ensembl also maintains a helpdesk facility that will answer any questions pertaining to the project. Either send an email to: helpdesk@ensembl.org, or use the form on the website. The Ensembl team is always interested in any feedback you might have on the site. Comments and suggestions are all carefully considered, and often make their way into the next version of the code. The web team can be contacted through the Ensembl helpdesk, or via webmaster@ensembl.org.

## Key References

Conway, D. (2000) Object-Oriented Perl: A Comprehensive Guide to Concepts and Programming Techniques.

Fernandez, X.M. and Schuster, M. (2007) Using the Ensembl Genome Server to Browse Genomic Sequence Data, Current Protocols in Bioinformatics, Supplement 16, Unit 1.15.

Hubbard, T.J.P. et al. (2007) Ensembl 2007, Nucleic Acids Res. 2007, Jan 1; Database Issue.

## Suggested Reading

### *Introduction*

1. Hubbard, T.J.P. et al. (2007) Ensembl 2007, Nucleic Acids Res. 2007, Jan 1; Database Issue.

### *Ensembl: An Open-Source Tool*

2. Raymond, E.S. (1999) The Cathedral and the Bazaar.

### *The Ensembl Code Base*

3. Conway, D., (2000) Object-Oriented Perl: A Comprehensive Guide to Concepts and Programming Techniques.
4. Havana (Human and Vertebrate Analysis and Annotation) http://www.sanger.ac.uk/HGP/havana/

### *Using the Ensembl Website*

5. Blast (Basic Local Alignment and Search Tool): http://www.ncbi.nlm.nih.gov/BLAST/
6. BioMart: http://www.biomart.org
7. Fernandez, X.M. and Schuster, M. (2007) Using the Ensembl Genome Server to Browse Genomic Sequence Data, Current Protocols in Bioinformatics, Supplement 16, Unit 1.15.
8. BioDAS: http://www.biodas.org

### *Using the Ensembl API*

9. BioPerl: http://www.bioperl.org

## Web Resources

www.ensembl.org
http://das.ensembl.org
http://apr2007. archive. ensembl.org
http://www.perC.com
http://ensembldb.ensembl.org
http://www.ensembl.org/info/data ft-links.html

## Chapter 25
# Management of Spatially Organized Biological Data using EMAGE

**Jeffrey H. Christiansen, Duncan R. Davidson, and Richard A Baldock**

**Abstract** Images are used to record the outcomes of many biological experiments, including those that employ *in situ* hybridization and immunohistochemistry assays. The most widespread method currently used to describe these images for archiving and analysis purposes is employing a text based description, whereby the pattern of staining observed in each image is described by a human using a standardized vocabulary of anatomical terms.

In this chapter, we describe a complementary approach, whereby, the spatially organized information inherently contained in data images is extracted and spatially integrated via a semi-automated process and then housed in a three-dimensional atlas of biological structure. Thus, spatial data from multiple assays can be interrogated directly by spatial based approaches to find patterns that display similarities or differences of patterns over local or entire regions of the specimen.

**Keywords** Biological images · Gene expression · Spatial · Database · *in situ* hybridization · Immunohistochemistry · Mouse · Embryo

## 25.1 Introduction

The practise of biology has always relied heavily on the use of images to record experimental observations. The format of these has changed over time according to available technologies: from hand-drawn images, to film photography, to digital photography and more recently onto 3D/4D digital image capture and beyond. Automation of laboratory procedures, advances in imaging technologies and the availability of affordable digital data storage have combined to allow for the production of this type of data on an unprecedented scale.

These images are sometimes used to document the gross morphology of a biopsy or a specimen (e.g., the appearance of a 'wild-type' vs. a mutant), or they may depict specimens that have been 'stained' to detect a specific biological process or entity. Staining procedures exists, for example, to detect cell proliferation, cell death, the output of distinct biological pathways or enzymatic reactions, particular tissue types or the presence of the expression of a single gene.

This chapter will focus on the images of the last type that depict the detection of specific mRNA or protein molecules from *in situ* hybridization or immunohistochemistry experiments. Topics covered will include a background of the theory of these experimental techniques and different approaches that have been utilized for the digital archiving and analysis of this data. As an illustration of a spatial-based approach to archiving and analysis of the image-based *in situ* gene expression data, the EMAGE database will be discussed in detail.

J.H. Christiansen
MRC Human Genetics Unit, Western General Hospital, Crewe Road, Edinburgh, EH4 2XU, United Kingdom
e-mail: jeffc@hgu.mrc.ac.uk

S. Krawetz (ed.), *Bioinformatics for Systems Biology*, DOI 10.1007/978-1-59745-440-7_25,

## 25.2 Methods to Detect Gene Expression

The genome of an organism such as a human or mouse contains approximately 20,000 genes. As every gene is not expressed in every cell throughout the body, if one wishes to gain an appreciation of the potential sites of action of a gene product, it is necessary to assess the levels and distribution of each gene's expression in different tissue and cell types. As gene expression involves transcription from genomic DNA to produce mRNA and then translation of these mRNAs to protein (see Fig 25.1A), assaying for the presence of either the mRNA or protein can be used as a read-out for the process of 'gene expression'.

### *25.2.1 Assaying for mRNA Presence*

The most common methods to assess the expression levels of the mRNA/transcript all rely on one of the central dogmas of molecular biology– that of base pairing between the sense and anti-sense complementary strands in nucleic acid hybrids. In these cases, a single-stranded nucleic acid *probe* is designed to be complementary to its intended single-stranded mRNA *target* and is synthesized *in vitro*, where it is labelled such that it can be subsequently detected. The probe and the target transcript are then mixed together under conditions permissive for the specific binding of the complementary strands to each other in a process called *hybridization*, and the excess, unbound probe is subsequently removed by extensive washing. Following this procedure, any remaining probe will be tightly bound to its intended target in a *hybrid*. Therefore, when the labelled probe is subsequently detected, the presence of its bound target hybrid complementary strand will also be detected.

This method has been used for many years to measure the presence of transcripts for individual genes, one at a time, within complex mixtures of thousands of transcripts, either from dissociated samples or to whole, fixed tissue biopsies (or sections of these on glass microscope slides). In the latter cases the target transcripts are permanently attached to adjacent molecules within the tissue at the site where they were naturally present by chemical fixation. Therefore, following the hybridization and signal detection process, signal is only detected at the sites of expression *in situ*. This technique is known as *in situ hybridization* and the output of the experiment is documented using an image (see Fig 25.1B).

### *25.2.2 Assaying for the Presence of a Protein*

One of the most common methods to detect the presence of proteins in biological samples is to use an antibody-based approach, where an *antibody* is generated against a single *target protein*. This can be achieved either by directly immunizing an animal with the target protein itself, or production of the antibody by other methods such as phage display. The antibody can be labelled by a number of methods. Because an antibody has the ability to recognize and tightly bind to the specific protein it was generated against, when bound, the presence of the antibody can be used to identify the presence of its target protein in a complex mixture of any number of proteins.

Similar to hybridization as discussed earlier, appropriate samples for antibody-detection analysis can also be either dissociated samples or whole fixed tissue biopsies (or sections of these, fixed onto glass microscope slides) that are subjected to the antibody binding and target detection procedure. When tissues or sections are used, the process is termed *immunohistochemistry*, and (as for *in situ* hybridization), the data output is also image based (see Fig 25.1B).

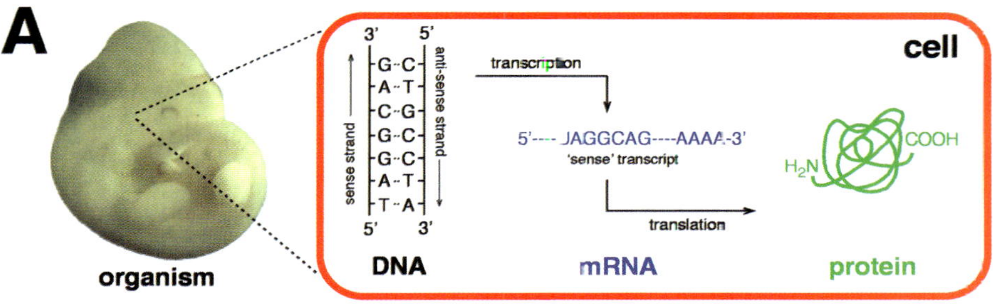


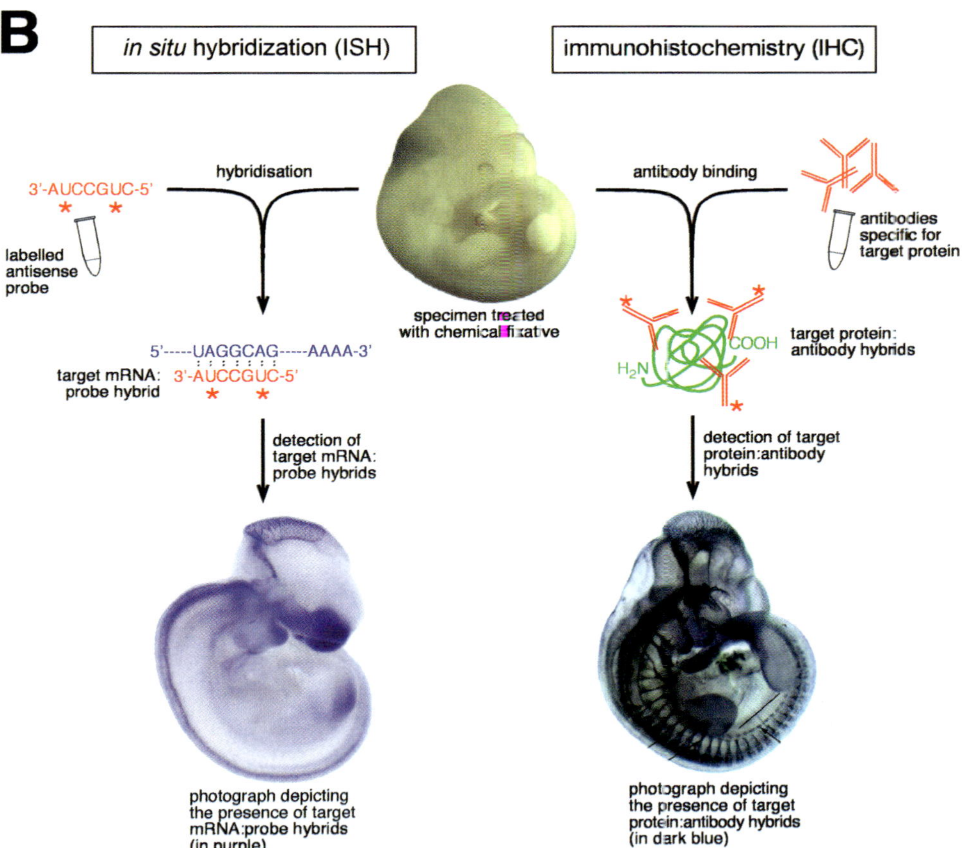


**Fig. 25.1** **(A)** Gene expression involves a two-step process where the DNA in the nucleus is first copied into an mRNA molecule. The mRNA copy is subsequently used as a template to instruct the building of a protein during translation. **(B)** The processes of *in situ* hybridization and immunohistochemistry are used to detect the presence of mRNAs and proteins within biological specimens. Both these images show whole, intact samples, yet the method can also be applied to tissue sections (Copies of figures including color copies, where applicable, are available in the accompanying CD)

### *25.2.3 Transgenic Techniques for Assessing Gene Expression*

Another powerful technique for assessing the expression profile of genes, particularly *in vivo*, is the use of transgenic reporter assays. The general method employed is to genetically modify an organism, such that it will express a detectable marker under the normal controls of the gene, whose expression one wants to examine. For example, the transgenic animal can be constructed such that a fusion protein is made between an endogenous protein and a fluorescent protein, so that monitoring the occurrence of fluorescing cells, points to the presence of the hybrid protein and therefore also to the endogenous protein (it is assumed that the hybrid protein is deployed in the cell in the same manner as the endogenous protein). When fluorescent markers are used, these techniques can offer the advantage that imaging of the target molecule can be achieved within intact, living specimens. 2D, 3D and 4D images can be captured to document these experimental outcomes.

### 25.2.4 The Pros and Cons of *in situ* Based Gene Expression Techniques

ISH, IHC and transgenic techniques, to assess gene expression patterns, all allow direct visualization of the signal within the organism *in situ*, to the single cell, and sometimes sub-cellular resolution. As such, it is possible to discern the precise positions of individual cells that express a gene of interest and their spatial relationships to other cells in a complex, spatially organized biological sample. Such an appreciation is generally not possible using non- *in situ* based gene expression profiling techniques like microarrays, mass spectrometry etc. because in these cases the sample is usually dissociated from a spatially organized tissue into a suspension prior to analysis.

The major drawback in the use of *in situ* based techniques, has until recently, been the amount of 'hands-on' time required to perform these procedures. However, the advent of automated robotic based methods to treat the samples, has now allowed these methods to be used in a high-throughput manner. For example, several *in situ* gene expression screening projects are currently in progress that will have collectively generated several millions of images documenting the sites of gene expression in the mouse or human upon completion e.g., *EURExpressII*; *BGEM*; *GENSAT*; *GUDMAP*; *GenePaint*; *Allen Brain Atlas* and the *Human Protein Atlas*.

These data-sets are all being produced to develop an understanding of the sites within the organism where the products of different genes are deployed. This will culminate in an appreciation of various molecular portfolios of different tissues, and is a biologically important question as it can point to possible relationships between sets of different genes. For example, if two or more gene products have similar spatio-temporal expression profiles within an organism (i.e., they are found in the same places at the same times), this can point to the existence of potential interactions between them, or a common regulatory mechanism governing their expression. Elucidating comprehensive gene expression profiles for each and every region of a species also has the potential to reveal an alternative view of the spatio-temporal organization of various tissues. This 'molecular' view of anatomy can be compared to classical anatomy that has been described based on visible morphological features.

## 25.3 Methods for Archiving and Describing Sites of Expression Identified from Image-based *in situ* Expression Profiling Techniques

### 25.3.1 Traditional Methods

Traditionally, the medium where *in situ* gene/protein profiling data has been archived, is the scientific literature. This has involved the publication of one or more photographs in a journal, documenting the outcomes of the experiment along with an accompanying text description – either in the figure legend or in the main text, outlining the author's interpretation of the results seen within the image(s). At least one journal: *Gene Expression Patterns* is entirely devoted to the publication of these types of data. While this method does allow for the archiving and distribution of this information, data published in this manner is often difficult to retrieve from the literature via database searching (e.g., PubMed), unless appropriate information pertaining to the expression pattern is included in the title, abstract, key words or MeSH terms of the paper. Unless the paper has been specifically published to describe the expression pattern itself, this is usually not the case. When this information is included in the title/abstract etc, it can still prove challenging to retrieve via literature citation database searching because of the variations in language and spelling used by the author to describe the sites of gene expression. For example, the 'stomatodaeum' (i.e., the precursor of the mouth during embryo development) could be described using the words 'stomatodaeum', 'stomatodeum', 'stomodaeum', 'stomodeum', 'oral pit' or 'mouth pit'. Therefore, literature databases would have to be searched using all of these words in order to retrieve all possible data.

### *25.3.2 Standardized Text-Based Descriptions of Gene Expression Patterns*

To help alleviate the problems associated with the use of non-standard text-based descriptions of sites of gene expression discussed above, various standardized vocabularies (or 'ontologies') to name anatomical structures present in an organism have been developed. These can be in a simple list format or hierarchically organized, with 'part-of' relationships, such that smaller structural units (e.g., 'eye') are part of a larger anatomical unit (e.g., 'head' or 'sensory system'). The hierarchical approach is usually adopted by databases because it allows greater flexibility in retrieval of the data. For example, if a gene expression is detected in the 'eye', it is therefore also detected in the 'head', and should be able to be retrieved from a database when either term is used for query. Some databases also offer greater search flexibility by a system of defining synonym relationships between certain words that have the same meaning. For example in the 'stomatodaeum' example mentioned earlier, the words 'stomatodaeum', 'stomatodeum', 'stomodaeum', 'stomodeum', 'oral pit' and 'mouth pit' could all be defined as synonyms and thus used interchangeably for query purposes.

Examples of databases that have adopted text as their main strategy to archive gene expression data from *in situ* gene expression experiments include all the major model organism databases, for e.g., MGI; ZFIN; XenBase; WormBase; FlyBase as well as a number of large screening projects currently underway e.g., GenePaint; Human Proteome Atlas; EURExpress, GUDMAP; EuRe-Gene; MEPD and GEISHA.

While this approach offers an excellent method to quickly describe the broad anatomical regions where expression is noted (e.g., "*GeneA* exhibits expression in the brain, the limbs and the eye") and is flexible for different data types (e.g., gene expression profiling data from diverse methods such as MS, SAGE, Microarrays and ISH/IHC can easily be integrated), it is very difficult to use effectively if one wants to describe the subtilities of an expression pattern as detected using an *in situ* based technique, as these need to include information on the sites, patterns and gradients of expression levels. For example, a text-based description of the expression pattern of a (hypothetical) gene with a reasonably complex expression pattern could be as follows: "*GeneA* has expression in the anterior-most third of the telencephalic part of the forebrain, on both sides of the midline. In this region, it is expressed in a ring that excludes the very anterior tip of the telencephalic vesicle. There is a gradient of expression in this region such that the outer part of the ring expresses the gene at a higher level than the inner part of the ring. Two other regions of expression are detected in the diencephalic part of the forebrain: one in a lateral stripe that extends in an anterior-posterior direction along the wall of the diencephalon, which sits just ventral of the dorso-ventral midline; and another that exhibits a spotted pattern with very high apparent expression levels in the dorsal most aspects of this region". This example illustrates that such text annotations can become unwieldy.

Due to the complexity of the text-based description that is required to describe a pattern as well as the associated time constraints that are usually available in which to perform the annotation, a drawback of this approach is that such annotations tend to be partial. Other factors that can contribute to the paucity of annotations made with this method can be a lack of anatomical knowledge that is required to perform the annotation in the first place, or 'selective vision' of the annotator. Text-based descriptions alone are also very difficult to use effectively if one wishes to convey information about variations of patterns in space or strength, or variations in morphology. Even if a careful full description is made using words, it is unlikely that the description could be used to describe a complex 3D expression pattern definitively and unambiguously. A parallel task would be attempting to definitively and unambiguously describe the face of one of your friends to a portrait artist using only words, and then expecting the artist to produce a photographic likeness of your friend's face from this information alone.

### *25.3.3 Spatial-Based Descriptions of Gene Expression Patterns*

A common feature of multi-cellular organisms is that the normal development is governed by a strict body-plan, which ensures that all members of a species develop with a similar and stereotypical morphology. Within each individual, the sites of the biological processes such as gene expression, cell division, cell death, enzyme production etc, are strictly spatially and temporally controlled to ensure correct development and function. Thus, one can envisage that developing a 3D computer model representation of an organism, and subsequently denoting regions within this model that exhibit each different biological process, would lead to an 'atlas' of biological functions for that organism. Subsequent spatial-based analyses of the spatial data itself could lead to an appreciation of those patterns that have similarities or differences in their spatial deployment and therefore regulation.

An example of this approach, used in practice to store and analyze gene expression patterns in a spatial format, is the EMAGE gene expression database. EMAGE contains data from *in situ* hybridization, immunohistochemistry and transgenic reporter assays that have been conducted on mouse embryo specimens at various stages of development.

## 25.4 EMAGE Database Structure

### *25.4.1 Concept*

The framework that houses the data contained in the EMAGE gene expression database is the EMAP Anatomy Atlas of Mouse Development (see Fig. 25.2). This interactive atlas consists of a set of reference 3D virtual models of the mouse embryos at different stages of embryonic development. The staging system utilized is that of Karl Theiler (1989), who defined 26 stages of mouse embryonic development, each characterized by a particular set of morphological features [7].

The reference embryo models are 3D digital images, made of voxels (volumetric pixels) stacked in 3D space. Voxel size in the models range from 2×2×2 m to 10×10×10m, depending on the method of model production. The reference embryo models can be viewed from any angle to assess external embryo morphology or can be virtually sliced in any plane to view internal embryo morphology. Within many of the reference model embryos, 3D anatomical regions have been delineated voxel by voxel.

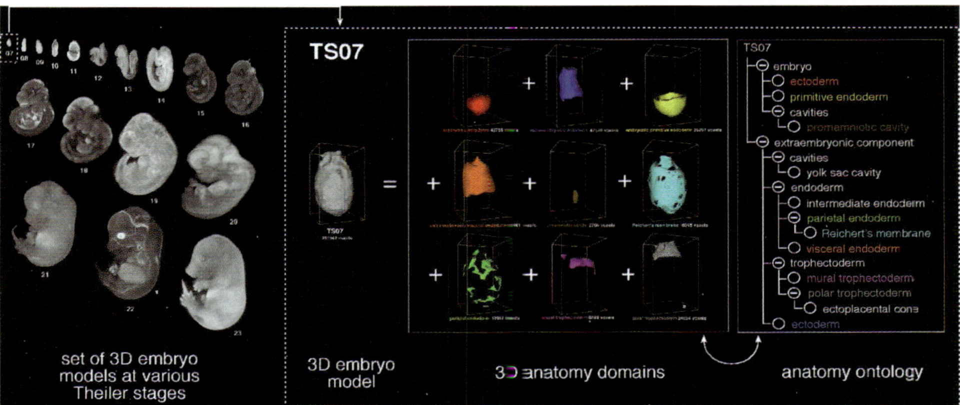


**Fig. 25.2** The EMAP Atlas of Mouse Development consists of a set of 3D digital images of mouse embryos at various stages of development and an accompanying ontology of anatomical terms at each stage. The models and anatomy ontology are linked by defining regions within the 3D models that correspond to various words in the ontology (Copies of figures including color copies, where applicable, are available in the accompanying CD)

The complementary part of the EMAP Atlas is an ontology of terms describing the anatomical parts visible at each Theiler Stage of development. This ontology contains approximately 17,500 anatomical terms across the 26 Theiler stages and includes common synonyms.

The 3D anatomical regions that have been defined in each model are linked to their corresponding names in the anatomy ontology. Thus, it is possible to either identify the name of a 3D anatomical structure found within a model, or to view the 3D structure within a model that corresponds to an anatomical name found within the ontology.

The EMAP anatomy atlas has subsequently been deployed to act as a scaffold for the EMAGE database, in which a combination of both standardized spatial representations of *in situ* gene expression data (used to show the sites of expression in the correct place) and standardized text-based descriptions (used to describe in words the sites of expression), are used to denote gene expression patterns.

### 25.4.2 *Software*

EMAGE has client-server architecture. The application-domain information is stored in a database. The database is currently a C + + object-oriented model but the intention is to change it to a relational model in the future. Client software includes a Java application delivered to a client computer using JavaWebStart, a web interface viewable through a normal HTML web browser, and web services delivered using the SOAP and WSDL standards.

## 25.5 EMAGE Data Sourcing and Entry

### 25.5.1 *Data Sourcing*

As a community resource, the data in EMAGE originates from a variety of sources. EMAGE provides an interface for data submissions designed to be used by individual laboratories or consortia. The interface allows creation of local, private databases for in-lab data management. This software also allows electronic submission to the EMAGE curators for entry to the public database. Alternatively, images and associated data can be provided in an ad hoc manner via post or electronic means to EMAGE staff for entry. EMAGE curators also obtain data from some collaborating projects by direct database access or through web site 'screen scraping'.

A proportion of data in EMAGE has been previously published in the literature, and incorporation of this data is achieved in collaboration with the text-based Gene Expression Database (GXD) at Mouse Genome Informatics (MGI), Maine, USA. GXD curators identify examples of in situ expression data in the literature and produce a standardized text description by annotating the EMAP text ontology to denote sites of gene expression in these assays, that is based on the author's non-standardized text description. GXD and EMAGE have secured global copyright agreements with the Company of Biologists Ltd and Elsevier BV to reproduce images that have been published in the journals *Development*, *Developmental Biology*, *Mechanisms of Development* and *Gene Expression Patterns*.

### *25.5.2 Data Entry*

In all the cases, the following information is obtained and entered in each database entry: [1] at least one image used to document the outcomes of the assay; [2] information on the reagent used to detect expression in the assay (i.e., nucleic acid probe, antibody or details of the construction of the transgenic reporter line); [3] information on the specimen (age, strain, sex etc.); [4] information about the experimental conditions used in the assay (e.g., fixation procedure of the specimen); [5] source of the data (e.g., screening consortia, submitter/lab details); and [6] sometimes a text-based description (either standardized or non-standardized) of the sites of gene expression.

#### 25.5.2.1 Text Annotation

In the case of non-standardized text annotations, EMAGE curation staff will transfer the given description across to equivalent terms in the standardized anatomy ontology.

### *25.5.3 Spatial Annotation*

EMAGE curators also assess all images for their suitability and for spatial annotation. When deemed to be suitable, specimens depicted in data images are assessed for their Theiler Stage of development, by visual inspection, for morphological features indicative of the stage. This information is then used to select an appropriately stage-matched target 'standard' embryo template from the EMAP Atlas for the annotation. If the data image depicts sectioned material, manual navigation through the 3D space of the target embryo model is used to find an appropriately morphologically matched arbitrary section plane. Determination of lateral views of WM specimens, as left- or right-hand side, is assessed by visual inspection and used to select the appropriate template model view. Subsequently, a bespoke in-house computer program, MAPaint, is used for data mapping. The process initially requires defining a set of anatomically equivalent points in both the data image and the template image. These are then used as anchor points in a further process where the data image is 'warped' onto the template embryo model view using a radial basis function transformation. Intensity of the signal in the data image is subsequently extracted by manual selection of a point on the image showing maximal signal intensity and then using an interactive thresholding procedure to extract the signal from the combined RGB color channels. Up to five regions of apparent signal intensity can be extracted. These include regions of 'strongest', 'moderate' and 'weakest' apparent signal intensity, regions where there is no apparent signal detected ('not detected'), and 'possible' where curators cannot confidently assess if the expression is either definitively present or absent. These domains are transferred to the template model using the warp parameters as a spatial guide, saved in Woolz digital image format and then housed in the EMAGE database (see Fig. 25.3). The color convention we have adopted to depict these regions of apparent signal intensity on the spatial templates are: red (strong); yellow (moderate); blue (weak); green (possible); black (detected) and cyan (not detected).

#### 25.5.3.1 Automatic Text Annotation

When 3D (e.g., section) data is spatially mapped into one of the 3D template models that has the 3D anatomical regions pre-defined, automatic text annotation is given for structures in the ontology

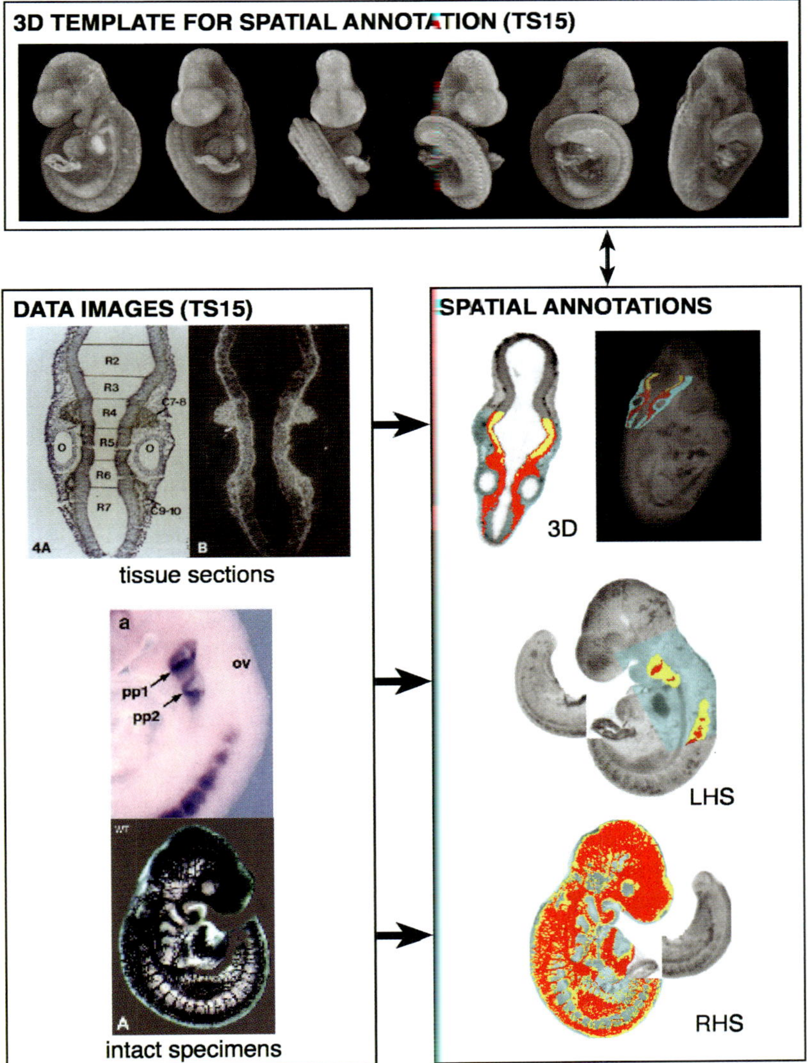


**Fig. 25.3** Spatial annotation in EMAGE is achieved by denoting gene expression in morphologically equivalent regions in the 3D embryo template that corresponds to the data image. This is performed for data that is from either a sectioned material or from left or right lateral views of intact specimens (Copies of figures including color copies, where applicable, are available in the accompanying CD)

where the gene expression domains intersect with each corresponding 3D anatomy region. This is articulated by denoting expression as 'present' in the structure/anatomical term, and also denoting the percentage of voxels of the underlying anatomy structure, in the regions examined, that show expression at each level of signal intensity. For example, think of a single section plane through the embryo that has been spatially annotated and contains 1000 voxels. If this plane intersects with 'structure X' (whose TOTAL 3D volume = 20,000 voxels) only over a region of 500 voxels, and out of these 30 have been annotated as having a 'strong' signal intensity, then the proportion of 'structure X' in this region, that has a signal intensity of 'strong', is 30/500 voxels or 6%.

## 25.6 EMAGE Data Querying

The modes for accessing EMAGE data are via a Java client interface, HTML pages and Web Services. All can be accessed via the EMAGE homepage.

### *25.6.1 Query via Java Client Interface*

Screencast movies showing each of the different searches outlined below being performed in the Java client interface are available on-line from http://genex.hgu.mrc.ac.uk/Emage/database/EMAGE_Docs/emagePreview.html.

#### 25.6.1.1 Performing Spatial Searches

By selecting the menu option, *Central Database>Search,* a user can formulate a query such as "What (genes) are (detected/possibly detected/not detected) in the following (region) at the following (Theiler stage) of development?" Two separate types of spatial data are housed in EMAGE – from wholemount embryos and from sectioned material, and for querying purposes these are treated independently during searches.

This process requires an initial selection of a left-or right-hand view (for wholemount data) or choosing an arbitrary section plane (for section data), within a template model by the user. The user subsequently defines a search region in one of the standard embryo models using a painting tool (search regions are denoted in magenta), and then clicks on 'Search'. The current underlying search algorithm, firstly demarcates a local region which will be used for the subsequent spatial comparison, and is defined as a region 30 voxels/pixels larger (in all directions) from the edges of the query domain. Subsequently, similarity values are calculated between the query domain ($d_1$) and all the domains (or parts thereof) ($d_2$) in the database that reside within the comparison region. The calculation is based on the Jaccard Index (V), which is defined as the ratio of the shared features of the two entities to the number of all features in both entities i.e., $V = (d_1 \cap d_2)/(d_1 \cup d_2)$. (see Fig. 25.4).

Spatial transformations from the right side of each template model and its left side have been defined, such that, the wholemount data from both sides of an embryo can be returned from a single spatial search. In addition, further spatial transformations *between* some template models and their next most morphologically similar/temporally adjacent template models have been added, such that, spatial searches *across* multiple stages of development can be performed. In these cases, the transformation parameters across the image have been pre-defined, and are based on prior definition of several anatomically equivalent points between each model. Under these circumstances, the pre-defined warp parameters are applied to the query region, and then used separately in Jaccard analyses for each relevant stage that was defined in the query.

In addition, the converse question can be asked: "What (regions) (express/possibly express/do not express) the following (genes) at the following (Theiler stage) of development?" The user defines

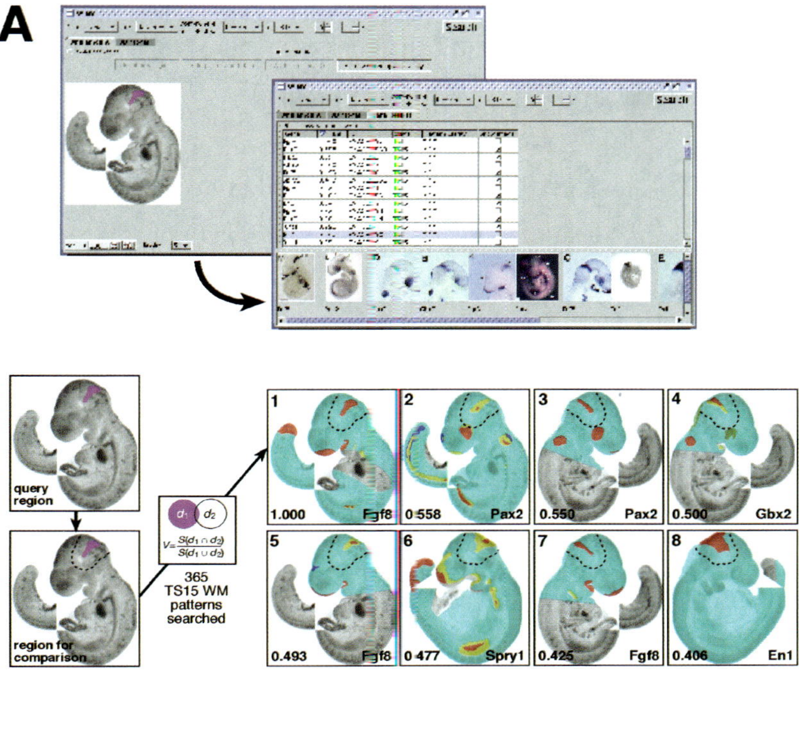


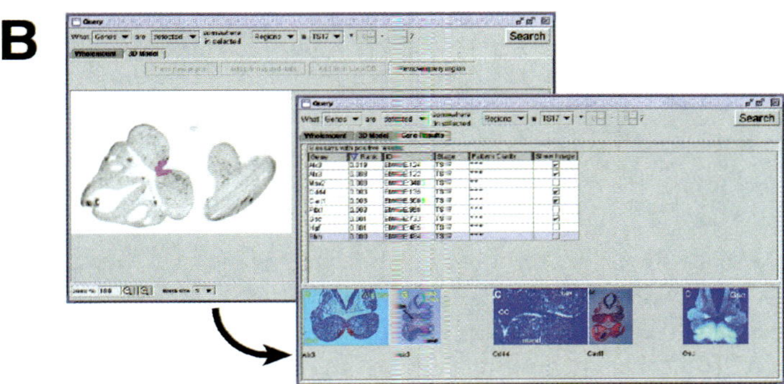


**Fig. 25.4** Spatial searches formulated using the EMAGE Java client interface. **(a)** Wholemount data searching. The interface, and an illustration of the underlying algorithm are shown with the 8 most similar query results. **(b)** 3D (section) data searching (Copies of figures including color copies, where applicable, are available in the accompanying CD)

their gene of interest and subsequently, a list of regions corresponding to the query result are returned. These are either 2D (wholemount data) or 3D (section data). The 2D results can be visualized directly in the results interface, and the 3D results (which cannot be visualized in the search interface) can be downloaded in Woolz image format for visualization in other suitable standalone software such as MAPaint (details on how to order a CD containing MAPaint) or JAtlasViewer. Any downloaded domains can also be used as an input domain for a subsequent spatial based search (as discussed previously) if desired.

In these searches, gene name/symbol synonym searches are supported. This is important as gene names and symbols may change over time. Official mouse gene names and symbols are decided by the Mouse Gene Nomenclature committee and the information outlining the current approved and withdrawn synonyms, as well as a non-changing 'gene ID' is stored in a database held at Mouse Genome Informatics. This data is utilized in EMAGE gene synonym searches. An example search containing the withdrawn gene symbol *Krox20*, would return data for this gene under its approved symbol *Egr2*.

#### 25.6.1.2 Performing Text-Based Searches in the EMAGE Java Client Interface

By selecting the menu option, *Central Database>Search,* a user can formulate a query such as "What (genes) are (detected/possibly detected/not detected) in the following (named anatomical components) at the following (Theiler stage) of development?" The user is presented with an hierarchically organized anatomy-ontology for that Theiler stage of development. The ontology tree can be browsed for a term by opening the branches, or searched by text for a specific structure. Common synonyms for anatomical structures (e.g., stomatodaeum = stomatodeum = stomodaeum = stomodeum = oral pit = mouth pit) are supported in these text searches, and are highlighted in the tree in yellow when found. Once a term is found, it can be selected (it will be highlighted in magenta) and used for query. Results are returned as a list of genes with accompanying original data images.

The converse question can also be asked: "What (named anatomical components) (express/possibly express/do not express) the following (genes) at the following (Theiler stage) of development?" The user defines their gene of interest and subsequently, a list of anatomical terms corresponding to the query result are returned. As discussed previously, gene name/symbol synonym searching is supported. Because of the hierarchical nature of the ontology, all terms including and higher than the search term (i.e., superstructures), are returned when querying for structures expressing a gene, and all terms including and lower than the search term (i.e., substructures), are returned when querying for structures that do not express the gene. For example, if a gene is *expressed* in the 'foot', by definition, it is also expressed in the 'limb'. Conversely, if a gene is *not expressed* anywhere in the 'limb', by definition, it is not expressed in the 'foot' either. All EMAGE text searches retrieve data that has been either manually or automatically text annotated.

### *25.6.2 Query via HTML*

EMAGE data is also accessible via a standard HTML web browser. Current search capabilities allow listing of all data entries in a tabular format, sorting of the various columns and performing simple text-based searches of the data values in the following columns: ID; Gene symbol (synonym searching is supported); Probe ID; Theiler Stage; Alternate Staging system value; Assay type (ISH/IHC/reporter); Specimen Type (wholemount/section); genotype (wild type/mutant). A thumbnail of the original data image is always shown (see Fig. 25.5).

Current re-factoring will move the Java Interface search capabilities, discussed previously, into a HTML-based web-browser interface in the near future. Under these circumstances, results will be returned in the same HTML list format as mentioned above. The capability to select one or more entries in these HTML lists and create a 'collection' will also be added (achieved by the use of cookies). The contents of the multiple 'collections' can then be compared for commonalities or differences, allowing Boolean-type searches. For example, if one wants to ask the question: "What genes are expressed in region X but not in the named anatomical structure Y?", this can achieved by performing the two searches "What genes are expressed in region X?" and "What gene are not expressed in the named anatomical structure Y?", independently and then finding the intersection of the two query result collections.

EMAGE data can also be accessed in HTML format from an Ensembl mouse Gene Report, via a distributed annotation system (DAS) server. In such cases, EMAGE is listed amongst the DAS sources for a mouse gene, and when selected, the results in EMAGE for that gene are displayed,

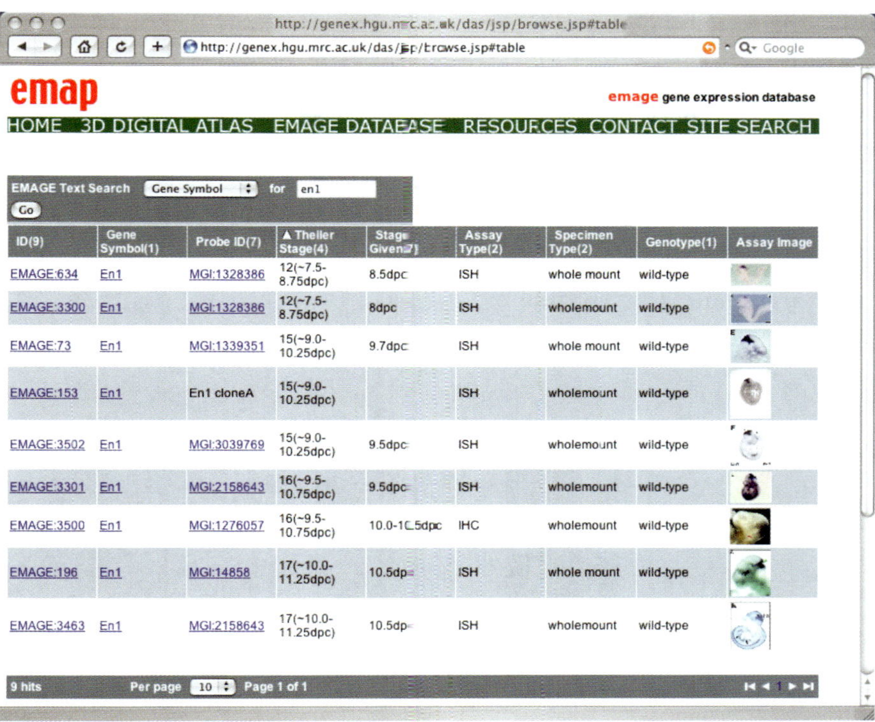


| ID(9) | Gene Symbol(1) | Probe ID(7) | ▲ Theiler Stage(4) | Stage Given(7) | Assay Type(2) | Specimen Type(2) | Genotype(1) | Assay Image |
|---|---|---|---|---|---|---|---|---|
| EMAGE:634 | En1 | MGI:1328386 | 12(~7.5-8.75dpc) | 8.5dpc | ISH | whole mount | wild-type | |
| EMAGE:3300 | En1 | MGI:1328386 | 12(~7.5-8.75dpc) | 8dpc | ISH | wholemount | wild-type | |
| EMAGE:73 | En1 | MGI:1339351 | 15(~9.0-10.25dpc) | 9.7dpc | ISH | whole mount | wild-type | |
| EMAGE:153 | En1 | En1 cloneA | 15(~9.0-10.25dpc) | | ISH | wholemount | wild-type | |
| EMAGE:3502 | En1 | MGI:3039769 | 15(~9.0-10.25dpc) | 9.5dpc | ISH | wholemount | wild-type | |
| EMAGE:3301 | En1 | MGI:2158643 | 16(~9.5-10.75dpc) | 9.5dpc | ISH | wholemount | wild-type | |
| EMAGE:3500 | En1 | MGI:1276057 | 16(~9.5-10.75dpc) | 10.0-10.5dpc | IHC | wholemount | wild-type | |
| EMAGE:196 | En1 | MGI:14858 | 17(~10.0-11.25dpc) | 10.5dpc | ISH | whole mount | wild-type | |
| EMAGE:3463 | En1 | MGI:2158643 | 17(~10.0-11.25dpc) | 10.5dpc | ISH | wholemount | wild-type | |

**Fig. 25.5** The HTML client interface for EMAGE data browsing. Simple text searches and sorting of the data by columns is supported (Copies of figures including color copies, where applicable, are available in the accompanying CD)

including the gene symbol and Theiler stages and the numbers of tissues in which expression annotations are present. Reciprocal links lead from EMAGE HTML entries to the relevant mouse gene entries in Ensembl. The number and type links from EMAGE HTML pages to other databases will be increasing over time.

### *25.6.3 Query via Web Services*

Web services are a way to access the EMAGE database (server), over the internet, using a software client. The interface to the web services is described by WSDL, a language that defines the services provided and the data structures involved. Programmers can use the WSDL description to design client software that makes requests to EMAGE and processes the results that are returned. The EMAGE database uses Apache Axis to deliver its web services.

### *25.6.4 Comparison Between Patterns in the Database for Spatial Similarity*

We have developed procedures to mine the data housed in EMAGE, to find examples of images of gene expression displaying overall spatial similarities. In these analyses, whole patterns housed in the database are compared for similarities with all the other whole patterns that have been spatially annotated to the same embryo model. This is achieved by calculating Jaccard similarity scores (as described previously) between all possible pairs of patterns in the database to give a matrix of numerical values that represents the spatial similarity of every pattern to all others in the data-set. Subsequent hierarchical clustering of these Jaccard values and visualization using an open source software that was originally developed for analysis of microarray data i.e., Cluster and TreeView, can be used to segregate patterns into groups that have overall similar values of spatial overlap with all the other patterns in the data set. Heat maps depicting the common regions of signal from constituent images on each branch can be generated to give a visual feedback to the user (see Fig. 25.6). In this manner, gene expression patterns that have overall global spatial similarities can be identified from large numbers of images.

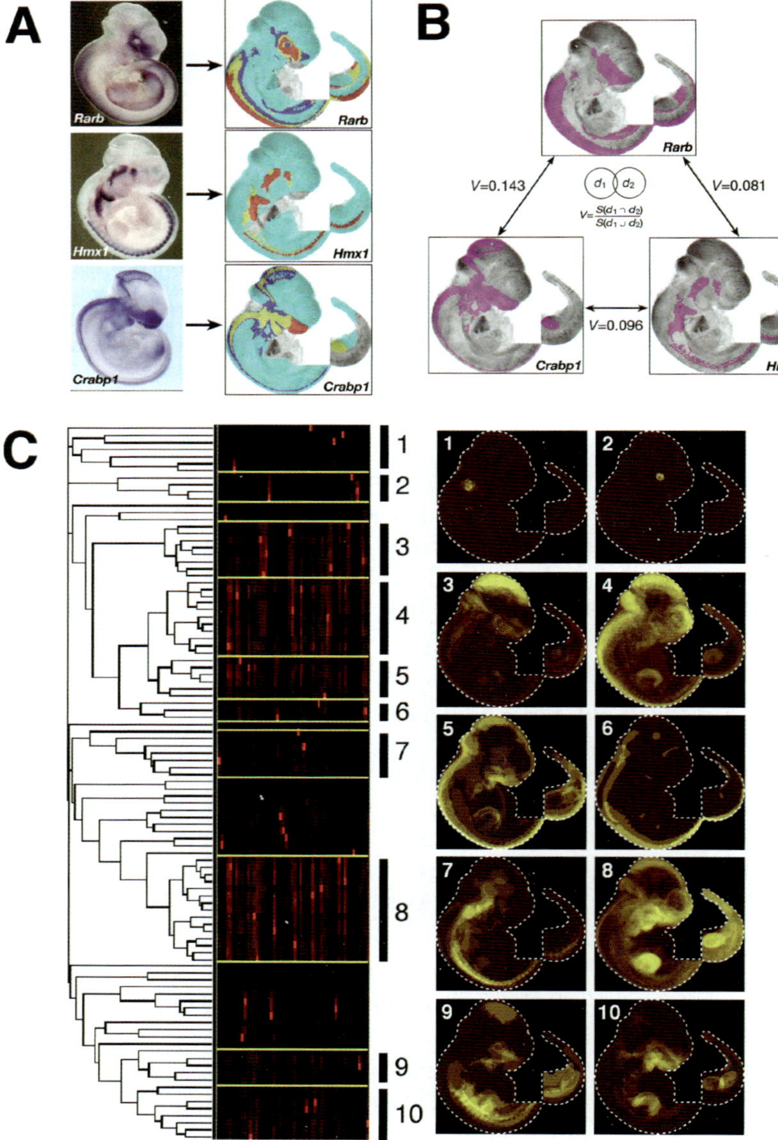


**Fig. 25.6** An example of data mining spatially-mapped biological data. **(A)** Three examples of spatially annotated TS18 gene expression data in EMAGE. **(B)** An illustration of the spatial comparison calculations between these three patterns using the Jaccard Index. **(C)** 101 TS18 spatially annotated patterns have been compared using the Jaccard Index in a pair-wise manner and the results hierarchically clustered. The method segregates the 101 gene expression patterns into smaller groups that display spatial similarities. 10 different branches of the tree with accompanying heat maps are shown. The heat maps show yellow regions in the embryo that correspond to regions of gene expression from the contributing images on each branch (Copies of figures including color copies, where applicable, are available in the accompanying CD)

Hierarchical clustering analyses of EMAGE data is currently performed offline and the pre-canned result files are made available through the EMAGE website, however these search capabilities will be moved to an interactive web-accessible format in the future.

### 25.6.5 *Future Developments in EMAGE*

EMAGE will continue to acquire and spatially annotate gene expression patterns in the developing mouse embryo. This includes data from the literature as well as data acquired from large scale screening projects. Planned developments apart from those discussed in the text, include the release of all data images prior to spatial annotation, and the inclusion of full 3D data images that are derived from stained whole mount embryos imaged using Optical Projection Tomography (OPT). A mailing list is used to announce major updates or a news release. Send emails requesting to join the list to MA-EMAGE-request@jiscmail.ac.uk

## 25.7 Perspective

Spatial-based data storage and query methods can allow for searching of large numbers of images of biological specimens for similarities or differences in staining patterns. This approach does not require an intermediate 'deconstruction' step where the image is described using a text-based language and importantly, allows for data access to users who are not, or who are only partially familiar with anatomy and its associated nomenclature. While the focus of this chapter has been on images depicting gene expression patterns *in situ*, the approaches discussed can be applied to any other *in situ* image-based data, for e.g., staining for cell death, cell division etc. As spatial annotation procedures are capable of being automated, the use of this approach in annotation of image-based biological data will increase over time. This will allow a powerful new approach for data analysis that is independent of the traditional text language-based anatomical descriptions.

## Glossary and Abbreviations

| | |
|---|---|
| DAS | Distributed annotation server |
| DNA | Deoxyribonucleic acid |
| HTML | Hypertext mark-up language |
| IHC | Immunohistochemistry |
| ISH | In situ hybridization |
| MeSH | Medical subject headings |
| MRNA | Messenger RNA |
| MS | Mass spectrometry |
| OPT | Optical projection tomography |
| RGB | Red-green-blue |
| RNA | Ribonucleic acid |
| SAGE | Serial analysis of gene expression |
| SOAP | Simple object access protocol |
| TS | Theiler stage |
| WSDL | Web service definition language |
| XML | Extensible mark-up language |

## References


1. Bard, JB. Anatomics: the intersection of anatomy and bioinformatics. J Anat 2005;206(1):1–16.
2. Davidson, D, Baldock, R. Bioinformatics beyond sequence: mapping gene function in the embryo. Nat Rev Genet 2001;2(6):409–417.
3. Baldock, RA, Bard, JB, Burger, A, et al. EMAP and EMAGE: a framework for understanding spatially organized data. Neuroinformatics 2003;1(4):309–325.
4. Sunkin, SM. Towards the integration of spatially and temporally resolved murine gene expression databases. Trends Genet 2006;22(4):211–217.
5. Wilkinson DG, Nieto, MA. Detection of messenger RNA by *in situ* hybridization to tissue sections and whole mounts. Methods Enzymol 1993;225:361–373.
6. Hsu, SM. Immunohistochemistry. Methods Enzymol 1990;184:357–363.
7. Theiler, K. The House Mouse: Atlas of Embryonic Development. New York: Springer-Verlag; 1989.


## Web Resources

http://www.eurexpress.org/
http://www.stjudebgem.org/web/mainPage/mainPage.php/
http://www.gensat.org/
http://www.gudmap.org/
http://www.genepaint.org/
http://www.brain-map.org/
http://www.proteinatlas.org/
http://www.nlm.nih.gov/mesh/
http://www.informatics.jax.org/
http://www.zfin.org/
http://www.xenbase.org/
http://www.wormbase.org/
http://www.flybase.bio.indiana.edu/
http://www.euregena.org/
http://www.ani.embl.de:8080/mepd/
http://www.geisha.arizona.edu/
http://www.informatics.jax.org/mgihome/GXD/aboutGXD.sthml/
http://www.genex.hgu.mrc.ac.uk/Emage/database/
http://www.genex.hgu.mrc.ac.uk/Emage/database/EMAGE-Docs/emagepreview.html/
http://www.genex.hgu.mrc.ac.uk/MouseAtlasCD/intro.html/
http://www.genex.hgu.mrc.ac.uk/Software/JavaTools/JAtlasViewer/intro.html/
http://www.biodas.org/
http://www.w3.org/TR/wsdl/
http://www.ws.apache.org/axis/
http://www.bonsai.ims-u-tokyo.ac.jp/~mdehoon/software/cluster/
http://www.jtreeview.sourceforge.net/
http://www.genex.hgu.mrc.ac.uk/Emage/cluster-analysis/
http://www.bioptonics.com/

# Chapter 26
# Equality of the Sexes? Parent-of-Origin Effects on Transcription and *de novo* Mutations*

**Rivka L. Glaser and Ian M. Morison**

**Abstract** Two main categories of the parent-of-origin effects are reviewed in this chapter: parent-of-origin effects on transcription, or genomic imprinting, and parent-of-origin effects on the development of *de novo* mutations. Each type of parent-of-origin effect is described, and the mechanisms that contribute to each discussed. The parent-of-origin effect database provides a catalog reports of genomic imprinting and related effects as well as reports of the parental origin of spontaneous mutations. This database provides a useful tool for finding genes, diseases, or traits that exhibit a parent-of-origin effect in humans and animals, conducting comparative analyses of the imprinted genes among different species, and examining the role of parent-of-origin effects for different types of spontaneous mutations in human genes.

**Keywords** Parental origin · Genomic imprinting · *de novo* mutations · Transcription

## 26.1 Introduction to the Range of Parent-of-Origin Effects

The term "parent-of-origin effects" describes two distinct phenomena: parent-of-origin effects on transcription and parent-of-origin effects on the development of spontaneous, or *de novo,* mutations. Parent-of-origin effects on transcription, also known as genomic imprinting, refer to unequal expression of parental alleles. For most genes, transcription occurs from both the alleles equally; however, for imprinted genes in mammals, one copy is exclusively or preferentially expressed. Expression from an allele is dependent upon the sex of the parent from which it was inherited. This phenomenon is caused by epigenetic modifications of the genome, not by changes to the *sequence* of the genome [1, 2].

Parent-of-origin effects on mutations refer to the preferential occurrence of some *de novo* mutations in either the paternal or maternal germline (Fig. 26.1). For example, base substitutions are more likely to be paternal in origin, whereas chromosomal abnormalities, such as trisomies, are more likely to be maternal in origin [3, 4]. The mechanisms that give rise to these types of spontaneous mutations differ depending upon the parental germline in which the mutation has occurred. Base substitutions caused by errors during DNA replication tend to be paternal in origin due to the greater number of cell divisions in spermatogenesis compared to oogenesis. Trisomies are mainly caused by maternal non-disjunction errors in meiosis. During oogenesis, oocytes are arrested in prophase of meiosis I until females reach sexual maturity, when one oocyte per month

* Portions of this chapter are reproduced with some modification from R. L. Glaser, E. W. Jabs, Dear Old Dad. Sci. Aging Knowl. Environ. 2004 (3), re1 (2004).

R.L. Glaser
Department of Biological Sciences, Stevenson University, 1525 Greenpring Valley Road, Stevenson, Maryland, 21153, USA
e-mail: f-glaser@mail.vjc.edu

S. Krawetz (ed.), *Bioinformatics for Systems Biology*, DOI 10.1007/978-1-59745-440-7_26,

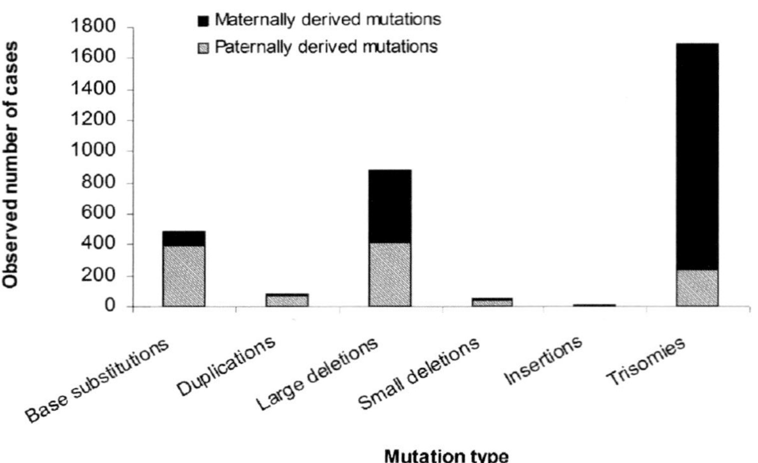


**Fig. 26.1** Parent-of-origin effects for different mutation types (Copies of figures including color copies, where applicable, are available in the accompanying CD)

is selected to resume the cell cycle. It is thought that the longer the oocytes are arrested in this phase of meiosis, the greater the chance for a non-disjunction event to occur [4]. Older parental age can influence the development of some types of mutations. When advanced paternal or maternal age is associated with the development of mutations, and therefore the development of genetic disorders, this phenomenon is referred to as the paternal or maternal age effect, respectively.

Both types of parent-of-origin effects play important roles in the inheritance of certain phenotypes in humans, mice, and other animals. Imprinted genes have been implicated in human disorders such as Prader-Willi, Angelman, and Beckwith-Weidemann syndromes [5–7], growth and maternal behavior in mice [8], and the callipyge "beautiful buttock" phenotype in sheep [9]. These phenotypes exhibit parent-of-origin effects since they only occur if there is loss of function of the expressing parental allele; deletion or mutation of the silent allele has no effect. Certain human genetic disorders such as achondroplasia [10] and Apert [11], Crouzon, and Pfeiffer syndromes [12] are caused by paternally-derived base substitutions and exhibit a paternal age effect, whereas Down syndrome and other trisomies are mainly caused by maternally-derived non-disjunction errors and exhibit a maternal age effect [13].

### *26.1.1 The Parent-of-Origin Effects Database (www.otago.ac.nz/IGC)*

The first catalogue of parent-of-origin effects was published in 1998 and it contained comprehensive lists of parent-of-origin effects on transcription in humans, mice, and other animals [14]. This published catalogue served as the basis for a more complete, searchable, online database, which was made publicly available in 1999. The original database included information on 41 imprinted genes and other parent-of-origin effects [15]. A comprehensive summary of the imprinting section was published in 2005 [16]. In 2006, a section on parent-of-origin effects on *de novo* mutations was added to the database [17]. Originally known as 'The imprinted gene and parent-of-origin effect database' or 'Imprinted Gene Catalogue', the database has been renamed to 'The catalogue of parent-of-origin effects'. The database provides unrestricted online access to this collation of parent-of-origin effects. Each section provides customizable searches and lists or tables of the parent-of-origin effects of interest.

#### 26.1.1.1 Introduction to the Imprinted Gene and Related Effects Catalogue

The genomic imprinting section of the database includes a wide variety of observations from animals. The user can search for entries according to taxon, chromosome, gene name, a text word from the description, and the category of entry (see below). The search results are displayed as a list of separate entries each of which, includes the taxon, chromosome, location, the gene name (or phenotype), a description of the parent-of-origin effect including PubMed links, the category of the entry and external links (most commonly to NCBI Gene or OMIM). The type and quality of evidence for the presence of genomic imprinting varies enormously, from well documented gene expression data showing parent-of-origin specific expression to poorly substantiated hypotheses regarding the possibility of imprinting for a specific phenotype.

The entries are divided into four main categories. The first category of entries describes genes for which genomic imprinting has been reported (182 entries). These entries include observations from human, mouse, sheep, cow, pig, rat, rabbit, and marsupial. To be included within the imprinted gene category, there has to be reasonable evidence of parental-specific imbalanced allelic expression from a specific gene transcript, but for some of the entries provisional or conflicting data are noted.

The second category includes 'parental effects not covered in other categories and includes 214 entries. Most of these are descriptions of human and animal diseases or traits for which a parent-of-origin effect has been reported.

The third category of entries describes the uni-parental disomies (UPD) and includes 32 entries from human and mouse. The usual state of any pair of chromosomes in a mammal is bi-parental heterodisomy, i.e., two different chromosomes, one from each parent. Rarely in humans, and experimentally in mice, an offspring inherits both copies of a chromosome from one parent (i.e., UPD). The occurrence of a consequent phenotype suggests that the two parental chromosomes provide an unequal genetic contribution to the offspring, often indicating the presence of imprinted genes within the region.

The fourth category, designated as 'others' (118 entries), includes a wide variety of records such as disputed observations, hypotheses, and the location of orthologs in taxons for which a gene is not imprinted or where the imprinting status is unknown.

#### 26.1.1.2 Introduction to the Parental Origin of *De Novo* Mutations Catalogue

This part of the database allows the user to search the entries in the catalogue according to the mutation type, disorder, chromosomal location, gene name, and inheritance pattern. Each entry in the database is hyperlinked to the relevant references in PubMed and Online Mendelian Inheritance in Man (OMIM). For example, if the user wanted to search the database for all reported cases of a parental origin for any base substitution (missense, nonsense, splice site mutations), the user would select "P" from the drop down menu for mutation type on the main search page (Fig. 26.2). The outcome of the search is presented in a table format, as shown in Fig. 26.3. Each entry is specific for one type of base substitution (missense, nonsense, or splice site). As some more current articles include a review of cases in the literature which may already be catalogued in the database, references are cross checked with information already in the database, in order to avoid cataloguing a mutation more than once. Other information included in the returned search is the disorder in which the mutations were found, inheritance pattern and incidence of the disorders, gene name, chromosomal location, evidence of a paternal or maternal age effect, recurrent mutations that have been found, number of paternal and maternal mutations, and the PubMed reference. In the case of base substitutions, data are further separated according to whether the mutation is a transition or transversion mutation, and whether the base substitution occurs within a CpG dinucleotide.

As shown in Fig. 26.3, a total of 11 paternally-derived base substitutions and no maternally-derived base substitutions have been reported for Crouzon syndrome. The three separate entries for this disorder show the user that four of the 11 mutations are transition missense mutations, five are transversion missense mutations, and two are transition splice-site mutations. The two transition splice-site mutations are recurrent mutations (A344A (G>A)). In achondroplasia, 40 paternally-derived missense mutations have been reported. No maternally-derived missense mutations have

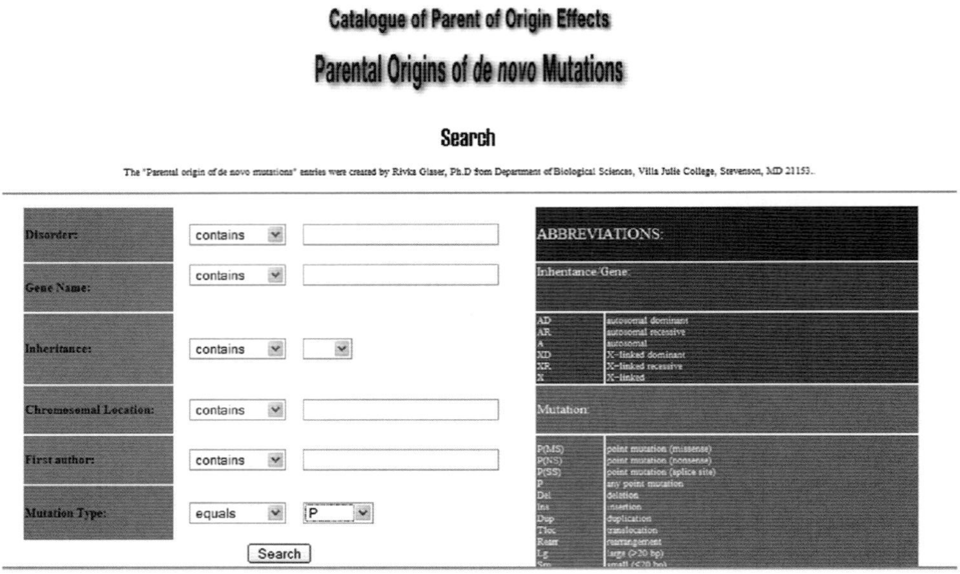

**Catalogue of Parent of Origin Effects**

**Parental Origins of de novo Mutations**

**Search**

The "Parental origin of de novo mutations" entries were created by Rivka Glaser, Ph.D from Department of Biological Sciences, Villa Julie College, Stevenson, MD 21153.

| | | |
|---|---|---|
| Disorder: | contains | |
| Gene Name: | contains | |
| Inheritance: | contains | |
| Chromosomal Location: | contains | |
| First author: | contains | |
| Mutation Type: | equals | P |

Search

ABBREVIATIONS:

Inheritance/Gene

| | |
|---|---|
| AD | autosomal dominant |
| AR | autosomal recessive |
| A | autosomal |
| XD | X-linked dominant |
| XR | X-linked recessive |
| X | X-linked |

Mutation

| | |
|---|---|
| P(MS) | point mutation (missense) |
| P(NS) | point mutation (nonsense) |
| P(SS) | point mutation (splice site) |
| P | any point mutation |
| Del | deletion |
| Ins | insertion |
| Dup | duplication |
| Tloc | translocation |
| Rearr | rearrangement |
| Lg | large (>20 bp) |
| Sm | small (<20 bp) |

**Fig. 26.2** The main search menu for the parental origin of *de novo* mutations catalogue. Any point mutation (P) was selected as the search criterion from the drop down menu for mutation type (Copies of figures including color copies, where applicable, are available in the accompanying CD)

**Catalogue of Parent of Origin Effects**

**Parental Origins of de novo Mutations**

ABBREVIATIONS: Inheritance/Gene — AD autosomal dominant; AR autosomal recessive; A autosomal; CGS Contiguous Gene Syndrome; XD X-linked dominant; XR X-linked recessive; X X-linked

ABBREVIATIONS: Mutation — P(MS) point mutation (missense); P(NS) point mutation (nonsense); P(SS) point mutation (splice site); P any point mutation; Del deletion; Ins insertion; Dup duplication; Tloc translocation; Rearr rearrangement; Lg large (>20 bp); Sm small (<20 bp); TS transition; TV transversion; CpG CpG mutation

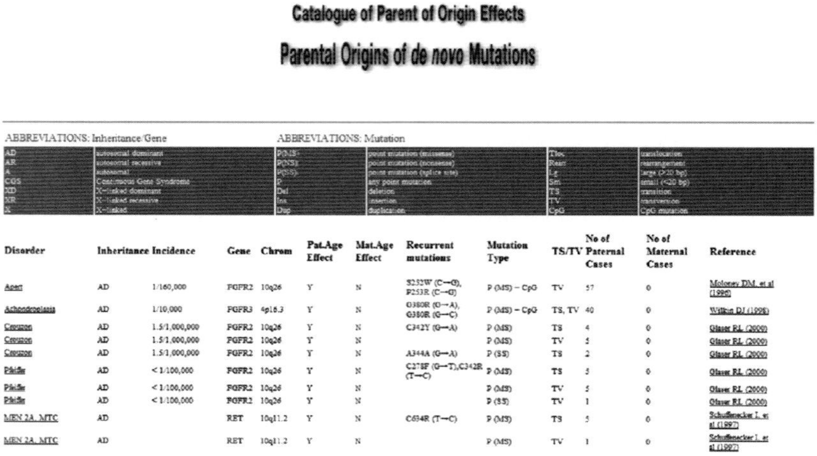

| Disorder | Inheritance | Incidence | Gene | Chrom | Pat.Age Effect | Mat.Age Effect | Recurrent mutations | Mutation Type | TS/TV | No of Paternal Cases | No of Maternal Cases | Reference |
|---|---|---|---|---|---|---|---|---|---|---|---|---|
| Apert | AD | 1/160,000 | FGFR2 | 10q26 | Y | N | S252W (C→G), P253R (C→G) | P (MS) – CpG | TV | 57 | 0 | Moloney DM. et al (1996) |
| Achondroplasia | AD | 1/10,000 | FGFR3 | 4p16.3 | Y | N | G380R (G→A), G380R (G→C) | P (MS) – CpG | TS, TV | 40 | 0 | Wilkin DJ (1998) |
| Crouzon | AD | 1.5/1,000,000 | FGFR2 | 10q26 | Y | N | C342Y (G→A) | P (MS) | TS | 4 | 0 | Glaser RL (2000) |
| Crouzon | AD | 1.5/1,000,000 | FGFR2 | 10q26 | Y | N | | P (MS) | TV | 5 | 0 | Glaser RL (2000) |
| Crouzon | AD | 1.5/1,000,000 | FGFR2 | 10q26 | Y | N | A344A (G→A) | P (SS) | TS | 2 | 0 | Glaser RL (2000) |
| Pfeiffer | AD | < 1/100,000 | FGFR2 | 10q26 | Y | N | C278F (G→T),C342R (T→C) | P (MS) | TS | 5 | 0 | Glaser RL (2000) |
| Pfeiffer | AD | < 1/100,000 | FGFR2 | 10q26 | Y | N | | P (MS) | TV | 5 | 0 | Glaser RL (2000) |
| Pfeiffer | AD | < 1/100,000 | FGFR2 | 10q26 | Y | N | | P (SS) | TV | 1 | 0 | Glaser RL (2000) |
| MEN 2A, MTC | AD | | RET | 10q11.2 | Y | N | C634R (T→C) | P (MS) | TS | 5 | 0 | Schuffenecker I. et al (1997) |
| MEN 2A, MTC | AD | | RET | 10q11.2 | Y | N | | P (MS) | TV | 1 | 0 | Schuffenecker I. et al (1997) |

**Fig. 26.3** A portion of the returned search for all points mutations (Copies of figures including color copies, where applicable, are available in the accompanying CD)

been reported for this disorder. The 40 reported mutations are comprised of both transition and transversion mutations. Achondroplasia is an excellent example of a disorder caused by recurrent mutations. As is shown in this entry, the 40 cases are caused by two recurrent missense mutations (G380R (G>A) and G380R (G>C)) in the *FGFR3* gene.

For deletions and insertions, the distinction is made between large deletions and insertions (>20 base pairs) and small deletions and insertions (< 20 base pairs). This size distinction is based upon the possibility of different mechanisms contributing to these different types of mutations, and therefore potentially different parental origins [3]. In general, large deletions do not appear to have a parent-of-origin effect, whereas small deletions are more often paternal in origin.

Currently, approximately 1800 mutations with a parent-of-origin effect found in 65 different disorders are catalogued in this database. Large deletions comprise the largest category in this database, with approximately 900 mutations catalogued. Base substitutions form the second largest category in the database, with approximately 500 mutations. Trisomies, other aneuploidies, and polyploidies are not currently catalogued in the database.

### *26.1.2 Other Imprinting and Parent-of-Origin Websites*

Two other databases provide specific information about the imprinted genes. The Mammalian Genetics Unit at the Medical Research Council in Harwell, United Kingdom provides an excellent resource for mouse imprinting. It contains maps, data and references relating to the location and function of imprinted genes in the mouse, along with descriptions of chromosome regions with imprinted phenotypes. Geneimprint, from the laboratory of Dr. R.L. Jirtle at Duke University in North Carolina, USA, provides well referenced commentary on the evolution, mechanisms and consequences of genomic imprinting.

In 2003, a list of 2100 candidate imprinted genes was published [18]. The data are available from CITE: Candidate Imprinted Transcript from gene Expression. However, our analysis of the genes within the list, focusing on those in mouse chromosome 7, revealed no enrichment of known imprinted genes [16]. The number of imprinted genes identified, by screening for differences in expression between parthenogenetic and androgenetic mouse embryos, was no more than would be expected by chance. Thus, it seems likely that the majority of candidate genes on this site do not reflect previously undetected imprinted genes.

## 26.2 Imprinting Effects

Imprinted genes are those genes for which the relative expression level of a parental allele is dependent upon the sex of the parent from which that allele was inherited. The 'strength' of the imprinting effect varies. For some imprinted genes, one parental allele is completely silenced in all tissues. For example, the *SNRPN* transcript is maternally silenced and paternally expressed in all expressing tissues. Other genes show weaker imprinting effects that are limited to specific tissues. For example, *Cd81* shows relative repression of the paternal allele, and only in extra-embryonic tissues.

The parental effect that establishes an imprinted locus is a 'mark' in an imprint control region (ICR) [19], or an imprint control element (ICE) [20]. Imprint control regions consists of differentially methylated domains (DMD) that are established during gametogenesis. The methylation marks that constitute the imprint control regions, usually exert primary regulatory control over one transcript, and subsequently through a variety of mechanisms bring about 'secondary imprinting' of a variable number of adjacent genes [19]. For example, at the mouse *Igf2r* locus, the ICR controls the imprinted expression of the *Air* transcript, whereas *Igf2r*, *Slc22a2* and *Slc222 a3* are secondarily imprinted through mechanisms that remain incompletely understood [21]. When considering the properties of imprinted genes and their adjacent sequences, it is important to distinguish those genes which are primarily imprinted from those which are secondarily imprinted. An example of a small but useful analysis is provided by Paoloni-Giacobino et al [22], who usefully compared the DMDs from three ICRs to determine features that are shared in common.

### *26.2.1 Summary of Known Imprinted Genes*

At the time of this writing, 101 transcriptional units (TU) in human, mouse, rat, cow, sheep, pig, rabbit, and marsupial were considered to be imprinted. Because of the presence of multiple-imprinted elements within single transcripts, we have used the TU concept when collating imprinted 'genes'.

A TU has been defined as a group of transcripts that contain a common core of genetic information having the same orientation, which does not necessarily correspond to protein-coding regions [23]. In addition to the 101 imprinted TUs, a further 12 TUs have provisional or conflicting data regarding their imprinting status. Apart from one example, *BEGAIN*, for which imprinting has only been reported in sheep, all the reported imprinted transcripts are represented by human or mouse orthologs. In mouse, 84 imprinted TUs have been reported whereas 50 have been reported for human, 11 for sheep, 8 for cow, and 5 each for pig, rat, and marsupial, and 1 for rabbit (Fig. 26.4).

In rodent extra-embryonic tissues such as the placenta [24] and in all female marsupial tissues [25], the paternal X chromosome is preferentially inactivated. Since several hundred X chromosome genes are therefore expressed in a parent-of-origin specific manner, it is reasonable to consider that these genes are imprinted. Although other secondarily imprinted genes, such as the genes surrounding *Kcnq1ot1*, are included in the lists of imprinted autosomal genes, traditionally *Xist* and *Tsix*, that control X-inactivation, are the only X chromosome genes that are listed to represent imprinted X-inactivation.

### 26.2.2 Imprinted Genes are Usually Clustered

The vast majority of mammalian imprinted genes aggregate together in clusters; 81 of the 101 TUs, being found within 16 clusters (Table 26.1). Indeed, about half [52] of the imprinted TUs are located within 5 clusters, these being the *Kcnq1ot1/Cdkn1c* cluster at mouse chromosome 7 F5 (human 11p15), the *Snrpn* cluster at mouse 7 B5 (human 15q11), the *Dlk1* cluster at 12 F1 (human 14q32), the *Peg3* cluster at 7 A2 (human 19q13) and the *Peg10* cluster at 6 A1 (human 7q21).

Eighteen imprinted genes appear to exist as singletons. Interestingly, seven (39%) of these genes are reported to have discordant imprinting status between mouse and human (AMPD3, DHCR7, DCN, GATM, IMPACT, HTR2A, L3mbtl), compared to seven (9%) discordances among the 82 clustered genes. For some of these singleton genes, independent confirmation of imprinting and more data about adjacent and overlapping genes would be desirable before concluding that these genes are indeed imprinted and, if so, not clustered.

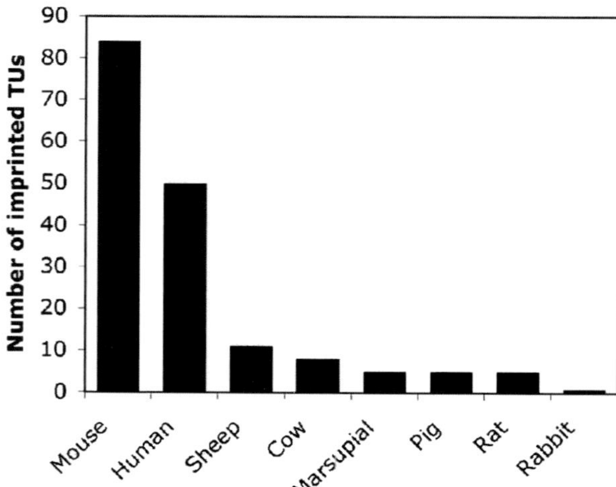


**Fig. 26.4** Frequency of reported imprinted transcriptional units by taxon (Copies of figures including color copies, where applicable, are available in the accompanying CD)

**Table 26.1** The majority of imprinted transcriptional units are located within 16 clusters (parentheses indicate clusters that are not imprinted)

| Mouse location | Human location | Number of TUs | Representative gene |
|---|---|---|---|
| 7 F5 | 11p15 | 15 | *CDKN1C* |
| 7 B5 | 15q11 | 13 | *SNRPN* |
| 12 F1 | 14q32 | 10 | *DLK1* |
| 7 A2 | 19q13 | 7 | *PEG3* |
| 6 A1 | 7q21 | 6 | *PEG10* |
| 6 A3 | 7q32 | 6 | *MEST* |
| 17 A1 | (6q25) | 4 | *Igf2r* |
| 7 F5 | 11p15 | 4 | *IGF2* |
| X A7 | (X) | 3 | *Xlr3b* |
| 11 A3 | (2p15) | 2 | *Commd1* |
| 10 A1 | 6q24 | 2 | *PLAGL1* |
| (2 E) | 11p13 | 2 | *WT1* |
| 9 E3 | (15q24) | 2 | *Rasgrf1* |
| 2 G3 | 20q11 | 2 | *MCTS2* |
| 2 E1 | 20q13 | 2 | *GNAS* |
| X D | (X) | 2 | *Xist* |
| | Total | 81 | |

(Copies of tables are available in the accompanying CD.)

### *26.2.3 Mechanisms of Imprinting*

The mechanisms by which the parental imprints are mediated into patterns of allelic expression differ. When a gene promoter includes a differentially methylated CpG island, the transcript is silenced on the methylated allele. Methylated DNA then recruits methyl-DNA binding domain proteins, which subsequently recruit histone modifying enzymes that result in repressive histone structures, compact chromatin, and gene silencing.

For many imprinted regions, the primarily imprinted transcript subsequently induces imprinting of adjacent genes. Examples include the non-coding transcripts *Air* and *Kcnq1ot1*. In mouse, the *Air* transcript itself, or the act of transcribing *Air* silences the expression of *Igf2r*, *Scl22a2* and *Slc22a3* in *cis*, i.e., on the paternal allele. In contrast, methylation-induced silencing of *Air* is associated with the expression of these three protein-encoding genes in *cis*. The mechanism by which transcription of *Air* silences the three adjacent genes is unclear, but sense/antisense transcriptional overlap can be excluded as the mechanism since *Air* does not overlap *Slc22a2* and *Slc22a3*. It has been proposed that the action of transcription interacts with a long-range *cis*-acting domain regulator [21,26]. Similarly, in mouse, transcription of the paternally-inherited allele of *Kcnq1ot1*, a non-coding RNA, results in the suppression of multiple protein-coding RNA transcripts. Suppression is associated with the acquisition of repressive histone modifications throughout the imprinted domain, but is also influenced by DNA methylation [27,28].

Antisense interference may be the mechanism by which *UBE3A* is imprinted. The long *SNURF/SNRPN* transcript extends for more than 460 kb to become *UBE3A-antisense*. Although it has been assumed that overlap from the *SNURF/SNRPN* transcript interferes with *UBE3A* transcription, this mechanism has not been proven and other plausible models exist [29].

Some ICRs mediate their imprinting effects via the binding of the insulator protein CTCF. Indeed, the differentially methylated ICR, upstream of *H19* contains multiple CTCF binding sites. Methylation on the paternal allele is not only associated with suppression of *H19* expression, but it also inhibits the binding of the insulator protein CTCF to its target site as well. Consequently, the absence of bound CTCF allows distant enhancers to access the paternal *IGF2* promoters. Conversely, the absence of methylation on the maternal allele allows CTCF to bind and exert its insulator activity, placing *IGF2* in an inaccessible chromatin loop [30].

### *26.2.4 Imprinted Non-coding RNAs*

Twenty-seven of the 101 imprinted TUs are non-coding RNAs. These non-coding RNAs include antisense transcripts, small nucleolar RNAs (snoRNAs), microRNAs, pseudo-genes, and other RNAs of unknown function. Nine imprinted non-coding genes appear to have a role in regulating the imprinting of adjacent protein-coding genes in *cis*. Five of these genes express overlapping antisense transcripts [*Air, KCNQ1OT1, UBE3A-AS, SANG, Tsix*], whereas in three, the regulatory effect occurs in *cis* from a distance [*H19, MEG3, Xist*] [31]. The maternally-expressed *anti-Rtl1* RNA plays a role in post-transcriptional regulation of paternally expressed *Rtl1* RNA [*PEG11*], via miRNAs that are processed from the *anti-Rtl1* transcript. These miRNAs regulate *Rtl1* in *trans* by guiding the RNA-induced silencing complex (RISC)-mediated cleavage of its mRNA [32]. In addition, there are six non-coding antisense RNA transcripts that are paired with the imprinted sense protein-coding transcripts, for which an imprint control function has yet to be determined [*MESTIT1, COPG2IT1, IGF2AS, WT1AS, Zfp127as* [antisense to *Mkrn3*], *AK014392* [antisense to *Ndn*]).

Small nucleolar (snoRNA) and microRNAs (miRNA) are commonly found within the imprinted transcripts. The extended human *SNURF/SNRPN* transcript is host to seven snoRNAs (four in mouse), the loss of some of which are believed to play a role in the pathogenesis of Prader-Willi syndrome [33]. Similarly, the mouse *Rian* transcript hosts seven snoRNAs that are expressed in the brain, but whose function is not known. Fourteen imprinted miRNAs have been reported within the three transcripts, *H19*, anti-*Rtl1*, and *Mirg*. As noted above, the miRNAs derived from anti-*Rtl1* guide RISC-mediated degradation of *Rtl1* [32], and the miRNAs within the *Mirg* transcript are thought to play a undefined role in mouse development [34]. The role of miR-625, derived from the highly abundant *H19* transcript, also remains to be established [35].

Twelve other non-coding RNAs have been reported. These include: *MEG3/Gtl2*, whose putative structure and function has been compared to *H19*; *Pec2*, *Pec3* and *PWRN1* from the mouse and human Prader-Willi locus; *ITUP1* which is upstream of, and oppositely oriented to, *PEG3*; *Zim3* and *Zfp264* which are both pseudogenes; *HYMAI* a transcript that overlaps *PLAGL1* in the same orientation; *Dlk* downstream transcripts; 4930524O08Ri*k* [*A19*]; *Peg13*; and *INPP5F_V2*.

### *26.2.5 Comparative Imprinting Data and Analysis*

The database provides a useful starting point for comparative analysis of imprinted genes. Such an analysis can be used to document the evolutionary history of an imprinted gene, or to explore the mechanism of the imprint. For example, a search for *IGF2* reveals the presence of imprinting in human, mouse, rat, cow, sheep, pig and marsupials, but not in monotremes, birds or fish, supporting the onset of *IGF2* imprinting in therian mammals [36]. Similarly, a search for *IGF2R* reveals the timing of evolutionary gain and loss of imprinting. It is imprinted in mouse, rat, cow, pig, sheep, and marsupials but not in monotremes, nor in euarchonta (tree shrew, flying lemur and human) [37]. In another example, the comparison of the imprinting status of the murine and human copper metabolism gene, *Commd1*, indicates that imprinting was acquired sometime after divergence of human and mouse. *Commd1* became imprinted in mouse after the insertion, by retrotransposition, of *Zrsr1* [*U2af1-rs1*] into its first intron [38].

### *26.2.6 Conflicting Data*

The database attempts to be inclusive rather than selective. The types of evidences that are used to judge whether a gene is imprinted vary enormously. For some transcripts, there are multiple independent reports showing exclusive expression of one parentally-derived allele in virtually all tissues at all developmental stages, whereas for other transcripts there may be a single report showing biased expression in a single tissue. Conflicting observations are noted when they are recognized

There are now many imprinted genes for which early reports of imprinting were probably incorrect. These include: human *MAS1*, mouse *Mas1*, human *COPG2*, human *HTR2A*, mouse *Ins1*, human *TRPM5*, human *SDHD*, mouse *Pira* and *Pirb*, and mouse *Mkrn1-ps1*. For each of these genes, the database discusses the basis for the early reports and subsequent evidence for lack of imprinting. For several other genes, there is an ongoing debate about the imprinting status. These include: human *IGF2R*, human GABA-A receptor subunit genes, mouse *Atp10a*, and human *WT1*.

Human *IGF2R* has been reported to be polymorphically imprinted by some groups [39,40], whereas other groups have found no evidence for its imprinting [41]. However human *IGF2R* maintains a differentially methylated region, and it has been speculated that the strong CpG island in intron 2 might allow the possibility of expression of a human *IGF2R* antisense [*AIR*] transcript (and thus *IGF2R* imprinting) in some situations [42].

The GABA-A receptor subunit genes were reported to be expressed in mouse A9 hybrids containing a single human paternal chromosome 15, but not in cells containing a maternal chromosome 15, leading to the conclusion that human *GABRB3*, *GABRA5*, *GABRG3* are imprinted and are exclusively paternally expressed [43]. Comparison of expression levels between lymphoblastoid cell lines derived from the Prader-Willi syndrome cases, with paternal deletions or maternal uni-parental disomy, led to the conclusion that *GABRA5* and *GABRB3* showed higher expression levels from the paternal allele [44]. However, expression data from the mouse brain provides no evidence for imprinting of *Gabrb3*, *Gabra5*, or *Gabrg3* [45], but in the heterozygous *Gabrb3* knockout mice, some phenotypes were associated with the parent-of-origin of the deletion, suggesting partial genomic imprinting [46].

Mouse *Atp10a* was reported to be predominantly expressed from the maternal allele in the brain, but others have reported that *Atp10a* is bi-allelically expressed in multiple tissues including the brain. It was suggested that imprinting may be strain-background dependent [47].

Polymorphic mono-allelic expression of the major transcript of human *WT1* has been reported in placental villous samples, preterm placentas, and in fetal brains (maternal expression was demonstrated in three of the placentas) [48,49]. In contrast, exclusive paternal expression of *WT1* was also shown in cultured fibroblasts and non-cultured lymphocytes from two of seven individuals [50]. Fetal kidney *WT1* showed no imprinting. However, the reported imprinting of an alternative *WT1* transcript and of an antisense transcript, suggest that some transcripts within the locus are imprinted [51].

For many other genes the evidence for imprinting should be regarded with caution, usually because of a combination of factors such as: a small number of cases studied; evidence derived solely from mono-chromosomal hybrids; the gene not being in a known imprinted cluster; or an ortholog not being imprinted. Examples of these genes include human *ZNF215*, *CTNNA3*, *TCEB3C*, *PON1*, *SLC22A1LS*, *ZNF331* and mouse *Pon2*, *Pon3*, *AK014392* [*Ndn-as*] and *F7*. Many non-imprinted gene transcripts show allelic imbalance [52] and the chance, for example, of 5 samples all showing repression of either the maternally or paternally-inherited allele is disturbingly high (12.5%).

## 26.3 Uni-parental Disomy

### *26.3.1 Human Uni-parental Disomy*

Recognition that human genes were imprinted came from early observations of uni-parental disomy, the situation in which both copies of a chromosome are derived from a single parent [53]. For example, the observation that Prader-Willi syndrome occurs when a child inherites both

**Table 26.2** Human developmental imprinting syndromes associated with uni-parental disomy

| Phenotype | Location | Gene |
|---|---|---|
| Beckwith-Wiedemann syndrome | Paternal UPD 11p15 | Double dose *IGF2*, loss of *CDKN1C* |
| Prader-Willi syndrome | Maternal UPD 15q11-q12 | Loss of *SNURF-SNRPN* transcript |
| Angelman syndrome | Paternal UPD 15q11-q12 | Loss of *UBE3A* expression |
| Silver-Russell syndrome | Maternal UPD 7p11-p13, 7q | Double dose of maternal growth repressor |
| Transient neonatal diabetes mellitus | Paternal UPD 6q24 | Double dose of *PLAGL1* |
| Maternal UPD 14 syndrome | 14 | Unknown |
| Paternal UPD 14 syndrome | 14 | Unknown |
| Maternal UPD 16 syndrome (possible) | 16 | Unknown |
| Maternal UPD 20 syndrome (possible) | 20 | Unknown |

(Copies of tables are available in the accompanying CD.)

copies of chromosome 15 from the mother, indicated that the respective parental copies of some chromosomes are not equivalent [54]. Uni-parental disomies that cause disease through the loss of gene expression or a doubling of gene expression are now recognized to cause at least seven developmental syndromes or diseases (Table 26.2).

### *26.3.2 Mouse Uni-parental Disomy*

In mice, Robertsonian and reciprocal translocations have been used to generate uni-parental disomies and uni-parental duplications of whole or selected chromosomal regions, respectively [55]. Abnormally imprinted phenotypes, including growth abnormalities, lethality, and behavioral changes, have been detected for 15 regions of maternal or paternal duplication [56]. For each of these regions, except the central chromosome 12, one or more imprinted genes have been mapped to the region of interest.

Some of the disomic mouse phenotypes correspond to those observed in human. For example, maternal disomy of central chromosome 11, a region that includes *Grb10*, causes fetal growth retardation in mice, comparable to the intra-uterine growth restriction observed in children who have maternal disomy of 7p, that contains human *GRB10*. However, in general, the lethality associated with most of the disomic mouse models precludes comparison with human phenotypes.

## 26.4 Parental Effects in Genetic Linkage

### *26.4.1 Inheritance Patterns of Human Familial Conditions*

A few of the imprinted genes have been discovered through study of loci that show parental effects on the transmission of inherited diseases or traits. For example, the rare familial cases of Beckwith-Wiedemann syndrome indicate that transmission almost always occurs maternally [57]. The phenotype of Albright hereditary osteodystrophy (AHO) depends on the sex of the parent transmitting the *GNAS* mutation; when inherited maternally, the phenotype includes end organ resistance to various hormones including the parathyroid (pseudo-hypoparathyroidism type 1a), but when the paternally transmitted offspring have the physical features of AHO but lacked hormone resistance (pseudopseudo-hypoparathyroidism) [58]. Familial paragangliomas type 1 (glomus tumors), provides an example of when inheritance patterns can be misleading. These tumors are always paternally inherited, suggesting that the mutated gene, *SDHD* on 11q23, was imprinted

[59, 60]. However, the results of subsequent studies have implied that the apparent imprinting of *SDHD* is attributable to the need to delete a maternally expressed tumor suppressor gene elsewhere on chromosome 11 [61].

### *26.4.2 Parental Effects and Quantitative Traits*

In humans and in livestock, there is considerable interest in the incorporation of parent-of-origin effects in models of genetically-complex diseases and quantitative trait loci (QTL) linkage analysis. When imprinted genes contribute to phenotypes, the power of tests for linkage can be enhanced if imprinting models are included (for example see [62]). In humans, parent-of-origin linkage analysis has detected putative imprinted loci involved in obesity, alcoholism, bipolar affective disorder, autism, type II diabetes mellitus, and others. If imprinted genes are indeed involved in genetically complex disorders, models that incorporate parent-of-origin effects may need to be used to unravel these conditions.

In pig, 63 QTLs, some of which overlap, have been reported to show parental effects Of particular interest is a QTL for lean meat that has been attributed to a nucleotide substitution in intron 3 of *IGF2*, a known imprinted gene. Many of the other QTLs with parent-of-origin effects may, however, reflect false discovery, a common problem for genome-wide association studies [63]. This assumption is supported by the reporting of QTLs with parent-of-origin effects in chickens [64, 65], a species that does not appear to exhibit genomic imprinting, and would not be expected to show parental effects. As for complex human diseases, the incorporation of imprinting models into animal QTL analysis is an important component for discovery of genetic association [66, 67].

### *26.4.3 Animal Phenotypes*

It has been known for a millennia that the breeding of mules requires the correct parent-of-origin in the horse-donkey cross. A mule is produced when a female horse is mated with a male donkey, whereas the less desired offspring, a hinny, is produced when the cross is reversed [68]. More recently, inter-species' crosses of zoo animals have revealed that the offspring from a tiger-lion mating have different phenotypes that depend on the direction of the cross. A female tiger crossed with a male lion produces a large animal (liger) while the reciprocal crosses (tigons) are smaller.

Given such blatant examples of imprinting, it is surprising that genomic imprinting was not recognized earlier, but the overzealous interpretation of Mendel's first law of heredity may have hindered the interpretation of these parental effects. Mendel, himself ,clearly described some parent-of-origin effects, noting that in certain crosses "The hybrids had the greatest similarity to the pollen parent. [69]".

## 26.5 Parent-of-Origin Effects on *De Novo* Mutations

### *26.5.1 The Influence of Parental Age on De Novo Mutations*

The parental age effect refers to the increased incidence of sporadic genetic disorders in children born to older men and women. The first observation of a relationship between age and the incidence of a disorder was in 1912, when Wilhelm Weinberg noticed that children with achondroplasia, an inherited skeletal disorder, were more likely to be born later in sibships. However, he could not distinguish between the effects of birth order, paternal age, and maternal age at this time [4, 70–72]. Three decades later, E.T. Mørch noted that parental age, not birth order, was responsible for the increased incidence of achondroplasia [71, 73, 74]. The distinction between the influences of paternal and maternal ages was made almost half a century after Weinberg's initial observation, when L.S. Penrose [75] observed that paternal age, not maternal age, correlated with the incidence of *de novo* achondroplasia.

Approximately 20 autosomal dominant disorders are now known to be associated with advanced paternal age (Table 26.3). Risch *et al.* [76] analyzed the effects of parental age on the *de novo* cases of these disorders by comparing the distribution of parental ages for the affected children (O) for each disorder, to the distribution of parental ages in the general population (E). A strong paternal age effect is evident in some disorders, such as achondroplasia and Apert, Crouzon, and Pfeiffer syndromes, where the birth frequency of the affected individuals increases rapidly with paternal age (Fig. 26.5A). However, for other disorders, such as neurofibromatosis, a much weaker paternal age effect is seen. The rate of increase in the frequency of sporadic cases of this disorder (O/E) is much less marked (Fig. 26.5B).

Clearly, paternal age influences the development of spontaneous mutations more in some disorders than in others. Penrose [83], suggested that one possible cause of spontaneous mutation was DNA copy errors in chromosomal replication during mitotic cell division. These mutations would most likely be base substitutions caused by DNA polymerase errors during DNA replication. Such errors would occur preferentially in the paternal germline, as opposed to the maternal germline, due to the larger number of cell divisions in spermatogenesis than in oogenesis. Furthermore, these mutations would be dependent upon paternal age and not maternal age, as the total number of cell divisions increases with age in spermatogenesis, but not oogenesis (Table 26.4) [71].

**Table 26.3** Disorders associated with advanced paternal age. Boldface type indicates the disorders with a reported preferential paternal origin of base substitutions. Italicized type indicates the disorders with a reported preferential paternal origin of deletions and other mutations. The parental origin of mutation has not yet been determined for the remaining disorders. Adapted from reference 3

| Disorder | Avg. paternal age (years) | Avg. paternal age in control population | Avg. maternal age (years) | Avg. maternal age in control population | Paternal age effect |
|---|---|---|---|---|---|
| Acrodysostosis | 33.0 | 28.7 | 28.1 | 25.9 | Strong |
| **Achondroplasia** | 36.8 | 30.4 | 31.0 | 27.5 | Strong |
| **Apert syndrome** | 35.8 | 27 | 30.8 | 27 | Strong |
| Basal cell nevus syndrome | 36.9 | 29.9 | 31.7 | 26.5 | Strong |
| Cleidocranial dysostosis | 33.1 | 28.7 | 29.8 | 25.9 | Strong |
| **Costello syndrome**[a] | 37.2[b] | 32.1[b] | 32.4[c] | ND[d] | Strong |
| **Crouzon syndrome** | 33.9 | 28.7 | 28.6 | 25.9 | Strong |
| Fibrodysplasia ossificans progressiva | 34.8 | 28.5 | 30.1 | 27.8 | Strong |
| Marfan syndrome | 36.6 | 29.9 | 29.3 | 26.5 | Strong |
| **Muenke syndrome**[a] | 34.7[e] | 30.6[e] | 31.2[e] | 28.0[e] | Strong |
| **Multiple endocrine neoplasia 2A**[a], **medullary thyroid carcinoma**[a] | 39.3[f] | 30.6 | 34.8[f] | 27.6 | Strong |
| **Multiple endocrine neoplasia 2B**[a] | 33[g] | 29.9 | 29.5[g] | 26.6 | Strong |
| **Noonan syndrome** | 35.6[h] | 29.5[h] | ND[d] | ND[d] | Strong |
| Oculodentodigital dysplasia | 32.3 | 28.7 | 28.1 | 25.9 | Strong |
| **Pfeiffer syndrome** | 35.9 | 28.2 | 29.6 | 26.0 | Strong |
| **Progeria** | 34.1 | 28.7 | 29.7 | 25.9 | Strong |
| Waardenburg syndrome | 34.8 | 29.9 | 30.5 | 26.5 | Strong |
| *Bilateral retinoblastoma* | 32.3 | 27 | 28.2 | 27 | Weak |
| Multiple extostoses | 30.6 | 28.7 | 26.1 | 25.9 | Weak |
| **Neurofibromatosis 1** | 32.5 | 29.7 | 29.8 | 26 | Weak |
| *Sotos syndrome* | 31.7 | 28.7 | 28.8 | 25.9 | Weak |
| *Treacher Collins syndrome* | 31.2 | 28.7 | 27.0 | 25.9 | Weak |

[a] Disorders not mentioned in ref. 76; [b] from ref. 77; [c] from ref. 78; [d] ND = not determined; [e] from ref. 79; [f] ref. 80; [g] from ref. 81; [h] from ref. 82.

(Copies of tables are available in the accompanying CD.)

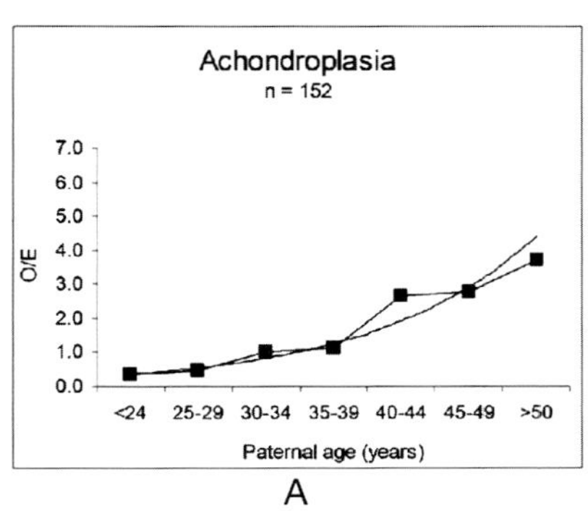


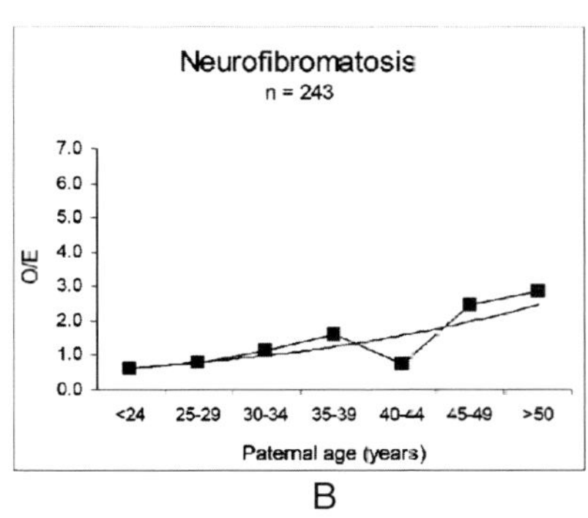


**Fig. 26.5** Ratio of the observed (O) to the expected (E) affected births in the general population, by age. **(A)** One example of a disorder with a strong paternal age effect. **(B)** One example of a disorder with a weak paternal age effect. Adapted from [3] (Copies of figures including color copies, where applicable, are available in the accompanying CD)

While increasing maternal age does not influence the incidence of base substitutions, it does influence the development of other spontaneous mutations, namely chromosomal abnormalities. Oocytes are arrested in the prophase of meiosis I, until females reach sexual maturity. During this stage of meiosis, homologous chromosomes pair up, or synapse [84]. When puberty begins, one oocyte per month is selected to come out of arrest and resume the cell cycle. Some oocytes, therefore, may be arrested for 10, 20, or more years. It is thought that the longer the oocytes are arrested, the harder it is for the chromosomes to separate properly when meiosis resumes, and therefore the greater the chance for the occurrence of a non-disjunction event that would lead to a chromosomal abnormality in the offspring [4, 85].

### *26.5.2 Parent-of-Origin Effects on Different Mutation Types*

In order to determine the parental origin of a mutation, the maternal and paternal alleles in the affected child must be distinguished from one another. Single nucleotide polymorphisms near the mutations of interest are used to identify the parental alleles in the child and thereby establish the parental origin of mutation.

#### 26.5.2.1 Base Substitutions

Using the database to conduct a search for the parental origin of base substitutions yields a list of mutations found in 26 different disorders. Several observations about the parental origin of base substitutions are immediately apparent. First, base substitutions in general, are paternally inherited. Of the 439 reported cases of base substitutions, where parental origin was established, 353, or 80%, are paternal in origin (Fig. 26.1). Secondly, point mutations in some disorders, such as achondroplasia, Apert syndrome, Crouzon syndrome, Pfeiffer syndrome, and progeria are exclusively paternal in origin (Table 26.5). Many of the disorders, with exclusive paternal origin of mutation, were previously shown to have a strong paternal age effect [76, Table 26.3).

**Table 26.4** Comparison between the number of cell divisions in spermatogenesis and oogenesis

| Age (years) | Spermatogenesis | Oogenesis |
|---|---|---|
| 20 | 150 | 24 |
| 30 | 380 | 24 |
| 40 | 610 | 24 |
| 50 | 840 | 24 |
| 60 | 1070 | 24 |
| 70 | 1300 | 24 |

(Copies of tables are available in the accompanying CD.)

**Table 26.5** Disorders with an exclusive paternal origin of base substitutions. Boldface type indicates the disorders that were previously identified as having a strong paternal age effect. Adapted from reference 3

| Disorder | Gene | No. of paternalcases | No. of maternal cases | $\alpha$[a] | Paternal age effect? | Recurrent mutations present? | Mutations in CpG dinucleotides? | Reference |
|---|---|---|---|---|---|---|---|---|
| **Achondroplasia** | *FGFR3* | 40 | 0 | ∞ | Y | Y | Y | 10 |
| Alport syndrome | *COL4A5* | 2 | 0 | ∞ | Unknown | N | N | 86 |
| **Apert syndrome** | *FGFR2* | 57 | 0 | ∞ | Y | Y | Y | 11 |
| Childhood hypophosphatemia | *ALPL* | 1 | 0 | ∞ | Unknown | N | N | 87 |
| Charcot-Marie-Tooth syndrome | *CMT* | 1 | 0 | ∞ | Unknown | N | N | 88 |
| **Crouzon syndrome** | *FGFR2* | 11 | 0 | ∞ | Y | Y | N | 12 |
| Diamond-Blackfan anemia | *RPS19* | 2 | 0 | ∞ | Unknown | N | N | 89 |
| MEN 2A[b], MTC[c] | *RET* | 11 | 0 | ∞ | Y | Y | N | 80, 90–92 |
| Muenke syndrome | *FGFR3* | 10 | 0 | ∞ | Y | Y | Y | 79 |
| Noonan syndrome | *PTPN11* | 14 | 0 | ∞ | Y | Y | N | 84 |
| **Pfeiffer syndrome** | *FGFR2* | 11 | 0 | ∞ | Y | Y | N | 12 |
| **Progeria** | *LMNA* | 8 | 0 | ∞ | Y | Y | Y | 93 |
| Total for all disorders | | 168 | 0 | ∞ | | | | |

[a] Ratio of the number of paternally derived mutations to maternally derived mutations for each disorder; [b] MEN 2A = multiple endocrine neoplasia type 2A; [c] MTC = medullary thyroid carcinoma.
(Copies of tables are available in the accompanying CD.)

There are several defining characteristics of the mutations found in disorders with exclusive paternal origin of mutation. In most of these disorders, there is a high prevalence of mutations at one or two loci in each gene, possibly due to these loci being mutational hotspots, bias of ascertainment, or selection for those mutations that cause a phenotype. Recurrent mutations are found in eight of the 13 (61%) disorders in this group (Table 26.5). In four of these nine disorders, one or two recurrent mutations account for the majority of all cases and are found within a CpG dinucleotide, a known mutational hotspot.

In achondroplasia, 98% of the cases are caused by one mutation within a CpG dinucleotide in the fibroblast growth factor receptor 3 gene [*FGFR3*]. In Apert syndrome, greater than 99% of all sporadic cases are caused by one of two mutations in the fibroblast growth factor receptor 2 gene [*FGFR2*]. One of these mutations occurs in a CpG dinucleotide [11]. More than 90% of cases of progeria are also caused by a recurrent mutation within a CpG dinucleotide in the lamin A gene [*LMNA*] [93, 94]. The most recent addition to this group of disorders is the Muenke syndrome. All the cases of this disorder are characterized by one recurrent missense mutation in the *FGFR3* gene, the same gene involved in achondroplasia. This recurrent mutation, like the recurrent mutation found in achondroplasia, also occurs within a CpG dinucleotide [83].

Recurrent missense mutations are also found in medullary thyroid cancer (MTC), multiple endocrine neoplasia 2A (MEN2A), Crouzon syndrome, Pfeiffer syndrome, and Noonan syndrome; however none of these mutations occur within a CpG dinucleotide. Mutations in the *RET* proto-oncogene, cause 30% and 85% of all cases of MTC and MEN2A, respectively [90, 95, 96]. The recurrent mutations found in Crouzon, Pfeiffer, and Noonan syndromes occur at much lower frequencies compared to the recurrent mutations found in the other disorders. The mutational heterogeneity is much greater in Crouzon, Pfeiffer, and Noonan syndromes as well. Over 25 and 30 different mutations have been reported in Pfeiffer and Crouzon syndromes, respectively, the majority of which are base substitutions [97]. Twenty-two different missense mutations have been reported in Noonan syndrome [98].

Perhaps the most intriguing observation about the disorders with an exclusive paternal origin of mutation is that five of the disorders are caused by gain-of-function mutations in the same two genes – *FGFR2* and *FGFR3*. These genes may be pre-disposed to mutational events and the high proportion of specific nucleotide changes may be due to a selective advantage of the mutated germ cell [11, 99, 100]. FGFRs are expressed in rat spermatogonial stem cells and seem to be important in maintaining spermatogenesis [100–103]. Dimerization of the receptors upon ligand binding results in auto-phosphorylation of the intracellular tyrosine kinase domains [104], which triggers downstream signaling pathways, resulting in changes in gene expression and other biological consequences. Phosphorylation of tyrosine kinases is also important in sperm motility and capacitation [105, 106]. The gain-of-function mutations in these genes may confer some selective advantage to the sperm, in terms of motility and capacitation, through constitutive or altered activation of the FGF receptors. In fact, many of the disorders with a strong paternal age effect and an exclusive paternal origin of mutation are caused by gain-of-function mutations in genes involved in similar or interacting signal transduction pathways [*FGFR2*, *FGFR3*, *RET*, and *PTPN11*] [84, 107–111].

For the 13 disorders in which some maternally-derived cases were reported (Table 26.6), approximately 70% of the reported base substitutions are paternal in origin. Recurrent mutations are found in just five (or 38%) of these disorders, compared to 8/13 (61%) disorders with exclusive paternal origin of mutation in which recurrent mutations were found. Mutations in CpG dinucleotides are found in three disorders: Rett syndrome, hemophilia B and von Hippel-Lindau syndrome. The disorders that have a paternal bias of mutation are caused by a variety of different mutation types and are generally not associated with advanced paternal age, whereas the disorders that have an exclusive paternal origin of mutation tend to be caused primarily, if not only, by point mutations and tend to be associated with advanced paternal age.

**Table 26.6** Parent-of-origin effects for base substitutions. Adapted from reference 3

| Disorder | Gene | No. of paternal cases (%) | No. of maternal cases (%) | αa | Paternal age effect? | Recurrent mutations present? | Mutations in CpG dinucleotides? | Reference |
|---|---|---|---|---|---|---|---|---|
| Alexander disease | *GFAP* | 24 (86) | 4 (14) | 6 | N | Y | Y | 112 |
| Costello syndrome | *HRAS* | 23 (92) | 2 (8) | 11.5 | Y | Y | Y | 77, 78 |
| Craniofrontal nasal syndrome | *EFNB1* | 5 (83) | 1 (17) | 5 | N | N | N | 113 |
| Familial adenomatous polyposis | *APC* | 5 (83) | 1 (17) | 5 | Y | N | N | 114 |
| Hemophilia A | *FVIII* | 8 (80) | 2 (20) | 4 | Y | N | N | 115 |
| Hemophilia B | *FIX* | 54 (51) | 51 (49) | 1.06 | Y | N | Y | 116 |
| Hirschsprung disease | *RET* | 0 | 3 | 0 | N | N | N | 117 |
| MEN 2B[b] | *RET* | 25 (93) | 2 (7) | 12.5 | Y | Y | N | 81 |
| Neonatal diabetes mellitus | *KCNJ11* | 13 (72) | 5 (28) | 2.6 | N | Y | Y | 118 |
| Neurofibromatosis 1 | *NF1 1* | 31 (89) | 4 (11) | 7.75 | N | N | N | 119–121 |
| Neurofibromatosis 2 | *NF2* | 13 (57) | 10 (43) | 1.3 | Unknown | Y | N | 122 |
| Pelizaeus-Merzbacher disease | *PLP* | 4 (80) | 1 (20) | 4 | N | N | N | 123 |
| Rett syndrome | *MECP2* | 29 (91) | 3 (9) | 9.67 | N | Y | Y | 124–126 |
| Townes-Brocks syndrome | *SALL1* | 14 (88) | 2 (12) | 7 | N | Y | Y | 127 |
| Tuberous sclerosis | *TSC2* | 2 (40) | 3 (60) | 0.67 | N | N | N | 128 |
| von Hippel-Lindau syndrome | *VHL* | 4 (57) | 3 (43) | 1.33 | N | N | Y | 129 |
| Total for all disorders | | 254 (72) | 97 (28) | 2.62 | | | | |
| Total for disorders with a paternal age effect | | 115 (66) | 58 (34) | 1.98 | | | | |

[a] Ratio of the number of paternally derived mutations to maternally derived mutations for each disorder;
[b] MEN 2B = multiple endocrine neoplasia type 2B.
(Copies of tables are available in the accompanying CD.)

There are some disorders that have been previously associated with a strong paternal age effect (Table 26.3), yet do not appear in this database. For some disorders in this group, like acrodysostosis and fibrodysplasia ossificans progressive, the parental origin of mutations has yet to be established because the gene is not known. For other disorders like cleidocranial dysostosis, oculodentodigital dysplasia (ODDD), basal cell nevus syndrome, Marfan syndrome, and Waardenburg syndrome, a range of mutation types are found [3], making the determination of the parental origin of mutation much harder, as a different experimental approach must be taken with each type of mutation. It remains to be seen whether these disorders will follow the pattern of other disorders with a strong paternal age effect, such as Apert syndrome and achondroplasia, and show a paternal bias of mutation.

#### 26.5.2.2 Deletions

Deletions are catalogued in the database as “large deletions” (involved >20 bp), “small deletions” (involving <20 bp), or simply “deletions”, if the size of the deletion was not reported in the literature. This size distinction is made based on the different mechanisms, and potentially different parental origins, that contribute to each type of deletion. In contrast to base substitutions, large deletions do not demonstrate a noticeable parent-of-origin effect or a strong paternal age effect, as 47% of all reported cases are paternal in origin (Fig. 26.1, Table 26.7).

**Table 26.7** Parent-of-origin effects for large deletions. Adapted from reference 3

| Disorder | Gene | No. of paternal cases (%) | No. of maternal cases (%) | α[a] | Paternal age effect? | Reference |
|---|---|---|---|---|---|---|
| Alagille syndrome | *JAG1* | 1 | 0 | 8 | Unknown | 130 |
| Bilateral retinoblastoma | *RB1* | 17 (89) | 2 (11) | 8.5 | N | 131–134 |
| Congenital deafness | *CGS*[b] | 1 | 0 | 8 | N | 135 |
| Craniofrontal nasal syndrome | *EFNB1* | 1 (33) | 2 (67) | 0.5 | N | 113 |
| Cri-du-chat syndrome | *CGS* | 85 (86) | 14 (14) | 6.07 | Unknown | 136–138 |
| Del 1p36 | *CGS* | 19 (36) | 34 (64) | 0.56 | Unknown | 139, 140 |
| Del 3p | *CGS* | 1 | 0 | 8 | Unknown | 141 |
| Del 9p | *CGS* | 10 (71) | 4 (29) | 2.5 | Unknown | 142, 143 |
| Del 14 | *CGS* | 5 (63) | 3 (37) | 1.67 | Y | 144–149 |
| Del 22q11 | *CGS* | 65 (30) | 153 (70) | 0.42 | Unknown | 150–157 |
| DiGeorge syndrome | *CGS* | 8 (53) | 7 (47) | 1.14 | Unknown | 152, 158 |
| Duchenne and Becker muscular dystrophies | *DMD* | 5 (14) | 30 (86) | 0.17 | Unknown | 159 |
| Familial adenomatous polyposis | *APC* | 3 | 0 | 8 | Unknown | 114 |
| Gorlin syndrome | | 1 | 0 | 8 | Unknown | 160 |
| Hemophilia A | *FVIII* | 2 (40) | 3 (60) | 0.67 | N | 161 |
| Hemophilia B | *FIX* | 3 (27) | 8 (73) | 0.38 | Y | 116, 162 |
| Hirschsprung disease | *RET* | 1 | 0 | 8 | N | 117 |
| HNPP[c] | *PMP22* | 4 (57) | 3 (43) | 1.33 | Unknown | 163 |
| Langer-Giedion syndrome (TRPSII) | *CGS* | 6 (67) | 3 (33) | 2 | Unknown | 164, 165 |
| Miller-Dieker syndrome | *CGS* | 5 (83) | 1 (17) | 5 | Unknown | 166, 167 |
| Neurofibromatosis 1 | *NF1* | 18 (23) | 60 (77) | 0.3 | N | 168–172 |
| Rett-like syndrome and autism | *CGS* | 1 | 0 | 8 | Unknown | 173 |
| Ornithine transcarbamylase deficiency | | 0 | 1 | 0 | Unknown | 174 |
| Severe myoclonic epilepsy of infancy | | 3 | 0 | 8 | Unknown | 175 |
| Sotos syndrome | *NSD1* | 18 (90) | 2 (10) | 9 | N | 176 |
| TRPS1[d] | *CGS* | 0 | 1 | 0 | Unknown | 177 |
| Tuberous sclerosis | *TSC2* | 0 | 2 | 0 | N | 128 |
| Unilateral retinoblastoma | *RB1* | 5 (45) | 6 (55) | 0.83 | N | 132–134 |
| von Hippel–Lindau | *VHL* | 3 (50) | 3 (50) | 1 | N | 129 |
| Williams syndrome | *CGS* | 93 (44) | 118 (56) | 0.79 | N | 178 |
| Wolf-Hirschhorn/Pitt-Rodgers-Danks syndrome | *CGS* | 27 (87) | 4 (13) | 6.75 | N | 179 |
| Gamma-delta-beta-thalassemia | *HBB* | 0 | 1 | 0 | Unknown | 180 |
| Total for all disorders | | 412 (47) | 464 (53) | 0.89 | | |

[a] Ratio of the number of paternally derived mutations to maternally derived mutations for each disorder; [b] CGS = contiguous gene syndrome; [c] HNPP = hereditary neuropathy with liability to pressure palsies; [d] TRPSI = trichorhinophalangeal syndrome type I.
(Copies of tables are available in the accompanying CD.)

While large deletions in general do not have a parent-of-origin effect, a clear parent-of-origin effect is observed in some individual disorders. For example, in bilateral retinoblastoma, cri-du-chat, Sotos, and Wolf-Hirschhorn syndromes, 86–90% of large deletions are paternal in origin (references in Table 26.7). In comparison, 64–86% of the large deletions in del 1p36 and del 22q11 syndromes, neurofibromatosis type 1 (NF1) and Duchenne and Becker muscular dystrophies are

**Table 26.8** Parent-of-origin effects for small deletions. Adapted from reference 3

| Disorder | Gene | No. of paternal cases (%) | No. of maternal cases (%) | $\alpha^a$ | Paternal age effect? | Reference |
|---|---|---|---|---|---|---|
| Alport syndrome | *COL4A5* | 0 | 1 | 0 | Unknown | 86 |
| Alexander disease | *GFAP* | 1 | 0 | 8 | N | 112 |
| Beta-thalassemia | *HBB* | 1 | 0 | 8 | Y | 181 |
| Familial adenomatous polyposis | *APC* | 8 (47) | 9 (53) | .89 | N | 114, 182 |
| Hirschsprung | *RET* | 1 | 0 | 8 | N | 117 |
| Lhermitte-Duclos syndrome | *PTEN* | 1 | 0 | 8 | Unknown | 183 |
| Neurofibromatosis 2 | *NF2* | 8 (89) | 1 (11) | 8 | Unknown | 122 |
| Rendu-Osler-Weber syndrome | *ENG* | 1 | 0 | 8 | Unknown | 184 |
| Rett syndrome | *MECP2* | 3 | 0 | 8 | N | 125, 126 |
| Townes-Brocks syndrome | *SALL1* | 5 | 0 | 8 | N | 127 |
| Treacher-Collins syndrome | *TCOF1* | 6 (67) | 3 (33) | 2 | N | 185 |
| Tuberous sclerosis | *TSC2* | 3 (60) | 2 (40) | 1.5 | N | 128 |
| Total for all disorders | | 38 (70) | 16 (30) | 2.38 | | |

[a] Ratio of the number of paternally derived mutations to maternally derived mutations for each disorder.
(Copies of tables are available in the accompanying CD.)

maternal in origin (references in Table 26.7). The nature of the repeated sequences that mediate deletions may influence the parental origin, as some sequences may be predisposed to recombination and deletion during oogenesis, whereas other deletions may be detrimental to mature sperm, and to other stages of spermatogonial development [155,156].

In contrast to large deletions, small deletions fewer than 20 base pairs tend to be paternal in origin (Table 26.8). This type of mutation is thought to arise during replication by mispairing and misalignment of the direct repeats or short runs of identical bases [186]. Because small deletions may occur during replication, it may explain the higher mutation frequency seen in males compared to females. However, small deletions are not associated with older paternal age, whereas point mutations, which also are thought to arise during replication, are associated.

#### 26.5.2.3 Duplications and Insertions

Duplications, like point mutations, also tend to be paternally inherited. In the literature, there are only 84 cases for which the parental origin is reported, 79 (94%) of which is paternally inherited. Duplications in Charcot-Marie-Tooth syndrome comprise the majority of the cases in the database.

There are very few reported cases of the parental origin of insertions in the literature. In some instances, the parental origins of several types of mutations have been reported together in one article and it is impossible to distinguish insertions from the other types of mutations.

#### 26.5.2.4 Parent-of-Origin Effects in Disorders Caused by Many Mutation Types

When a variety of mutations are observed in one disorder, a paternal age effect for one type of mutation (e.g., point mutations) may be diluted by the presence of other types of mutations (e.g., deletions) that are not associated with age [4]. Some of these disorders, such as NF1, Sotos syndrome, bilateral retinoblastoma, multiple exostoses, and Treacher Collins syndrome, were previously shown to have a weak paternal age effect (76, Table 26.3). It would be interesting to see if a strong paternal age effect is observed for only the point mutations in these disorders.

Interestingly, in NF1, 89% of base substitutions are paternal in origin, whereas only 23% of large deletions are paternal in origin (references in Tables 26.6 and 26.7). Conversely, in NF2, base substitutions do not show a bias of parental origin, whereas deletions do; 90% of the deletions are paternal in origin (Tables 26.6 and 26.8) [122]. The fact that ratios of paternally-derived to maternally-derived mutations differ depending on the mutation type and that some mutations are not strongly associated with parental age, argue for the existence of more than one mechanism underlying the origin of these mutations.

#### 26.5.2.5 Trisomies

As can be seen from Fig. 26.1, the majority of trisomies are maternal in origin. Observations of an association between trisomies and advanced maternal age have been made since the mid–1900's [187]. Maternal non-disjunction in meiosis I is responsible for the majority of cases of trisomy 13, 16, 21, and 22. Maternal non-disjunction in meiosis II has been shown to be responsible for most cases of trisomy 18 [188].

### 26.5.3 *Mechanisms of Mutation*

#### 26.5.3.1 Higher Frequency of Base Substitutions in Males

In his study of X-linked hemophilia, Haldane suggested that there is a higher mutation frequency in the germ cells of human males as compared to females [189]. He based this hypothesis on his observation that affected males were more likely to be born to heterozygous carrier mothers than unaffected mothers, meaning that the mutation must have occurred in an earlier generation. While other studies have found a higher mutation frequency in human males and males of other species, there is a disagreement about how much higher the mutation frequency is, in males compared with in females (reviewed in 190).

The ratio of the mutation frequency in males compared to females ($\alpha$) provides a measure of the extent of parental bias for each disorder. As discussed earlier, point mutations in several disorders were found to be exclusively paternal in origin and associated with older age. These disorders have the highest $\alpha$, as no maternally-derived cases have been observed yet (Table 26.5). Since these disorders also have a strong paternal age effect, there is an association between the disorders with a strong paternal age effect and disorders with the highest $\alpha$. Is this the case for base substitutions in other disorders? For the disorders listed in Table 26.6, the average $\alpha$ is 2.62, although the value varies greatly between the individual disorders. Combining the data from all base substitutions (Tables 26.5 and 26.6), yields 422 paternally-derived base substitutions compared with 97 cases of maternally-derived base substitutions ($\alpha = 4.35$).

While point mutations in general may be more prone to occurring in the male germline, it should be noted that $\alpha$ is close to 1 (i.e., there is no parent-of-origin effect) in four disorders, Hemophilia B, neurofibromatosis type 2 (NF2), tuberous sclerosis, and von Hippel-Lindau syndrome. In tuberous sclerosis and von Hippel-Lindau syndrome, however, the estimated $\alpha$ may not be correct due to the small number of cases reported for each of these disorders. Interestingly, there were no disorders in which a significantly greater number of maternally-derived base substitutions was observed.

This higher prevalence of mutations in males may explain the observed lack of affected males and reduced fitness of females with any of the several dominant X-linked disorders, once thought to be lethal in males [191]. For these disorders, the higher frequency of mutation in males would result in paternally-derived *de novo* mutations on the X chromosome, which would manifest only in heterozygous (affected) females, due to lack of X chromosome transmission from father to son. In the absence of maternally-derived mutations, affected males would be born only to affected mothers, but because of the reduced fitness of such females, familial cases are very rare.

#### 26.5.3.2 The Role of Methylation

Differences in the methylation status of the DNA in male and female germ cells may play a role in the observed higher frequency of mutation in males than females. CpG dinucleotides are considered hypermutable because in most of these dinucleotides, the cytosine is methylated, which predisposes it to spontaneous deamination to thymine. In humans and mice, sperm DNA has been shown to be more methylated than oocyte DNA, and this has been suggested to account for the greater number of paternally-derived mutations occurring within a CpG dinucleotide [192,193]. Indeed, the ratio of paternally-derived to maternally-derived base substitutions in a CpG dinucleotide is 1.5–2 times higher than the ratio for base substitutions at all locations (Table 26.9). This mechanism could explain the increased incidence of paternally-derived base substitutions, in CpG sites, in disorders with no age effect, such as Rett syndrome [124–126]. This mechanism would not, however, explain the association of the frequency of CpG mutations with older paternal age, as seen in achondroplasia, Apert syndrome, Muenke syndrome, and progeria. Few studies have examined the relationship between age and the extent of DNA methylation in the sperm as an explanation for the paternal age effect seen in these disorders [195]. One such study found that methylation was stable with increasing age at nucleotide 1138 of *FGFR3*, the locus implicated in most cases of achondroplasia. The existence of CpG mutations that are mostly paternal in origin and not associated with older paternal age; such as those in Rett syndrome, suggests that the mechanisms that generate base substitutions with both parent-of-origin and paternal age effects differ from the mechanisms that generate base substitutions with only a parent-of-origin effect.

#### 26.5.3.3 Mutations Arising in Meiosis

Mutations arising in meiosis would not be expected to show a parent-of-origin or a parental age effect because a limited number of meiotic divisions occur in both spermatogenesis and oogenesis

**Table 26.9** Parent-of-origin effects for CpG mutations. Adapted from reference 3

| Disorder | Gene | No. of paternal cases (%) | No. of maternal cases (%) | $\alpha$[a] | Paternal age effect? | Reference |
|---|---|---|---|---|---|---|
| Achondroplasia | *FGFR3* | 40 (100) | 0 | $\infty$ | Y | 10 |
| Alexander disease | *GFAP* | 14 (93) | 1 (7) | 14 | N | 112 |
| Apert syndrome | *FGFR2* | 38 (100) | 0 | $\infty$ | Y | 11 |
| Costello syndrome | *HRAS* | 22 (92) | 2 (8) | 11 | Y | 77, 78 |
| Hemophilia B | *FIX* | 14 (48) | 15 (5) | 0.93 | Y | 116, 162 |
| Muenke syndrome | *FGFR3* | 10 (100) | 0 | $\infty$ | Y | 79 |
| Neonatal diabetes mellitus | *KCNJ11* | 11 (79) | 3 (21) | 3.7 | N | 118 |
| Progeria | *LMNA* | 7 (100) | 0 | $\infty$ | Y | 93, 194 |
| Rett syndrome | *MECP2* | 26 (90) | 3 (10) | 8.67 | N | 124-126 |
| Townes-Brocks syndrome | *SALL1* | 6 (86) | 1 (14) | 6 | N | 127 |
| von Hippel-Lindau syndrome | *VHL* | 2 | 0 | $\infty$ | N | 129 |
| Total for all disorders | | 190 (88) | 25 (12) | 7.6 | | |
| Total for disorders with a paternal age effect | | 131 (89) | 17 (11) | 7.7 | | |

[a] Ratio of the number of paternally derived mutations to maternally derived mutations for each disorder.
(Copies of tables are available in the accompanying CD.)

[4,196]. Recombination between repeated sequences, repetitive sequence elements (like Alu elements), or in some cases, non-homologous sequences during meiosis have been hypothesized to account for many of the large deletions listed in Table 26.7 (Refs from table). As discussed previously, large deletions do not show either a parent-of-origin or a parental age effect. This same mechanism may generate other chromosomal abnormalities such as large insertions and duplications [196], for which we would not expect to see parent-of-origin or parental age effects either.

### 26.5.4 Challenging Penrose's Copy Error Hypothesis

A number of different mechanisms have been mentioned that could give rise to mutations associated with a parent-of-origin effect either alone or in conjunction with a paternal age effect. One such mechanism, copy error during DNA replication, predicts a paternal origin of base substitutions associated with older paternal age. However, several studies and observations have provided evidence that challenge this hypothesis. First, there are several disorders that show a parent-of-origin effect for base substitutions, but not a parental age effect (Table 26.6). Secondly, the linear increase in the number of chromosomal replications in sperm, with age alone, is expected to produce a linear increase in the number of base substitutions with age. However, the increase in the frequency of affected children born to older fathers actually is exponential [4,71,76] (Fig. 26.5a). This suggests that mechanisms in addition to replication errors contribute to the paternal age effect. Regardless of the mechanism by which these mutations are generated, the increase in the incidence of disorders with a paternal age effect should be a reflection of the increasing number of mutations present in sperm as men age.

Over the last several years, various studies have attempted to quantify the number and type of mutations in sperm with age. Age-related increases in the frequency of the most common achondroplasia mutation (1138 G>A in *FGFR3*) and several Apert syndrome mutations in *FGFR2*, in the sperm of unaffected men have been observed [195,197,198]. No age-related increases in the frequency of sperm with aneuploidies or diploidies were observed [199].

Most recently, "protein-driven selection" of mutations was proposed to explain the accumulation of *FGFR2* mutations in the male germline. These mutations are thought to arise in low frequency in mitotic spermatogonial stem cells and confer the cells with a selective advantage, allowing them to clonally expand over time [100].

## 26.6 Closing Remarks

The range of parent-of-origin effects is wide – from epigenetic modifications of the genome that affects the expression of parental alleles to parent-of-origin effects on the development of *de novo* mutations. Genomic imprinting, that results in parentally-derived suppression of allelic expression, has been documented for approximately 100 genes in mammals. The total number of genes that are affected by genomic imprinting remains unclear. Given that most imprinted genes occur in clusters, which have now been well characterized, the number of additional imprinted genes may be small. Alternatively, subtle or undiscovered imprinting effects may contribute to a large number of complex phenotypes in humans and other mammals.

*De novo* base substitutions, regardless of the context in which they occur, are generally paternally-derived. Not all base substitutions, however, are associated with older paternal age, suggesting that there are a number of different mechanisms that contribute to this type of mutation. Large deletions do not show a parent-of-origin effect or a paternal age effect, whereas small deletions are more often paternally derived, but also do not show a paternal age effect. Certain genes, such as *FGFR2* and *FGFR3*, are frequently mutated at specific nucleotides. These mutations are exclusively paternal in origin and have a strong paternal age effect, suggesting that the sequence of these genes may predispose them to mutation, or that the mutations themselves may confer a selective advantage to the sperm cells.

## Key References

Crow JF. The origins, patterns and implications of human spontaneous mutation. Nature Rev Genet. 2000;1:40–47.

Edwards CA, Ferguson-Smith AC. Mechanisms regulating imprinted genes in clusters. Curr Opin Cell Biol. 2007;19:281–289.

Glaser RL, Jabs EW. Dear Old Dad. Sci Aging Knowl Environ. 2004;3 re1.

Morison IM, Ramsay JP, Spencer HG. A census of mammalian imprinting. Trends Genet. 2005;21:457–465.

Wood AJ, Oakey RJ. Genomic Imprinting in Mammals: Emerging Themes and Established Theories. PLoS Genet. 2006;2:e147

## Suggested Reading

### *Introduction to the Range of Parent-of-Origin Effects*

1. Reik W, Walter J. Genomic imprinting: parental influence on the genome. Nature Rev Genet. 2001;2:21–32.
2. Wood AJ, Oakey RJ. Genomic imprinting in mammals: emerging themes and established theories. PLoS Genet. 2006;2: e147.
3. Glaser RL, Jabs EW. Dear old dad. Sci Aging Knowl Environ. 2004;3 re1.
4. Crow JF. The origins, patterns and implications of human spontaneous mutation. Nature Rev Genet. 2000;1:40–47.
5. Bittel DC, Butler MG. Prader-Willi syndrome: clinical genetics, cytogenetics and molecular biology. Expert Rev Mol Med. 2005;7:1–20.
6. Lalande M, Calciano MA. Molecular epigenetics of Angelman syndrome. Cell Mol Life Sci. 2007;64:947–960.
7. Enklaar T, Zabel BU, Prawitt D. Beckwith-Wiedemann syndrome: multiple molecular mechanisms. Expert Rev Mol Med. 2006;8:1–19.
8. Tycko B, Morison IM. Physiological functions of imprinted genes. J Cell Phys. 2002; 192:245–258.
9. Georges M, Charlier C, Cockett N. The callipyge locus: evidence for the trans interaction of reciprocally imprinted genes. Trends Genet. 2003;19:248–252.
10. Wilkin DJ, Szabo R, Cameron S, et al. Mutations in fibroblast growth factor receptor 3 in sporadic cases of achondroplasia occur exclusively on the paternally-derived chromosome. Am J Hum Genet. 1998;63:711–716.
11. Moloney DM, Slaney DF, Oldridge M, et al. Exclusive paternal origin of new mutations in Apert syndrome. Nat Genet. 1996;13:48–53.
12. Glaser RL, Jiang W, Boyadjiev S, et al. Paternal origin of FGFR2 mutations in sporadic cases of Crouzon syndrome and Pfeiffer syndrome. Am J Hum Genet. 2000;66:768–777.
13. Petersen MB, Mikkelsen M. Nondisjunction in trisomy 21: Origin and mechanisms. Cytogenet Cell Genet. 2000;91:199–203.
14. Morison IM, Reeve AE. A catalogue of imprinted genes and parent-of-origin effects in humans and animals. Hum Mol Genet. 1998;7:1599–1609.
15. Morison IM, Paton CJ, Cleverly SD. The imprinted gene and parent-of-origin effect database. Nucleic Acids Res. 2001;29:275–276.
16. Morison IM, Ramsay JP, Spencer HG. A census of mammalian imprinting. Trends Genet. 2005;21:457–465.
17. Glaser RL, Ramsay JP, Morison IM. The imprinted gene and parent-of-origin effect database now includes parental origin of *de novo* mutations. Nucleic Acids Res Database Issue. 2006;34:D29-D31.
18. Nikaido, I, Saito, C, Mizuno, Y, Meguro, M, Bono, H, Kadomura, M, Kono, T, Morris, GA, Lyons, PA, Oshimura, M, Hayashizaki, Y, and Okazaki, Y. (2003) Discovery of imprinted transcripts in the mouse transcriptome using large-scale expression profiling. Genome Res, 13, 1402–1409.

## *Imprinting Effects*

19. Edwards CA, Ferguson-Smith AC. Mechanisms regulating imprinted genes in clusters. Curr Opin Cell Biol. 2007;19:281–289.
20. Spahn L, Barlow DP. An ICE pattern crystallizes. Nat Genet. 2003;35:11–12.
21. Seidl CI, Stricker SH, Barlow DP. The imprinted air ncRNA is an atypical RNAPII transcript that evades splicing and escapes nuclear export. Embo J. 2006;25:3565–3575.
22. Paoloni-Giacobino A, D'Aiuto L, Cirio MC, Reinhart B, Chaillet JR. Conserved features of imprinted differentially methylated domains. Gene. 2007; Epub ahead of print.
23. Okazaki Y, Furuno M, Kasukawa T, et al. Analysis of the mouse transcriptome based on functional annotation of 60,770 full-length cDNAs. Nature. 2002;420:563–573.
24. Takagi N, Sasaki M. Preferential inactivation of the paternal derived X chromosome in the extraembryonic membranes of the mouse. Nature. 1975;256:640–642.
25. Graves JA. Mammals that break the rules: genetics of marsupials and monotremes. Ann Rev Genet. 1996; 30:233–260.
26. Pauler FM, Koerner MV, Barlow DP. Silencing by imprinted noncoding RNAs: is transcription the answer? Trends Genet. 2007;23:284–292.
27. Lewis A, Mitsuya K, Umlauf D, et al. Imprinting on distal chromosome 7 in the placenta involves repressive histone methylation independent of DNA methylation. Nat Genet. 2004;36:1291–1295.
28. Umlauf D, Goto Y, Cao R, et al. Imprinting along the Kcnq1 domain on mouse chromosome 7 involves repressive histone methylation and recruitment of Polycomb group complexes. Nat Genet. 2004;36: 1296–1300.
29. Le Meur E, Watrin F, Landers M, Sturny R, Lalande M, Muscatelli F. Dynamic developmental regulation of the large non-coding RNA associated with the mouse 7C imprinted chromosomal region. Dev Biol. 2005;286:587–600.
30. Kurukuti S, Tiwari VK, Tavoosidana G, et al. CTCF binding at the H19 imprinting control region mediates maternally inherited higher-order chromatin conformation to restrict enhancer access to Igf2. Proc Natl Acad Sci USA. 2006;103:10684–10689.
31. O'Neill MJ. The influence of non-coding RNAs on allele-specific gene expression in mammals. Hum Mol Genet. 2005;14 Suppl 1:R113–120.
32. Davis E, Caiment F, Tordoir X, et al. RNAi-mediated allelic trans-interaction at the imprinted Rtl1/Peg11 locus. Curr Biol. 2005;15:743–749.
33. Schule B, Albalwi M, Northrop E, et al. Molecular breakpoint cloning and gene expression studies of a novel translocation t(4;15)(q27; q11.2) associated with Prader-Willi syndrome. BMC Med Genet. 2005;6:18.
34. Seitz H, Royo H, Bortolin ML, Lin SP, Ferguson-Smith AC, Cavaille J. A large imprinted microRNA gene cluster at the mouse Dlk1-Gtl2 domain. Genome Res. 2004;14:1741–1748.
35. Cai X, Cullen BR. The imprinted H19 noncoding RNA is a primary microRNA precursor. Rna. 2007; 13:313–316.
36. Killian JK, Nolan CM, Stewart N, et al. Monotreme IGF2 expression and ancestral origin of genomic imprinting. J Exp Zool. 2001;291:205–212.
37. Killian JK, Byrd JC, Jirtle JV, et al. M6P/IGF2R imprinting evolution in mammals. Mol Cell. 2000;5:707–716.
38. Wang Y, Joh K, Masuko S, et al. The mouse Murr1 gene is imprinted in the adult brain, presumably due to transcriptional interference by the antisense-oriented U2af1-rs1 gene. Mol Cell Biol. 2004;24:270–279.
39. Xu Y, Goodyer CG, Deal C, Polychronakos C. Functional polymorphism in the parental imprinting of the human *IGF2R* gene. Biochem Biophys Res Comm. 1993;197:747–754.
40. Monk D, Arnaud P, Apostolidou S, et al. Limited evolutionary conservation of imprinting in the human placenta. Proc Natl Acad Sci USA. 2006;103:6623–6628.
41. Killian JK, Nolan CM, Wylie AA, et al. Divergent evolution in M6P/IGF2R imprinting from the Jurassic to the Quaternary. Hum Mol Genet. 2001;10:1721–1728.
42. Braidotti G, Baubec T, Pauler F, et al. The air noncoding RNA: an imprinted cis-silencing transcript. Cold Spring Harb Symp Quant Biol. 2004;69:55–66.
43. Meguro M, Mitsuya K, Sui H, et al. Evidence for uni-parental, paternal expression of the human GABA(A) Receptor subunit genes, using microcell-mediated chromosome transfer. Hum Mol Genet.1997;6:2127–2133.
44. Bittel DC, Kibiryeva N, Talebizadeh Z, Butler MG. Microarray analysis of gene/transcript expression in Prader-Willi syndrome: deletion versus UPD. J Med Genet, 2003;40:568–574.
45. Buettner VL, Longmate JA, Barish ME, Mann JR, Singer-Sam J. Analysis of imprinting in mice with uni-parental duplication of proximal chromosomes 7 and 15 by use of a custom oligonucleotide microarray. Mamm Genome. 2004;15:199–209.

46. Liljelund P, Handforth A, Homanics GE, Olsen RW. GABAA receptor beta3 subunit gene-deficient heterozygous mice show parent-of-origin and gender-related differences in beta3 subunit levels, EEG, and behavior. Brain Res Dev Brain Res. 2005;157:150–161.
47. Kayashima T, Ohta T, Niikawa N, Kishino T. On the conflicting reports of imprinting status of mouse ATP10a in the adult brain: strain-background-dependent imprinting? J Hum Genet. 2003;48:492–493; author reply 494.
48. Jinno Y, Yun K, Nishiwaki K, et al. Mosaic and polymorphic imprinting of the WT1 gene in humans. Nat Genet. 1994;6:305–309.
49. Nishiwaki K, Niikawa N, Ishikawa M. Polymorphic and tissue-specific imprinting of the human Wilms tumor gene, WT1. Jpn J Hum Genet. 1997;42:205–211.
50. Mitsuya K, Sui, Meguro M, et al. Paternal expression of WT1 in human fibroblasts and lymphocytes. Hum Mol Genet.1997;6:2243–2246.
51. Dallosso AR, Hancock AL, Brown KW, Williams AC, Jackson S, Malik K. Genomic imprinting at the WT1 gene involves a novel coding transcript (AWT1) that shows deregulation in Wilms' tumours. Hum Mol Genet. 2004;13:405–415.
52. Lo HS, Wang Z, Hu Y, et al. Allelic variation in gene expression is common in the human genome. Genome Res. 2003;13:1855–1862.

## *Uni-parental Disomy*

53. Engel E. Uni-parental disomy revisited: The first twelve years. Am J Med Genet. 1993; 46:670–674.
54. Nicholls RD, Knoll JHM, Butler MG, Karam S, Lalande M. Genetic imprinting suggested by maternal heterodisomy in non-deletion Prader-Willi syndrome. Nature. 1989;342:281–285.
55. Cattanach BM, Beechey CV. Genomic imprinting in the mouse: possible final analysis. In: Reik W, Surani A, eds. Genomic Imprinting. Oxford: New York: OUP/IRL Press, 1997:118–145.
56. Beechey CV, Cattanach BM, Blake A, Peters J. MRC Mammalian Genetics Unit, Harwell, Oxfordshire. World Wide Web Site — Mouse Imprinting Data and References 2005. cited; Available from: http://www.mgu.har.mrc.ac.uk/research/imprinting/

## *Parental Effects in Genetic Linkage*

57. Niikawa N, Ishikiriyama S, Takahashi S, et al. The Wiedemann-Beckwith syndrome: pedigree studies on five families with evidence for autosomal dominant inheritance with variable expressivity. Am J Med Genet. 1986;24:41–55.
58. Davies SJ, Hughes HE. Imprinting in Albright's hereditary osteodystrophy. J Med Genet. 1993;30:101–103.
59. Baysal BE, Farr JE, Rubinstein WS, et al. Fine mapping of an imprinted gene for familial nonchromaffin paragangliomas, on chromosome 11q23. Am J Hum Genet. 1997;60:121–132.
60. Milunsky J, DeStefano AL, Huang XL, et al. Familial paragangliomas: linkage to chromosome 11q23 and clinical implications. Am J Med Genet. 1997;72:66–70.
61. Hensen EF, Jordanova ES, Van Minderhout IJ, et al. Somatic loss of maternal chromosome 11 causes parent-of-origin-dependent inheritance in SDHD-linked paraganglioma and phaeochromocytoma families. Oncogene. 2004;23:4076–4083.
62. Shete S, Zhou X. Parametric approach to genomic imprinting analysis with applications to Angelman's syndrome. Hum Hered. 2005;59:26–33.
63. Dudbridge F, Gusnanto A, Koeleman BP. Detecting multiple associations in genome-wide studies. Hum Genom. 2006;2:310–317.
64. Siwek M, Cornelissen SJ, Nieuwland MG, et al. Detection of QTL for immune response to sheep red blood cells in laying hens. Anim Genet. 2003;34:422–428.
65. Buitenhuis AJ, Rodenburg TB, van Hierden YM, et al. Mapping quantitative trait loci affecting feather pecking behavior and stress response in laying hens. Poult Sci. 2003;82:1215–1222.
66. Gorlova OY, Lei L, Zhu D, et al. Imprinting detection by extending a regression-based QTL analysis method. Hum Genet. 2007;Epub ahead of print.
67. Cui Y. A statistical framework for genome-wide scanning and testing of imprinted quantitative trait loci. J Theor Biol. 2007;244:115–126.
68. Savory TH. The mule. Sci Amer. 1970;223:102–109.
69. Mendel GJ. Experiments in plant hybridisation. Edinburgh: Oliver & Boyd. 1965;7–51.

## *Parent-of-Origin Effects on* De Novo *Mutations*


70. Weinberg W. Zur Vererbung des Zwergwuchses. Arch Rassen-u Gesell. Biol. 1912;9:710–718.
71. Vogel F and Rathenberg R. Spontaneous mutation in man. Adv Hum Genet. 1975;5:223–318.
72. Vogel F and Motulsky AG. Mutation: Spontaneous mutation in germ cells. In: Vogel F and Motulsky, eds. Human Genetics: Problems and Approaches, 3rd ed., New York: Springer, 1997:385–430.
73. Mørch ET. Chondrodystrophic dwards in Denmark. (Opera ex Domo Biol Hered Hum Univ Hafn Munskgaard, Copenhagen). 1941:3.
74. Crow JF. Spontaneous mutation in man. Mutat Res. 1999;43:5–9.
75. Penrose LS. Parental age in achondroplasia and mongolism. Am J Hum Genet. 1957;9:167–169.
76. Risch R, Reich EW, Wishnick MW and McCarthy JG. Spontaneous mutation and parental age in humans. Am J Hum Genet. 1987;41:218–248.
77. Zampino G, Pantaleoni F, Carta C, et al. Diversity, parental germline origin, and phenotypic spectrum of de novo HRAS missense changes in Costello syndrome. Hum Mutat. 2007;28:265–272.
78. Sol-Church K, Stabley DL, Nicholson L, Gonzalez IL, Gripp KW. Paternal bias in parental origin of HRAS mutations in Costello syndrome. Hum Mutat. 2006;27:736–741.
79. Rannan-Eliya SV, Taylor IB, de Heer IM, van den Ouweland AMW, Wall SA, Wilkie AOM. Paternal origin of FGFR3 mutations in Muenke-type craniosynostosis. Hum Genet. 2004;115:200–207.
80. Schuffenecker I, Ginet N, Goldgar D, et al. Prevalence and parental origin of de novo RET mutations in multiple endocrine neoplasia type 2A and familial medullary thyroid carcinoma. Le Groupe d'Etude des Tumeurs a Calcitonine. Am J Hum Genet. 1997;60:233–237.
81. Carlson KM, Bracamontes J, Jackson CE, et al. Parent-of-origin effects in multiple endocrine neoplasia type 2B. Am J Hum Genet. 1994;55:1076–1082.
82. Tartaglia M, Cordeddu V, Change H, et al. Paternal germline origin and sex-ratio distortion in transmission of PTPN11 mutations in Noonan syndrome. Am J Hum Genet. 2004;75:492–497.
83. Penrose LS. Parental age and mutation. Lancet. 1955;269:312–313.
84. Eppig JJ, Vivieros MM, Marin-Bivens C, De La Fuente R. Regulation of mammalian oocyte maturation. In: Leung PCK and EY Adashi, eds. The Ovary. Amsterdam: Elsevier Academic Press, 2004:113–129.
85. Muller F, Rebiffé M, Taillandier A, Oury JF, Mornet E. Parental origin of the extra chromosome in prenatally diagnosed fetal trisomy 21. Hum Genet. 2000;106:340–344.
86. Hertz JM, Juncker I, Persson U, et al. Detection of mutations in the COL4A5 gene by SSCP in X-linked Alport syndrome. Hum Mutat. 2001;18:141–148.
87. Taillandier A, Sallinen SL, Brun-Heath I, De Mazancourt P, Serre JL, Mornet E. Childhood hypophosphatasia due to a de novo missense mutation in the tissue-nonspecific alkaline phosphatase gene. J Clin Endocrinol Metab. 2005;90:2436–2439.
88. Roa BB, Garcia CA, Suter U, et al. Charcot-Marie-Tooth disease type 1A. Association with a spontaneous point mutation in the PMP22 gene. N Engl J Med. 1993;329:96–101.
89. Orfali KA, Ohene-Abuakwa Y, Ball SE. Diamond Blackfan anaemia in the UK: clinical and genetic heterogeneity. Br J Haematol. 2004;125:243–252.
90. Mulligan LM, Eng C, Healey CS, et al. A de novo mutation of the RET proto-oncogene in a patient with MEN 2A. Hum Mol Genet. 1994;3:1007–1008.
91. Zedenius J, Wallin G, Hamberger B, Nordenskjöld M, Weber G, Larsson C. Somatic and MEN 2A de novo mutations identified in the RET proto-oncogene by screening of sporadic MTC:s. Hum Mol Genet. 1994;3:1259–1262.
92. Wohllk N, Cote GJ, Bugalho MM, Ordonez N, Evans DB, Goepfert H, Khorana S, et al. Relevance of RET proto-oncogene mutations in sporadic medullary thyroid carcinoma. J Clin Endocrinol Metab. 1996;81:3740–3745
93. Eriksson M, Brown WT, Gordon LB. Recurrent de novo point mutation sin lamin A case Hutchinson-Guilford progeria syndrome. Nature. 2003;423:293–298.
94. Cao H, Hegel A. LMNA is mutated in Hutchinson-Guilford progeria (MIM 176670) but not in Wiedemann-Rautenstrach progeroid syndrome (MIM 264090). J Hum Genet. 2003; 48:271–274.
95. Mulligan LM, Kwok JB, Healey CS, et al. Germ-line mutations of the RET proto-oncogene in multiple endocrine neoplasia type 2A. Nature. 1993;363:458–460.
96. Eng C, Clayton D, Schuffenecker I. The relationship between specific RET proto-oncogene mutations and disease phenotype in multiple endocrine neoplasia type 2. International RET mutation consortium analysis. JAMA. 1996;276:1575–1579.
97. Passos-Bueno MR, Wilcox WR, Jabs EW, Sertie AL, Alonso LG, Kitoh H. Clinical spectrum of fibroblast growth factor receptor mutations. Hum Mutat. 1999;14:115–125.
98. Tartaglia M, Kalidas K, Shaw A, et al. *PTPN11* Mutations in Noonan Syndrome: Molecular Spectrum, Genotype-Phenotype Correlation, and Phenotypic Heterogeneity. Am J Hum Genet. 2002;70:1555–1563.

99. Oldridge M, Lunt PW, Zackai EH, et al. Genotype-phenotype correlation for nucleotide substitutions in the IgII-IgIII linker of FGFR2. Hum Mol Genet. 1997;6:137–143.
100. Goriely A, McVean GA, van Pelt AM, et al. Gain-of-function amino acid substitutions drive positive selection of FGFR2 mutations in human spermatogonia. Proc Natl Acad Sci USA. 2005;102:6051–6056.
101. Van Dissel-Emiliani FM, De Boer-Brouwer M, de Rooij DG. Effect of fibroblast growth factor-2 on Sertoli cells and gonocytes in coculture during the perinatal period. Endocrinology. 1996;137:647–654.
102. Cancilla B, Risbridger GP, Differential localization of fibroblast growth factor receptor-1, -2, -3, and -4 in fetal, immature, and adult rat testes. Biol Reprod. 1998;58:1138–1145.
103. Cancilla B, Davies A, Ford-Perriss M, Risbridger GP. Discrete cell- and stage-specific localisation of fibroblast growth factors and receptor expression during testis development. J Endocrinol. 2000;64;149–159.
104. Lemmon MA, Schlessinger J. Regulation of signal transduction and signal diversity by receptor oligomerization. Trends Biochem Sci. 1994;19:459–463.
105. Bajpai M, Asin S, Doncel GF. Effect of tyrosine kinase inhibitors on tyrosine phosphorylation and motility parameters in human sperm. Arch Androl. 2003;49:229–246.
106. Urner F, Sakkas D. Protein phosphorylation in mammalian spermatozoa. Reproduction. 2003;125:17–26.
107. Asai N, Iwashita T, Matsuyama M, Takahashi M. Mechanism of activation of the ret proto-oncogene by multiple endocrine neoplasia 2A mutations. Mol Cell. Biol. 1995; 15:1613–1619.
108. Wang Y, Spatz MK, Kannan K, et al. A mouse model for achondroplasia produced by targeting fibroblast growth factor receptor 3. Proc Natl Acad Sci. USA. 1999;96: 4455–4460.
109. Kannan K, Givol D. FGF receptor mutations: dimerization syndromes, cell growth suppression, and animal models. IUBMB Life. 2000;49:197–205.
110. Yu K, Herr AB, Waksman A, Ornitz DM. Loss of fibroblast growth factor receptor 2 ligand-binding specificity in Apert syndrome. Proc Natl Acad Sci. USA. 2000;97:14536–14541.
111. Ibrahimi OA, Eliseenkova AV, Plotnikov AN, Yu K, Ornitz DM, Mohammadi M. Structural basis for fibroblast growth factor receptor 2 activation in Apert syndrome. Proc Natl Acad Sci. USA. 2001;98:7182–7187.
112. Li R, Johnson AB, Salomons GS, et al. Propensity for paternal inheritance of de novo mutations in Alexander disease. Hum Genet. 2006;119:137–144.
113. Twigg SR, Matsumoto K, Kidd AM, et al. The origin of EFNB1 mutations in craniofrontonasal syndrome: frequent somatic mosaicism and explanation of the paucity of carrier males. Am J Hum Genet. 2006;78:999–1010.
114. Aretz S, Uhlhaas S, Caspari R, et al. Frequency and parental origin of de novo APC mutations in familial adenomatous polyposis. Eur J Hum Genet. 2004;12:52–58.
115. Becker J, Schwaab R, Möller-Taube A, et al. Characterization of the factor VIII defect in 147 patients with sporadic hemophilia A: family studies indicate a mutation type-dependent sex ratio of mutation frequencies. Am J Hum Genet. 1996;58:657–670.
116. Ketterling RP, Vielhaber E, Li X, et al. Germline origins in the human F9 gene: frequent G:C->A:T mosaicism and increased mutations with advanced maternal age. Hum Genet. 1999;105:629–640.
117. Yin L, Seri M, Barone V, Tocco T, Scaranari M, Romeo G. Prevalence and parental origin of de novo RET mutations in Hirschsprung's disease. Eur J Hum Genet. 1996;4:356–358.
118. Edghill EL, Gloyn AL, Goriely A, et al. Origin of de novo KCNJ11 mutations and risk of neonatal diabetes for subsequent siblings. J Clin Endocrinol Metab. 2007;92:1773–1777.
119. Stephens K, Kayes L, Riccardi VM, Rising M, Sybert VP, Pagon RA. Preferential mutation of the neurofibromatosis type 1 gene in paternally-derived chromosomes. Hum Genet. 1992;88:279–282.
120. Lazaro C, Gaona A, Ainsworth P, et al. Sex differences in mutational rate and mutational mechanism in the NF1 gene in neurofibromatosis type 1 patients. Hum Genet. 1996; 98:696–699.
121. Jadayel D, Fain P, Upadhyaya M, et al. Paternal origin of new mutations in von Recklinghausen neurofibromatosis. Nature. 1990;343:558–559.
122. Kluwe L, Mautner V, Parry DM, et al. The parental origin of new mutations in neurofibromatosis 2. Neurogenetics. 2000;3:17–24.
123. Mimault C, Giraud G, Courtois V, et al. Proteolipoprotein gene analysis in 82 patients with sporadic Pelizaeus-Merzbacher Disease: duplications, the major cause of the disease, originate more frequently in male germ cells, but point mutations do not. The Clinical European Network on Brain Demyelinating Disease. Am J Hum Genet. 1999;65:360–369.
124. Amir RE, Van den Veyver IB, Schultz R, et al. Influence of mutation type and X chromosome inactivation on Rett syndrome phenotypes. Ann Neurol. 2000;47:670–679.
125. Girard M, Couvert P, Carrie A, et al. Parental origin of de novo MECP2 mutations in Rett syndrome. Eur J Hum Genet. 2001;9:231–236.
126. Trappe R, Laccone F, Cobilanschi J, et al. MECP2 mutations in sporadic cases of Rett syndrome are almost exclusively of paternal origin. Am J Hum Genet. 2001;68:1093–1101.

127. Böhm J, Munk-Schulenburg S, Felscher S, Kohlhase J. SALL1 mutations in sporadic Townes-Brocks syndrome are of predominantly paternal origin without obvious paternal age effect. Am J Med Genet A. 2006;140:1904–1908.
128. Roberts PS, Chung J, Jozwiak S, et al. SNP identification, haplotype analysis, and parental origin of mutations in TSC2. Hum Genet. 2002;111:96–101.
129. Richards FM, Payne SJ, Zbar B, Affara NA, Ferguson-Smith MA, Maher ER, Molecular analysis of de novo germline mutations in the von Hippel-Lindau disease gene. Hum Mol Genet. 1995;4:2139–2143.
130. Deleuze JF, Hazan J, Dhorne S, Weissenbach J, Hadchouel M. Mapping of microsatellite markers in the Alagille region and screening of microdeletions by genotyping 23 patients. Eur J Hum Genet. 1994;2(3):185–190.
131. Horsthemke B, Greger V, Barnert HJ, Hopping W, Passarge E. Detection of submicroscopic deletions and a DNA polymorphism at the retinoblastoma locus. Hum Genet. 1987;76:257–261.
132. Ejima Y, Sasaki MS, Kaneko A, Tanooka H. Types, rates, origin and expressivity of chromosome mutations involving 13q14 in retinoblastoma patients. Hum Genet. 1988;79:118–123.
133. Dryja TP, Mukai S, Petersen R, Rapaport JM, Walton D, Yandell DW. Parental origin of mutations of the retinoblastoma gene. Nature. 1989;339:556–558.
134. Zhu XP, Dunn JM, Phillips RA. et al. Preferential germline mutation of the paternal allele in retinoblastoma. Nature. 1989;340:312–313.
135. Petek E, Windpassinger C, Mach M, et al. Molecular characterization of a 12q22-q24 deletion associated with congenital deafness: confirmation and refinement of the DFNA25 locus. Am J Med Genet A. 2003;117:122–126.
136. Overhauser J, McMahon J, Oberlender S, et al. Parental origin of chromosome 5 deletions in the cri-du-chat syndrome. Am J Med Genet. 1990;37:83–86.
137. Church DM, Bengtsson U, Nielsen KV, Wasmuth JJ, Niebuhr E. Molecular definition of deletions of different segments of distal 5p that result in distinct phenotypic features. Am J Hum Genet. 1995;56:1162–1172.
138. Mainardi PC, Perfumo C, Cali A, et al. Clinical and molecular characterisation of 80 patients with 5p deletion: genotype-phenotype correlation. J Med Genet. 2001;38:151–158.
139. Shapira SK, McCaskill C, Northrup H, et al. Chromosome 1p36 deletions: the clinical phenotype and molecular characterization of a common newly delineated syndrome. Am J Hum Genet. 1997;61:642–650.
140. Heilstedt HA, Ballif BC, Howard LA, et al. Physical map of 1p36, placement of breakpoints in monosomy 1p36, and clinical characterization of the syndrome. Am J Hum Genet. 2003;72:1200–1212.
141. Petek E, Windpassinger C, Simma B, Mueller T, Wagner K, Kroisel PM. Molecular characterisation of a 15 Mb constitutional de novo interstitial deletion of chromosome 3p in a boy with developmental delay and congenital anomalies. J Hum Genet. 2003; 48:283–287.
142. Micale MA, Haren JM, Conroy JM, Crowe CA, Schwartz S. Parental origin of De Novo chromosome 9 deletions in del(9p) syndrome. Am J Med Genet. 1995;57:79–81.
143. Olivieri C, Maraschio P, Caselli D, et al. Interstitial deletion of chromosome 9, int del(9)(9q22.31–q31.2), including the genes causing multiple basal cell nevus syndrome and Robinow/brachydactyly 1 syndrome. Eur J Pediatr. 2003;162:100–103.
144. Hreidarsson SJ, Stamberg J. Distal monosomy 14 not associated with ring formation. J Med Genet. 1983;20:147–149.
145. Telford N, Thomson DA, Griffiths MJ, Ilett S, Watt JL. Terminal deletion (14)(q32.3): a new case. J Med Genet. 1990;27:261–263.
146. Elliott J, Maltby EL, Reynolds B, A case of deletion 14(q22.1→q22.3) associated with anophthalmia and pituitary abnormalities. J Med Genet. 1993;30:251–252.
147. Shapira SK, Anderson KL, Orr-Urtregar A, Craigen WJ, Lupski JR, Shaffer LG. De novo proximal interstitial deletions of 14q: cytogenetic and molecular investigations. Am J Med Genet. 1994;52:44–50.
148. Byth BC, Costa MT, Teshima IE, Wilson WG, Carter NP, Cox DW. Molecular analysis of three patients with interstitial deletions of chromosome band 14q31. J Med Genet. 1995;32:564–567.
149. Petek E, Plecko-Startinig B, Windpassinger C, Egger H, Wagner K, Kroisel PM. Molecular characterisation of a 3.5 Mb interstitial 14q deletion in a child with several phenotypic anomalies. J Med Genet. 2003;40:e47.
150. Demczuk S, Levy A, Aubry M, et al. Excess of deletions of maternal origin in the DiGeorge/velo-cardio-facial syndromes. A study of 22 new patients and review of the literature. Hum Genet. 1995;96:9–13.
151. Morrow B, Goldberg R, Carlson C, et al. Molecular definition of the 22q11 deletions in velo-cardio-facial syndrome. Am J Hum Genet. 1995;56:1391–1403.
152. Ryan AK, Goodship JA, Wilson DI, et al. Spectrum of clinical features associated with interstitial chromosome 22q11 deletions: a European collaborative study. J Med Genet. 1997;34:798–804.
153. Fokstuen S, Arbenz U, Artan S, et al. 22q11.2 deletions in a series of patients with non-selective congenital heart defects: incidence, type of defects and parental origin. Clin Genet. 1998;53:63–69.

154. Matsuoka R, Kimura M, Scambler PJ, et al. Molecular and clinical study of 183 patients with conotruncal anomaly face syndrome. Hum Genet. 1998;103:70–80.
155. Lu JH, Chung MY, Hwang B, Chien HP. Prevalence and parental origin in Tetralogy of Fallot associated with chromosome 22q11 microdeletion. Pediatrics. 1999;104:87–90.
156. Chung MY, Lu JH, Chien HP, Hwang B. Chromosome 22q11 microdeletion in conotruncal heart defects: clinical presentation, parental origin and de novo mutations. Int J Mol Med. 2001;7:501–505.
157. Eliez S, Antonarakis SE, Morris MA, Dahoun SP, Reiss AL. Parental origin of the deletion 22q11.2 and brain development in velocardiofacial syndrome: a preliminary study. Arch Gen Psychiatry. 2001;58:64–68.
158. Saitta SC, Harris SE, McDonald-McGinn DM, et al. Independent de novo 22q11.2 deletions in first cousins with DiGeorge/velocardiofacial syndrome. Am J Med Genet A. 2004;124:313–317.
159. Bakker E, Veenema H, Den Dunnen JT, et al. Germinal mosaicism increases the recurrence risk for 'new' Duchenne muscular dystrophy mutations. J Med Genet. 1989;26:553–559.
160. Chen CP, Lin SP, Wang TH, Chen YJ, Chen M, Wang W. Perinatal findings and molecular cytogenetic analyses of de novo interstitial deletion of 9q (9q22.3→q31.3) associated with Gorlin syndrome. Prenat Diagn. 2006;26:725–729.
161. Youssoufian H, Kasper CK, Phillips DG, Kazazian HH, Antonarakis SE. Restriction endonuclease mapping of six novel deletions of the factor VIII gene in hemophilia A. Hum Genet. 1988;80:143–148.
162. Green PM, Saad S, Lewis CM, Giannelli F. Mutation rates in humans. I. Overall and sex-specific rates obtained from a population study of hemophilia B. Am J Hum Genet. 1999;65:1572–1579.
163. LeGuern E, Gouider R, Ravisé N, et al. A de novo case of hereditary neuropathy with liability to pressure palsies (HNPP) of maternal origin: a new mechanism for deletion in 17p11.2? Hum Mol Genet. 1996;5:103–106.
164. Ludecke HJ, Burdiek R, Senger G, Claussen U, Passarge E, Horsthemke B. Maternal origin of a de novo chromosome 8 deletion in a patient with Langer-Giedion syndrome. Hum. Genet. 1989;82:327–329.
165. Lopes J, Ravisé N, Vandenberghe A, et al. Fine mapping of de novo CMT1A and HNPP rearrangements within CMT1A-REPs evidences two distinct sex-dependent mechanisms and candidate sequences involved in recombination.Hum Mol Genet. 1998;7:141–148.
166. Schwartz CE, Johnson JP, Holycross B, et al. Detection of submicroscopic deletions in band 17p13 in patients with the Miller-Dieker syndrome. Am J Hum Genet. 1988; 43:597–604.
167. van Tuinen P, Dobyns WB, Rich DC, et al. Molecular detection of microscopic and submicroscopic deletions associated with Miller-Dieker syndrome. Am J Hum Genet. 1988;43:587–596.
168. Kayes LM, Burke W, Riccardi VM, et al. Deletions spanning the neurofibromatosis 1 gene: identification and phenotype of five patients. Am J Hum Genet. 1994;54:424–436.
169. Upadhyaya M, Maynard J, Osborn M, et al. Characterisation of germline mutations in the neurofibromatosis type 1 (NF1) gene. J Med Genet. 1995;32:706–710.
170. Upadhyaya M, Ruggieri M, Maynard J, et al. Gross deletions of the neurofibromatosis type 1 (NF1) gene are predominantly of maternal origin and commonly associated with a learning disability, dysmorphic features and developmental delay. Hum Genet. 1998;102:591–597.
171. Lopez Correa C, Brems H, Lazaro C, et al. Molecular studies in 20 submicroscopic neurofibromatosis type 1 gene deletions. Hum Mutat. 1999;14:387–393.
172. Lopez Correa C, Brems H, Lazaro C, Marynen P, Legius E. Unequal meiotic crossover: a frequent cause of NF1 microdeletions. Am J Hum Genet. 2000;66:1969–1974.
173. Pescucci C, Meloni I, Bruttini M, et al. Chromosome 2 deletion encompassing the MAP2 gene in a patient with autism and Rett-like features. Clin Genet. 2003;64:497–501.
174. Azevedo L, Soares PA, Quental R, et al. Mutational spectrum and linkage disequilibrium patterns at the ornithine transcarbamylase gene (OTC).Ann Hum Genet. 2006; 70:797–801.
175. Madia FStriano P, Gennaro E, et al. Cryptic chromosome deletions involving SCN1A in severe myoclonic epilepsy of infancy. Neurology. 2006;10;67:1230–1235.
176. Miyake N, Kurotaki N, Sugawara H, et al. Preferential paternal origin of microdeletions caused by prezygotic chromosome or chromatid rearrangements in Sotos syndrome. Am J Hum Genet. 2003;72:1331–1337.
177. Nardmann J, Tranebjaerg L, Horsthemke B, Ludecke HJ. The tricho-rhino-phalangeal syndromes: frequency and parental origin of 8q deletions. Hum Genet. 1997; 99:638–643.
178. Wang MS, Schinzel A, Kotzot D, et al. Molecular and clinical correlation study of Williams-Beuren syndrome: No evidence of molecular factors in the deletion region or imprinting affecting clinical outcome. Am J Med Genet. 1999;86:34–43.
179. Wieczorek D, Krause M, Majewski F, et al. Unexpected high frequency of de novo unbalanced translocations in patients with Wolf-Hirschhorn syndrome (WHS). J Med Genet. 2000;37:798–804.
180. Driscoll MC, Dobkin CS, Alter BP. Gamma delta beta-thalassemia due to a de novo mutation deleting the 5' beta-globin gene activation-region hypersensitive sites. Proc Natl Acad Sci U S A. 1989;86:7470–7474.
181. Chehab FF, Winterhalter KH, Kan YW. Characterization of a spontaneous mutation in beta-thalassemia associated with advanced paternal age. Blood. 1989;74:852–854.

182. Ripa R, Bisgaard ML, Bülow S, Nielsen FC. De novo mutations in familial adenomatous polyposis (FAP). Eur J Hum Genet. 2002;10:631–637.
183. Delatycki MB, Danks A, Churchyard A, Zhou XP, Eng C. De novo germline PTEN mutation in a man with Lhermitte-Duclos disease which arose on the paternal chromosome and was transmitted to his child with polydactyly and Wormian bones. J Med Genet. 200340:e92.
184. Lastella P, Sabbà C, Lenato GM, et al. Endoglin gene mutations and polymorphisms in Italian patients with hereditary haemorrhagic telangiectasia. Clin Genet. 2003;63:536–540.
185. Splendore A, Jabs EW, Félix TM, Passos-Bueno MR. Parental origin of mutations in sporadic cases of Treacher Collins syndrome. Eur J Hum Genet. 2003;11:718–722.
186. Antonarakis SE, Krawczak M, Cooper DN. Disease causing mutations in the human genome. Eur J Pediatr. 2000;159:S173–S178.
187. Penrose LS. Parental age in achondroplasia and mongolism. Am J Hum Genet. 1957;9:167–169.
188. Chen CP, Chern SR, Tsai FJ. A comparison of maternal age, sex ratio and associated major anomalies among fetal trisomy 18 cases with different cell division of error. Prenat diagn. 2005;25:327–330.

## *Mechanisms of Mutation*

189. Haldane JBS. The rate of spontaneous mutation of a human gene. J Genet. 1935;31:317–326.
190. Li WH, Yi S, Makova K. Male-driven evolution. Curr Opin Genet Dev. 2002;12:650–656.
191. Thomas GH. High male:female ratio of germ-line mutations: an alternative explanation for postulated gestational lethality in males in X-linked dominant disorders. Am J Hum Genet. 1996;58:1364–1368.
192. Monk M, Boubelik M, Lehnert S. Temporal and regional changes in DNA methylation in the embryonic, extraembryonic and germ cell lineages during mouse embryo development. Development. 1987;99:371–382.
193. Driscoll DJ, Migeon BR. Sex difference in methylation of single-copy genes in human meiotic germ cells: implications for X chromosome inactivation, parental imprinting, and origin of CpG mutations. Somatic Cell Mol Genet. 1990;16:267–282.
194. D'Apice MR, Tenconi R, Mammi I, van den Ende J, Novelli G. Paternal origin of LMNA mutations in Hutchinson-Gilford progeria. Clin Genet. 2004;65:52–54.
195. Tiemann-Boege I, Navidi W, Grewal R, et al. The observed human sperm mutation frequency cannot explain the achondroplasia paternal age effect. Proc Natl Acad Sci. USA. 2002;99:14952–14957.
196. Cooper DN, Krawczak M, Antonarakis SE. The nature and mechanisms of human gene mutation. In: Scriver CS, Beaudet AL, Sly WS, Valle D, eds. The Metabolic and Molecular Basis of Inherited Disease, 7th ed. New York: McGraw-Hill, 1995:65–94.

## *Challenging Penrose's Copy Error Hypothesis*

197. Glaser RL, Broman KW, Schulman RL, Eskenazi B, Wyrobek AJ, Jabs EW. The paternal age effect in Apert syndrome is due, in part, to the increased frequency of mutations in sperm. Am J Hum Genet. 2003;73:939–947.
198. Goriely A, McVean GAT, Röjmyr M, Ingemarsson B, Wilkie AOM. Evidence for selective advantage of pathogenic mutations in the male germline. Science. 2003;301:643–646.
199. Wyrobek AJ, Eskenazi B, Young S, et al. Advancing age has differential effects on DNA damage, chromatin integrity, gene mutations, and aneuploidies in sperm. Proc Natl Acad Sci. USA. 2006;103:9601–9606.

## Web Resources

www.otago.ac.nz/IGC
http://www.mgu.har.mrc.ac.uk/research/imprinting/index.html
http://www.geneimprint.com/site/home
http://fantom2.gsc.riken.go.jp/imprinting/

# Part VII
# Biological Networks

# Chapter 27
# Methods for Structural Inference and Functional Module Identification in Intracellular Networks

**Maria Manioudaki, Eleftheria Tzamali, Martin Reczko, and Panayiota Poirazi**

**Abstract** The ways in which intracellular components interact in order to produce certain functions remains a mystery to the scientific community. Although parts of biological systems become more and more characterized, a more global understanding of the structure, dynamics and functionalities of complex intracellular networks is currently lacking. Systems Biology approaches aim at providing such a global picture by combining analytical and experimental techniques across several multi-disciplinary fields. In this chapter, we provide an overview of the analytical approaches and computational tools that have been applied to biological systems in order to describe them at different levels of abstraction. We start by reviewing methods that model or infer a topological map of complex biological networks (structural inference) and move on to discuss ways of discovering the functionalities of sub-network entities that comprise these networks (functional module inference). Although clearly not exclusive, this chapter aims at providing a representative overview of the currently available methods that have been successfully used to characterize complex biological networks and reveal their structure and function.

**Keywords** Modelling · Mathematical methods · Cellular networks · Modules · Motifs · Inference

## 27.1 Introduction

The central dogma of biology states a rather simple principle: DNA is the carrier of the genetic information and it is transcribed to mRNA, which is thereafter translated into proteins. Behind this simple idea of genetic information flow, lays a dramatically complex cascade of regulatory events. A large number of biological components influence each other in a selective, functionally diverse, and non-linear manner to effectively produce stable behaviors. Understanding how cellular components interact in time and space is crucial for deciphering the functions inside a living cell. Towards this goal, Systems Biology approaches are frequently used to identify the components of the system and their relationships, to build *in silico* models that explain complex cellular processes and make testable predictions, as well as to represent these interactions graphically so that the complexity becomes more comprehensible.

Technological advances in the last decade allow simultaneous detection of thousands of biological variables, resulting in the ongoing accumulation of massive amounts of experimental data. Microarrays, the epitome of these high-throughput techniques, are used to measure the expression

P. Poirazi
Computational Biology Laboratory, Institute of Molecular Biology and Biotechnology (IMBB), Foundation of Research and Technology-Hellas (FORTH) Vassilika Vouton, P.O. Box 1385, GR 711 10 Heraklion, Crete, Greece
e-mail: poirazi@imbb.forth.gr

S. Krawetz (ed.), *Bioinformatics for Systems Biology*, DOI 10.1007/978-1-59745-440-7_27, 

levels of thousands of genes simultaneously, yeast-two-hybrid assays are used to locate large-scale protein-DNA interactions and ChIP-chip methods are used to identify interacting proteins, just to name a few. The challenge in this post-genomic era is to properly integrate all the available information so as to reconstruct, as accurately as possible, the complex network of interactions inside a cell, beginning from the identification of its static structure and moving on to characterize its dynamic behavior and functional role. Towards this goal, the main properties of the system, such as complexity, robustness and adaptation must first be derived from available data, in order to design a model that not only describes faithfully its structure but also makes useful predictions about its behavior. In order to cope with the massive amount of data that need to be analyzed and synthesized for modeling a biological network, mathematical and computational methods are inevitably required.

A biological system can be viewed as a set of diverse and multi-functional components (genes, gene products, and metabolites), whose population levels change over time in response to internal interactions and external signals. A snapshot of this dynamic evolution, referred to as the system *state*, describes the values of each component of the system at a given time. In gene regulatory networks, for instance, the system components are usually genes and gene products, which produce directly (protein profiles) or indirectly (mRNA profiles) observation values that vary over time. The interactions among the system components reflect the influence that the value of a component has on the values of other components. These interactions are usually governed by a set of biophysical laws, most of which are only partially known. Modeling concerns inference of both the interaction map (*structural inference*) of the system and the mathematical formalisms that approximate the dynamic biophysical laws the system follows (*dynamical inference*). Both of these approaches aim at characterizing the systems function at different levels of abstraction. As cellular components often interact selectively in order to accomplish certain functions, identification of the sub-networks responsible for such tasks (*functional module inference*) provides an alternative approach towards this goal.

Identifying the mathematical framework that better describes a biological system is not trivial. While the general suggestion is that, *things should be made as simple as possible, but not simpler* – a razor that places a trade off between simplicity and complexity – available data and computational complexity issues comprise a hurdle that can't be ignored. Demanding a low level of abstraction – that is a 'fine' model to describe the biological system – requires large amounts of experimental data but the risk of overfitting is particularly high in the absence of such information. Alternatively, a high level of abstraction – that is a 'coarse' model – requires less data but presents the challenge of building a model, capable of capturing the biological behaviors of the system without losing generality. The information we have for the system under study and the type of information we wish to infer from the model, define more or less the golden section for method selection, keeping in mind that the model will eventually be as good as the assumptions made for it.

Boolean and Bayesian networks comprise the main methods of a broader class of algorithms, the *network-based algorithms*, that qualitatively describe the topology of a biological system (*structural inference*). Based on various assumptions, these algorithms search for a network, among all possible structures, which best describes the information given for the underlying system.

Another approach used to reduce the complexity of this search and extend structural to functional inference, is to decompose the problem, for the identification of interconnected functional sub-networks (*modules*) within the biological system (*functional module inference*). Functional modules are discrete entities whose task is separable from those of other modules [1]. Modules interact with each other, in order to form larger networks, thus allowing for an orchestrated regulation of intracellular processes. Searching for functional modules is thus a critical step in illuminating intracellular networks and their functionalities. Computational methods for inferring both structural and functional characteristics of intracellular networks will be discussed in this chapter.

At the other end of the modeling spectrum stand highly specified models that strive to capture, in detail, the fundamental biophysical dynamics of the system (*dynamical inference*). These models

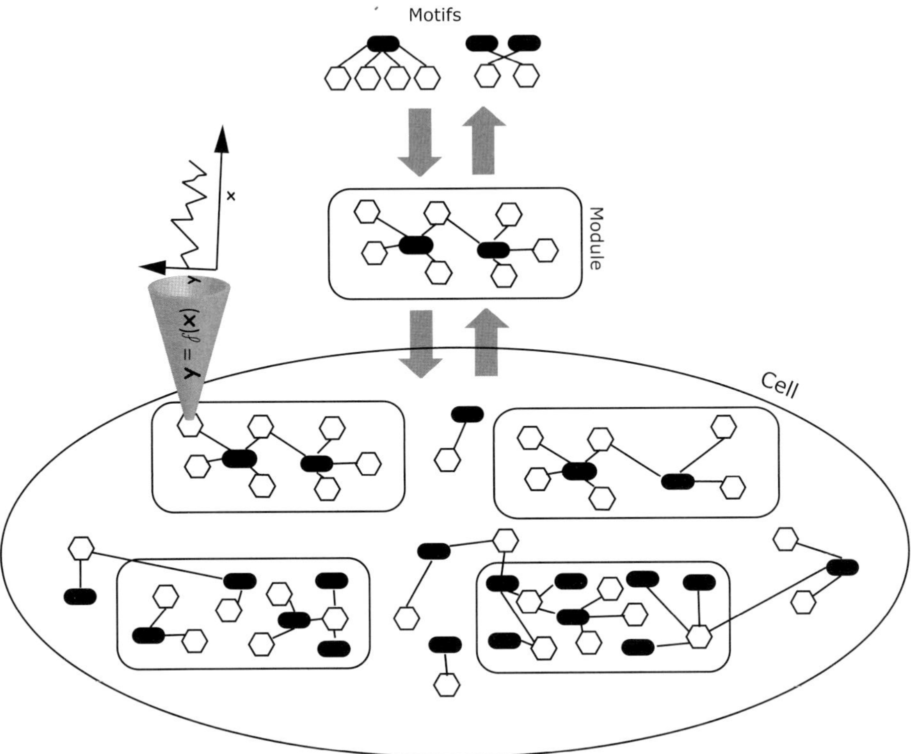


**Fig. 27.1** A general drawing summarizing intracellular network organization, which is the main focus of this chapter (Copies of figures including color copies, where applicable, are available in the accompanying CD)

(discussed in Chapter 28) use various types of equations to describe the interactions between the system components (*equation-based algorithms*). According to the assumptions made about the nature of the underlying system, these equations can be linear, non-linear, differential equations and so on. The dynamical equations are often applied on a network structure that is either known *a priori* or derived from network-based algorithms.

In-depth analysis of each of the available modeling techniques is beyond the scope of this chapter. Our main goal is to provide the reader with the intuition behind the methods and highlight the capacity of a given method to describe the real biological phenomena, as well as the possibility to infer a method based model from the available data. A graphical representation of the biological components and their relations discussed in this chapter is shown in Fig. 27.1.

## 27.2 Structural Inference

A biological system consists of a large set of diverse and multi-functional molecular components, that interact selectively, obeying specific biophysical laws in response to certain environmental conditions and signals. The selective interactions among the components of the system define a connectivity map (*network*) which is what structural inference approaches try to reveal. Specifically, the *network inference* problem is defined as follows:

"Given a set of interacting components as well as any prior knowledge and any set of observations that the system components produce, find the network connectivity that satisfies a given set of constraints and assumptions (*model*)."

To understand the variety of the modeling approaches used for inference, explanatory definitions of what a *network* and a *state* of a network represent are needed.

*Network:* A network is a representation of the interactions among the components of a system. The components of interest are depicted as nodes in the network, while the interactions between these components are depicted as edges. An edge can either be directed (arc), representing the flow of information or undirected representing simply the interaction between the components. Each component can also be seen as a variable, whose values describe the component's molecular population levels.

Examples of network nodes include genes, proteins, and the metabolite variables. Interactions can be physical, representing the potential of physical contact between two components (such as protein-DNA binding) or temporal, describing the influence a component will have on another component in a posterior time, thus tracking the temporal transition of each component of the system, or they may even depict dependencies among components under certain environmental conditions. Any kind of meaningful integration of the above is also possible.

Metabolism, for example, is commonly represented as a flow network where the nodes are metabolites and the edges establish the metabolic reactions that are catalyzed by certain enzymes. In an alternative representation, the nodes of the network correspond to metabolic genes, while an edge among two components exists if and only if there is a metabolite that is catalyzed by enzymes encoded by both components. In gene regulatory networks, genes and gene products comprise the nodes of the networks, while the edges represent the assumed influences between them as shown in the Fig. 27.2.

*State:* A state of a component is a description of a certain property (e.g., concentration) of that component at a given time point. Similarly, the state of a system describes the properties of each component at a given time.

Overall, several different representations are usually possible for the same system of components, depending on the algorithms used and their underlying assumptions. As a result, the kind of influence we want to represent (physical, temporal, conditional), the underlying assumptions regarding the system behavior (deterministic, stochastic), the level of detail in the system's observations (from Boolean to continuous variables) and the algorithms used, define the qualitative and quantitative information that can be incorporated into and/or inferred from a model.

### 27.2.1 Boolean Networks

While Boolean networks were mainly designed for studying in a simplified manner the dynamical properties of a system, they are frequently used to infer structural properties in intracellular networks by identifying causal relationships between the network components under the Boolean assumptions. In these networks, the nodes are reduced to binary (0 or 1) variables and the interactions between them are described by logical functions.

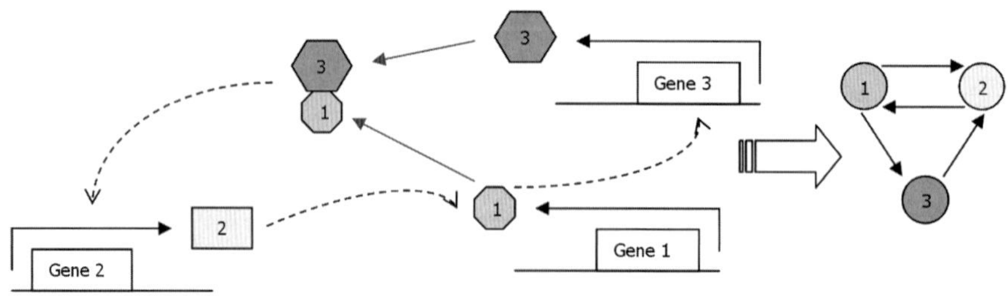


**Fig. 27.2** A small gene regulatory system and its corresponding network representation. Genes are translated into proteins which may form complexes as the gene products 1 and 3; these proteins or complexes regulate in turn the genes, thus forming a network of interactions. Discontinuous arrows indicate regulations. (Copies of figures including color copies, where applicable, are available in the accompanying CD)

Boolean networks were first used to model gene regulation by Kauffman [2], who was convinced that despite their inherent simplicity, these algorithms can capture the complex dynamics of gene networks and give a qualitative interpretation of their behavior. In Kauffman networks, the activation of a gene (node state) was assumed to be binary, i.e., to have only two states, *on* (expressed or over-regulated) or *off* (not expressed or down-regulated). All intermediate levels of activation were ignored. Each node was randomly connected with K nodes that remain its regulators throughout the system's temporal evolution. The interactions between genes were described by non-linear deterministic Boolean logic rules such as AND, OR and NOT, combined properly to enable a dynamic network behavior, which resembles that of the biological system. In essence, having the state of each network component (gene expression) at a given time, step $t$ and a set of rules that approximate the regulation functions among components, the state of any network component at a next time, step $t+1$ could be determined. The network updates its values in a deterministic parallel fashion. An example of a Boolean network is shown in Fig. 27.3.

This early work was able to describe certain behaviors of gene regulatory networks with relatively high accuracy. A limitation of the Kauffman Boolean network is that the model exhibits

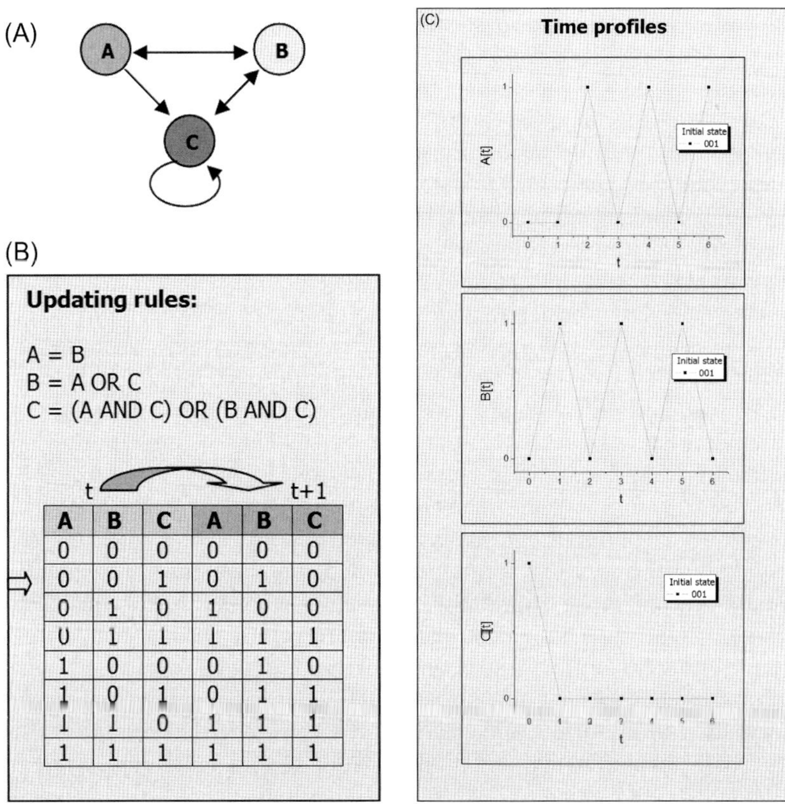


| A | B | C | A | B | C |
|---|---|---|---|---|---|
| 0 | 0 | 0 | 0 | 0 | 0 |
| 0 | 0 | 1 | 0 | 1 | 0 |
| 0 | 1 | 0 | 1 | 0 | 0 |
| 0 | 1 | 1 | 1 | 1 | 1 |
| 1 | 0 | 0 | 0 | 1 | 0 |
| 1 | 0 | 1 | 0 | 1 | 1 |
| 1 | 1 | 0 | 1 | 1 | 1 |
| 1 | 1 | 1 | 1 | 1 | 1 |

**Fig. 27.3** **(A)** A regulatory Boolean network of 3 components A, B and C with their corresponding interactions. Edge directions indicate flow of causality. The regulators of the component B, for example, are the components A and C. Thus, B changes its values in respect to the updating rule between A and C. **(B)** Each component can take a value of 1 or 0. The updating of logical rules for each node of the network are shown. The rules are applied to the nodes under a synchronous updating scheme and the table of the corresponding temporal transitions of the system for each possible state is presented in table **(C)** An example of the temporal profiles of each component of the network when the initial state corresponds to the values 0 for A, 0 for B and 1 for C. The states of the components A and B change periodically in time. The component C reaches a steady state (Copies of figures including color copies, where applicable, are available in the accompanying CD)

a robust behavior similar to that of biological systems, only if the number K of connections per component is less or equal to two [3, 4]. In reality, this constraint is rarely satisfied. Individual genes, for example, are known to interact with several –often with up to 20– regulatory factors. Fortunately, the homogeneous network topology and the 'match in step' updating scheme of Kauffman networks that impose this restriction can be modified to better reflect the natural, intrinsic properties of the biological systems. The dynamic behavior of the system is strongly dependent on those intrinsic properties, thus both issues should be addressed. The choice of Boolean functions, the updating schema from synchronous to asynchronous and the global network properties are some of the issues currently being revised in the field. For the interested reader, C. Gershenson discusses many of the state-of-the-art topics in the field of Boolean networks [5].

The amount of theoretical work done on Boolean networks [6–11] and their respective biological applications [12–16], show that these algorithms are capable of qualitatively describing the complex, dynamic phenomena present in biological systems. More importantly, computational methods have been developed to provide further insights towards understanding the regulatory mechanisms within a biological system, by enabling the inference of the biological network structure from observations of its constituents. Specifically, the value of any given node in the Boolean network is derived by a set of K parent nodes, which are sufficient to determine it fully. Thus, the problem of revealing causal relationships in the data, also known as a consistency problem, amounts to finding the parent nodes for each network component subject to a constraint: a Boolean function applied on the parent nodes of any component should determine its two possible values and reproduce its observation profile.

Given unlimited computational recourses, an exhaustive search over all possible tuples of the system components could be performed to determine the parent set for each target component that fulfils the observation consistency constraint. However, as this is computationally very expensive a more efficient search algorithm is required. Furthermore, given the high content of noise in most types of biological data, it is likely that not even a single consistent tuple is identified. Thus, the consistency search problem has been reformulated to search for a Boolean formula that minimizes the number of inconsistencies, known as the *best-fit extension approach*. Several theoretical studies have analyzed the pros and cons of these two different approaches [17–22]. Since the number of examples required to identify the parent set is exponentially dependent on the size of parent nodes [23], when limited amounts of data are available, it is prudent to constrain the size of candidate tuples by sacrificing real representations.

For both theoretical and practical reasons it is interesting to know whether the network under construction is unique. The inference problem does not necessarily imply a single structure for the system under study. On the contrary, a number of Boolean networks are frequently suggested as optimal for the data given. A key question, however, is whether one should demand a unique solution. Keeping in mind that biological networks are robust and adaptable to environmental (internal or external) changes, different structures corresponding to a variety of internal alterations in the interactions may be equally good representations, as long as the dynamic behavior of the system does not change dramatically. Given additional knowledge concerning the structure of the system, the set of all possible networks could be reduced to a more representative subset [24]. Additionally, it could be interesting to investigate the dynamic behavior of all possible structures. Networks with similar dynamics may depict various manifestations of the same biological system that ensure robustness against environmental changes [25].

Boolean networks have frequently been used to infer and model the transcriptional regulatory network from sets of time-course gene expression data. A variety of algorithmic strategies for inference have been proposed in the literature [17, 19, 20] and applied for the Boolean network reconstruction and analysis of the *Saccharomyces cerevisiae* cell cycle [22, 24], the immunology microarray data-sets [25] and the metastatic melanoma related gene expression data [12, 26], among others.

### 27.2.2 Generalized Logical Networks

The generalized logical method was introduced by Thomas and colleagues in 1995 [27] and has been extended thereafter [28, 29]. Like Boolean networks, the method assumes discrete values for the system components, with the exception that these values can be more than two. It also allows a more sophisticated definition of logical interactions as well as the asynchronous temporal transition of system components.

In the Boolean formalism, a threshold is needed to convert the continuous values of the system observations to binary values. Only when a component's value surpasses the threshold an effect on the other components is generated. Multivariate logic values offer a more realistic representation of biological reality, where single elements interact with multiple elements in different ways, thus requiring multiple thresholds to be defined. The molecular population bandwidth of a node variable, in that sense, may have to be quantized to as many degrees needed to reflect the different levels of activation of each node that it regulates.

In the generalized logical networks shown in Fig. 27.4, each edge is labeled with a sign that expresses inhibition ('–') or activation ('+') and a rank number of the threshold above which, the regulatory effect occurs. The logical functions, used to determine the component values, are generalizations of the Boolean functions specified in a way that is consistent with threshold restrictions and biological considerations. To achieve an asynchronous temporal transition of the system from one state to another, the logical functions map the current states to all potential states the system can undergo when no simultaneous update of the components is allowed. Taking into account the time delays in information flow from one component to another, the transition path that the system will follow given an initial state, can be determined. Under the above considerations, Rene Thomas observed that the dynamics of such systems correspond to discrete states of the original system and can be approximated by differential equations [30].

The generalized logic approach is primarily used to analyze systems of known interactions. It has been successfully applied to a number of small-scale, well studied regulatory systems, such as the flower morphogenesis gene network in Arabidopsis [31], the control of cell determination during early Drosophila development [32] and the core network, controlling the mammalian cell cycle [15]. Since the method is not particularly suited for identifying a structure that best represents the real system under certain assumptions, inference is a major drawback, that results from the difficulty to find a natural scoring function for all the possible networks and thus identify the one that best describes the data.

### 27.2.3 Bayesian Networks

The main characteristic of these models is the introduction of stochasticity in the system states; the system components are allowed to take on values that depend on the inter-relations between them

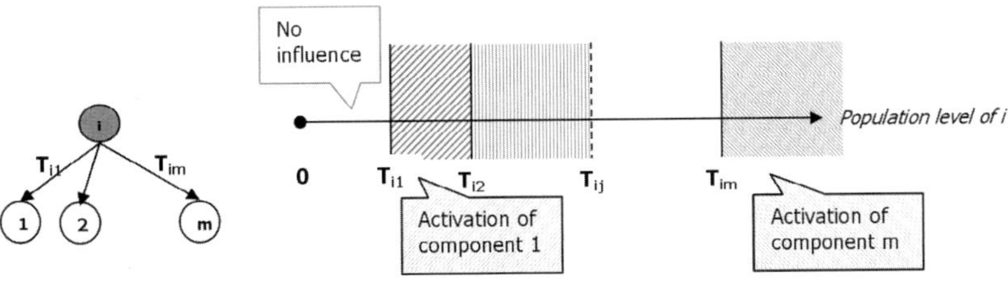


**Fig. 27.4** The node *i* influences m other nodes (*left*). *m* distinct thresholds ($T_{i1}$, $T_{i2}$, . . ., $T_{im}$) on the population level of the *i*th component may hold for its influence to be present in each of the *m* components (*right*) (Copies of figures including color copies, where applicable, are available in the accompanying CD)

and can be described in a probabilistic manner by what is called a joint probability distribution *joint probability distribution*. These inter-relations are characterized by two main principles: dependency and conditional independency. Two components are termed dependent if the knowledge of one provides predictive knowledge of the other. However, if knowledge of a set of components is enough to determine another component, the latter is conditionally independent of any other component of the system, given that set.

A Bayesian network is a graphical representation of the joint probability distribution of the underlying system, which incorporates faithfully the dependencies and conditional independencies among the system components. Thus, it is minimal with respect to the connections it represents. A Bayesian network is a directed acyclic graph, which means that feedback loops are not accommodated in the model and only static regulations can be described. In this graphical representation, the nodes represent the system components and the edges the direct dependencies among those components. Each component is associated with a conditional probability distribution that quantifies the strength of its dependencies. The value of a system component can be either discrete or continuous, and the Bayesian models have the potential to include any or both of these representations, describing the system at different levels of detail.

Consider a very simple example of three genes in a Bayesian representation as shown in Fig. 27.5. How is information transferred in that network? Gene 1 and Gene 2 depend directly on Gene 3. This means that a change in the value of Gene 3 will affect the values of the others. Moreover, Gene 1 and Gene 2 are indirectly dependent if the change of Gene 3 is not given. However, when the value of Gene 3 is known, Gene 1 and Gene 2 are no longer dependent. The probability of the system being in a certain state (joint probability) where each gene takes values $g_1$, $g_2$, and $g_3$, respectively is given by:

$$P(g_1, g_2, g_3) = P(g_3)P(g_2|g_3)P(g_1|g_2, g_3) \tag{27a}$$

$$P(g_1, g_2, g_3) = P(g_3)P(g_2|g_3)P(g_1|g_3) = \prod_i (g_i|P_a(g_i)) \tag{27b}$$

The joint probability distribution can always be factorized according to the chain rule as shown in equation (27a). The conditional independences of the system, depicted in the structure of the Bayesian network (Fig. 27.5), can be used to further simplify the factorized form as shown in equation (27b). Equation (27b) shows the *Markov condition* implied in the Bayesian networks according to which the probability distribution of any node ($g_i$) in the network, given its parents $P_a(g_i)$, are independent of any subset of its non-descendent nodes. The state of the system is thus probabilistically determined by the knowledge of the factor terms.

Two important issues should be mentioned about Bayesian networks. The first deals with the fact that there is no unique graphical representation of the joint probability distributions and the set

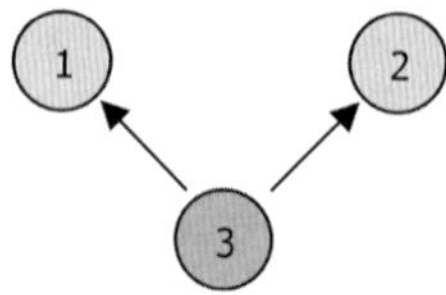


**Fig. 27.5** A Bayesian network with 3 components (e.g., genes) depicts the independencies and conditional dependencies among its constituents. Knowledge of the value of gene 3 implies that the value of gene 1 can be determined independently of the value of gene 2. However, if the value of gene 3 is not known then there is an information flow among genes 1 and 2 (Copies of figures including color copies, where applicable, are available in the accompanying CD)

of independencies the system comprises. More than one graph may imply exactly the same set of independencies. This ambiguity is crucial when trying to infer a network from the data and understand its properties. However, it has been shown that all equivalent graphs, with respect to the joint probability distribution they represent, have the same underlying undirected structure[1]. The second issue concerns the flow of causality among the system components. A directed edge in the graph is not necessarily an edge from a cause to its effect. Inferring a causal Bayesian network from observations alone is usually not possible and prior knowledge – especially knock out experiments – have to be taken into account. In general, Bayesian networks have several advantages: they can handle missing data and unobserved values of system components by allowing latent components to be present in the network; they are robust for noisy data, as they are probabilistic in nature, and are inherently optimal for describing processes composed of locally interacting components.

Inferring the most probable Bayesian network from the data is generally a hard problem. The difficulty comes from the number of system components that the network consists of and the complexity of their interactions. Current state-of-the-art algorithms struggle to scale up to thousands of components in a reasonable time while learning remains a challenge. Incorporation of prior biological knowledge and integration of different sources of data (binding data, protein expression data, knock-out experiments, etc.) for the system under investigation, make the problem more tractable. Bayesian methodology provides a principled way of incorporating additional information as prior knowledge, but the challenge is to assign a weight of trust in these different sources of information.

The Bayesian framework was first used to represent gene interactions Friedman et al. [33] was applied. A learning method to infer the cell cycle regulatory system from gene expression data. In [34], the method was used to reconstruct the galactose regulatory network in *Saccharomyces cerevisiae*, while Jansen, et al. [35] used the Bayesian framework to predict genome-wide protein-protein interactions in yeast and Le Phillip, et al. [36] studied the impact of incorporating prior structural knowledge on learning.

### 27.2.4 Dynamic Bayesian Networks

Bayesian networks can only describe static regulatory systems due to their acyclic structural constraint property. They are capable of describing causal interactions, but the dynamic evolution of the system is not taken into account. As dynamic data are richer in information than static data, their use should further reduce the ambiguities concerning the underlying network structure. Temporal or dynamic Bayesian networks are extensions of the Bayesian networks that explicitly model the stochastic evolution of the system components through discrete time. Dynamic Bayesian networks can be seen as a stream of static pictures/frames of the system spanning across time, which are connected through temporal arcs explaining the time interaction among the components. The connections that describe each frame do not change over time so that the network is stable. It is also assumed that the state of the system at a given time point depends only on the state of the system at a previous time (Markov property). The assumption is satisfied when the regulatory process and the experimental sampling rate[2] are in agreement. The whole dynamic phenomenon is thus

[1] A directed graph is comprised of a set of nodes and a set of directed edges that express the direction-flow of interaction among nodes. The undirected structure of a directed graph is comprised of the same set of nodes and edges but the edges have no direction.

[2] The sampling rate is defined as the frequency of samples (e.g. gene expression levels/gene) taken over the temporal evolution of the phenomenon.

described in a probabilistic manner and the joint probability distribution tracks the temporal evolution of each of the components of the system.

Theoretical work on how to learn Dynamic Bayesian networks from time course data was first done by Murphy [37]. Since prior knowledge improves the quality of inferred networks, a lot of research work has been done towards the integration of such knowledge in the Bayesian framework and the resulting algorithms have been applied in several biological systems [38–41].

### *27.2.5 Supervised Structural Inference*

As more and more knowledge about biological systems is acquired, not only on the level of different sources of observations for the system components, but also on the level of the sub-structures that comprise those systems, a multi-level integrated model can be more beneficial for structural inference of the whole system. Methods that take advantage of the knowledge provided by experimentally verified sub-structures (such as functional modules) within a network, in addition to the known observations, are referred to as supervised approaches and will be discussed in the following section. The intuition is that these sub-networks offer structural information about the system and the goal is to complete the whole network by adding, under supervision, the rest of the components and their potential connections to better match the observations [42, 43].

## 27.3 Identifying Functional Modules in Biological Networks

In our efforts towards identifying the structure and function of intracellular networks, it is important to remember that cells are well-organized systems having their components strategically positioned and regulated in a functionally independent, modular manner. This form of internal organization was selected throughout evolution as a 'divide and conquer' method to successfully manage the increasing complexity and also as a 'safety switch' where diverse reactions could be carried out without cross-talk that might harm the cell. Connectivity among such functional modules is the key feature that makes the cell operate as an integrated system, allowing internal functions to influence one another [1]. Identifying functional modules is thus crucial for understanding intracellular functions. In this section, we discuss two approaches that try to reach this goal: the bottom-up approach (Section 27.3.1) where raw data are being analyzed and grouped based on a common property, thus revealing discrete blocks that would comprise the functional modules and the top-down approach (Section 27.3.2) where modular organization is derived from already structured networks.

### *27.3.1 From Modules to Networks: A Bottom-up Approach*

The ongoing accumulation of large amounts of multi-faceted biological data and the need for handling and interpreting the available information led to the development of many novel computational methods. Initially, the algorithms used aimed at identifying similarities or extremes among the components of a data-set, in an effort to group together those sharing a similar property – thus creating the first modules – or to identify the components with a particularly different behavior compared to the rest [44]. In these early attempts, two main types of data were used to derive functional modules: genomic data (primarily from microarray experiments) and proteomic data (such as protein expression and protein-protein interactions). Several algorithms were applied based on different similarity measures, resulting in the partitioning of the data into groups. The degree to which these groups formed 'functional modules' was estimated by comparing the computational predictions to known functional entities or designing novel experiments to support (or reject) the biological significance of each group. Later on, in an attempt to delineate the biological mechanisms that govern the regulation of these modules, more sophisticated algorithms were developed, capable of combining different kinds of experimental data including sequence data, annotation data, protein-DNA interactions etc [45, 46]. In the following paragraphs, we will briefly discuss the pioneering or most widely used algorithms as well as some examples of their application to biological data.

Many of the algorithms used to analyze and integrate large-scale biological information focus on grouping similar data together. Various clustering (e.g., Hierarchical, Self Organizing Maps, k-Means, k-Nearest Neighbor) and dimensionality reduction (e.g., Principal Component Analysis, Independent Component Analysis, Multi-Dimensional Scaling, Partial Least Squares) algorithms are some of the methods most frequently used to identify functional modules.

#### 27.3.1.1 Clustering Algorithms

Clustering techniques are used to organize the data in groups (clusters), based on their similarities (or differences), in an attempt to reveal well-defined sub-structures that may be of biological significance. Three main issues need to be addressed when using a clustering algorithm. First, what is the specific property or combination of properties of the system under consideration whose similarity should be measured, i.e., what can be clustered?. Then, a measure of similarity (metric) has to be defined. This is the critical step for all clustering methods as different measures can result in different clusters [47]. Finally, a way for evaluating the clustering results needs to be selected, the best being experimental verification.

Since there are many clustering algorithms that can be applied to the same data-set, each one imposing different biases on the structure of the clusters [48], the selection of an optimal algorithm is not-trivial. However, several criteria can be used as a measure of clustering 'correctness', such as cluster accuracy (how biologically meaningful the derived clusters are compared to existing knowledge) and cluster stability (if the same clusters are derived when the algorithm is applied to synthetic data) [49]. However, since clustering is the answer to the question, 'what is the best partitioning of objects given specific similarity criteria', a measure of 'good' clustering from the mathematical point of view is how accurately an algorithm performs this task. Therefore, a technically optimal algorithm may not necessarily provide the best answer to a biological question unless special care is paid on posing the question in a mathematically sound way.

The simplest method for arranging genes in groups is called *Guilt-By-Association* (GBA) and was initially used to find genes with similar expression patterns compared to that of a known cancer-associated gene [50]. The algorithm calculates the pair-wise association between the genes in the data-set and the gene of interest, using the hyper-geometric distribution, and computes a p-value showing the probability that the observed co-expression is due to chance. This method has also been used to search for modules of co-expressed genes where, nearby genes in the expression space were grouped together using a pre-defined p-value cut-off and the results were then confirmed experimentally [51].

Clustering methods are divided in *hierarchical* and *non-hierarchical*. Hierarchical algorithms result in a tree-shaped structure called dendrogram, where each cluster connects to others, thus forming larger groups, as shown in Fig. 27.6. Hierarchical clustering can be either agglomerative or divisive, that is use top-down or bottom-up approaches. Agglomerative clustering starts with each vector forming its own cluster, and merges similar clusters together until there is one single super-cluster remaining or a pre-defined termination condition satisfied. Divisive clustering starts with the entire data-set as a single cluster, which is iteratively split into smaller clusters according to some dissimilarity criterion. Since the computational cost for dividing a cluster into two, given all possible choices is high, divisive clustering is less commonly used. In one of the first applications of hierarchical clustering on gene expression data, Eisen et al. [52] derived clusters that contained

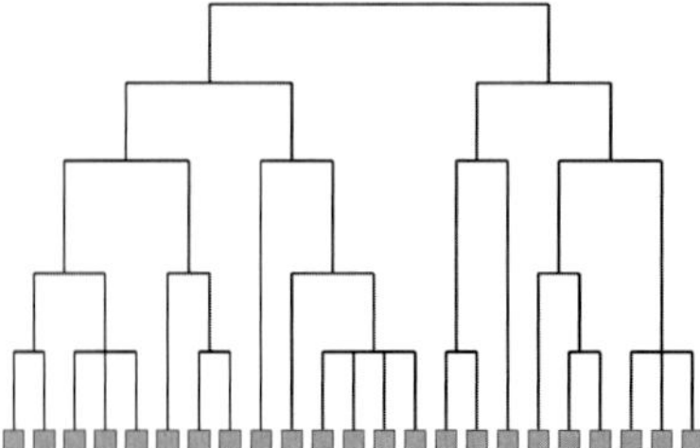

**Fig. 27.6** Dendrogram representing the output of a hierarchical algorithm. At different stages of the algorithm there are different numbers of clusters, ranging from a single cluster on the top merging all the components, to the maximum number of clusters at the bottom, representing each component as an individual cluster (Copies of figures including color copies, where applicable, are available in the accompanying CD)

genes known to belong to the same biological function using microarray data from the budding yeast *Saccharomyces cerevisiae* and human fibroblasts. This work was based on the assumption that co-expressed genes are most likely to be co-regulated under particular conditions and may have the same biological function, which was also the case in several subsequent studies [53–57].

Non-hierarchical clustering methods, like *k-Means* and *Self-Organizing Maps* (SOMs), divide the data into *k* different clusters, such that the objects in a cluster are more similar to each other than to objects in different clusters. The *k-Means* algorithm starts by randomly assigning the input points (e.g., genes) into *k*-pre-determined initial groups and then calculates the mean point called centroid for each group. The data are then reassigned so that they belong to the group whose centroid is closer, and this procedure iterates until the means converge. Different number of clusters (*k*) as well as different initial centroid positions may yield different results. Therefore, it is important to run the algorithm several times with different random initial conditions [48]. The *k-Means* algorithm has been used to analyze microarray cell cycle data from *Saccharomyces cerevisiae*, resulting in the partitioning of gene expression profiles in 30 clusters, with 49–186 genes per cluster. A significant number of clusters contained genes that belonged to the same functional class [53]. Other applications of this algorithm for functional module identification can be found in [58–60].

A variation of the centroids approach is also used by the *Self Organizing Maps* algorithm, which was developed to model the way in which neurons in the brain become tuned to specific receptive fields [61]. It is a neural-network based divisive clustering technique that projects high-dimensional data to a low-dimensional, usually 2D or 3D grid. The nodes in the grid represent neurons, which, during an iterative unsupervised learning process becomes specifically tuned to various input signal patterns or classes of patterns in an orderly fashion. The algorithm starts by defining the dimensions of the grid whose nodes are subsequently initialized with random values. Each sample in the initial data-set is compared, using a specified metric, to each node and assigned to the most similar one. For each assigned sample, the centroids of the 'winner' node along with the adjacent nodes are re-calculated to incorporate the new sample. During this learning process, individual changes may be contradictory, but after a large number of iteration steps, the net process is a mapping of the initial data-set, so that similar samples are clustered together, as well as nearby clusters in the topology mapping contain samples that are near each other in the input space. SOMs can thus serve as a clustering tool as well as a tool for visualizing high-dimensional data. An example of a SOM algorithm output generating a $5 \times 5$ grid is illustrated in Fig. 27.7. Several studies used SOMs to search for functional modules in biological data [62, 63]. A good example is the work of Tamayo et al., who used SOMs to recognize and classify gene expression features from complex multi-dimensional data: a cell cycle microarray data-set from *Saccharomyces cerevisiae* and a human hematopoietic differentiation data-set. In both cases, SOMs identified biologically significant clusters capturing cell cycle periodicity and known pathways, respectively [64].

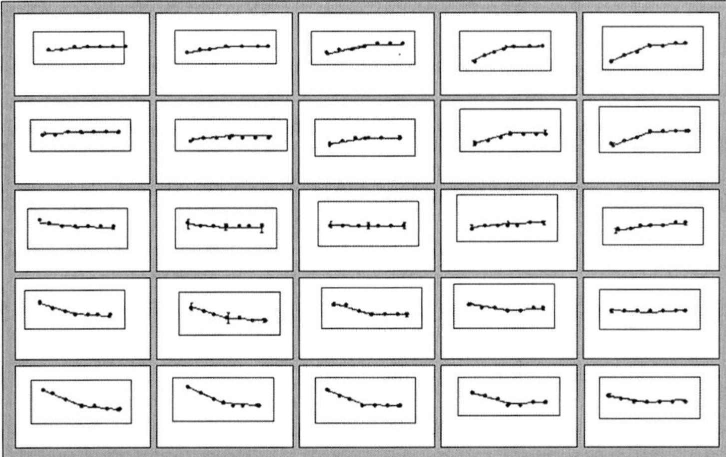

**Fig. 27.7** Output of a 2-dimensional 5×5 Self-Organizing Map. Each node represents a cluster of samples while neighboring clusters include the samples which are located close to each other in the original space (Copies of figures including color copies, where applicable, are available in the accompanying CD)

#### 27.3.1.2 Dimensionality Reduction Methods

Clustering is a quantitative way to gain first order knowledge from the data by partitioning data points into disjoint groups and maximizing intra-group similarity. However, clustering techniques cannot simplify the large number of dimensions in the data space, and dimensionality reduction methods are utilized to perform this task. *Principal Component Analysis* (PCA) achieves dimensionality reduction by finding new axes, called principal components, which identify the linear combinations of signaling axes most tightly connected with one another. Instances are thus assigned to these real-valued vectors in a lower dimensional space (often just two or three dimensions) that capture the most important information in each of the original signaling axes. Briefly, in a data-set with $N$ vectors from $k$-dimensions, the algorithm finds $c \leq k$ orthogonal vectors that can be used to represent the data without significant loss of information. The original data-set is thus reduced to one consisting of $N$ data vectors on $c$ principal components, where each data vector is the linear combination of the $c$ principal component vectors. The PCA method has the advantage of radically reducing dimensionality while preserving most of the variance in the original data space. Clustering approaches like those described in the previous paragraphs can then be applied to the reduced data-set. Although some scientists argue against the quality of the clusters after using PCA, this method has proved to be valuable for a number of biological applications for viewing high-dimensional data spaces compactly [47]. For example, PCA is frequently used on microarray data to improve visualization and facilitate clustering, classification or modeling efforts by filtering out the components that contribute to noise [65, 66].

*Partial Least Squares* (PLS) is another method used for dimensionality reduction, similar to PCA. The main difference is that while PCA factors the entire data set into principal components in an unsupervised manner, PLS predicts a set of dependent variables from a larger set of independent variables based on a proposed relationship [47]. PLS places strong emphasis on predicting the system responses without explaining the underlying relationship between the interacting variables. For example, it aims to predict future signaling responses from events at earlier times [47]. However, experimental observations are frequently not adequate for calculating a unique solution that estimates the contribution of each signaling variable in the independent block to those in the dependent block. It is in this context that principal components and PLS are most useful. Rather than performing the regression in the original data-set, PLS reduces the dimensions to a principal-component space and then regresses the independent and dependent principal components [47].

PLS methods are characterized by high computational efficiency, flexibility, and versatility which make them particularly attractive for analyzing biological data. Using mRNA expression data and protein-DNA binding information, Boulesteix and Strimmer estimated transcription factor activities as well as the functional interactions of regulators [67]. Following the analysis of microarray data on sporulation of budding yeast, the gene expression levels of representative genes were regressed against the expression levels of the remaining genes [68]. Finally, using amino acid composition data, Clementi et al. attempted to predict secondary structure [69] using PLS.

#### 27.3.1.3 Integration Methods

Most of the computational approaches that have been used to identify functional gene modules, describe the regulatory events in a cell utilizing a single type of biological data, e.g., gene expression or protein-DNA interaction data and so on. However, expression data based approaches, often assume that co-expression of a gene implies co-regulation and that the expression levels of regulators and their target genes are somehow related. This is not always true as post-translational modifications of transcription factors, for example, are well documented [45, 70]. Algorithms based solely on protein-DNA interaction data cannot distinguish between positive and negative regulation as such data are often very noisy due to non-specific binding [46].

Algorithms that integrate multiple data types are currently a fast developing area in the bioinformatics field [71]. The goal of these algorithms is to integrate diverse, high-throughput data to discover patterns of combinatorial regulation and to understand how the activity of genes, involved in similar biological processes is coordinated and interrelated. Examples of such algorithms that have been extensively used to identify functional gene regulatory modules by combining various types of data (such as gene expression, protein interaction, growth phenotype, transcription factor binding data etc) include the GRAM [46], the ReMoDiscovery method [72], GeneXpress [45] and SAMBA [73], to name a few. Next, we briefly describe GRAM as a representative example of this class of algorithms.

The GRAM algorithm [46] combines gene expression with protein-DNA interaction and annotation data to identify modules of genes that are bound by the same transcription factors and share similar expression profiles. Using annotation data the algorithm also pin points the most likely biological function of the genes in the identified modules. The method starts by finding groups of genes that share the same set of regulators based on a p-value criterion. Within each group, the genes that are co-expressed form the core of a module. Additional genes with similar expression profiles, which are also bound by the same transcription factors but with a more relaxed binding criterion, are then added to the module. The algorithm also correlates the expression profile of each regulator to that of the average gene expression pattern in the module, to infer whether it acts as a positive or a negative regulator. The GRAM algorithm was first applied to *Saccharomyces cerevisiae* using binding data for 106 transcription factors and over 500 microarray experiments [46]. 106 gene modules containing 655 distinct genes regulated by 68 transcription factors were identified and visualized as a network of transcriptional interactions. Comparing derived modules with experimental results, for example, by performing independent chromatin-IP experiments and literature research, it was verified that GRAM identifies biologically meaningful regulatory modules with high accuracy.

### *27.3.2 From Networks to Modules: A Top-down Approach*

Complex biological networks, often consisting of hundreds of interconnected components, are best visualized as tree-like structures, where nodes represent biological entities and edges represent a relationship (interaction, regulation, reaction and activation/inhibition) among pairs of the components. A plethora of biological networks are present in the literature, including protein-interaction networks [74–76], metabolic networks [77, 78], gene regulatory networks [79, 80], pathway networks [81] etc. The ongoing accumulation of experimental data further increases the complexity of such networks making them hard to visualize and interpret. Since 'modularity' is an intrinsic property of biological systems [1], a number of algorithms have been developed that try to isolate smaller functional groups (modules) inside large biological networks, in order to better visualize and interpret their functions. We will discuss three examples of such approaches, each one implementing different criteria for specifying modularity.

#### 27.3.2.1 Modules in Metabolic Networks

Cellular metabolism can be represented graphically by the metabolic network, a fully connected biochemical network where the metabolic substrates (nodes) are densely inter-connected through biochemical reactions (edges). Jeong and colleagues [77] analyzed the metabolic pathways from 43 organisms, and came to the conclusion that metabolic networks exhibit scale-free topology, as the probability $P(k)$, that a certain substrate is connected with $k$-reactions follows a power-law distribution, i.e., $P(k) \sim k^{-y}$ where $y \cong 2.2$ for all organisms. Metabolic networks are therefore characterized by the presence of a few, highly connected, substrates (hubs) participating in a large number of metabolic reactions; a feature that provides robustness against random mutations or environmental fluctuations of non-hub substrates [77]. This observation led Ravasz et al. [82] to pose the dilemma whether modularity is indeed a key feature of such networks, since the appearance of hubs integrates all the components in a single web, in which the existence of fully separated modules is, by definition, prohibited. On one hand, evidence from the analysis of metabolic networks indicates the existence of functionally separated entities (for example the biochemical pathways) [83, 84] ,while exhibiting high clustering coefficients, indicative of a modular structure. On the other hand, scale-free topology implies a highly non-uniform distribution of the interconnections, opposing the idea of modularity. This dichotomy was overcome by describing the metabolic network with a hierarchical architecture keeping both the modular and the scale-free properties. A synthetic generation of such an architecture would be the iterative quadruplication of an initial cluster, leading to an inherently hierarchical organization of the nodes with different degrees of modularity [82].

Application of different algorithms to cluster metabolic networks based on topological properties has often resulted in the identification of functional modules representing known biological pathways [82], [78], [85, 86]. Based on the assumption that neighboring components in a biochemical network may belong to the same pathway, one approach used a topological matrix representing the degree to which two substrates share common connections with other substrates and applied a clustering algorithm to identify such topological modules. When applied to the metabolic network of *Escherichia coli*, this method resulted in modules with a major overlap with already characterized pathways such as the pyrimidine metabolism [82]. Another approach for the partitioning of a network into modules is to find the partition that maximizes the within-module links and minimizes the between-module links, using appropriate algorithms like simulated annealing [78]. Application of this algorithm to the metabolic networks of twelve organisms including bacteria, eukaryotes and archaea resulted in an average of 15 different modules in each network with high inter-module density of links and module content that attributes to a specified pathway according to information from the KEGG database. An alternative computational approach is the application of flux coupling analysis to metabolic pathways to identify functional groups of reactions [85, 86].

#### 27.3.2.2 Modules in Gene Regulatory Networks

A gene can affect the expression of another gene by binding its product on the control region of the target gene. At the level of the full genome, we refer to regulatory networks as the regulatory interactions between genes.

Carter et al. [87] built a gene network using expression profiles and by considering two genes as linked if their expression profiles were correlated to a significant degree. The authors confirmed a power-law distribution model indicative of a hub-oriented network and also a high-level clustering coefficient that, similar to metabolic networks [82], is characteristic of a hierarchical organization of modularity. Using deletion experiments it became apparent that highly connected nodes represent essential genes and that gene-classes enriched in connectivity are process specific, thus revealing the modular structure of the network.

In an attempt to unravel the structural organization of the transcriptional network of *Escherichia coli,* the presence of modules as well as their regulators emerged. The transcriptional regulatory network appeared to have a five-layer hierarchical structure with a feed-forward regulatory relationship between the layers. Each gene in a top layer could directly regulate any of the downstream genes (but not the upstream) with an average path length of 1.36. Based on the connections in this hierarchy, groups of common-bound genes were considered as functional modules and their regulators were identified [88].

#### 27.3.2.3 Modules in Protein Interaction Networks

A 'functional module' is an entity with a separable function that can act independently through chemical, functional, or spatial isolation. In that concept, protein complexes, whether long-lived (like the F1Fo ATP synthase) or transient (like CDK-cyclin complexes), can be regarded as modules having spatial and functional isolation with strong interactions within the module and weak interactions with the external components [89]. However, when considering a protein as part of a larger cellular network, where each protein can interact with numerous other proteins, clusters of such interactions could represent functional modules.

Protein-interaction networks also form a highly in homogeneous scale-free structure [74]. There are a few proteins with many connections (the 'hubs'), while the vast majority of the proteins have a very small number of links in the network. The biological role of hubs is considered highly significant as provend by deletion experiments, where the removal of hubs is usually lethal for the cell. In a study regarding the analysis of protein interactions in the yeast *Saccharomyces cerevisiae* [76], the authors further arranged the hubs in two categories, based on the way they interact with their partners according to mRNA expression data. They identified 'party' hubs that interact with their partners simultaneously and 'date' hubs that bind their different partners at different times or locations. The authors showed that 'date' hubs are, in fact, the connectors between modules, while the 'party' hubs provide the within-module connections.

Traditional clustering algorithms have also been applied to protein-interaction networks to identify functional modules. A common technique is to describe the interactions with some form of a weight network. For example, networks whereby weight values are analogous to the reliability of the particular interaction [90, 91], or to the shortest path length reveal modules that contain proteins with biologically significant function [92, 93].

#### 27.3.2.4 On a Smaller Scale: Network Motifs

Analyzing networks on a global scale helps uncover the relatively large groups of components that are somehow related, and play a specific and distinct role in the orchestrated regulation of interactions. Complementarily, the search for repeated local graph properties has received increasing interest over the last decade and the identification of network motifs is now an important field in Systems Biology. Network motifs are structural elements (patterns) that occur in statistically significant quantities compared to the respective random structures. Their presence has been documented in many kinds of networks, from social to biological ones, indicating their significance in offering a functional advantage to the system [94]. Interestingly, analysis of protein and genetic networks revealed that network motifs are dominant within these network structures [81, 94]. Different types of networks are characterized by their distinct motifs, for example, the three-node feed-forward motif is regularly seen in biological networks [81, 94–97] while the four-node feedback loop is basically apparent in electric circuits [94]. To identify the motifs present in a given network, all substructures of $n$ nodes are determined [94]. Motifs appear to be evolutionary conserved in the yeast protein interaction network [98] as well as in genetic networks, indicating a key role in the optimal design of biological networks [99].

## 27.4 Evolution of Functional Modules

The revelation of the presence and role of functional modules in numerous complex biological networks, raises the question of whether a module is evolutionarily conserved. In other words, does evolution play a role in shaping inter- or intra-module interactions? The answer to this question will be given by the efforts of future workers. Initial findings suggest that modules are, in general, evolutionarily conserved entities; the extent of this conservation depends on the type of the functional module. For example, protein complexes appear to be more evolutionary conserved than co-expressed gene clusters. Evolutionary changes are more frequently observed in the module linkage than in the intra-module composition [89, 100].

## 27.5 Conclusion

In this chapter, we discussed some of the most popular analytical and computational approaches used to identify the structure and infer the function of intracellular networks at various levels of abstraction. Since complex intracellular systems are often composed of smaller, functionally independent sub-network structures, we discussed different approaches that partition a system into functional modules or reconstruct it based on the interactions between these entities. Different algorithms may result in different compartmentalization of the underlying structure of the whole, but when combined effectively, these approaches can – at least in principle – provide a global view of the coordinated functionalities inside complex biological systems.

Limitations induced by the amount of biological knowledge and the complexity bounds of certain methods can be overcome by assuming well-defined constraints in the computational algorithms, while maintaining a high level of modeling accuracy. Nice examples include the application of Boolean methods, in which the network nodes are reduced to binary variables and the interactions between them are described by logical functions, in modeling the *Saccharomyces cerevisiae* cell cycle [22, 24]. Alternatively, methods that reconstruct a cellular system by first breaking it down into functional modules which can be studied in more detail and then connecting the modules into a more complex network are also good examples. The work of Bar-Joseph et al., [46] in *Saccharomyces cerevisiae* showed that combining binding data with expression information can reveal biologically plausible, functionally distinct sub-network structures with very high accuracy. Overall, the many examples discussed in this chapter illustrate the far-reaching capabilities of available computational methods, especially when combined with experimental techniques, in extending our understanding of complex biological networks. Some of the advantages and disadvantages of the methods discussed in this chapter are shown in Table 27.1 while several implementations of these methods that are available on the internet are listed in Table 27.2.

**Table 27.1** Summary of the strengths and weaknesses for some of the computational methods discussed in the chapter. A similar comparison for some of these methods can be found in [47]

| Method class | Method | Optimal dimensions | Strengths | Weaknesses |
|---|---|---|---|---|
| Clustering | Hierarchical | Dendrogram 'branches' | Simple and unbiased; entire dendrogram can be scanned for assembly of clusters | Clusters must be assembled pairwise; some clusters might lack biological relevance; dendrogram does not simplify the data set |
| Clustering | k-means/ SOMs | Centroids | Clusters are assembled in groups; allows user to specify an expected number of biological classes; centroids provide a simplified representation of the data set (for SOMs, vicinity of partitions is a measure of similarity) | Requires user to specify initial number of centroids and their starting positions; some centroids might lack biological relevance |
| Principal components analysis (PCA) | | Principal components | Simple and unbiased; scores and loadings vectors provide simplified representations of the data set | Cannot pose a hypothetical relationship within the data set; some principal components might lack biological relevance |
| Partial least squares (PLS) | Classification | Principal components | Allows user to specify an expected set of biological classes without the need for additional data | Class predictions are inherently qualitative; principal components might lack biological relevance when classes are too distantly related to the independent variables |
| PLS | Prediction | Principal components | Allows user to pose a biological hypothesis; predictions are quantitative | Often requires an additional data set of dependent measurements; assumes a linear relationship between independent and dependent variables |

(Copies of tables are available in the accompanying CD.)

**Table 27.2** A list of available software tools implementing computational methods discussed in this chapter

| Tool name | Description | Link |
|---|---|---|
| Cluster | Performs hierarchical clustering, self-organizing maps | http://rana.lbl.gov/ |
| GenePattern | Contains clustering, classification, cluster analysis, and preprocessing algorithms | http://www.broad.mit.edu/cancer/software/software.html |
| GEPAS | A suite of gene expression pattern analysis tools (k-means, SOMs, SOTA etc) | http://gepas.bioinfo.cipf.es/cgi-bin/clus |
| caGEDA | A web application for the integrated analysis of global gene expression patterns in cancer | http://bioinformatics.upmc.edu/GE2/GEDA.html |
| GRAM | Identifies modules, collections of genes that share common regulators as well as expression profiles. | http://psrg.lcs.mit.edu/GRAM/Index.html |
| SAMBA | A novel biclustering algorithm for the identification of modules of genes that exhibit similar behavior under a subset of the examined biological conditions. | http://www.cs.tau.ac.il/~samba/ |
| Expander | Performs clustering, visualizing, biclustering and downstream analysis of clusters and biclusters such as functional enrichment and promoter analysis. | http://www.cs.tau.ac.il/%7Ershamir/expander/expander.html |
| GeneXpress | A visualization and analysis tool for gene expression data, integrating clustering, gene annotation, and sequence information. | http://genexpress.stanford.edu/ |
| ReMoDiscovery | An intuitive algorithm that correlates regulatory programs with regulators and corresponding motifs to a set of co-expressed genes. | http://homes.esat.kuleuven.be/~kmarchal/Supplementary_Information_Lemmens_2006/Index.html#SW |
| Cytoscape | Software platform for *visualizing* molecular interaction networks and *integrating* these interactions with gene expression profiles and other state data. | http://www.cytoscape.org/ |
| RBN MATLAB Toolbox | Random Boolean network | http://www.teuscher.ch/rbntoolbox/ |
| BN/PBN MATLAB Toolbox | Boolean and Probabilistic Boolean toolbox | http://personal.systemsbiology.net/ilya/PBN/PBN.htm |
| BNSOFT | Bayesian Network Software packages | http://www.cs.ubc.ca/~murphyk/Bayes/bnsoft.html |
| BAGEL | Bayesian Analysis of Gene Expression Levels: a program for the statistical analysis of spotted microarray data. | http://web.uconn.edu/townsend/software.html |
| Bayes Net MATLAB Toolbox | Bayesian Network | http://bnt.sourceforge.net/ |
| Causal Explorer (CE) | Local and Global Causal discovery | http://discover1.mc.vanderbilt.edu/discover/public/ |

(Copies of tables are available in the accompanying CD.)

**Acknowledgments** The authors involved in this chapter were supported by the GSRT, Hellas (project PENED 03ED842), the EMBO Young Investigator Program, and the European Commission (project INFOBIOMED, FP6-IST-507585). We acknowledge the authors of the tools listed in this chapter for placing them in the public domain. We thank all members of our lab for fruitful discussions and comments on the manuscript.

## Key References

Hartwell, L.H., et al., *From molecular to modular cell biology*. Nature, 1999. **402**(6761 Suppl): p. C47–52.

Bar-Joseph, Z., et al., *Computational discovery of gene modules and regulatory networks*. Nat Biotechnol, 2003. **21**(11): p. 1337–1342.

D'Haeseleer, P., *How does gene expression clustering work?* Nat Biotechnol, 2005. **23**(12): p. 1499–1501.

Martin, S., et al., *Boolean Dynamics of Genetic Regulatory Networks Inferred from Microarray Time Series Data*. Bioinformatics, 2007.

Dojer, N., et al., *Applying dynamic Bayesian networks to perturbed gene expression data*. BMC Bioinformatics, 2006. **7**: p. 249.

## Suggested Reading

1. Hartwell, L.H., et al., *From molecular to modular cell biology*. Nature, 1999. **402**(6761 Suppl): p. C47–52.
2. Kauffman, S.A., *Metabolic stability and epigenesis in randomly constructed genetic nets*. J Theor. Biol., 1969. **22**(3): p. 437–467.
3. Drossel, V.K.a.B., *Relevant components in critical random Boolean networks*. New J. Phys., 2006. **8**(228).
4. Derrida, B.a.P., Y., *Random networks of automata: A simple annealed approximation*. Europhys. Lett., 1(2): 45–49., 1986. **1**(2): p. 45–49.
5. Gershenson, C., *Introduction to Random Boolean Networks*. Workshop and Tutorial Proceedings, Ninth International Conference on the Simulation and Synthesis of Living Systems 2004: p. 160–173.
6. Greil, F. and B. Drossel, *Dynamics of critical Kauffman networks under asynchronous stochastic update*. Phys Rev Lett, 2005. **95**(4): p. 048701.
7. Klemm, K. and S. Bornholdt, *Stable and unstable attractors in Boolean networks*. Phys Rev E Stat Nonlin Soft Matter Phys, 2005. **72**(5 Pt 2): p. 055101.
8. Raeymaekers, L., *Dynamics of Boolean networks controlled by biologically meaningful functions*. J Theor Biol, 2002. **218**(3): p. 331–341.
9. Shmulevich, I. and S.A. Kauffman, *Activities and sensitivities in boolean network models*. Phys Rev Lett, 2004. **93**(4): p. 048701.
10. Aldana, M., et al., *Robustness and evolvability in genetic regulatory networks*. J Theor Biol, 2007. **245**(3): p. 433–448.
11. Wuensche, A., *Genomic regulation modeled as a network with basins of attraction*. Pac Symp Biocomput, 1998: p. 89–102.
12. Datta, A., et al., *External control in Markovian genetic regulatory networks: the imperfect information case*. Bioinformatics, 2004. **20**(6): p. 924–930.
13. Szallasi, Z. and S. Liang, *Modeling the normal and neoplastic cell cycle with "realistic Boolean genetic networks": their application for understanding carcinogenesis and assessing therapeutic strategies*. Pac Symp Biocomput, 1998: p. 66–76.
14. Shmulevich, I., et al., *Probabilistic Boolean Networks: a rule-based uncertainty model for gene regulatory networks*. Bioinformatics, 2002. **18**(2): p. 261–274.
15. Faure, A., et al., *Dynamical analysis of a generic Boolean model for the control of the mammalian cell cycle*. Bioinformatics, 2006. **22**(14): p. e124–131.
16. Herrgard, M.J., et al., *Integrated analysis of regulatory and metabolic networks reveals novel regulatory mechanisms in Saccharomyces cerevisiae*. Genome Res, 2006. **16**(5): p. 627–635.
17. Akutsu, T., S. Miyano, and S. Kuhara, *Identification of genetic networks from a small number of gene expression patterns under the Boolean network model*. Pac Symp Biocomput, 1999: p. 17–28.
18. Boros, E. and T. Ibaraki, *Error-free and Best-fit extensions of partially defined boolean functions*. Information and Computation 1998. **140**(2): p. 254–283.
19. Nam, D.S., Seunghyun; Kim, Sangsoo, *An efficient top-down search algorithm for learning Boolean networks of gene expression*. Machine Learning, 2006. **65**(1): p. 229–245.

20. Liang, S., S. Fuhrman, and R. Somogyi, *Reveal, a general reverse engineering algorithm for inference of genetic network architectures*. Pac Symp Biocomput, 1998: p. 18–29.
21. Kim, H., J.K. Lee, and T. Park, *Boolean networks using the chi-square test for inferring large-scale gene regulatory networks*. BMC Bioinformatics, 2007. **8**: p. 37.
22. Lähdesmäki, H., *On Learning Gene Regulatory Networks under the Boolean Network Model*. Machine Learning, 2002. **52**: p. 147–163.
23. Akutsu, T., et al., *A System for Identifying Genetic Networks from Gene Expression Patterns Produced by Gene Disruptions and Overexpressions*. Genome Inform Ser Workshop Genome Inform, 1998. **9**: p. 151–160.
24. Osamu Hirose, N.N., Yoshinori Tamada, Hideo Bannai, Seiya Imoto and Satoru Miyano, *Estimating Gene Networks from Expression Data and Binding Location Data via Boolean Networks*, in Computational Science and Its Applications – ICCSA 2005. 2005, Springer Berlin / Heidelberg. p. 349–356.
25. Martin, S., et al., *Boolean Dynamics of Genetic Regulatory Networks Inferred from Microarray Time Series Data*. Bioinformatics, 2007.
26. Pal, R., et al., *Intervention in context-sensitive probabilistic Boolean networks*. Bioinformatics, 2005. **21**(7): p. 1211–1218.
27. Thomas, R., D. Thieffry, and M. Kaufman, *Dynamical behaviour of biological regulatory networks—I. Biological role of feedback loops and practical use of the concept of the loop-characteristic state*. Bull Math Biol, 1995. **57**(2): p. 247–276.
28. Thomas, R. and M. Kaufman, *Multistationarity, the basis of cell differentiation and memory. I. Structural conditions of multistationarity and other nontrivial behavior*. Chaos, 2001. **11**(1): p. 170–179.
29. Thomas, R. and M. Kaufman, *Multistationarity, the basis of cell differentiation and memory. II. Logical analysis of regulatory networks in terms of feedback circuits*. Chaos, 2001. **11**(1): p. 180–195.
30. Thomas, R., *Boolean formalization of genetic control circuits*. J Theor Biol, 1973. **42**(3): p. 563–585.
31. Mendoza, L., D. Thieffry, and E.R. Alvarez-Buylla, *Genetic control of flower morphogenesis in Arabidopsis thaliana: a logical analysis*. Bioinformatics, 1999. **15**(7–8): p. 593–606.
32. Thieffry, D. and L. Sanchez, *Dynamical modelling of pattern formation during embryonic development*. Curr Opin Genet Dev, 2003. **13**(4): p. 326–330.
33. Friedman, N., et al., *Using Bayesian networks to analyze expression data*. J Comput Biol, 2000. **7**(3–4): p. 601–620.
34. Hartemink, A.J., et al., *Using graphical models and genomic expression data to statistically validate models of genetic regulatory networks*. Pac Symp Biocomput, 2001: p. 422–433.
35. Jansen, R., et al., *A Bayesian networks approach for predicting protein-protein interactions from genomic data*. Science, 2003. **302**(5644): p. 449–453.
36. Le Phillip, P., A. Bahl, and L.H. Ungar, *Using prior knowledge to improve genetic network reconstruction from microarray data*. In Silico Biol, 2004. **4**(3): p. 335–353.
37. Murphy K, M.S., *Modelling gene expression data using dynamic Bayesian networks*. 1999, Computer Science Division, University of California, Berkeley.
38. Beal, M.J., et al., *A Bayesian approach to reconstructing genetic regulatory networks with hidden factors*. Bioinformatics, 2005. **21**(3): p. 349–356.
39. Bernard, A. and A.J. Hartemink, *Informative structure priors: joint learning of dynamic regulatory networks from multiple types of data*. Pac Symp Biocomput, 2005: p. 459–470.
40. Dojer, N., et al., *Applying dynamic Bayesian networks to perturbed gene expression data*. BMC Bioinformatics, 2006. **7**: p. 249.
41. Zou, M. and S.D. Conzen, *A new dynamic Bayesian network (DBN) approach for identifying gene regulatory networks from time course microarray data*. Bioinformatics, 2005. **21**(1): p. 71–79.
42. Kato, T., K. Tsuda, and K. Asai, *Selective integration of multiple biological data for supervised network inference*. Bioinformatics, 2005. **21**(10): p. 2488–2495.
43. Yamanishi, Y., J.P. Vert, and M. Kanehisa, *Supervised enzyme network inference from the integration of genomic data and chemical information*. Bioinformatics, 2005. **21 Suppl 1**: p. 2468i477.
44. D'Haeseleer, P., S. Liang, and R. Somogyi, *Genetic network inference: from co-expression clustering to reverse engineering*. Bioinformatics, 2000. **16**(8): p. 707–726.
45. Segal, E., et al., *Module networks: identifying regulatory modules and their condition-specific regulators from gene expression data*. Nat Genet, 2003. **34**(2): p. 166–176.
46. Bar-Joseph, Z., et al., *Computational discovery of gene modules and regulatory networks*. Nat Biotechnol, 2003. **21**(11): p. 1337–1342.
47. Janes, K.A. and M.B. Yaffe, *Data-driven modelling of signal-transduction networks*. Nat Rev Mol Cell Biol, 2006. **7**(11): p. 820–828.
48. D'Haeseleer, P., *How does gene expression clustering work?* Nat Biotechnol, 2005. **23**(12): p. 1499–1501.
49. Yeung, K.Y., M. Medvedovic, and R.E. Bumgarner, *Clustering gene-expression data with repeated measurements*. Genome Biol, 2003. **4**(5): p. R34.

50. Walker, M.G., et al., *Prediction of gene function by genome-scale expression analysis: prostate cancer-associated genes*. Genome Res, 1999. **9**(12): p. 1198–1203.
51. Thompson, H.G., et al., *Identification and confirmation of a module of coexpressed genes*. Genome Res, 2002. **12**(10): p. 1517–1522.
52. Eisen, M.B., et al., *Cluster analysis and display of genome-wide expression patterns*. Proc Natl Acad Sci U S A, 1998. **95**(25): p. 14863–14868.
53. Tavazoie, S., et al., *Systematic determination of genetic network architecture*. Nat Genet, 1999. **22**(3): p. 281–285.
54. Claverie, J.M., *Computational methods for the identification of differential and coordinated gene expression*. Hum Mol Genet, 1999. **8**(10): p. 1821–1832.
55. Michaels, G.S., et al., *Cluster analysis and data visualization of large-scale gene expression data*. Pac Symp Biocomput, 1998: p. 42–53.
56. Ben-Dor, A., R. Shamir, and Z. Yakhini, *Clustering gene expression patterns*. J Comput Biol, 1999. **6**(3–4): p. 281–297.
57. Alon, U., et al., *Broad patterns of gene expression revealed by clustering analysis of tumor and normal colon tissues probed by oligonucleotide arrays*. Proc Natl Acad Sci U S A, 1999. **96**(12): p. 6745–6750.
58. Kim, K., et al., *Measuring similarities between gene expression profiles through new data transformations*. BMC Bioinformatics, 2007. **8**: p. 29.
59. Kuncheva, L.I. and D.P. Vetrov, *Evaluation of stability of k-means cluster ensembles with respect to random initialization*. IEEE Trans Pattern Anal Mach Intell, 2006. **28**(11): p. 1798–1808.
60. Lu, Y., et al., *Incremental genetic k-means algorithm and its application in gene expression data analysis*. BMC Bioinformatics, 2004. **5**: p. 172.
61. Kohonen, T., *Self Organizing Maps*. 1995, Berlin: Springer.
62. Wang, J., et al., *Clustering of the SOM easily reveals distinct gene expression patterns: results of a reanalysis of lymphoma study*. BMC Bioinformatics, 2002. **3**: p. 36.
63. Toronen, P., et al., *Analysis of gene expression data using self-organizing maps*. FEBS Lett, 1999. **451**(2): p. 142–146.
64. Tamayo, P., et al., *Interpreting patterns of gene expression with self-organizing maps: methods and application to hematopoietic differentiation*. Proc Natl Acad Sci U S A, 1999. **96**(6): p. 2907–2912.
65. Misra, J., et al., *Interactive exploration of microarray gene expression patterns in a reduced dimensional space*. Genome Res, 2002. **12**(7): p. 1112–1120.
66. Alter, O., P.O. Brown, and D. Botstein, *Singular value decomposition for genome-wide expression data processing and modeling*. Proc Natl Acad Sci USA, 2000. **97**(18): p. 10101–10106.
67. Boulesteix, A.L. and K. Strimmer, *Predicting transcription factor activities from combined analysis of microarray and ChIP data: a partial least squares approach*. Theor Biol Med Model, 2005. **2**: p. 23.
68. Datta, S., *Exploring relationships in gene expressions: a partial least squares approach*. Gene Expr, 2001. **9**(6): p. 249–255.
69. Clementi, M., et al., *Robust multivariate statistics and the prediction of protein secondary structure content*. Protein Eng, 1997. **10**(7): p. 747–749.
70. Zhang, Z. and M. Gerstein, *Reconstructing genetic networks in yeast*. Nat Biotechnol, 2003. **21**(11): p. 1295–1297.
71. Cassman, M., *Barriers to progress in Systems Biology*. Nature, 2005. **438**(7071): p. 1079.
72. Lemmens, K., et al., *Inferring transcriptional modules from ChIP-chip, motif and microarray data*. Genome Biol, 2006. **7**(5): p. R37.
73. Tanay, A., et al., *Revealing modularity and organization in the yeast molecular network by integrated analysis of highly heterogeneous genomewide data*. Proc Natl Acad Sci U S A, 2004. **101**(9): p. 2981–2986.
74. Jeong, H., et al., *Lethality and centrality in protein networks*. Nature, 2001. **411**(6833): p. 41–42.
75. Tong, A.H., et al., *A combined experimental and computational strategy to define protein interaction networks for peptide recognition modules*. Science, 2002. **295**(5553): p. 321–324.
76. Han, J.D., et al., *Evidence for dynamically organized modularity in the yeast protein-protein interaction network*. Nature, 2004. **430**(6995): p. 88–93.
77. Jeong, H., et al., *The large-scale organization of metabolic networks*. Nature, 2000. **407**(6804): p. 651–654.
78. Guimera, R. and L.A. Nunes Amaral, *Functional cartography of complex metabolic networks*. Nature, 2005. **433**(7028): p. 895–900.
79. Stuart, J.M., et al., *A gene-coexpression network for global discovery of conserved genetic modules*. Science, 2003. **302**(5643): p. 249–255.
80. Tong, A.H., et al., *Global mapping of the yeast genetic interaction network*. Science, 2004. **303**(5659): p. 808–813.
81. Lee, T.I., et al., *Transcriptional regulatory networks in Saccharomyces cerevisiae*. Science, 2002. **298**(5594): p. 799–804.
82. Ravasz, E., et al., *Hierarchical organization of modularity in metabolic networks*. Science, 2002. **297**(5586): p. 15511555.

83. Schilling, C.H., D. Letscher, and B.O. Palsson, *Theory for the systemic definition of metabolic pathways and their use in interpreting metabolic function from a pathway-oriented perspective*. J Theor Biol, 2000. **203**(3): p. 229–248.
84. Schuster, S., D.A. Fell, and T. Dandekar, *A general definition of metabolic pathways useful for systematic organization and analysis of complex metabolic networks*. Nat Biotechnol, 2000. **18**(3): p. 326–332.
85. Burgard, A.P., et al., *Flux coupling analysis of genome-scale metabolic network reconstructions*. Genome Res, 2004. **14**(2): p. 301–312.
86. Kuepfer, L., U. Sauer, and L.M. Blank, *Metabolic functions of duplicate genes in Saccharomyces cerevisiae*. Genome Res, 2005. **15**(10): p. 1421–1430.
87. Carter, S.L., et al., *Gene co-expression network topology provides a framework for molecular characterization of cellular state*. Bioinformatics, 2004. **20**(14): p. 2242–2250.
88. Ma, H.W., J. Buer, and A.P. Zeng, *Hierarchical structure and modules in the Escherichia coli transcriptional regulatory network revealed by a new top-down approach*. BMC Bioinformatics, 2004. **5**: p. 199.
89. Pereira-Leal, J.B., E.D. Levy, and S.A. Teichmann, *The origins and evolution of functional modules: lessons from protein complexes*. Philos Trans R Soc Lond B Biol Sci, 2006. **361**(1467): p. 507–517.
90. Cho, Y.-R., W. Hwang, and A. Zhang, *Efficient Modularization of Weighted Protein Interaction Networks using k-Hop Graph Reduction*, in *Sixth IEEE Symposium on BioInformatics and BioEngineering (BIBE'06)*. 2006.
91. Pereira-Leal, J.B., A.J. Enright, and C.A. Ouzounis, *Detection of functional modules from protein interaction networks*. Proteins, 2004. **54**(1): p. 49–57.
92. Rives, A.W. and T. Galitski, *Modular organization of cellular networks*. Proc Natl Acad Sci U S A, 2003. **100**(3): p. 1128–1133.
93. Arnau, V., S. Mars, and I. Marin, *Iterative cluster analysis of protein interaction data*. Bioinformatics, 2005. **21**(3): p. 364–378.
94. Milo, R., et al., *Network motifs: simple building blocks of complex networks*. Science, 2002. **298**(5594): p. 824–827.
95. Shen-Orr, S.S., et al., *Network motifs in the transcriptional regulation network of Escherichia coli*. Nat Genet, 2002. **31**(1): p. 64–68.
96. Ma, H.W., et al., *An extended transcriptional regulatory network of Escherichia coli and analysis of its hierarchical structure and network motifs*. Nucleic Acids Res, 2004. **32**(22): p. 6643–6649.
97. Zhang, L.V., et al., *Motifs, themes and thematic maps of an integrated Saccharomyces cerevisiae interaction network*. J Biol, 2005. **4**(2): p. 6.
98. Wuchty, S., Z.N. Oltvai, and A.L. Barabasi, *Evolutionary conservation of motif constituents in the yeast protein interaction network*. Nat Genet, 2003. **35**(2): p. 176–179.
99. Conant, G.C. and A. Wagner, *Convergent evolution of gene circuits*. Nat Genet, 2003. **34**(3): p. 264–266.
100. Snel, B. and M.A. Huynen, *Quantifying modularity in the evolution of biomolecular systems*. Genome Res, 2004. **14**(3): p. 391–397.

# Chapter 28
# Methods for Dynamical Inference in Intracellular Networks

**Eleftheria Tzamali, Panayiota Poirazi, and Martin Reczko**

**Abstract** Equation-based algorithms make hypotheses regarding the biophysical dynamical laws that govern a biological system and in the form of a mathematical expression, aiming to interrelate the system components, in an effort to explain and verify the experimental observations. This approach is what we mainly regard as dynamical inference. Assumptions such as the deterministic or stochastic laws that govern the system dynamics, the degree of modeling spatial phenomena, the exact mathematical representations of these biophysical laws and constraints, comprise some of the main issues of the dynamical inference problem. Another class of algorithms considers the cell as a whole system that orchestrates its components under physio-chemical constraints towards the accomplishment of certain cellular functions. These approaches avoid the search of detailed equation forms as well as the demand of knowledge of the parameters involved in the kinetics, and produce a steady state dynamic picture of the complex, genome-scale metabolic network of chemical reactions at the flux level. The constraint-based methods are essential for the analysis of the metabolic capabilities of organisms as well as the elucidation of systemic properties that cannot be described by descriptions of individual components or sub-systems.

The current biological knowledge, the available data and the computer power, are the issues that actually determine the upper limit for the system size and its complexity that can be simulated, thus defining our level of understanding.

**Keywords** Differential equations · Stochastic simulation · Spatial organization · Constraint-based methods · Flux balance analysis

## 28.1 Introduction

Cells are complex biological systems that consist of components that interact with each other, under regulatory strategies, in response to internal and environmental signals. These components can be also seen as players in a game with short and long-term goals which follow certain, finite and evolutionary sub or fully optimized strategies with the potentiality of updating and self adjusting. The goals can be interpreted as the accomplishment of certain functional tasks towards maintenance of life. A strategy involves the temporal succession of the interactions that the components of the system will undergo during the progress of the dynamic phenomenon. The biophysical laws the system obeys as well as the evolution process that has driven the system towards a better survival and source utilization, are the constraints that shape the regulatory strategies under which the system efficiently orchestrates its components.

M. Reczko
Institute of Computer Science (ICS), Foundation of Research and Technology-Hellas (FORTH), Vassilika Vouton, P.O. Box 1385, GR 711 10, Heraklion, Crete, Greece
e-mail: mreczko@gmail.com

S. Krawetz (ed.), *Bioinformatics for Systems Biology*, DOI 10.1007/978-1-59745-440-7_28, 

Identification of the system strategies in a qualitative manner is achieved by the network-based class of algorithms described analytically in the previous chapter. In network representations, either a dynamics-free-picture of the system or an interpretation-free or a simplified dynamic interpretation of the interactions is mainly depicted. Besides network inference, explaining and formulating those strategies is a further step towards understanding the dynamic phenomena that take place in the system. Equation-based algorithms make hypotheses regarding the biophysical dynamical laws that govern the system and in the form of a parametric equation aim to interrelate the system components, in an effort to explain and verify the experimental observations. This approach is what we call dynamical inference and will be discussed in the following.

Dynamical inference can be seen as a two-fold problem. The first part concerns the determination of the biophysical description of the system, including assumptions and decisions such as the deterministic or stochastic nature of the system dynamics, the influence of the in homogeneity of the actual reacting volume in the dynamics, and the linear or non-linear dynamic interrelation of the system components. The optimal model is selected based on the available data, and the biological knowledge and assumptions that are most proper to fit to the biophysical dynamics of the system under study. The second part attempts to make the biophysical description, selected in the first step specific, by identifying the missing parameters involved and incorporating them to the pre-defined mathematical framework. Actually many of the kinetic parameters (reaction rates, diffusion or transport rates) involved in the governing equations are usually missing. Therefore, methods for estimating those parameters from experimental data should be applied. Optimization of the parameters involved is a problem of its own and usually a large computational effort is needed.

An expressional rich model (many degrees of freedom in the parameters involved), is more probable to include a representation close to the real dynamics. However, the limited amount of data and constraints make the identification problem hard and a unique solution is unlikely to be found. At that point, one may also wonder whether a unique solution is indeed, what describes a biological system. A plethora of solutions may soundly depict the inherent robustness of the system. Discrimination between modeling weaknesses and the actual biological system properties is essential when aiming to understand the real phenomena and predict system behaviors. Chen et al. (2005) and Mahadevan and Schilling (2003) [1, 2], among others, investigate the robustness reflected, either as variation in the kinetic parameters or as alternate optimal fluxes with respect to steady state the maintenance.

Current knowledge certainly places limits to the level of description, and the challenge is to build mathematical models with the available data at hand that will be able to reveal new properties and understand the system functioning to further guide new experiments. When a valid and biologically plausible unified description of the system is found, the iterative model refinement procedure will come to an end. For gene regulatory systems, the biophysical reaction mechanisms are not yet fully understood. The individual steps involved in the gene expression pathway, which includes the information transferred from DNA to RNA and then to protein, are thoroughly described [3]. A theory of modeling the reaction steps and placing them in a network of interactions remains, however, an open problem in Systems Biology. Recent research attempts that work on the molecular levels of understanding the underlying regulatory mechanisms, include references [4–6]. Quantitative information on kinetic parameters and molecular populations is also limited, making the model reconstruction harder. In metabolic networks, on the other hand, for organisms like *E. Coli* or yeast, more knowledge is available but the data is far from being fully complete. An exceptional case comprises the human red cell for which all the kinetic parameters are available [7, 8].

The chapter is divided into the equation-based and the constraint-based modeling approaches. The breadth of the description of each approach is proportional to the research works on the fields and the aspects that we consider educationally essential in understanding.

## 28.2 Equation-based Modeling

*Problem Definition*: Given sequential snapshots of the time evolution of the components of a system (time-course data), a function that explains the dynamic change of the values of each component with respect to the system's past values is asked. The solution for the problem should explain and predict the population levels of the species at any time, given the molecular populations of the biochemical reacting system at an initial time.

Population levels can either be continuous or discrete variables depending on the resolution, that better describes the system. In principle, molecular concentrations can be assumed continuous variables only in the thermodynamic limit of a large number of molecules. Quantization of the continuous variables is also possible to simplify modeling.

The deterministic or stochastic description of the dynamics of the biochemical reacting system under study is the first decision to be made. The second, concerns the form of the temporal transition function, which reflects the further assumptions (e.g., linearity, non-linearity, reaction time evolution rules) made on the appropriate deterministic or stochastic description of the real biophysical laws that govern the system, and comes with the cost of the parameter estimations from the current knowledge.

A short overview of the main aspects that concern the dynamic modeling will be presented first while a deeper analysis of the particular methods that each approach uses will follow.

### *28.2.1 Overview*

#### 28.2.1.1 Deterministic Approach

The deterministic dynamic evolution of a biochemical reacting system, of $N$ species, can be described by a set of $N$ coupled time-dependent equations of the form presented in Fig. 28.1.

$v_i$[t] expresses the population level of the $i$th species at time $t$, and $f_i$ is a transition function that quantifies the temporal relation between the $i$th species and all the species of the system. $V$[t] represents the population levels of each species in a vector form (system state).

In the above representation, the biological system is described by a set of coupled difference equations. However, a dynamical system can also be represented by a set of coupled, ordinary differential equations where both the time domain and the molecular concentrations of the components are considered continuous. If these assumptions hold, each of the above equations takes the differential form:

$$\frac{dv_i}{dt} = f_i(v_1(t), v_2(t), \ldots, v_N(t)) \tag{28.1}$$

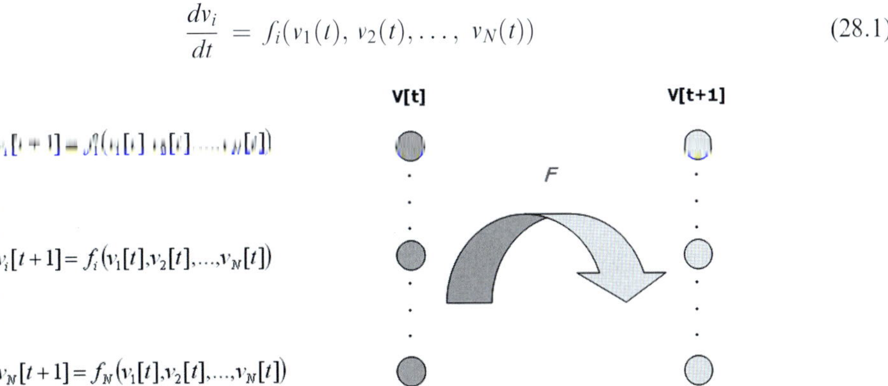


**Fig. 28.1** The temporal transition of a system. $V$[t] and $V$[t+1] depict the state vectors of the system components at time $t$ and $t$+1 respectively (Copies of figures including color copies, where applicable, are available in the accompanying CD)

Differential equations are commonly used to model biochemical systems [9, 10], particularly the chemical reactions involved in metabolic networks [11, 12]. However, since they require thorough biochemical detail of the system and because of their deterministic description of the dynamics, they are generally difficult to apply to modeling gene regulatory networks [13]. Analytical solutions exists for a few simple systems, thus numerical, computational methods are used to solve the equations.

The representation of system dynamics in the above formulas implies that the information propagates quickly in the reacting medium, in a sense that for a certain sampling rate, a component has the time to respond to a change. In that sense, the population levels of each species at time $t$ can be determined by those at a previous time step. When this is not the case and time delays arise from the possible time required to complete processes such as transcription, translation or/and transportation, a modification is needed in the dynamic description. Time delays, which are not modeled explicitly, can be accommodated in the above (deference or differential) representation [14]. An example of how a difference equation is reformulated is:

$$v_i[t+1] = f_i(v_1[t-\tau_{i1}], v_2[t-\tau_{i2}], \ldots, v_N[t-\tau_{iN}]) \tag{28.2}$$

where $\tau_{i1}, \ldots, \tau_{iN}$ are discrete time delays

Later we shall discuss how spatial models reflect on these phenomena.

The dynamical representations described, additionally imply that biological systems are closed systems. However, rarely is this the case. Biological systems are open systems where external signals and environmental changes also control the dynamics. In order for the proposed function to consistently describe the dynamic phenomenon, the control property should be also accommodated in the formalism as follows:

$$v_i[t+1] = f_i(v_1[t], v_2[t], \ldots, v_N[t], \quad \vec{u}[t]) \tag{28.3}$$

where $\vec{u}$ is the vector of input signals

A decomposed form of the dynamics to a rate of production ($f_i$) and a rate of consumption or degradation ($g_i$) is a common representation and is more convenient since it further allows the expression of different sub-networks and laws to govern each phenomenon if necessary. The temporal transition for the $i^{\text{th}}$ component in the decomposed formula now takes the form:

$$v_i[t+1] = f_i(v_1[t], v_2[t], \ldots, v_N[t]) \quad - \quad g_i(v_1[t], v_2[t], \ldots, v_N[t]) \tag{28.4}$$

The degradation effect is usually modeled by a first order chemical kinetic reaction and the above equation becomes:

$$v_i[t+1] = f_i(v_1[t], v_2[t], \ldots, v_N[t]) \quad - \quad \gamma_i v_i[t], \tag{28.5}$$

where the parameter $\gamma_i$ denotes the degradation rate of the $i^{\text{th}}$ component.

Within each time step, usually, more than one interaction-reaction takes place. Hence, we have expressed the population level of each component in the system as a function of all the system components of which a certain component may interact within that time interval. Usually, the total regulatory effect on each component is assumed additive, in a sense that the values of each component are combined additively, in a proper manner to change the value of a certain component. The additive effect can be expressed either as a function of a weighed sum of the values of the system components (linear models, linear weight matrices models), or as a sum of the regulation functions of each system component (piece-wise linear models). The approaches will be discussed separately in following sections.

$$
\begin{aligned}
v_i[t+1] &= f_i\Big(\sum_j w_{ij} v_j[t]\Big), \text{function of a weighted sum} \\
v_i[t+1] &= \sum_j w_{ij} f_i\big(v_j[t]\big), \text{weighted sum of regulation functions}
\end{aligned} \tag{28.6}
$$

Using the additive expression alone to describe the dynamic change of a component, implies a system where components behave independently (non-cooperatively). There are cases, however, where components act cooperatively for an influence to be apparent as a change in the population level of a certain component. In gene regulation this synergistic effect is particularly common [15] and expresses tough regulation of some biological tasks. The synergism effect adds (further) non-linearities to the dynamic expression and different formalisms, have been proposed which are analyzed further in the paragraphs to follow.

#### 28.2.1.2 Stochastic Approach

The physical basis of the approaches we have mentioned counts on the deterministic dynamic behavior of the chemically reacting system. Recalling our problem definition of finding a dynamical expression to explain and predict the time evolution of a biophysical reacting system, Gillespie first in 1977 placed the problem differently asking how predictable this system actually is. Adopting the collision theory to describe the chemical reactions, he proposed that reactions take place at discrete time intervals of random length and that the observed temporal change in the populations of a species is the consequence of this random process. In that sense, describing the system in its full momentum-position space and under the assumption that classical mechanics hold, the time evolution of the system can be considered a deterministic process. However, in the N-dimensional subspace that we want to describe it, it is not.

Stochastic phenomena have been observed in biological systems. Even in cloned cell populations and under the same (as possible) experimental conditions, significant phenotypic variations have been reported in each cell including variations in the rates of development, morphology and population levels of each species [16–18].

In systems where the populations of the species are large enough (e.g., metabolites), random effects are averaged out. Nevertheless, if the population of a system component is low enough (e.g., mRNA copies, transcription factors), the fluctuations in the molecular levels (e.g., protein levels) could not be predicted by a deterministic approach. When the microscopic fluctuations produce macroscopic effects [6] that the deterministic reaction rate equations are unable to predict, microscopic stochastic simulation approaches described in a later section are required [19].

#### 28.2.1.3 Spatial Consideration

All organisms, even bacteria, show a spatial organization into cellular compartments. In the modeling perspectives discussed so far we assumed a homogeneous distribution of the cellular components. Including a spatial constituent into the biochemical dynamic expressions is essential when different reactions evolve differently in separate compartments and the molecular mobility causes significant non-linear delays to the dynamic system. Spatial phenomena arise when reaction rates are shown to be comparatively faster than diffusion rates.

In a simple case where the compartments can be assumed homogeneous inside, we could apply the dynamic equations separately for the species involved in each compartment and use an additional term to include the transportation of molecules between the compartments. Alternatively, we can consider space, like time in ordinary differential equations, a continuous variable and express the dynamic evolution of the system spatio-temporal using partial differential equations [20].

Spatial phenomena are also apparent when the reacting volume is of high molecular density (molecular crowding). Signaling pathways have reported suffering from phenomena such this [21–23].

The following present give a more detailed description of the equation-based dynamical inference methods we have sententiously mentioned above. The deterministic models, from linear to non-linear assumptions on their dynamics, are firstly discussed. Afterwards, approaches inspired by the inherent stochastic nature of some biological systems are analyzed. And lastly, a section describing the additional consideration of spatial resolution to the models is included.

### *28.2.2 Deterministic Approach – Methods*

Under the deterministic framework, the biophysical laws that describe the temporal evolution of a system produce precisely and consistently the future states of the system given its initial condition. A variety of mathematical expressions have been proposed in the literature to specify those laws and explain the observations.

#### 28.2.2.1 Linear Models

Linear models assume that the response of a component $v_i[t+1]$ depends linearly on its input $\vec{V}[t]$, which consists of the values of all of the components of the system at a previous time. The response of each component of the system is fully determined in deterministic manner by weighting properly the input. The basic mathematical form is:

$$v_i[t+1] = \sum_j w_{ij} v_j[t] = \vec{W}_i \vec{V}[t], \text{which is commonly extended to}$$

$$v_i[t+1] = \sum_j w_{ij} v_j[t] + b_i. \tag{28.7}$$

The terms $w_{ij}$ express the influence of the component $j$ to the component $i$. The $b_i$ is a bias term to the model that expresses the activation of a component in the absence of any other regulatory inputs. Under the above considerations, the dynamical inference problem is defined as a problem of finding the parameters $w_{ij}$ and $b_i$ for each component of the system, given a time sequence of finite, usually, too few time points. Usually, clustering methods are used to make the inference tractable. Linear models have been used to describe several gene regulatory systems [24–29]. Chen et al. [29] extended the linear differential equation to explicitly include degradation terms of mRNAs and proteins. Wu et al. [27] viewed gene expression levels as observation variables, whose expressions depend on current internal state variables, and assumed a linear model to describe the dynamic evolution of those internal variables.

#### 28.2.2.2 Piece-wise Linear Models

Biochemical reactions are non-linear in nature; therefore, linear models are not an optimal design strategy to describe the dynamics of the biological systems. Without loosing the descriptive power and keeping simplicity in the formalism of the biophysical laws, locally linear models, known as piece-wise linear models have been proposed originally by Glass and Kauffman [30]. Thus, piece-wise linear systems of differential equations are also known as Glass networks.

The method is applicable in interaction systems of switch-like character. Gene regulatory networks, for instance, are assumed to usually approximate, that kind of behavior well. Genes are activated when the concentration of a regulator (transcription factor) reaches a certain level. Activation, here, can either mean a positive or negative effect on the amount of a gene. Furthermore, the activation effect is saturated and the whole regulatory effect is similar to a sigmoidal shape function which can be approximated by a piece-wise linear function (step or logoid function) (Fig. 28.2). Numerical simulation studies have shown that there are no qualitative differences when approximating a sigmoid regulation function with a step function. Each regulator shows its own characteristic region of activation that depends on the gene it activates. Eventually, all the regulators act independently and additively, in a total regulatory effect that remains piece-wise linear:

$$f_i(\vec{V}(t)) = \sum_j w_{ij} r_{ij}(v_j(t)) \tag{28.8}$$

It is possible for the regulator function to further express the synergism effect, in which two or more regulators form a complex and then activate a gene. This non-linearity can be described in a piece-wise linear manner as a product of step functions each with its own characteristics (firing threshold), corresponding to the different regulators that form the complex. The note here is that the product of step functions is still a piece-wise linear function, a property which is not generally true.

Switching thresholds play a key role to the dynamic behavior of the system. Determining those thresholds from data is generally not a difficult problem. Methods applied to linear systems can be used. However, we must always keep in mind that the system works cooperatively with reference to its components, and generating a single model independently for each component is probably not the optimal way [31]. A lot of theoretic work that investigates the properties of the piece-wise dynamic systems has been conducted [32–38]. Radde et al. [33] has adopted the formalism to investigate the DNA repair system of *Mycobacterium tuberculosis*.

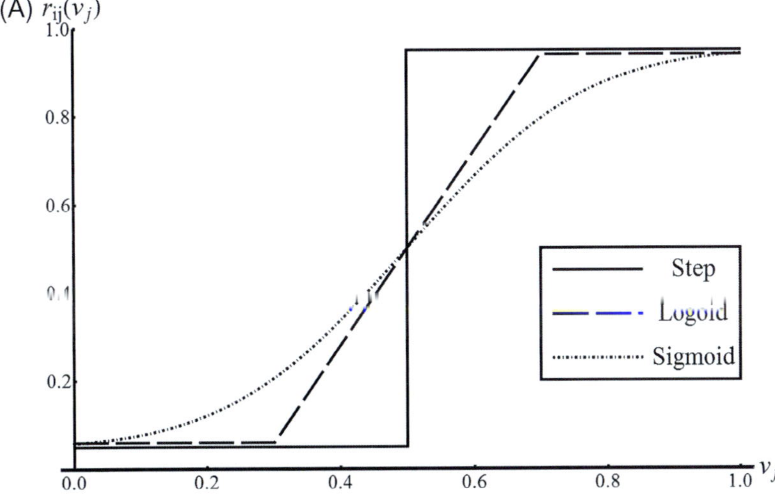


**Fig. 28.2** (**A**) The un-weighted regulatory contribution $r_{ij}$ of the component $v_j$ to the component $v_i$. Step, logoid and sigmoid functions are commonly used to approximate switch-like regulatory effects. (**B**) A network representation of the temporal transition of states, under the assumption that an additive regulatory effect is valid. The next state of a component is determined by the weighted sum of regulation functions (Copies of figures including color copies, where applicable, are available in the accompanying CD)

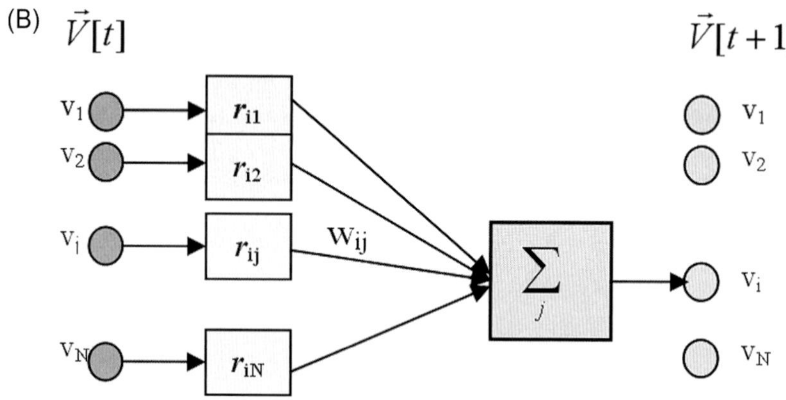


**Fig. 28.2** (Continued)

#### 28.2.2.3 Non-linear Models

Various non-linear functions have been proposed to describe the dynamic evolution of biological systems. The difficulty that these models present is that analytical solutions are often not possible and numerical techniques have to be applied to solve the underlying equations. The lack of measurements of the kinetic parameters involved in those equations comprises a significant hurdle in numerical approaches. As the system complexity grows, the parameter space becomes huge and a global solution requires a vast amount of data to be found.

Piece-wise linear models were used to replace the non-linear sigmoidal shape updating functions as shown in the previous paragraph. Artificial neural networks have been used to model the regulatory effect among genes using sigmoidal transfer functions to simulate the real phenomenon. In that case, the level of activation of a gene is determined by:

$$\frac{dv_i}{dt} = a_i f_i(\sum_j w_{ij} v_j(t) + b_i) - \gamma_i v_i(t) \qquad (28.9)$$

The factors $\alpha_i$ and $\gamma_i$ depict the activation and degradation rates respectively and the $f_i$ represents the sigmoidal transfer function. Each gene-regulator contributes additively to the expression level of a certain gene *i*. The total contribution is given by the weighted sum of all these individual expression levels causing the gene to fire according to the sigmoidal function $f_i$. The rate constant, $\alpha_{i,}$ is applied to the total activation effect. There is no additive sigmoid contribution, as in piece-wise linear models, but a sigmoidal additive contribution instead.

Each node represents a particular gene and each layer of the network represents the level of expression of genes at a given time t. In general, all genes can control all others and full connectivity is assumed if no evidences of the dependencies are available. In reality, a gene is regulated by a subset of genes imposing a certain structure in the network (weight matrix). The training algorithms of neural networks are subject to reach a local instead of global minima and conclude with a system wiring not representative of the biological systems. When limited data are available proper constraints should be added to the model. Recurrent neural networks have been used in several gene regulatory networks [39–44].

Examples of nonlinear functions commonly used to describe biochemical reaction systems are presented below.

H-system

Represents an enhancement of the deterministic linear additive (weight matrix) models by the addition of a non-linear term to the dynamic formula.

$$\frac{dv_i(t)}{dt} = \left[\sum_j w_{ij} v_j + b_i\right] + \left[v_i \sum_j k_{ij} v_j\right] \tag{28.10}$$

S-System

Dynamical and structural rich formalism that captures the dynamics of the complex, non-linear biological systems such as gene regulatory and metabolic networks are s-sytem. They produce a number of parameters estimated from data and optimization algorithms that can search effectively the solution space. Usually evolutionary algorithms are used as optimization strategy. S-systems are a type of power low formalism:

$$\frac{dv_i(t)}{dt} = a_i \prod_{j=1}^{N} v_j(t)^{G_{ij}} - \beta_i \prod_{j=1}^{N} v_j(t)^{H_{ij}} \tag{28.11}$$

The first term represents the production effect (synthesis)– all influences that increase the value of the $i^{th}$ component which can either be gene, protein or metabolite, while the second term corresponds to the consumption effect (degradation) – $\alpha_i$ and $\beta_i$ are non-negative parameters that express the rate constants of the two processes accordingly. The terms $G_{ij}$ and $H_{ij,}$ on the other hand, are called kinetic exponents and they express the effectiveness of the influence among the components *i* and *j* of the system. The kinetic exponents determine the structure of the regulatory network and may have either positive or negative values, inducing or inhibiting correspondingly the effect of the process they participate.

S-systems have been used widely to explore the dynamic behavior of a fermentation pathway in *Saccharomyces cerevisiae* in different experimental conditions [45], in a biochemical system of yeast glycolysis [46] in an attempt to provide a rational basis for the optimization of citric acid production by *Aspergillus niger* [47], on the yeast galactose utilization pathway [48], to analyze cDNA micro-array data of *Thermus thermophilus* HB8 strains [49], to evaluate the interactions among genes in the SOS signaling pathway in *Escherichia Coli* [50], to model the cadBA system in *E. Coli* [51], on many case-study models to study optimization methods and dynamic behavior [14, 48, 51–56].

Law of Mass Action and Michaelis-Menten Kinetics

The law of mass action comprises the fundamental, empirical law governing reaction rates in biochemical systems. This "law" states that for a reaction in a homogeneous, free medium the rate of a process is proportional to the concentrations of the molecular species involved.

$$A + B \xrightarrow{k_+} C \qquad \frac{dC}{dt} = k_+[A][B] \tag{28.12}$$

Most of the chemical reactions that take place in biological systems are enzymatic. Michaelis and Menten proposed the formation of an enzyme-substrate complex as an intermediate step to product formation. Taking the decomposed form into account and applying the law of mass action in each elementary reaction step, the system kinetics can be found. Under the above considerations assumptions, any chemical reacting system can be expressed as a collection of coupled non-linear first order differential equations.

## 28.2.3 Stochastic Approach – Methods

Molecular reactions are inherently random processes. According to the collision theory, reactants collide first and if the collision is efficient enough then the reaction takes place. When exactly the next collision will take place within a volume is not known because molecules undergo Brownian motion and the exact position-momentum phase space of the system is not provided. Consequently, the exact time of the next reaction cannot be estimated and the time evolution of the system can thus only be described under a probabilistic manner. The reaction rates, we have previously seen, are now replaced by the corresponding reaction probabilities per unit time.

As the number, $N$, of the reactants increases, the randomness phenomenon smoothes out and the dynamic behavior of the system becomes more deterministic ($noise \approx 1/N^{1/2}$). In the thermodynamic limit, the system can be viewed macroscopically. Metabolic, signaling and gene networks are characteristic examples of systems with components of populations in different scales. The number of each metabolite within a cell is of the order of millions. Thus, fluctuations in metabolic systems are usually negligible. In gene networks, though, the number of mRNA molecules produced per transcription is usually of the order of tens. Metabolic systems are usually comprised of millions of molecules. Fluctuations can be considered negligible and macroscopic descriptions are approximations. In gene networks, mRNA copies are of the order of 10. Fluctuations are extremely significant (finite number effect), and a microscopic view of those systems, which "sees" and models these fluctuations is required.

In consequence of the above discussion, the physics a system obeys and correspondingly the appropriate mathematical formalism selection depends on the certain characteristics of the system. In some cases, it is possible to describe stochasticity by adding a noise term in the deterministic formalism (Langevin equation). In other cases, the whole dynamics must change description (chemical master equation).

### 28.2.3.1 Chemical Master Equation

If we can consider that $N$ distinct molecular species are spatially and uniformly distributed in a certain volume, $V$, and can chemically react through M finite reaction channels, then the chemical master equation describes the time evolution of the molecular populations of the species under a probabilistic manner. There are, actually $M+1$ different ways in which the chemically reacting system can evolve in the next infinitesimal time interval, either through accomplishing one of the $M$ reaction channels or none of them.

The probability that a certain reaction, $m$, will occur in the volume, $V$, in the time interval $(t, t+dt)$, given the system state at time t, depends on the propensity function $\alpha_m$ and comprises the fundamental hypothesis of the stochastic chemical kinetics.

The chemical master equation is of no practical use since, in general it cannot be solved analytically. Gillespie then placed the problem differently asking when and which reaction probably occurs next, given the system's current state. The simulation method he proposed was in accordance to the fundamental premise of the stochastic chemical kinetics and the realizations of the simulated Markov process describe precisely the chemical master equation. Gillespie's simulation algorithm is exact in that sense, though it remains computationally intensive especially when applied to large and highly reactive biochemical systems. Several approximation algorithms have been proposed to accelerate the exact Gillespie's algorithm. Instead of leaving the system to proceed according to the probable next time of the next reaction, thus simulating one reaction per step, the methods propose one step to leap over many reactions. As long as the propensity function remains constant within the leap, the leap approximations stay close to the exact algorithm [57–61].

#### 28.2.3.2 Chemical Langevin Equation

The chemical Langevin equation describes the time evolution of the chemically reacting system through the determination of the number of times each reaction channel occurs within a time interval, given the system's current state. The number of occurrences of each reaction is a random variable that depends on the time interval given, the current state of the system and the propensity function. In the Langevin equation, this random variable is approximated by a Gaussian random variable. As a result of this approximation, the system dynamics are simplified to the corresponding known deterministic reaction dynamics with an additional temporally uncorrelated and statistically independent Gaussian white noise term. When the approximation holds, Gillespie has shown that the Gaussian assumption is valid when two conditions are satisfied. The first condition requires the time interval to be short enough that any propensity function suffers noticeable change. The second condition needs the time interval to be long enough for any reaction channel to fire several times. In most systems, these two conditions are hardly satisfied simultaneously. Simpson et al. 2004 [62] presented a modification of the Langevin approach to overcome the unsound conditions. Furthermore, hybrid stochastic-deterministic methods (multiscale algorithms), able to partition the system into subsystems where different assumptions may hold in each partition have been proposed [63, 64].

Stochastic methods can be solved for small systems [4, 6, 65]. However, as the system size increases, the number of reaction events scales up and stochastic simulations face a computational burden. Methods that either improve the stochastic simulation performance or combine micro-macroscopic scales in the system or even work on both directions, making minor sacrifices in accuracy are indispensable.

The simulations of the stochastic models we have discussed, thus far, attempt to solve the forward problem. The stochastic kinetic parameters involved in modeling are assumed to be given. However, rarely are these parameters known. With the aim of modeling real large-scale systems, besides the fruitful theoretical work, the inverse problem should also be considered [66].

Other formalisms to describe the dynamics of a system as the combined effect of a deterministic driven system and Gaussian noise have also been proposed in the literature. An example is a model for the glycolytic yeast genes experiencing a diauxic transition upon glucose depletion [67].

#### 28.2.3.3 Hidden Markov Models

Hidden Markov models (HMMs), are observation driven approaches aiming to explain the observation data the system produces. The system being modeled is assumed to be a stochastic process, where the Markov property stands. In HMMs, the state of the system is assumed not to be directly visible, yet variables influenced by the system state are visible. In gene expression experiments for example, observations can be the mRNA measured values over time, while the exact truth behind, which causes the chemical reacting-interacting system to produce those observations is actually hidden. In microscopic scale, we could consider the number of molecules of each species involved on the process as the observation state vector and the frequency of each reaction channel at a given time interval as, the hidden state vector and the actual reason of the observations [68]. In another consideration, the hidden state vector can represent the expression-molecular population of gene modules. The transition probability, expresses the interactions among the modules in a probabilistic manner and the observation state vector represents the gene expression of all the genes involved in the process as the outcome of these interactions [27, 69].

The temporal transition of the system from one (hidden) state to the next is described probabilistically. The observation state at a given time depends on the hidden state of the same time and is also given under a probabilistic manner. When linear dynamics drive the system from the one (hidden) state to the next and both the hidden transition and emission probabilities are assumed to be Gaussian, the model is called Kalman filter representing the Gaussian equivalent of a HMM.

## *28.2.4 Spatial Approach – Method*

The cell is far from a homogeneous medium! As the volume of the medium in which molecular chemical reactions and interactions take place increases, well-stirred assumptions are no longer valid in modeling of the system dynamics. Similar to the importance of the temporal behavior, the spatial organization should also be considered in modeling to describe the molecular locomotion to places of action. A recent interesting investigation of the chemical reaction laws describing the dynamics in intracellular environments is presented in [70]. Furthermore, [71] proposes a criterion that determines whether the biological system under study needs spatial models to be described.

Spatio-temporal models of the biological processes have recently been proposed including from coarse-grained to realistic geometries (Fig. 28.3). Virtual Cell [72, 73], is an example of a computational tool that provides spatial simulations of biological systems at several spatial resolutions and geometries. The tool allows digital microscopic images to be used for the system space to be segmented appropriately.

### 28.2.4.1 Compartmental Modeling

The least computationally expensive and the first more plausible assumption, is to consider cell a multi-compartment system. Compartments usually reflect the physical entities of the cell (cytoplasm, nucleus), where the within distributions of species can be considered time-persistent homogeneous. Components present in different compartments are considered as different entities in modeling, thus the system size increases. Surfaces (e.g., cellular membranes) allow the formation of gradients and support molecular fluxes across them. The dynamics within the cellular compartments can be described with any of the ways we have seen so far. Additionally, terms for molecular transportation from one compartment to the other should also be added in whole cell simulations.

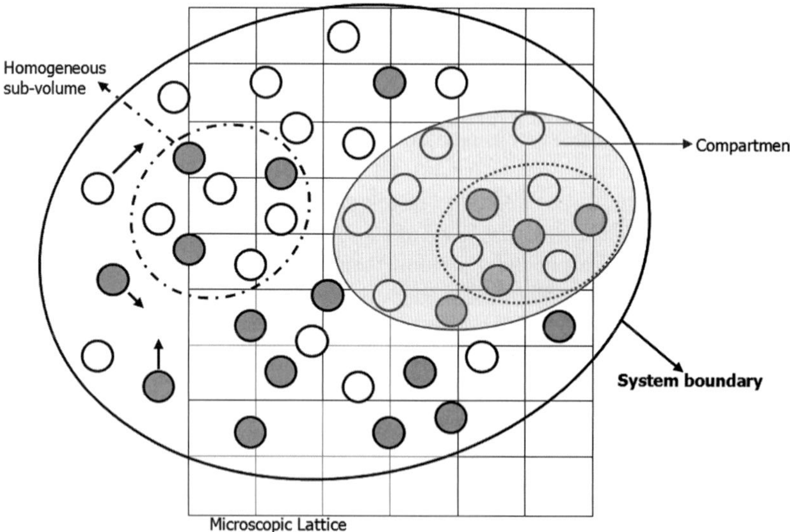


**Fig. 28.3** Different spatial resolution approaches. The system boundary defines the volume of the interacting molecular populations. A coarse segmentation of system space is into compartments (*grey area*). Homogeneity can be assumed within compartments, however, further segmentation of the space into sub-volumes might be essential. A microscopic lattice is also shown in the figure (Copies of figures including color copies, where applicable, are available in the accompanying CD)

Hirschberg et al. [74] have used compartmental models to describe, kinetically, the VSVG-GFP protein-traffic through the various compartments of the secretory pathway. Compartmental modeling has been also used, in modeling of the segment polarity network in early developmental stages of *Drosophila melanogaster* [75].

#### 28.2.4.2 Mesh space and Mean-field Approximation Modeling

In order for models to depict the heterogeneous distributions within compartments and their apparent stationary diffusion gradients, compartments are further divided into sub-volumes. Finer sub-volumes produce more accurate solutions, yet make simulations computationally expensive. In general, the partitioning of space comprises a challenging problem. Sub-volumes should be small enough for the spatial phenomena not to arise in them and for the partition to be assumed homogeneous. Thus, the partition should be such that diffusion within the sub-volume is faster than the rate of reactions. Partial differential equations have been proposed to describe system dynamics, representing together, molecular populations and the spatio-temporal evolution of the system in a continuous manner. In that way, the intracellular kinetics is approached at a macroscopic scale under the mean-field approximation where molecular concentrations become a continuous function of space and time. Partial differential equations require more parameters than their corresponding ordinary differential equations and are solved numerically, which means that both the spatial and temporal region should properly be divided to sub-regions. Thus, numerical integrations face the partitioning problem described above. Partial differential equations have been used to describe several systems [76–79]. When stochastic effects become significant giving rise to fluctuations around the mean field, molecular populations should be treated as system of discrete entities.

#### 28.2.4.3 Discrete Space Modeling

Discrete space models consider the chemically reacting space as a 2D or 3D microscopic lattice. Each site of the lattice may be occupied by one or several molecules. The molecules propagate from one site to another according to their velocities or the diffusion rate and then collide and react to other molecules. The sites may have different shapes or sizes according to the size and the number of molecules they are supposed to accommodate. Multi-scale segmentation of the space with high and low-resolution regions is also possible. In accordance to the microscopic or macroscopic assumptions of the kinetics, the evolution of the system is determined under that prism (cellular automata). Discrete space modeling provides a framework with significant advantages in computational efficiency over the even finer particle space models presented below that have been widely used [80–85].

#### 28.2.4.4 Particle Space Modeling

Particle space is the natural space of the system. In the molecular level of cell description, neither species populations are considered continuous variables nor is the reacting space artificially partitioned. Molecules are single entities moving within the cell volume and interacting according to the bio-physical forces. This perspective actually transfers the problem to a particle space definition regardless of the kinetic assumptions. The finest many-body molecular dynamics, the Gillespie stochastic model and the Brownian dynamics, all consider molecules as distinct entities nicely embedded in a particle space [86–89]. NAMD comprises the state- of- the- art computational framework for bio-molecular dynamics simulation of large-scale systems of the size of even multi-million atoms [90, 91].

## 28.3 Constraint-Based Modeling

Equation based models attempt to reveal the real biophysical forces, and describe the dynamic laws that drive the system to its functional properties by going deep into even microscopic scales. In a complex system of such a variety of components and interactions, robust and highly regulated, as the biological systems are, the more plausible way to understand the system, at that high resolution, was to restrict the studies to small parts of the puzzle and then follow the bottom-up approach towards whole picture reconstruction.

From a different perspective, holistic models consider the cell as a complete system that orchestrates its components under physio-chemical constraints towards the accomplishment of certain functions. The constraints actually describe the biophysical laws, the components that the system obey, as also the evolution process that has driven the system towards a better survival and source utilization, in a systemic fashion. By taking into account the holistic consideration, constraint-based models aim to integrate knowledge ("omics" data) at different levels in the cascade, from genes to proteins and further to metabolic fluxes in a genome-scale metabolic network to describe and understand the overall cellular functions [92, 93]. Genome-scale metabolic networks allow the direct correlation between the genomic information and metabolic activity at the flux level and the elucidation of properties (such as network robustness, product yield, and metabolic versatility, environmental and genetic phenotypic effects) that cannot be described by descriptions of individual components. In that sense, constraint-based models are also known as metabolic models.

How the holistic consideration is actually adopted in modeling? The core assumption of constraint-based models is that the system reaches a steady state (intracellular flux balancing) that satisfies the physio-chemical constraints under any given environmental condition. The hypothesis is based on the fact that the time-constants which describe metabolic transients are fast as compared to the constants associated with cell growth experiments. Under that macroscopic time scale, the cell can be considered as being in a quasi-steady state.

The constraints, the system inevitably obeys, include mass balance, energy balance, flux limitations; constraints which are not fully known though. Originally, the models assumed that the metabolic system reaches a steady state constrained by its stoichiometry (mass balance). Nevertheless, the stoichiometry alone only bounds the solution space, determining the region of the possible fluxes the system can reach. The metabolic networks always have more fluxes than metabolites; thus the system is undetermined under the stoichiometry constraint alone. Thermodynamic constraints that determine the reversibility of the reactions involved and enzymatic capacity constraints were also included to place limits on the range of possible fluxes. Figure 28.4 abstractly depicts the successive restrictions of the flux space the constraints of the system impose.

What information may the feasible flux space provide for the system? Convex analysis was used to enumerate the unique set of all distinct metabolic routes (elementary modes) of the network [94–96], as also to determine the minimal set of convex vectors (extreme pathways) needed to describe all allowable steady state flux distributions [97–99]. Elementary mode analysis and extreme pathways were an essential attempt to characterize the solution space. What remains unanswered is the solution the cell chooses under the given conditions.

Constraint-based models further assume that the cell performs the way it performs following an optimization strategy to accomplish certain cellular tasks such as the maximization of biomass production, the minimization of nutrient utilization, the maximization of ATP, under certain conditions. With the insertion of an optimization function, the problem is formulated to a combinatorial optimization problem. If the optimization function is linear, with respect to the fluxes involved, then the optimization problem is a linear programming problem and can be solved exactly by providing the flux distributions of the system. Flux balance analysis (FBA) models estimate the *optimal* flux distribution of the entire biochemical reacting system, providing a

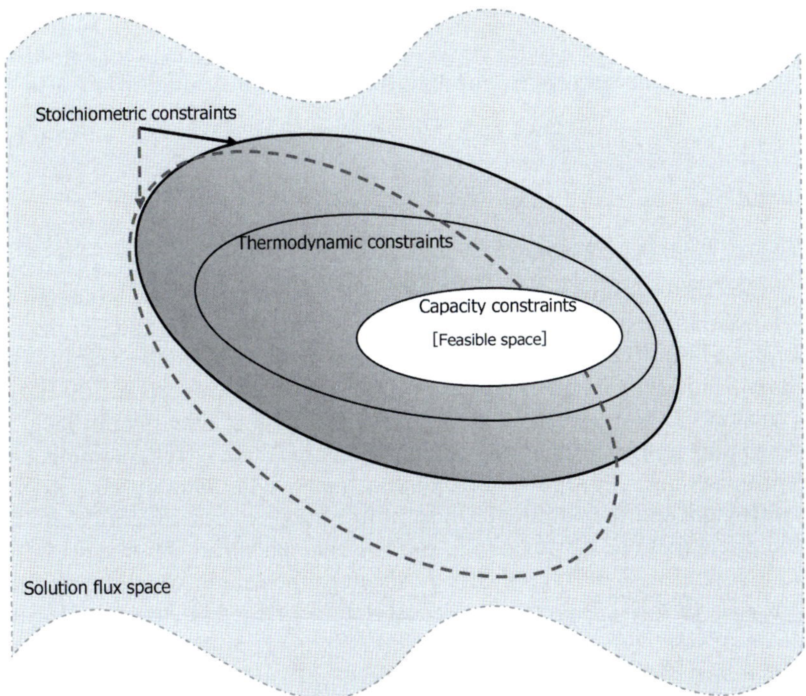


**Fig. 28.4** Abstract representation of the flux space. Feasible space is the flux subspace that satisfies all the constraints the system obeys. FBA methods search for the optimal flux distribution of the system within the feasible space, and assume that the system's phenotype is the reflection of this flux optimality. The set of the enzymatic interactions present in the system under study, together with the corresponding stoichiometric coefficients determine the stoichiometric constraints of the flux space. Different sets of enzymatic interactions defines different constraints resulting in different segmentation of the solution space (dotted external curve). The thermodynamic and capacity constraints are applied on the present set of enzymatic reactions, thus the feasible space will possibly be different for the different active networks (Copies of figures including color copies, where applicable, are available in the accompanying CD)

quantitative description of the system when the intracellular fluxes are in balance. Flux balance methods search the feasible space (Fig. 28.4) for the optimal flux distribution.

For the model to better represent real biological systems and accurately predict the experimental evidence, further improvements concerning both the constraints and the optimization functions have been suggested. The thermodynamic constraints were developed to take into account the intracellular and environmental conditions of the system [100]. Flux limitations with respect to the way the system responds and adapts to genetic perturbations were also investigated [101]. Furthermore, the incorporation of the temporal constraints to the model who also studied. The stoichiometry constraint implies that all the metabolic reactions present in an organism do actually participate, and all gene products are available to contribute to the certain task-optimal solution. However, the system may utilize different metabolic pathways to respond to the various conditions throughout its temporal evolution. Consequently, an 'active' subset among all the possible metabolic reactions, actually plays a role in the system dynamics [92, 102–104]. Moreover, researchers have considered the possibility of having multiple optimal solutions in the flux space [2, 105]. It is interesting to mention that flux data have also been used to determine inversely the optimization strategy that the system followed to measure the flux distribution [106].

The method has proved successful in analyzing the metabolic capabilities of several organisms, including its ability to predict deletion phenotypes, to determine the relative flux values of the metabolic reactions, to identify alternate optimal growth states [107–109], to guide the iterative refinement of the model and to validate metabolic network reconstruction [104, 110–112], and to identify a group of reactions being active under all environmental conditions [113].

## 28.4 Conclusions

In principle, with enough computing power, an initial picture of the system at the molecular level and a proper unified theory of the biophysical laws that govern the system's dynamics, which takes into account all peculiarities of the biological system (in homogeneity, randomness, redundancy, robustness) we have discussed, a whole cell simulation at that molecular-reaction level of detail could be achieved. However, even though a massive amount of experimental data is currently available and substantial biological knowledge has been gained, they remain insufficient for the inference of the missing knowledge, in order to simulate large scale systems at molecular resolution. There are compromises that, if properly applied, may improve the simulation speed and reduce the dimensionality problem and the parameter space, while making minor sacrifices in the description accuracy of the phenomenon. For example, models that partition the system into subsystems, where different assumptions can be applied, have been proposed including the stochastic-deterministic hybrid models [63, 64]. A recent multi-level integrated software tool for simulation of complex biochemical systems is COPASI [114]. COPASI incorporates deterministic and stochastic approaches to simulate biochemical reactions and also provides a package of optimization algorithms to estimate the unknown parameters involved. To keep the problem tractable, the cellular system has also been partitioned into functional modules (previous chapter) where detailed kinetic models can be constructed for each. Those *divide-conquer* (bottom-up) approaches face two difficulties. The partitioning of the system into functional or mathematical parts, in addition to the integration which has to be followed is not always a trivial task. Furthermore, when validation or optimization is needed for the sub-models, we must have in mind that the data are usually referred to the complete system and not to the parts which are indeed not independent of the rest system. Based on the modular approach, nice examples of integrative dynamic models are presented by Snoep et al. and Klipp et al. [115, 116]. Alternative models, which simulate large scale systems as a whole by incorporating information and data from genes to proteins and enzymes,

**Table 28.1** Integrated dynamic systems

| Description | Name | Webpage |
|---|---|---|
| Rule-based differential equations for metabolism | E-Cell | http://www.e-cell.org/software/e-cell-system |
| Simulation of metabolic pathways & optimization tool | GEPASI | http://www.gepasi.org/ |
| Stochastic-deterministic simulation of complex pathways & optimization tool | COPASI | http://www.copasi.org/tiki-index.php |
| Compartmental & Spatial model | VCell | http://www.vcell.org/ www.nrcam.uchc.edu/login/login.html |
| Experimentally based simulation | SiC | http://www.siliconcell.net/ |
| Stochastic Simulator | SGNsim | http://bioinformatics.oxfordjournals.org/cgi/content/abstract/btm004v2 |
| Stochastic Spatial Simulator | SSS | http://www.webpages.uidaho.edu/~krone/winsss/sss-main.html |
| Reaction-Diffusion network | SmartCell | http://smartcell.embl.de/introduction.html |
| Molecular dynamics simulation | NAMD | http://www.ks.uiuc.edu/Research/namd/ |
| GRN inference & simulation | JCell | http://www-ra.informatik.uni-tuebingen.de/software/JCell/tutorial/index.html |

(Copies of tables are available in the accompanying CD.)

have also been proposed [93], sacrificing dynamic description resolution. Constraint-based models are widely used as top-down models, for the investigation of the metabolic capabilities of certain organisms under specific environmental conditions and perturbations. Dynamic phenomena can be approximated by changing the constraints to shape according to the feasible flux space. The temporal path of flux distributions that the system undergoes throughout its dynamic evolution remains an open problem. Additionally, a better way to incorporate other interacting systems such as signal pathways, and gene regulatory networks to the complex metabolic network leaves room for improvement towards a multi-level integrated system. Some representative implementations of the methods described in this chapter available on the internet are listed in Table 28.1.

**Acknowledgments** The authors involved in this chapter were supported by the GSRT, Hellas (project PENED 03ED842), the EMBO Young Investigator Program, and the European Commission (project INFOBIOMED, FP6-IST-507585). We acknowledge the authors of the tools listed in this chapter for placing them in the public domain. We thank all the members of our lab for fruitful discussions and comments on the manuscript.

## Key References

Vitaly V. Gursky, J.J., Konstantin N. Kozlov, John Reinitz, Alexander M. Samsonova, *Pattern formation and nuclear divisions are uncoupled in Drosophila segmentation: comparison of spatially discrete and continuous models*. Physica D, 2004. 197: p. 286–302.

Herrgard, M.J., et al., *Integrated analysis of regulatory and metabolic networks reveals novel regulatory mechanisms in Saccharomyces cerevisiae*. Genome Res, 2006. 16(5): p. 627–635.

Hoops, S., et al., *COPASI—a COmplex PAthway SImulator*. Bioinformatics, 2006. 22(24): p. 3067–3074.

Snoep, J.L., et al., *Towards building the silicon cell: a modular approach*. Biosystems, 2006. 83(2–3): p. 207–216.

## Suggested Reading

1. Chen, B.S., et al., *A new measure of the robustness of biochemical networks*. Bioinformatics, 2005. 21(11): p. 2698–2705.
2. Mahadevan, R. and C.H. Schilling, *The effects of alternate optimal solutions in constraint-based genome-scale metabolic models*. Metab Eng, 2003. 5(4): p. 264–276.
3. Orphanides, G. and D. Reinberg, *A unified theory of gene expression*. Cell, 2002. 108(4): p. 439–451.
4. Ribeiro, A., R. Zhu, and S.A. Kauffman, *A general modeling strategy for gene regulatory networks with stochastic dynamics*. J Comput Biol, 2006. 13(9): p. 1630–1639.
5. Chen, K.C., et al., *A stochastic differential equation model for quantifying transcriptional regulatory network in Saccharomyces cerevisiae*. Bioinformatics, 2005. 21(12): p. 2883–2890.
6. McAdams, H.H. and A. Arkin, *Stochastic mechanisms in gene expression*. Proc Natl Acad Sci U S A, 1997. 94(3): p. 814–819.
7. Joshi, A. and B.O. Palsson, *Metabolic dynamics in the human red cell. Part I—A comprehensive kinetic model*. J Theor Biol, 1989. 141(4): p. 515–528.
8. Nakayama, Y., A. Kinoshita, and M. Tomita, *Dynamic simulation of red blood cell metabolism and its application to the analysis of a pathological condition*. Theor Biol Med Model, 2005. 2(1): p. 18.
9. Chen, K.C., et al., *Integrative analysis of cell cycle control in budding yeast*. Mol Biol Cell, 2004. 15(8): p. 3841–3862.
10. Ingram, P.J., M.P. Stumpf, and J. Stark, *Network motifs: structure does not determine function*. BMC Genomics, 2006. 7: p. 108.
11. Yang, C.R., et al., *A mathematical model for the branched chain amino acid biosynthetic pathways of Escherichia coli K12*. J Biol Chem, 2005. 280(12): p. 11224–11232.
12. Goryanin, I., T.C. Hodgman, and E. Selkov, *Mathematical simulation and analysis of cellular metabolism and regulation*. Bioinformatics, 1999. 15(9): p. 749–758.
13. Widder, S., J. Schicho, and P. Schuster, *Dynamic patterns of gene regulation I: Simple two-gene systems*. J Theor Biol, 2007.

14. Mocek, W.T., R. Rudnicki, and E.O. Voit, *Approximation of delays in biochemical systems.* Math Biosci, 2005. 198(2): p. 190–216.
15. Carey, M., *The enhanceosome and transcriptional synergy.* Cell, 1998. 92(1): p. 5–8.
16. Spudich, J.L. and D.E. Koshland, Jr., *Non-genetic individuality: chance in the single cell.* Nature, 1976. 262(5568): p. 467–471.
17. Hasty, J., et al., *Noise-based switches and amplifiers for gene expression.* Proc Natl Acad Sci U S A, 2000. 97(5): p. 2075–2080.
18. Elowitz, M.B., et al., *Stochastic gene expression in a single cell.* Science, 2002. 297(5584): p. 1183–1186.
19. Gillespie, D.T., *Stochastic Simulation of Chemical Kinetics.* Annu Rev Phys Chem, 2006.
20. Lacalli, T.C., *Modeling the Drosophila pair-rule pattern by reaction-diffusion: gap input and pattern control in a 4-morphogen system.* J Theor Biol, 1990. 144(2): p. 171–194.
21. Aranda, J.S., E. Salgado, and A. Munoz-Diosdado, *Multifractality in intracellular enzymatic reactions.* J Theor Biol, 2006. 240(2): p. 209–217.
22. Schnell, S. and T.E. Turner, *Reaction kinetics in intracellular environments with macromolecular crowding: simulations and rate laws.* Prog Biophys Mol Biol, 2004. 85(2–3): p. 235–260.
23. Weiss, M., et al., *Anomalous subdiffusion is a measure for cytoplasmic crowding in living cells.* Biophys J, 2004. 87(5): p. 3518–3524.
24. de Hoon, M.J., et al., *Inferring gene regulatory networks from time-ordered gene expression data of Bacillus subtilis using differential equations.* Pac Symp Biocomput, 2003: p. 17–28.
25. D'Haeseleer, P., et al., *Linear modeling of mRNA expression levels during CNS development and injury.* Pac Symp Biocomput, 1999: p. 41–52.
26. van Someren, E.P., L.F. Wessels, and M.J. Reinders, *Linear modeling of genetic networks from experimental data.* Proc Int Conf Intell Syst Mol Biol, 2000. 8: p. 355–366.
27. Wu, F.X., W.J. Zhang, and A.J. Kusalik, *Modeling gene expression from microarray expression data with state-space equations.* Pac Symp Biocomput, 2004: p. 581–592.
28. Gustafsson, M., M. Hornquist, and A. Lombardi, *Constructing and analyzing a large-scale gene-to-gene regulatory network–lasso-constrained inference and biological validation.* IEEE/ACM Trans Comput Biol Bioinform, 2005. 2(3): p. 254–261.
29. Chen, T., H.L. He, and G.M. Church, *Modeling gene expression with differential equations.* Pac Symp Biocomput, 1999: p. 29–40.
30. Glass, L. and S.A. Kauffman, *The logical analysis of continuous, non-linear biochemical control networks.* J Theor Biol, 1973. 39(1): p. 103–129.
31. Drulhe, S., Ferrari-Trecate, G., H. de Jong, and A. Viari, *Reconstruction of Switching Thresholds in Piece-wise-Affine Models of Genetic Regulatory Networks.* LECTURE NOTES IN COMPUTER SCIENCE, 2006(3927): p. 184–199.
32. Vercruysse, S. and M. Kuiper, *Simulating genetic networks made easy: network construction with simple building blocks.* Bioinformatics, 2005. 21(2): p. 269–271.
33. Radde, N., J. Gebert, and C.V. Forst, *Systematic component selection for gene-network refinement.* Bioinformatics, 2006. 22(21): p. 2674–2680.
34. Mason, J., et al., *Evolving complex dynamics in electronic models of genetic networks.* Chaos, 2004. 14(3): p. 707–715.
35. Edwards, R., P. van den Driessche, and L. Wang, *Periodicity in piece-wise-linear switching networks with delay.* J Math Biol, 2007.
36. Casey, R., H. de Jong, and J.L. Gouze, *Piece-wise-linear models of genetic regulatory networks: equilibria and their stability.* J Math Biol, 2006. 52(1): p. 27–56.
37. Ben-Hur, A. and H.T. Siegelmann, *Computation in gene networks.* Chaos, 2004. 14(1): p. 145–151.
38. Mestl, T., E. Plahte, and S.W. Omholt, *A mathematical framework for describing and analysing gene regulatory networks.* J Theor Biol, 1995. 176(2): p. 291–300.
39. Hu, X., A. Maglia, and D. Wunsch, *A general recurrent neural network approach to model genetic regulatory networks.* Conf Proc IEEE Eng Med Biol Soc, 2005. 5: p. 4735–4738.
40. Vohradsky, J., *Neural network model of gene expression.* Faseb J, 2001. 15(3): p. 846–854.
41. Wahde, M. and J. Hertz, *Modeling genetic regulatory dynamics in neural development.* J Comput Biol, 2001. 8(4): p. 429–442.
42. Xu, R., X. Hu, and D. Wunsch Ii, *Inference of genetic regulatory networks with recurrent neural network models.* Conf Proc IEEE Eng Med Biol Soc, 2004. 4: p. 2905–2908.
43. Weaver, D.C., C.T. Workman, and G.D. Stormo, *Modeling regulatory networks with weight matrices.* Pac Symp Biocomput, 1999: p. 112–123.
44. Vohradsky, J., *Neural model of the genetic network.* J Biol Chem, 2001. 276(39): p. 36168–36173.
45. Sorribas, A., R. Curto, and M. Cascante, *Comparative characterization of the fermentation pathway of Saccharomyces cerevisiae using biochemical systems theory and metabolic control analysis: model validation and dynamic behavior.* Math Biosci, 1995. 130(1): p. 71–84.

46. Voit, E.O. and T. Radivoyevitch, *Biochemical systems analysis of genome-wide expression data*. Bioinformatics, 2000. 16(11): p. 1023–1037.
47. Alvarez-Vasquez, F., C. Gonzalez-Alcon, and N.V. Torres, *Metabolism of citric acid production by Aspergillus niger: model definition, steady-state analysis and constrained optimization of citric acid production rate*. Biotechnol Bioeng, 2000. 70(1): p. 82–108.
48. Kitayama, T., et al., *A simplified method for power-law modelling of metabolic pathways from time-course data and steady-state flux profiles*. Theor Biol Med Model, 2006. 3: p. 24.
49. Kimura, S., et al., *Inference of S-system models of genetic networks using a cooperative coevolutionary algorithm*. Bioinformatics, 2005. 21(7): p. 1154–1163.
50. Noman, N. and H. Iba, *Reverse engineering genetic networks using evolutionary computation*. Genome Inform, 2005. 16(2): p. 205–214.
51. Gonzalez, O.R., et al., *Parameter estimation using Simulated Annealing for S-system models of biochemical networks*. Bioinformatics, 2007. 23(4): p. 480–486.
52. Voit, E.O., *Smooth bistable S-systems*. Syst Biol (Stevenage), 2005. 152(4): p. 207–213.
53. Marino, S. and E.O. Voit, *An automated procedure for the extraction of metabolic network information from time series data*. J Bioinform Comput Biol, 2006. 4(3): p. 665–691.
54. Chou, I.C., H. Martens, and E.O. Voit, *Parameter estimation in biochemical systems models with alternating regression*. Theor Biol Med Model, 2006. 3: p. 25.
55. Hernandez-Bermejo, B., V. Fairen, and A. Sorribas, *Power-law modeling based on least-squares minimization criteria*. Math Biosci, 1999. 161(1–2): p. 83–94.
56. Savageau, M.A., *A theory of alternative designs for biochemical control systems*. Biomed Biochim Acta, 1985. 44(6): p. 875–80.
57. Cai, X. and Z. Xu, *K-leap method for accelerating stochastic simulation of coupled chemical reactions*. J Chem Phys, 2007. 126(7): p. 074102.
58. Cao, Y., D.T. Gillespie, and L.R. Petzold, *Efficient step size selection for the tau-leaping simulation method*. J Chem Phys, 2006. 124(4): p. 044109.
59. Chatterjee, A., et al., *Time accelerated Monte Carlo simulations of biological networks using the binomial tau-leap method*. Bioinformatics, 2005. 21(9): p. 2136–2137.
60. Tian, T. and K. Burrage, *Binomial leap methods for simulating stochastic chemical kinetics*. J Chem Phys, 2004. 121(21): p. 10356–10364.
61. Puchalka, J. and A.M. Kierzek, *Bridging the gap between stochastic and deterministic regimes in the kinetic simulations of the biochemical reaction networks*. Biophys J, 2004. 86(3): p. 1357–1372.
62. Simpson, M.L., C.D. Cox, and G.S. Sayler, *Frequency domain chemical Langevin analysis of stochasticity in gene transcriptional regulation*. J Theor Biol, 2004. 229(3): p. 383–394.
63. Haseltine, E.L. and J.B. Rawlings, *On the origins of approximations for stochastic chemical kinetics*. J Chem Phys, 2005. 123(16): p. 164115.
64. Salis, H. and Y. Kaznessis, *Accurate hybrid stochastic simulation of a system of coupled chemical or biochemical reactions*. J Chem Phys, 2005. 122(5): p. 54103.
65. Achimescu, S. and O. Lipan, *Signal propagation in non-linear stochastic gene regulatory networks*. Syst Biol (Stevenage), 2006. 153(3): p. 120–134.
66. Reinker, S., R.M. Altman, and J. Timmer, *Parameter estimation in stochastic biochemical reactions*. Syst Biol (Stevenage), 2006. 153(4): p. 168–178.
67. Wang, S.C., *Reconstructing genetic networks from time ordered gene expression data using Bayesian method with global search algorithm*. J Bioinform Comput Biol, 2004. 2(3): p. 441–458.
68. Goutsias, J., *A hidden Markov model for transcriptional regulation in single cells*. IEEE/ACM Trans Comput Biol Bioinform, 2006. 3(1): p. 57–71.
69. Inoue, L.Y., et al., *Cluster-based network model for time-course gene expression data*. Biostatistics, 2006.
70. Grima, R. and S. Schnell, *A systematic investigation of the rate laws valid in intracellular environments*. Biophys Chem, 2006. 124(1): p. 1–10.
71. Mayawala, K., D.G. Vlachos, and J.S. Edwards, *Spatial modeling of dimerization reaction dynamics in the plasma membrane: Monte Carlo vs. continuum differential equations*. Biophys Chem, 2006. 121(3): p. 194–208.
72. Loew, L.M. and J.C. Schaff, *The virtual cell: a software environment for computational cell biology*. Trends Biotechnol, 2001. 19(10): p. 401–406.
73. Slepchenko, B.M., et al., *Quantitative cell biology with the virtual cell*. Trends Cell Biol, 2003. 13(11): p. 570–576.
74. Hirschberg, K., et al., *Kinetic analysis of secretory protein traffic and characterization of golgi to plasma membrane transport intermediates in living cells*. J Cell Biol, 1998. 143(6): p. 1485–1503.
75. Von Dassow, G. and G.M. Odell, *Design and constraints of the Drosophila segment polarity module: robust spatial patterning emerges from intertwined cell state switches*. J Exp Zool, 2002. 294(3): p. 179–215.

76. Schaff, J., et al., *A general computational framework for modeling cellular structure and function*. Biophys J, 1997. 73(3): p. 1135–1346.
77. Wylie, D.C., et al., *A hybrid deterministic-stochastic algorithm for modeling cell signaling dynamics in spatially inhomogeneous environments and under the influence of external fields*. J Phys Chem B Condens Matter Mater Surf Interfaces Biophys, 2006. 110(25): p. 12749–12765.
78. Vitaly V. Gursky, J.J., Konstantin N. Kozlov, John Reinitz, Alexander M. Samsonova, *Pattern formation and nuclear divisions are uncoupled in Drosophila segmentation: comparison of spatially discrete and continuous models*. Physica D, 2004. 197: p. 286–302.
79. Smith, A.E., et al., *Systems analysis of Ran transport*. Science, 2002. 295(5554): p. 488–491.
80. Broderick, G., et al., *A life-like virtual cell membrane using discrete automata*. In Silico Biol, 2005. 5(2): p. 163–178.
81. Weimar, J.R. and J.P. Boon, *Class of cellular automata for reaction-diffusion systems*. Physical Review. E. Statistical Physics, Plasmas, Fluids, and Related Interdisciplinary Topics, 1994. 49(2): p. 1749–1752.
82. Shimizu, T.S., S.V. Aksenov, and D. Bray, *A spatially extended stochastic model of the bacterial chemotaxis signalling pathway*. J Mol Biol, 2003. 329(2): p. 291–309.
83. Dab, D., et al., *Cellular-automaton model for reactive systems*. Physical Review Letters, 1990. 64(20): p. 2462–2465.
84. Wishart, D.S., et al., *Dynamic cellular automata: an alternative approach to cellular simulation*. In Silico Biol, 2005. 5(2): p. 139–161.
85. Kier, L.B., et al., *A cellular automata model of enzyme kinetics*. J Mol Graph, 1996. 14(4): p. 227–231, 226.
86. Andrews, S.S. and D. Bray, *Stochastic simulation of chemical reactions with spatial resolution and single molecule detail*. Phys Biol, 2004. 1(3–4): p. 137–151.
87. Erban, R. and S.J. Chapman, *Reactive boundary conditions for stochastic simulations of reaction-diffusion processes*. Phys Biol, 2007. 4(1): p. 16–28.
88. Elf, J. and M. Ehrenberg, *Spontaneous separation of bi-stable biochemical systems into spatial domains of opposite phases*. Syst Biol (Stevenage), 2004. 1(2): p. 230–236.
89. Chiam, K.H., et al., *Hybrid simulations of stochastic reaction-diffusion processes for modeling intracellular signaling pathways*. Phys Rev E Stat Nonlin Soft Matter Phys, 2006. 74(5 Pt 1): p. 051910.
90. Sanbonmatsu, K.Y. and C.S. Tung, *High performance computing in biology: multimillion atom simulations of nanoscale systems*. J Struct Biol, 2007. 157(3): p. 470–480.
91. Phillips, J.C., et al., *Scalable molecular dynamics with NAMD*. J Comput Chem, 2005. 26(16): p. 1781–1802.
92. Covert Markus, Schilling Christophe, and P. Bernhard, *Regulation of Gene Expression in Flux Balance Models of Metabolism*. 2001: p. 73–78.
93. Varma, A. and B.O. Palsson, *Stoichiometric flux balance models quantitatively predict growth and metabolic by-product secretion in wild-type Escherichia coli W3110*. Appl Environ Microbiol, 1994. 60(10): p. 3724–3731.
94. Cakir, T., B. Kirdar, and K.O. Ulgen, *Metabolic pathway analysis of yeast strengthens the bridge between transcriptomics and metabolic networks*. Biotechnol Bioeng, 2004. 86(3): p. 251–260.
95. Klamt, S. and J. Stelling, *Combinatorial complexity of pathway analysis in metabolic networks*. Mol Biol Rep, 2002. 29(1–2): p. 233–236.
96. Carlson, R., D. Fell, and F. Srienc, *Metabolic pathway analysis of a recombinant yeast for rational strain development*. Biotechnol Bioeng, 2002. 79(2): p. 121–134.
97. Wiback, S.J. and B.O. Palsson, *Extreme pathway analysis of human red blood cell metabolism*. Biophys J, 2002. 83(2): p. 808–818.
98. Papin, J.A., et al., *The genome-scale metabolic extreme pathway structure in Haemophilus influenzae shows significant network redundancy*. J Theor Biol, 2002. 215(1): p. 67–82.
99. Schilling, C.H., et al., *Combining pathway analysis with flux balance analysis for the comprehensive study of metabolic systems*. Biotechnol Bioeng, 2000. 71(4): p. 286–306.
100. Henry, C.S., L.J. Broadbelt, and V. Hatzimanikatis, *Thermodynamics-based metabolic flux analysis*. Biophys J, 2007. 92(5): p. 1792–1805.
101. Shlomi, T., O. Berkman, and E. Ruppin, *Regulatory on/off minimization of metabolic flux changes after genetic perturbations*. Proc Natl Acad Sci U S A, 2005. 102(21): p. 7695–7700.
102. Herrgard, M.J., S.S. Fong, and B.O. Palsson, *Identification of genome-scale metabolic network models using experimentally measured flux profiles*. PLoS Comput Biol, 2006. 2(7): p. e72.
103. Mahadevan, R., J.S. Edwards, and F.J. Doyle, 3rd, *Dynamic flux balance analysis of diauxic growth in Escherichia coli*. Biophys J, 2002. 83(3): p. 1331–1340.
104. Herrgard, M.J., et al., *Integrated analysis of regulatory and metabolic networks reveals novel regulatory mechanisms in Saccharomyces cerevisiae*. Genome Res, 2006. 16(5): p. 627–635.
105. Patil, K.R., et al., *Evolutionary programming as a platform for in silico metabolic engineering*. BMC Bioinformatics, 2005. 6: p. 308.

106. Knorr, A.L., R. Jain, and R. Srivastava, *Bayesian-based selection of metabolic objective functions*. Bioinformatics, 2007. 23(3): p. 351–357.
107. Forster, J., et al., *Genome-scale reconstruction of the Saccharomyces cerevisiae metabolic network*. Genome Res, 2003. 13(2): p. 244–253.
108. Borodina, I., P. Krabben, and J. Nielsen, *Genome-scale analysis of Streptomyces coelicolor A3(2) metabolism*. Genome Res, 2005. 15(6): p. 820–829.
109. Edwards, J.S., R.U. Ibarra, and B.O. Palsson, *In silico predictions of Escherichia coli metabolic capabilities are consistent with experimental data*. Nat Biotechnol, 2001. 19(2): p. 125–130.
110. Feist, A.M., et al., *Modeling methanogenesis with a genome-scale metabolic reconstruction of Methanosarcina barkeri*. Mol Syst Biol, 2006. 2: p. 2006 0004.
111. Becker, S.A. and B.O. Palsson, *Genome-scale reconstruction of the metabolic network in Staphylococcus aureus N315: an initial draft to the two-dimensional annotation*. BMC Microbiol, 2005. 5(1): p. 8.
112. Duarte, N.C., M.J. Herrgard, and B.O. Palsson, *Reconstruction and validation of Saccharomyces cerevisiae iND750, a fully compartmentalized genome-scale metabolic model*. Genome Res, 2004. 14(7): p. 1298–1309.
113. Almaas, E., Z. Oltvai, and A. Barabasi, *The Activity Reaction Core and Plasticity of Metabolic Networks*. PloS Computational Biology, 2005. 1(7).
114. Hoops, S., et al., *COPASI–a COmplex PAthway SImulator*. Bioinformatics, 2006. 22(24): p. 3067–3674.
115. Snoep, J.L., et al., *Towards building the silicon cell: a modular approach*. Biosystems, 2006. 83(2–3): p. 207–216.
116. Klipp, E., et al., *Integrative model of the response of yeast to osmotic shock*. Nat Biotechnol, 2005. 23(8): p. 975–982.

## Chapter 29
# ASIAN: Network Inference Web Server

**Sachiyo Aburatani, Shigeru Saito, and Katsuhisa Horimoto**

**Abstract** Network inference in a living cell is one of the main themes in high-throughput assays, and some computational methods have been developed to deduce the relationships between the genes and proteins. Recently, we developed a method for inferring a network from a large amount of numerical data obtained from high-throughput analyses. Our method is now open to the public on the web, named as ASIAN (http://eureka.cbrc.jp/asian/).

In this chapter, we describe the ASIAN (Automatic System for Inferring A Network) web server for inferring a network from a large amount of numerical data, based on graphical Gaussian modeling (GGM) in combination with hierarchical clustering. GGM is based on a simple mathematical structure, which is the calculation of the inverse of the correlation coefficient matrix between variables. The ASIAN web server can analyze a wide variety of data within a reasonable computational time. The server allows users to input the numerical data, and it outputs the dendrogram of the objects by several hierarchical clustering techniques, the cluster number is estimated by a stopping rule for hierarchical clustering, and the network between the clusters by GGM, with the respective graphical presentations. ASIAN is useful for inferring the framework of networks from redundant empirical data, in addition to the clustering, concomitant with the estimation of the cluster number. In particular, the visual presentation of the result provides an intuitive means for understanding the putative network between the objects.

**Keywords:** Network inference · Statistical analysis · Hierarchical clustering · Microarray data · Partial correlation coefficient

## 29.1 Introduction

The elucidation of the complete genome sequences of model organisms has facilitated the high-throughput assays of living cells on a genome-wide scale. By the high-throughput assays, molecular networks and their components in the living cells are measured at multiple levels, such as mRNA transcript quantities, protein-protein and protein-DNA interactions, chromatin structure, and protein quantities, localization, and modifications. Among the many genome-wide assays, whole-genome expression profiling, facilitated by the development of DNA microarrays, represents a major advance in genome-wide functional analysis. The advances in microarray techniques have enabled us to monitor the expression levels of thousands or tens of thousands of genes simultaneously, under multiple conditions. The data obtained by microarray assays are rich

S. Aburatani
Biological Network Team, Computational Biology Research Center, National Institute of Advanced Industrial Science and Technology, AIST Tokyo Waterfront Bio-IT Research Building, 2-42 Aomi, Koto-ku, Tokyo, 135-0064, Japan
e-mail: s.aburatani@aist.go.jp

S. Krawetz (ed.), *Bioinformatics for Systems Biology*, DOI 10.1007/978-1-59745-440-7_29,

numerical data that reflect the cellular processes at work. Therefore, to transform the numerical data into biological insights about the underlying mechanisms, the computational analysis of the numerical data is required. To reveal the underlying information in the complex expression data, various types of methods have been proposed, in terms of two analytical approaches. One is gene classification by detecting similar profile patterns, and the other is network inference by modeling the expression changes of the genes over the measured conditions.

To classify the genes, cluster analyses and other approaches are usually adopted to the expression profiles. The genes are classified into some groups by various computational methods, as the first step toward identifying the gene function. Based on the gene classifications, for example, genes are allocated into functional categories, and searches for regulatory sequences are performed in the upstream regions among the genes belonging to each cluster. Thus, classification methods, such as clustering, have been established as a pre-requisite for the identification of gene function from gene expression profiles, and several web servers have been developed to perform the clustering of profiles integrated from different resources. One of the important issues in cluster analysis is the estimation of gene groups.

As a further challenging investigation, the network of regulatory relationships is inferred by various approaches directly from the expression profiles. For example, the Boolean and Bayesian networks have been successfully applied to infer the regulatory networks, and some improvements and modifications have been reported in the application of Boolean and Bayesian networks. However, since the two approaches require specific techniques and large amounts of computational time, it would be difficult to develop a web server, based on the two approaches, to analyze large numbers of gene expression profiles. Furthermore, the difficulty in inferring regulatory networks is affected by the composition of the expression profiles. In many profile data-sets, the expression changes of thousands of genes have been measured under all most two hundred conditions. For example, the expression changes of about 6200 genes in yeast have been measured under about 200 different conditions. The results indicated that many genes show quite similar patterns of expression profiles under a restricted number of measured conditions. Thus, the features of the present profile data on a genomic scale essentially prevent us from inferring the explicit relationship between each gene. In other words, the pre-processing of the profiles on a genomic scale may be expected to be the first step towards revealing the whole regulatory relationship.

To enhance the investigations, we developed two methods for expression profile analysis: one is a method for the automatic estimation of cluster boundaries in a hierarchical clustering, and the other is a method for the inference of a genetic network by the application of graphical Gaussian modeling (GGM). The two newly developed methods have been synthesized into a system, named Automatic System for Inferring A Network (ASIAN), for automatically deducing the framework of a genetic network, with only the input of a large amount of expression profile data. GGM is one of the graphical models that include the Boolean and Bayesian models. Among the graphical models, GGM is the simplest structure in a mathematical sense; only the inverse of the correlation coefficient between the variables is needed. Therefore, GGM can be easily applied to a wide variety of data. To apply GGM to expression profiles, ASIAN is devised as a procedure combination with hierarchical clustering, since the expression profiles often share a similar pattern.

Although ASIAN was developed for inferring the regulatory network from gene expression profiles, this procedure can also be applied to other biological data. Usually, the numerical data obtained by high-throughput assays, include some redundancy. ASIAN classifies the large number of numerical data into groups, according to the usual hierarchical clustering method, as a pre-processing for network inference. To avoid the generation of a non-regular correlation coefficient matrix from the redundant data, a stopping rule is adopted in the hierarchical clustering. Then, the relationship between the clusters is inferred by GGM. Thus, ASIAN provides a framework for inferring the network between the clusters. The new version provides a quick analysis by ASIAN, a step-by-step analysis by ASIAN, and graphical presentations of the clustering and the cluster boundary estimation.

## 29.2 The Overview of ASIAN

The overview of ASIAN is shown in Fig. 29.1. ASIAN is composed of four main parts: (i) the calculation of a correlation coefficient matrix for the input data, (ii) the hierarchical clustering, (iii) the estimation of cluster boundaries, and (iv) the application of GGM to the clusters.

On the front page, users can select either a high-throughput analysis or a partial analysis. In the partial analysis, the user can independently perform the four parts of ASIAN. Thus, the present ASIAN web site is able to perform network inferences and various statistical analyses based on the user's interests.

### *29.2.1 The Calculation of a Correlation Coefficient Matrix*

In GGM, the network is inferred by the calculation of a partial correlation coefficient matrix from the correlation coefficient matrix. Thus, the calculation of correlation coefficients between all pairs of given objects is the first step for inferring the network from the data. Actually, the correlation coefficient is recognized as one of the criteria to measure the strength of the relationship between the two objects. This part will generate the correlation coefficient matrix between all pairs of the objects.

### *29.2.2 The Hierarchical Clustering*

The partial correlation coefficient matrix can only be obtained if the correlation coefficient matrix is regular. Since, the empirical data from a high-throughput analysis often include some redundancy; the correlation coefficient matrix between the objects is not always regular. Thus, the classification of the objects according to the similarity of the redundant data is needed for network inference by GGM. As the classification method, the hierarchical clustering is performed with the data as the second step in the ASIAN system.

Furthermore, the hierarchical clustering is one of the major classification methods from the observed data. By the hierarchical clustering, the distance of similarity between each pair of the objects is calculated from their numerical data. On the ASIAN system, 3 metrics for measuring the distance between the objects are prepared, and the Euclidean distance between the Pearson correlation coefficients is set as a default metric:

$$d_{jk} = \sqrt{\sum_{l=1}^{n} (r_{il} - r_{jl})^2} \tag{29.1}$$

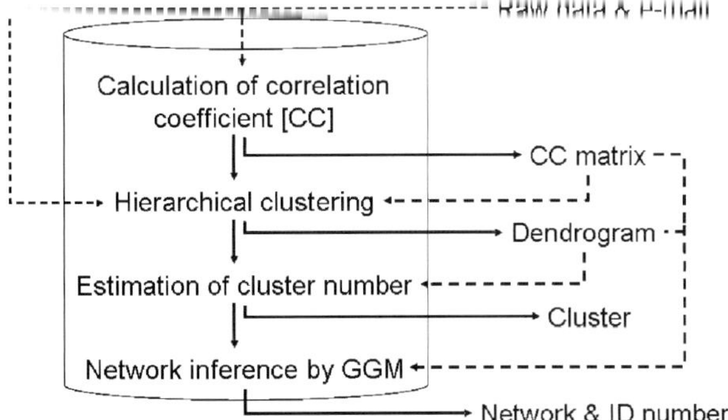


**Fig. 29.1** Organization of the ASIAN System. The ASIAN website is indicated by a cylinder, and the ASIAN system is composed of four parts. The input by a user and the output from the ASIAN web site are denoted by a broken and solid arrows, respectively. (Copies of figures including color copies are available in the accompanying CD)

where $n$ is the total number of genes, and $r_{ij}$ is the Pearson correlation coefficient between the $i$ and $j$ objects of the numerical data that are measured under $n_c$ conditions, $p_{ik}$, ($k = 1,2,\dots n_c$ ):

$$r_{ij} = \frac{\sum_{k=1}^{n_c} (p_{ik} - \overline{p_i})(p_{jk} - \overline{p_j})}{\sqrt{\sum_{k=1}^{n_c} (p_{ik} - \overline{p_i})^2 \cdot \sum_{k=1}^{n_c} (p_{jk} - \overline{p_j})^2}} \tag{29.2}$$

where $\overline{p_i}$ is the arithmetic average of $P_{ik}$ over $n_c$ conditions. Thus, the similarity between the numerical data is defined based on the correlation coefficient on the default system. This metric is robust for both missing and abnormal values. Then, the metric defined in equation 29.1 is analyzed by a standard hierarchical clustering technique, the unweighted pair-group method using arithmetic averages (UPGMA), as a default on the ASIAN system.

### *29.2.3 The Estimation of Cluster Boundaries*

In the ASIAN system, the cluster boundaries along the dendrogram, obtained in the former part, are estimated based on the statistical properties of the data. In this way, the redundancy of the empirical data is summarized as a group. In addition, the estimation of the object groups is one of the important issues in the classification methods. In hierarchical clustering, the cluster numbers are determined by visual inspection and consideration of the other information, but some ambiguity exists in determining the object groups. For estimating the cluster boundaries automatically, the variance inflation factor (VIF) is utilized in the ASIAN system.

The variance inflation factor (VIF) is employed to estimate the linear relationship between the numerical data in the clusters along the dendrogram. In the multiple regression analysis, the existence of a high linear relationship among the explanatory variables is known as multi-co-linearity, and the variables that are involved in the multi-co-linearity are diagnosed by the VIF, as follows :

$$\mathrm{VIF}_i = r_{ii}^{-1}, \tag{29.3}$$

where $r_{ii}^{-1}$ is the $i$th diagonal element of the inverse of the correlation coefficient matrix (CCM) between explanatory variables. Thus, $m$ VIFs are calculated in the correlation coefficient matrix among the $m$ exploratory variables. The magnitude of the VIF increases when the correlation among any independent variables increases. When the explanatory variables in equation 29, correspond to the empirical data, the VIF expresses the degree of linear relationship between the data. Given that the threshold in the diagnosis of multi-co-linearity is set to be $v_c$, the linear relationships of the $i$th variable exist when $\mathrm{VIF}_i$ is larger than $v_c$; on the ASIAN system, the $v_c$ value is set to the popular threshold of 10.0 as the default. The $m$ VIFs are assessed with the following conditions:

$$\max\{\mathrm{VIF}_i\} < v_c \quad \text{for } i = 1, 2, \dots .m. \tag{29.4}$$

If the condition is satisfied, then no linear relationship exists in the m sets of data, and in contrast, the linear relationship still exists in the data, if the condition is not satisfied. Thus, the maximum number of clusters with no linear relationship is searched for along the dendrogram.

### 29.2.4 The Application of GGM to the Clusters

The last part of ASIAN is the network inference by GGM. In the high-throughput analysis on the ASIAN system, the first three parts are prerequisites for analyzing the redundant data, including many similar numerical patterns. In high-throughput analysis on the ASIAN system, the average of the numerical data is calculated for each cluster obtained in the former part, as the pre-processing for the application of GGM. The averaged data of the $j$th condition for each cluster are calculated, i.e.,

$$\bar{p}_j^{(k)} = \frac{1}{n_k} \sum_{i \in clusterk}^{n_k} p_{ij}, \tag{29.5}$$

where $n_k$ is the number of members in the $k$th cluster. Then, the averaged data over $n_c$ conditions in $k$ clusters are subjected to GGM. In the independent analysis, the input data are utilized without modification. Then, the averaged or input numerical data are subjected to GGM by a step-wise and interactive algorithm, in order to evaluate which pair of the variables is conditionally independent.

The conditional independency is recognized when the element of a partial correlation coefficient matrix (PCCM) is zero in GGM. A partial correlation coefficient matrix (PCCM) is calculated from the inverse of CCM between the averaged or input data, and an element in PCCM with a minimum absolute value is replaced with zero. Subsequently, the CCM restored from the PCCM is tested with the original CCM by calculating the deviance, which is expressed as follows:

$$dev = n_c \ln \frac{|R_{i+1}|}{|R_i|}, \tag{29.6}$$

where $R_{i+1}$ and $R_i$ are the determinants of the CCMs at the $i+1$ th and $i$th iteration steps, respectively. The dev is known approximately to follow a $\chi^2$distribution with one degree of freedom. The iteration of the replacement of minimum values with 0 values in PCCM is stopped on the ASIAN system, when the probability of deviance is less than 0.05. Finally, a model of the PCCM is obtained. When the partial correlation coefficient for a pair is equal to 0, the pair is conditionally independent, indicating no connection between the objects in the network.

## 29.3 ASIAN Website

ASIAN (see Website: http://eureka.cbrc.jp/asian/) is one of the useful servers, for inferring the framework of a network, from a large amount of numerical data, in addition to clustering concomitant with the automatic estimation of cluster number. The visual presentation of the results provides an intuitive means for understanding the putative relationships between genes, proteins, and other objects. Furthermore, the ASIAN system now provides an independent statistical analysis. The front page of the ASIAN website, shown in Fig. 29.2, describes the analysis procedure and the updated news. A new analysis window will open, when the "Analyses" button on the front page is clicked.

## 29.4 Website Tour

To use the ASIAN system, the user first uploads the raw data from their machine, and then inputs some parameters and their e-mail address. Immediately after receiving the data, the ASIAN system successively performs the four steps. Concomitantly, the machine will send an ID number and the details of the input data by e-mail. The user can view the results analyzed on the website, with security, by the user's ID number and e-mail address.

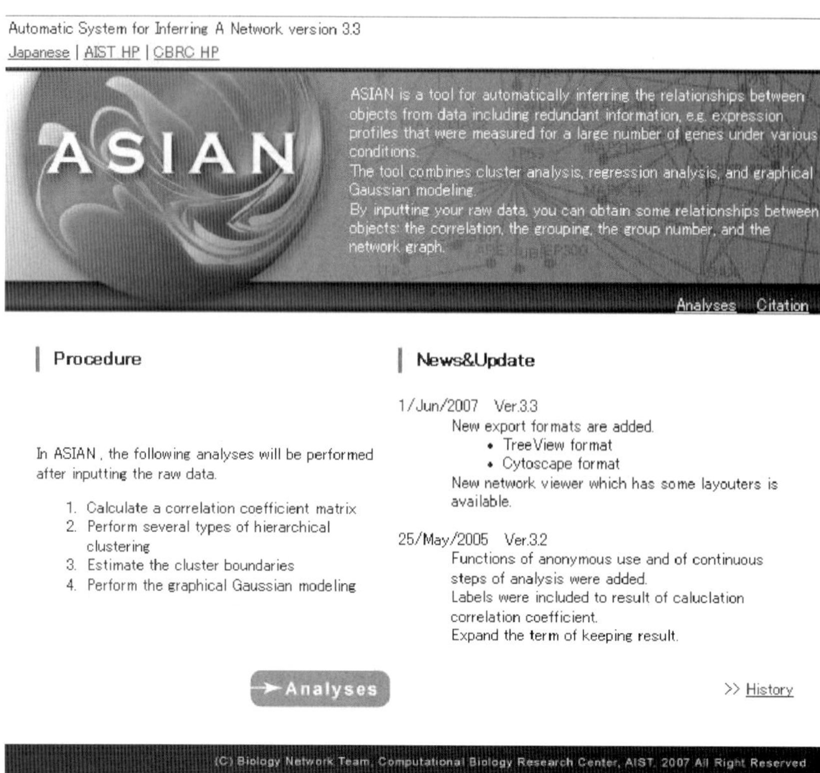


**Fig. 29.2** The front page of the ASIAN web site. On the front page of the ASIAN website, the outline of the procedure and the information about the upgrades are displayed on the left and right sides, respectively. The ASIAN system has been improved since the first version, with the current ver 3.3. The analysis page will open when the "Analyses" button is clicked (Copies of figures including color copies, where applicable, are available in the accompanying CD)

### *29.4.1 Data Import*

The clickable button 'Analyses' opens the analysis page. The top part of an analysis page is shown in Fig. 29.3A. The ASIAN system can analyze the uploaded data in two ways: one is a batch process that can successively perform the aforementioned four parts with the default parameters, and the other is a process that can allow the user to input the parameter values in each analysis.

In the batch process, only two steps are needed. First, the program runs by uploading the numerical data to be analyzed, and then the user selects one of two ways to receive the results: one is an anonymous use to display the results simultaneously with the processing, and the other is a signed use to receive the results after all the processes are finished, through a web site that can be accessed by inputting the user's email address. The format of the numerical data, which the user prepares, is assumed to be csv or tab-delimited text files. Information about the input data is available at the help page, which the user can open by clicking the help button. Immediately after receiving the data and selecting the method for receiving the results, the server successively performs the four calculation parts with the default values.

### *29.4.2 Parameter Setting*

On the ASIAN System, all parameters are set as default, but the user can set some parameters for network inference according to their interests. In this case, the user can input some parameters for

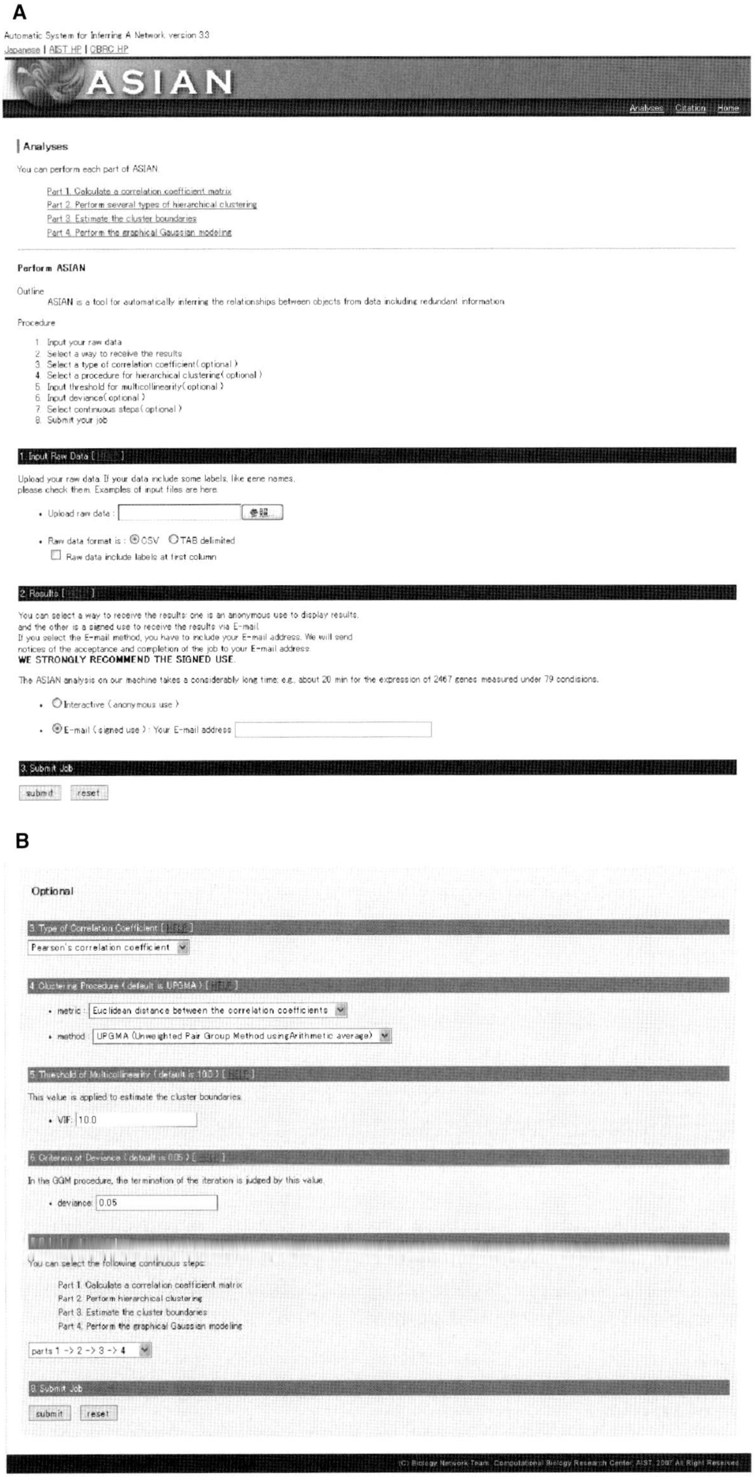

**Fig. 29.3** The analysis page of ASIAN. The analysis page is the interface of the ASIAN system. **(A)** The upper part of the analysis page. The data and the e-mail address are imported by the user. Furthermore, the user can select an independent part of the analysis on the ASIAN system. (**B**) The lower part of the analysis page. The user can set each of the parameters for network inference according to their interest, as an option. Usually, the parameters are set as default (Copies of figures including color copies, where applicable, are available in the accompanying CD)

each step. The user can set these parameters at the option part on the analysis page. The option part, which exists at the lower part on the analysis page, is shown in Fig. 29.3B.

#### 29.4.2.1 Types of Correlation Coefficient

For the calculation of the correlation coefficient matrix, the user can select one type of correlation coefficient from the three different types: (i) the Pearson's correlation coefficient (the default type), which is a representative correlation coefficient for a continuous variable, (ii) the Kendall's rank-correlation coefficient, which is a representative of a categorical variable, and (iii) the Eisen's correlation coefficient for the gene expression profile data. In general, the Pearson's correlation coefficient is suitable for data obtained from a bi-variate population according to the normal distribution, while the Kendall's rank-correlation coefficient is for data that are far from normal. The Eisen's correlation coefficient is devised to consider the experimental conditions, by setting the reference state as a term that corresponds to the average of the Pearson's correlation coefficient.

#### 29.4.2.2 Clustering Procedure

The user can select a pair of metric and clustering techniques in hierarchical clustering. Since the metrics and the techniques in the clustering depend on the user's data and interests, ASIAN allows the user to select one metric and technique pair from three metrics and seven techniques. The three metrics, the Euclidian distance between a pair of objects, the Euclidian distance between correlation coefficients and Eisen's distance, especially for gene expression analyses, are available in the present version of ASIAN.

Based on one of the metrics, the profiles are subjected to a hierarchical clustering analysis by one of these seven techniques: Single Linkage (nearest neighbor), Complete Linkage (furthest neighbor), Unweighted Pair-Group Method using Arithmetic average (UPGMA), Unweighted Pair-Group Method using Centroid average (UPGMC), weighted pair-group method using arithmetic average (WPGMA), Weighted Pair-Group Method using Centroid average (WPGMC) and Ward's method. The default metric and technique pair is the Euclidian distance between correlation coefficients and the UPGMA.

#### 29.4.2.3 Threshold of Multi-co-linearity

One of the remarkable features of the ASIAN system is that it can allow users to estimate the cluster number by a stopping rule for the hierarchical clustering. As mentioned above, in the cluster number estimation, the variance inflation factor (VIF) is utilized as a measure for the degree of separation between the clusters. Empirically, 10.0 is used as a cut-off value of VIF in various statistical analyses, and the cluster numbers estimated by the empirical value have been quite consistent with the previous numbers, as assessed by visual inspection and consideration of the biological function in the expression profile analyses. Although the default value of VIF is set at 10.0, the user can set any VIF value in this system.

#### 29.4.2.4 Criterion of Deviance

In the network inference, the average correlation coefficient matrix is calculated from the average profiles calculated within the members of each cluster. Then, the average correlation coefficient matrix between the clusters is subjected to the GGM. In the GGM, the co-variance selection is adopted, and the server allows the users to set the significance probability for the deviance in the modeling. The default significance probability is set to 0.05.

#### 29.4.2.5 Select Continuous Step

Furthermore, apart from the high-throughput inference of the network, the system can provide a step-by-step approach to ASIAN. The user can select several continuous steps, such as parts (1) and (2), to submit the numerical data as input and to receive the correlation coefficient matrix and the clustering results as output. The user can select one of the four types of continuous steps in the box, and then the server performs the checked steps. The default of the step is set to the four continuous parts.

## 29.5 ASIAN Outputs

Figure 29.4 shows the top page of the result. The results analyzed can be presented on the display immediately after each process is finished, if the user selects the anonymous use setting. If the user inputs their email address, then an email notice with the ID number and the URL will be sent, when the analyses are complete. In the latter case, the user can view the results analyzed on the web site, with security, by the ID number and the email address. In the automatic inference of the network, the analyzed results are composed of the correlation coefficient matrix, the dendrogram of hierarchical clustering with the cluster boundary in both text and graphic forms, the results of hierarchical clustering with two other format files (.gtr and .atr) for TreeView software, the average correlation coefficient matrix, the network between clusters in text and graphic forms, and the other

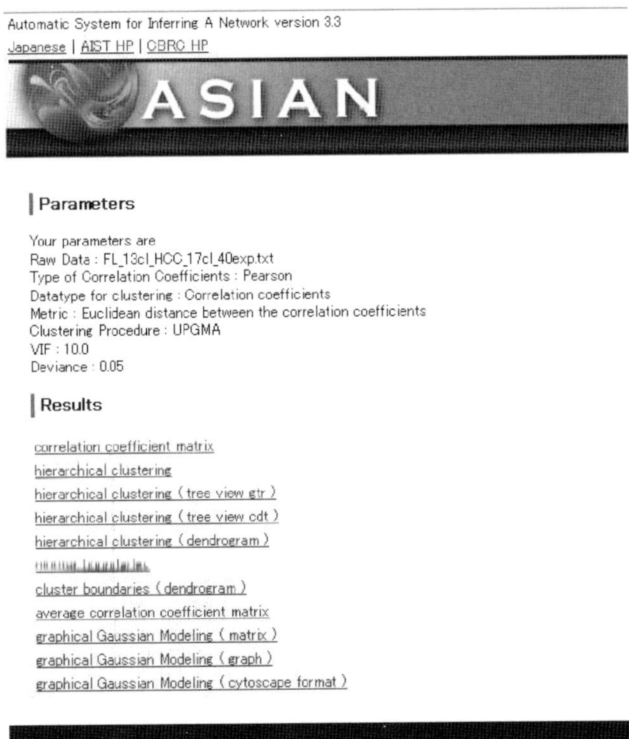


**Fig. 29.4** The top page of the results in ASIAN. In both the interactive form and e-mail form, this page is returned to the user as the top page of the results. On the upper part of the page, ASIAN displays the set parameters of the analysis. By the high-throughput network analysis, 11 types of results are returned. After the user clicks on each sentence, a new window opens to indicate the numerical result or graphical result (Copies of figures including color copies, where applicable, are available in the accompanying CD)

format results of the inferred network for displaying the network by the Cytoscape software. In the independent analysis, the user can obtain each result, according to the selected analysis.

All of the above results are kept in the user's web site for 30 days after the analysis is completed. If the uses wish the analyzed results to be deleted or to be kept for more than 30 days, they can request their wishes by email (asian@cbrc.jp).

### 29.5.1 Correlation Coefficient Matrix

If the user clicks the "correlation coefficient matrix" on the result front page, then the correlation coefficient matrix calculated from the input data is displayed. Each numerical element of the matrix indicates the correlation coefficient values between each pair of objects. In ASIAN, the object included in the input data is not displayed on this matrix, and only the numerical data are displayed. Since the order of the rows and columns on this matrix is according to the user's input data, the user can easily distinguish the correlation coefficient value between the two objects of interest.

### 29.5.2 Hierarchical Clustering

ASIAN provides four types of results as the hierarchical clustering. When the user clicks "hierarchical clustering", the text type result is displayed. On this page, the third column indicates the distance, which is calculated from the input data, between the two objects displayed in the first and second columns on the same line. Thus, the user can obtain the numerical distance between the two objects from this txt format file.

The graphical result of hierarchical clustering is displayed when the user clicks the words "hierarchical clustering (dendrogram)". The dendrogram provides a visual inspection of the hierarchical clustering. Furthermore, ASIAN has been improved to return two other types of format files, .gtr and .atr, after the hierarchical clustering. According to the TreeView software. The TreeView software was developed by Eisen et al., and it is the one of the major graphical interfaces of hierarchical clustering in microarray data analysis. If the TreeView software is used for displaying the hierarchical clustering, then the user imports the .gtr and .atr files obtained by the ASIAN hierarchical clustering.

The ASIAN system provides 21 types of metric and technique pairs in the hierarchical clustering module, and thus the user can obtain the results of the hierarchical clustering with the metric and technique pair of interest.

### 29.5.3 Cluster Boundaries

The remarkable feature of the hierarchical clustering in the ASIAN system is that the cluster boundaries are automatically estimated by VIF. The results of these automatically estimated cluster boundaries are displayed as both txt type and graphic type.

If "cluster boundaries" is clicked on the front page of the results, then the txt type result of cluster boundaries is displayed in the other window. This file includes two columns. The first column indicates the estimated cluster numbers, and the second column indicates the objects that are classified in the cluster. The estimated cluster numbers are also indicated at the headline of this txt type result, and the first line of each cluster indicates the number of members included in the clusters.

The "cluster boundaries (dendrogram)" has a hyperlink to a graphical display of the results of the estimation of cluster boundaries. Figure 29.5 shows an example of a dendrogram with the

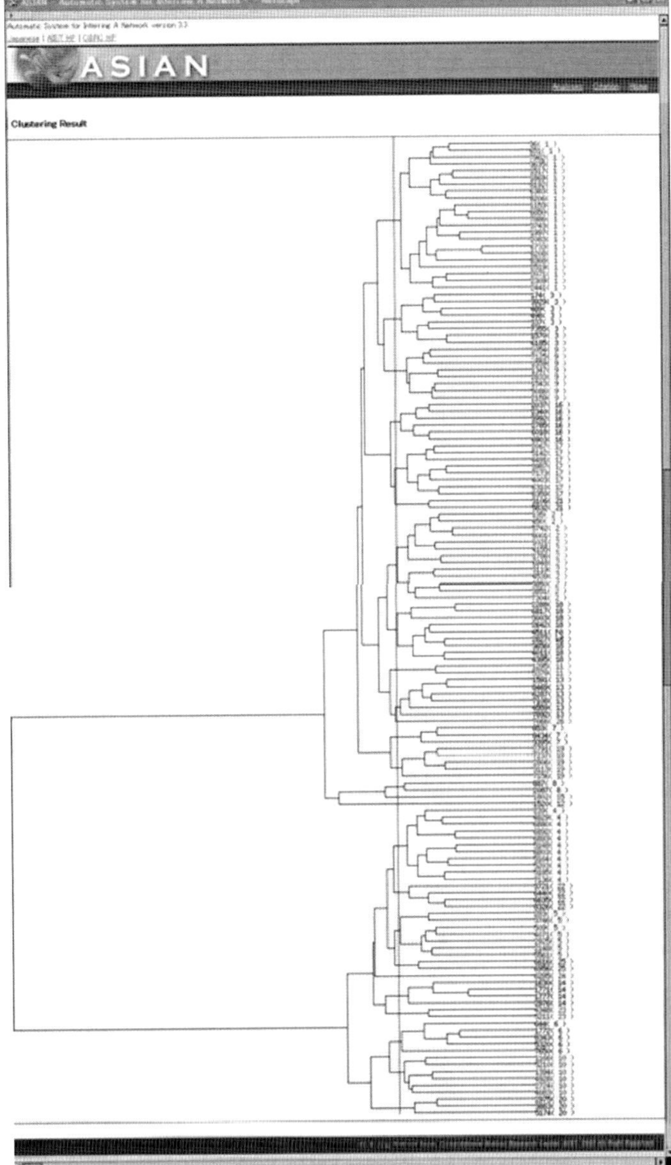


**Fig. 29.5** Example of the results of hierarchical clustering and estimated cluster boundaries. The hierarchical clustering is performed with default parameters, such as setting the distance as the Euclidian distance between correlation coefficients and the metric as UPGMA. The VIF is set as 10.0. The estimated cluster boundaries are displayed by the line on the dendrogram. The objects are classified into 26 clusters. The members in the neighbouring clusters are discriminated by object names with other colors (Copies of figures including color copies, where applicable, are available in the accompanying CD)

cluster boundary estimated by the default value of VIF. The cluster boundary is indicated by a verticle line on the dendrogram, and the members in the neighboring clusters are discriminated by object names, colored blue and red.

### 29.5.4 Average Correlation Coefficient Matrix

In the high-throughput analysis for the network inference by the ASIAN system, the calculation of the average numerical data in each cluster is performed. Thus, the correlation coefficient matrix of the averaged numerical data is calculated again. ASIAN displays the results of the re-calculated correlation coefficient matrix between the estimated clusters.

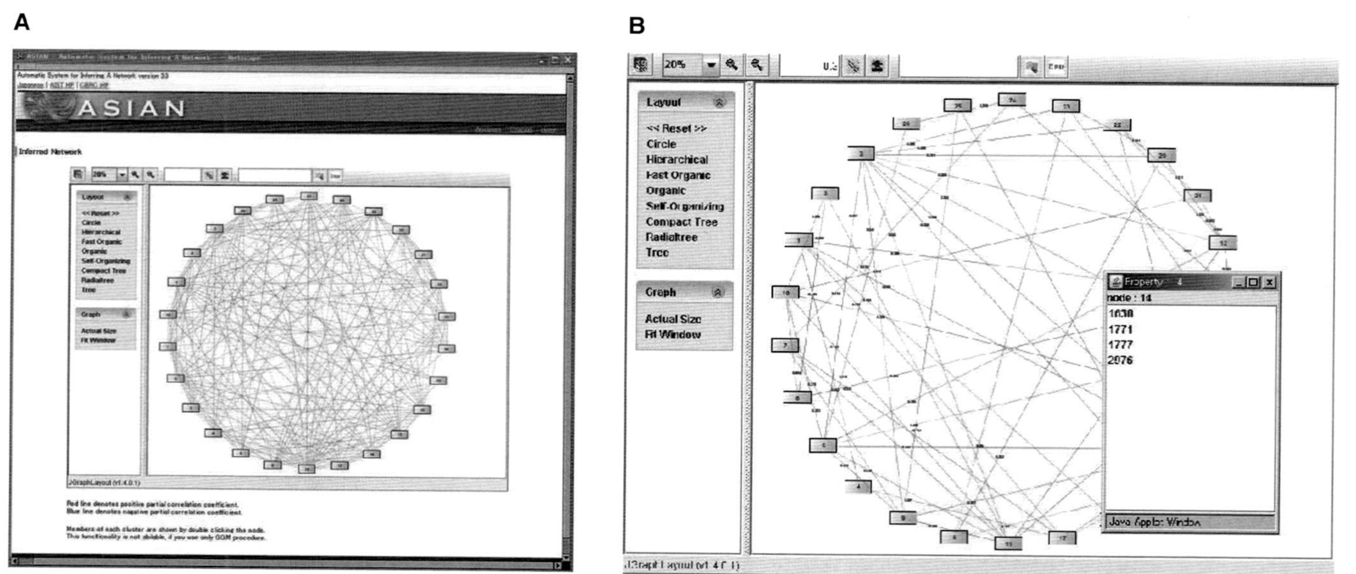


**Fig. 29.6** Example of an inferred network. **(A)** Graphic view of the inferred network between the 24 clusters. The clusters that are automatically estimated by the hierarchical clustering are indicated by rectangles. The inferred relationships between the clusters are indicated by the edges between the clusters. On the web-site, positive relationships and negative relationships are displayed by red and blue lines, respectively. **(B)** The summarized network and the member list of a cluster. The edges, which are those PCC values over 0.3, remain from the raw inferred network. The member list is indicated as the other window on the network (Copies of figures including color copies, where applicable, are available in the accompanying CD)

The result page of the average correlation coefficient matrix only includes the numerical elements of the matrix. Although the cluster numbers are not indicated, the order of the rows and columns in the matrix is according to the cluster number. Thus, the user can easily recognize the correlation coefficient value between the two clusters.

### 29.5.5 Graphical Gaussian Modeling

The inferred network by the GGM is displayed at the "graphical Gaussian Modeling" part on the front page of the results. Three types of files are displayed in ASIAN: matrix, graph and Cytoscape format.

The "graphical Gaussian Modeling (matrix)" is linked to the modeled partial correlation coefficient matrix estimated by GGM. In the GGM, the modeling is performed based on the partial correlation coefficient matrix, which is calculated from the inverse correlation coefficient matrix. In the high-throughput network analysis in the ASIAN system, the partial correlation coefficient matrix is calculated from the correlation coefficient matrix between the estimated clusters. Thus, the indicated partial correlation coefficient matrix, as the result of GGM, includes the same number of elements as the average correlation coefficient matrix. If the user selects the independent analysis for network inference by GGM, on the analysis front page, then the modeled partial correlation coefficient includes the elements according to the input numerical data. The modeled partial correlation coefficient matrix also indicates only the numerical data without the object name. However, the order of this matrix is according to the cluster number or the order of the input data, and thus, the user can easily recognize the value of the modeled partial correlation coefficient between the pair of two objects.

In the partial correlation coefficient, the absolute value of an element indicates the strength of the direct interaction between the two objects. The diagonal element is –1. The sign of an element indicates that the estimated interaction between the two objects is a positive or negative interaction. Therefore, the user can obtain the details of the inferred network from this partial correlation coefficient matrix as the numerical data.

If the user clicks the "graphical Gaussian Modeling (model)", a new graphic page is opened. A graphic example of the network inferred by the present ASIAN is shown in Fig. 29.6A. In the default graph, the nodes that indicate the clusters are connected at the edges, if the partial correlation coefficient between the corresponding clusters is estimated as non-zero by GGM. When the user clicks the node, the members that are included in the cluster are displayed on a new page as shown in Fig. 29.6B. In the network graph, the positive and negative partial correlation coefficients are discriminated by the red and blue lines in the graph, respectively. Furthermore, the user can set the threshold of the partial correlation coefficient for visualizing the edges. When the partial correlation coefficient between the clusters is greater than the threshold defined by the user, the nodes are connected by the edges between the corresponding clusters. This option facilitates the interpretation of the network, especially that of a complex network with many edges and nodes.

Recently, other options on this graphic page have been implemented. The layout of the inferred network can be changed to various patterns. In the default mode, the inferred network is displayed as a circle. On the left side of the window, the user can select from seven other types of layouts for the inferred network, corresponding with each layout algorithm, Hierarchical, Fast Organic, Organic, Self-Organizing, Compact Tree, Redialtree and Tree. The layout of the inferred network is reset by clicking of "Reset (Circle)". The other improved option of the graphic page is that the user can search each object included in a cluster. When the user clicks the search button on the top of page, the text field is opened. After the object name is given in this text field, the search is performed with all of the cluster members. Thus, the user can find the object easily.

The last result of graphical Gaussian modeling is the inferred network with the Cytoscape format. The user can retrieve this file by the clicking of the "graphical Gaussian Modeling (cytoscape format)". This is the hyperlink word to download the result file in the cytoscape format. This result file is composed of three columns, and the first and third columns indicate the connected cluster numbers. Cytoscape is freeware for a graphical view of the network between some objects. If Cytoscape is used for displaying the inferred network by ASIAN, then the cytoscape format file should be chosen.

## 29.6 Concluding Remarks

By using the ASIAN system, an automatic-throughput for object classification and network inference between the estimated clusters is realized from a large amount of redundant data. Empirical data usually includes redundancy, but the ASIAN system summarizes the redundancy by the combination of hierarchical clustering. The improvement of the object classification, by setting a user-defined threshold for VIF, allows the user to compare frameworks between different clusters. Of course, if a small amount of data, which has no redundancy is prepared, the user can utilize the part of the ASIAN analysis for network inference and the other statistical analysis. The visual presentation of the inferred network provides an intuitive means for understanding the putative relationships between the objects of interest.

## 29.7 ASIAN on a Personal Computer

A package of *ASIAN* on PC (Windows and LINUX), "Auto Net Finder", is commercially distributed by INFOCOM CORPORATION.

## 29.8 Contact List

ASIAN is supported by INFOCOM CORPORATION. If you have any questions or comments about ASIAN, send an e-mail to: asian@cbrc.jp. The ASIAN support team will answer your questions or respond to your comments as soon as possible. The ASIAN support team is always interested in any feedback you might have about the site. Comments and suggestions are all carefully considered, and often make their way into the next version of the site.

**Acknowledgments** Sachiyo Aburatani was supported by a Grant-in-Aid for Scientific Research (grant 18681031) from the Ministry of Education, Culture, Sports, Science and Technology of Japan, and Katsuhisa Horimoto was partly supported by a Grant-in-Aid for Scientific Research on Priority Areas "Systems Genomics" (grant 18016008) and by a Grant-in-Aid for Scientific Research by the Ministry of Health, Labour and Welfare of Japan. This study was supported in part by the New Energy and Industrial Technology Development Organization (NEDO) of Japan, and by a Grant-in-Aid for Scientific Research (grant 19201039) from the Ministry of Education, Culture, Sports, Science and Technology of Japan.

## Suggested Reading

### *Automatic estimation of cluster boundaries*

Horimoto K, Toh H, Statistical estimation of cluster boundaries in gene expression profile data. Bioinformatics; 2001; 17:1143–1151.

## *Graphical Gaussian modeling*

Whittaker J, Graphical Models in Applied Multivariate Statistics. John Wiley, NY; 1990
Edwards D, Introduction to Graphical Modelling. 2nd ed., Springer-Verlag New York, Inc.; 2000

## *Network Inference by GGM*

Toh H, Horimoto K, Inference of a genetic network by a combined approach of cluster analysis and graphical Gaussian modeling. Bioinformatics; 2002; 18:287–297.
Aburatani S, Kuhara S, Toh H, Horimoto K, Deduction of a gene regulatory relationship framework from gene expression data by the application of graphical Gaussian modeling. Signal Processing; 2003; 83:777–788.

## *ASIAN Website*

Aburatani S, Goto K, Saito S, Fumoto M, Imaizumi A, Sugaya N, Murakami H, Sato M, Toh H, Horimoto K, ASIAN: a web site for network inference. Bioinformatics; 2004; 20:2853–2856.
Aburatani S, Goto K, Saito S, Toh H, Horimoto K, ASIAN: a Web Server for Inferring a Regulatory Network Framework from Gene Expression Profiles. Nucl. Acid. Res.; 2005; 33: W659-W664

## Web Resource

http://www.infocom.jp//bio/datamine/autonetfinder_en.html

# Part VIII
# Bridging the Gap

# Chapter 30
# Bioinformatics for Metabolomics

**David S. Wishart**

**Abstract** This chapter is intended to familiarize readers with the field of metabolomics and how it relates to Systems Biology. It also describes how bioinformatics and metabolic modeling are being used to facilitate metabolomics research, which in turn, is being used to enable important developments in Systems Biology. Specifically, this chapter summarizes four areas where computational biology is playing a key role in both Systems Biology and metabolomics. These include: (1) metabolic pathway databases; (2) metabolomics databases; (3) spectral and statistical analysis tools for metabolomics; and (4) metabolic modeling. The needs, challenges and recent progress being made in the four areas are also discussed.

**Keywords** Metabolomics · Databases · Bioinformatics · Modeling · Cheminformatics

## 30.1 Introduction

Systems Biology is an integrated discipline that combines high-throughput experimental techniques such as genomics, proteomics and metabolomics with computational techniques such as bioinformatics and computer simulations in an attempt to fully understand or mechanistically model a biological system, such as a cell, organ or organism [1]. While most readers are at least modestly familiar with genomics, proteomics and/or bioinformatics, relatively few are likely to be familiar with the other "omics" term — metabolomics. Metabolomics (or metabonomics as it is sometimes called) is a newly emerging field of "omics" research, concerned with the high-throughput identification and quantification of the small molecule metabolites in the metabolome [2]. The metabolome can be defined as the complete collection of all small molecule (<1500 Da) metabolites found in a specific cell, organ, or organism. It is a close counterpart to the genome, the transcriptome, and the proteome.

Interestingly, some of the first successes in Systems Biology involved the deciphering and modeling of metabolic pathways [3,4]. Indeed, there are many experts who view metabolomics as having a central role in Systems Biology [5–7], and still others who see metabolomics seamlessly blending in with the other "omic" sciences under the banner of Systems Biology (Fig. 30.1). Metabolomics not only serves as a cornerstone to Systems Biology, it is beginning to serve as a cornerstone to other fields as well. In particular, because of its unique focus on small molecules and small molecule interactions, metabolomics is finding widespread applications in drug discovery and development [5–8], clinical toxicology [9,10], clinical chemistry [11] and nutritional genomics [12,13].

---

D.S. Wishart
Departments of Computing Science and Biological Sciences, University of Alberta, 2-21 Athabasca Hall, Edmonton, AB, T6G 2E8, Canada
National Institute for Nanotechnology, 11421 Saskatchewan Drive, Edmonton, AB, T6G 2M9, Canada
e-mail: david.wishart@ualberta.ca

S. Krawetz (ed.), *Bioinformatics for Systems Biology*, DOI 10.1007/978-1-59745-440-7_30,

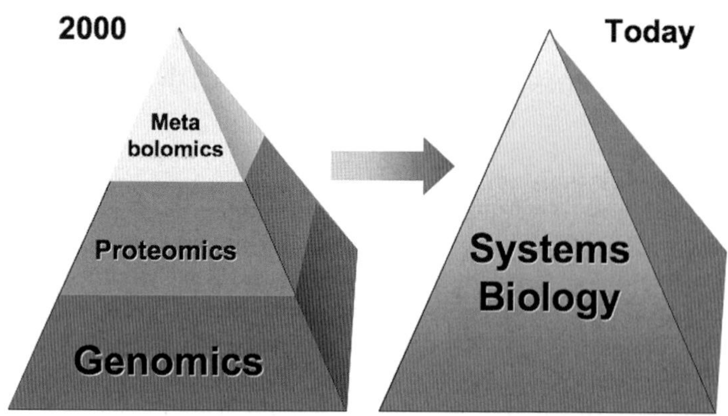


**Fig. 30.1** The blending of genomics, proteomics and metabolomics into Systems Biology (Copies of figures including color copies, where applicable, are available in the accompanying CD)

Metabolomics, being a relatively new addition to the "omics" sciences, is still evolving some of its basic computational infrastructure [14]. Whereas most data in the field of proteomics, genomics, or transcriptomics is readily available and easily analyzed through on-line electronic databases, most metabolomic data is still housed in books, journals and other paper archives. Metabolomics also differs from the other "omics" sciences because of its strong emphasis on chemicals and analytical chemistry techniques such as nuclear magnetic resonance (NMR) spectroscopy, mass spectrometry (MS) and chromatography. As a result, the analytical software used in metabolomics is often quite different than most of the software used in genomics, proteomics or transcriptomics [14]. The field of metabolomics is not only concerned with the identification and quantification of metabolites, it is also concerned with relating metabolite data to genes, proteins, pathways, physiology and phenotypes. As a result, metabolomics requires that whatever chemical information it generates must be linked to both biochemical causes and physiological consequences. This means that metabolomics must combine two very different fields of informatics: bioinformatics and cheminformatics.

Despite these differences, metabolomics still shares many of the same computational needs with genomics, proteomics, and transcriptomics. All the four "omics" techniques require electronically accessible and searchable databases, all of them require software to interpret or process data from their own high-throughput instruments and all require software tools to predict or model properties, pathways and processes. These shared computational needs are the common thread that links metabolomics with all of the other "omics" sciences, and ultimately to Systems Biology.

This chapter is intended to familiarize readers with the field of metabolomics and how it relates to Systems Biology. It also describes how bioinformatics and metabolic modeling are being used to facilitate metabolomics research, which in turn, is enabling important developments in Systems Biology. Four key areas of computational metabolomics are discussed in detail. These include: (i) metabolic pathway databases; (ii) metabolomics databases; (iii) spectral and statistical analysis tools; and (iv) metabolic modeling.

## 30.2 Metabolic Pathway Databases

Electronic databases lie at the heart of almost every "omics" field. Genomics would not exist without GenBank [15] while proteomics would be bereft without SwissProt [16] or the PDB [16]. These electronic databases serve as the repository of the most current information about genes, proteins, and protein structures. The shear volume of data contained in these databases (100's of gigabytes) makes it impractical to assemble or distribute the information via books, journals, or even electronic media such as CDs or DVDs. While the volume of data concerning metabolism and metabolic pathways is not quite as large, the same issues of currency and distribution are still crucial to metabolomics. So, it is fair to say that databases also lie at the heart of metabolomics.

There are two types of metabolomic databases: (i) general metabolic pathway databases, and (ii) dedicated metabolomic databases. Metabolic pathway databases are designed to house and display biochemical pathways or metabolite-gene-protein interactions. They are fundamentally visual aids, designed to facilitate the exploration of metabolism and metabolites across many different species. Metabolomic databases, on the other hand, contain much more information about metabolites and their physical, chemical, or spectroscopic properties. They also include physiological, biological, or chemical information, but this information is typically restricted to a single type of organism. In other words, metabolic pathway databases can be characterized as being broad in coverage but shallow in content, while metabolomic databases can be characterized as narrow in coverage but deep in content.

Metabolic pathway databases are generally better known than their metabolomic database counterparts and include such familiar resources as the Roche Biochemical Pathways Chart [17], KEGG [18], MetaCyc [19], Reactome [20], BRENDA [21], PUMA2 [22] and others. Table 30.1 lists the names, web addresses, and general features for these and other notable databases. A more detailed description of some of these databases (excluding the KEGG database which is described in chapter 25 of this book) is given below.

### *30.2.1 The Roche Biochemical Pathways Database*

Perhaps the first metabolic pathway database to gain wide popularity was not really a database, but rather a paper wall chart developed by Dr. Gerhard Michal, a staff scientist at Boehringer Mannheim (now Roche Applied Sciences) and published in 1968 [17]. This richly annotated and compactly illustrated poster depicts most of the known metabolic pathways, metabolites, and enzymes associated primarily with human metabolism. It has been revised and expanded many times since its first release and has proven to be very popular, with more than 1 million copies now in print. To facilitate its distribution the last published version of the Biochemical Pathways chart was digitized in 1997, given a queryable web interface and posted on the EXPASY website. To access the chart, users may type in compound names (or parts of names) or they may select zoomable thumbnail/subsection views of the chart. In this way it is possible to interactively explore metabolic pathways or cellular and molecular processes on a desktop computer. All compound queries provide hyperlinked lists to EC numbers (Enzyme Classification codes and the corresponding enzyme information) and to the specific location of these compound(s) on the Biochemical Pathways wall chart. However, the web version of the Roche Biochemical Pathways chart does have some limitations. For instance, the presentation format is now a little outdated and it lacks the interactivity and extensive hyperlinking to other biologically relevant databases typically found in most other bioinformatics resources.

### *30.2.2 The MetaCyc Database*

The MetaCyc database or "Metabolic Encyclopedia" has been maintained by Peter Karp's group at the Stanford Research Institute since 1999 [19]. MetaCyc is a web-accessible database containing more than 966 experimentally elucidated metabolic pathways from more than 1,000 different species gathered from the scientific literature or inferred through a variety of computational methods. At last count (Version 11.1) the MetaCyc staff had catalogued 4271 different types of enzymes, 6,354 different metabolites, 6,464 reactions and compiled nearly 15,000 references. MetaCyc contains extensively hyperlinked metabolic pathway diagrams, enzyme reactions, enzyme data, chemical structures, chemical data, and gene information. Likewise, users can query MetaCyc by the name of a protein, gene, reaction, pathway, chemical compound, or EC (enzyme classification number). Just as with KEGG, most MetaCyc queries or browsing operations return a rich and colorful collection of hyperlinked figures, pathways, chemical structures, reactions, enzyme names, references and protein/gene sequence data.

**Table 30.1** Alphabetical Summary of Metabolite or Metabolic Pathway Databases

| Database Name | URL or Web Address | Comments |
|---|---|---|
| BRENDA (BRaunschweig ENzyme Database) | *http://www.brenda.enzymes.info/* | -Enzyme database containing rate constants and some metabolic pathway data |
| Chemicals Entities of Biological Interest (ChEBI) | *http://www.ebi.ac.uk/chebi/* | -Covers metabolites and drugs<br>-Focus on ontology and nomenclature not biol. |
| HumanCyc (Encylopedia of Human Metabolic Pathways) | *http://humancyc.org/* | -MetaCyc adopted to human metabolism |
| KEGG (Kyoto Encyclopedia of Genes and Genomes) | *http://www.genome.jp/kegg/* | -Best known and most complete metabolic pathway database<br>-Covers many organisms<br>-Small (<15) number of data fields, no biomedical data |
| LipidMaps | *http://www.lipidmaps.org/* | -Limited to lipids only (not species specific)<br>-Nomenclature standard |
| MetaCyc (Encyclopedia of Metabolic Pathways) | *http://metacyc.org/* | -Similar to KEGG in coverage, but with different emphasis<br>-Well referenced<br>-Small (<15) number of data fields, no biomedical data |
| Nicholson's Metabolic Minimaps | *http://www.iubmb.nicholson.org/* | -Used for teaching (limited coverage) |
| PUMA2 (Evolutionary Analysis of Metabolism) | *http://compbio.mcs.anl.gov/puma2/cgi-bin/index.cgi* | -Used for metabolic pathway comparison and genome annotation<br>-Requires registration |
| Reactome (A Curated Knowledgebase of Pathways) | *http://www.reactome.org/* | -Pathway database with more advanced query features<br>-Not as complete as KEGG or MetaCyc |
| Roche Applied Sciences Biochemical Pathways Chart | *http://www.expasy.org/cgi- bin/search-biochem-index* | -The old metabolism standard (on line) |

(Copies of tables are available in the accompanying CD.)

Unlike most other metabolic pathway databases, MetaCyc provides much more detailed enzyme information including data on substrate specificity, kinetic properties, activators, inhibitors, cofactor requirements and links to sequence/structure databases. Additionally, MetaCyc supports sophisticated relational queries, allowing complex searches to be performed and more detailed information to be displayed. These search utilities are supplemented with a very impressive "Omics Viewer" that allows gene expression and metabolite profiling data to be painted onto any organism's metabolic network. MetaCyc also displays metabolic pathway information at varying degrees of resolution, allowing users to interactively zoom into a reaction diagram for more detailed views and more detailed pathway annotations. Because it covers metabolic data for so many organisms, the MetaCyc database supports organism-specific queries and displays this taxonomic information in its illustrated pathways through gene and enzyme names. One particularly interesting organism-specific database contained within MetaCyc (or the BioCyc family) is called HumanCyc [23]. HumanCyc (Release 10.6) supports many of the same features found in MetaCyc. It also contains information on 28,782 human genes, 2,594 human enzymes, 1,253 reactions, 974 human metabolites, and 178 human-specific metabolic pathways.

### *30.2.3 The Reactome Database*

A much more recent addition to the collection of metabolic pathway databases is the Reactome database [20]. The Reactome project was started in 2002 to develop a curated resource of core pathways and reactions in human biology. The "reactome" is defined as the complete set of possible reactions or pathways that can be found in a living organism, including the reactions involved in intermediary metabolism, regulatory pathways, signal transduction and cell cycle processes. The Reactome database is a curated resource authored by biological researchers with expertise in their fields. Unlike KEGG or MetaCyc, the Reactome database takes a much more liberal view of what constitutes metabolism (or biochemical reactions) by including such processes as mitosis, DNA-repair, insulin mediated signaling, translation, transcription and mRNA processing in addition to the standard metabolic pathways involving amino acids, carbohydrates, nucleotides, and lipids. Furthermore, unlike KEGG or MetaCyc, the primary focus in the Reactome database is on *Homo sapiens* (although additional data from 20 model organisms are now included).

The Reactome database (Version 22), currently has 691 human-associated pathways assembled from 1,845 reactions involving 1,473 proteins or protein complexes. Central to the Reactome database is a schematic "Reaction Map" which graphically summarizes all the high-level reactions contained in the Reactome database. This map allows users to navigate through the database in an interactive and progressively more detailed fashion. Users may also browse through the database by selecting topics from a table of contents, or they may query the database using a variety of text and keyword searches. The Reactome database also supports complex Boolean text queries for different combinations of reactions, reaction products, organisms, and enzymes. The results from these queries include higher-resolution pathway maps (in PDF, PNG and SVG formats), SBML (Systems Biology mark-up language) descriptions, and synoptic Reactome web "cards" on specific proteins or metabolites, with hyperlinks to many external databases.

One of the most useful and innovative features of the Reactome database is a tool called the Reactome "skypainter". This allows users to paste in a list of genes or gene identifiers (GenBank, UniProt, RefSeq, EntrezGene, OMIM, InterPro, Affymetrix, Agilent and Ensembl formats) and to "paint" the Reactome reaction map in a variety of ways. In fact it is even possible to generate "movies" which can track gene expression changes over different time periods – as might be obtained from a time series gene or protein expression study. This tool is particularly useful for analyzing microarray data, but it is also useful for visualizing disease genes (say from OMIM) and mapping the roles they play and the pathways in which they participate. In general, the central concepts behind the Reactome database are quite innovative and it certainly appears that this resource could play an increasingly important role in many areas of biology, biochemistry and Systems Biology.

### *30.2.4 Miscellaneous Pathway Databases*

In addition to the four major pathway resources just described, there are a number of smaller or lesser-known metabolic pathway databases (Table 30.1). Some, like BRENDA [21] and PUMA2 [22], are actually very comprehensive resources and contain quantities of information comparable to MetaCyc or KEGG. For instance PUMA2, which is maintained at the Los Alamos National Laboratory, is both a database and a genomic annotation system. PUMA2, which stands for "Phylogeny-Metabolism-Alignments" is primarily designed for high-throughput genetic sequence analysis and metabolic reconstructions from sequence data. PUMA2 currently contains pre-computed analysis of >1,000 genomes and automated metabolic reconstructions for >200 organisms covering more than 3,000 pathways.

BRENDA (BRaunschweig ENzyme DAtabase) is primarily an enzyme database with considerable supplementary information on metabolism. Under development since 1987, BRENDA includes data on 3,500 different enzymes from 7,500 different organisms. BRENDA is essentially a web-accessible relational database system consisting of 46 interlinked tables containing EC numbers, source organism, enzyme names, references, substrate/product information, inhibitors, enzyme mutants, Km values, measurement conditions, enzyme structure data, tissue or organ sources, subcellular localization, isolation and separation information, and disease information. Registered BRENDA users may query any combination of these fields through a half dozen different query tools. Originally, BRENDA was developed with a focus on organism and/or EC data, but now it is more aligned with other databases that focus on protein or compound searches. The information in BRENDA is quite unique and can be exploited to model metabolic pathways, to explore tissue differences in metabolism and to understand the function of potential drugs and inhibitors.

Table 30.1 also lists several other metabolite-only or somewhat more modest metabolic pathway databases designed more for educational, ontological or illustrative purposes. These include LipidMaps [24], ChEBI [25] and Donald Nicholson's Metabolic Charts (Table 30.1). Of particular interest to clinicians and biomedical researchers is Dr. Nicholson's beautifully illustrated Inborn Errors of Metabolism chart which is now available as a downloadable PDF file. This chart identifies the sites of over 120 metabolic disorders that are listed along the sides of the chart.

## 30.3 Metabolomics Databases

Metabolic pathway databases such as KEGG, MetaCyc, and Reactome are designed to facilitate the exploration of metabolism and metabolites across many different species. This broad, multi-organism perspective has been critical to enhancing our basic understanding of metabolism and our appreciation of biological diversity. Metabolic pathway databases also serve as the backbone to facilitate many practical applications in biology including comparative genomics and targeted genome annotation. However, the information contained in these "traditional" databases does not meet the unique data requirements for most metabolomics researchers.

This is because metabolomics is concerned with rapidly characterizing dozens of metabolites at a time and then using these metabolites or combinations of metabolites to identify disease biomarkers or model large scale metabolic processes. As a result, metabolomics researchers need databases that can be searched not just by pathways or compound names, but also by Nuclear Magnetic

Resonance (NMR) spectra, mass spectra (MS), GC-MS retention indices, chemical structures, or chemical concentrations. In addition to these query requirements, metabolomics researchers routinely need to search for metabolite properties, tissue/organ locations, or metabolite-disease associations. Therefore metabolomics databases require information not only about compounds and reaction diagrams, but also data about physio-chemical properties, compound concentrations, bio-fluid or tissue locations, subcellular locations, known disease associations, nomenclature, descriptions, enzyme data, mutation data and characteristic MS or NMR spectra. These data need to be readily available, experimentally validated, fully referenced, easily searched, readily interpreted and they need to cover as much of a given organism's metabolome as possible.

These are very tall orders, but there are now a number of newly emerging metabolomics databases that are beginning to address these needs, either in whole or in part. These include the Human Metabolome Database or HMDB [26], the METLIN database [27], the BioMagResBank or BMRB [28], the Golm Metabolome database [29], the BiGG metabolic reconstruction database [30] and others (see Table 30.2 for a more complete list). Some, like the HMDB, attempt to address all of the above-mentioned database needs, while others, such as the BMRB tend to focus on the specific need for creating spectral reference libraries. Following is a brief summary of five of these metabolomic databases.

### 30.3.1 The Human Metabolome Database

The HMDB [26] currently contains more than 2,650 human metabolite entries that are linked to more than 24,800 different synonyms. These metabolites are further connected to some 70 non-redundant pathways, 3,360 distinct enzymes, 103,000 SNPs as well as 862 metabolic diseases (genetic and acquired). Much of this information is gathered manually or semi-automatically from thousands of books, journal articles, and electronic databases. In addition to its comprehensive literature-derived data, the HMDB also contains an extensive collection of experimental metabolite concentration data for plasma, urine, CSF and/or other bio-fluids for a total of 883 compounds. The HMDB also has more than 8,600 compounds for which experimentally acquired "reference" $^{1}$H and $^{13}$C NMR and MS/MS spectra have been acquired.

The HMDB is fully searchable with many built-in tools for viewing, sorting and extracting metabolites, biofluid concentrations, enzymes, genes, NMR or MS spectra and disease information. Each metabolite entry in the HMDB contains an average of 90 separate data fields including a comprehensive compound description, names and synonyms, structural information, physico-chemical data, reference NMR and MS spectra, biofluid concentrations, disease associations, pathway information, enzyme data, gene sequence data, SNP and mutation data as well as extensive links to images, references and other public databases. A screen shot montage of the HMDB and some of its data content is given in Fig. 30.2. A key feature that distinguishes the HMDB from other metabolic resources is its extensive support for higher level database searching and selecting functions. In particular, the HMDB offers a chemical structure search utility, a local BLAST search [31] that supports both single and multiple sequence queries, a boolean text search, a relational data extraction tool, an MS spectral matching tool and an NMR spectral search tool. These spectral query tools are particularly useful for identifying compounds via MS or NMR data from other metabolomic studies.

### 30.3.2 The METLIN Metabolomic Database

The METLIN database, which is currently housed at the Scripps Research Institute, is designed primarily as a human metabolome resource [27]. It contains an annotated list of 15,000 known and "predicted" metabolites (primarily 8,400 di and tripeptides) including their mass, chemical formula,

**Table 30.2.** Alphabetical Summary of Metabolomic Databases

| Database Name | URL or Web Address | Comments |
|---|---|---|
| BiGG (Database of Biochemical, Genetic and Genomic metabolic network reconstructions) | *http://bigg.ucsd.edu/home.pl* | -Database of human, yeast and bacterial metabolites, pathways and reactions as well as SBML reconstructions for metabolic modeling |
| BioMagResBank (BMRB – Metabolimics) | *http://www.bmrb.wisc.edu/metabolomics/* | -Emphasis on NMR data, no biological or biochemical data |
| Fiehn Metabolome Database | *http://fiehnlab.ucdavis.edu/compounds/* | -Specific to plants (Arabadopsis)<br>-Tabular list of ID'd metabolites with images, synonyms and KEGG links |
| Golm Metabolome Database | *http://csbdb.mpimp-golm.mpg.de/csbdb/gmd/gmd.html* | -Emphasis on MS or GC-MS data only<br>-No biological data<br>-Few data fields<br>-Specific to plants |
| Human Metabolome Database | *http://www.hmdb.ca* | -Largest and most complete of its kind. -Specific to humans only |
| METLIN Metabolite Database | *http://metlin.scripps.edu/* | -Human specific<br>-Mixes drugs, drug metabolites together<br>-Name, structure, ID only |
| NIST Spectral Database | *http://webbook.nist.gov/chemistry/* | -Spectral database only (NMR, MS, IR)<br>-No biological data, little chemical data<br>-Not limited to metabolites |
| Spectral Database for Organic Compounds (SDBS) | *http://www.aist.go.jp/RIODB/SDBS/cgi-bin/direct_frame_top. cgi?lang=eng* | -Spectral database only (NMR, MS, IR)<br>-No biological data, little chemical data<br>-Not limited to metabolites |

(Copies of tables are available in the accompanying CD.)

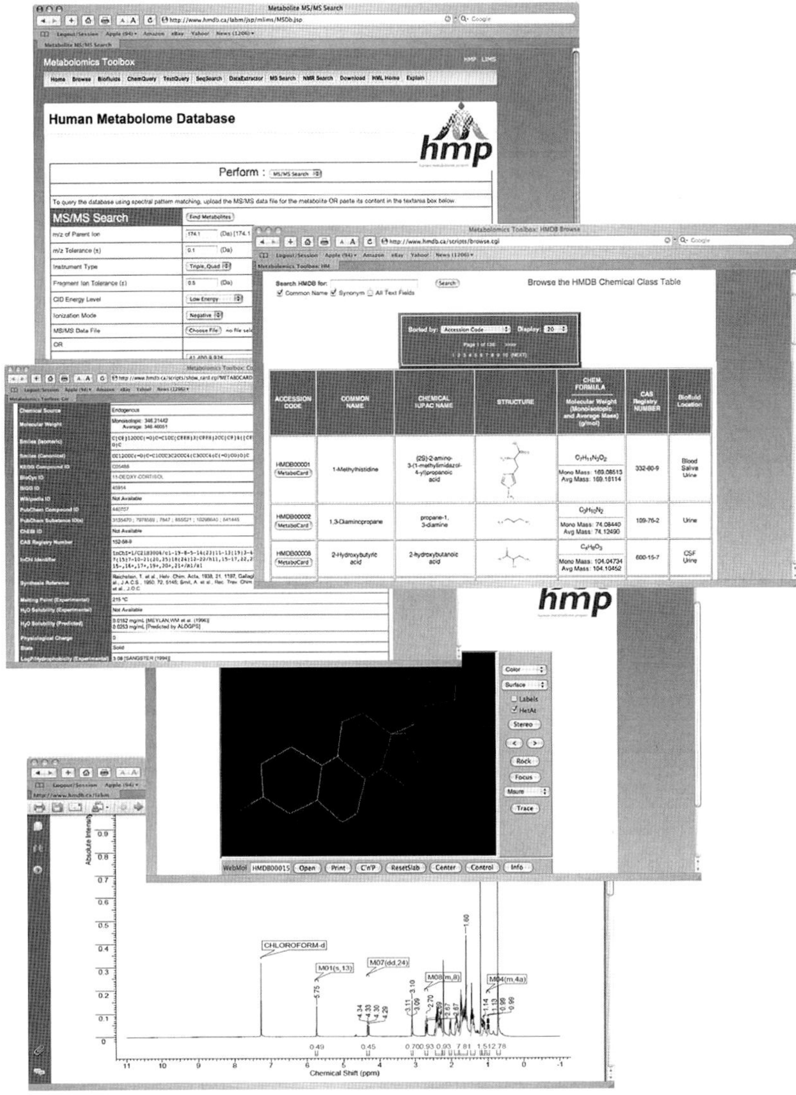


**Fig 30.2** A montage of screen shots from the Human Metabolome Database HMDB, illustrating some of the data content and query capabilities of the database (Copies of figures including color copies, where applicable, are available in the accompanying CD)

structure image and associated KEGG hyperlink. While the METLIN database contains a very large number of molecules, many of the listed compounds are actually drugs or drug metabolites rather than endogenous metabolites. In addition to this metabolite list, METLIN also contains a database of FT-MS spectra from a number of chromatographically separated human serum fractions, as well as a smaller number of spectra from other bio-fluids and tissues. In addition, a set of 216 reference LC/MS profiles and 79 MS/MS data from a variety of tissues and bio-fluids in various disease states are also present. The METLIN database is one of only a small number of metabolome databases dedicated to human metabolites and human metabolite profiles. Certainly, as this database grows it could evolve to be a resource of particular interest to clinicians and biomedical researchers.

### 30.3.3 The Golm Metabolome Database

The Golm Metabolome Database (GMD) contains mass spectral and retention index (MSRI) libraries obtained from purified tissue of bio-fluid samples derived from mammals, plants, yeast, and bacteria (corynebacterium) [29]. These libraries include more than 2,000 mass spectral data, collected from two different kinds of capillary GC-MS systems – a quadrupole GC-MS and a GC-TOF (time-of-flight)-MS. To date the GMD has assembled 1,089 non-redundant peaks or mass spectral tags of which 360 have been identified. The GMD supports both compound searches and mass spectral queries. The compound search utility allows filtered searches, by compound name, and provides access to the linked mass spectral information housed at the GMD. The GMD mass spectral search supports both NIST02 and AMDIS formatted input files. The GC-MS searches are performed by computing the fragment-intensity agreement based on a normalized Euclidean, Hamming and Jaccard distance. The resulting matches are presented as a sorted table containing the similarity score, the retention index (RI), the spectral identifier and the compound name (if known).

### 30.3.4 The BiGG Database

The BiGG database [30] is the one metabolomics database that is, perhaps, most intimately connected to Systems Biology. Developed by the members of Bernhard Palsson's lab at UCSD, the BiGG database is actually a database of databases containing metabolic reconstruction data on 6 organisms including yeast, *Escherichia coli* and *Homo sapiens*. These carefully curated databases contain information about metabolites, their pathways, their reactions, and their cellular compartments. The Human entry in the BiGG database contains data on 1,805 distinct metabolites involved in 3,311 metabolic and transport reactions. SBML representations of all of the reactions are also provided. Particular care has gone into assembling BiGG's reaction databases, since these require mass and charge balance as well as internal self-consistency, especially for the Flux Balance Analyses that these databases were specifically developed.

### 30.3.5 The BioMagResBank

The BioMagResBank [28], which was originally developed for deposition of macromolecular NMR data, has recently been expanded to include NMR data from small molecule metabolites. Currently, the BMRB (under its "Metabolomics" link) contains one-dimensional $^{1}$H and $^{13}$C data as well as two-dimensional $^{1}$H TOCSY and $^{1}$H,$^{13}$C HSQC data from $\sim$250 metabolites, many of which are common to both plants and animals. Each compound is listed alphabetically and users may click on the name to bring up a metabolite page giving compound names, synonyms, chemical formulas, SMILES strings, CAS numbers, structural details, interactive 3D renderings, hyperlinks to PubChem, KEGG, ChEBI and other databases, molecular weights (monoisotopic mass), predicted chemical shifts, NMR data collection conditions and — most importantly — hyperlinks to all of the experimental NMR spectra and associated peak lists. All the reference NMR spectra are collected in water so this makes the BMRB data quite compatible with standard metabolomic measurements (which are also done in water). Like the HMDB's MS/MS and NMR spectral libraries, the BMRB's libraries also supports NMR-based peak searches and spectral comparisons. These can be particularly useful in identifying or confirming the identity of compounds found in high-throughput metabolomics experiments.

The need for spectral (MS/MS, NMR, GC-MS) databases and spectral processing tools, is as fundamental, to metabolomics, as are image repositories and image processing tools for microarrays (in genomics) and 2D gels (in proteomics). In the next section we will discuss these spectral tools and the methods used for spectral analysis in more detail.

## 30.4 Spectral and Statistical Analysis Tools for Metabolomics

Metabolomics shares more than a few similarities with clinical chemistry. Both fields are focused on identifying or quantifying small molecules from tissues, cells, or bio-fluids. However, what distinguishes metabolomics from clinical chemistry is the fact that in metabolomics one is measuring not just 1 or 2 compounds at a time, but literally hundreds at a time. Furthermore, in clinical chemistry, most metabolites are typically identified and quantified using colorimetric chemical assays. In contrast, metabolomics attempts to rapidly measure large numbers (tens to hundreds) of metabolites using non-chemical, non-colorimetric methods such as GC-MS (gas chromatography – mass spectrometry), LC-MS (liquid chromatography – mass spectrometry), CE (capillary electrophoresis), FT-MS (Fourier transform mass spectrometry) or NMR spectroscopy [32].

The fact that most metabolomic data contains hundreds of peaks or spectral "features" means that it must be analyzed quite differently than conventional clinical chemistry data. In fact, there are two very distinct schools-of-thought about how metabolomic data should be collected, processed, and interpreted. In one version (the chemometric or non-targeted approach) the compounds are not formally identified – only their spectral patterns and intensities are recorded, compared and used to make diagnoses, identify phenotypes or draw conclusions [33]. In the other version (targeted profiling or quantitative metabolomics), the compounds are actually identified and quantified. The resulting list of compounds and concentrations (a metabolic profile) is then used to make diagnoses, identify phenotypes or draw conclusions [11,34]. Both methods have their advantages and both have their advocates.

### *30.4.1 Chemometrics and Metabolomic Data Analysis*

Chemometrics can be defined as the application of mathematical, statistical, graphical or symbolic methods to maximize the information which can be extracted from chemical or spectral data. Chemometric approaches for spectral analysis emerged in the 1980's and are primarily used to extract useful information from complex spectra consisting of many hard-to-identify or unknown components [35,36]. Chemometric approaches can also be used to identify statistically significant differences between large groups of spectra collected on different samples or under different conditions.

To facilitate the spectral analysis process, each input spectrum is usually divided up into smaller regions or bins. This spectral partitioning process is called "binning", and it allows specific features, peaks, or peak clusters in a multi-peak spectrum to be isolated or highlighted. Once binned, the peak intensities (or total area under the curve) in each bin are tabulated and analyzed using multivariate statistical analysis. This "divide-and-conquer" approach allows spectral components to be quantitatively compared within a single spectrum or between multiple spectra. Of course, the number of components or "dimensions" that a binned spectrum may represent could number in hundreds or even thousands. To reduce the complexity or the number of parameters, chemometricians use dimensional reduction to identify the key components that seem to contain the maximum amount of information or which yield the greatest differences. The most common form of dimensional reduction is known as Principal Component Analysis or PCA.

PCA is not a classification technique, rather it is an unsupervised clustering or data reduction technique. Formally, principal component analysis determines an optimal linear transformation for a collection of data points such that the properties of that sample are most clearly displayed along the coordinate (or principal) axes. A more simplified explanation of PCA is given in Fig. 30.3, where we use the analogy of the projecting shadows on a wall using a flashlight to find a "maximally informative projection" for a particular object. If the object of interest is a doughnut (but the observer can only look at the projected shadow), then by shining the flashlight directly over the doughnut hole one would generate the tell-tale "O" shadow. On the other hand, if the flashlight was directed at the edge of the doughnut, the resulting shadow would be a less informative "hot-dog

**Fig 30.3** A simplified picture of Principle Component Analysis (PCA), where a three dimensional object is reduced to a 2 dimensional representation by prudent selection of the projection plane (Copies of figures including color copies, where applicable, are available in the accompanying CD)

bun" shape. This kind of shadow would, likely, lead the observer to the wrong conclusion about what the object was. While this example shows how a 3D object can be projected or have its key components reduced to two dimensions, the strength of PCA is that it can do the same with a hyperdimensional object just as easily.

In practice, PCA is most commonly used to identify how one sample is different from another, which variables contribute most to this difference, and whether those variables contribute in the same way (i.e., are correlated) or independently (i.e., uncorrelated) from each other. PCA also quantifies the amount of useful information or signal that is contained in the data. As a data reduction technique, PCA is particularly appealing as it allows one to easily detect, visually or graphically, sample patterns or groupings.

PCA is not the only chemometric or statistical approach that can be applied to spectral analysis in metabolomics. A second class of chemometric method is known as supervised learning or supervised classification. Supervised classifiers require that information about the class identities have to be provided by the user in advance of running the analysis. Examples of supervised classifiers include SIMCA (Soft Independent Modeling of Class Analogy), PLS-DA (Partial Least Squares – Discriminant Analysis) and K-means clustering. All of these techniques have been used to interpret NMR, MS/MS and FTIR spectral patterns in a variety of metabolomic or metabonomic applications [37–39].

PLS-DA or Partial Least Squares – Discriminant Analysis can be used to enhance the separation between groups of observations by rotating PCA components, such that a maximum separation among classes is obtained. In doing so, it is hoped that one can better understand which variables are most responsible for separating the observed (or apparent) classes. The basic principles behind PLS (partial least squares) are similar to that of PCA. However, in PLS a second piece of information is used, namely, the labeled set of class identities. PLS-DA, which is a particular form of PLS, is a regression or categorical extension of PCA that takes advantage of a priori or user-assigned class information to attempt to maximize the covariance between the "test" or predictor variables and the training variable(s). PLS-DA is typically used after a relatively clear separation between two or more groups has been obtained through an unsupervised (PCA) analysis. Care must be taken in using PLS-DA methods as it is easy to create convincing clusters or classes that have no statistical meaning (i.e., they over-fit the data). The best way of avoiding these problems is to use N-fold cross-validation methods, boot-strapping or re-substitution approaches to ensure that the data clusters derived by PLS-DA or other supervised methods are real and robust [40].

The intent in using pattern classification for spectral analysis in metabolomics is not to identify any specific compound but, rather, to look at the spectral profiles of bio-fluids or tissues and to classify them in specific categories, conditions or disease states. This trend to pattern classification represents a significant break from the classical methods of analytical chemistry or traditional clinical chemistry which historically have depended on identifying and quantifying specific compounds. With chemometric profiling methods one is not so interested in quantifying known metabolites, but rather in trying to look at all the metabolites (known and unknown) at once [33,37]. The strength of this holistic approach lies in the fact that one is not selectively ignoring or including key metabolic data in making a phenotypic classification or diagnosis. These pattern classification methods can perform quite impressively and a number of groups have reported success in diagnosing certain diseases such as colon cancer [38], in monitoring organ rejection [41] and in classifying different strains of mice and rats [42,43].

### *30.4.2 Targeted or Quantitative Metabolic Profiling*

Targeted (or quantitative) metabolic profiling is fundamentally different than most chemometric approaches. In targeted metabolic profiling, the compounds in a given bio-fluid or tissue extract are actually identified and quantified by comparing the bio-fluid spectrum of interest to a library of reference spectra of pure compounds [11,34,44,45]. The basic assumption in targeted profiling is that the spectra obtained for the bio-fluid (which is a mixture of metabolites) is the sum of individual spectra for each of the pure metabolites in the mixture (see Fig 30.4). This approach to compound identification is somewhat similar to the approach historically taken by GC-MS methods and, to a much more limited extent, LC-MS methods [46,47]. For NMR, this particular approach requires that the sample pH be precisely known or precisely controlled. It also requires the use of sophisticated curve-fitting software and specially prepared databases of NMR spectra of pure metabolites collected at different pH values and at multiple spectrometer frequencies (400, 500, 600, 700 and 800 MHz) [34].

One of the strengths of the NMR-curve fitting approaches is the fact that the NMR spectra for many individual metabolites are often composed of multiple peaks covering a wide range of chemical shifts. This means that most metabolites have unique or characteristic "chemical shift" fingerprints. This particular characteristic of NMR spectra helps reduce the problem of spectral (or chromatographic) redundancy as it is unlikely that any two compounds will have identical numbers of peaks with identical chemical shifts, identical intensities, identical spin couplings or identical peak shapes. Likewise, with higher magnetic fields (>600 MHz) the chemical shift separation among different peaks and different compounds is often good enough to allow the unambiguous identification of up to 100 compounds at a time, through simple curve fitting [11,34, 45].

Targeted metabolic profiling is not restricted to NMR or GC-MS. It is also possible to apply the same techniques to LC-MS systems [47]. In the case of MS spectroscopy, the sample MS/MS spectra must be collected at reasonably similar collision energies and on similar kinds of instruments [48]. Quantification of metabolites by LC-MS is somewhat more difficult than GC-MS or by NMR. Typically, quantification requires the addition or spiking of isotopically labeled derivatives

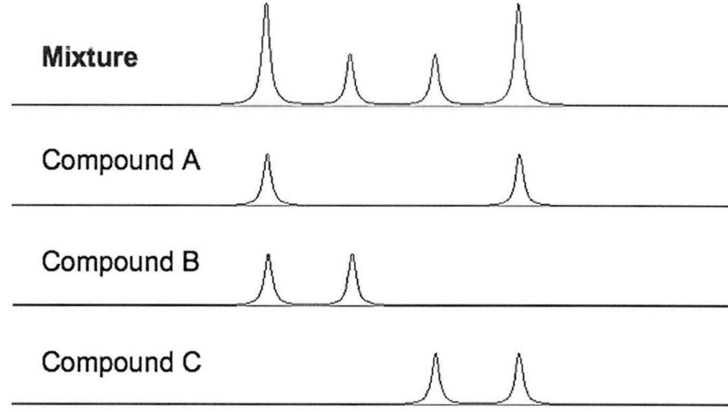


**Fig. 30.4.** The principle of spectral fitting from a spectral library to deconvolute (i.e., identify and quantify) components in a complex mixture (Copies of figures including color copies, where applicable, are available in the accompanying CD)

of the metabolites of interest to the bio-fluid or tissue sample. The intensity of the isotopic derivative can then be used to quantify the metabolite of interest.

A key advantage targeted metabolic profiling is that it does not require the collection of identical sets of cells, tissues or lab animals and so it is more amenable to human studies or studies that require less day-to-day monitoring. In other words, there is no need for specially designed metabolic chambers. A key disadvantage of this approach is the relatively limited size of most current spectral libraries (~250 compounds). Such a small library of identifiable compounds may bias metabolite identification and interpretation. Both the targeted and chemometric approaches have their advocates. However, it appears that there is a growing trend towards combining the best features of both methods.

Because targeted metabolic profiling yields information about both the identity and concentration of compounds, it is possible to use a large range of statistical and machine learning approaches to interpret the data. In fact, the same statistical techniques used in chemometric or non-targeted studies – PCA, SIMCA, PLS-DA, K-means clustering – can also be used with targeted profiling. However, instead of using binned spectra or arbitrary peak clusters as input to these algorithms, the actual names of the compounds and their concentrations are used as input. This added specificity seems to significantly improve the discriminatory capabilities of most statistical techniques, over what is possible for unlabeled or binned spectral data [34]. Targeted profiling also seems to be particularly amenable to other, more powerful, classification techniques such as artificial neural networks (ANNs), support vector machines (SVMs) and Decision Trees (DTs).

## 30.5 Metabolic Modeling and the Interpretation of Metabolomic Data

As we have already seen, the statistical and computational methods described in section 4 are particularly useful at identifying metabolic differences or finding interesting bio-markers. However, these approaches are not designed to provide a great deal of biological insight nor can they provide a clear perspective on the underlying biological causes for the metabolic profiles that are seen. To gain this sort of insight, it is often necessary to either mine the literature or to turn to metabolic modeling. Metabolic modeling is where metabolomics probably begins to align itself most closely with Systems Biology.

Metabolic modeling or metabolic simulation can be done in a variety of ways. Traditionally, it is done by writing down and solving systems of time-dependent ordinary differential equations (ODEs) that describe the chemical reactions and reaction rates of the metabolic system of interest. There are now a host of metabolic simulation programs that allows very complex, multi-component simulations to be performed [49,50]. These include programs such as GEPASI [51], CellDesigner [52], SCAMP [53] and Cellerator [54]. GEPASI is a good example of a typical metabolic or biochemical pathway simulation package. This program, which has been under development for almost 15 years, uses a simple interface to allow one to build models of metabolic pathways and simulate their dynamics and steady state behaviour for the given sets of parameters. GEPASI also generates the coefficients of Metabolic Control Analysis for steady states. In addition, it allows one to study the effects of several parameters on the properties of the model pathway. GEPASI allows users to enter the kinetic equations of interest and their parameters (Km, reaction velocity, starting concentrations), solves the ODEs using an ODE solver, and generates plots that can be easily visualized by the user. GEPASI has been used in a wide variety of metabolic studies such as bacterial glucose/galactose metabolism [55] and glutathione/phytochelitin metabolism [56] and continues to be used in many metabolomic or kinetic analyses.

An alternative to solving large systems of time-dependent rate equations is a technique known as constraint-based modeling [57,58]. Constraint-based modeling uses physico-chemical constraints such as mass balance, energy balance and flux limitations to describe the potential behavior of a large metabolic system (a cell, an organ, an organism). In this type of modeling, the time dependence and rate constants can be ignored, as one is only interested in finding the steady state conditions that satisfy the physico-chemical constraints. Because cells and organs are so inherently complex and because it is almost impossible to know all the rate constants or instantaneous metabolite concentrations at a given time, constraint-based modeling is particularly appealing to those involved in large-scale metabolomic studies. In particular, through constraint-based modeling, models and experimental data can be more easily reconciled and studied on a whole-cell or genome-scale level [57,58]. Furthermore, experimental data sets can be examined for their consistency against the underlying biology and chemistry represented in the models.

### 30.5.1 Flux Balance Analysis

One of the most popular approaches to constraint-based metabolic modeling is known as flux-balance analysis or FBA [59,60]. FBA requires knowledge of the stoichiometry of most of reactions and transport processes that are thought to occur in the metabolic system of interest. This collection of reactions defines the metabolic network. FBA assumes that the metabolic network will reach a steady state constrained by the stoichiometry of the reactions. Normally the stoichiometric constraints are too few and this leads to more unknowns than equations (i.e., an underdetermined system). However, possible sets of solutions can be found by including information about all feasible metabolite fluxes (metabolites added or excreted) and by specifying maximum and minimum fluxes through any particular reaction. The model can also be refined or further constrained by adding experimental data, from known physiological or biochemical data obtained from specific metabolomic studies. Once the solution space is defined, the model is refined and its behavior can be studied by optimizing the steady state behavior with respect to some objective function. Typically, the objective function optimization involves the maximization of biomass, the maximization of growth rate, the maximization of ATP production, the maximization of the production of a particular product or the maximization of reducing power. Once the model is fully optimized, it is possible to use that FBA model to create predictive models of cellular, organ or whole organism metabolism. These predictions can be done by changing the network parameters or flux balance, changing the reactants, adding new components to the model or changing the objective function to be maximized.

Critical to the success of any FBA model is the derivation or compilation of appropriate mass and charge balance [59,60]. Mass balance is defined in terms of both the flux of metabolites through each reaction, the stoichiometry of that reaction and the conservation of mass and charge. Mass and charge balance considerations give rise to a set of coupled differential equations. This set of equations is often expressed as a matrix equation which can be solved through simple linear algebra and optimized through linear programming. The goal of FBA is to identify the metabolic fluxes in the steady state (i.e., where the net flux is 0). Because there are always more reactions than metabolites, the steady state solution is always underdetermined. As a result, additional constraints must be added to determine a unique solution. These constraints can be fluxes measured through metabolomics experiments (such as isotope labeling experiments) or through estimated ranges of allowable (feasible) flux values.

FBA methods have been used in a variety of metabolomic studies. In particular, they have been used in the genome-scale modeling of many bacterial metabolic systems including, *Lactococcus lactis, Corynebacterium glutamicum, Streptomyces coliecolor, Helicobacter pylori and E. coli* [61–65]. Flux balance analysis has also been used to look at yeast metabolism [66,67], erythrocyte metabolism [68], myocardial metabolism [69] and most impressively the entire human metabolomic network [30]. Certainly, as more detailed flux data is acquired through isotope tracer analysis and more information is obtained from quantitative, targeted metabolic profiling it is likely that flux balance analysis and other kinds of constraint-based modeling will play an increasingly important role in the interpretation of metabolomic data – and in Systems Biology.

## 30.6 Conclusions

This chapter has attempted to highlight the areas where most of the activity in computational metabolomics has occurred and where most of the interest (at least among the current practitioners) seems to lie. These areas include: (i) the creation of comprehensive metabolomics databases; (ii) the refinement of data analysis tools; and (iii) the development of metabolic modeling tools and methods. While good progress is being made in most areas, it is also clear that there are still many opportunities for algorithmic development and bio/chemo-informatics innovation. One of the most obvious trends in computational metabolomics is the growing alignment or integration of metabolomics with Systems Biology [5,6,30]. This integration will require that metabolomics methods and data reduction techniques will have to become much more quantitative. While chemometric methods for spectral analysis will likely continue to be popular among some groups for certain types of applications, the long term trend in metabolomics seems to be towards rapid/high-throughput compound identification and quantification. These so-called targeted or quantitative methods will require greater reliance on spectral libraries and spectral standards and will no-doubt lead to the appearance of organism-specific metabolite databases. This trend towards large-scale metabolite identification and quantification will likely encourage metabolomics researchers to adopt many of the analytical approaches commonly used in transcriptomics and proteomics, where transcript and protein levels are routinely quantified, compared and analyzed. Given the importance that bioinformatics has played in establishing genomics and proteomics, it is likely that continuing developments in bioinformatics will have an equally profound impact on metabolomics and, ultimately, in its role in Systems Biology.

**Acknowledgments** I would like to acknowledge Genome Canada, Genome Alberta and the Alberta Ingenuity Centre for Machine Learning (AICML) for their financial support.

## References

1. Ideker T, Galitski T, Hood L. A new approach to decoding life: Systems Biology. Annu Rev Genomics Hum Genet 2001; 2:343–372.
2. German JB, Hammock BD, Watkins SM. Metabolomics: building on a century of biochemistry to guide human health. Metabolomics 2005; 1:3–9.
3. Ideker T, Thorsson V, Ranish JA, Christmas R, Buhler J, Eng JK, Bumgarner R, Goodlett DR, Aebersold R, Hood L. Integrated genomic and proteomic analyses of a systematically perturbed metabolic network. Science 2001; 292:929–934.
4. Edwards JS, Ibarra RU, Palsson BO. In silico predictions of Escherichia coli metabolic capabilities are consistent with experimental data. Nat Biotechnol 2001; 19:125–130.
5. Lindon JC, Holmes E, Nicholson JK. Metabonomics: Systems Biology in pharmaceutical research and development. Curr Opin Mol Ther 2004; 6:265–272.
6. Kell DB. Systems Biology, metabolic modelling and metabolomics in drug discovery and development. Drug Discov Today 2006; 11:1085–1092.
7. Schnackenberg LK, Beger RD. Monitoring the health to disease continuum with global metabolic profiling and Systems Biology. Pharmacogenomics 2006; 7:1077–1086.
8. Watkins SM, German JB. Metabolomics and biochemical profiling in drug discovery and development. Curr Opin Mol Ther 2002; 4:224–228.
9. Bugrim A, Nikolskaya T, Nikolsky Y. Early prediction of drug metabolism and toxicity: Systems Biology approach and modeling. Drug Discov Today 2004; 9:127–135.
10. Griffin JL, Bollard ME. Metabonomics: its potential as a tool in toxicology for safety assessment and data integration. Curr Drug Metab 2004; 5:389–398.
11. Wishart DS, Querengesser LMM, Lefebvre BA et al. Magnetic resonance diagnostics: a new technology for high-throughput clinical diagnostics. Clin Chemistry 2001; 47:1918–1921.
12. Trujillo E, Davis C, Milner J. (2006) Nutrigenomics, proteomics, metabolomics, and the practice of dietetics. J Am Diet Assoc 2006; 106:403–413.
13. Gibney MJ, Walsh M, Brennan L et al. Metabolomics in human nutrition: opportunities and challenges. Am J Clin Nutr. 2005; 82:497–503.

14. Shulaev V. Metabolomics technology and bioinformatics. Brief Bioinform 2006; 7:128–139.
15. Benson DA, Karsch-Mizrachi I, Lipman DJ, Ostell J, Wheeler DL. Nucleic Acids Res 2007; 35(Database issue):D21–25.
16. Wu CH, Apweiler R, Bairoch A, Natale DA, Barker WC, Boeckmann B, Ferro S, Gasteiger E, Huang H, Lopez R, Magrane M, Martin MJ, Mazumder R, O'Donovan C, Redaschi N, Suzek B. The Universal Protein Resource (UniProt): an expanding universe of protein information. Nucleic Acids Res 2006; 34(Database issue): D187–191.
16. Kouranov A, Xie L, de la Cruz J, Chen L, Westbrook J, Bourne PE, Berman HM. The RCSB PDB information portal for structural genomics. Nucleic Acids Res 2006 1;34(Database issue):D302–305.
17. Michal G. Biochemical pathways wall chart. Mannheim, Boehringer Mannheim, 1968.
18. Kanehisa M, Goto S, Hattori M, et al. From genomics to chemical genomics: new developments in KEGG. Nucleic Acids Res 2006; 34(Database issue):D354–357.
19. Caspi R, Foerster H, Fulcher CA, et al. MetaCyc: a multiorganism database of metabolic pathways and enzymes. Nucleic Acids Res 2006; 34(Database issue):D511–516.
20. Joshi-Tope G, Gillespie M, Vastrik I, et al. Reactome: a knowledgebase of biological pathways. Nucleic Acids Res 2005; 33(Database issue):D428–432.
21. Barthelmes J, Ebeling C, Chang A, Schomburg I, Schomburg D. BRENDA, AMENDA and FRENDA: the enzyme information system in 2007. Nucleic Acids Res 2007; 35(Database issue):D511–514.
22. Maltsev N, Glass E, Sulakhe D, Rodriguez A, Syed MH, Bompada T, Zhang Y, D'Souza M. PUMA2—grid-based high-throughput analysis of genomes and metabolic pathways. Nucleic Acids Res 2006; 34(Database issue):D369–372.
23. Romero P, Wagg J, Green ML, Kaiser D, Krummenacker M, Karp PD. Computational prediction of human metabolic pathways from the complete human genome. Genome Biol 2005; 6:R2.
24. Cotter D, Maer A, Guda C, Saunders B, Subramaniam S. LMPD: LIPID MAPS proteome database. Nucleic Acids Res 2006; 34(Database issue):D507–510.
25. Brooksbank C, Cameron G, Thornton J. The European Bioinformatics Institute's data resources: towards Systems Biology. Nucleic Acids Res 2005; 33(Database issue):D46–53.
26. Wishart DS, Tzur D, nox C, Eisner R, Guo AC, Young N, Cheng D, Jewell K, Arndt D, Sawhney S, Fung C, Nikolai L, Lewis M, Coutouly MA, Forsythe I, Tang P, Shrivastava S, Jeroncic K, Stothard P, Amegbey G, Block D, Hau DD, Wagner J, Miniaci J, Clements M, Gebremedhin M, Guo N, Zhang Y, Duggan GE, Macinnis GD, Weljie AM, Dowlatabadi R, Bamforth F, Clive D, Greiner R, Li L, Marrie T, Sykes BD, Vogel HJ, Querengesser L. HMDB: the Human Metabolome Database. Nucleic Acids Res 2007; 35(Database issue):D521–526.
27. Smith CA, O'Maille G, Want EJ, Qin C, Trauger SA, Brandon TR, Custodio DE, Abagyan R, Siuzdak G. METLIN: a metabolite mass spectral database. Ther Drug Monit 2005; 27:747–751.
28. Seavey BR, Farr EA, Westler WM Markley JL. A relational database for sequence-specific protein NMR data. J Biomol NMR 1991; 1:217–236.
29. Kopka J, Schauer N, Krueger S, Birkemeyer C, Usadel B, Bergmuller E, Dormann P, Weckwerth W, Gibon Y, Stitt M, Willmitzer L, Fernie AR, Steinhauser D. GMD@CSB.DB: the Golm Metabolome Database. Bioinformatics 2005; 21:1635–1638.
30. Duarte NC, Becker SA, Jamshidi N, Thiele I, Mo ML, Vo TD, Srivas R, Palsson BO. Global reconstruction of the human metabolic network based on genomic and bibliomic data. Proc Nat Acad Sci 2007; 104:1777–1782.
31. Altschul SF, Madden TL, Schaffer AA, Zhang J, Zhang Z, Miller W, Lipman DJ. Gapped BLAST and PSI-BLAST: a new generation of protein database search programs. Nucleic Acids Res 1997; 25:3389–3402.
32. Dunn WB, Bailey NJ, Johnson HE. Measuring the metabolome: current analytical technologies. Analyst 2005; 130:606–625.
33. Trygg J, Holmes E, Lundstedt T. Chemometrics in metabonomics. J Proteome Res 2007; 6:469–479.
34. Weljie AM, Newton J, Mercier P, Carlson E, Slupsky CM. Targeted profiling: quantitative analysis of 1H NMR metabolomics data. Anal Chem 2006; 78:4430–4442.
35. Lavine B, Workman JJ Jr. Chemometrics. Anal Chem 2004; 76:3365–3371.
36. Lindon JC, Holmes E, Nicholson JK. Metabonomics and its role in drug development and disease diagnosis. Expert Rev Mol Diagn 2004; 4:189–199.
37. Holmes E, Nicholls AW, Lindon JC, Connor SC, Connelly JC, Haselden JN, Damment SJ, Spraul M, Neidig P, Nicholson JK. Chemometric models for toxicity classification based on NMR spectra of bio-fluids. Chem Res Toxicol 2000; 13:471–478.
38. Smith IC, Baert R. Medical diagnosis by high resolution NMR of human specimens. IUBMB Life 2003; 55:273–277.
39. Wilson ID, Plumb R, Granger J, Major H, Williams R, Lenz EM. HPLC-MS-based methods for the study of metabonomics. J Chromatogr B Analyt Technol Biomed Life Sci 2005; 817:67–76.

40. Molinaro AM, Simon R, Pfeiffer RM. Prediction error estimation: a comparison of resampling methods. Bioinformatics 2005; 21:3301–3307.
41. Wishart DS. Metabolomics: the principles and potential applications to transplantation. Am J Transplant 2005; 5:2814–2820.
42. Robosky LC, Wells DF, Egnash LA, Manning ML, Reily MD, Robertson DG. Metabonomic identification of two distinct phenotypes in Sprague-Dawley (Crl:CD(SD)) rats. Toxicol Sci 2005; 87:277–284.
43. Gavaghan McKee CL, Wilson ID, Nicholson JK. Metabolic phenotyping of nude and normal (Alpk:ApfCD, C57BL10J) mice. J Proteome Res 2006; 5:378–384.
44. Serkova NJ, Rose JC, Epperson LE, Carey HV, Martin SL. Quantitative analysis of liver metabolites in three stages of the circannual hibernation cycle in 13-lined ground squirrels by NMR. Physiol Genomics 2007 May 29; [Epub ahead of print]
45. Serkova NJ, Zhang Y, Coatney JL, Hunter L, Wachs ME, Niemann CU, Mandell MS. Early detection of graft failure using the blood metabolic profile of a liver recipient. Transplantation 2007; 83:517–521.
46. Niwa T. Metabolic profiling with gas chromatography-mass spectrometry and its application to clinical medicine. J Chromatogr. 1986; 379:313–345.
47. la Marca G, Casetta B, Malvagia S, Pasquini E, Innocenti M, Donati MA, Zammarchi E. Implementing tandem mass spectrometry as a routine tool for characterizing the complete purine and pyrimidine metabolic profile in urine samples. J Mass Spectrom 2006; 41:1442–1452.
48. Jiang H, Somogyi A, Timmermann BN, Gang DR. Instrument dependence of electrospray ionization and tandem mass spectrometric fragmentation of the gingerols. Rapid Commun Mass Spectrom 2006; 20:3089–3100.
49. Alves R, Antunes F, Salvador A. Tools for kinetic modeling of biochemical networks. Nat Biotechnol 2006 Jun;24:667–672.
50. Materi W, Wishart DS. Computational Systems Biology in drug discovery and development: methods and applications. Drug Discov Today 2007; 12:295–303.
51. Mendes P. GEPASI: a software package for modelling the dynamics, steady states and control of biochemical and other systems. Comput Appl Biosci 1993; 9:563–571.
52. Kitano H, Funahashi A, Matsuoka Y, Oda K. Using process diagrams for the graphical representation of biological networks. Nat Biotechnol 2005; 23:961–966.
53. Sauro HM. SCAMP: a general-purpose simulator and metabolic control analysis program. Comput Appl Biosci 1993; 9:441–450.
54. Shapiro BE, Levchenko A, Meyerowitz EM, Wold BJ, Mjolsness ED. Cellerator: extending a computer algebra system to include biochemical arrows for signal transduction simulations. Bioinformatics 2003; 19:677–678.
55. Demir O, Aksan Kurnaz I. An integrated model of glucose and galactose metabolism regulated by the GAL genetic switch. Comput Biol Chem 2006; 30:179–192.
56. Mendoza-Cozatl DG, Moreno-Sanchez R. Control of glutathione and phytochelatin synthesis under cadmium stress. Pathway modeling for plants. J Theor Biol 2006; 238:919–936.
57. Gagneur J, Casari G. From molecular networks to qualitative cell behavior. FEBS Lett 2005; 579:1867–1871.
58. Joyce AR, Palsson BO. Toward whole cell modeling and simulation: comprehensive functional genomics through the constraint-based approach. Prog Drug Res 2007; 64:267–309.
59. Kauffman KJ, Prakash P, Edwards JS. Advances in flux balance analysis. Curr Opin Biotechnol 2003; 14:491–496.
60. Lee JM, Gianchandani EP, Papin JA. Flux balance analysis in the era of metabolomics. Brief Bioinform 2006; 7:140–150.
61. Marx A, Eikmanns BJ, Sahm H, de Graaf AA, Eggeling L. Response of the central metabolism in Corynebacterium glutamicum to the use of an NADH-dependent glutamate dehydrogenase. Metab Eng 1999; 1:35–48.
62. Oliveira AP, Nielsen J, Forster J. Modeling Lactococcus lactis using a genome-scale flux model. BMC Microbiol 2005; 5:39.
63. Price ND, Thiele I, Palsson BO. Candidate states of Helicobacter pylori's genome-scale metabolic network upon application of "loop law" thermodynamic constraints. Biophys J 2006; 90:3919–3928.
64. Borodina I, Krabben P, Nielsen J. Genome-scale analysis of Streptomyces coelicolor A3(2) metabolism. Genome Res 2005; 15:820–829.
65. Edwards JS, Palsson BO. Metabolic flux balance analysis and the in silico analysis of Escherichia coli K-12 gene deletions. BMC Bioinformatics 2000; 1:1.
66. Segre D, Deluna A, Church GM, Kishony R. Modular epistasis in yeast metabolism. Nat Genet 2005; 37:77–83.
67. Jin YS, Jeffries TW. Stoichiometric network constraints on xylose metabolism by recombinant Saccharomyces cerevisiae. Metab Eng 2004; 6:229–238.
68. Durmus Tekir S, Cakir T, Ulgen KO. Analysis of enzymopathies in the human red blood cells by constraint-based stoichiometric modeling approaches. Comput Biol Chem 2006; 30:327–338.
69. Luo RY, Liao S, Tao GY Li YY, Zeng S, Li YX, Luo Q. Dynamic analysis of optimality in myocardial energy metabolism under normal and ischemic conditions. Mol Syst Biol 2006; 2:2006.0031.

## Key References

Shulaev V. Metabolomics technology and bioinformatics. Brief Bioinform 2006; 7:128–139.

Kell DB. Systems Biology, metabolic modelling and metabolomics in drug discovery and development. Drug Discov Today 2006; 11:1085–1092.

Dunn WB, Bailey NJ, Johnson HE. Measuring the metabolome: current analytical technologies. Analyst 2005; 130:606–625.

Wishart DS, Tzur D, Knox C, Eisner R, uo AC, Young N, Cheng D, Jewell K, Arndt D, Sawhney S, Fung C, Nikolai L, Lewis M, Coutouly MA, Forsythe I, Tang P, Shrivastava S, Jeroncic K, Stothard P, Amegbey G, Block D, Hau DD, Wagner J, Miniaci J, Clements M, Gebremedhin M, Guo N, Zhang Y, Duggan GE, Macinnis GD, Weljie AM, Dowlatabadi R, Bamforth F, Clive D, Greiner R, Li L, Marrie T, Sykes BD, Vogel HJ, Querengesser L. HMDB: the Human Metabolome Database. Nucleic Acids Res 2007; 35(Database issue):D521–526.

# Chapter 31
# Virtual Reality Meets Functional Genomics

**Andrei L. Turinsky and Christoph W. Sensen**

**Abstract** The emergence of the post-genomic era requires new approaches to biomedical data analysis. To meet this challenge, we are building technologies that combine information from molecular biology, anatomy and advanced visualization. Our top-down approach to integrative data analysis utilizes existing software tools and life science knowledge bases. It allows various data components to be visually manipulated and cross-referenced with other data sources via standard interfaces. Such techniques are useful for a range of tasks, from gene expression studies to pharmacokinetic explorations, within the visual model of an organism.

**Keywords** Bioinformatics · Virtual reality · Java 3D · Ontologies · Atlas

## 31.1 Introduction

Since the release of the first microbial genome in 1996 [1], the field of genome research has exploded. To date, hundreds of genomes, from viruses to the human genome has been completely sequenced and submitted to the public databanks. In addition, new large-scale, high-throughput experiments have been designed, which allow the characterization of the messenger RNA levels and the protein contents of an organism. Other fields, such as metabolomics, are rapidly emerging, leading to complete inventories of all biologically active compounds in an organism.

The rapidly growing data space has posed major challenges for the field of bioinformatics. The initial task of organizing and annotating the genome information has more or less been solved, but it has turned out that, unfortunately, genome analysis and annotation through sequence comparison is not sufficient to make much sense of the blueprint of life [2]. Without the additional information derived from gene expression studies, proteomics and metabolomics experiments, many organismal systems cannot be understood. For example, the human genome contains approximately 30,000 genes, but the number of proteins, including all modifications, is estimated to be between ten and one hundred times larger than this [3].

Aside from the sheer numbers, the distribution of molecules within a cell and their presence over time in an organ or organism is a spatio-temporal problem and can only be understood if four-dimensional models are built, which integrate all known facts. Currently, most experiments only measure the difference between a "normal" population and one with a certain condition, but we expect that very shortly, time series experiments, which can follow the development of a condition will become the norm. Spatio temporal experiments are very hard to understand and their results can become very complex.

C.W. Sensen
Sun Center of Excellence for Visual Genomics, Faculty of Medicine, University of Calgary, 1150 HS, 3330 Hospital Drive NW, Calgary, AB, T2N 4N1, Canada
e-mail: csensen@ucalgary.ca; Homepage: http://www.visualgenomics.ca/sensencw

S. Krawetz (ed.), *Bioinformatics for Systems Biology*, DOI 10.1007/978-1-59745-440-7_31, 

We have begun to create a new bioinformatics (and medical informatics) environment, which is capable of handling this kind of information in the context of the organism. Our approach strives to combine all known facts with a surface model of the organism and provide the scientists, who use this model, with tools that allow them to manipulate the view of the results in ways that they already know. At no time do we require the end user to write programs in order to explore the data space.

This top-down approach is in contrast to many approaches, which try to understand the functional networks within cells through the reconstruction of the interaction of proteins with other proteins and/or small molecules. Unlike our approach, which allows the existence of knowledge gaps, the bottom-up approach needs complete information in order to succeed. Unfortunately, to date we have studied very few biological systems to the point where we have enough knowledge to describe the overall-system completely.

Our approach can be divided into four components. The usual two-dimensional display, which is currently in use for the visualization of most Bioinformatics results, is not sufficient for this kind of work. We have moved to the use of *virtual reality displays*, which provide a much higher resolution (billions of three-dimensional pixels, also called voxels). Using these displays requires the development of a new *software environment*, which allows users to manipulate the virtual objects they experience, with the tools that they ideally are already used to. For each organism which we study, we need to develop an *anatomical atlas*, onto which additional information can be mapped. The last component is the *functionality*, that the atlas needs to be equipped with in order to accurately reflect the biological or medical phenomenon studied.

## 31.2 Hardware

In 2002, we established the world's first Java 3D™-enabled CAVE automated virtual reality environment in our laboratory [4]. While CAVE systems, operating based on proprietary software environments, were in use approximately 10 years before we installed our machine [5], our system was the first one to operate based on a generic programming environment, as Java 3D™ can be used on almost any Java-enabled computer, from Windows and Macintosh personal computers, to Linux servers and mainframes. In addition, Java is a generic programming environment that is often taught as the first programming language to Computer Science undergraduate students.

CAVE automated environments consist of multiple stereo displays, from two to six walls; round displays have been described as well. To achieve a cubed display, which provides the full CAVE functionality, at least four walls are necessary, with three of them forming a U-shaped enclosure and the fourth wall being a floor display. Each wall is operated through a separate graphics unit (graphics card or processor on a multi-processor graphics card), ultimately providing the illusion of a three-dimensional view with the help of stereo goggles, which allow each eye to experience a slightly different perspective.

Our initial setup had four display units, as described above, and was driven by a mainframe computer (Sun Enterprise 6900 class). This setup is quite expensive (more than one million US dollars) and therefore only suitable for use at large organizations. We have since installed a much more modest system, which uses a single wall, a laptop and two passive stereo (polarized light) projectors. This system, which only costs approximately 35,000 US dollars, allows users to "look at" rather than immerse themselves within the models, while its portability and affordability allow the deployment in much smaller settings, such as an individual research laboratory or a rural hospital.

The interaction with the models displayed in the CAVE is facilitated through a tracking system, whose sensor is attached to the users glasses, which allows the computer to determine and monitor the position of the user, and the magic wand (similar to a three-dimensional joystick), which is held in the user's hand and can be equipped with various manipulation modes.

## 31.3 Software

The manipulation of objects in virtual reality requires a middleware layer, with which the end users interact. Within Java 3D, virtual objects are organized in a hierarchical fashion. This is quite beneficial for handling biological and medical information, as almost all of the data can be organized using controlled vocabularies (also called ontologies), which are essentially hierarchical expressions of the knowledge base.

Given that all interactions within the CAVE environment are mediated by the tracking device attached to the stereo glasses (which monitors the user's position with six degrees of freedom) and the wand (which looks like a joystick, but of course has six degrees of freedom), we had to create an entirely new approach to the interaction with objects. Our software system is called JABIRU, or *J*ava 3D *A*pplication *B*ehavior for *I*mmersive Virtual *R*eality *U*tilities [6]. JABIRU is a suite of interactions available to users in a virtual reality environment, as well as a mechanism for creating new interactions for Java 3D software tools. Through this middleware layer, users can manipulate virtual objects in many ways: (i) zooming in and out; (ii) rotating objects and positioning them after picking them up with the wand pointer; (iii) making parts of the model transparent or (iv) deconstructing the model piece by piece.

The importance of JABIRU lies in its ability to separate 3D interactive behaviors from the main Java 3D applications that use such behaviors. For example, the same mechanism of zooming into a 3D visual object and increasing its size ten-fold may be re-used by diverse bioformatics applications: from the studies of 3D structure of RNA and protein molecules [7], to tissue-level comparisons between anatomical samples [8], to very complex scenes involving hundreds and even thousands of visual components in the course of surgical simulations [9]. Furthermore, the separation between a Java application and the suite of available interactive behaviors allows one to "rewire" any or all such behaviors to different combinations of 3D joystick buttons, mouse movements, and/or keyboard key strokes.

Consequently, the users are able to transfer their bioinformatics Java 3D applications to different visualization systems or device configurations. This offers tremendous improvement over other existing approaches to virtual reality development, where tools are commonly created for only one device setup and require recompilation and/or code adjustment for another setup, hindering exchange of methods and ideas between researchers.

To achieve this functionality, JABIRU uses the Java 3D *ConfiguredUniverse* package, which enables the runtime configuration of Java 3D tools to the desired computational environment. The viewing environment and the program executables are separated, allowing for complete portability of the Java 3D code, regardless of the viewing environment that was initially used for development. It therefore allows the users to port existing molecular biology software packages, such as tools for viewing protein 3D structure in Protein Data Bank format [10], between an immersive CAVE, stereo projection system (active or passive), UNIX thin client or personal computer, which is just equipped with a regular two-dimensional display. Users can install and use such software locally, which is both faster and more secure than exploring remote data over the Web. Making prototypes like JABIRU, applicable to a wider range of systems, is a prominent component of our research.

As an example of a JABIRU-enabled functionality, we adopted several standard 2D graphical interactions into the CAVE virtual reality environment. To achieve this, we created a 3D wand mode that controls the mouse movement and generates mouse events from within the immersive CAVE. As a result, mouse interactions are now available through a regular CAVE wand device, which opens the way for using common Java 2D Swing GUI objects [11], without leaving the CAVE immersion. Most 3D bioinformatics applications contain 2D GUI elements, such as various menus, buttons, check boxes and radio buttons (because they were often created for use on a standard 2D desktop setup), and therefore require mouse-based interactions, which we have now made available.

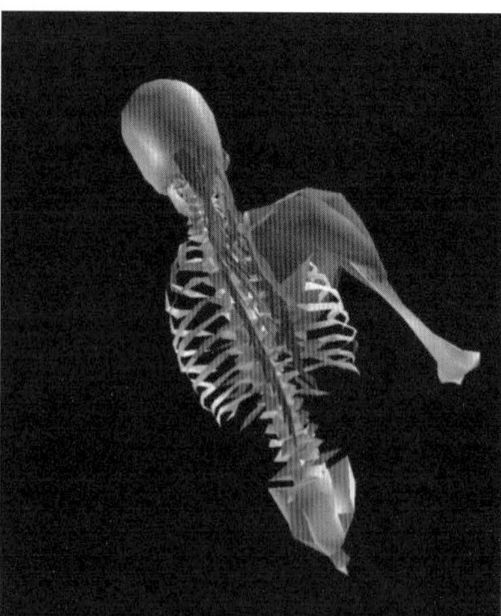
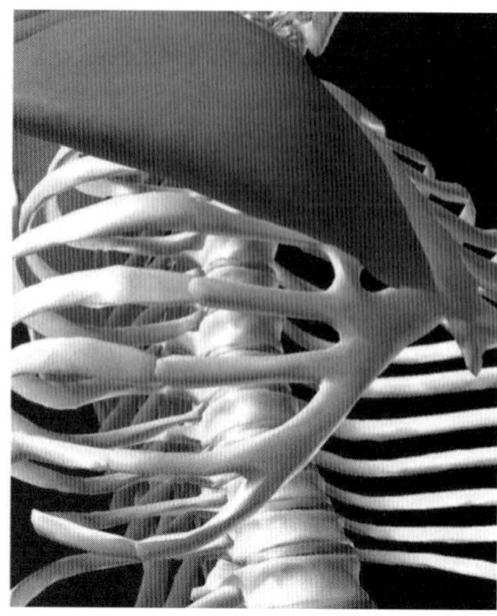

**Fig. 31.1** Level-of-detail management mechanism in a Java 3D environment. A model with a smaller polygon count (*left*) is used at a larger distance from the viewer, and is automatically replaced by a more detailed version (*right*) when the user zooms into the 3D scene (Copies of figures including color copies, where applicable, are available in the accompanying CD)

Our software environment allows the users to interact with objects in-natural ways, by pointing at them and subsequently using the wand controls for the manipulation. The key to the seamless operation is the level of detail management, which is a major component of our middleware software layer. Our complete model of the human body is constructed from hundreds of objects, each of which is rendered in very high resolution. To manipulate the displayed objects in real time, the displayed images need to be reduced to a total number of less than one million triangles at any given time. In addition, objects or parts of objects not visible to the users need to be deleted from the rendition and re-loaded into the graphics memory if they are becoming visible. This requires a level of detail management system, which we have implemented. Figure 31.1 shows two views of the same object, each time from a different angle, clearly showing the changing level of granularity, as objects move to the forefront and back, respectively.

## 31.4 Models

Most current anatomical models are derived directly from three-dimensional data such as the Visible Human [12] or from a combination of MRI and CAT scans. Typically, these kinds of models are volumetric, i.e., they are three-dimensional pictures, not object-oriented models. While it is relatively simple to create volumetric models, they are intrinsically meaningless, as they are not automatically indexed, and need experts for their interpretation.

We have chosen to create a surface-based atlas model, instead of a volumetric model, in order to be able to automatically use it for the mapping of spatio-temporal data [13]. The hierarchy of objects within the human atlas is derived from the *Terminologia Anatomica* [14]. The model is constructed in a way that allows the addition of new features seamlessly, as leaf nodes are added to the branches of the hierarchical tree. When using a surface model, automated indexing becomes possible. This will eventually allow the automated and unattended mapping of volumetric patient data onto the atlas.

The entire set of objects that now constitutes the complete atlas of the adult male human body was rendered by two artists. They used a number of sources for the model construction, from anatomy books, to the virtual human data, to information provided by experts for particular organs and the inspection of human cadavers at the University of Calgary's anatomy laboratory. Special emphasis was given to rendering of a body that reflects a live human, rather than a deceased individual (as in the case of the Virtual Human). Table 31.1 lists the major components of the overall system.

**Table 31.1** The structure of the *Realism* anatomical atlas

| Organ systems | Surface polygons | Separate 3D organs |
|---|---|---|
| Skeletal / articular systems | 385,107 | 214 |
| Muscular system | 853,199 | 1365 |
| Alimentary system | 712,628 | 26 |
| Respiratory system | 272,830 | 68 |
| Urinary system | 262,920 | 32 |
| Genital system (male) | 81,182 | 24 |
| Endocrine glands | 218,243 | 7 |
| Cardiovascular system | 1,490,485 | 154 |
| Lymphoid system | 127,563 | 29 |
| Nervous system | 1,673,322 | 338 |
| Sense organs | 265,681 | 46 |
| Integument | 84,408 | 32 |
| **Total** | 6,427,568 | 2,335 |

(Copies of tables are available in the accompanying CD.)

Each component of the atlas is broken down into subcomponents, which can be loaded into the computer separately. The level of granularity that is currently included in the objects varies. It was determined by the anatomical experts, who oversaw the creation of the respective objects. As the atlas is constructed "top down", we will be able to add additional granularity if the need arises. Figure 31.2 shows a close rendition of what users can expect to experience when displaying objects in the CAVE automated virtual reality environment.

We have begun to explore how automated indexing can be implemented within the context of our model. The hierarchical surface model provides a connection between a structure, which can be detected in a volumetric data set and the function of this structure, which is usually provided through the analysis of a volumetric data-set by an expert. The extraction of hierarchical objects from volumetric data, without the assistance of a surface model, is usually done through the use of an equidensity approach. As an example, an expert identifies a set of gray shades within TIFF file (most of these are 8-bit files with a total of 256 gray shades), which represents the bone, and the computer then uses this information to extract all the parts of the image with the same gray shades.

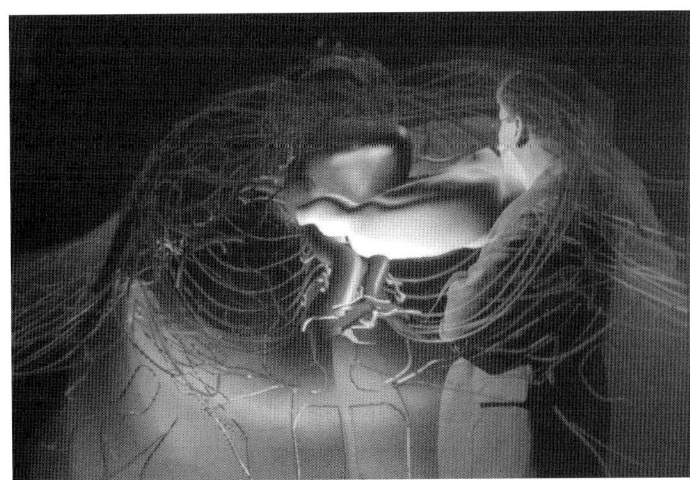

**Fig. 31.2** Typical user session in a Java 3D-enabled CAVE virtual reality environment, where users can immerse themselves into the spatio-temporal models of the studied organisms (Copies of figures including color copies, where applicable, are available in the accompanying CD)

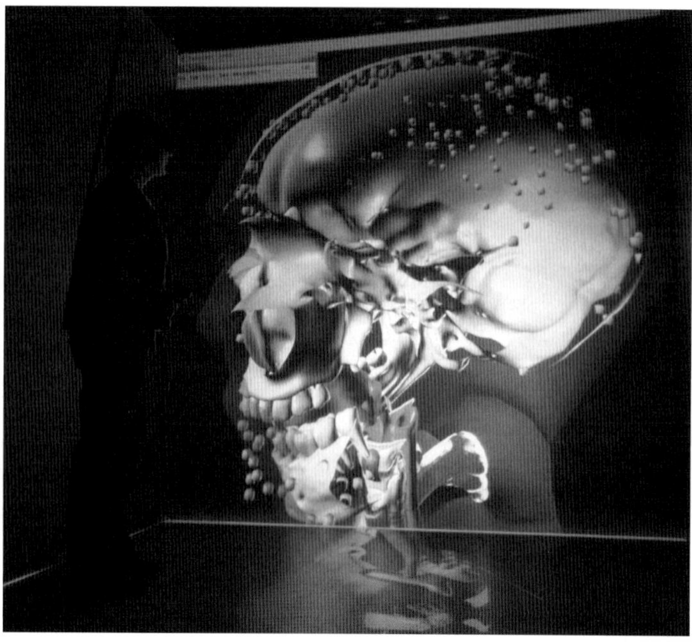

**Fig. 31.3** Working in the CAVE on the 3D image registration. The virtual scene contains the generic 3D skull model from the human body atlas, and a set of landmarks that outline the shape of a patient's skull. The landmarks are used for the registration of the two skulls, with the exception of the jaws; the atlas jaw is open whereas the patient jaw is closed (Copies of figures including color copies, where applicable, are available in the accompanying CD)

Typically, this approach tends to over-predict, and thus the resulting model is quite crude. In the case of bones, parts of the surrounding tissues, which in the set of TIFF files have the same gray shade values, are often included in the extracted model [15].

Using our approach of combining a surface model with the volumetric data, very few gray shades can be used to identify "landmarks", rather than attempting to extract the complete surface model from the TIFF files. The landmarks can subsequently be used to morph the surface model into the patient's shape. This approach is more likely to yield a high-quality model of the patient's phenotype than the direct and unguided extraction from volumetric data sets. A warping of the rendered 3D space will be required between the atlas and the individual patient's scan, which will have to be non-linear to achieve a satisfactory match. Figure 31.3 shows some initial results of our land marking strategy.

## 31.5 Modular Design

The atlas-based software system for functional genomics currently consists of several essential components: the virtual reality atlas of the organism's anatomy, data analysis modules for each type of biomedical data, whose results can be mapped onto the atlas (such as data from genomics, metabolomics, pharmaco dynamics, and medical imaging), and a component for the visual 3D scene management. The underlying data indexing mechanism ensures that the anatomical atlas can be cross-referenced with other biomedical data types. The atlas then serves as a common context for visual correlation and queries of the bioinformatics and medical data. The use of Java 3D makes the atlas-based system portable and platform independent. It also allows the incorporation of existing Java tools for biomedical data analysis and data mining into the system, as shown below.

To maintain the biomedical data, we use a modular design similar to a database federation [16,17,18]. Heterogeneous data from different sources remain in separate databases and an additional meta-component provides unified data access for the user. This approach presents a number of advantages: (i) easy integration of additional data sources; (ii) easy integration of additional data analysis tools, which are typically designed for a specific type of data and can therefore operate autonomously on a corresponding database component; (iii) support of conceptually different data types: images, numerical values, persistent objects, textual annotations; (iv) support of established data formats, which makes the source data immediately available to a user; (v) minimal need for data pre-processing, which is often subjective and error prone.

The usability of the system critically depends on both the quality and the efficiency that the meta-component can relate heterogeneous data to each other. We are building an XML-based cross-referencing mechanism, which utilizes the existing standards for XML links and the XML query language. Other benefits of using XML for metadata include the abundance of open source XML processing tools and easy web access for remote users. Our team has already used XML extensively in the development of the MAGPIE genome annotation system [19] and the Bluejay genomic browser [20]. We have recently published our XML-based strategy and it's connection to Web-based ontologies [21]. Ontology support provides a standard mechanism for the semantic annotation between data sets from different domains of knowledge [22].

The existing modules can be complemented easily through the addition of existing Java-based applications as well as new modules for the integration of additional data types or for the exploration of the integrated results.

## 31.6 Applications

The main outcome of the virtual reality approach to functional genomics is a software system for visual data analysis. The system focuses on two core areas: the exploration of the effects of biochemical variation on phenotypical outcomes, and the creation of standard mechanisms of cross-associations between diverse data types. Molecular processes are captured through the so-called *omics* data, such as data from genomics, proteomics, and metabolomics. The organism's phenotype is reflected in physical measurements, medical imaging, patient profiles and demographics data.

It is our goal to overlay the anatomical model with additional information, such as gene expression patterns, protein distributions or metabolomic information. For many reasons, this poses quite a challenge. Most of the additional data is multi-dimensional and efforts need to be made to reduce the dimensionality of the visualization. From the user's standpoint, the 3D virtual body becomes a convenient visual portal to biomedical data. However, given the complexity of the molecular data sets (typically with thousands of attributes), it is often not feasible to bring an entire original data set into the 3D scene at once. Handling of large volumes of data, such as gene expression data, may require the use of massive computing resources. The situation is further complicated when multiple molecular data sets are retrieved simultaneously.

Our solution is to data mine the *omics* data sets on the fly and retrieve only the resulting visual models. The models are created by domain-specific data mining tools in response to user activity. Data mining helps to reduce the size of massive life science data sets, prior to their integration into the visual context, providing a substantial performance gain, while preserving the data topology. As the user explores multiple organs, the system automatically locates and processes relevant *omics* data, merges new models into the 3D scene, and removes the models for organs that are no longer selected or visible to the user. This technology allows us to work with limited computing resources.

When starting a new case study in functional genomics, we initially limit the scope of our work to conducting visual exploration of existing data, finding associations between various data types, and identifying correlations evident in the collected data. Visualization aspects include not only the data per se, but also visual models derived from the data, e.g., 3D graphs, visualization of patterns, abstract maps, and connectivity networks. A typical example of such an exploration includes filtering down the population and animating several data types simultaneously. Filtering may require setting the ranges for multiple parameters such as age, gender, acceptable lymphocyte concentration, blood type, and disease subtype. Once the data volume is reduced through the filtering process, the collected molecular and imaging data can be visualized simultaneously. For example, the user may observe that a group of similar genomic profiles (as identified by clustering gene expression microarray data) comes from a group of patients of the predominantly relapsing-remitting subtype of multiple sclerosis, which may indicate a pattern worthy of further studies.

After the preliminary patterns are visually identified, pattern discovery, data mining, and predictive modeling tasks should follow. Building on the visual associations found in the visualization phase, the goal is to establish statistically significant correlations and hypothesise about the cause-effect links between the *omics* and the phenotype data types. Naturally, correlation does not necessarily mean causation, but may nevertheless lead to valuable hypotheses. As with any pattern-discovery initiatives, the simulation phase is inherently unpredictable: it is generally not known *a priori*: (a) to what extent the correlation patterns within the data exist and are learnable via standard methods; (b) how complex the simulation models should be, in order to match accurately the data already collected; or (c) to what extent the simulated results will be validated by future experiments with sufficient statistical significance.

The necessary steps to model such hypotheses involve the recovery of relevant genetic pathways via genomic data analysis and literature searches; and the enabling of visual manipulations of the pathways, annotations, and other components. For example, the user should be able to manipulate a pathway model by visually "knocking out" genes or changing their expression, and then observing the *predicted* effects of these manipulations on the dynamics of a disease in question. Subsequent steps then involve the validation of predictive accuracy with respect to real experimental data.

One of the advantages of using Java for our system is that other Java-based applications can be rapidly bound into the toolkit. For example, the TIGR Multi-Experiment Viewer (MeV) now functions from within our system, allowing users to combine gene expression data with the anatomical model [23]. TIGR MeV implements normalization, clustering, classification, statistical analysis, and visualization of gene expression data. We provide a visual mapping between the TIGR MeV output and the Java 3D visual scene, so that data mining results can be piped directly into the displayed anatomical structures (Fig. 31.4).

We primarily use unsupervised machine learning methods [24] to model gene expression patterns. Generally speaking, unsupervised methods separate massive data sets into much smaller groups. By extracting a small descriptive model for each group, the user can instantly reduce the

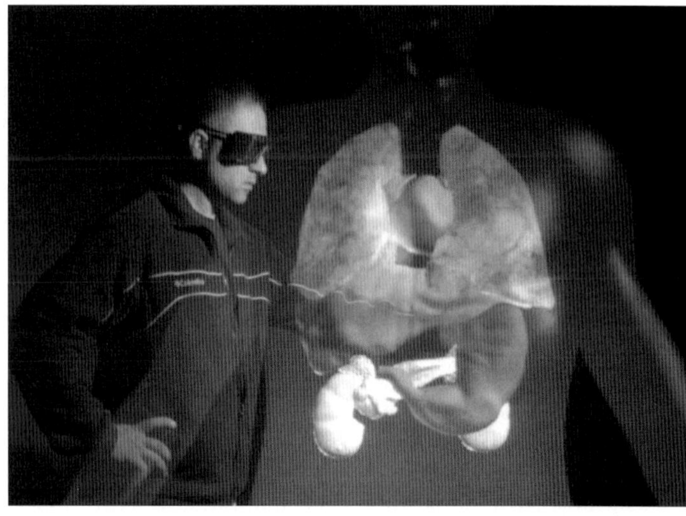

**Fig. 31.4** Gene expression profiles are analyzed in the TIGR Multi-Experiment Viewer. Patterns of interest are visualized as animated color maps directly on the chosen organs of the virtual body (Copies of figures including color copies, where applicable, are available in the accompanying CD)

amount of visual input by several orders of magnitude, e.g., from tens of thousands of gene expression log-ratios to just a dozen prominent trends. We believe that this approach to the reduction-based retrieval is more appropriate than, for example, data sampling; it avoids the user-introduced bias usually associated with sampling, and therefore provides access to a more standardized interpretation of the results. The data mining model creates a high-level overview, but detailed features are also available for drill-down explorations.

Various combinations of selected organs give rise to different data mining results. Although local sets are relatively static (unless the user uploads new data into the system), data mining models built on various combinations of organs may differ substantially. For example, gene expression in the urinary system differs drastically from gene expression in the central nervous system, and so do the respective statistical models. Depending on user activity, the system derives high-level representations of the relevant data on the fly, which are then integrated dynamically into the virtual reality scene.

Similar to the gene expression studies, we can map the distribution of small molecules within organs. Using the Java 3D atlas of the human body, we can visualize chemical reactions between proteins, enzymes, and other chemicals in an anatomical context. Most importantly, we are moving from modeling generic molecular processes in a human body towards modeling these processes for each individual patient.

As a first example, we have developed a prototype visualization system for biochemical data mapping using pharmaco dynamics data. Using aspirin (acetylsalicylic acid) as an example, we were interested in following the events after a patient takes it to relieve a headache (Fig. 31.5). After the initial uptake, it takes about four hours for each 500 mg pill to be digested. The dissolved aspirin first appears in the stomach, subsequently migrates through the blood vessels and ultimately reaches the brain. The concentration changes can be visualized using coloration time series, changing textures, and other visual cues as part of the virtual patient body. Once in the digestive system, the main content of the aspirin pill is gradually converted into salicylic acid [25]. As the aspirin is being digested, the concentration of the salicylic acid rapidly rises, as shown by the various visual time series. At the same time, both chemicals are absorbed from the digestive organs into the blood stream. Once in the blood, the salicylic acid breaks into four additional chemical by-products: salicyl acyl glucuronide, salicyluric acid, gentisic acid, and salicyl phenolic glucuronide. By this time, the patient's pain subsides, as the substances finally reach the receptors in the brain. Eventually, all chemicals are filtered from the blood into the urinary system, which excretes them from the body.

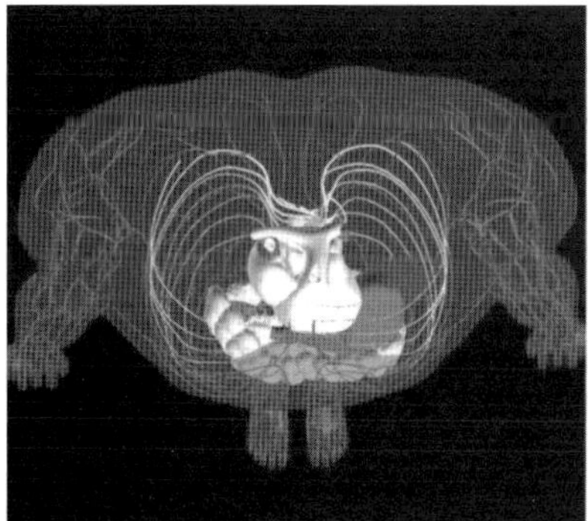
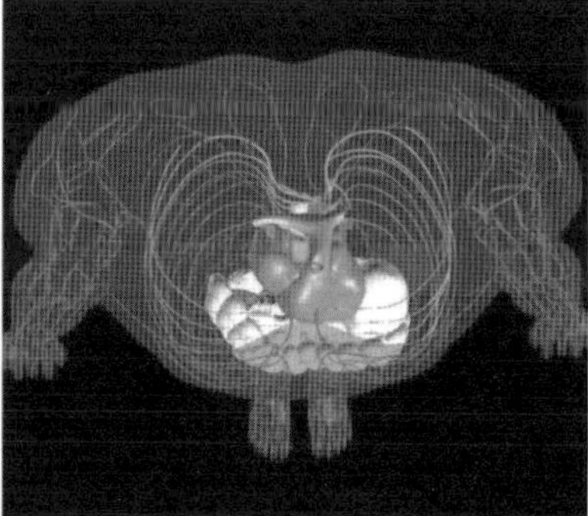

**Fig. 31.5** Aspirin metabolism in selected organ systems of a human body atlas. Initially, the concentrations of the acetylsalicylic acid and the salicylic acid are high in the digestive organs, especially in the stomach, as shown by its deeper shade (*left*). After about 30 minutes, the two acids are concentrated predominantly in the blood, and are visualized using deeper shades of the heart and the central arteries (*right*) (Copies of figures including color copies, where applicable, are available in the accompanying CD)

Visualizing the metabolic activity allows us to observe the biochemical changes directly within a (virtual) patient's body. If advanced imaging information is coupled with this information, we are even able to observe the morphological and histological changes, which occur at the same time. We are now collecting data on complex drugs, in order to advance to more elaborate pharmaco dynamic models.

The analysis of the differences between the predicted and the observed metabolite concentrations may: (a) allow validation and course correction in pharmaceutical experiments, (b) help assess the applicability of theoretical models in a given line of research, and (c) possibly lead to other insights. Thanks to the Java language portability, even the smallest hospital, patient clinic, or classroom and will be able to use our Java 3D system to model the effects of drug therapies.

Similarly, we use Java 3D technology to model some of the main anatomical manifestations of genetic diseases and other pathologies. For example, we have begun to visualize the behavior of various lymphocytes and macrophages, which has important applications to many diseases of the human immune system. In particular, our collaborators use this technology in their exploration of multiple sclerosis [26], where changes in blood cell dynamics are important indicators of the on-going nerve cell damage, such as demyelination and axonal loss. In contrast, physical evidence may only be available after substantial damage to nerve tissues has already occurred, as evident from MRI scans.

Our goal is to make a time-series model of molecular and cell proliferation for each pathway associated with a genetic condition we explore, based on the clinical data. The result of this research activity will be a mechanism of visualizing spatio temporal patterns of cell proliferation in a patient's organ or body fluid. In a simple scenario, this virtual model can then be correlated with the additional clinical data on drug dosage, represented as one of the controllable parameters in our visual model. We will then focus on the discovery of the associations between different drug dosages on one hand (such as beta interferon for multiple sclerosis), and different disease patterns on the other hand. Similarly, genomics data collected through RT-PCR, Western Blot, and microarray gene expression analysis will be incorporated into the system.

Naturally, personalized medicine does not stop with molecular and biochemical therapies but also includes macro-scale treatments, such as surgeries. Through automated indexing, we will soon be able to re-shape the idealized atlas of the human body into the form of an individual patient. The possibilities of the combination of volumetric data with a surface model are almost endless. They include the study of a particular disease, the normalization of information from large patient cohorts, the planning of surgical procedures and maybe in the future even new methods to assist the execution of robotic surgery.

In summary, we will soon be able to upload any individual's advanced imaging information, merge it with the biomedical information and patient history and explore the resulting model in real time, with a very high resolution and in a way is familiar to physicians. We hope that this will overcome the fragmented medical landscape, which is reality today. If someone is unfortunate enough to develop cancer, they are currently forced to see several specialists, who may give them conflicting advice on how to treat their condition.

In the future, we hope that all specialists will see the same model, which includes all the data collected from the patient and be able to discuss the treatment jointly. This will lead to the personalized medicine that we are currently forecasting and hopefully will provide us with the insights, which are necessary to understand the mechanism of complex genetic diseases and other ailments for which no treatment is currently available.

## 31.7 Challenges for the Future

Thus far, the number of true spatio temporal experiments that could be used in connection with our models is scarce. While we expect this to change over time, this currently poses a major challenge for the usefulness of our work. Most experiments to date have only attempted to characterize the difference between "normal" and "diseased", or the very beginning and the end status of a phenomenon. This is, of course, insufficient for the understanding of the mechanisms and networks that are underlying the observed characteristics.

To date, we have created a model of the male adult human anatomy. We are planning to expand this collection to include the female body, as well as models of children to reflect the entire process of human development. This naturally requires much more work, and eventually will lead to a system that can seamlessly morph between the fetus and the adult stage.

Most biological and medical research uses model organisms. Similar to the creation of surface models for the various stages of the human body, we need to create surface models for the animals that we work with. Naturally, this includes rats and mice, and other vertebrates. Eventually, we will have to continue our work to also include renditions of other models, from bacteria to plant models such as *Arabidopsis*, *Brassica* or rice. Thankfully, the software environment that we have developed thus far is generic and can rapidly be expanded to include the models proposed above.

The greatest challenge though is presented by the question on how to expand the current models to the cellular and sub cellular level. Given that a single human finger has more than 1 billion cells, it is quite clear that no current computer system would be capable of rendering a model with this kind of resolution. Level-of-detail (LOD) management is one possible direction, in which we create only one elaborate model of a cell, but replace it with highly simplified, symbolic "placeholder" models for higher zoom levels. The data is thus separated into a hierarchy of layers and the system allows the user to switch between the layers, refining both the visual precision and the behavioral veracity of the 4D component. The LOD mechanism is used with great success, for example, in most of the geographical information systems [27]. In Java 3D, a visual model is internally represented by a "scene graph", a tree-like hierarchical structure. The nodes of the scene graph correspond to visible objects, light sources, geometric transformations, and interactive behaviors that define the 3D model.

To represent cellular processes, the data layering mechanism would need to use runtime pruning and re-growth of the data within this hierarchical structure, based on a system of semantic rules. The incoming requests for cellular-level data and models are placed within the atlas using the semantic attributes, such as the anatomical term from the standard *Anatomica Teminologia* ontology (e.g., *corpus callosum*), the time stamp for time series data, and the required level-of-detail zoom level. A semantic 4D zoom [28] will retrieve and show high-resolution details only within a small region of interest, thus balancing the depth of the detail against the breadth of the region of interest.

The reduction of massive life science data sets prior to their integration into the visual context also poses a serious challenge. Our current approach is to use data mining in order to reduce the data sets to a small number of groups, descriptors, or trends. A promising complementary approach is outlier detection. Instead of integrating an entire molecular data-set into the atlas, the users may focus on extracting only a small "interesting" sample of the data that represents pathology or an unusual behavior. This approach may provide substantial performance gains during data integration into the organism's atlas. Outlier detection is a prominent and rapidly developing area of data mining. However, one must take into account the potentially very high dimensionality of the data (e.g., thousands of attributes in microarray gene expression data) and develop the algorithms that are resistant to the curse of dimensionality [29]. The main challenge is in detecting "interesting" rare anomalies, and filtering out what can be called "normal" data, such as housekeeping genes in gene expression experiments. We envisage many situations where only a few outliers are of interest to the investigators.

Finally, haptic technology is emerging as one of the most intriguing new developments with a huge application potential [30]. By simulating touch and force feedback, haptic devices delve into a completely new realm of the user's senses, which has until now been under-utilized. As just one example of what will likely become possible in the coming years, haptic applications may allow surgeons not only to see but also to feel the response of different anatomical tissues under the scalpel, and therefore allow more precise surgical simulations, and perhaps even remote surgeries. This is an exciting new frontier in bioinformatics.

**Acknowledgments** The authors would like to thank Quang Trinh, Benedikt Hallgrimsson, David Wishart, George Robertson, Jane Shearer, Jana Rieger and Johan Wolfaardt for helpful discussions, as well as Masumi Yajima for her art work. This research has been supported by the Government of Canada and the Government of Alberta through the Western Economic Partnership Agreement; by Genome Alberta, in part through Genome Canada, a not-for-profit corporation which is leading a national strategy on genomics with $375 million in funding from the Government of Canada; by the iCORE/Sun Microsystems Industrial Research Chair program; by the National Research Council of Canada's Industrial Research Assistance Program; as well as by the Alberta Science and Research Authority, Western Economic Diversification, the Alberta Network for Proteomics Innovation, and the Canada Foundation for Innovation. The Kasterstener Realism 3D atlas of human anatomy is available from Kasterstener Publications, Inc.

## References

### *Introduction*

1. Fleischmann RD, Adams MD, White O, Clayton RA, Kirkness EF, Kerlavage AR, Bult CJ, Tomb JF, Dougherty BA, Merrick JM, McKenney K, Sutton G, Fitzhugh W, Fields C, Gocayne JD, Scott J, Shirley R, Liu LI, Glodek A, Kelley JM, Weidman JF, Phillips CA, Spriggs T, Hedblom E, Cotton MD, Utterback MR, Hanna MC, Nguyen DT, Saudek DM, Brandon RC, Fine LD, Fritchman JL, Fuhrmann JL, Geoghagen NSM, Gnehm CL, McDonald LA, Small KV, Fraser CM, Smith HO, Venter JC. Whole-genome random sequencing and assembly of Haemophilus influenzae Rd. Science 1995;269:496–512.
2. Ideker T, Galitski T, Hood L. A new approach to decoding life: Systems Biology. Annu Rev Genomics Hum Genet 2001;2:343–372.
3. Godovac-Zimmermann J, Kleiner O, Brown LR, Drukier AK. Perspectives in spicing up proteomics with splicing. Proteomics 2005;5(3):699–709.

### *Hardware*

4. Sensen CW. Using CAVE technology for functional genomics studies. Diabetes Technology and Theraputics 2002;4:867–871.
5. DeFanti T, Sandin DJ, Cruz-Neira C. A "Room" with a "View". IEEE Spectrum, 1993;30(10):30–33.

### *Software*

6. Stromer JN, Quon GT, Gordon PMK, Turinsky AL, Sensen CW. Jabiru: harnessing Java 3D behaviors for device and display portability. IEEE Computer Graphics & Applications 2005;25:70–80.
7. Quon GT, Gordon P, Sensen CW. 4D bioinformatics: a new look at the ribosome as an example. IUBMB Life 2003;55(4-5):279–283.
8. Cooper DM, Turinsky AL, Sensen CW, Hallgrimsson B. Quantitative 3D analysis of the canal network in cortical bone by micro-computed tomography. Anat Rec B New Anat 2003;274(1):169–179.
9. Liu A, Tendick F, Cleary K, Kaufmann C. A survey of surgical simulation: applications, technology, and education. Presence: Teleoperators and Virtual Environment 2003;12:599–614.
10. Berman HM, Westbrook J, Feng Z, Gilliland G, Bhat TN, Weissig H, Shindyalov IN, Bourne PE. The Protein Data Bank. Nucleic Acids Res 2000;28(1):235–242.
11. Elliott J, Eckstein R, Loy M, Wood D, Cole B. Java Swing. O'Reilly Media, Inc.; 2002

### *Models*

12. Ackerman MJ. The Visible Human Project. J Biocommunications 1991;18:14.
13. Lajeunesse D, Edwards C, Grosenick B. Realism: A study in human structural anatomy. Red Deer: Kasterstener Publications Inc.; 2003.

14. Federative Committee on Anatomical Terminology. Terminologia Anatomica – international anatomical terminology. Stuttgart, New York: Thieme; 1998.
15. Turinsky A, Sensen CW. On the way to building an integrated computational environment for the study of developmental patterns and genetic diseases. Int J Nanomedicine 2006;1(1):89–96.

## *Modular Design*

16. Wiederhold G. Mediators in the architecture of future information systems. IEEE Computer, 1992;25(3):38–49.
17. Sheth AP, Larson JA. Federated database systems for managing distributed, heterogeneous, and autonomous databases. ACM Computing Surveys 1990;22:183–236.
18. Haas LM, Lin ET, Roth MA. Data Integration through database federation, IBM Systems Journal 2002;41(4):578–596.
19. Gaasterland T, Sensen CW. Fully automated genome analysis that reflects user needs and preferences — A detailed introduction to the MAGPIE system architecture. Biochimie 1996;78:302–310.
20. Turinsky AL, Ah-Seng AC, Gordon PMK, Stromer JN, Taschuk ML, Xu EW, Sensen CW. Bioinformatics visualization and integration with open standards: The Bluejay genomic browser. In Silico Biology. 2005;5(2):187–198.
21. Horrocks I, Patel-Schneider P, van Harmelen F. From SHIQ and RDF to OWL: The making of a web ontology language. J. Web Semantics 2003;1:7–26.
22. Noy NF. Semantic integration: A survey of ontology-based approaches. SIGMOD Record 2004;33:65–70.

## *Applications*

23. Saeed AI, Sharov V, White J, Li J, Liang W, Bhagabati N, Braisted J.M.K, Currier T, Thiagarajan M, Sturn A, Snuffin M, Rezantsev A, Popov D, Ryltsov A, Kostukovich E, Borisovsky I, Liu Z, Vinsavich A, Trush V, Quackenbush J. TM4: a free, open-source system for microarray data management and analysis. Biotechniques 2003;34:374–378.
24. Quackenbush J. Computational analysis of microarray data. Nature Reviews 2001;2:418–427.
25. Needs CJ, Brooks PM. Clinical pharmacokinetics of the salicylates. Clin Pharmacokinet 1985;10(2):164–177.
26. Robertson GS, Crocker SJ, Nicholoson DW, Schulz JB. Neuroprotection by the Inhibition of Apoptosis. Brain Pathology 2000;10:283–292.

## *Challenges for the Future*

27. Burrough PA, McDonnell RA. Principles of Geographical Information Systems (Spatial Information Systems). Oxford University Press; 1998.
28. Loraine AE, Helt GA. Visualizing the genome: techniques for presenting human genome data and annotations. BMC Bioinformatics 2002;3(1):19.
29. Aggarwal C, Yu P. Outlier detection for high dimensional data. Proc SIGMOD'2001, 2001;30(2):37–46.
30. Ernst MO, Banks MS. Humans integrate visual and haptic information in a statistically optimal fashion. Nature 2002;415(6870):429–433.

## Key References

DeFanti T, Sandin DJ, Cruz-Neira C. A "Room" with a "View". IEEE Spectrum, 1993;30(10):30–33.
Federative Committee on Anatomical Terminology. Terminologia Anatomica – international anatomical terminology. Stuttgart, New York: Thieme; 1998.
Quackenbush J. Computational analysis of microarray data. Nature Reviews 2001;2:418–427.

## Web Resources

www.kasterstener.com

# Chapter 32
# Systems Biology of Personalized Medicine

**Craig Paul Webb and David Michael Cherba**

**Abstract** To achieve the promise of individualized molecular-based medicine, the application of informatic tools to the entire hierarchy of biological system interactions and dynamics will be required in order to promote the effective discovery, validation and application of new diagnostic and treatment strategies in a real-time environment. As the field of biomarker discovery continues to unravel the underlying molecular mechanisms of diseases, the utility of the acquired and expanding knowledge lags far behind. Systems Biology represents a pivotal component of the personalized medicine workflow through its ability to consolidate complex data and knowledge into definable networks, and reproducibly identify key convergence/divergence points representing the biomarkers of interest. In this chapter, we will provide a brief update in the field of personalized medicine, and how Systems Biology tools can be used to support biomarker discovery. This chapter emphasizes the potential utility of Systems Biology for the prediction of network-based treatments in oncology, using empirical biomarker data sets in conjunction with knowledge and pre-existing drug resources.

**Keywords** Pharmacogenomics · Systems Biology · Cancer · Informatics · Biomarkers · Individualized therapy

## 32.1 Introduction: The Challenge – The Complexity of Biological Systems

Before providing an overview relating to the potential utility of Systems Biology in personalized medicine, it is essential to clearly define some terms within this framework. Physicians, after all, have personalized medicine for decades using their continued education and acquisition of knowledge in the field to optimize care for their individual patients. For this purpose, they use various objective and subjective inputs such as signs, symptoms and test results. Systems Biology is an equally ill-defined term derived from an emerging discipline. According to the latest in a long string of classifications, Systems Biology characterizes an integrated approach by which scientists study pathways and networks which touch all areas of biology. While the semantic definitions continue to represent an area of debate, it is far more important to consider how the discipline of Systems Biology can enable individualized medicine at the molecular level, thereby transforming healthcare into a more predictive and integrated discipline. Essentially, through Systems Biology it should be possible to leverage recent improvements in technology and the vast expansion of acquired knowledge to redefine the wiring of diseases, and design network-targeted therapies based upon the provided input of comprehensive and standardized molecular data. When applied appropriately,

C.P. Webb
Program of Translational Medicine, Van Andel Institute, 333 Bostwick Avenue, Grand Rapids, MI, 49503, USA
e-mail: craig.webb@vai.org

S. Krawetz (ed.), *Bioinformatics for Systems Biology*, DOI 10.1007/978-1-59745-440-7_32, 

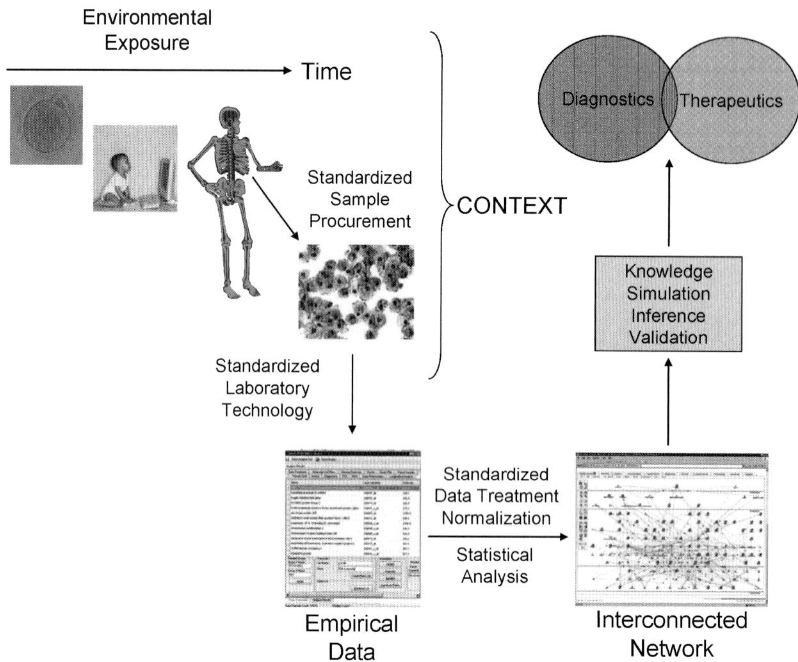


**Fig. 32.1** A simplified overview of the role of Systems Biology in the area of individualized molecular-based medicine. As described in the text, the cellular systems are determined by intrinsic molecular determinants (including the genetics of the cell) in conjunction with external environmental factors that provide a fluctuating temporal and spatial context. Systems Biology can assist in the conversion of standardized empirical data to network information that, in conjunction with pre-existing drug-target knowledge, can be used to predict pathway-directed treatments (Copies of figures including color copies, where applicable, are available in the accompanying CD)

Systems Biology converts empirical data to network information and, coupled with the knowledge of biomarker-drug associations, is highly relevant to optimization of targeted therapeutics (Fig. 32.1).

After years of extensive basic research focused on furthering our understanding of the etiology and progression of disease at the molecular level, it is now generally accepted that virtually all phenotypes, which may ultimately manifest themselves as signs or symptoms of a disease, are governed by the molecular constituents of individual cells. These cells interact in the form of tissues, organs, and the physiological organism as a whole. Molecular constituents increase in complexity, from the DNA sequence and chromosomal organization, through gene regulation and transcription at the DNA–RNA interface, to protein translation at the RNA-protein boundary. This classical transfer of genetic code to functional protein is further complicated by protein modifications, protein-protein interactions, and protein-chemical (metabolomic) reactions. This diversity can provide the capacity to form essentially infinite "systems" within an individual cell or collection of cells working in concert; i.e., the system is highly dependent on the context, which includes both temporal (time dependent) and spatial (location dependent) factors. It is unlikely that any cellular system will be exposed to the identical external environment due to contextual fluctuations. A high degree of intracellular complexity is required to provide the ultimate plasticity and allow the individual cell to adapt to or exploit extracellular cues and ensure the long-term survival of itself and/or the multi-cellular organism. The multicellular organism requires a highly coupled coordination between its cellular sub-systems, which cannot be excessively rigid in the face of a selective environment that drives micro and macro-evolutionary processes (1). As multi-cellular organisms evolved from unicellular precursors, the degree of complexity within the intracellular networks increased to permit adaptation to, or exploitation of, the changing micro and macro environment

(2). The charge of Systems Biology is to understand how this structured complexity can be simulated to derive predictive models for future healthcare applications (3). In conjunction with classical control system modeling, the development of predictive models is feasible using system topology and transfer functions. For example, a regulatory NF-κβ module was constructed using 15 molecular variables and a classical control transfer model (4). In the new era of individualized molecular medicine, the task is to reduce the complexity to practical applications, whereby the molecular data in conjunction with traditional assessments of physiological systems, can be used together to predict, detect, diagnose, treat and/or monitor disease.

## 32.2 The Added Complexity of Cancer

Within oncology, the system complexity is further exacerbated due to the inherent genomic instability of a tumor cell. Alterations in DNA repair mechanisms at the onset of tumorogenesis, essentially instigate an accelerated micro-evolutionary process, which naturally results in further heterogeneity within the intrinsic micro-systems (5). Furthermore, tumors represent a complex interaction between malignant cells and their host micro-environment, which can include a multitude of proteomic and metabolomic constituents and other cellular components such as endothelial, stromal and inflammatory cells (6,7). Thus, even if one were able to identify the complete molecular-chemical systems within the individual cells of a tumor mass, no two cellular systems would be identical at a given point in time. Indeed, the same cell would likely exhibit differences at multiple time points, and variation would also be prevalent between spatially separated tumor cells within the same population (such as the primary tumor mass), and between side populations (such as metastases within different anatomical sites). The various organ or tissue sites hosting primary tumors and their resulting metastases provide contextual complexity, which coupled with the underlying genetics of the tumor, drive the interaction between the tumor-host cellular systems that determines the path of disease progression. Thus, while the molecular genetics of a tumor cell may promote an accelerated micro-evolutionary process within a population of tumor cells (8), the interplay between each tumor cell and its micro-environment provides a constantly shifting context and selective milieu that resembles a high dimensional form of conditional probability. Current efforts to sequence the cancer genome will naturally identify the predominant genetic changes within the tumor system as a whole, and may provide some candidate genes, proteins and/or networks that are frequently utilized by specific tumor types (9,10). However, while in other diseases the genome is relatively stable, the genetic instability in conjunction with the invariable changes in environmental context within a tumor, provides arguably the most significant challenge for individualizing treatments in oncology towards a cure or chronic control of the disease. The ability of Systems Biology to model interactions between multivariate molecular and chemical components at any level, and provide a means to identify the driving network aberrations associated with the malignant phenotype, is pivotal to these efforts.

## 32.3 Personalized Medicine — The Objectives

The advent of pharmaco-genomics (a subset of personalized medicine), has stemmed from the realization that individuals and their disease(s) demonstrate a differential response to various medications that are governed by heterogeneity at the genomic level. The level of interest in this field has been driven by a number of factors including public demand (11), regulatory agencies (12), and the possible financial incentives associated with biomarker-drug co-development within the pharmaceutical and biotechnology industry (13). Coupled with enhanced computational power and the explosion in new technologies, allowing us to measure the multivariate components of cellular systems (DNA, RNA, proteins and metabolites) at an unprecedented density, personalized medicine is now a reality with many highly active areas of research. However, the lack of expertise and data integration across medical and scientific disciplines, including those engaged in Systems Biology, represents a significant challenge to achieving the practical objectives of individualized, molecular-based medicine.

## 32.4 Biomarkers in Practice – The Role of Systems Biology

A biomarker is generally defined as a biological feature (DNA, RNA, protein or metabolite) that can be objectively measured and used to predict or detect the progression of disease or the effects of treatment. From an idealistic standpoint, biomarkers could be used to (i) predict the onset of a condition prior to occurrence; (ii) detect a condition in its infancy (so called occult disease); (iii) accurately diagnose the condition at presentation and predict prognosis; (iv) predict the appropriate treatment for a condition; (v) detect target modification within the disease tissue or an appropriate surrogate tissue; and/or (vi) monitor clinical response (efficacy and/or toxicity). Within this natural continuum of care, a number of univariate biomarkers and in some cases multivariate composites, have been successfully translated into clinical practice within oncology. For example, genetic screening of BRCA1/BRCA2 mutations in cellular DNA, from females with a family history of breast or ovarian cancer, can now be performed to identify those at high risk of developing tumors, thereby providing options for prophylactic surgery (14). Various isoforms of the liver cytochrome p450 enzymes are known to be involved in drug metabolism, and ultimately govern how individual patients respond to a number of drugs prescribed for various indications including cancer (15). The outlier patients, who either over- or under-metabolize some drugs often experience serious adverse events or exhibit poor drug efficacy due to differential rates of drug or pro-drug metabolism. The first FDA-approved multivariate DNA microarray that assesses the mutations in several cytochrome p450 genes, is now available to physicians and their patients, although the medical reimbursement for this test remains a challenge (16). Recently, the FDA changed the label for the drug irinotecan, commonly used in metastatic colon and lung cancer, to include genetic testing for UGT1A1 DNA mutations (17). Some genetic isoforms of UGT1A1 confer an increased risk of drug-induced toxicity. These and other such genetic tests are performed on normal cellular DNA, and hence provide a life-time risk assessment for physicians to consider when determining the required frequency of patient assessments and/or therapeutic selections.

Several biomarkers associated with occult and early progressive disease are also in mainstream clinical use, and include genomic, proteomic and metabolomic markers. These are typically assayed from readily attainable physiological fluids such as blood, urine, saliva, or stool and may be measured repeatedly over time to determine the rate of change in the level of the biomarker. The rate of change is typically a better predictor of early disease onset and/or progression, when compared with the absolute biomarker level at any given time point (18). Perhaps, the most renowned protein marker in clinical oncology is Prostate Specific Antigen (PSA), which represents a relatively non-specific blood-based marker for prostate cancer (19). In some cases, potential protein markers of early disease detection can be inferred from genome-wide microarray experiments, comparing normal and diseased tissue using some rudimentary aspects of Systems Biology. For example, in a comparative analysis of the gene expression profiles of malignant and normal mesothelium from patients with mesothelioma, osteopontin (OPN) was identified as a known extracellular and plasma protein that was increased in aggressive tumors, and appeared to represent a common terminal output node within a highly enriched molecular network (20). Screening of blood OPN levels in patients at a relatively high risk of developing this disease may hold some promise for identifying those with occult disease (21).

At the time of disease presentation, it is critical to provide a comprehensive work-up of the disease to provide a definitive diagnosis and prognosis, which are in turn used to determine the treatment course. Typically, pathologists utilize microscopic tissue morphology or specific protein/RNA stains, to characterize the disease and these may provide some prognostic information in conjunction with macroscopic observations. More recently, molecular diagnostics that simultaneously assess multiple biomarkers have been developed to improve on the traditional methods of diagnosis and prognostication. The most advanced of these is arguably in breast cancer, where gene expression signatures within tumors have been shown to characterize those at high risk of recurrence post surgery (22). Commercial tests are now available that can identify those patients at the lowest and highest risk of recurrent disease (23). There are numerous studies in other tumor types that have identified putative molecular signatures associated with differential phenotypes and/or clinical outcomes. While the standardization of the various genome-wide technologies under evaluation is critical to ensure raw data accuracy (24), Systems Biology tools can greatly enhance the data analysis component that ultimately plays a fundamental role in defining tests that will perform with the necessary accuracy in the clinical setting. Network-based diagnostic optimizations, in which components of the same pathway are assessed collectively and not treated as independent data elements, represent an exciting opportunity for Systems Biology. Analysis of disparate data-sets at the network level has revealed consistencies across studies that were not evident from the analysis of the individual network components (25). In this form, Systems Biology can represent a means of data reconciliation, and can be used to select pivotal biomarkers for further validation and potential diagnostic development. While individual elements within a network are often redundant and sensitive to disparity in experimental techniques, the network as a whole represents a unifying theme that is representative of the underlying biology of the disease.

Once an accurate disease diagnosis and associated prognosis has been established, the selection of an optimal therapeutic is the next natural step. Currently, the vast majority of treatments are selected based upon a large cohort of clinical trial data in which a portion of patients with the "same" disease phenotype were shown to benefit in some fashion after administration of the treatment. The degree of benefit in the mythical average patient is a relative term, and treatment approval very much depends on the "gold standard" for the given disease population at the time of the pivotal trial(s). Drugs are currently approved with a specific set of phenotypic indications based upon the diseases studied in the clinical trial(s). Physicians are able to use their best judgment as well as several other key factors including, known drug-drug interactions or contraindications to personalize the patients' treatment. As it has become clear that drugs, which after all target molecular entities (typically proteins) with varying degrees of selectivity display differential patient toxicity and efficacy depending on the molecular makeup of the patient and/or their disease, there has been a shift in thinking towards utilizing molecular-based indications and diagnostics to improve the probability of the treatment success.

Historically, disease and drug-specific diagnostic tests may be ordered prior to the administration of a restricted class of drugs. Perhaps, the best example of this remains trastuzumab, which is recommended for the treatment of tumors that are positive for the ERBB2 molecular target (26). While plainly intuitive, a series of clinical trials have now demonstrated that ERBB2 positive tumors showed an improved response to trastuzumab relative to ERBB2 negative tumors, as determined using assays that measure gene amplification or protein over-expression (27). Indeed, the drug would not have likely been approved for general use in breast cancer without this companion diagnostic test, which enriches for a sub-population that is more likely to achieve therapeutic benefit. In another landmark study, the drug imatinib was developed to target the ABL kinase due to its constitutive activity in patients with chronic myelogenous leukemia, in which tumors harbor a BCR-ABL genetic translocation (28). Of critical importance to the field of Systems Biology, resistant tumors have recently emerged that utilize alternative pathways downstream of the drug target to circumvent the absolute requirement for the ABL kinase (28). Similar results with other targeted agents used in single or minimal combinational modalities are emerging, and would indeed be predicted from analysis of the target network that demonstrates the level of redundancy and robustness within signaling pathways (29). Essentially, the plasticity within these networks reduces the dependency on single nodes within the pathways. Systems Biology will play a critical role in modeling the pathways to *de novo* and acquired resistance. This knowledge can be used in the development of targeted approaches to minimize the probability that a tumor will develop resistance, and/or target key network nodes/hubs to reverse the resistant phenotype.

An important area of focus for Systems Biology in the field of oncology, will be its utility in the modeling of the inactivation status of tumor suppressor genes (30) and/or the activation of oncogenes (31), within complex systems from multivariate molecular data. The latter, in particular, represents therapeutic targets against which a number of drugs are currently available or are in active development. The classic definition of the oncogene should be stretched to include any molecular aberration that is causative with respect to the etiology and/or progression of the disease at the network level. As discussed above, it is well established that there is a high degree of redundancy within molecular networks, and that tumor cells will ultimately evolve to fully utilize different means to achieve the same end-result at the sub-system level. For example, inactivation of a canonical pathway can occur through a number of redundant mechanisms which would be predicted based upon a pre-requisite knowledge of the pathway (32). A key aspect of future studies will be to determine the utility of multiplexed "omics" technologies and network reconstruction, to accurately predict the *activation status* of potential targets within complex disease systems. The molecular portrait of a tumor likely reflects, at some level, the consequence of persistent activation/ inactivation of key network inputs (divergent points) and outputs (convergent points). These in turn, represent excellent targets, and in conjunction with pre-requisite drug-target knowledge, can be used to infer optimal therapeutic strategies with existing agents. As highlighted below, this approach represents a key component of our current predictive therapeutics protocol that remains in its rudimentary stages. The ability to systematically predict potential drugs using an integration of standardized technology, Systems Biology, and knowledge of biomarker-drug associations, for a given input remains a fundamental challenge.

Biomarkers that detect drug efficacy or toxicity, typically referred to as surrogate markers of response, represent another active area of biomarker research. Such biomarkers, could logically be used to determine the viability of a therapy, and/or be used in an adaptive trial design to optimize the dosing schedule for individual patients (33). In the case of validated markers of toxicity, these could be used to terminate treatment in advance, of a serious adverse event. Recently, a Predictive Safety Testing Consortium, sponsored by the FDA and the Critical Path Institute was formed to validate the utility of predictive biomarkers of organ toxicity that have been identified in pre-clinical models. Analogous to early detection biomarkers, these analytes must be assayed from readily accessible physiological fluids, and often require a pre-treatment baseline to determine the relative change following drug exposure. Current efforts are predominantly focused on the measurement of specific proteins within the blood or urine, or gene expression of circulating blood lymphocytes that may represent an excellent surrogate cell type for toxicity assessment (34).

Pharmacodynamic markers are also being investigated within physiological fluid and/or diseased tissue. Early markers of target modification would ideally be evaluated rapidly after administration of the treatment to determine the relative drug effect on the proposed molecular target and its associated network. This would be dependent on the pharmaco-kinetic and pharmaco-dynamic properties of the agent, as well as its molecular mechanism of action. These analyses would ideally be performed from diseased tissue, but repetitive sampling of a tumor represents a challenge for most tumor sites. As such, differential protein profiles within blood serum or plasma that reflect changes in the tumor-host micro-environment provide a possible opportunity for biomarker development, although the sensitivity of detection remains challenging

(35). Changes in the molecular target within the circulating lymphocytes may also represent a surrogate of target modulation within the disease (34), but cannot necessarily be used to ascertain the adequate delivery of the agent to the target tissue. Molecular-based imaging, and the ability to detect and characterize low levels of circulating tumor or endothelial cells from blood represent attractive areas of investigation, since they may permit the detection of changes in the molecular target within the tumor (36,37). If validated, these methods could be used to optimize the dosing schedule based upon the relative effect on the molecular target. They may also be used to triage ineffective drugs earlier in the drug discovery pipeline (38). Systems Biology can play a significant role in establishing the optimal biomarker panel for evaluation. For example, the constituents of the network downstream of the molecular target (the primary network input), typically represent an amplified signal, and this may provide a more sensitive and systematic means to detect target modulation. As described above, changes in possible extracellular plasma proteins can be inferred from a pre-requisite knowledge of the molecular target and its associated network (20). Coupled with Systems Biology tools, empirical experiments *in vitro* using gene-targeting technologies in comparison with a therapeutic agent could be used to identify a network associated with early target modulation, and this could conceptually be assayed *in vivo* to assess the degree of target activation. Reverse engineering of the measured changes in molecular events downstream of the primary input is also possible, and has been used to reveal off-target effects and/or identify the target for drugs with an unknown molecular mechanism of action (39) (CPW, unpublished observations).

The same considerations apply to the detection of biomarkers that are associated with long term therapeutic efficacy (typically termed surrogate markers of clinical response). Validated biomarkers that detect early clinical response would logically shorten the clinical development time considerably, and provide a more rapid assessment of the ineffective treatments in subjects. These late onset markers reflect a hallmark of cascading molecular events, many of which result from contextual changes in the tumor-host system that arise following long-term drug exposure (40). The fundamental cause of the observed changes within a treated tumor may be difficult to reveal in the noise of an evolved system that has been modified over time, in the backdrop of a changing tumor-host micro-environment and fluctuating local drug concentrations (the primary input). Unlike the persistent activation of oncogenes that can cause constant system deregulation, variations in target activity following intermittent drug exposure may be more difficult to model back to the target. In this context, Systems Biology is probably more appropriately used to infer biomarkers that could be used to monitor clinical response, investigate the molecular mechanisms of acquired drug resistance, and develop therapeutic strategies at the network level that may overcome resistance. The same principals related to Systems Biology essentially apply to all levels of biomarker discovery; the association between a primary input (the target) and the downstream molecular consequences can be used reciprocally to simulate and infer target-biomarker associations.

## 32.5 Challenges to Current Biomarker Development

While there is palpable excitement relating to the development of molecularly-targeted agents in conjunction with specific biomarkers that may indicate their selectivity in subsets of patients, the co-development of the necessary companion diagnostic test adds significant time to the cycle of drug development and clinical implementation. Typically, unless the biomarker positively impacts the drug approval process itself, drug manufacturers are unlikely to invest in diagnostic development after the successful marketing of their agent unless post-market forces, such as value-based reimbursement, encourage their subsequent development and use (41,42). Major obstacles to the accelerated adoption of biomarkers in medicine include the persistent development of new technologies for biomarker detection (the "me too" diagnostic), the development of biomarkers that are restricted to a specific manufacturer's drug, and the current uncertainty in the precise workflow towards regulated approval for *in vitro* diagnostics or laboratory tests (43). In principle, a standardized technology that reproducibly determines the absolute or relative levels of a comprehensive set of biomarkers (ideally with full genomic/proteomic coverage), in concert with standard operating procedures (SOP) for tissue procurement and processing, could provide a consistent measurement of the molecular constituents of a biological system. Systems Biology tools could be applied to these standardized data-sets to generate network-based mathematical models that are robust in the face of some missing data elements, including those measurements with a lower degree of confidence. Validated algorithms could be logically applied to new data-sets to predict the desired phenotype based upon the standardized input, and/or infer likely molecular-based intervention strategies. The new wave of molecularly-targeted drugs, which target subsets of molecular entities with increased specificity, provide an excellent opportunity for investigators in the discipline of Systems Biology. In theory, it should be possible to achieve a more specific disruption of the identified network nodes or hubs with tailored combinations of these agents. From the view point of the drug manufacturer, these efforts could readily expand the indications portfolio of existing agents by identifying molecular networks within additional disease subtypes, although economic and regulatory barriers to combinational therapies remain a significant challenge. In the remainder of this chapter, we will describe how we have begun to utilize Systems Biology in parallel with standardized genomic technologies and the current biomarker-drug knowledge to implement a strategy for the delivery of personalized and combinational therapeutics.

## 32.6 Personalized Molecular Medicine – The Future of Oncology?

The preceding well-established examples illustrate some success stories relating to the implementation of specific sets of biomarkers into the clinical setting, where they are now being used to improve the prediction, detection, diagnosis, treatments and/or monitoring of patients with different diseases including cancer. Systems Biology approaches are frequently used for the discovery of novel targets in conjunction with gene targeting technologies, and/or to predict the mechanism of action of a new agent (44). The development of inducible vectors for regulating small interfering RNA molecules, for example, will allow for more temporal control of target activity and assessment of the molecular consequences at the system level. Perturbations in the system could then be systematically extracted from cellular "noise" at the empirical level, and the defined sub-systems fully described and used to detect target activity within tumors. The degree of network similarity can also be used to compare pre-clinical models with the clinical disease. While homologous molecular components of the disease may differ across species, similarities in the overall modular network often provide a measure of the relevance between model systems (45). Thus, network information can be used at every step within a drug research and development program. The remainder of this chapter, however, briefly describes the potential use of Systems Biology for the optimization of existing drugs in various combinations, based upon the identification of the specific sub-systems at play within the tumor of the individual cancer patient. Whether this approach is used to expand the indications of compounds within a drug company's pipeline, or improve the probability of tumor response in individual patients, the overall theme is consistent. Despite differences in opinion regarding methods for predicting optimal treatments, few in the field of oncology now dispute the clear need for combinational treatment strategies that are designed to target the Achilles heal of the tumor to maximize efficacy while sparing normal cell function to minimize toxicity. Given the genetic instability within a tumor and the redundancy within molecular networks, the probability that a cell within the average tumor mass will acquire resistance to a single monotherapy, particularly targeted agents, during the life span of a tumor population would be expected to be very high (46).

## 32.7 A Paradigm Shift – Molecular-Based Indications

Currently, the most advanced diagnostics that accompany the prescription of a drug (known as companion diagnostics or theranostics) are developed for a single drug or class of drugs in a relatively small subset of site-specific diseases. Current approaches fail to expedite the utility of molecular biomarkers, Systems Biology and existing knowledge of biomarker-drug associations in the selection of patients for potential treatment with any drug or drug combination irrespective of the patient's disease classification. Consider the following hypothetical example: a patient with stage-IV glioblastoma who has failed the standard-of-care therapy visits his/her medical oncologist, who enters him/her into a research study that generates a molecular profile of the tumor. Coupled with Systems Biology inference, the profile suggests that the tumor expresses a high level of activated EGFR, providing some evidence that the glioblastoma may respond to anti-EGFR therapy (47). While a single validated assay to measure active EGFR and/or its down-stream signaling molecules can be developed for future screening (48), the development cycle for such a companion diagnostic is arduous. In addition, the treating physician also acquires knowledge regarding an association between increased osteonectin gene expression and Abraxane efficacy (49), and reduced levels of the ERCC1 nucleotide excision repair gene and enhanced sensitivity to the DNA cross-linking platinum-based therapy (50). Osteonectin is shown to be increased in this patient's tumor, while ERCC1 is decreased relative to a series of other well-characterized tumors. Logically, the gene expression profile in conjunction with informatics, drug-biomarker knowledge, and Systems Biology may predict a non-tested combination of an EGFR inhibitor, Abraxane and a platinum-based agent. Should the physician administer this unproven yet logically-inferred prediction? How do we systematically evaluate the simulation methodologies that can model optimal combinational strategies? As the FDA considers the future labeling of drugs to include molecular-based indications, there is an increasing list of agents with known molecular mechanisms of action (targets) or biomarker associations that have been implicated as markers of drug sensitivity and/or resistance (51); the number of drugs within our knowledge base with a reported biomarker association at the time of this writing is 1,526. A major challenge remains how to optimize treatments with the currently available drug resources and biomarker knowledge.

## 32.8 Optimizing the Translational Research Workflow – Development of the Critical Infrastructure

At the onset, it is apparent that there are several infrastructural components that are required to realize the full potential of evidence-based individualized therapy that may be able to benefit patients in an accelerated and real time fashion. These include full integration of medical and bioinformatics to permit synchronized data acquisition, consistent sample collection and processing, assay validation, standardized and documented data processing, a knowledge base that can be readily expanded and applied in a systematic fashion, and a means to develop and test predictive models and triage ineffective methodologies. Computational power alone is insufficient, and the full integration of empirical multivariate data-sets, coupled with knowledge and some theory (logic) of how to utilize these resources is essential. An efficient means to test the hypothesis remains critical as with any other scientific discipline. It is important to state that despite some anecdotal successes in-patient and mouse test cases, our genome and drug-wide approach that is disease agnostic remains unproven. Nonetheless, our systematic method may significantly advance the adoption of a predictive therapeutics philosophy, in which molecular-based evidence in conjunction with Systems Biology tools are used to identify drugs for consideration. At the very least, this approach could be used to serve as a clearing-house for patient enrichment on drug-specific protocols.

While not the focus of this chapter, it is important to briefly discuss the practicalities of a multi-disciplinary translational research project. Translational medicine requires coordinated multi-disciplinary teams and an efficient means to communicate the multi-faceted data and results. We have focused a large extent of our development efforts on building and establishing the necessary relationships and infrastructure that enable more efficient discovery, validation, and application of new molecular-based diagnostic and therapeutic strategies. This includes the development of the necessary informatics (XB-Bio Integration Suite, XB-BIS) that becomes the central hub of any effective translational research project.

In brief, XB-BIS permits the collection, management, analysis and reporting of data within our translational research protocols. On the front end, XB-BIS interfaces with existing data bases including clinical electronic medical record (EMR) data through an IRB-HIPPA portal. This aspect alone allows us to collect clinical data in a highly efficient and compliant fashion that includes key date-time stamped data elements such as patient demographics, diagnoses, test results, treatments, and response variables. These data describe in part the context of the physiological system under study (Fig. 32.1). The samples that are collected from the subject are also tracked within XB-BIS, and include descriptions of microscopic pathology that further define histological context (Fig. 32.1). Standardized experiments are performed that measure the absolute or relative amount of molecular constituents within the defined sample at a specific date and time (the time of sample acquisition). The derived molecular (DNA, RNA and/or protein) data are statistically analyzed within the same system, and the results of the analyses subjected to further network analysis. The identified networks are further analyzed to identify enriched hubs and/or input/output nodes that could represent key points for intervention. The results are overlaid with a knowledge of drug-biomarker associations, that is maintained within the knowledge base of XB-BIS. Each target (and hence each associated drug) is weighted based upon the inferred significance at the expression and network level. While we typically utilize transcriptional profiling to infer the aberrant networks within a given sample, it is critical to reiterate that gene or protein expression do not necessarily equate to target activation. Within complex biological systems, there is a large extent of negative and positive feedback and feed-forward loops that can disrupt linear relationships between target expression and activity. For example, one of the first molecular events that occurs in some cells following stimulation with an extracellular ligand can be the down-regulation of the cell surface receptor (reduced protein expression) and reduced receptor transcription (52). These and other negative feedback loops within environmental sensors (receptors) have likely evolved to maintain the cells adaptivity and sensitivity to environmental cues. In tumors, some of these feed back loops are deregulated, permitting the cancer cell to fully leverage the microscopic environment for survival and/or expansion. Regardless, the waves of transcriptional events that occur within a tumor cell likely represent a hallmark of upstream chronic signaling cascades on which the tumor more likely depends (53). As such, we are currently utilizing standardized gene expression profiling as a surrogate for the prediction of network activity.

## 32.9 Implementation of Systems Biology into a Predictive Therapeutics Protocol

Systems Biology/network analysis tools utilize public and/or proprietary knowledge to associate molecules via various interaction types, and reconstruct possible molecular networks based upon the input. While this is an effective method to maximize the visualization of existing knowledge in the form of connectivity maps, misinterpretation of the results is frequent due to the inherent bias (enrichment) in public knowledge. In conjunction with GeneGo, Inc. we have developed an algorithm that evaluates the topological significance within reconstructed networks. This tool determines the number of paths traversing each node within an identified shortest path network, and compares this to the total number of paths going via the same node in the complete global network. The significance for each node within the identified network is calculated based upon its probability for providing network connectivity. A result of a network topology analysis from the tumor of a patient with late stage non-small cell lung carcinoma is shown in Fig 32.2. In this specific tumor, gene expression data was normalized to a relative Z-score by comparison with a large reference set of tumors processed using standardized methods within a CLIA/CAP-accredited laboratory that implements critical sample and process quality control. Over-expressed genes were identified and submitted to network topological analysis. A major divergence point was identified as EGFR, which was also over-expressed at the transcript level relative to other tumors. Based upon this inference, the medical oncologist confirmed EGFR gene amplification in the tumor using traditional FISH analysis. This patient was treated with a combination of erlotinib, cisplatin (reduced expression of the ERCC1 gene), and bevacizumab (inferred constitutive activation of the VEGF-VEGFR pathway). The patient exhibited a partial response to this network targeted, combinational treatment.

In the first testing phase on 50 late stage cancer patients, a handful of anecdotal responses were observed in patients whose tumors exhibited a strong network signal, providing an impetus to further develop this general approach. As with all scientific disciplines, Systems Biology

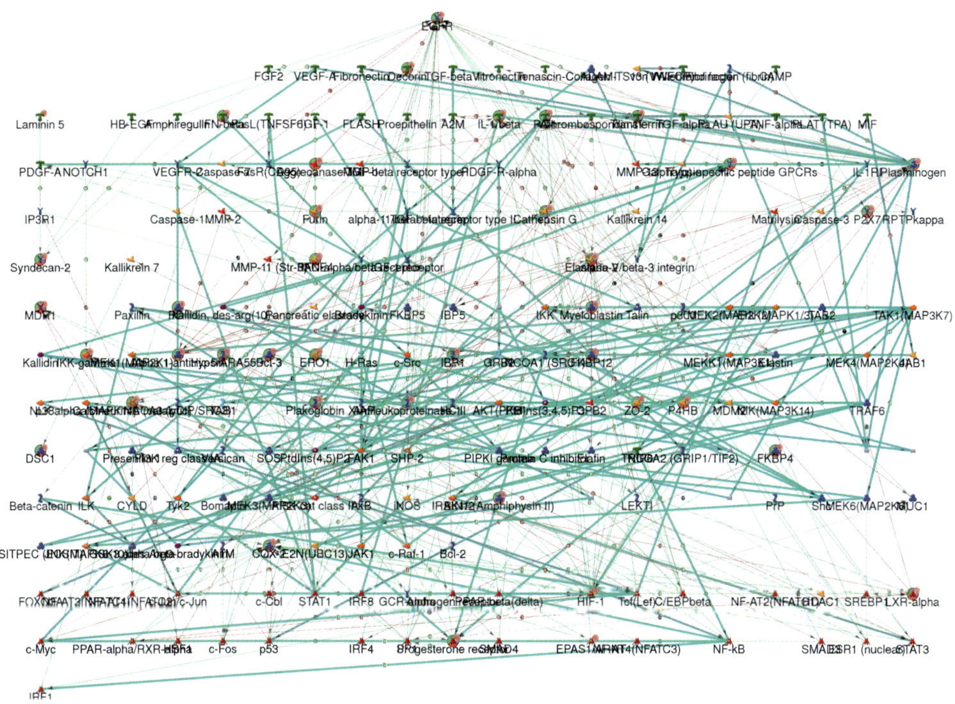


**Fig 32.2** Topological network analysis of gene expression data from a non-small cell lung carcinoma, inferred a key input node at the level of EGFR. The results of these analyses are displayed using MetaCore™, a Systems Biology network tool produced by GeneGo (www.genego.com). A knowledge base of drug-target interactions within XB-BIS was applied to these results, and the inferred significance of system connectivity used to select a corresponding EGFR inhibitor. The patient's tumor showed a partial response to erlotinib in combination with cisplatin and bevacizumab, which were also indicated from the molecular profiling data (low ERCC1 gene expression and inference of constitutive VEGF-VEGFR network signaling respectively). Among other applications, this type of systems approach can be readily applied for the discovery of new disease targets, prioritization and/or validation of the existing targets, and/or the identification of new indications for compounds that have a known or associated molecular mechanism of action. A key aspect is to successfully identify the significant convergence or divergence hubs or nodes within the identified networks (Copies of figures including color copies, where applicable, are available in the accompanying CD)

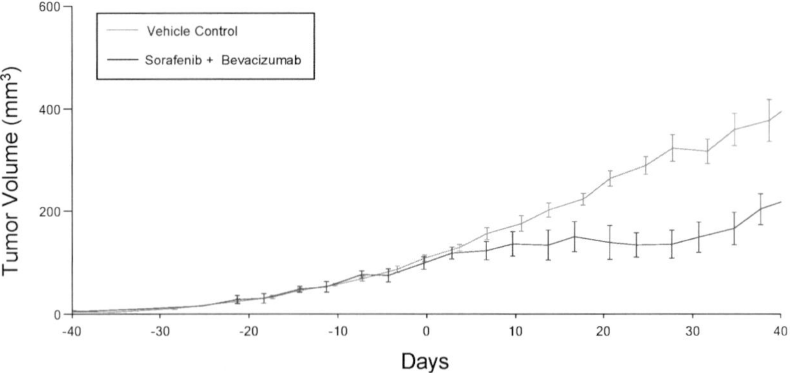


**Fig. 32.3** An example of a tumor response to a network-targeted combinational treatment strategy using pre-clinical models to validate the simulation methodologies. In this example, tumor-grafts from an adult patient with metastatic melanoma were established in immune compromised mice, and processed using gene expression profiling. After data normalization and comparison to the patient's original tumor, the network activity was inferred using a Systems Biology tool that identifies topologically significant nodes and hubs within networks (see Fig. 32.2). In this example, Sorafenib and Bevacizumab were selected based upon their targets inferred causative role (key input nodes) within the identified networks. Mice treated with a combination of these drugs showed partial tumor response relative to vehicle controls. The mean volumes of the measured tumors are shown over time relative to the start of treatment (day 0) (Copies of figures including color copies, where applicable, are available in the accompanying CD)

approaches that infer target activation and hence, drug selection, require rigorous testing in independent sample sets. Because of the anecdotal nature of patient responses in these case studies, we developed a pre-clinical component in the protocol in which subsets of the patient tumors were expanded as tumor-grafts in immune compromised mice. We have shown that these tumor-grafts closely resemble the patient's tumor at the histopathological, and more importantly, the molecular level. However, it is also important to identify that the environmental context for a tumor cell is evidently different in the mouse and human host. Nonetheless, the expansion of a single patient tumor to a large cohort of tumor-grafts permits the evaluation of Systems Biology tools for identifying the active molecular networks in this pre-clinical context. This in turn, provides for the cyclic simulation and testing environment that is critical to evaluate the different computational methodologies of Systems Biology in a more high-throughput fashion. Although this protocol remains in its infancy, the preliminary data from early pre-clinical experiments are encouraging (Fig. 32.3).

## 32.10 Summary and Future Challenges

In this chapter, we have summarized the potential role that Systems Biology tools may play in biomarker discovery and the realization of personalized medicine. The strength of Systems Biology is its ability to model standardized molecular data input, and provide consistent algorithm-based methods for predicting a phenotype of interest. Our current approach uses standardized genomic technologies, network reconstruction, and pre-requisite target-drug knowledge to infer treatment strategies for aggressive tumors. This represents only one approach to the realization of individualized treatment that is based upon the identified molecular systems within the tumor. The application of this general approach to sub-populations of tumor cells known as tumor stem cells (54) represents a further area of interest, and may permit the targeted disruption of stem cell networks that are believed to represent a critical component within a tumor. The relative system complexity within this proposed side-population of tumor originating cells remains to be determined, but the key divergence or convergence points within this subsystem represent additional therapeutic targets for consideration. The challenges for the mainstream adoption of Systems Biology in the area of diagnostics and/or predictive therapeutics remain prevalent, and require bio-analytical technologies that provide the standardized data input, as well as simulation methodologies that provide for consistent outputs. The complete translational workflow from human disease, to data acquisition and analysis, computational modeling, validation and clinical application, must be fully integrated to provide the efficient and practical foundation for personalized medicine initiatives. Relevant and scalable pre-clinical models that can test the accuracy of the systems approach are needed to select the most promising computational methodologies, before the general clinical acceptance of the approach can be expected. While the regulatory hurdles facing the adoption of multiplexed biomarkers and off-label drug indications remain uncertain, existing knowledge and associated logic, coupled with the complexity of the tumor system, provide unique opportunities for those in the Systems Biology field. Systems Biology represents the interface between the genotypic complexity and phenotypic manifestation, and when integrated with other scientific and medical disciplines, will likely represent the hub of translational medicine in the 21st century.

## Glossary and Abbreviations

| | |
|---|---|
| ABL | v-abl Abelson Murine Leukemia Viral Oncogene |
| BCR | Breakpoint Cluster Region |
| BRCA1 | Breast Cancer 1, Early Onset |
| BRCA2 | Breast Cancer 2, Early Onset |
| CAP | College of American Pathologists |
| CLIA | Clinical Laboratory Improvement Amendments |
| DNA | Deoxyribonucleic Acid |
| EGFR | Epidermal Growth Factor Receptor |
| EMR | Electronic Medical Record |
| ERBB2 | v-erb-b2 Erythroblastic Leukemia Viral Oncogene Homolog 2 |
| ERCC1 | Excision Repair Cross-Complementing Rodent Repair Deficiency, Complementation Group 1 |
| FDA | Food and Drug Administration |
| FISH | Fluorescence In Situ Hybridization |
| IRB | Institutional Review Board |
| mRNA | Messenger Ribonucleic Acid |
| OPN | Osteopontin |
| PSA | Prostate Specific Antigen |
| RNA | Ribonucleic Acid |
| SOP | Standard Operating Procedure |
| UGT1A1 | UDP Glucuronosyltransferase 1 Family, Polypeptide A1 |
| VEGF | Vascular Endothelial Growth Factor |
| VEGFR | Vascular Endothelial Growth Factor Receptor |
| XB-BIS | XenoBase Bio-Integration Suite |

## References

### *The Challenge – The Complexity of Biological Systems*

1. Stumpf MP, Kelly WP, Thorne T, Wiuf C. Evolution at the system level: the natural history of protein interaction networks. Trends Ecol Evol 2007;22(7):366–373.
2. Huang S. Back to the biology in Systems Biology: what can we learn from biomolecular networks? Brief Funct Genomic Proteomic 2004;2(4):279–297.

3. van der Greef J, Martin S, Juhasz P, et al. The art and practice of Systems Biology in medicine: mapping patterns of relationships. J Proteome Res 2007;6(4):1540–1559.
4. Fujarewicz K, Kimmel M, Lipniacki T, Swierniak A. Adjoint systems for models of cell signaling pathways and their application to parameter fitting. IEEE/ACM transactions on computational biology and bioinformatics/ IEEE, ACM 2007;4(3):322–335.

## *The Added Complexity of Cancer*

5. Wang E, Lenferink A, O'Connor-McCourt M. Cancer Systems Biology: exploring cancer-associated genes on cellular networks. Cell Mol Life Sci 2007;64(14):1752–1762.
6. Hanahan D, Weinberg RA. The hallmarks of cancer. In: Cell; 2000:57–70.
7. Webb CP, Vande Woude GF. Genes that regulate metastasis and angiogenesis. J Neurooncol 2000; 50(1-2):71–87.
8. Aranda-Anzaldo A. Cancer development and progression: a non-adaptive process driven by genetic drift. Acta Biotheor 2001;49(2):89–108.
9. Balakrishnan A, Bleeker FE, Lamba S, et al. Novel somatic and germline mutations in cancer candidate genes in glioblastoma, melanoma, and pancreatic carcinoma. Cancer Res 2007;67(8):3545–50.
10. Sjoblom T, Jones S, Wood LD, et al. The consensus coding sequences of human breast and colorectal cancers. Science 2006;314(5797):268–274.

## *Personalized Medicine – The Objectives*

11. Maron BJ, Hauser RG. Perspectives on the failure of pharmaceutical and medical device industries to fully protect public health interests. Am J Cardiol 2007;100(1):147–151.
12. Goodsaid F, Frueh FW. Implementing the U.S. FDA guidance on pharmacogenomic data submissions. Environ Mol Mutagen 2007;48(5):354–358.
13. Jain KK. Challenges of drug discovery for personalized medicine. Curr Opin Mol Ther 2006;8(6):487–492.

## *Biomarkers in Practice – The Role of Systems Biology*

14. Nusbaum R, Isaacs C. Management updates for women with a BRCA1 or BRCA2 mutation. Mol Diagn Ther 2007;11(3):133–44.
15. Daly AK. Individualized drug therapy. Curr Opin Drug Discov Devel 2007;10(1):29–36.
16. de Leon J, Susce MT, Murray-Carmichael E. The AmpliChip CYP450 genotyping test: Integrating a new clinical tool. Mol Diagn Ther 2006;10(3):135–151.
17. Marsh S. Impact of pharmacogenomics on clinical practice in oncology. Mol Diagn Ther 2007;11(2):79–82.
18. Duffy MJ. Role of tumor markers in patients with solid cancers: A critical review. Eur J Intern Med 2007;18(3):175–184.
19. Thompson IM, Ankerst DP. Prostate-specific antigen in the early detection of prostate cancer. Cmaj 2007;176(13):1853–1858.
20. Webb CP, Pass HI. Translation research: from accurate diagnosis to appropriate treatment. J Transl Med 2004;2(1):35.
21. Pass HI, Lott D, Lonardo F, et al. Asbestos exposure, pleural mesothelioma, and serum osteopontin levels. N Engl J Med 2005;353(15):1564–1573.
22. Miller LD, Liu ET. Expression genomics in breast cancer research: microarrays at the crossroads of biology and medicine. Breast Cancer Res 2007;9(2):206.
23. Kaklamani VG, Gradishar WJ. Gene expression in breast cancer. Curr Treat Options Oncol 2006;7(2):123–128.
24. Canales RD, Luo Y, Willey JC, et al. Evaluation of DNA microarray results with quantitative gene expression platforms. Nat Biotechnol 2006;24(9):1115–1122.
25. Rhodes DR, Kalyana-Sundaram S, Tomlins SA, et al. Molecular concepts analyzes links tumors, pathways, mechanisms, and drugs. Neoplasia 2007;9(5):443–454.
26. O'Donovan N, Crown J. EGFR and HER-2 antagonists in breast cancer. Anticancer Res 2007;27(3A):1285–1294.
27. Tsuda H. HER-2 (c-erbB-2) test update: present status and problems. Breast Cancer 2006;13(3):236–248.

28. Hochhaus A, Erben P, Ernst T, Mueller MC. Resistance to targeted therapy in chronic myelogenous leukemia. Semin Hematol 2007;44(1 Suppl 1):S15–24.
29. Bublil EM, Yarden Y. The EGF receptor family: spearheading a merger of signaling and therapeutics. Curr Opin Cell Biol 2007;19(2):124–134.
30. Wen L, Li W, Sobel M, Feng JA. Computational exploration of the activated pathways associated with DNA damage response in breast cancer. Proteins 2006;65(1):103–110.
31. Schafer R, Schramme A, Tchernitsa OI, Sers C. Oncogenic signaling pathways and deregulated target genes. Recent Results Cancer Res 2007;176:7–24.
32. Lin J, Gan CM, Zhang X, et al. A multidimensional analyzes of genes mutated in breast and colorectal cancers. Genome Res 2007;17(9):1304–1318.
33. Lassere MN, Johnson KR, Boers M, et al. Definitions and validation criteria for biomarkers and surrogate endpoints: development and testing of a quantitative hierarchical levels of evidence schema. J Rheumatol 2007;34(3):607–615.
34. Burczynski ME, Dorner AJ. Transcriptional profiling of peripheral blood cells in clinical pharmacogenomic studies. Pharmacogenomics 2006;7(2):187–202.
35. Floyd E, McShane TM. Development and use of biomarkers in oncology drug development. Toxicol Pathol 2004;32 Suppl 1:106–115.
36. Kelloff GJ, Sigman CC. New science-based endpoints to accelerate oncology drug development. Eur J Cancer 2005;41(4):491–501.
37. Bertolini F, Shaked Y, Mancuso P, Kerbel RS. The multifaceted circulating endothelial cell in cancer: towards marker and target identification. Nat Rev Cancer 2006;6(11):835–845.
38. Graul AI. Promoting, improving and accelerating the drug development and approval processes. Drug news & perspectives 2007;20(1):45–55.
39. Goutsias J, Lee NH. Computational and experimental approaches for modeling gene regulatory networks. Curr Pharm Des 2007;13(14):1415–1436.
40. Kuick R, Misek DE, Monsma DJ, et al. Discovery of cancer biomarkers through the use of mouse models. Cancer Lett 2007;249(1):40–48.

## *Challenges to Current Biomarker Development*

41. Logue LJ. Genetic testing coverage and reimbursement: a provider's dilemma. Clin Leadersh Manag Rev 2003;17(6):346–350.
42. Roberts TG, Jr., Chabner BA. Beyond fast track for drug approvals. N Engl J Med 2004;351(5):501–555.
43. Swanson BN. Delivery of high-quality biomarker assays. Dis Markers 2002;18(2):47–56.

## *Personalized Molecular Medicine – The Future of Oncology?*

44. Caldwell JS. Cancer cell-based genomic and small molecule screens. Adv Cancer Res 2007;96:145–173.
45. Watters JW, Roberts CJ. Developing gene expression signatures of pathway deregulation in tumors. Mol Cancer Ther 2006;5(10):2444–2449.
46. Michor F, Nowak MA, Iwasa Y. Evolution of resistance to cancer therapy. Curr Pharm Des 2006; 12(3):261–271.

## *A Paradigm Shift – Molecular-Based Indications*

47. Sarkaria JN, Yang L, Grogan PT, et al. Identification of molecular characteristics correlated with glioblastoma sensitivity to EGFR kinase inhibition through use of an intracranial xenograft test panel. Mol Cancer Ther 2007;6(3):1167–1174.
48. Sequist LV, Bell DW, Lynch TJ, Haber DA. Molecular predictors of response to epidermal growth factor receptor antagonists in non-small-cell lung cancer. J Clin Oncol 2007;25(5):587–595.
49. Gradishar WJ. Albumin-bound paclitaxel: a next-generation taxane. Expert Opin Pharmacother 2006;7(8):1041–1053.
50. Gossage L, Madhusudan S. Current status of excision repair cross complementing-group 1 (ERCC1) in cancer. Cancer Treat Rev 2007;33(6):565–577.
51. Lee JK, Havaleshko DM, Cho H, et al. A strategy for predicting the chemosensitivity of human cancers and its application to drug discovery. Proceedings of the National Academy of Sciences of the United States of America 2007;104(32):13086–13091.

### *Optimizing the Translational Research Workflow – Development of the Critical Infrastructure*

52. Shtiegman K, Kochupurakkal BS, Zwang Y, et al. Defective ubiquitinylation of EGFR mutants of lung cancer confers prolonged signaling. Oncogene 2007;26(49):6968–6978.
53. Weinstein IB, Joe AK. Mechanisms of disease: Oncogene addiction—a rationale for molecular targeting in cancer therapy. Nat Clin Pract Oncol 2006;3(8):448–457.

### *Summary and Future Challenges*

54. Schulenburg A, Ulrich-Pur H, Thurnher D, et al. Neoplastic stem cells: a novel therapeutic target in clinical oncology. Cancer 2006;107(10):2512–2520.

## Key References

Huang S. Back to the biology in Systems Biology: what can we learn from biomolecular networks? Briefings in functional genomics & proteomics 2004;2(4):279–297.

Sjoblom T, Jones S, Wood LD, et al. The consensus coding sequences of human breast and colorectal cancers. Science (New York, NY 2006;314(5797):268–274.

Rhodes DR, Kalyana-Sundaram S, Tomlins SA, et al. Molecular concepts analyzes links tumors, pathways, mechanisms, and drugs. Neoplasia (New York, NY 2007;9(5):443–454.

Michor F, Nowak MA, Iwasa Y. Evolution of resistance to cancer therapy. Current pharmaceutical design 2006;12(3):261–271.

Weinstein IB, Joe AK. Mechanisms of disease: Oncogene addiction—a rationale for molecular targeting in cancer therapy. Nature clinical practice 2006;3(8):448–457.

## Web Resources

www.fda.gov/cber/gdlns/ichel5term.htm
www.fda.gov/bbs/topics/news/2006/NEW01337.html
www.fda.gov/cdrh/oivd/guidance/1610.html
www.xbtransmed.com
www.genego.com

# Index

## E



## F



## G

**Q**



**R**



**S**

Printed in the United States of America

Springer